3rd International Conference of Planning in the Era of Uncertainty 2017

IOP Conference Series: Earth and Environmental Science
Volume 70

Malang, Indonesia
6 – 7 March 2017

ISBN: 978-1-5108-4293-9
ISSN: 1755-1307

Printed from e-media with permission by:

Curran Associates, Inc.
57 Morehouse Lane
Red Hook, NY 12571

Some format issues inherent in the e-media version may also appear in this print version.

Printed by Curran Associates, Inc. (2017)

For permission requests, please contact the Institute of Physics at the address below.

Institute of Physics
Dirac House, Temple Back
Bristol BS1 6BE UK

Phone: 44 1 17 929 7481
Fax: 44 1 17 920 0979

techtracking@iop.org

Additional copies of this publication are available from:

Curran Associates, Inc.
57 Morehouse Lane
Red Hook, NY 12571 USA
Phone: 845-758-0400
Fax: 845-758-2633
Email: curran@proceedings.com
Web: www.proceedings.com

Table of contents

Volume 70

3rd International Conference of Planning in the Era of Uncertainty

6–7 March 2017, Malang, Indonesia

Accepted papers received: 6 June 2017
Published online: 22 June 2017

Preface

Environment, Infrastructure, and Information System

Regional Development and Public Policy

Urban Planning and Design

IOP Conf. Series: Earth and Environmental Science **70** (2017) 011001 doi:10.1088/1755-1315/70/1/011001

PREFACE: The 3rd International Conference in The Era of Uncertainty (ICPEU)

In the spirit of achieving Sustainable Development Goals declared by the United Nations, the 3rd ICPEU looks at harmonizing contextual perspective associated with the innovation and creativity that still considers the economic, ecology, and social value to create sustainable nations. The journey towards achieving a sustainable nation requires recommendations and synchronization from a wide array of stakeholders including academics, private sectors as well as government officials to create inclusive, comfortable and sustainable places for living.

The Urban and Regional Planning Department at Universitas Brawijaya is one of the institutions which has been actively promoting sustainability in all its educational obligations. In the past years, the Department has established cooperation with international institutions, one of which is Kyoto University, to better resonate the awareness to a wider scale. The cooperation with Kyoto University has been extremely crucial for the organization of ICPEU.

For two days, copious data and hypotheses on urban and regional planning issues were discussed and exchanged. More than 200 outstanding intellectuals of the world participated in aiding those involved in such ventures solving any problem that occur. It is only natural that a considerable number of schemes as well as projections were constructed as the result, all assembled in this conference proceeding. It is to our highest expectation that all ideas expressed in this conference could participate and contribute in the global movement toward achieving sustainability.

The committee of the 3rd ICPEU would like to thank the Sixth Southeast Asia Network Forum for the collaboration and to all keynote speakers who participated:

- Prof. Kiyoshi Kobayashi of Kyoto University, Japan,
- Prof. Hans WestLund of Kungliga Tekniska Hogskolan (KTH) Royal Institute of Technology, Sweden,
- Prof. Masaaki Okamoto of Kyoto University, Japan
- Dr. Surjono of Universitas Brawijaya, Indonesia,
- Dr (HC) Tri Rismaharini (Mayor of Surabaya),
- Dr. Emil Elestianto Dardak (Mayor of Trenggalek), and
- Mr. John Taylor as the founder of "Kota Kita".

See you at the next ICPEU in 2019.

Malang, 3-4 March 2017
Universitas Brawijaya
Indonesia

IOP Conf. Series: Earth and Environmental Science **70** (2017) 011001 doi:10.1088/1755-1315/70/1/011001

LIST OF COMMITTEES

Steering Commitee (SC) : Ir. Ismu Rini Dwi Ari, MT., Ph.D
Dr. Eng. I Nyoman Suluh Wijaya, ST., MT
Dr. Ir. Budi Sugiarto W, MSP
Dr. Ir. Agus Dwi Wicaksono, Lic. Rer.Reg

Organizing Committer (OC) :
Chairman : Aris Subagiyo, ST., MT
Technical Chairman : Eddi Basuki Kurniawan, ST., MT

Scientific Committee : Dr. Eng. Turniningtyas Ayu R, ST., MT *(Universitas Brawijaya, Indonesia)*
Wawargita Permata Wijayanti, ST., MT *(Universitas Brawijaya, Indonesia)*
Dadang Meru Utomo, ST., MURP *(Universitas Brawijaya, Indonesia)*
Dr. techn. Christia Meidiana, ST., M.Eng *(Universitas Brawijaya, Indonesia)*
Imma Widyawati Agustin, ST., MT., Ph.D *(Universitas Brawijaya, Indonesia)*
Dr. Septiana Hariyani, ST., MT *(Universitas Brawijaya, Indonesia)*
Wulan Dwi Purnamasari, ST., MT *(Universitas Brawijaya, Indonesia)*
AR. Rohman Taufiq H. ST., M.Agr.Sc *(Universitas Brawijaya, Indonesia)*
Dian Indah Shofarini, ST., MT *(Universitas Brawijaya, Indonesia)*
Dwi Maulidatuz Zakiyah, ST., MT., M.Sc *(Universitas Brawijaya, Indonesia)*
Gunawan Prayitno, SP., MT., Ph.D *(Universitas Brawijaya, Indonesia)*
Dr. Eng. Fadly Usman, ST., MT *(Universitas Brawijaya, Indonesia)*
Dian Dinanti, ST., MT *(Universitas Brawijaya, Indonesia)*
Kartika Eka Sari ST., MT *(Universitas Brawijaya, Indonesia)*
Nindya Sari, ST., MT *(Universitas Brawijaya, Indonesia)*
Chairul Maulidi, ST., MT *(Universitas Brawijaya, Indonesia)*
Deni Agus Setiyono, ST., M.Eng *(Universitas Brawijaya, Indonesia)*
Prof. Kiyoshi Kobayashi *(Kyoto University, Japan)*
Prof. Hans Westlund *(Kungliga Tekniska Hogskolan – KTH Royal Institute of Technology, Sweden)*
Dr. Ir. Surjono, MTP *(Universitas Brawijaya, Indonesia)*
Prof. Masaaki Okamoto *(Kyoto University, Japan)*
Dr.Md. Hafizur Rahman *(Independent University, Bangladesh)*
Dr. Sharina Farihah Hasan *(International Islamic University Malaysia – IIUM, Malaysia)*

Technical Assistance : Edo Dwi Imawan
Dimas Prasetya
Rendiana Satya Pangestika
Utari Salsabila Adam
Fadhilatus Shoimah
Rofifah Yusadi
Tubagus Wahyu Jatmikko
Patricia Emanuella Tambunan

Hanifa Muslimah
Ahmad Beny Zamroni
Bimar Jalu Arditama
Muammal

IOP Conf. Series: Earth and Environmental Science **70** (2017) 011002 doi:10.1088/1755-1315/70/1/011002

Peer review statement

All papers published in this volume of *IOP Conference Series: Earth and Environmental Science* have been peer reviewed through processes administered by the proceedings Editors. Reviews were conducted by expert referees to the professional and scientific standards expected of a proceedings journal published by IOP Publishing.

IOP Conf. Series: Earth and Environmental Science **70** (2017) 012001 doi:10.1088/1755-1315/70/1/012001

Distribution of green open space in Malang City based on multispectral data

A W Hasyim, F P Hernawan

Department of Urban and Regional Planning, Faculty of Engineering, Universitas Brawijaya, Malang, Indonesia

awhasyim@yahoo.com

Abstract. Green open space is one of the land that its existence is quite important in urban areas where the minimum area is set to reach 30% of the total area of the city. Malang which has an area of 110,6 square kilometers, is one of the major cities in East Java Province that is prone to over-land conversion due to development needs. In support of the green space program, calculation of green space is needed precisely so that remote sensing which has high accuracy is now used for measurement of green space. This study aims to analyze the area of green open space in Malang by using Landsat 8 image in 2015. The method used was the vegetation index that is Normalized Difference Vegetation Index (NDVI). From the study obtained the calculation of green open space was better to use the vegetation index method to avoid the occurrence of misclassification of other types of land use. The results of the calculation of green open space using NDVI found that the area of green open space in Malang City in 2015 reached 39% of the total area.

1. Introduction

Malang is one of the urban areas in East Java Province with 110,06 square kilometer administrative area and 851,298 of total population in 2015 (Malang City In Figures, 2016). Urban areas with a high population tendency make the need for land increasing every year. Conversion of green land into built land directly related to the needs of land which will eventually lead to a decline in the area of green land which is one of the supporting elements for sustainability of a city.

Green open space in urban areas according to law no.26 of 2007 consists of puclic green open space and private green open space. The proportion of green open spaces in urban areas shall be at least 30 (thirty) percent of total city area. The proportion of public green open spaces in urban areas shall be at least 20 (twenty) percent total area of the city. The green open space is an elongated / lane and / or clumped area, which is open to use, where crops grow, either naturally or intentionally planted [1].

The presence of green open space in an urban area has the benefit of an air circulation control system (urban lung). Indirectly, green open space functions in watershed protection and biodiversity conservation / for biodiversity. In addition, the presence of green space in an urban area can be utilized by the population to perform various social activities in a city where its existence produces various social, economic, and environmental impacts.

This study aimed to determine the distribution of green open space in Malang City in 2015 using Landsat 8 image. The calculation of the green open space was done by interpretation of Landsat 8 image using ArcGIS 10.3 software.

2. Methods

This study was based on the phenomenon of land conversion rate is quite high in Malang because of the high demand for land. Malang was classified in a large urban classification with a population above 500,000 who also accumulated by the existence of students and commuters that make the need for land to be high and at risk of green land conversion function.

The data used in this research was Landsat 8-OLI image with date of acquisition 16 June 2015 for scene scene number / row 118/066. Radiometric correction was performed to calculate the reflectance value of the image. The NDVI value (Normalized Difference Vegetation Index) is calculated by adopting a calculation method determined by NASA. NDVI value used to separate between the green area and built area for calculated of green area/ green space in Malang.

2.1. Remote sensing

Remote sensing is a technique of collecting information about objects and the environment remotely without physical touch. This technique produces some form of image that further processing and interpretation to produce data related needs in the field of geography, geology, planning, forestry and so forth [2].

The landsat 8 satellite has an OLI sensor (Onboard Operational Land Imager) and TIRS (Thermal Infrared Sensor) with a total of 11 units. Among the canals, 9 channels (bands 1-9) are in OLI and 2 others (bands 10 and 11) on the TIRS. Most channels have specifications similar to Landsat-7.

2.2. Radiometric correction

Radiometric correction was used to correct the pixel value to fit the supposedly normally considering atmospheric disturbance factor as the main fault source and also to eliminate or minimize radiometric error due to external aspect of atmospheric disturbance during recording [3].

Radiometric correction was performed using the ToA (Top of Atmosphere) correction which includes reflectance ToA and sun correction. Digital number (DN) is a digital value that describes an object's brightness level in satellite data. Digital numbers are stored in a pixel, whereas pixels are a point that is the smallest element of a satellite image. Spectral reflectance is the amount of energy reflected by an object per unit area and a certain wavelength. Conversion from digital number to spectral rerflectance is done to change the value of the reflection contained in the digital number that previously did not have a unit into the energy value that already has a unit. Here is the equation for the correction of reflectance ToA:

$$\rho\lambda' = M\rho * Qcal + A\rho \tag{1}$$

Where:

$\rho\lambda'$ = ToA reflectance, without sun elevation correction
$M\rho$ = Reflectance_MULT_BAND_x, which x is band number
$Qcal$ = Digital Number (DN)
$A\rho$ = Reflectance_ADD_BAND_x, which x is band number

Furthermore, the image was corrected to the sun angle to eliminate the difference in DN values caused by the position of the sun. The position of the sun against the earth changes depending on the recording time and the location of the recorded object. The equations for correction with the angle of the sun are:

$$\rho\lambda = \rho\lambda'/(\cos(\theta SZ)) = \rho\lambda'/(\sin(\theta SE)) \tag{2}$$

Where:

$\rho\lambda$ = ToA reflectance
θSZ = Sun elevation
θSE = Angle zenith sun, $\theta SZ = 90° - \theta SE$

2.3. NDVI (Normalized Difference Vegetation Index)

NDVI (Normalized Difference Vegetation Index) is an image calculation used to determine the greenish level. NDVI can show parameters related to vegetation parameters, among others, green foliage biomass, green foliage area that is predictable for vegetation division.

The absorption of red light by chlorophyll and the reflection of infrared light close by the mesophyll tissue in the leaves will make the brightness that receives the satellite sensors through these channels will be much different. On land without vegetation, including areas and settlements, open land, or damaged vegetation conditions, can not provide a high ratio to the comparison of the canals. In contrast, on land with dense vegetation and healthy conditions, the comparison of the value of the two channels provides a very high value.

The value of the vegetation index is built on the difference between maximum absorption in red channels as a result of chlorophyll pigment and maximum reflectance in near infrared (near infrared / NIR) infrared spectral channel due to the effect of leaf cellular structure [4].

NDVI has a range between -1.0 to +1.0. Values greater than 0.1 usually indicate an increase in the degree of greenness and intensity of vegetation. The value between 0 and 0.1 is characteristic of rocks and vacant land. A value less than 0 applies cloud, ice, and snow. The vegetation surface has a range of NDVI 0.1 values for grassland and shrublands up to 0.8 for tropical rainforest areas.

$$NDVI = (NIR - VIS)/(NIR + VIS) \quad (3)$$

Where:
NIR = Reflectant value of near infra red channel
VIS = Reflectant value infra-red channel

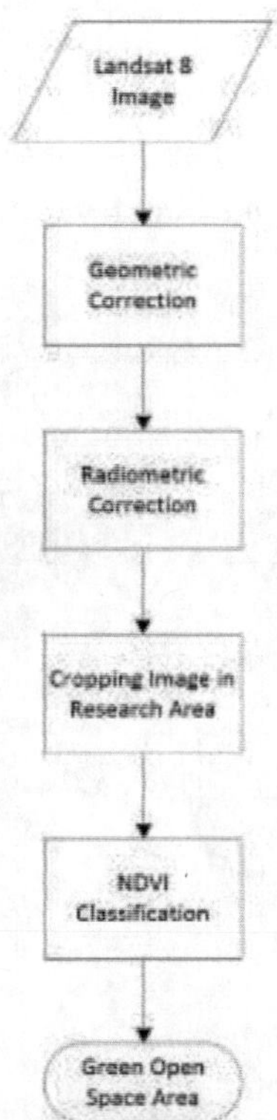

Figure 1. Flow chart of Landsat image processing.

IOP Conf. Series: Earth and Environmental Science **70** (2017) 012001 doi:10.1088/1755-1315/70/1/012001

3. Results

Image processing of Landsat 8 was done on the administrative area of Malang City. The result of Landsat 8 data before and after the ToA correction can be seen in Figure 2. The result of band composite on Landsat 8 image of Malang city using combination between band 4,3 and 2. The Landsat 8 data has been corrected ToA, the value of the Digital Number was turned into a reflectance value. To support the NDVI classification, an image calculation was performed to determine the greenness level.

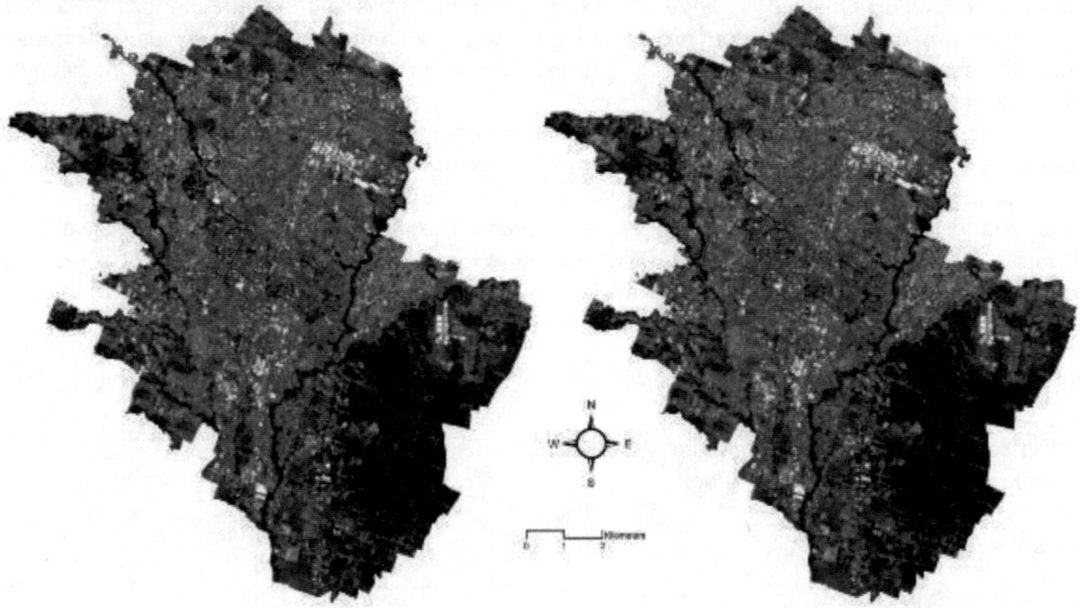

Figure 2. Landsat 8 Image Using Composite Band 432 (a) Before ToA Correction; (b) After ToA Correction

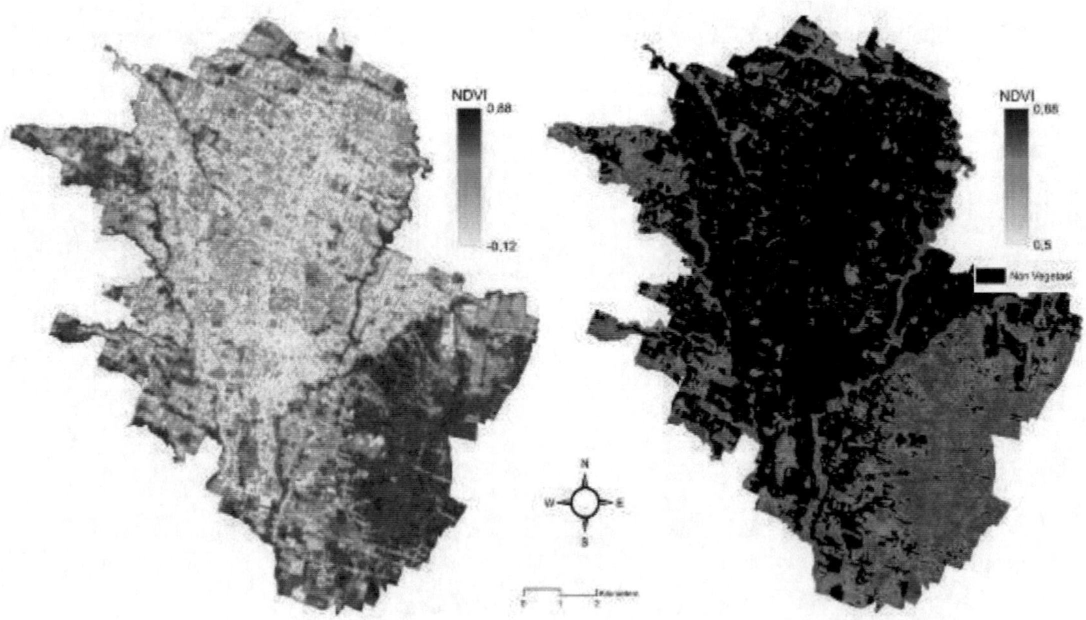

Figure 3. Vegetation Indeks (a) NDVI; (b) Reclassification NDVI

IOP Conf. Series: Earth and Environmental Science **70** (2017) 012001 doi:10.1088/1755-1315/70/1/012001

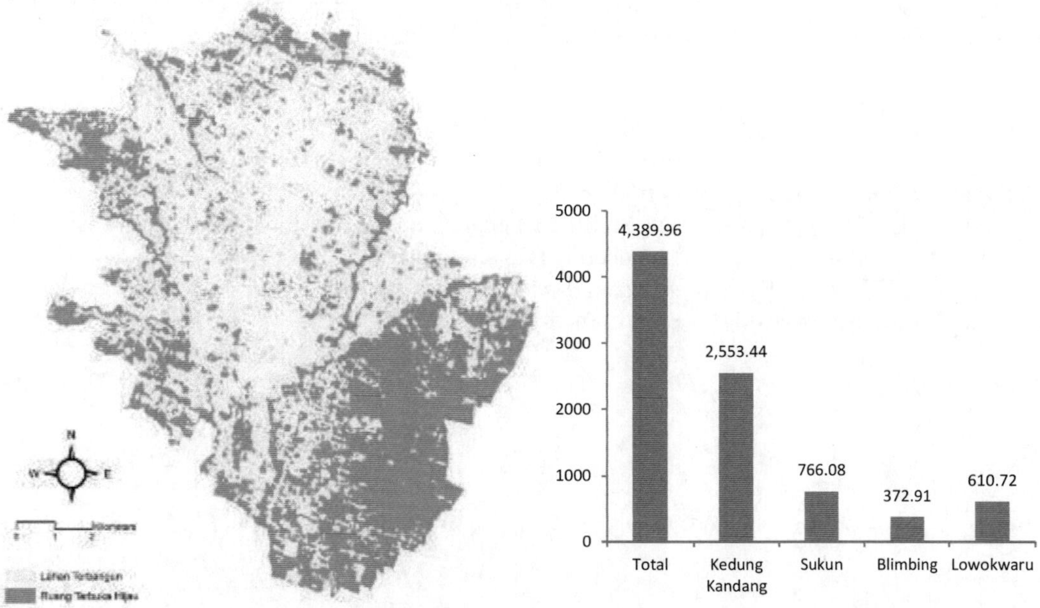

Figure. 4. (a) Land Cover; (b) Green Open Space Width of Each Sub-District

3.1. The Vegetation Index (NDVI)
From the results of NDVI classification in Malang, can be seen the spread of vegetation in Malang tends to be located in the suburb of Kedung Kandang and surrounding areas. The calculation results obtained NDVI for Landsat 8 data 2015 as shown in Figure 3.

In the downtown area, the NDVI value tends to be low as it is dominated by the area built in the area. In the southeastern region is still found plants that have high NDVI value (blue) where the area is a hill that is still dominated by tall trees and forests. After the classification stage of NDVI value, calculation of green open space in Malang is done by converting the image of NDVI processing into polygon with minimum limit (classification) to determine the built area and green area (green open space). After the classification of each land, then calculated the area between land and green area. Classification for green open space used value is more than 0,5. This classification value was used to classify forage on trees, excluding grass, rice and other low plants. Thus the obtained distribution green open space in Malang has a value between 0,5 to 0,88 as shown in Figure 3.

From the results of the calculation area, it is known that the total green open space in Malang reached 4389,959 hectares. While the bulit up area was 6816,3913 hectares. According to its own percentage, the area of green open space of Malang City reaches 39,7% and the land area is 60,3%. The width of green open space in each sub-district can be seen in Figure 4.

4. Conclusion
From the results of the study was found that the calculation of green open space was better to use vegetation index method to avoid the error of other types of land use. By using the classification of NDVI values from 0,5 to 0,88 as green open space, it is known that the green open space in Malang reaches 4389,959 hectares. Green Open Space in each district respectively, Klojen District of 86,80 hectare, Lowokwaru District of 610,72 hectares, Blimbing District with 372,91 hectares, Sukun District covering 766,08 hectares and Kedung Kandang District with 2553,44 Hectare. The total area of green

open space in Malang reaches 39.7% of the total area in which it exceeds the minimum limit for the green open space area in a city that is equal to 30% of the total administrative area.

References

[1] Undang-undang No.26 Tahun 2007 Tentang Penataan Ruang.
[2] Wolf, P. R. (1993). Elemen Fotogrametri: Dengan Intepretasi Foto Udara dan Penginderaan Jauh. Yogyakarta: Gajah Mada University Press.
[3] Soenarmo, S. H., 2009. Penginderaan Jauh dan Pengenalan Sistem Informasi Geografis untuk Bidang Ilmu Kebumian. Bandung: Penerbit ITB Bandung.
[4] Tucker, C. J. (1979, Vol. 8). *Red and Photographic Infrared Linear Combinations for Monitoring Vegetation*. Remote Sensing of Environment, p. 127–150.

IOP Conf. Series: Earth and Environmental Science **70** (2017) 012002 doi:10.1088/1755-1315/70/1/012002

Towards democracy in spatial planning through spatial information built by communities: The investigation of spatial information built by citizens from participatory mapping to volunteered geographic information in Indonesia

Adipandang Yudono[1, 2]

[1] Department of Urban and Regional Planning, Faculty of Engineering, Universitas Brawijaya, Malang, Indonesia
[2] Department of Urban Studies and Planning, University of Sheffield, S10 2TN, Sheffield, UK

adipandangyudono@ub.ac.id; ayudono1@sheffield.ac.uk

Abstract. Recently, crowd-sourced information is used to produce and improve collective knowledge and community capacity building. Triggered by broadening and expanding access to the Internet and cellular telephones, the utilisation of crowd-sourcing for policy advocacy, e-government and e-participation has increased globally [1]. Crowd-sourced information can conceivably support government's or general social initiatives to inform, counsel, and cooperate, by engaging subjects and empowering decentralisation and democratization [2]. Crowd-sourcing has turned into a major technique for interactive mapping initiatives by urban or rural community because of its capability to incorporate a wide range of data. Continuously accumulated spatial data can be sorted, layered, and envisioned in ways that even beginners can comprehend with ease. Interactive spatial visualization has the possibility to be a useful democratic planning tool to empower citizens participating in spatial data provision and sharing in government programmes. Since the global emergence of World Wide Web (WWW) technology, the interaction between information providers and users has increased. Local communities are able to produce and share spatial data to produce web interfaces with territorial information in mapping application programming interfaces (APIs) public, such as Google maps, OSM and Wikimapia [3][4][5]. In terms of the democratic spatial planning action, Volunteered Geographic Information (VGI) is considered an effective voluntary method of helping people feel comfortable with the technology and other co-participants in order to shape coalitions of local knowledge. This paper has aim to investigate 'How is spatial data created by citizens used in Indonesia?' by discussing the characteristics of spatial data usage by citizens to support spatial policy formulation, starting with the history of participatory mapping to current VGI development in Indonesia.

1. Introduction

Initiation concept of participatory research emerged in the era of 1960s, with the proliferation of widespread social repression as a result of the Second World War in the West (viz. Europe). It triggered social studies to involve the community in overcoming the problems of urban and regional issues at the time. Furthermore, the emergence of Marxist influence to overcome the influx of social justice issues in the social sciences in the 1970s encourage the development of research methods aligned to marginalized communities. One of the breakthrough social research methods was a participatory research initiated by Freire, who use the method of dialogue between elites and citizens in overcoming problems in marginalized communities.

Participatory research is one of the breakthroughs in social research that see knowing subjects as capable of making their own knowledge about themselves [6]. The implication of this approach is the use of dialogue in research methods to support communities to take control of their own lives and change repressive conditions. In this sense, the method of citizen participation from participatory mapping to Volunteered Geographic Information (VGI) can be understood as a process of political dialogue between the communities being affected by government planning policies and the government making the policies.

Currently, the growth of the internet has supported various human activities, including demands for information related to spatial analysis and more specifically, provides opportunities for spatial policy formulation in cooperation with publics wanting more control over their own spaces. The needs of citizens and governments increasingly demand accurate data and information.

Citizens as actors in governance for the implementation of spatial planning has a role in developing, using and managing selected target planning regions in the government agenda. In Indonesia, the Geospatial Information Act of Law No. 4 of 2011, Article 23, Paragraphs 1 and 4, and Law No.26 of 2007, Article 65 on spatial planning provide an opportunity for participatory-based approaches to the provision of spatial data and information to achieve spatial planning policy goals.

A variety of community participatory-based activities have direct effects on government policies and programmes. In 2004, the Asian Development Bank (ADB) presented a comprehensive report on public participation in decision-making processes and argued that the participatory approach in the implementation of government programmes is efficient and accommodates social aspirations at low-cost but with positive outcomes for communities [7].

There have been many studies by GIS and planning scholars of community-based spatial data management and usage for regional planning purposes [8][9][10][11]. In general, their research shows that communities affected by government planning and development agendas are helped immensely by participatory mapping approaches that assist in creating political dialogue with decision-makers. At the same time, from the perspective of the political elites, governments receive valuable input to align community needs with the realisation of the government development agenda.

In Indonesia, a variety of community-based research approaches to spatial data usage, from paper-based mapping to GIS, GPS and WebGIS digital-based methods, have been applied to various social problems [12][13]. This paper has aim to investigate 'How is spatial data created by citizens used in Indonesia?' by discussing the characteristics of spatial data usage by citizens to support spatial policy formulation, starting with the history of participatory mapping to current VGI development in Indonesia.

2. Methods

This research predominantly used qualitative methods with specifically approach was discourse analysis using legal document analysis and in-depth interview to participatory mapping and VGI coordinators. The legal document collections were needed to gain a comprehensive understanding of spatial information built by citizens in spatial planning process. Furthermore, the interviews were used to understand the perspective of the government and planning stakeholders in terms of their current strategy and initiatives for promoting spatial planning process from a political and social background.

An in-depth interview activity is undertaken to get information about the acquainted and substantial spatial information built by citizens the spatial planning process. In this case, the respondents explained their current working activities and experiences using Geographical Information Systems (GIS), including their expectancy and threat towards spatial data sharing amongst institutions. In order to facilitate the process of interview and to ensure all information are secured, note taking with the help of digital voice recording has been applied and later transcribed. Furthermore, the legal document collections were used to serve the purpose of providing a fundamental background of the policy context to gain a comprehensive understanding of the spatial data development and sharing in spatial planning formulation that has been done and will do in the future.

3. Research Results

This section discusses the characteristics and development of spatial data usage by citizens to support any activities related to spatial planning processes in Indonesia. The discussion begins by exploring the history of spatial data usage through the printed spatial visualization (viz. map) by indigenous people, known as participatory mapping. Understanding the historical background of participatory mapping provides knowledge of the beginnings of spatial data use by citizens. The discussion is continued by examining the role of participatory mapping in spatial planning advocacy in Indonesia.

Entering the open data and internet era, spatial mapping methods have changed from paper-based to digital spatial data formats. Now many people produce and share spatial data voluntarily to achieve specific goals. Thus, participatory mapping that was originally limited to communities living in a particular region, now through digital mapping, can be created and used by communities from other areas as well. This is called crowd-sourcing geographic information, which in this research involves VGI. Research into VGI for this paper will be discussed after exploring the history of participatory mapping in Indonesia

3.1. Historical background of participatory mapping in Indonesia

Participatory mapping initiation in Indonesia to be carried out by indigenous people, communities, and the Non-Governmental Organisations (NGOs) for the advocacy in spatial planning for their land when it was proposed for commercial development. Many indigenous people and communities have used participatory mapping methods to negotiate with planning actors, such as the government and businesses. Planning advocacy through participatory mapping in Indonesia has been strengthened, especially in relation to the increasing number of spatial impact of policies contained in local spatial plans (RTRW), either at the province or regency/municipality levels.

The participatory mapping activities related to spatial planning in Indonesia can be identified as two types; firstly, mapping activity for advocacy zone management between communities; and secondly, mapping activity for advocacy zone management amongst indigenous or rural communities with over development agendas in their region.

The mapping activity for advocacy zone management can be defined that citizens participates and engage to keep their claim territories regarding natural resource management through participatory mapping [14]. The map from participatory mapping result can be used to as a media to presents issues regarding their areas to other parties (e.g. government, investor, developer) for land right advocation purposes [14].

An example of mapping in relation to advocacy zoning mapping can be seen in a case study of participatory mapping on Pahawang Island (See Figure 1). In this area, the local community used the map to build consensus between traditional fishermen and modern fishermen in relation to fishing activity. The fishing activity areas mapping presented an agreed upon between traditional and modern fishermen only allowed to use the conventional method (e.g. using a fish hook) and not using trawlers, bombs and potassium cyanide concerning fishing activity in maintaining the sustainability of natural resources [14].

The impact of participatory mapping was that the use of trawlers, bombs and potassium cyanide by modern fishermen were vanished. As a result, fish numbers in the areas surrounding Pahawang island increased, and traditional fishermen are able to continue to fish. In addition, Indigenous society on Pahawang Island has used maps as a basis for the formulation of village regulations, peraturan desa (Perdes) formulation [14].

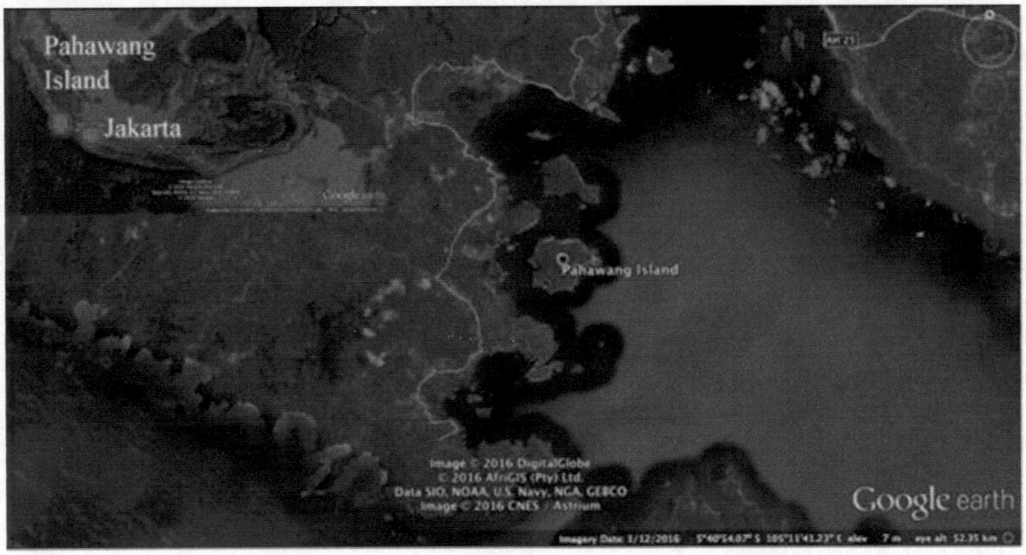

Figure 1. Pahawang Island geographical location.

An example of participatory mapping activities in relation to the management of territory in local community relations with the government is Katu village territory (See Figure 2). The map created by the Katu Village community through participatory mapping was used as a basis for proposed activities in the National Programme for Community Empowerment, *Program Nasional Pemberdayaan Masyarakat* (PNPM) and the Accelerating Development of Disadvantaged Areas Programme, Program Percepatan Pembangunan Daerah Tertinggal dan Khusus (P2DTK) [15].

Figure 2. Katu Village geographical location.

The participatory mapping provides an inventory of indigenous peoples' resources and living territories, so that, local spatial knowledge can be recorded. According to Sabu (2009), the implementation of participatory mapping in Indonesia has changed local community perceptions and patterns of natural resource management in four aspects, namely political, social, economic and cultural aspects [16].

3.1.1. The social aspect. Participatory mapping can changed the way a local society thinks about the area where they live in some of the following ways:

- Indigenous people have local spatial knowledge about their local resources and conditions within their areas and understand the various threats faced;
- Maps have the ability to identify the community involved in territorial management;
- Participatory mapping can contribute to creating solidarity within communities / indigenous peoples [16].

3.1.2. The economic aspect. Because indigenous people can produce their own daily food needs from harvesting their own agricultural products, maps produced by local communities are able to identify local natural resources including those valuable to that particular community. As a result, the fulfilment of their daily needs becomes more secure [16].

3.1.3. The cultural aspect. Participatory mapping helps indigenous people to identify, preserve and develop traditional customs and habits in managing the area handed down by their parents from their ancestors. Area management customs and organisations can be explained and illustrated through participatory maps. For example, zones for indigenous forests can be identified, areas considered to have religious purposes can be replanted with plants having significant symbolic value, such as the banyan tree. Participation in the determination of the zones in question allows local wisdom to contribute to the organisation and management of their territory [16]. However, participatory mapping for spatial planning activities does not always run successfully. Problems have occurred due to indigenous people being sceptical of the participatory mapping method. The next section discussed some of these impediments to participatory mapping section.

3.2. The Impediments of Participatory Mapping Activities Mapping in Indonesia

At the time of the introduction of participatory mapping approaches to solving social problems, local societies had at first been sceptical of this method for the management of their territories. It took a long time and commitment in supporting local society for participatory mapping to become accepted. Sometimes, the benefits can appear only after mapping activities have been completed.

In practice, participatory mapping approaches, from promotion to the ratification of the map, are still not optimal in implementation. Derived from Jaringan Kerja Pemetaan Partisipatif (JKPP), The Indonesian Participatory mapping network organisation study in 2009, constraints on the implementation of participatory mapping in Indonesia can be summarized as follows:

- Participatory mapping was of only temporary interest;
- Participatory mapping budget was still predominantly derived from NGOs;
- The idea of participatory mapping was only known by certain local community actors;
- Maps were only put into use when indigenous peoples were in conflict. [17]

Generally, this research indicates that participatory mapping in Indonesia came about because of the background of agrarian conflict in indigenous communities living in forest regions that intersected with government and commercial interets. Maps are used to claim ownership of an area: for instance, land certificates are issued by the Indonesian National Land Agency (BPN) with maps attached that illustrate individual's tenures and timber concession to companies issued by the Ministry of Forestry [17].

But many forest concessions issued by the Ministry of Forestry were generally located on land that was controlled and managed by indigenous societies who were then able to get bans placed on businesses in areas claimed by timber companies. The situation got worse when the companies claimed land ownership with an attached a map of the forest concession from the government and threatened indigenous groups with intimidation and violence. In other words, spatial information visualization has economic and political functions that can have serious effects on the socio- cultural aspects of the lives of indigenous people.

Participatory mapping relying on paper-based methods has created problems for management and storage. With the fast development of digital technology, participatory mapping methods have changed from paper-based to digital format, and terminologies changed with Participatory GIS (PGIS).

Participatory mapping and PGIS limited to the community level led to limited opportunity to integrate with other spatial data producers, thus spatial data management by indigenous societies, urban communities and government is difficult to integrate. In the light of this situation, in the 2010s, Indonesia introduced a digital-based participatory mapping system, or VGI. The system has capabilities to produce spatial data voluntarily with extensive community outreach beyond their living regions. The metamorphosis of participatory mapping from paper-based, to a VGI system is discussed in the next session.

3.3. From Participatory Mapping to VGI Activities in Indonesia

Interactive internet usage is growing rapidly with the next-generation Web 3.0 which allows not only the interaction between two or more internet users and information providers, but also sharing files. These developments along with facilities for data and information sharing have changed the outlook on the production and use of data, since now the user an also be involved as a data and information producer and provider (see Figure 3).

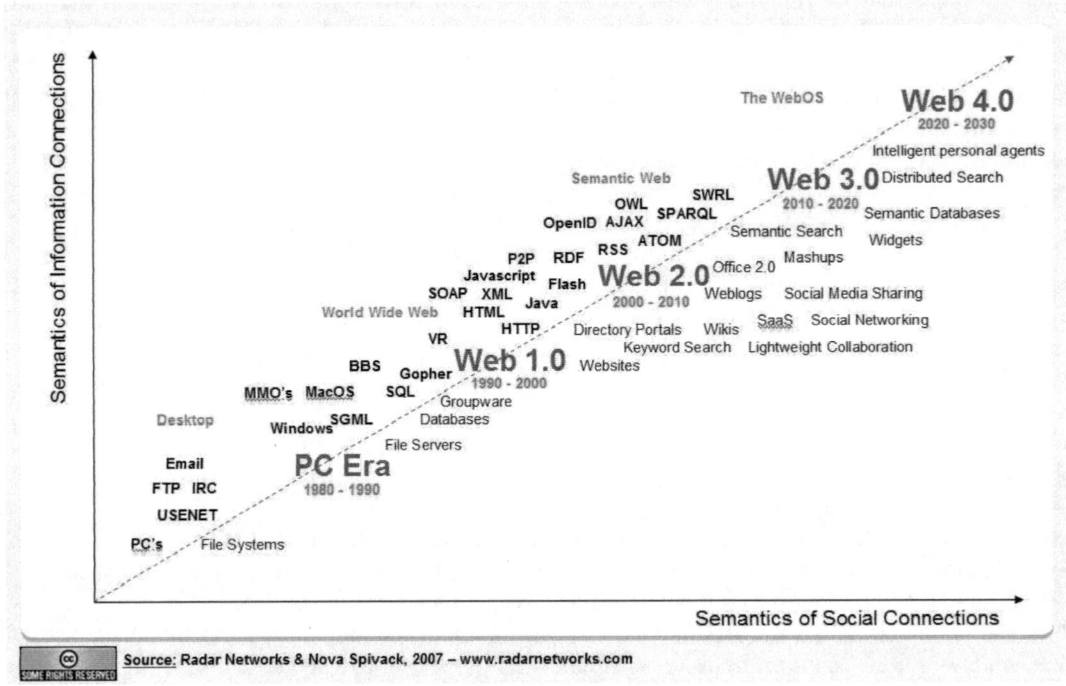

Figure 3. Internet generation era.

In this context, citizens can create and share spatial data and place their territorial observations on public web mapping application programming interfaces (APIs) such as Google Maps, WikiMapia, OSM and Microsoft's Virtual Earth [3][4][5]. Globally, in the 1990s, spatial data production and provision was dominated by national mapping agencies. However, advances in mapping technology and support of the internet provide GIS open source data for the production and procurement of spatial data or information. Since the 2000s, spatial data providers outside government agencies have been widely established to produce and provide spatial data development and sharing services amongst their communities. The development of GIS from a stand-alone application to crowd-sourced geographic information be seen in Figure 4

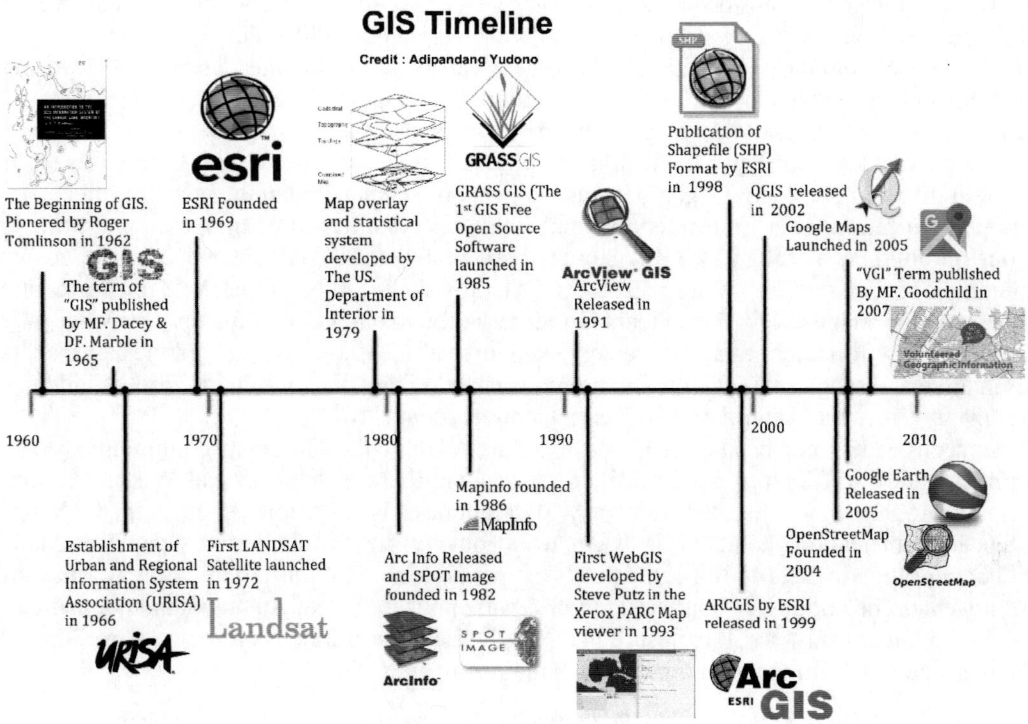

Figure 4. GIS Timeline.

Generally, spatial data providers outside government can be divided into two groups: commercial spatial data providers and open public spatial data providers. The first group produces spatial data to be sold to the public. This data is produced by persons with cartographic expertise and having high quality products. The spatial data generated are encrypted and cannot be re-processed. Spatial data providers in this category include Google maps, Navtech, TomTom and Tele Atlas.

The second category of spatial data provider provides a platform for building spatial data which can be shared amongst communities. Unlike the first group, this group consists of people from various professions with no cartographical abilities but who contribute to the building of spatial databases. The open source nature means there is no charge for downloading data and data can be re-edited and re-added. In other words, this category of spatial data provider is VGI. And include groups like Wikimap, OSM and Waze.

3.4. VGI initiatives in Indonesia

Global voluntary spatial data production has been implemented since 2006 after participatory GIS open source platform released, such as OpenStreetMap (OSM) which were established in 2004 (See Figure 4). One of the initial VGI activities in Indonesia was a Green Map Program which was a map generated by local communities to map the potential of natural and cultural resources of the communities' living places. The Green Map charted all places and phenomena, whether positive or negative and aimed to help people see, judge, connect, and care about the environment where they are located. The Indonesian Green Map is a part of the World Green Map organisation and can be viewed at: http://www.opengreenmap.org/greenmap.

Participatory mapping was used for informal or recreational pursuits such as mapping restaurants at: https://www.google.com/maps/d/viewer?mid=1Ok41gwEcc0oW0Nu-D0LKF5wApPQ; outdoor activities at: http://www.navigasi.net and tourist attractions at:

https://www.google.com/maps/d/viewer?mid=19cYg4A1S0xD-dXENVYbQ-pnP1d0&hl=en; and traffic conditions using Waze platform at: https://www.waze.com/id/livemap.

These kinds of voluntary GIS activities have been carried out in Indonesia since OSM was released in 2004. Spatial data created by earlier VGI communities was more aimed at displaying targeted spatial objects on the WebGIS platform or putting up geo-tagged objects for personal interest rather than building a geo-database to re-use for various purposes. This section discusses VGI focuses on the GIS platform in the form of building a geo-database to support the spatial planning application.

Drawing on information from interview and field observations, an open source platform with the function of building a voluntary GIS geo-database and spatial data created and shared amongst communities concentrated on Wikimap and OSM providers. Study of the VGI characteristics and developments on both these VGI providers in Indonesia found that OSM is the only provider which has management based in Indonesia. Conversely, Wikimapia does not have an Indonesian management representative and spatial data created by the Indonesian Wikimapia community directly interacts with central management that control by two Russian entrepreneurs [18].

This means it has not been possible to examine VGI in the Wikimap community because the researcher found it difficult to assess official reports and there was no official Wikimapia Indonesia management to interview. Thus, the study of VGI in Indonesia is only represented by the OSM provider.

OSM implementation in Indonesia was carried out by the Humanitarian OpenStreetMap Team (HOT) organisation in 2011 to fulfill the needs of spatial data and information before, during and after disaster events. The initiation of Indonesian OSM was sponsored from Australia and the main aim was to carry out natural disaster risk exposure mapping in the eastern Indonesia region (See Figure 5). As one of the senior HOT Indonesian organisers explained in an interview:

> *"The entry of VGI activities in Indonesia through OpenStreetMap started from HOT International which aims to help humanitarian issues. Viewing Indonesia's geographical position in disaster prone areas and supported by limited spatial data availability, HOT International entered Indonesia in 2011 to build a spatial database base map with an OSM platform. In May 2011, HOT Indonesia began a pilot project in the form of Community Mapping for risk exposure of natural disasters in Indonesia under the Australian Community Development and Civil Society Strengthening Scheme (ACCESS) project. Social mapping was first focused on eastern Indonesia."*

(HOT Indonesia Senior Management member: Interview at 6th February 2015)

Figure 5. Indonesian Western and Eastern Regional Divisions.

The initial activities implementing VGI in Indonesia and carried out in June 2011 include:

- Conducting trial-error experiments using the OSM platform to be adopted in rural and urban areas in Indonesia
- Identifying the need for tools and materials for subsequent mapping of spatial objects transferred to the OSM platform

- Ensuring OSM application can be useful for the assessment of contingency planning and emergency situations. [19]

Indonesian OSM activities were based on annual projects in which initial activity was to conduct digitization of spatial data on buildings, roads, bridges and other spatial infrastructure objects in urban and rural areas, to support the activities of the government in disaster management. The evaluation of OSM activities to build geo-database for disaster management by the central government showed it could improve central government performance in relation to identifying catastrophe victims. Since 2012, the Indonesian OSM has worked in cooperation with the Indonesian National Disaster Management Agency, Badan Nasional Penanggulangan Bencana (BNPB) because BNPB lacked spatial data management experience in identifying and quantifying the impact of disaster events.

In supporting operationalization in the field, BNPB required spatial information visualisation in large-scale maps (1: 500, 1: 1000; 1; 5000; 1: 10,000). But Badan Informasi Geospasial (BIG), as the official government institution for creating and providing spatial data, could not meet the needs of the large-scale spatial data. But at the time of the tsunami and in the post-disaster period, BNPB needed spatial data for disaster management quickly. Constraints on supplying large scale spatial data were further studied and followed up by foreign aid from Australia through the Australia-Indonesia Facility for Disaster Reduction (AIFDR). AIFDR work with HOT international because of their experience in managing community building geo-databases using open source GIS platform [19].

Having outlined the historical background to the initiation of VGI in Indonesia, the following discussion will examine current VGI development in Indonesia. This is followed by a discussion of the experience of spatial data usage built by Indonesian VGI communities in relation to activities that interact with spatial planning.

3.5. Current VGI Development in Indonesia

Based on field observations for this research, the development of VGI activities in Indonesia growing rapidly, especially for geo-tagging activities for hobbies like travel, restaurants, locating traffic congestion, hiking and cycling route tracking. VGI activities building geo-databases for updating the latest spatial data and information with GIS-based open source platforms such as WikiMapia and OSM are slowly but surely showing an increase in the number of objects mapped across the globe. As stated previously, due to the limited access to information from WikiMapia provider, this research has taken the development of OSM as representative of current VGI development in Indonesia.

According to the HOT Indonesia (2012) project report, in the OSM pilot study of implementation between June 2011 to March 2012, 163.912 buildings were mapped. HOT Indonesian's 2013 Annual Report records that 1.3 million buildings had been mapped. In other words, nearly four times the number of buildings were mapped in the period 2012-2013, then at the beginning of OSM operations (Figure 6). The number of buildings mapped by May 9, 2016 was 4,238,869 (Figure 7)

Figure 6. Time series of the number of buildings mapped in Indonesia through the openstreetmap platform.

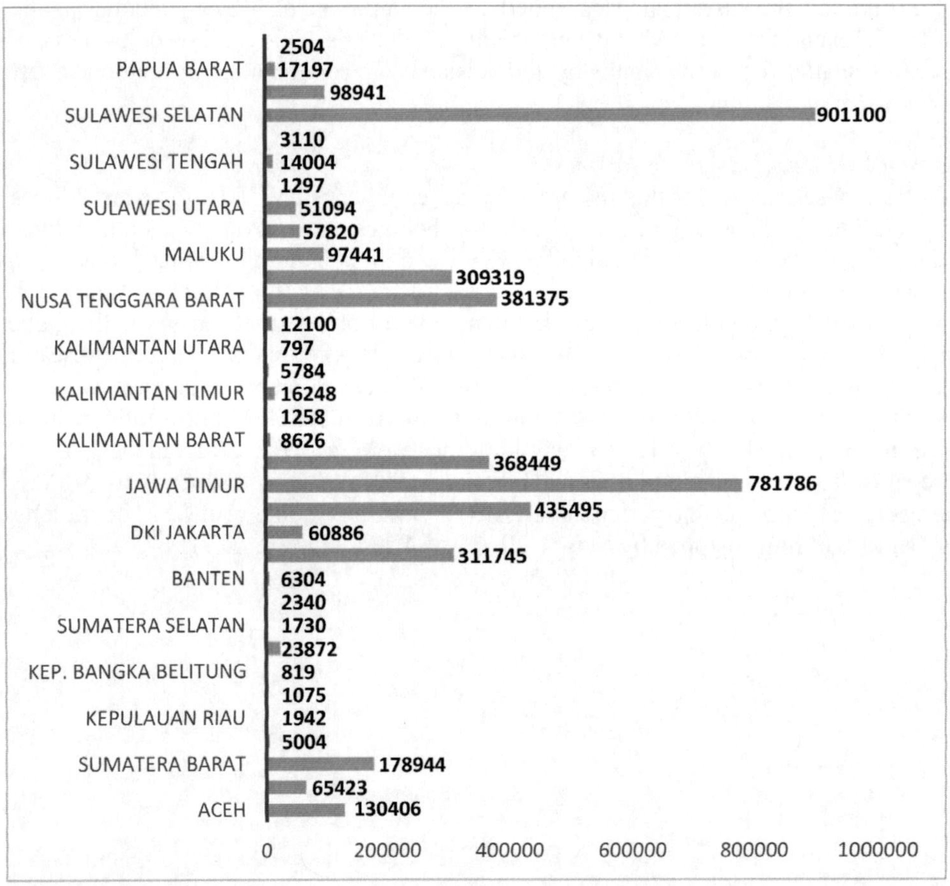

Figure 7. Statistics of the number of buildings mapped in the territory of Indonesia by 9th May 2016.

Strategies have been undertaken by HOT Indonesia for training OSM communities, including: basic and advanced training in building geo-databses in OSM; mapathons (face-to-face groups for building spatial object mapping); and OSM mapping and consultations through social media such as Facebook and Twitter. In an effort to increase the digitization of spatial objects mapped, HOT Indonesia held a mapping competition with rewards for members who mapped the greatest number of spatial objects.

In addition, other efforts at OSM community training were undertaken by HOT Indonesia. These included promotions to recruit new members by cooperating with government officials, schools and universities as well as civil society organisations such as the Scouts (Pramuka) and the Indonesian Red Cross (Palang Merah Indonesi (PMI)).

Building the geo-database and training and recruiting new members of the OSM community continue to be effective, but according to interview respondents, voluntary digital mapping in Indonesia is still largely project-based even though it is the communities' own initiatives to identify the objects to be mapped in their areas. When the project was over, slow or stop participatory digital mapping progress occurred, due to no supporting budget for participatory mapping community empowerement. Also, less or no supporting budget will affect the awareness of local community to keep producing and updating spatial data, exception for particular persons who interest to produce and update spatial data and information. This is an existing deficiency the appropriate pattern in building awareness of digital mapping for communities of their own territory has not found yet.

3.6. The significant VGI actors to achieve success building interactive digital mapping community empowerment in Indonesia

The significant VGI contribution to build spatial database in Indonesia is owing to the substantial VGI actors role in encouraging communities to map particular areas on the open source GIS platform. Shkabatur (2014) argues that there are four groups of actors needed to achieve successful of interactive community mapping action globally: external interactive community mapping experts, local civil society organisations, local community members; and local public officials[21]. However, drawing on observations, interview and HOT internal reports, the researcher identified five main actors for creating successful VGI activities in Indonesia: international funders; HOT administrators/VGI organisers; local government; local organisation partners; and local mapping contributors.

3.6.1. International funders. Literally, Shkabatur (2014) mentions that international donors give attention regarding financing interactive community mapping operationalisation. However, she was not explicitly say that international donors become one of significant actors to get the success of digital community mapping activities. The different argument with Shkabatur's idea, the researcher argues that international funders has a substantial role regarding VGI programmes in Indonesia. According to fieldwork and interview to one of the senior HOT Indonesian members, International funders or donors do not necessarily contribute to VGI initiatives, but their primary contribution is significant in creating a good atmosphere between governments and citizens.The case study of early OSM projects in Indonesia, the Australian Agency for International Development (AusAID) and the World Bank have supported local community mapping activities to fill the lack of official spatial data and information. International funders have a critical part in bringing government on board, guaranteeing active backing for the undertaking, and organising and influencing the performance of all relevant partners. As the active engagement of government authorities bolsters the long-term producing and provision of official spatial data or information, it is imperative to guarantee coordinated efforts from the beginning of the VGI venture.

3.6.2. VGI organisers. The generation of spatial data by a VGI community is ordinarily encouraged by citizens and VGI organisers. These organisers regularly have extensive involvement in configuring and creating VGI. HOT is one of the VGI organisers that works with the OSM platform and frequently initiates VGI procedures, endeavouring to extend their aptitudes and abilities in new areas. They normally contact a local citizens partner to gain an understanding of the needs and capacities of local communities and work with them on the outline and usage of the VGI procedure. These organisers then lead the VGI procedure, preparing group mappers to utilize mapping innovations, helping them to collect spatial data and information, and delivering relevant maps with appropriate symbols (legends) on the basis of the data and information collected. In collecting spatial data in particular areas, VGI organisers will work in a structured way to obtain high spatial data quality. In general, VGI organisers include data entry specialists, quality assurance specialists and team supervisors. Data entry specialists are selected by recruiting local people who understand local geographic conditions of the area being mapped. Communities in this category do not require high skills in mapping or cartography, but their contributions will be subject to monitoring by Quality Assurance Specialist. A quality assurance specialist is a VGI organiser responsible for selecting the quality of data that is mapped from a group of data entries. Members of this group are usually selected from local public figures who have knowledge of mapping and cartography. Lastly, the team supervisor is the highest position in the mapping structure, and has the role of final quality assurance before submitting digital spatial data to the global VGI portal (See Figure 8).

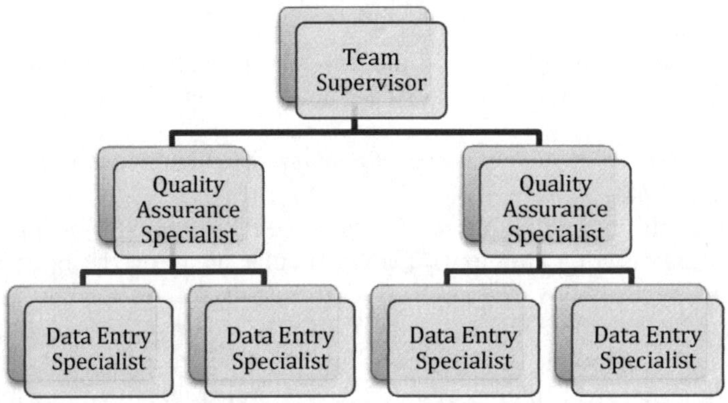

Figure 8. VGI organisational structure for spatial data collection in OSM procedure.

3.6.3. Local government. Government advocacy of the VGI procedure and the cooperation of local authorities with VGI organisers, local organisation partners, and local community mappers are vital to secure the achievement and effects of VGI activities. Active government engagement to community increase opportunity that the resulting interactive mapping will be consistently guided to develop public services and other government programmes to community. Government responsibility for guiding public participatory in spatial planning may likewise guarantee support for the mapping procedure, increase the motivations of local inhabitants to take part in it, and enhance the financing of VGI. However, while the three partners— VGI organisers, local organisation partners, and local community mappers —are consistent elements of all VGI activities, the part of neighbourhood government and legislators shifts markedly from one VGI activity to the next. Organisation, leadership and political settings play significant parts in achieving VGI outcomes.

3.6.4. Local organisation partners. Usually, independent VGI organisers work with local communities for a limited period, helping them to make volunteered geographic data and then leaving after the project contract is over. As VGI organisers are not embedded into the life of the areas being mapped, they have to work together closely with local organisation partners who are usually social groups and social activists who live and work in the area, for example, the Scouts (Pramuka), Red Cross (PMI) or local youth associations (Karang Taruna)—who serve as the contact point for VGI organisers with the mapping group. Participation between VGI organisers and local organisation partners is imperative in all phases of the VGI procedure. To start with, local organisation partners, local governments, or citizens activists can identify data needs and requests of the community and offer direction as to implementation in the specific local setting. At that point, local organiser partners can help by connecting with, and assembling, the community to participate in the VGI procedure, sorting out community discussions, activating enthusiasm to create spatial data or information on the open source GIS platform, recruiting mapping contributors, and supporting them through the mapping procedure. After fulfillment of the procedure, local organisation partners can serve as its "hosts," guaranteeing the utilization and further improvement of voluntary digital mapping guides.

3.6.5. Local mapping contributors. Like conventional community mapping, the centre of the volunteered geographic data procedure is the engagement of local inhabitants. The VGI procedure should give local inhabitants important specialized aptitudes, help them to speak to their communities and to the outside their areas, and increase their voice in regions that matter to them. While local occupants sometimes initiate the mapping procedure, more frequently, VGI is a supply-driven process, presented and supported by VGI organisers and local organisation partners. As already discussed, finding the right motivations is a testing assignment, as poor groups frequently do not directly benefit from the mapping exercise and cannot afford to volunteer for the undertaking without getting paid. Facing the digital mapping problem, the researcher argues that building awareness of the benefits of geo-database built by community participation can be based on the theory of 'communities of practice' [22]. The community is aware of the benefits of community mapping for others outside the region and this contributes to establishing an external identity for the region. Each community member has a psychological connection to their homes, community and place, and benefits when these are recognized externally. In other words, public awareness of the importance of updating spatial data voluntarily through digital mapping occurs automatically when a community is able to show that mapping spatial data provides learning and helps form community identity. In the long term, community willingness to engage with updating spatial data. The researcher argues that if a community is homogeneous – for example, an indigenous community in a particular area – then it is more likely to rapidly accept participatory mapping in a VGI context. However, if a community in a particular place is heterogeneous – for example, indigenous and immigrants in the same territory - it needs intensive communication over a relatively long time period to achieve cooperation in performing consistent updating of digital mapping. VGI provides great opportunities in describing spatial visualization for spatial planning applications through the provision of detailed and up-to-date spatial data for particular areas. However, as noted previously VGI activity is still being conducting on a project-by-project basis. It is feared after the project is over, the consistency to perform spatial data updating does not happen. In addition, citizen awareness of the ability to contribute to spatial data production still lacking. As one of the senior HOT Indonesian organizers explained in an interview:

> *"Digital community mapping in Indonesia so far, is still on a project basis, and local community awareness of spatial data production participation is still very low. Until now, HOT Indonesia still has not found the right pattern in building a minimum awareness of digital mapping to a minimum for its territory. It could be expected,in the future, awareness of mapping will be embedded in each individual in the community so that they will engage in intense mapping or spatial data production."*

(HOT Indonesia Senior Management member: Interviewed at 6th February, 2015)

According to the interview, HOT Indonesia still has problems in finding a suitable method for building awareness of community digital mapping. With this in mind, the researcher attempts to explore the willingness of individuals in particular communities to participate in specific activities with the purpose of achieving community goals. The discussion in the next section explores community empowerment theory.

3.7. The role of community empowerment through VGI activities to support spatial planning formulation

One of the local community involvement methods in spatial policy processes is collecting data and information about their local areas. In this context, VGI becomes one these methods for data collection.

With the limited resources available, communities need decision-making based on understanding the importance of sustainability and the conflicts of interest that exist between the quality of life of a community (neighborhood); social, economic, environmental assets; and the potential benefits for different stakeholders in the decision taken. In principle, there must be a less distance bureaucracy between government and citizens at the level of participatory decision-making, but it also increases the intensive communications amongst stakeholders, including researchers, experts and policy makers.

Academicians can act as facilitators helping to identify local community needs as well as mapping local social issues through collecting data and analyzing the information on the VGI platform. At a later stage, their knowledge of local communities, the kinds of additional data needed and their ability to access data is a crucial contribution in a community.

In terms of citizen participation, the following kinds of spatial data and information can be collected to support the spatial planning process:

- Community inventory and asset evaluation;
- Social resources;
- Infrastructure resources;
- Environmental resources;
- Culture / history;

Currently, GIS applications are used extensively by citizens, individually or at local community level for involvement in spatial planning processes (these activities include collecting data, mapping, analysing and decision-making).

Participation in digitalized spatial data can be categorized in four degrees or intensities: information sharing, consultation, involvement, action.

1. Information Sharing
   One or two-way communication between 'outside' and local community involvement
2. Consultation
   Consultation is more like mentoring, directed to the key problems and the kinds of training needed by the community.
3. Involvement in decision-making by all actors
   Interaction between internal and external actors to identify priorities, analyze conditions, select alternatives and tools.
4. initiating Action
   Independent initiatives driven by local communities who mobilize themselves to support relevant activities. [23]

This does not imply that participation must always be of maximum intensity, but the intensity should be appropriate to the task, competence and the specific relationships between the actors participating in the spatial planning process.

The role of VGI in the spatial planning process has been examined by McCall (2004) who argues that the role of VGI as as a voluntarily mapping activity depends on the type of VGI community itself. The researcher argues that a local VGI community does have a direct impact on the policies made by the government, so spatial information visualization products created by a VGI community are more like the sharing information and consultation level identified by McCall. Communities directly affected by planning policies decided by the government are involved at Levels 3 and 4 - involvement in decision-making by all actors; and initiate action respectively.

4. Conclusion

This paper has aim to investigate 'How is spatial data created by citizens used in Indonesia?' by discussing the initiation, characteristics and development of public participation involved in the spatial data production and provision in Indonesia. Community participation in spatial data provision in Indonesia has various advantages as: an advocacy tool to gain recognition for local areas through negotiation with government; activities to identify spatial objects for the purposes of disaster management; and the fulfillment of the requirements for informal or recreational information such as cycling, tourist sites or restaurants.

A variety of participatory community-based activities with direct impacts on government policies and programmes in the planning of specific areas are the objects of current participatory research. In 2004, the Asian Development Bank (ADB) presented a comprehensive report on public participation in the decision-making processes which showed that participatory approaches in the spatial policy process provides efficiency and can accommodate all local community aspirations at low cost but with highly positive outcomes.

Many people are now voluntarily creating and sharing spatial data for specific purposes. So participatory mapping that was originally only accessible to limited communities, currently through interactive digital mapping, is also available to other communities outside of their areas so that they can map other regions.

References

[1] Shirky, C., 2008, Here comes everybody: The power of organizing without organizations. Penguin Press, New York.

[2] Bott, M., Gigler, BS., Young, G., 2014, The Role of Crowdsourcing for Better Governance in Fragile State Contexts In Gigler, BS., and Bailur, S., (eds) *Closing the Feedback Loop – Can Technology Bridge the Accountability Gap?*, The World Bank, Washington, D.C., The USA, pp.107-146

[3] Rouse, L. J., Bergeron, S. J., and Harris, T. M., 2007, Participating in the geospatial Web: Collaborative mapping, social networks and participatory GIS. In Scharl, A and Tochtermann, K. (eds.), *The Geospatial Web: How Geo-Browsers, Social Software and the Web 2.0 are Shaping the Network Society,* Springer-Verlag, London. pp. 153–158.

[4] Goodchild, M.F., 2007, Citizens as Sensors: The World of Volunteered Geography, in *VGI Specialist Meeting Position Papers*. 2007: Santa Barbara, CA.

[5] Tulloch, D. L., (2007), Many many maps: Empowerment and online participatory mapping. *First Monday*, Vol.12, No.2 at http://firstmonday.org/ojs/index.php/fm/article/view/1620/1535 accessed 26th January, 2016.

[6] Freire, P., 1971, *Pedagogy of the Oppressed*, Continuum, New York

[7] Asian Development Bank (ADB), 2004, Effectiveness of Participatory Approaches: Do the New Approaches Offer an Effective Solution to the Conventional Problems in Rural Development Projects? *Special Evaluation Study by the Asian Development Bank,* December 2004.

[8] Roche, S., 2014, Geographic Information Science I: Why does a smart city need to be spatially enabled?, *Progress in Human Geography*, Vol. 38, No.5, pp. 703-711.

[9] Roche, S., Zimmermann, EP., Merickskay, B., 2013, GeoWeb and crisis management: Issues and perspectives of volunteered geographic information, *GeoJournal*, Vol. 78, pp.21-40.

[10] Heipke, C., 2010, Crowdsourcing geospatial data, *ISPRS Journal of Photogrammetry and Remote Sensing* Vol. 65, pp. 550- 557.

[11] Haklay, M., and Weber, P., 2008, OpenStreetMap: User-Generated Street Maps, *IEEE Pervasive Computing*, Vol. 7, No. 4. pp. 12-18.

[12] Aditya, T., 2010, Usability Issues in Applying Participatory Mapping for Neighbourhood Infrastructure Planning, *Transactions in GIS*, Vol. 14 (S1), pp. 119-147.

[13] Mustofa, FC., Aditya, T., *Sutanta, H., 2014,* Sisten Informasi Geografis Partisipatif (PGIS) untuk menuntaskan pemetaan bidang tanah: peluang dan tantangan (Participatory Geographic

Information Systems (PGIS) to Complete the Mapping of Land Cadaster: Opportunities and Challenges, in Sutanta (ed), *Menuju Pengelolaan Informasi Secara Spasial: Fondasi, Evaluasi dan Implementasi* (Towards Spatial Information Management: Foundations, Evaluation and Implementation), Jurusan Teknik Geodesi, Fakultas Teknik, Universitas Gadjah Mada

[14] Rizani and Karim, M.S., 2009, Pemetaan Partisipatif: Sebagai upaya masyarakat melindungi dan mengelola sumberdaya alam pesisir laut di Pulau Pahawang (Participatory Mapping: As the community's efforts to protect and manage the natural resources of marine coasts on the Pahawang island) in Pramono, AH., Samperante, F., Safitri, H., Achmaliadi, R., (eds) *Menuju Demokratisasi Pemetaan: Refleksi Gerakan Pemetaan Partisipatif di Indonesia* (Toward Mapping Democratisation: Reflectinos on the Participatory Mapping Act in Indonesia , Jaringan Kerja Pemetaan Partisipatif (JKPP), Bogor, Indonesia

[15] Azar, 2009, Pemetaan Partisipatif: Mempertahankan Hak atas tanah leluhur, pengalaman orang Katu (Participatory Mapping: Retain Right to ancestral land, the experience of Katu). In Pramono, AH., Samperante, F., Safitri, H., Achmaliadi, R., (eds) *Menuju Demokratisasi Pemetaan: Refleksi Gerakan Pemetaan Partisipatif di Indonesia* (Toward Mapping Democratisation: Reflections on Participatory Mapping Act in Indonesia , Jaringan Kerja Pemetaan Partisipatif (JKPP), Bogor, Indonesia

[16] Sabu, A.S.S.S., 2009, Mengambil alih hutan lindung: Kasus Komunitas Adat Nian Uwe Wari Tana Kera PU Hikong-Boru Kedang, Nusa Tenggara Timur (Taking Over the Protected Forest: the Case of Indigenous Communities Nian Uwe Wari Tana Kera Hikong PU-Boru Kedang, East Nusa Tenggara). In Pramono, AH., Samperante, F., Safitri, H., Achmaliadi, R., (eds) *Menuju Demokratisasi Pemetaan: Refleksi Gerakan Pemetaan Partisipatif di Indonesia* (Toward Mapping Democratisation: Reflections on Participatory Mapping Act in Indonesia , Jaringan Kerja Pemetaan Partisipatif (JKPP), Bogor, Indonesia

[17] Safitri, H., and Pramono, AH., 2009, Refleksi atas gerakan pemetaan partisipatif dan tantangan dimasa depan (Reflection on the Participatory Mapping Act and Future Challenges). In Pramono, AH., Samperante, F., Safitri, H., Achmaliadi, R., (eds) *Menuju Demokratisasi Pemetaan: Refleksi Gerakan Pemetaan Partisipatif di Indonesia* (Toward Mapping Democratisation: Reflections on the Participatory Mapping Act in Indonesia , Jaringan Kerja Pemetaan Partisipatif (JKPP), Bogor, Indonesia

[18] Kersky, J.J., and Clark, J., 2012, The GIS Guide to Public Domain Data, ESRI Press, Redlands-California, USA

[19] Australia-Indonesia Facility for Disaster (AIFDR), 2014, *Expansion of InaSAFE and OpenStreetMap in Indonesia*, January 2014 – February 2015 Report, Jakarta, Indonesia

[20] Putten, K., and Akhadi, Y., 2013, *Pengembangan skenario untuk rencana kontijensi* (Scenario Development for contingency plans) Presentation in AIFDR Project on 16th November 2013.

[21] Shkabatur, J., 2014, Interactive Community Mapping: Between Empowerment and Effectiveness. In Gigler, BS., and Bailur, S., (eds) *Closing the Feedback Loop – Can Technology Bridge the Accountability Gap?*, The World Bank, Washington, D.C., The USA

[22] Wenger, E., 1998, *Communities of Practice: Learning, Meaning, and Identity*, Cambridge University Press, Cambrdige, UK

[23] McCall, M.K., 2004, Can Participatory-GIS Strengthen Local-level Spatial Planning? Suggestions for Better Practice, Paper presented *GISDECO 2004* on 10-12th May 2004 , Johor, Malaysia.

IOP Conf. Series: Earth and Environmental Science **70** (2017) 012003 doi:10.1088/1755-1315/70/1/012003

The study of environmental carrying capacity for sustainable tourism in Telaga Warna Telaga Pengilon Nature Park, Dieng Plateu, Central Java

Alexander Melat Aryasa[1] , Azis Nur Bambang[1,2] , Fuad Muhammad[1,3]

[1]Master Program of Environmental Science School of Studies Diponegoro University, Semarang, Indonesia
[2]Departement of Fisheries and Marine, Diponegoro University, Indonesia
[3]Departenemt of Science and Mathematics, Diponegoro University, Semarang, Indonesia

ale_melat@yahoo.com

Abstract. The increasing in quantity of the tourists visiting Telaga Warna Telaga Pengilon Nature Park, Dieng Plateau, Central Java, can cause a potential threat toward the conservation sustainability of the tourist attraction and the surrounding area. The utilization of conservation area for tourist attraction has to be carried out based on the principal of Environmental Carrying Capacity so that it will not affect the ecosystem. This study aims to determine the value of Telaga Warna Telaga Pengilon Nature Park environmental carrying capacity as a conservation area used for tourism activities. The environmental carrying capacities calculated in this study were physical carrying capacity, real carrying capacity, and effective carrying capacity. Results of this research show that the physical carrying capacity of The Telaga Warna Telaga Pengilon Nature Park was 31,302 visitors, the real capacity was 869 visitors/day and the effective carrying capacity was 579 visitors/day. Thus, the sustainable tourism development strategy is needed to manage the everlasting natural resources.

1. Introduction

Forest ecosystems in protected areas has the uniqueness of the landscape as well as the biodiversity [1]. As an attempt to preserve the nature, sustainability activities in tourism sector is needed, which is called sustainability tourism. The sustainability tourism will balance the positive and negative impact from tourism activities [2].

Tourism activities in a protected area are related to the characteristic and the congruence of the area as a destination for ecotourism. The right ecotourism planning and management are needed in order to decrease the negative impact caused by the tourism activities and to maximize its positive impact [3]. To preserve the environmental quality of the tourist object, integrated management effort has to be applied [4]. The ability of the environment to support human life, living creatures, and the balance between the two is called the environmental carrying capacity. Environmental carrying capacity studies are very crucial to be conducted. These studies are the preventive precaution measures to avoid the environmental destruction [5].

2. Literature Review

The basic principles of Sustainable Tourism is optimally utilize environmental resources while maintaining the ecology and conservation, respect for cultural authenticity and communities and ensure the long term sustainability [6]. Environmental Carrying Capacity of natural attractions is the ability of natural attractions in the area and a certain time unit to accommodate the number of tourists [7]. Tourism carrying capacity has an important role in the management of the area because it was considered as a systematic, strategic policy tool in the planning process [8].

Management of protected areas as a tourist spot is one effort to build a sustainable management. Development of tourism in the conservation area occurred from mass tourism to an individual or small group tourism. The development of this travel pattern focuses on the experience with the motivation to get the expansion of life [9].

Physical carrying capacity is the maximum number of travellers who are physically fulfilled by the space provided at a certain time. The real carrying capacity was the number of visitors allowed to visit a tourist attraction with a correction factor derived from the characteristics of the object that is applied to the Physical carrying capacity [10]. The effective carrying capacity was the maximum number of visitors received at a tourist attraction, was obtained from the value of the real carrying capacity that related to the capacity menagemen attractions manager [8].

3. Methods

The method used in this study was qualitative. Information was obtained through questionnaires. The carrying capacity of environment was determined by using the calculation of physical carrying capacity, real carrying capacity and effective carrying capacity. The questionnaires was filled out by visitors and the manager of the Telaga Warna Telaga Pengilon. The number of visitors was determined using the formula slovin with accuracy boundary of 10%. The research was located at Jojogan Village, Kejajar District, Wonosobo Regency, Central Java province. The location was one of the tourist attractions at the Dieng Plateau which managed by the Central Java BKSDA, the Ministry of Environment and Forestry.

3.1. Physical Carrying Capacity/PCC

Physical carrying capacity intended to find out the limit of maximum excursions by considering the satisfaction in travelling in a unit of time [10].

$$PCC = A.\frac{1}{B}.Rf \qquad (1)$$

Note:
PCC = Physical Carrying Capacity
A = Area for tourism activities
B = The required tourist area to obtain satisfaction in tourism activities

3.2. Real Carrying Capacity/PCC

Real carrying capacity was the result value of physical carrying capacity based correction factors that have been determined because it is considered influential in tourism activities. Some elements into the correction factor was vegetation, slope, landscape potential, rate of erosion and rainfall.

$$RCC = PCC.\frac{100-cf1}{100}.\frac{100-cf2}{100}.\frac{100-cfn}{100} \qquad (2)$$

Note:
RCC = Real Carrying Capacity
Cf = Correction Factor

3.3. Effective Carrying Capacity/ECC

The effective carrying capacity was the maximum value for the number of visits at certain locations by considering the management capacity and the real carrying capacity.

$$MC = \frac{Rn}{Rt}.100\% \tag{3}$$

Note:
MC = Management capacity
Rn = The Officer on duty
Rt = The officer required for on duty

The capacity management will affect the real carrying capacity, it's called the effective carrying capacity

$$ECC = MC.RCC \tag{4}$$

Note:
MC = Management Capacity
ECC = Effective Carrying Capacity
RCC = Real Carrying Capacity

4. Results

Telaga Warna Telaga Pengilon was located at an altitude 1965-2300 meters above sea level. Telaga Warna and Pengilon was geographically located between 7° 12 '3 "and 7° 13' 3" LS and the 109° 54 '47 "and 109° 55' 10" BT. The area was composed of ecosystems covered by plants such as *Acacia Sp.*, *Podocarous* sp., *Melaleuca* Sp., *Schima Sp, Casuarina junghuniana*. Relative Frequency value, relative density and relative dominance was as follows:

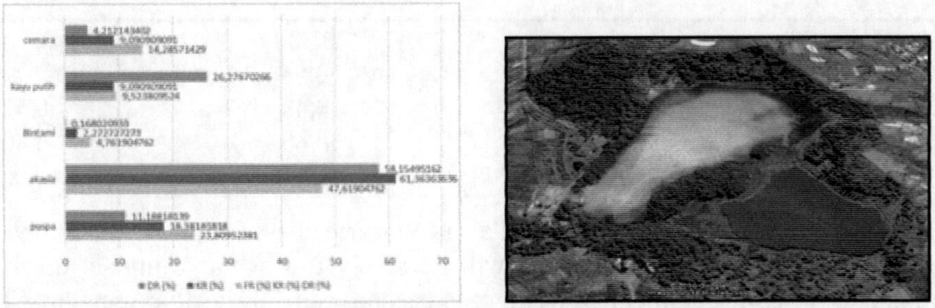

Figure 1. (a) Relative frequency, relative density and relative dominance; (b) Location of Telaga Warna Telaga Pengilon.

Table 1. The calculation of INP

Name	Attendance	Total individu	Total basal area [cm^2]	INP [%]
Schima Sp..	5	8	13381.53	53.18
Acacia Sp	10	27	69555.73	167.14
Podocarous Sp.	1	1	200.96	7.20
Melaleuca Sp.	2	4	31428.03	44.89

Casuarina junghuniana	3	4	5037.90	27.59
Total	21	44	119604.14	300

According to the table 1. Telaga Warna Pengilon has 21 types of trees that dominated by *Acacia* Sp. were classified as type mountain forests. *Podocarous* Sp. has the smallest attendance since rarely encountered in the research. The composition of mountain forests was one of the main factors Telaga Warna Telaga Pengilon was a nature-themed recreation. Visitors motivation in tourism activities were based on the physical motivation among others, to refresh their body and soul, rest for health, sports, recreation for fun [11]

Tourism activities were increased the number of visitors over the last few years. These trends were influenced by the nature-themed tourism, accommodation and the ease of access information such as internet information, newspaper on the objects that they visit. Telaga Warna Telaga Pengilon area was a challenging locations as destination activities and nature tourism activities in accordance with the natural surroundings. In the conservation area it was possible to obtain the knowledge so that it can be packaged into various forms of tourism activities [9].

Table 2. Total visitors the last 4 years

Year	Total Visitor [visitor/year]
2012	155.516
2013	210.372
2014	345.393
2015	388.521

The value of physical capacity shows that customer satisfaction will remain awake even if the visitors of Telaga Warna Telaga Pengilon raise into 31.302 people. The value of the physical carrying capacity was based solely on the area of conservation without considering other environmental factors. Biotic and abiotic environmental conditions are the limiting factor in the physical carrying capacity so that it can be regarded as the real carrying capacity of a tourist area.

Table 3. The correction factor value of environmental carrying capacity

Correction Factor	value Index	Value Factor Correcting
Simpson diversity index	0,586	0,414
Bureau Of Land Management index	0,56	0,44
Slope Index	0,48	0,52
The type of soil against erosion sensitivity	0,60	0,40
Rainfall index	0,267	0,733

Calculation of the real carrying capacity of Telaga Warna Telaga Pengilon was 869 visitor/ day. It means that the Telaga Warna Telaga Pengilon will be received a negative impact on the biotic and abiotic environments when the number of visitors exceeded into 869 visitors per day. The negative impact on the abiotic environment can be seen in the area, such as degradation of water quality due garbage, fandalisme action, etc. While the negative impact on the biotic environment also has been happpen such as disruption of the distribution of birds in the area

The value of the effective carrying capacity of the Telaga Warna Telaga Pengilon was 579 visitor/ day. That means if one day the tourist area can accommodate 579 visitors in accordance with the capacity management. It can be said that if the number of visitors is not more than the value of ECC, the ecosystem will not be negatively affected and the visitors will still get good service Management Capacity with a value of 66.67% has an effect on the level of visitor satisfaction. Management capacity less than optimal, it can affect the maintenance and monitoring of the tourist area. The value of Management Capacity can be increased 33.33% by adding the officers in accordance with the necessary requirements so that it will correspond to the real carrying capacity calculation. Adding a place to enjoy the lake it is possible to break the concentration of visitors. It was intended so that visitors do not accumulates at one point. Carrying capacity is flexible, because the physical carrying capacity of the region would change which is influenced by other factors including the pattern of a visits, the experience of the visit, the climate, and policy manager. Flexible carrying capacity can be maximized by adding attractions and supporting facilities in the area provided [12].

The value of the real carrying capacity was not exceeded in comparison with the average amount of visitors/day in the last four years, namely 753 visitors/day. However, when it was compared to the effective carrying capacity, the value had exceeded the average amount of visitors per day in the last 4 years. An ecosystem had the ability to recover against a disturbance, on condition that the disorder does not take place continuously, and requires a certain time [4].

5. Conclusions

The environmental carrying capacity in the area is not constant but dynamic, so the ability of a system can be increased or decreased. Sustainable development strategy tourism is necessary for the management of natural resources sustainability. Sustainable tourism aims to manage all natural resources as best as possible. Management of tourism is determined based on the ability of the manager to manage resources and potential, with attention to the impact on the environment.

References

[1] C. M. Pickering and W. Hill, "Impacts of recreation and tourism on plant biodiversity and vegetation in protected areas in Australia," *J. Environ. Manage.*, vol. 85, pp. 791–800, 2007.

[2] J. Zuzana and L. Zuzana, "Monitoring System of Sustainable Development in Cultural and Mountain Tourism Destinations," *J. Compet.*, vol. 7, no. 1, pp. 35–52, 2015.

[3] J. S. Jeong, L. García-Moruno, J. Hernández-Blanco, and F. J. Jaraíz-Cabanillas, "An operational method to supporting siting decisions for sustainable rural second home planning in ecotourism sites," *Land use policy*, vol. 41, pp. 550–560, 2014.

[4] F. Muhammad, S. Basuni, A. Munandar, and H. Purnomo, "Kajian Daya Dukung Ekowisata Hutan Mangrove Blanakan , Subang , Jawa Barat The Study of Carrying Capacity in Mangrove Ecotourism Blanakan , Subang , West Java," *J. BIOMA*, vol. 14, no. 2, pp. 64–72, 2012.

[5] J. Damanik, *Pariwisata Indonesia Antara Peluang dan Tantangan*, Cetakan I. Yogyakarta: Pustaka Pelajar, 2013.

[6] Kementerian Pariwisata dan Ekonomi Kreatif Republik Indonesia, "Rencana Strategis 2012-2014." Jakarta, 2014.

[7] O. Soemarwoto, *Ekologi, Lingkungan Hidup dan Pembangunan*, 10th ed. Jakarta: Djambatan, 2004.

[8] F. Salerno, G. Viviano, E. C. Manfredi, P. Caroli, S. Thakuri, and G. Tartari, "Multiple Carrying Capacities from a management-oriented perspective to operationalize sustainable tourism in protected areas," *J. Environ. Manage.*, vol. 128, pp. 116–125, 2013.

[9] C. Fandeli and M. Nurdin, "Pengembangan Ekowisata Berbasis Konservsi di Taman Nasional," Yogyakarta: Pusat Studi Pariwisata, Kantor Kementrian Lingkungan Hidup. Fakultas Kehutanan Universitas Gajah Mada, 2005, p. 265.

[10] M. Cifuentes, *Determinacion De Capacidad De Carga Turistica En Areas Protegidas*, Serie Tecn. Turrialba, Costa Rica: Publicacion Patrocinada Por el Fondo Mundial para la Naturaleza-WWF, 1992.

[11] C. Fandeli, *Bisnis Konservasi, Pendekatan Baru dalam Pengelolaan Sumberdaya Alam dan Lingkungan Hidup*. Yogyakarta: Gadjah Mada University Press., 2012.

[12] F. Muhammad, "Model Ekowisata Kawasan Hutan Mangrove Resiliensi Ekologi (Kasus Ekowisata Mangrove Blanakan , Subang , Jawa Barat)." Bogor, 2012.

IOP Conf. Series: Earth and Environmental Science **70** (2017) 012004 doi:10.1088/1755-1315/70/1/012004

Effectiveness of Waste Management in Mataram City

B H Widayanti[1], F P Hirsan, and A Kurniawan

Department of Urban and Regional Planning, Universitas Muhammadiyah Mataram, Mataram 83114, Indonesia

efpit_inter@yahoo.co.id

Abstract. Mataram city as National Activity Center (NAC) led to increased of activity that occurs in this region. This condition impacted the increasing of population and the amount of waste. The amount of waste in Mataram City currently reaches 1,444 m3/day and that has been transported by the Sanitation Department as much as 1,033.82 m3 or 71.59%. This research aims to analyze the effectiveness of community-based waste or waste management. The method that was used is quantitative descriptive analysis of waste heaps and analysis of waste management. The results of the analysis of waste heaps is that in the next 10 years (2026) the amount of waste will reach 2,019 m3/day. By using the analysis of waste management, if there are 25 units machines today and 48 waste management groups are effectively utilized, then 948 m3 amount of waste could be processed in a day or as much as 65.65% of the waste is managed by the community. So that, in order to get over this waste problems, collaboration between government and the community in Mataram City is needed.

1. Introduction

According to the constitution number 18 / 2008, waste is the remnant of everyday human activities or natural processes in the solid form. Generally waste management technique in Indonesia are; solid waste generated, onsite waste processing, waste collection, waste treatment and landfill (Faizah, 2008 in Setiadi, 2015:2)

Mataram city is the capital of West Nusa Tenggara, which became the National Activity Center (NAC) that has function as the central government, public service center in regional and national scale, trade and services center in regional and national scale, as well as a major ground transportation node in national scale. As National Activity Center (NAC) and as the central provincial capital city, Mataram conduced as the first choice in the move and reside. These conditions resulted annual increase of the number of people in the city of Mataram. The population of the Mataram city in 2015 as many as 441,064 inhabitants which is the number increased from the year 2010 as many as 402,843 inhabitants.

Table 1. Total Population of Mataram 2014

Districts	Area (Km^2)	Population (inhabitant)
Mataram	10,76	81.450
Ampenan	9,46	86.052
Cakranegara	9,67	66.516
Sekarbela	10,32	62.508
Selaparang	10,77	74.513
Sandubaya	10,32	70.025

Districts	Area (Km^2)	Population (inhabitant)
Jumlah	61,30	441.064

Source: Mataram in Figures, 2015

Increasing population and activities in Mataram have an impact on the amount of waste productions, which are increased the amount of waste up to 441,650 m3 in 2010 and increased up to 446,790 m^3 in 2011. This is in line with the average increase of solid waste every day, which reached 1,444 m^3/day in 2016. The currently available infrastructures in Sanitation Department of Mataram City will be handling waste every day as much as 1,033.82 m^3 (71.59%). Moreover, sanitary landfill is also a very important issue to be addressed in Mataram. Currently, the sanitary landfill site in Mataram City residents is at Kebon Kongok landfill, located in District Gerung, West Lombok Regency, within ± 29 km from the Mataram city. Mataram city has limitations in the provision of sanitary landfill in its jurisdiction because there is no decent location as sanitary landfill.

Mataram City has 61.30 km^2 of area, which is dominated by the use of plots in residential areas with amenities that the existences spread throughout the city of Mataram. At the same time, the requirements that must be met in building landfills are:

1. Landfills are not built near the sources of drinking water or other water sources used by humans
2. Not in a place often hit by floods
3. The place far from human residence

The distances that are often used to measure the minimum distance of the location of landfills from residential areas is 2 km, 15 km from the sea, and 200 m of water resources, so that, there is no single location in the area of Mataram city which meet those requirements. The problem of waste in Mataram city requires a serious attention from either local governments himself or for environmentalists. Then this research will assess the potential of existing waste management groups and waste management infrastructure.

2. Literature Review

2.1. Waste definition

Waste is a garbage in the form solid or semi-solid that comes from urban activity or daily activity of human, animals, and plants (Kodoatie, R. J, 2005: 216). The source of a municipal solid waste are mostly from residential, traditional market, trading estate, industrial park, animal husbandry and other public facilities. There are two types of solid waste (Kodoatie, R. J, 2005 : 216) :

1. Organic waste
   Organic waste is classified as biodegradable garbage, which is easy to decompose by bacteria. For instance: vegetables, leaves, timbers, etc.
2. Inorganic waste
   Inorganic waste also called non-biodegradable because it is hard to decay, for examples plastics, cans, irons, glasses, etc.

2.2. Waste Management

Waste management is defined as every action which conducted to dispose the waste from its inception to its final disposal. In general, this action involving monitoring solid waste generation, waste collecting, transferring and transporting waste, treatment, and its final disposal (Kartikawan, 2007) :

1. Solid waste generation
   As it seen from its definition, the solid waste is generated not produced. Hence, to decide the perfect disposal management, the number of subjects and their activity are mattered the most.
2. On-site handling
   On-site handling is all waste management before it transferred to the landfill. This activity is encouraged by the fact that many wastes are possessed good value in an economic way. On-site handling will give significant influence to the next management step. This activity varies depend

on the types of waste, includes shorting, reuse and recycle. The aim of on-site handling is to reduce the amount of waste.

3. Collecting
   Collecting is a process to pile up the waste from the inception to the waste tanks or to the landfill.
4. Transfer and transport
   Transfer and transport are an activity to convey the waste and its tanks to the final disposal.
5. Treatment
   Based on the waste type and its compositions, waste management are divided into:
   a. Physical transformation
      It includes sorting and compacting to simplify the storage and the transport.
   b. Incineration
      Incineration is a technique that conducted to transform the waste into a flue gas, thus, it will reduce the waste volume up to 90-95%. Inceneration means the act of burning something until there is nothing left but only the ashes. An incinerator is a unit or facility that is used in other to burn used trash and some other different kind of waste until it is finally reduced to only ashes (Hickmann and Lanier Jr. 2003). An incenerator is built with strong and well-insulated material so that during combusting the extreme heat is not lost but contained. The heat is left inside the furnace so that all the waste inside the incenerator plant could be burn very quickly and as well effecient. But when the heat is not well contained the waste will not be completely burn with level of effeciency expected of it. Incinerator is classified based on its utility and its feed system. There are two types incinerator based on its utilization, (Bagus T, 2002 : 19) :
      1) Incinerator without heat combustion, this system usually constructed to burn small scale waste (0.2-1 ton/h).
      2) Incinerator with heat and its conversion. The capacity of this incinerator is around 40 ton/h.
   c. Composting
      Compost is an organic fertilizer that made of a green compound and other organic matter which added to speed up the decomposing, for instance, livestock manure and urea (if it is needed).
   d. Energy recovery
      Energy recovery is a transformation from wasted into energy, either heat or electrical energy.
6. Final disposal
   In principle, final disposal must follow the health rules and environmental sustainability. Nowadays, the technique that has been used is open dumping, in which waste is placed on the specific place until its overload. This phenomenon has high probability in creating an environmental disturbance. To prevent the problem, another technique is recommended, namely sanitary landfill, where the waste stockpile is managed.

2.3. Effectiveness
Effectiveness is a measurement to measure how far the target has been achieved (quantity, quality, and time) (Hidayat, 1986). Strees in Tangkilisan (2005:141) stated 5 (five) variables in measuring effectiveness, there are: productivity, working adaptability, job satisfaction, profitability, and resources capabilities. Thus, measurement is a value in target achievement that has been determined in the first place. If the aim is achieved, so the activity is effective.

3. Research Methodology

3.1. Types of Research
This research uses a case study approach and field research, where the goal is to intensively study the background, the last status and environmental interactions that occur in a social unit. Case study investigates many variables and many conditions in a small sample (Azwar S., 2014 : 8).

3.2. Research Focus
The aim of this research is to determine the effectiveness of waste management in Mataram with a focus on community empowerment groups that have been formed in Mataram City. This study focuses on the ability of a group of communities in managing organic and inorganic waste.
This research uses descriptive quantitative approach that the populations are the entire waste management groups in Mataram as many as 48 groups and the samples are as many as 10 groups with the classification of processing of organic waste and inorganic waste. The selected groups that are still active in waste management are: Komunitas Hijau Biru, Kimzeefarm, Muda Kratif, Pokja Lisan, Kampung Hijau, KSM Cinta Bersih, KSM Terampil Pagutan, Peduli Lingkungan, Banjar and TPST Sandubaya. To determine the effectiveness of 5 (five) concept of effectiveness criteria proposed by Sterss in Tangkilisan (2005), this research only uses the productivity criteria, because this criteria can be measured.

3.3. Techniques of Data Collection and Data Analysis
This research uses data collection techniques such as: direct observation and interviews. Primary data collected in the form of information related to government policy about waste management in Mataram, while the methods of documentation are used to obtain secondary data like the number of population, waste infrastructure and the amount of midden.
The analytical methods that is used are midden analysis and analysis of midden management. The measurement method and composition of midden are based on SNI M 36-1991-03(02). To predict the midden, the following equation can be used

$$Qn = Qt(1 + Cs)^n \tag{1}$$

Which

$$Cs = \frac{[1+(Ci+Cp+Cqn)/3]}{[1+p]} \tag{2}$$

Which are:

Qn = n midden in coming years
Qt = midden in the early years of calculation
Cs = increase / growth of the city
Ci = the growth rate of the industrial sector
Cp = the growth rate of the agricultural sector
Cqn = the rate of increase in per capita income
P = population growth rate

The analysis of the midden management is based on the results of a field survey carried out on samples that have been specified above. The result of the survey later will be accumulated as an average capability of processing waste per day.

4. Discussion

4.1. A General Overview
Waste generated in Mataram city is derived from residential area, offices, commerce, education, markets, roads and others. Based on data from the Sanitation Department in Mataram city, average

volume of waste per day is 1,444 m3. For more details, the volume of waste generated per day from each district in the Mataram City can be seen in the table 2.

Table 2. Waste Volume in Mataram city

No	Districts	Volume (m^3/day)
1	Ampenan	283
2	Sekarbela	190
3	Mataram	263
4	Selaparang	260
5	Cakranegara	229
6	Sandubaya	219
	Total	1.444

Source : Sanitation Department of Mataram City, 2016

The amount of waste is currently processed by the Sanitation Department in Mataram by dispose it straight to TPA Kebon Kongok. The transportation mediums and waste transportation personnel that are owned by the Mataram City can be seen in the table 3.

Table 3. Waste Transportation Infrastruktur im Mataram City

No	Infrastructure	Total	Capacity
1	Dump truck	46 unit	8 m3
2	Armroll	19 unit	8 m3
3	Open pick up (without a basket)	9 unit	1,5 m3
4	Open pick up (with baskets)	6 unit	8 m3
5	Sweeping 300 day (25 day/month)	3 unit	8 m3
	Total	83 unit	

Source : Sanitation Department of Mataram City, 2016

Table 4. Waste Transporter Officer in Mataram city

No	Transportation Type	Status (person)		
		Governent employees	Honorary	Contract
1	Dump truck	17	15	14
2	Armroll	18	1	-
3	Open pick up	14	1	9
	Total	49	17	23

Source : Sanitation Department of Mataram City, 2016

With the current existing facilities and infrastructure conditions, Sanitation Department has been able to transport the waste every day as much as 1033.82 m^3 (71.59%) that are disposed in TPA Kebun Kongok, West Lombok regency. TPA Kebon Kongok is a regional landfill in accordance to memorandum of understanding between the West Lombok regent and the Mayor of Mataram in 2006. Kebon Kongok landfill has an area of 8.6 ha including facilities and infrastructure, with a controlled landfill system. Based on existing conditions and field observations, there are some problems in waste management in Mataram City, those are:

1. Limitations of waste transportation infrastructure
2. Community behavior that often throw garbage into the river and lack of attention to the environment
3. The emergence of temporary landfills or dumped in vacant lots
4. Waste management is still relying on the conventional scheme: stacks – transport – throw away
5. Lack of separation of organic waste - inorganic at the community level
6. Minimizing waste by 3R (reduce, reuse and recycle) is not rooted in the community. 3R movement has been implemented by most communities in each neighborhood. Recycling bins

are currently relying on informal waste processing subjects like scavengers, both located at the source, TPS and TPA

7. Not optimal market organic waste processing

4.2. Midden Analysis

The midden in the city of Mataram is very closely related to the amount of waste generated by the sources of waste. Sources of waste may come from settlements, markets, streets, shops, offices, restaurants, hotels and hospitals. The largest source of midden in Mataram city derived from settlements (households), which are, based on existing documents, the amount of waste currently reaches 1.444m3/day. Total midden will be higher due to increased population and the number of activities undertaken by the community.

By the availability of current transportation waste infrastructure, the number of trash capable of being transported to the landfill every day can reach 1.033,82 m3 (71,59 %) and the remaining still cannot be transported to the landfill. To forecast the amount of midden for the next 10 years by using the approach of measuring the composition and waste SNI 36-1991-03 M (02), the predicted amount of waste in the Mataram city as follows:

Based on this formula, then:

1. The amount of midden in Mataram city on 2016 are = 1,444 m3 / day
2. The growth rate of the industrial sector = 7.22%
3. The growth rate of the agriculture = 2.67%
4. The rate of increase in income per capita = 7.67%
5. Population growth rate = 1.01%

Based on the data above, the obtained increase / growth of the city of Mataram is:

$$Cs = \frac{[1+(7{,}22\%+2{,}67\%+7{,}67)/3]}{[1+1{,}01]} = 3{,}41\% \quad (3)$$

So that, with the growth of Mataram City as much as 3.41 %, and using formulas $Qn = Qt(1 + Cs)^n$, then obtained a projection of the amount of midden for next 10 years are :

Table 5. Projections Of Midden

Years	The Volume of Midden (m^3/day)
2017	1.493
2018	1.544
2019	1.597
2020	1.651
2021	1.707
2022	1.766
2023	1.826
2024	1.888
2025	1.953
2026	2.019

Source : Analiysis, 2016

4.3. Community-Based Analysis Of Waste Management

Government of Mataram City through various programs has been providing facilities and infrastructures to facilitate the public in waste processing. The infrastructure that has been available is in a form of a thrasher machine that converts organic waste into a smaller size. Those organic wastes that are processed and converted by the community become more useful materials, for example, that has developed today is the manufacture of compost. The number of waste processing machine in Mataram is 25 units of machines which are distributed around groups of waste processing community. Generally, the waste processing machine are in good condition, but the fact is, the machines are not optimally used by the waste processing communities. The waste processing activities in Mataram city can be seen in the picture bellow:

Figure 1. (a) Waste Processing Mechine (b) Waste Processing Activity

Besides that, Mataram City Government has also trained several waste processing communities. There are some communities formed because of social concerns in the waste processing, aside of the governmental trained communities. On the figure bellow we can see the activity of waste processing of inorganic waste:

Figure 2. (a) and (b) Inorganic Waste Processing by Waste Community

Waste processing in Mataram City not only uses 3R concept, but also currently tries on Integrated Waste Processing. The location of it in Mataram City is at Sandubaya. At this location, besides of processing the waste into compost, this group is also decreasing the midden using the incinerator. The facilities that are provided at the location are waste conveyances with waste processing operator that can be seen in the table below.

Table 6. Waste Conveyance Facilities at Integrated Waste Processing in Sandubaya

No	Facilities	Number of Facilities	Capacity
1	Open pick up	2 unit	1,5 m3
2	Waste conveyance (motorcycle)	7 unit	1 m3
3	Waste conveyance (cart)	8 unit	1 m3
	Total	17 unit	

Source : Primary Survey, 2016

According to the table above, waste conveyance facilities (motorcycle and cart) are the most effective conveyances because those could pass through the narrow roads (less than 3 meters wide).

Table 7. Operator at Waste Processing Location in Sandubaya

No	Employment Status	Number of Employee (people)
1	Government employee	12
2	Honorary employee	7
	Total	19

Source : Primary Survey, 2016

The number of operating employees at integrated waste processing in Sundabaya is 19 people which consist of 12 people of government employees and 7 people of honorary employees. In every fireplace of incinerator, there are 3 people is needed which their tasks are dry the waste and put it into the fireplace of the incinerator.

Figure 3. Fireplace of Incinerator at Sandubaya

According to the result of the field survey towards determined samples, the total of midden that is processed by each group can be estimated as in the table below.

Table 8. Volume of Waste Processing

No	Group	Number of Members	Processing Form	Average Volume of Waste Prcessing (kg/day)
1	Hijau Biru	10	Organic into compost	80
2	Kimzeefarm	5	Organic into compost	150
3	Muda Kreatif	12	Inorganic into craft	1
4	Pokja Lisan	20	Inorganic into craft	1
5	Kampung Hijau	5	Organic into compost	100
6	KSM Cinta Bersih	14	Organic into compost	100
7	KSM Terampil	15	Organic into compost Inorganic into craft	50
8	Peduli Lingkungan	5	Organic into compost	50
9	Banjar	8	Organic into compost Inorganic into craft	150
10	TPST Sandubaya	25	Organic into compost Incinerator	3.000

Source : Primary Survey, 2016

The findings in the field for waste processing from inorganic forms into crafts did not done daily; this depends on the craft's raw materials that have to be sought, collected, and cleaned so that it can be processed into a craft. Besides, the activity of making a craft from scrap papers would be high when the groups got orders.

While for the waste processing using incinerator can be operated using 3 (three) fireplaces daily with the size is 2 x 2 x 1.5 m which has capacity of 5 tons/2h with the operational duration is 8 hours/day and now has been processing 3,000 kg waste daily.

For the organic waste processing into compost, the processing activity in some surveyed groups has gone down. This condition is caused by the heap of the compost that has not been marketed. The main cause of bad marketing is the quality of the compost that is not standardized by each group. The 3R processing machines in Mataram City are 3.5-5 pk machines, with the capacity of 1.5 tons/h (4.5 m3/h) and some of it are 9-11.5 pk machines with capacity of 3 tons/h (9 m3/h) with operational duration is 8 hours/day.

With the potential that has been owned by Mataram City in waste processing, the number of groups that are concerned about waste are as many as 48 groups that are scattered throughout the Mataram City territory and as many as 25 waste processing machines, the calculation of each machine capacity that can be achieved if every machine can processing 1.5 ton waste hourly in average is 36 m^3/day.

So that, according to the data available and the size of capacity of the machine with assumption that all of the machines available in Mataram City is working optimally, those machines can process as much as 36 m3 waste x 25 machines = 900 m3 waste daily. For the waste processing groups available in Mataram City, if there is assumed that all of the groups are active in processing the waste from inorganic waste into crafts or processing the waste becomes more valuable stuffs with a group is consisted of 5 people, so for each group, they can process 1 m3 waste daily. So that, according to the currently number of groups that are available, as many as 48 groups, the inorganic waste that can be processed is as much as 48 m3 daily.

If we accumulate it, the waste in Mataram City can be processed daily by the social groups is 948 m3 (65.65%) compared to the total of current midden. But, by looking at how poorly the society in processing the waste in Mataram City, the alternative method of processing the waste is using incinerator.

The capacity of incinerators that are currently available in Sandubaya District can be calculated is 180 m3/day. Based on that calculation, if all of the fireplaces that are available in Sandubaya District are used optimally, with current situation (total of midden reach 219 m^3/day), the volume that can be cut is as much as 180 m^3 (82.19%) waste a day.

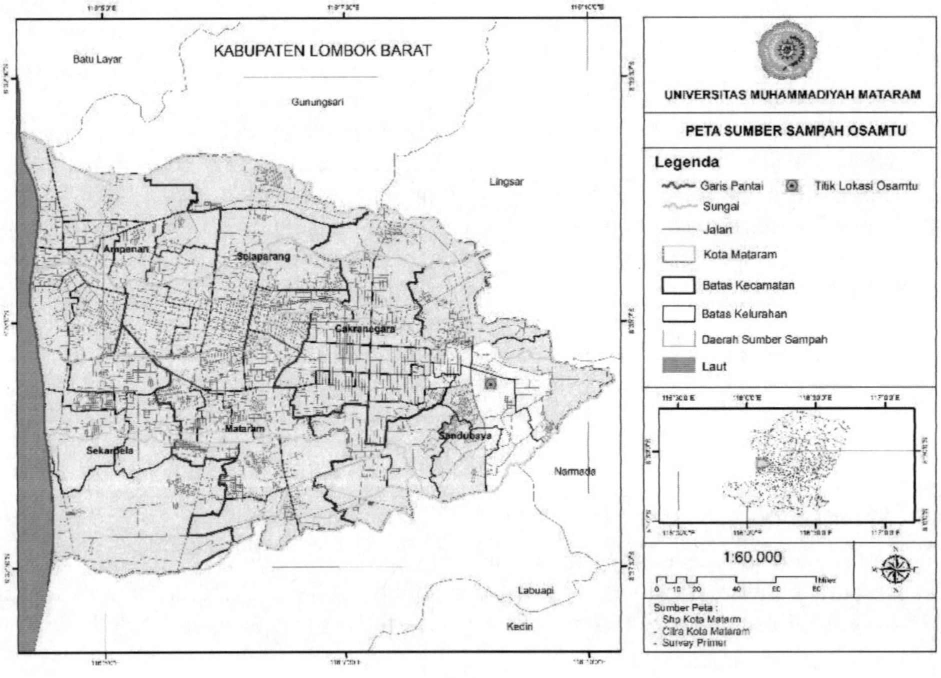

Figure 4. Osamtu's Waste Sources Map.

5. Conclusion

According to the discussion result above, can be concluded that:

1. Number of midden in Mataram City can be managed optimally by the waste concerned groups – if there is an incentive supervising either from the government, private, or from the university – is as much as 948 m3 waste a day or 65.65% of the total midden currently. While by using Integrated Waste Processing in Sandubaya District can reduce 82.19% of the midden there.
2. Total of the midden will increase in 10 years becomes 2,019 m3 midden in a day, so that, addition of either machine facilities or capacity reinforcement of waste concerned group are needed.

IOP Conf. Series: Earth and Environmental Science **70** (2017) 012004 doi:10.1088/1755-1315/70/1/012004

6. Recommendation

1. If Integrated Waste Processing will be replicated in Mataram City, the Integrated Waste Processing in Sandubaya District as the trial should be successful in keeping the temperature above 800^0C so that perfect combustion can be achieved and poisonous gas that is harmful for human's health will not be produced.
2. In waste processing with Integrated Waste Processing technology, there should be waste separation first, so that the waste that will be put in to the fireplace is dry waste. For plastics and glasses waste can be separated from the waste sources, so that in the implementation of it, the clear and understandable SOP is needed for all of the operators.
3. The training and reinforcement of the capacity for the operators of Integrated Waste Processing technology are needed so that they can understand the ways of working and safety system in the implementation of waste processing with Integrated Waste Processing technology.

References

[1] Anonim. Mataram in Figures. Statistics of Mataram City. Mataram ; 2015
[2] Anonim. Rencana Aksi Komunitas Hijau, Kota Mataram ; 2013
[3] Azwar Saifuddin. Metode Penelitian. Yogyakarta : Penerbit Pustaka Pelajar ; 2014
[4] Setiadi Amos. Studi Pengelolaan Sampah Berbasis Komunitas pada Kawasan Permukiman Perkotaan di Yogyakarta. Jurnal Wilayah dan Lingkungan Volume 3 Nomor 1 April 2015, p. 27-38
[5] SK SNI-M-36-1991-03. Metoda Pengambilan dan Pengukuran Contoh Timbulan Komposis Sampah Perkotaan.
[6] Tangkilisan, Hesel Nogi S. Manajemen Publik. Jakarta : Gramedia Widia Sarana ; 2005
[7] Undang-undang No 18 Tahun 2008 tentang Pengelolaan Sampah

Road network – land use interaction model: Malang City in Indonesian case

B S Waloejo

Department of Urban and Regional Planning, Faculty of Engineering, Universitas Brawijaya, Malang, Indonesia

budieswe@yahoo.co.id

Abstract. Urban population in Indonesia is significantly increasing from 44% of total population in 2002 to approximately 60% in 2015. Rapid population growth has resulted in rapid urban land use changes. The problems became more complicated since the changes created mixed use development along the main urban corridors that resulted in higher trip generation and attraction while urban land very limited that made road widening and creation of new road less possible. This led to an accumulation of movement, worse road's level of service and congestion in the main urban corridor. The aims of the research are to analyse trip generation/attraction of the mixed- land uses of the main corridors Malang City; and to formulate road network – land use interaction model in the case of Malang (a nearly 1 million population city in 2015). The selected corridors in Malang City a main road in the west district of Malang City - Indonesia. Correlational method (Pearson Product Moment) and regression method (stepwise, anova for land use's trip generation/attraction), and analysis of road's level of service (LOS), using Indonesian Road Capacity Manual, are employed in this research. The research formulated the interaction model as:

$V_{total} = \sum V_{internal} + \sum V_{external}$ *Where*

$\sum V_{internal} = e_{residential} Y_{residential} + e_{schools} Y_{schools} + e_{unis} Y_{unis} + e_{offices} Y_{offices} + e_{hospital} Y_{hospital} + e_{chemists} Y_{chemists} + e_{commercials} Y_{commercials} + e_{market} Y_{market} + e_{fuel\ station} Y_{fuel\ station} + e_{bus\ station} Y_{bus\ station}$ *and*

$\sum V_{external} = V_{local\ roads} + V_{continous\ traffic\ flow}$

The research showed that V_total=23,033 car unit/day (internal) + 32,746 car unit/day (external) = 55,779 car unit/day. The trip higher than the road's capacity (C) which 40.695 car unit/day. The levels of services of road's segments dominantly F (congested). The simulation of the model showed that traffic engineering can be used to increase road's capacity while land use changes (reduction of the variety of mixed land uses), particularly mix land uses at frontage area, can improve level of services (from F to C).

1. Introduction

Most of recent literature of urban land use in the last two decades, such as Cervero [1] [2] promoted mixed uses development (MXD) to increase compactness of the city and therefore reduce trip generation of the whole city. However, compactness is claimed to be negative for several aspects of urban life, related to transport, is likely to reduce level of walking and cycling [3] and to increase congestion [4]. Mixed land uses are common characteristic of urban development in Indonesia. In most cases, mixed land uses are flourished in the main road of big cities and they tend to convert housing to commercial uses [5]. In the case of Malang City, East Java Province, this change was even legitimated by the Local Government which approved commercial uses along the main corridors in its Evaluation of Malang City Spatial Plan.

Especially in the field of transport planning and infrastructure provision the so called four-stage models were applied. These models consist of a sequential algorithm, which includes the estimation of transport demand, transport distribution, modal split and assignment [6]. These models mainly applied to identify bottlenecks in existing road networks and based on that new, additional infrastructure is justified. But these models do not include all means of transport nor long term effects on land use caused by this additional supply and so wrong decisions are made. Newer developments in the domain of transport planning realised this shortcoming of four-stage models and took a more holistic approach. They combined land use models with transport models to display the interaction of urban transport systems and land use development over time in a combined model. An overview of existing Land Use Transport Interaction (LUTI) model. An advantage of this approach is, that the exogenous input data which are needed e.g. the transport model can be calculated in the land use part and vice versa [7].

An important step to this quest is to examine to what extent the mixed land-uses can be developed on urban corridor area based on the maximum capacity of road's network within the area. The hypothesis if the traffic volume indicates to reach the capacity threshold of the road, then land uses should be controlled by, e.g. preventing conversion of low trip to high trip generator/attractor land uses through zoning ordinance. Therefore it is significance to examine the character/type of each land use. Therefore, land use planning for areas along the urban main corridor and its correlation to transportation in Indonesian city, which also representing most cities in South East Asia, is relevant to explore to formulate a more well-planned and efficient regional transportation system. Promotion of a model of interaction between land-uses and capacity of the road is intended to measure the desired threshold of the capacity of the land-uses and their interaction to road's service.

2. Research Method

The approach of the researh is a mix of qualitative and quantitative research. Qualitative approach used to describe the condition of the study area. The study area located on areas of a main road in the west district of Malang City i.e. a main road that connects Malang to Batu City, a tourist destination city in the Province of East Java. The characteristics of trip attraction and generation along the main road qualitatively explored. Quantitative approach used to formulate regression model of the interaction between land uses and trip generation/attraction [8]. The analyses included stepwise method, correlation, ANOVA, road capacity, and road's level of services.

Based on trip attraction and generation, we proposed a mathematical model using correlation between response variable (the sum of trip generation/attraction) and independent variables. We proposed 10 (ten) independent variables as follows:

1. Residentials; X_1= building area, X_2= number of bedrooms, X_3= number of household's members, X_4= number of motorized vehicles possessed by the family, X_5 = family income.
2. Schools (kinder garten to high school); X_6= number of students, X_7= number of theachers/employees, X_8= number of classrooms, X_9= land area, X_{10}= building area.
3. Universities; X_{11}= building area, X_{12}= number of students, X_{13}= number of lecturers/employee, X_{14}= number of courses.
4. Offices; X_{15}= number of employees, X_{16}= number of visitors/guests, X_{17}= land area, X_{18}= building area.
5. Hospital; X_{19}= number of doctors, X_{20}= number of daily patients, X_{21}= number of rooms, X_{22}= parking area, X_{23}= building area.
6. Chemists; X_{24}= number of employees, X_{25}= number of customers, X_{26}= building area, X_{27}= parking area.
7. Commercials; X_{28}= parking area, X_{29}= building area, X_{30}= number of employee, X_{31}= number of visitors.

8. Market; X_{32}= number of sellers, X_{33}= floor area for meat and fish, X_{34}= floor area for food and beverage, X_{35}= floor area for daily basic need, X_{36}= floor area for fruits, X_{37}= floor area for vegetables, X_{38}= floor area for clothing/fashion, X_{39}= floor area for other goods.
9. Fuel station; X_{40}= number of employees, X_{41}= built up area.
10. Bus station; X_{42}= land area, X_{43}= number of buses/minibuses, X_{44}= number of employee and passangers.

Using Borland Delphi 7 and Microsoft Access 2010 we developed an application program (B.eswe2013) to simulate data based on the regression equation. This software can be used to recommend the extent of mixed land use that can be accommodated in the area based on the capacity of the road.

3. Result and Discussion

The road in the study area is a two way road, un-separated, with 8 m pavement width of hot mix asphalt surface. The road had about 0.5 to 1 m shoulder/curb width. The curb/shoulder occupied by many street vendors. The condition of the road as follows: C_O = 2,900 veh/h; FC_W = 1.14, FC_{SP} = 0.97; FC_{SF} = 0.90, and FC_{CS} = 0.94, so the actual capacity of the road 2,712.97 veh/h. Continuous movement during the weekdays 27,523 veh/day, higher than local movement (10,763 veh/day).

3.1. The characteristic of mixed land use in the study area:

1. Residential
   Most the the buildings (60%) are single story, consisted of maximum four rooms with 3 persons per household in the average. The majority of the resident's employees with the destination to city center riding motorbike (50%). This land use generated 75 veh/day and attracted 75 veh/day, so the total 150 veh/day. Highest trip generation (30 veh/hr) from 7.00 to 8.00 am while the highest trip attraction from 4.00 – 5.00 pm. The ratio (e) 20%. The lowest movement at 10.00 – 11.00 am (6 veh/hr) with ratio (e) at 4%.
2. Schools (kindergarten – high school)
   Most of the building single storey, except one high school that consist two – three storeys. Origins of the movement to these land uses from inside the district (45%) and the rest form other districts. Transportation modes are motorbike (47%) with travel time less than 25 minutes (74%). Total trip from and to these education facilities 140 veh/day. The highest trip attraction (15 veh/hr) at 7.00 – 8.00 am while trip generation at 3.00 – 4.00 pm. The ratio at peak hour 15.19% and the lowest at 5.00 – 6.00 pm with the ratio of 2.79%.
3. Universities
   Origins of the movement to this land use 50% form other districts and the transportation mode is motorbike (50%), with travel time less than 5 minutes (54%). There are three Universities along the corridor. Average movement 648 veh/day, a total of 324 veh/day trip attraction with peak hours at 12.00 am – 1.00 pm (56 veh/hr) and 324 veh/day trip generation with peak hour at 3.00 – 4.00 pm (44 veh/hr). Peak of trip attraction and generation at 12.00 am – 1.00 pm with 88 veh/hr and e ratio at 13.50%, while the lowest movement at 8.00 – 9.00 pm with only 4 veh/hr and e ratio at 0.62%.
4. Offices
   Most (57%) of the origin of movement from other district and 56% used motorbike. Travel times varied from 10-25 minutes (65%) to 6-10 minutes (35%). Trip attraction and generation for offices only 16 veh/day for each movement, then the total movement 32 veh/day. The peak for trip attraction at 8.00-9.00 am with 2.75 veh/hr, while the peak for trip generation at 1.00-2.00 pm with 2.58 veh/hr. The peak of total movement at 2.00-3.00 with 4.25 veh/hr and its e ratio at 13.28%, while the lowest at 5.00 – 6.00 pm with only 0.67 veh/hr and e ratio at 2.08%.

5. Health facilities (Hospitals and chemists)
   The building varied 1 – 2 floors. Most of the movements from outside the district (53%). Popular mode of transport is motorbike (57%) which took 11-15 minutes (31%) and 15-25 minutes (69%). There one hospital and five chemists. Trip attraction and generation of the hospital 192 veh/day for each movement, thus the total movement 384 veh/day. Peak hour of total movement 68 veh/hr with e ratio 17.71% at 5.00-6.00 pm. Trip attraction for chemists 20 veh/hr equal to trip generation. The peak hour for trip generation 2.25 veh/hr at 9.00-10.00 am, 10.00-11.00 am, 5.00-6.00 pm, while the peak for trip attraction 4.75 veh/hr at 9.00-10.00 am with e ratio 11.88%.
6. Commercials
   Origins of the movement from outside the neighbouring districts (54%). Similar to other land uses, the favourite mode is motor bike (64%) with travel times varied from 6-10 minutes (33%) to 10-25 minute (67%). Land uses for commercials consisted of trades and commercial services, markets, and fuel station. Trades and commercial services generated and attracted 28 veh/day. The peak for trip attraction 1.53 veh/hr at 12.00 am-1.00 pm. The peak for trip generation occured at the same time with 1.51 veh/hr then the total movement at the peak hour 3 veh/hr with e ratio 10.71%.
   Trip generation and attraction for Dinoyo market 841 veh/day, with the peak of attraction 105 veh/hr at 7.00 – 8.00 am and the peak of trip generation at 68 veh/hr at 10.00-11.00 am. The peak of total movement occured at 7.00-8.00 with 145.25 veh/hr with e ratio 17.27%.
   The fuel station generated and attracted the second highest trip along the corridor with total trip 3,548 veh/day. The peak hour at 4.00-5.00 pm with 175 veh/hr (attraction) and 175 veh/hr (generation) then the total 350 veh/hr with e ratio at 9.86%.
7. Bus station
   The bus station covered an area of 4,770 sqm which allocated for bus parking space (2,290 sqm). The terminal operated 337 minibuses, 48 buses. The terminal accommodated 150 motor bikes in its parking space. Landungsari transit station generated and attracted the highest trip with 4.381 veh/day (in + out). The peak hour is at 4.00-5.00 pm with 210 veh/hr (attraction) and 210 veh/hr (generation). The e ratio at the peak hour was 9.59%

3.2. Formulation of the model
We used linier regression (stepwise method) for each land use. The result of each land use is as in the following table (Table 1):

Table 1. Equation model for each land use.

No	Land Use	Equation
1	Residential	$Y_{residential} = 1.012 + 0.004\ (X_1) + 0.264\ (X_3) + 0.251\ (X_4)$
2	Schools	$Y_{schools} = 2.498 + 0.141\ (X_6) + 0.024\ (X_{10})$
3	Universities	$Y_{unis} = -3.555 + 0.008\ (X_{11}) + 0.149\ (X_{12})$
4	Offices	$Y_{offices} = 9.456 + 0.564\ (X_{16}) + 0.089\ (X_{18})$
5	Hospital	$Y_{hospital} = 13.715 + 0.291\ (X_{20}) + 0.055\ (X_{21})$
6	Chemists	$Y_{chemists} = 25.323 + 0.322\ (X_{25}) + 0.084\ (X_{26})$
7	Commercials	$Y_{commercials} = 4.639 + 0.180\ (X_{29}) + 0.189\ (X_{31})$
8	Market	$Y_{market} = 1.648 + 0.500\ (X_{33}) + 0.979\ (X_{35}) + 0.505\ (X_{36}) + 1.127\ (X_{37})$
9	Fuel station	$Y_{fuel\ station} = -3.255 + 12.867\ (X_{41})$
10	Bus station	$Y_{bus\ station} = 49.435 + 8.097\ (X_{43})$

3.3. The influence land uses to vehicle trip at the corridor
By comparing the results of land use model analyses at the corridor (Mayjen Haryono Street – Raya Tlogomas Street) we found that the peak volume of trip at 12.00 am – 1.00 pm with 2,040 veh/hr while

the capacity of the street 2,713 veh/hr, therefore the influence 75.21%. The influence for each hour is described in Table 2 and Figure 1.

Table 2. The influence of land uses to trip volume.

Time	Vehicle volume (veh/hr)	Road capacity	The influence (%)
06.00-07.00	608.67	2,712.97	22.44
07.00-08.00	1,377.47	2,712.97	50.77
08.00-09.00	1,758.03	2,712.97	64.80
09.00-10.00	1,710.22	2,712.97	63.04
10.00-11.00	1,775.22	2,712.97	65.43
11.00-12.00	1,853.70	2,712.97	68.33
12.00-13.00	2,040.34	2,712.97	75.21
13.00-14.00	1,885.56	2,712.97	69.50
14.00-15.00	1,757.06	2,712.97	64.77
15.00-16.00	1,840.05	2,712.97	67.82
16.00-17.00	1,871.90	2,712.97	69.00
17.00-18.00	1,583.31	2,712.97	58.36
18.00-19.00	1,169.54	2,712.97	43.11
19.00-20.00	1,091.66	2,712.97	40.24
20.00-21.00	710.15	2,712.97	26.18
Total	**23,032.90**		

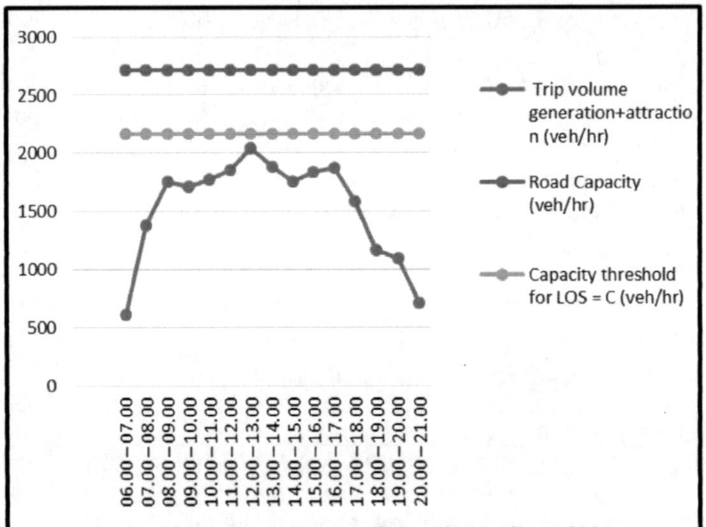

Figure 1 Comparison between the road capacity and trip volume contributed by the mixed land use at the corridor

Figure 1. Comparison between the road capacity and trip volume contributed by the mixed land use at the corridor.

The total volume of vehicle trip interacted with road capacity to provide road's performance or level of service. The road performance is based on the following formula:

$$VCR = \frac{V_{total}}{C} \tag{1}$$

$$VCR = \frac{\sum Vi + \sum V_{external}}{C} \tag{2}$$

$$\sum Vland\ uses = 23{,}033\ veh/day \tag{3}$$

$$\sum Vexternal = V\ local\ roads + Vcontinuous\ trip = 32{,}746\ veh/day \tag{4}$$

$$\sum VTotal = 23{,}033\ veh/day + 32{,}746\ veh/day = 55{,}779\ veh/day \tag{5}$$

The output of the interaction between trip generation/attraction of the mixed land use and road network system is described in Table 3.

Table 3. The interaction between trip volume of the mixed land use and road network system.

Time	Trip volume resulted from land uses (veh/hr)	Trip volume resulted from local road heading to/from the corridor + continuous trip (veh/hr)	Total trip volume (veh/hr)	Road capacity	Ratio V/C	LOS
06-08	608.67	1,708.90	2,317.57	2,712.97	0.85	D
07-08	1,377.47	2,421.55	3,799.02	2,712.97	1.57	F
08-09	1,758.03	2,427.05	4,185.08	2,712.97	1.55	F
09-10	1,710.22	2,334.75	4,044.97	2,712.97	1.51	F
10-11	1,775.22	2,295.45	4,070.67	2,712.97	1.49	F
11-12	1,853.70	2,300.40	4,154.10	2,712.97	1.54	F
12-13	2,040.34	2,371.05	4,411.39	2,712.97	1.65	F
13-14	1,885.56	2,434.90	4,320.46	2,712.97	1.61	F
14-15	1,757.06	2,346.05	4,103.11	2,712.97	1.51	F
15-16	1,840.05	2,186.70	4,026.75	2,712.97	1.51	F
16-17	1,871.90	2,250.60	4,122.50	2,712.97	1.52	F
17-18	1,583.31	2,179.00	3,762.31	2,712.97	1.36	F
18-19	1,169.54	2,126.65	3,296.19	2,712.97	1.22	F
19-20	1,091.66	1,914.25	3,005.91	2,712.97	1.11	F
20-21	710.15	1,448.30	2,158.45	2,712.97	0.78	C
Total	23,032.90	32,745.60	55,778.50	40,694.55		

Table 3 shows that from 7.00 am to 8.00 pm the road's LOS is F. This proves that the model can be used to estimate the influence of mixed land use at the corridor to road performance in Malang City. This also indicates that the level of mixed land use in Malang (we argue that this is the typical land use in most South East Asian Countries) generated/attracted vehicle trip that override road's capacity. The interaction model, therefore, can measure the influence of mixed land use to trip generation and attraction along a main road. The model can also provide a comparison between continuous trip (regional trip) with local trip (generated/attracted by the activities of the mixed land use system and by movement from interconnected local roads to the main road.

We use the equation models to develop user friendly application program software (B.eswe2013) with the help of Borland Delphi 7 and Microsoft Access 2010. This program accelerates equation by executing the interaction of mixed land use with the road system and estimates the road performance. (The program is still developed in Indonesian language). The program consisted of four parts:

1. Data for response variable:
   This part has two tables: response variable and ratio. Response variable table consisted of response variable data (read only) resulted from execution using SPSS (version 17.0) by inputting data of independent variables for each land use, thus the trip volume can be calculated. Ratio table consists of the list of each ratio of response variable from 6.00 am to 9.00 pm. This table is for inputting e ratio for each column corresponding to each land use (figure 2).

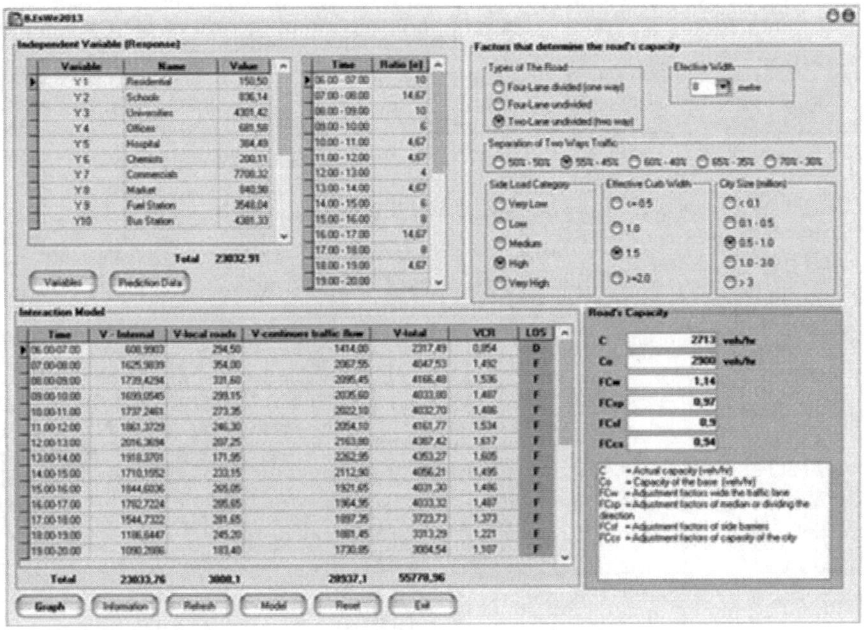

Figure 2. The view of program showing interaction process of land use and road network.

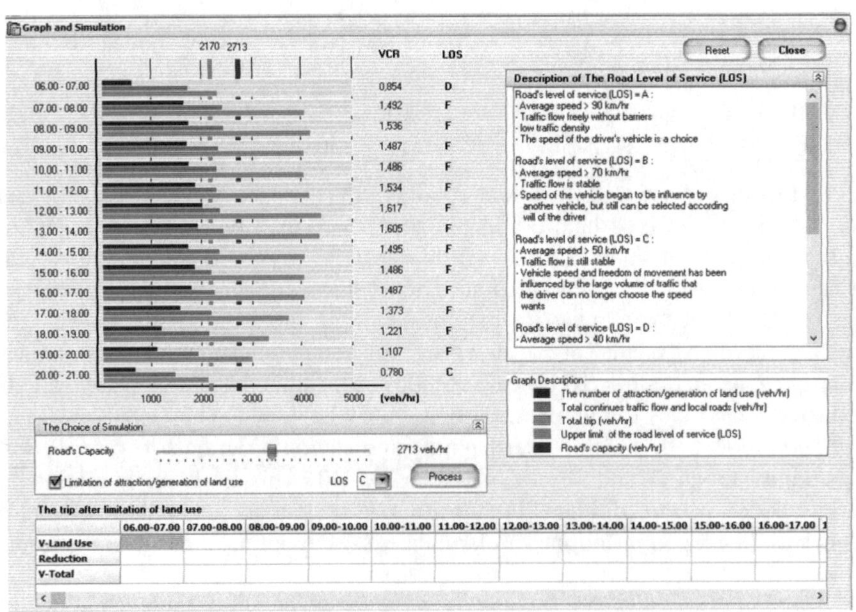

Figure 3. Graph showing total trip volume and road's capacity.

2. Factors that determine the road's capacity
   Several setting options, including type of the road, effective width, separation of two ways traffic, side load category, effective curb width, and city size can be modified in this part.
3. Road's capacity
   This part executes C value by inputting CO, FCW, FCSP, FCSF, FCCS data, the values will change when the previous data is modified.
4. Interaction model
   Execution of the interaction model of land use and road network is calculated in this part. The value resulted from the analysis can be simulated to provide road's performance following the interaction process (figure 3).

4. Conclusion

The character of movement identified from 6.00 am to 9.00 pm. The primary observation showed that CO = 2,900 veh/hr; FCW =1.14; FCSP = 0.97; FCSF = 0.90; and FCCS = 0.94. Based on these values, the actual capacity of the corridor is 2,713 veh/hr. The continuous trip week days 27,523 veh/day while the local trip only 10,763 veh/day. The majorities (60%) of the building permanent and one storey building. 58% of trips heading to/from other districts and 50% of trip used motorbikes.

The research promotes an interaction model between activity system (mixed land use) and road network system using a serie of equations incorporating $\sum$ Vinternal, $\sum$Vexternal, e ratio and Y. V obtained from the observation, while e ratio, which is the ratio of trip volume of each land use at particular time and the total trip volume (veh/day), provided from the trip in each land use. The Y obtained from the regression analyses for each land use (see Table 1). The analysis and estimation can be executed using B.eswe2013, an application that formulated from the interaction process based on the research on the main corridor in Malang City, representing typical intermediate/big city of Indonesia (or East Asian Countries).

The notion, promoted by, among other, Cervero [2] [1], who stated that mixed land uses contribute to reduce trip volume, is not applicable to the study area. It also means the notion is not applicable to Indonesian Cities which are already 'very mixed'. There is a need to measure the extent of mix use, since the typology of East Asian Countries is different to American (Western) Cities.

References

[1] R. Cervero, Land use mixing and suburban mobility, UC Berkeley: University of California Transportation Center, 1988, pp. 429-446.
[2] R. Cervero, "Mixed land-uses and commuting: Evidence from the American History Survey," Transportation Research Part A: Policy and Practice, vol. 30, no. 5, pp. 251-260, 1996.
[3] E. Burton, "The compact city: Just or just compact? A preliminary analysis," Urban Studies, vol. 37, no. 11, pp. 1969-2006, January 2000.
[4] M. Breheny, "The compact city and transport energy consumption," Transactions of the institute of British geographers, vol. 20, no. 1, pp. 81-101, 1995.
[5] B. S. Waloejo, Surjono and H. Sulistio, "The influence of trip attraction on the road's level of service (LOS) at traditional market land use," Journal of Applied Environmental and Biological Sciences (JAEBS), pp. 92-96, 2012.
[6] J. d. D. Ortuzar and L. G. Willumsen, Modelling transport, New York: Wiley, 1994.
[7] G. Emberger, P. C. Pfaffenbichler and S. Jaensirisak, "Application of the European Landuse-Transport interation model MARS to Asian cities," in CUPUM 05, London, 2005.
[8] P. M. Torrens, "How land-use-transportation models work," Centre for Advanced Spatial Analysis University College London, London, 2000.

IOP Conf. Series: Earth and Environmental Science **70** (2017) 012006 doi:10.1088/1755-1315/70/1/012006

Street-service-level approach towards the calculation of CO emission in Malang City, Indonesia

D M Utomo, A Bakkara and K E Sari

Department of Urban and Regional Planning, Faculty of Engineering, Universitas Brawijaya, Malang, Indonesia

dadang.utomo@ub.ac.id

Abstract. Malang has shown an annual vehicle growth of 15%. However, it is an umfortunate fact that 32% of 44 main streets are identified as having low service level, according to a local transportation ranking report. Such condition results in the decline of average vehicle velocity, approaching to the level of velocity (v) = 0; or in other words, street saturation degree could reach >1. The condition is not proportional to the increase of CO concentration in Malang in 2013-2014 as shown in the result of Evaluation of City Air Quality in 2014 which jumped from 3000 $\mu m/m^3$ in 2013 to almost 5000 $\mu m/m^3$ in 2014. This study was aimed at evaluating the extend to which street-service-level variables influence the production of CO emission from motorized transportation activity in an urban street in Malang. Gatot Subroto Street is chosen as a case study according to Multi Criteria Analysis. Furthermore, the street-service-level variables being evaluated include vehicle volume, velocity, side friction, effective roadside width and effective street width. Through a qualitative statistical analysis approach using a multiple linear regression analysis, the result suggests that vehicle volume and side friction are the most dominant factors (Xi) that significantly influence CO emission loads (Y).

1. Introduction

The number of motor vehicle registered in City of Malang in 2015 reached 512,072 vehicles and based on Document of Local Transportation Ranking in City of Malang in 2015-2035, the number of vehicle increases 15% annually. The number excludes vehicles that come from other regions as a result of regional movement. The geographical location of Malang is considered strategic in East Java because City of Malang is surrounded by Regency of Malang and has regional connection with City of Surabaya, Blitar, Lumajang, Pasuruan, Kediri and with some tourism objects. Such condition becomes the main factor of the intensity of external-external and external-internal movements in City of Malang.

Unless there is an effective control, the traffic growth in city of Malang as mentioned above will likely result in the decline of average vehicle velocity, approaching to velocity (v) = 0 or in other words, street saturation degree (volume/capacity) could reach >1 [1]. From the data, it can be seen that 32% of 44 main streets in City of Malang have low serviceability level [2]. Beside the growth of vehicle as its prominent cause, traffic jam is also caused by lack of existing street facility capacity or the activities that cause the deterioration of existing capacity. Tamin (2000) explained that activities such as vehicles parking along the roadside or street vendors using roadside for selling-buying activities force pedestrians to use the street and such condition will reduce street capacity in and as its consequence the street does not have capacity as initially planned [3]. The condition of street serviceability as determined by factors such as vehicle volume, velocity and street network characteristics such as effective roadside width, effective street width and side friction as determinants for capacity greatly influences efficiency and effectiveness of vehicle movement [4]. Consequently, if it is not planned optimally or if disruption is still present, vehicle, and volume shows linear growth, and saturation degree will reach >1.

Traffic jam causes environmental issues such as air pollution whose intensity keeps increasing especially for CO gas. Fardiaz (2010) described that 60% emission produced by vehicle is carbon monoxide [1]. Furthermore, transportation contributes 90% CO gas in air. The study tries to analyze how serviceability level influences the production of CO emission load from motor vehicle through the evaluation of determinant factors of street serviceability level. The study will evaluate whether factors that so far have been regarded as the base of estimation for street serviceability level can significantly influence the existing CO emission production. The focus of this study is limited only on the production of emission gas. Its background builds upon the evaluation result of Indonesia's National Department of Environment (2014) that showed the concentration increased in 2013-2014 in City of Malang as well as CO gas concentration that is much more dominant in comparison to other emission gas from total emission produced by motor vehicle [5]. Street corridor that would be analyzed was street corridor considered a prioritized artery that had been previously determined through the use of Multi Criteria Analysis method.

2. Methodology

2.1. Multi Criteria Analysis

We carried out Multi Criteria Analysis to determine 1 prioritized street out of 44 artery streets in City of Malang. The decision was made by at least 6 experts who have expertise in research area [6]. The applied criteria to determine prioritized street in this study consisted of pavement quality, street dimension, traffic pattern, saturation degree, and area usage. The determination of those criteria was based on past studies.

2.2. Characteristic Analysis of Determinant Factors of Street Serviceability Level

In this study we applied characteristic analysis of determinant factors of street serviceability level to describe or illustrate effective street width, velocity, effective roadside, vehicle's volume and side friction. For vehicle volume, vehicle unit was converted into passenger car unit (pcu). We carried out the analysis of side friction by summing up all present weight based on MKJI 1997 with weighting categories such as the number of pedestrians walking and crossing along street segment (Pedestrian/PED) with weight of 0.5, the number of stationary vehicles with weight of 1.0, the number of vehicles moving in and out parking area (Parking and Slow Vehicles/PSV) with weight of 0.7 and the number of slow moving vehicles (Slow Moving Vehicles/SMV) with weight of 0.4.

2.3. Analysis of Vehicle CO Emission Load

We conducted the analysis of vehicle emission load to reveal the production of CO gas emission in street under our study by previously collecting data of vehicle volume in accordance with vehicle type, total street length and CO emission factor [7]. The applied formula was

$$E_p = \sum_{i=1}^{n} L \times Ni \times Fpi \tag{1}$$

Where:

L	= Street length under current study
Ni	= Number of type i vehicles that pass on the street (vehicle/hour)
Fpi	= Emission factor of type i vehicle (g/Km)
i	= Vehicle type (1 ñn)
Ep	= Emission intensity from certain street (g/hour/km)
P	= Estimated pollutant type

2.4. Multiple Linear Regression Analysis

In this study, we applied Multiple Linear Regression Analysis to know the influence of determinant factors of street serviceability level on the emission production (CO) of motor vehicle. Subsequently some tests were also conducted to ensure whether the resulted model was feasible or not. The applied test would be Model Feasibility Test that comprised of Model Reliability Test (F Test), Determination Coefficient Test, Regression Coefficient Test (T Test), and Classical Assumption Test that consisted of Normality Test, Multicolinearity Test and Heteroscedasticity Test.

2.5. Research Variables

Variables used in the study were based on previous studies as shown in the following table:

Table 1. Research variables.

No	Variable	Sub Variables
1	Street Condition	Street Pavement Quality Street Dimension
2	Activity	Area Usage
3	Accessibility	Traffic pattern Saturation Degree
4	CO emission load from motor vehicle	Street length CO gas emission factor Vehicle volume
5	Determinant Factors of Street Serviceability Level	Effective street width Velocity Effective roadside Vehicle volume Side Friction

3. Results

3.1. Prioritized Street

There were 19 primary artery streets and 23 secondary artery streets included into multi criteria analyses. Based on secondary data and the result of 5 criteria weighting by 10 experts from Transportation Department of City of Malang, Environment Department of City of Malang, Regional Planning and Development of City of Malang and Street Construction and Maintenance Division in Public Works Department, it was recognized that criterion with the highest weight was saturation degree. Moreover, Gatot Subroto Street showed the highest total score out of other 41 streets with total weight of 2933.33 or normalized score of 3.25%. It showed that Gatot Subroto Street had AN existing condition that was assumed to be able to produce emission load from motor vehicle that was higher than that of other streets. General Gatot Subroto Street is located in Kelurahan Jodipan, Kecamatan Blimbing that spans 0.713 km long. This street is A primary artery street that functions as national road to connect City of Malang with Malang Regency and Pasuruan Regency. The location of Gatot Subroto Street and the division of research segments can be seen in the following figure:

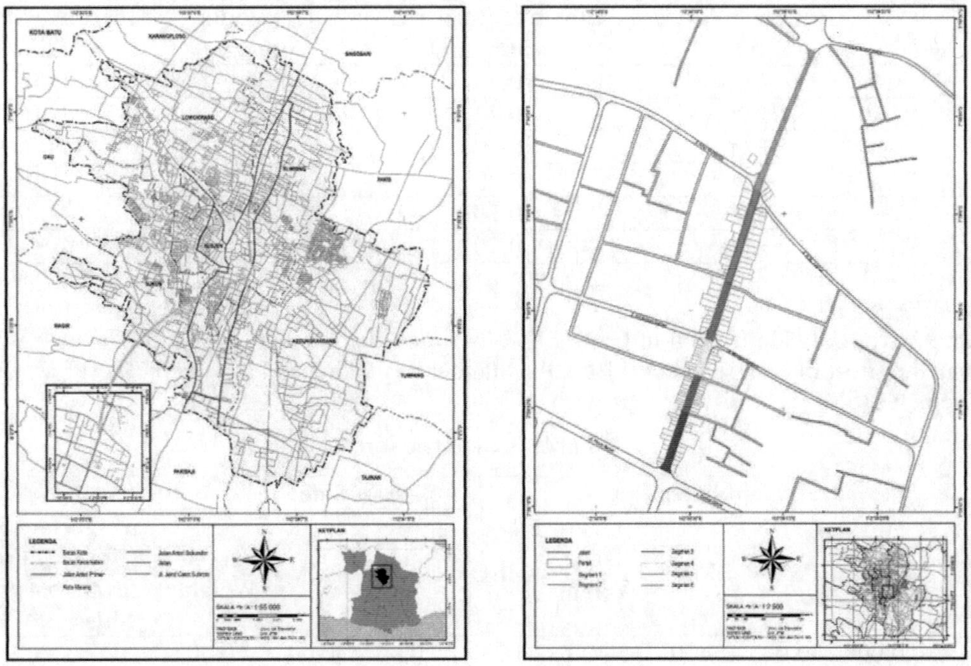

Figure 1. (a) The location of General Gatot Subroto Street; (b) The division of Gatot Subroto Street segments.

3.2. Characteristic of Determinant Factors of Gatot Subroto Street Serviceability Level

3.2.1. Vehicle volume and average velocity

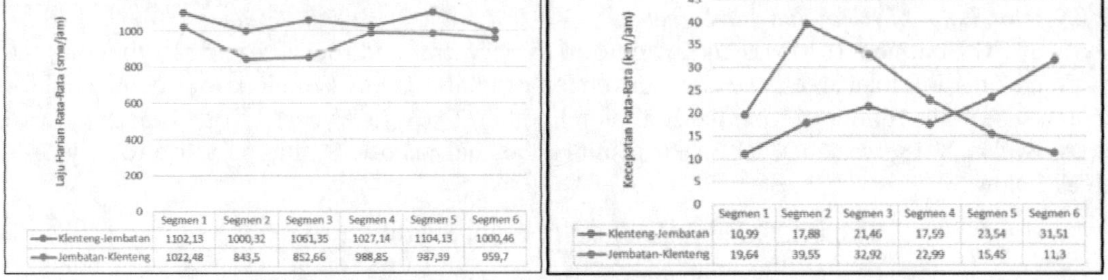

Figure 2. (a) Average daily speed; (b) Average velocity

3.2.2. Street Width and Effective Roadside

Table 2. Characteristic of street width and effective roadside width.

Segment	Side	Roadside Width (m)	Effective Roadside Width (m)	Street Width (m)	Effective Street Width (m)
1	East	0.30	0.30	4.00	4.00
	West	0.35	0.35	4.00	4.00
2	East	1.43	0.72	5.40	5.00
	West	1.60	0.75	5.40	5.00
3	East	1.43	0.72	5.40	5.40
	West	1.60	0.75	5.40	5.00

Segment	Side	Roadside Width (m)	Effective Roadside Width (m)	Street Width (m)	Effective Street Width (m)
4	East	3.06	0.70	4.00	4.00
	West	2.20	0.70	4.50	4.50
5	East	3.06	0.70	4.00	4.00
	West	2.20	0.70	4.50	4.50
6	East	3.06	0.70	4.00	4.00
	West	2.20	0.70	4.50	4.50

3.2.3. Side Friction
There were 3 types of side friction in Gatot Subroto Street; the number of pedestrians who walk and cross along the street in each segment (PED), the number of parking and slow vehicles (PSV) and slow moving vehicles (SMV).

Table 3. Side friction.

Segment	West Side				East Side				Total Weight for Both Sides	Side Friction Class
	PED	PSV	SMV	Total Weight	PED	PSV	SMV	Total Weight		
1	0.00	30.00	24.00	54.00	0.00	20.00	30.00	50.00	104.00	Low
2	18.00	26.00	25.20	69.20	13.50	30.00	23.20	66.70	135.90	Low
3	10.00	25.00	29.20	64.20	12.50	22.00	28.00	62.50	126.70	Low
4	24.50	13.00	18..00	55.50	18.50	9.00	17.20	44.70	100.20	Low
5	15.50	10.00	18.80	44.30	11.00	12.00	19.20	42.20	86.50	Very Low
6	12.00	17.00	26.00	55.00	16.50	8.00	17.20	41.70	96.70	Very Low

3.3. *CO Emission Load of Gatot Subroto Street*
The analysis of CO Emission Load was carried out for six segments and routes. After carrying out the analysis of load emission for 1 week, the data conversion of total CO emission load was conducted for 1 year with the use of the following formula: Emission Load in 1 year (ton/year) = Total Emission Load in 1 week (gr/week) X 48 (weeks) X 10-6. The result of the analysis of CO emission load for 1 year is presented as follows:

Table 4. CO emission load of Gatot Subroto street.

Segment	Flow Direction	Total Emission Load for 1 week (gram/week)	Emission Load in 1 year (ton/year)	Emission Load in 1 year per segment (ton/year)
1	Klenteng-Jembatan	1289797.22	61.91	115.90
	Jembatan-Klenteng	1124763.74	53.99	
2	Klenteng-Jembatan	772740.87	37.09	66.36
	Jembatan-Klenteng	609712.84	29.27	
3	Klenteng-Jembatan	482994.81	23.18	40.76

Segment	Flow Direction	Total Emission Load for 1 week (gram/week)	Emission Load in 1 year (ton/year)	Emission Load in 1 year per segment (ton/year)
	Jembatan-Klenteng	366137.66	17.57	
4	Klenteng-Jembatan	290305.32	13.93	26.43
	Jembatan-Klenteng	260253.97	12.49	
5	Klenteng-Jembatan	345804.07	16.60	30.26
	Jembatan-Klenteng	284574.54	13.66	
6	Klenteng-Jembatan	518328.72	24.88	46.63
	Jembatan-Klenteng	453030.03	21.75	
Total		**6798443.80**	**326.33**	**326.34**

3.4. Multiple Linear Regression Analysis

There were 3 times OF modeling with the different number of independent variables. However 12 samples (6 segments and 2 directions) and 1 dependent variable (CO emission load) were still used. Model 1 showed that velocity variable and effective street width variable had coefficient value that positively influenced Y so the result could not be accepted because it didn't conform to the existing theory. In model 2, those two variables were excluded and the result showed that only volume variable and side friction variable had sig value <0.05. Consequently, in model 3 vehicle volume variable and friction side variable were included with the following result:

3.4.1. Feasibility Test Model

1) Model Reliability Test (F test)

Count prob. F values (sig) in model 1, model 2 and model 3 were 0.000. Three models had significant values smaller than error (alpha) that was 0.05. As a result, we concluded that the estimated linear regression model was feasible to apply in order to describe the influence of Side Friction, Effective Roadside Width, Volume, Velocity and Effective Street Width on Emission Load (dependent variable).

2) Determination Coefficient

R value in model 3 was 0.986 so it means that the influence of vehicle volume and side friction on Load Emission was 98.6% (0.986 x 100%). Moreover, R square value of 0.973 showed that the influence proportion of vehicle volume variable and side friction variable on Emission Load was 97.3% while the rest, 2.7% (100%-97.3%) was influenced by other variables that were not included in this model.

3) Regression Coefficient Test (T Test)

Table 5. Regression coefficient test

Model		t	Sig.
1	(Constant)	-3.327	.016
	Volume	12.114	.000
	Effective Roadside Width	-.096	.926
	Velocity	.388	.712
	Effective Street Width	.208	.842
	Side Friction	.772	.469

Model		t	Sig.
2	(Constant)	-4.156	.003
	Volume	15.773	.000
	Effective Roadside Width	.126	.903
	Side Friction	2.330	.048
3	(Constant)	-4.906	.001
	Volume	17.596	.000
	Side Friction	2.494	.034

The previous table describes that in model 1 the variable that significantly influenced dependent variable was vehicle volume while in model 2 in which velocity and effective street width were excluded the result showed that the variables that significantly influenced dependent variable were vehicle volume and side friction so in model 3 the included variables were volume and side friction and the result showed that both variables were significant in the model.

3.4.2. Classical Assumption Test

1. Normality Test

In model 1 skewness ratio was 0.278/0,637 = 0.436; while kurtosis ratio was -0.522/1,232 = -0.424. In model 2 skewness ratio was 0.426/0.637 = 0.669; while kurtosis ratio was -0.007/1.232 = -0.006. In model 3 skewness ratio was 0.448/0.637 = 0.703; while kurtosis ratio was 0.025/1.232 = 0.020. Because skewness ratio and kurtosis ratio in all models were between -2 and +2, so it could be concluded that data distribution was normal.

2. Multicollinearity Test

From the result of multicollinearity test we knew that all independent variables either in model 1, 2, OR 3 had VIF value smaller than 10 and tolerance value bigger than 0.1. Consequently, we concluded that those variables didn't have any multicollinearity issue.

3. Heteroscedasticity Test

After carrying out heteroscedasticity test with glejser method, the result showed that t-statistic values from all independent variables either in model 1, 2 OR 3 were not significant so we concluded that this model did not have any heteroscedasticity issue.

3.4.3. Model Interpretation

Table 6. Coefficients

Model		Unstandardized Coefficients		Standardized Coefficients	t	Sig.
		B	Std. Error	Beta		
3	(Constant)	-17.857	3.639		-4.906	.001
	Volume	.052	.003	1.025	17.596	.000
	Side Friction	.062	.025	.145	2.494	.034

The formed equilibrium is as follows:

$$Y = -17.857 + 0.052\, X_1 + 0.062\, X_5 \qquad (2)$$

Where:
Y = Emmision Load (gr/hour per meter of street length)
a = -17.857 (Constant)
$X1$ = Vehicle volume (pcu/hour)
$X5$ = Side Friction (present weight/hour)
$b1$ = 0.052 (Variable coefficient of vehicle volume)
$b5$ = 0,062 (Variable coefficient of side friction)
Model interpretation can be carried out as follows:

1. b1 value = +0.052

When side friction (X5) is constant, each increase in vehicle volume value (X1) of one pcu/hour will increase emission load (Y) to 0.050 gr/hour per meter of street length.

2. b5 value = +0.062
   When vehicle volume (X1) is constant, each increase in side friction value (X5) of one present weight/hour will increase emission load (Y) to 0.062 gr/hour per meter of street length.

4. Conclusion

The result of Multiple Linear Regression analysis shows that from 5 variables, variables that significantly influence dependent variable are vehicle volume and side friction. When volume and side friction increase, CO emission load from vehicle will also increase. And the comparison of two coefficients shows that coefficient of side friction variable has higher value than coefficient of vehicle volume variable. It shows that side friction gives more influence on CO emission production than vehicle volume.

References

[1] Panjaitan, T. P. M., 2010. Model Pengelolaan Transportasi Berkelanjutan Di Kawasan Pinggiran Metropolitan. Disertasi. Postgraduate Program Institute of Agriculture Bogor. Bogor
[2] Tataran Transportasi Lokal Kota Malang Tahun 2015-2035. Dinas Perhubungan Kota Malang. Malang.
[3] Tamin, O. Z. Perencanaan Dan Permodelan Transportasi. 2 ed. Bandung: Penerbit ITB; 2000.
[4] Manual Kapasitas Jalan Indonesia. 1997. Directorate General Bina Marga. South Jakarta.
[5] Evaluasi Kualitas Udara Perkotaan Tahun 2014. Kementerian Lingkungan Hidup. East Jakarta.
[6] Mendoza, G. A and Macoun, P. Panduan Untuk Menerapkan Analisis Multikriteria Dalam Menilai Kriteria Dan Indikator. 1 ed. Jakarta: Center For International Forestry Research (CIFOR); 1999.
[7] Minister of Environment Regulation No. 12 Year 2010. Pelaksanaa Pengendalian Pencemaran Udara Di Daerah. 26 March 2010. Minister of Agriculture in Environmental Planning. Jakarta.

IOP Conf. Series: Earth and Environmental Science **70** (2017) 012007 doi:10.1088/1755-1315/70/1/012007

Societal perception toward transportation modes based on online (Go-Jek) In Malang City

D M Buamona

Department of Urban and Regional Planning, Faculty of Engineering, Universitas Brawijaya, Malang, Indonesia

dedibuamona@gmail.com

Abstract. Driver online which is renowned as Go-jek is one of alternative transportation exist in Malang city based on IT. This service is supportive enough for all societies and easy to do serving, namely it is enough by using Go-jek application which is exist in Smartphone. Some services served by Go-jek including Go-Ride, Go-Car, Go-Food, Go-Send dan Go- Massage. By the development of technology and the emergence of alternative transportation, it will not close any chance to get competition.In its development some Go-Jek users started complaining about the services provided. This study aims is to assess the level of importance and performance and measure the level of community satisfaction in using Go-Jek with the method of Importance Performance Analysis (IPA) and Customer Satisfaction Index (CSI), Importance Performance Analysis (IPA) is used to determine the attributes of importance and level of satisfaction of performance while the Customer Satisfaction Index (CSI) is used to view the percentage of satisfaction level. Accidental Sampling technique is used to determine the sample based on spontaneity factor as well as for data collection using questionnaires with 100 respondents. The results of this study indicate that services considered important by the community but less attention namely waiting time for short reservations, ability to reach a specific location. Drivers deliver you to the right destination and reliability. For percentage of respondent satisfaction level, it is seen from several criteria: a) 0,81-1,00 Very satisfied, b) 0,66-0,80 Satisfied, c) 0,51-0,65 Simply satisfied, d) 0,35 -0.50 Less satisfied, and d) 0.00-0.34 Very dissatisfied. From the results of this study, it shows that people feel very satisfied with the service given today with a value of 0.84 CSI, this value is at the interval of 0.81-1.00. Based on the results of the analysis, although the percentage of this study shows high level of satisfaction of the respondents but the service improvements should continue if Go-Jek wants to make this transport as a public transportation alternative choice for people that is not to move on to other alternative transportation.

1. Introduction

Malang city is a city located in East Java Province. The city is located at 90 Km south of Surabaya City and is the largest city in Indonesia by population [1]. The development of Malang City does not mean that the city is free from various problems. With limited job opportunities coupled with economic demands, then the motor that is used as a daily activity personally has been converted in function as a service provider / transportation tool that transports people and goods.

Recently the motorcycle taxi transformed into a professional means of transportation supported by technological sophistication called go-jek. Go-jek is a service company which is offering speed, security and convenience. One of the companies that carries the concept of business sharing economy is now present in Indonesian [2], especially in Malang. Because Go-jek uses applications that can be used only temporarily, this is what distinguishes between Go-jek with a conventional motorcycle taxi. A service quality can be measured using five dimensions, namely, Reliability, Responsiveness, Assurance, Empathy, and Tangibility [3]. The dimensions which set out above is to assess and measure the quality of services to the company. The quality of services provided by Go-Jek will affect people's perception differently on the use of such services.

2. Methods

2.1. Importance Performance Analysis (IPA)

Importance Performance Analysis (IPA) is an application method to measure attributes according to importance and performance level or satisfaction level, it is to develop effective strategies for the company. This technique is based on four quadrant matrix that is bounded by two line sections intersecting perpendicular to point X (Y satisfaction level) and Y (average satisfaction level) and identifying the strengths and weaknesses of the service and determining improvement opportunities to develop strategic planning [4]. Martilla and Jams in [5] suggest the use of Importance Performance Analysis method in measuring service satisfaction level [5]. The main function is to display information relating to service facts which, according to consumers, greatly affect their satisfaction and loyalty. To find out how big customers feel satisfied with the performance level, use the following formula:

$$T_{ki} = \frac{X_i}{Y_i} \times 100\% \tag{1}$$

Where:
T_{ki} = Respondent appropriateness value
X_i = Evaluation score based on satisfaction
Y_i = Evaluation score based on importance

After doing appropriateness level measurement, for the next is to make map position of importance performance which constitutes a building consisted of four quadrants.

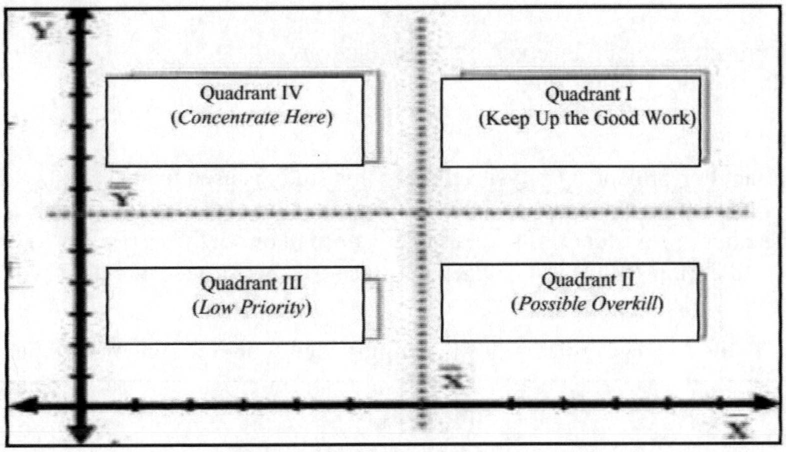

Picture 1. Cartesius diagram of IPA.

Strategies that can be done in regard to the position of each variable from the four quadrants can be explained as follows:

a. Quadrant I (Keep up the Good Work), factors that are considered important & have high performance.
b. Quadrant 2 (Possible Overkill), Factors that are considered less important to customers, but have a performance which is quite satisfactory.
c. Quadrant 3 (Low Priority), the factors that are considered less important and not very special

d. Quadrant 4 (Concentrate Here), Factors that are considered important by the customer but have not met expectations.

2.2. Customers satisfactory methods
Measurement of the Customer Satisfaction Index is necessary because the results of these measurements can be used as a reference to determine the target next year. Without any customer satisfaction index, it is impossible to determine the management in improving customer satisfaction. Assessment of satisfaction is an evaluation for management to improve service and win the competition [6]. The Customer Satisfaction Index (CSI) is used to view the overall customer satisfaction index by considering the importance of service attributes [7]. The steps to calculate the customer satisfaction index is explained such as follows:

a. Calculating Weighting Factor (WF)
The function of the median importance score or median score of importance for each variable is in the form of percent from the total median score of importance for all attributes tested.
b. Counting Weighted Score (WS)
The function of the median satisfaction score or median score of satisfaction level of each attribute is multiplied by the Weighting Factor (WF) of each variable.
c. Calculating the Total Weighed Median (WMT)
Total of the overall Weighted Score (WS) score
d. Customer Satisfaction Index
Calculations from Weighed Median Total (WMT) are divided into the maximum scale or high test scale multiplied by 100%. Overall satisfaction level of respondents can be seen from the criteria of customer satisfaction or consumer level with the following criteria: a. 0.81-1.00 (Very satisfied); B. 0.66-0.80 (Satisfied); C. 0.51-0.65 (Simply satisfied); D. 0.35-0.50 (Satisfied); E. 0,00-0,34 (Very dissatisfied).

Here is the formula to score the Customer Satisfaction Index (CSI) value such as follow:

$$IKP = \frac{T}{5Y} \times 100\% \quad (2)$$

Named with:
T = Multiplication result of importance value by satisfaction
Y = the result of importance accumulated

3. Results
Analysis of Important Performance Analysis (IPA) in this study is used to analyze the factors of interest and community satisfaction by using go-jek. The importance level of the variable service can be measured from whether the existence of the variable is needed or not. Variable services with high value should get serious attention from the management because this service variable will have dominant influence at the level of performance.

Table 1. The evaluation toward importance score on each of variables.

No	Statement	Importance Average
A	**Physical proof (Tangibility)**	
1.	Physical condition (Motor is in a good condition)	4,4
2.	Equipment (Driver uses uniform, helm and etcetera)	4,3
3.	The amount of vehicle (The available vehicle in appropriate condition)	4,2
B	**Empathy**	
4.	Kecakapan operator craftmanship (Fast in understanding your problems)	4,1
5.	Driver gives attention toward you	4,2
6.	Driver cares with collective safety	4,4
C	**Responsiveness**	

No	Statement	Importance Average
7.	The capability to give fast serving	4,3
8.	The fast waiting time from consumer	4,3
9.	Driver understand your needs	4,1
D	**Reability**	
10	The capability to reach some particular areas	4,4
11.	Driver will deliver you into the right place in a right time	4,2
12.	Driver deliver you in place faster	4,2
E	**Assurance**	
13.	Driver can be relied upon	4,1
14.	Driver obey the traffic	4,1
15.	You feels comfortable with any crime	4,1
	Total	63,4

Table 2. The evaluation toward satisfactory score on each of variables

No	Statement	Importance Average
A	**Physical proof (Tangibility)**	
1.	Physical condition (Motor is in a good condition)	4,3
2.	Equipment (Driver uses uniform, helm and etcetera)	4,4
3.	The amount of vehicle (The available vehicle in appropriate condition)	4,2
B	**Empathy**	
4.	Kecakapan operator craftmanship (Fast in understanding your problems)	4,1
5.	Driver gives attention toward you	4,2
6.	Driver cares with collective safety	4,3
C	**Responsiveness**	
7.	The capability to give fast serving	4,2
8.	The fast waiting time from consumer	3,9
9.	Driver understand your needs	4,1
D	**Reability**	
10.	The capability to reach some particular areas	4,0
11.	Driver will deliver you into the right place in a right time	4,1
12.	Driver deliver you in place faster	4,4
E	**Assurance**	
13.	Driver can be relied upon	4,1
14.	Driver obey the traffic	4,2
15.	You feels comfortable with any crime	4,0
	Total	62,4

Based on the calculation by using the IPA method, the variables that need to get the attention are the variables contained in quadrant IV. The results of the analysis show that quadrant IV is a factor that is considered important and located and it is in bad condition so that it needs attention and improvement. The degree of importance and satisfaction in this study is presented in the form of a Cartesius diagram (Figure 1). The variables that are in quadrant IV consist of waiting time for short order, ability to reach certain location and can deliver you to the right destination which is surely reliable drive.

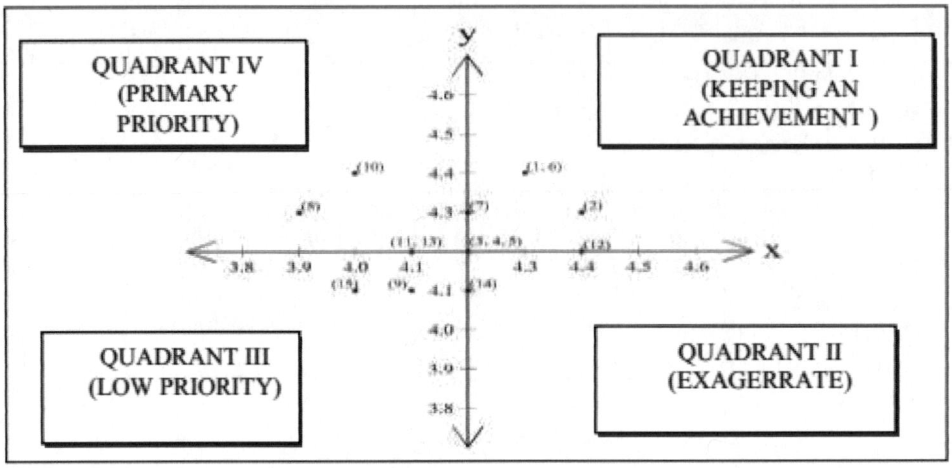

Picture 2. Cartesius Diagram.

For the calculation result with Customer Satisfaction Indeks (CSI), it shows the satisfaction of Go-Jek consumer which can be seen in table 3.

Table 3. Customer satisfaction index.

No	Statement	Satisfaction (P) Scale 1-5	Understanding (I) Scale 1-5	Score (S) S = (P) x (I)
A	**Physical proof (Tangibility)**			
1	Physical condition (Motor is in a good condition)	4,3	4,3	18,2
2	Equipment (Driver uses uniform, helm and etcetera)	4,4	4,3	18,6
3	The amount of vehicle (The available vehicle in appropriate condition)	4,2	4,2	17,6
B	**Empathy**			
4	Kecakapan operator craftmanship (Fast in understanding your problems)	4,1	4,1	16,7
5	Driver gives attention toward you	4,2	4,2	17,7
6	Driver cares with collective safety	4,3	4,4	18,7
C	**Responsiveness**			
7	The capability to give fast serving	4,2	4,3	18,2
8	The fast waiting time from consumer	3,9	4,3	16,8
9	Driver understand your needs	4,1	4,1	16,6
D	**Reability**			
10	The capability to reach some particular areas	4,0	4,4	17,6
11	Driver will deliver you into the right place in a right time	4,1	4,2	17,1
12	Driver deliver you in place faster	4,4	4,2	18,6
E	**Assurance**			
13	Driver can be relied upon	4,1	4,1	16,9
14	Driver obey the traffic	4,2	4,1	17,2
15	You feels comfortable with any crime	4,0	4,1	17,2
	Total		63,3	263,44
	CSI			83,22

From recapitulation score on the table above, so it can be calculated such as follow:

$$CSI = \frac{263{,}44}{5\times 63{,}3}\times 100\% = 83.33 \sim 0{,}83\% \quad (2)$$

From the calculation results using CSI it gets score up to 0.83. The value is on value interval of 0.81-1.00 which means that customers / consumers feel "very satisfied" with Go-Jek's current performance. This satisfaction does not mean that Go-jek management does not need to improve the level of service. Improvement of service level is necessary as in Cartesian Quadrant IV diagram.

4. Conclusion

From the result of calculation by using IPA, there are four variables which is stated as important for the costumer but the work performance is low and it is needed to pay the attention including, the time for waiting the to get instant serving, the skill to reach particular location, the driver will deliver you to the exact purpose, and driver can be relied upon meanwhile the work performance is assumed as good and need to be protected which consist of , vehicle physical condition, driver uses more detailed equipment, the total vehicle are well prepared, the operator capability in answering the problem, driver gives attention toward you, the driver cares about your safety. The capability to give the fast serving, and driver deliver you in brief time. Meanwhile for the satisfactory percentage, it shows that CSI calculation reach score up to 0,83. Those value is in 0,81 – 1,00 interval score which means that the consumer is very satisfied with Go-jek performance up to day.

According to the analysis result done , eventhough the result of presentation on this research shows the satisfactory level of highly motivated respondent but the service improvement must be done in keeping if the Go-Jek wants to make this vehicle as alternative transportation societal choice in order for the society is not transported into the other alternative

References

[1] Kota Malang dalam angka tahun 2008, (Malang City in Figures)

[2] Aghia Ghassani Harish, Aditya Wardhana.(2013). Analisis Faktor-Faktor Pembentukan Preferensi Konsumen Go-jek di Kota Jakarta. Universitas Telkom

[3] Jasfar, Farida. (2009). Manajemen Jasa. Ciawi, Bogor. Ghalia Indonesia.

[4] Mohebifar, Rafat, et.el (2016). Evaluating Service Quality from Patients Perceptions : Application of Importance Performance Analysis Method. Journal Osong Public Health and Research Perspective. Volume 7, Issue 4, Pages 233-238

[5] Zeithaml, V.A., Parasuraman, A. dan Berry, L.L., (1990), Delivering Quality Service, New York.

[6] Siti, Husna. (2014). Penerapan Customer Satisfaction Index (CSI) dan Analisis GAP Pada Kualitas Pelayanan Trans Jogja. Jurnal Ilmiah Teknik Industri, Volume 13, No 2.

[7] Maiyanti, S. I., Irmeliyana, dan Verawati, 2009, Aplied Customer Satisfaction Index (CSI) and Importance Performance Analysis (IPA) to Know Student Satisfaction Level of Sriwijaya University Library Services, Universitas Sriwijaya, Palembang.

[8] Hair, JF, RE Anderson, RL Tatham, and WC Black. Multivariate Data Analysis, Prentice Hall International, New Jersey; 1998.

[9] Ridwan, Cetakan kedua. Skala Pengukuran Variabel-Variabel Penelitian, Bandung. CV Alfabetta; 2003

IOP Conf. Series: Earth and Environmental Science **70** (2017) 012008 doi:10.1088/1755-1315/70/1/012008

Factor weighting in DRASTIC modelling for assessing the groundwater vulnerability in Salatiga groundwater basin, Central Java Province, Indonesia

D A Kesuma[1, 5], P Purwanto[1, 2], T T Putranto[1, 3], and T P D Rahmani[4]

[1]Master Program in Environmental Science, School of Postgraduate Studies, Diponegoro University, Semarang 50241, Indonesia
[2]Department of Chemical Engineering, Faculty of Engineering, Diponegoro University, Semarang 50275, Indonesia
[3]Department of Geological Engineering, Faculty of Engineering, Diponegoro University, Semarang 50275, Indonesia
[4]Department of Biological Science, Faculty of Science and Engineering, Islamic State University of Walisongo, Semarang 50185, IndonesiaType the author addresses here

da.kesuma@gmail.com

Abstract. The increase in human population as well as area development in Salatiga Groundwater Basin, Central Java Province, will increase the potency of groundwater contamination in that area. Groundwater quality, especially the shallow groundwater, is very vulnerable to the contamination from industrial waste, fertilizer/agricultural waste, and domestic waste. The first step in the conservation of groundwater quality is by conducting the mapping of the groundwater vulnerability zonation against the contamination. The result of this research was groundwater vulnerability map which showed the areas vulnerable to the groundwater contamination. In this study, groundwater vulnerability map was assessed based on the DRASTIC Method and was processed spatially using Geographic Information System. The DRASTIC method is used to assess the level of groundwater vulnerability based on weighting on seven parameters, which are: depth to the water table (D), recharge (R), aquifer material (A), soil media (S), topography (T), impact of vadose zone (I), and hydraulic conductivity (C). The higher the DRASTIC Index will result in the higher vulnerability level of groundwater contamination in that area. The DRASTIC Indexes in the researched area were 85 – 100 (low vulnerability level), 101 -120 (low to moderate vulnerability level), 121 – 140 (moderate vulnerability level), 141 – 150, (moderate to high vulnerability level), and 151 – 159 (high vulnerability level). The output of this study can be used by local authority as a tool for consideration to arrange the policy for sustainable area development, especially the development in an area affecting the quality of Salatiga Groundwater Basin.

1. Introduction

Water on earth, either surface or groundwater, is very sensitive towards the changing in land utilization, especially the groundwater. The freshwater groundwater can be polluted by contaminants from human activity waste. The contaminants can be originated from agricultural, industrial, as well as domestic waste from the human settlement. Therefore, the decreasing quality of groundwater caused by

IOP Conf. Series: Earth and Environmental Science **70** (2017) 012008 doi:10.1088/1755-1315/70/1/012008

contaminations needed to be urgently treated since almost half of the world's populations are still consuming the groundwater for drinking and other activities [1].

Salatiga groundwater basin is mostly located in the KEDUNGSEPUR (Kendal, Ungaran, Semarang, Salatiga, Purwodadi) area, which is one of the eight strategic areas included in the Central Java Province Spatial Plans (Rencana Tata Ruang Wilayah (RTRW)) Number 21 Year 2003 juncto Central Java Province Local Regulation (PERDA) Number 6 Year 2010, regarding the Central Java Province Spatial Plans. This area is planned to be the center of national economic growth for the local income equalization as well as to decrease the economic imbalance [2]. The Salatiga Groundwater Basin Area is also located on the strategic track which connecting the cities of Jogjakarta, Solo, and Semarang (JOGLOSEMAR). The strategic location of the Salatiga Groundwater Basin Area will open the opportunities to the development of the area, especially for the agricultural, trade and service, industrial as well as tourism sectors development [3].

The establishment of that strategic area was aimed to create the safe, comfortable, productive, and sustainable area so that the social, economic, and environmental aspects could be harmonized. However, its existence can provoke the uncontrollable population growth and land conversion, and thus will give rise to the environmental issues. As a picture, the land utilization in the City of Salatiga for the five-year range from 2010 to 2014, had undergone settlement sector development (increased for 120,40 Ha), trade sector development (increased for 4,48 Ha), industrial sector development (increased for 14,55 Ha), and non-agricultural area utilization. In general, the alteration of above land utilization was the conversion from non-irrigated agricultural land to mixed plantation land [4]. The alteration of land utilization to support the human activities, such as the settlement and industrial development, the usage of fertilizer for agriculture, the construction of the sanitary landfill, are the potential hazards which can influence the groundwater quality and increase the potency of groundwater contamination [5].

Thus, the assessment of aquifer vulnerability is very crucial to be conducted as the early attempt to protect the quality of groundwater. Zabet (2002) described vulnerability as the aquifer system sensitivity to be deteriorated caused by external influence [6]. The groundwater vulnerability concept is assuming that physical environment has the ability to protect the groundwater against the impact caused by the nature and human activities. The Earth materials can be acted a natural filter for the flowing through contaminants [7, 8].

The term ''vulnerability of groundwater to contamination'' was first used by Margat (1968). The groundwater vulnerability is used to the term natural protection against contamination. Groundwater vulnerability to contamination was defined by the National Research Council (1993) as ''the tendency or likelihood for contaminants to reach a specified position in the groundwater system after introduction at some location above the uppermost aquifer'' [9].

Many groundwater vulnerability assessment techniques had been developed in the last few decades, such as overlay and index method, statistics, as well as computer based process. Groundwater vulnerability assessment method used in this study was a DRASTIC Method. The DRASTIC Method is one of the overlay and index methods which can be broadly applied for several aquifers, such as basaltic rock, sedimentary rock, carbonaceous rock, hard rock, and coastal aquifers [1].

The objective of this research was to determine the aquifer vulnerability level against the contamination using the DRASTIC Method as well Geographic Information System to map the contamination-vulnerable areas in Salatiga Groundwater Basin.

2. Study Area

Salatiga groundwater basin is one of 19 groundwater basins across administrative areas regencies / cities in Central Java province. Administratively, most of Salatiga Groundwater Basin is included in the administrative area of Salatiga and Semarang regency as well as a small portion of Boyolali regency. Geographically, Salatiga groundwater basin is located approximately between 110° 27'56.81"-110°32'4.64" East longitude and 7°17'-7°17.23 'South Latitude with an altitude of 220 to 1,450 meters above Mean Sea Level. Salatiga groundwater basin lies mostly in the slopes of Mount Merbabu and small mountains, among others: Gajah Mungkur, Telomoyo, and Payung Rong [10]. The average annual Rainfall of the last 5 years is about 2,198.33 mm/ year [11, 12, 13].

Aquifer system at Salatiga groundwater basin is unconfined and confined aquifer. The rocks that make up the unconfined aquifer usually consisting of tuffaceous sandstone originating from volcanic activity of Mount Merbabu. In the southern part of the study area, the depth of the lower unconfined aquifer is at a depth of about 15-20 meters below local ground level surface, ie starting from the slopes of Mount Merbabu up toward the northeast. In the eastern part of the study area, the position of the bottom of the unconfined aquifer reaches a depth of 30 meters below local ground level surface.

The confined aquifer layer is composed of tuffaceous sandstone layers alternating with lahar and lava. Based on observations from some locations of spring, there are aquifer through the cracks. The confined aquifer is generally located at 43 up to 81 meters below local ground level surface on the slopes of Mount Merbabu and increasingly shallow eastward around the District of Wonosegoro. In the northern part of the research area, the upper confined aquifer depth is ranging from 18 meters below local ground level surface.

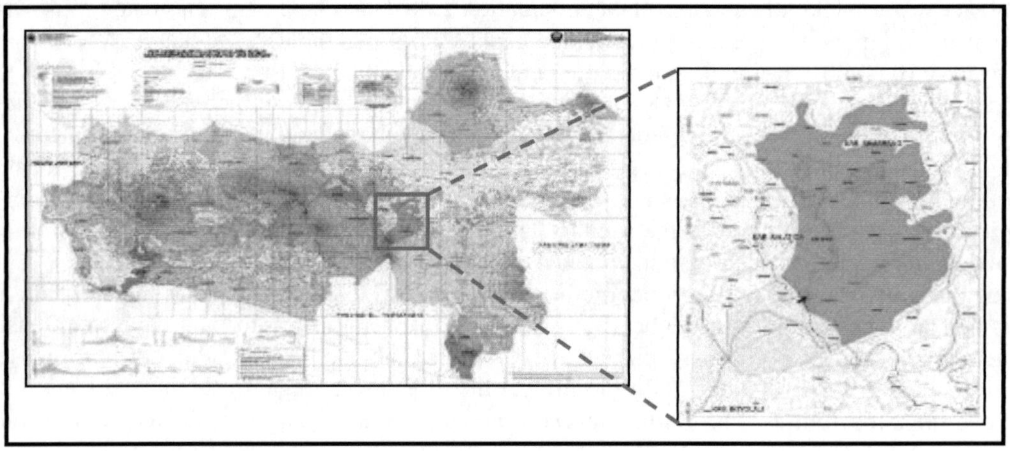

Figure 1. The location of study area.

3. Methodology

The DRASTIC Method of groundwater vulnerability level assessment uses seven parameters, which are: depth to the water level (D), recharge (R), aquifer material (A), soil media (S), topography (T), impact of vadose zone (I), and hydraulic conductivity (C). Depth to the water level data was taken from the depth measurements of the resident's dug wells in October-December 2016. Recharge Value was determined by considering the precipitation, evapotranspiration, and runoff levels in studied area between 2011 until 2015. Recharge level was calculated using meteoric water balance analysis. The aquifer materials, soil media, and impact of vadose zone parameters were obtained from the borehole correlation and interpretation log results as well as from geoelectrical survey results data taken from Energy and Mineral Resources Department of Central Java Province [10]. The topography of study area was prepared based on the Shuttle Radar Topographical Mission (SRTM) satellite data. While the hydraulic-conductivity data was obtained from the rock permeability tests in the study area. Each of the DRASTIC Parameters was weighted according to its level of importance, whereas each of the components from every parameters was rated based of its level of importance. Depth to the water table is weighted 5, recharge is weighted 4, aquifer material is weighted 3, soil media is weighted 2, topography is weighted 1, impact of vadose zone is weighted 5, and hydraulic conductivity is weighted 3. All seven parameters were then converted into digital format via Geographic Information System. By using ArcGIS 10.2, each of the components of the DRASTIC parameters were rated and weighted according to Aller,et.al., 1987 [14].

The next step was to calculate the DRASTIC Index using raster calculator with the formula below:

IOP Conf. Series: Earth and Environmental Science **70** (2017) 012008 doi:10.1088/1755-1315/70/1/012008

$$DI = Dr\,Dw + Rr\,Rw + Ar\,Aw + Sr\,Sw + Tr\,Tw + Ir\,Iw + Cr\,Cw \tag{1}$$

With r = rating and w = weight.

The DRASTIC Index were classified into:

Table 1. Vulnerability Ranges Classification [15]

No	Vulnerability Level	DRASTIC Index
1	Very low	< 80
2	Low	80 - 100
3	Low to moderate	101 - 120
4	Moderate	121 - 140
5	Moderate to high	141 - 150
6	High	151 - 180
7	Very high	181 - 200
8	Extremely high	> 200

4. Result and discussion

4.1. Hydrogeologic Parameters

4.1.1. Depth to groundwater level. Depth to groundwater level is the minimum distance of ground surface from the water table. This is a crucial factor, since the condition of the depth to groundwater level will affect the time needed for the contaminants to reach the water table. The depth to groundwater level data was obtained from the measurements of 70 dug wells from October to December 2016. Depth to groundwater level was the shallowest at 1 meter below ground surface until up to > 30 meter below the local ground surface. The parameter of depth to groundwater level has a weight of 5 (Figure 2 and Table 2).

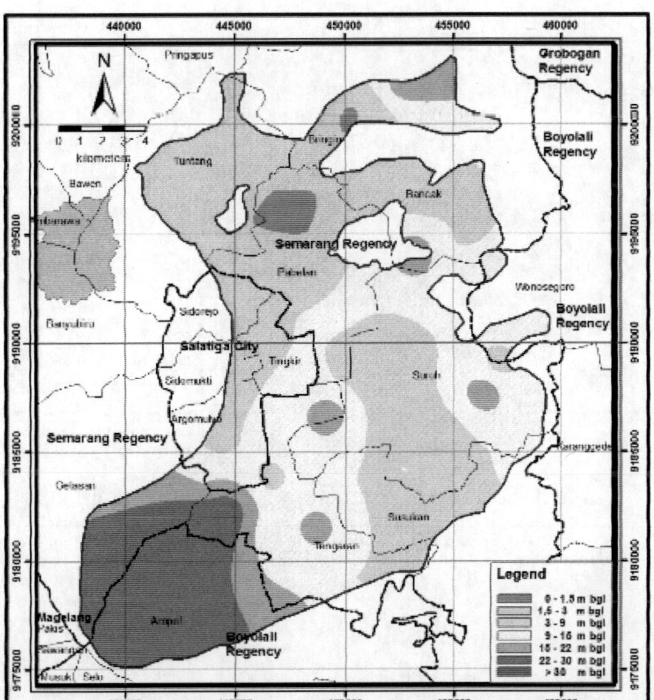

Figure 2. Depth to groundwater level.

Table 2. Value of depth to groundwater level.

No	Depth Interval (m)	Rating	Weight	Index
1	0 - 1.5	10	5	50
2	1.5 - 3	9	5	45
3	3 - 9	7	5	35
4	9 - 15	5	5	25
5	15 - 22	3	5	15
6	22 - 30	2	5	10
7	> 30	1	5	5

4.1.2. Net Recharge. The net recharge describes the amount of rainwater absorbed into the ground and reached the aquifer. The rainwater absorbed in the ground through the soil pores was assumed to be able to help transporting the contaminants vertically toward the groundwater level and horizontally inside the aquifer. Based on the calculation results, the recharge values of the studied areas, which were the recharge zone, were 0-51 mm/year, 52-102 mm/year, 103-178 mm/year, 179-254 mm/year, and 254-1,092 mm/year. The recharge parameter was weighted 4. The parameter score measurement, recharge amount and recharge parameter map can be viewed at Table 3 and Figure 3.

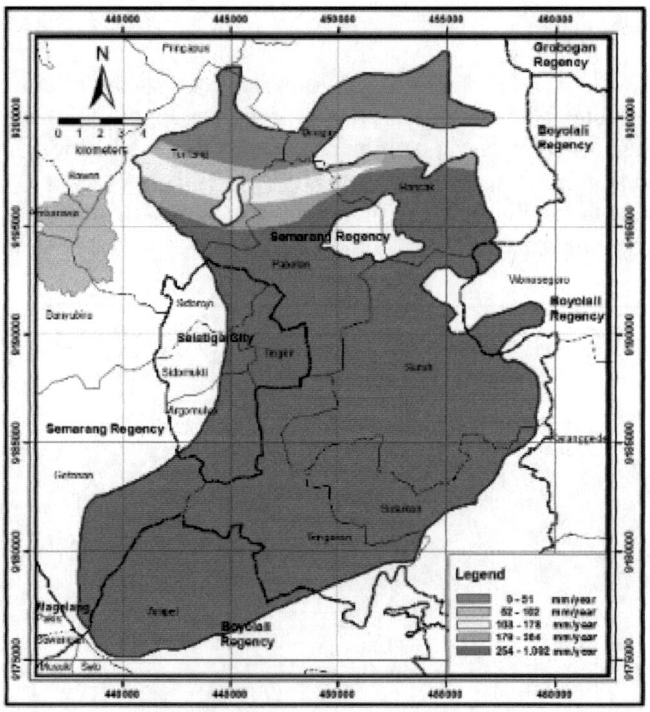

Figure 3. Net recharge.

Table 3. Score Measurement for Recharge Parameter.

No	Recharge (mm/year)	Rating	Weight	Index
1	0-51	1	4	4
2	52-102	3	4	12
3	103-178	6	4	24
4	103-178	8	4	32
5	254-1092	9	4	36

4.1.3. Aquifer Media. Aquifer media type has a role in controlling the movements of the contaminants inside the water saturated rocks (aquifer). The rocks assembling the aquifer in the study area were from Merbabu volcano. Based on the correllation results via drilling data and subsurface estimation with geoelectrical method, the type of rocks functioned as an aquifer was interpreted as Tuffaceous Sandstone Alternate with Volcanic Breccia and Fractured Andesite Lava. Aquifer media parameter weighted 3. The aquifer media measurement and the distribution maps can be seen in Table 4 and Figure 4.

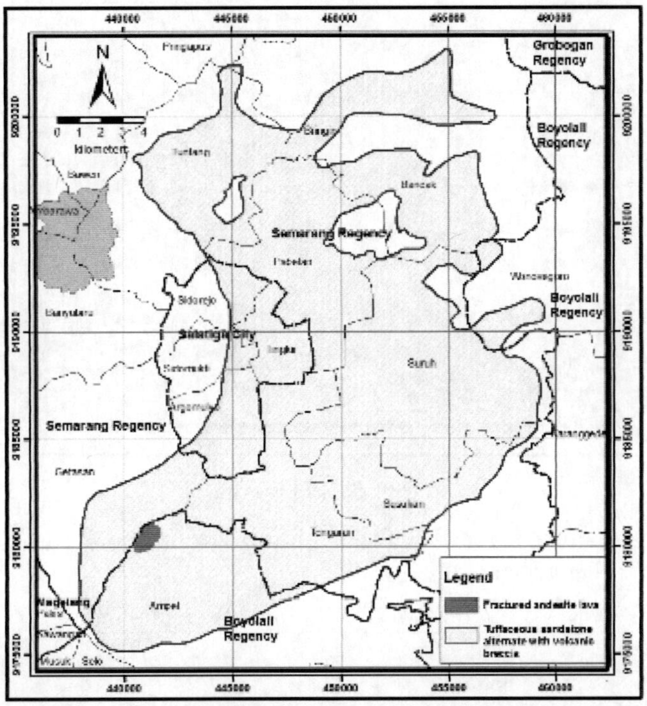

Figure 4. Aquifer Media

Table 4. Score Calculation for the Aquifer Media Parameter

No	Aquifer Media	Rating	Weight	Index
1	Tuffaceous Sandstone Alternate with Volcanic Breccia	6	3	18
2	Fractured Andesite Lava	3	3	9

4.1.4. Soil Media. Soil media is the uppermost of the unsaturated water zone which not undergone weathering. Based on the field observation and the soil permeability value of some soil samples in the study area, it can be interpreted that the soil media consisted of sandy loam and sand. Soil media parameter was weighted 2 points. The calculation of soil media parameter score as well as the distribution maps of soil media parameter can be viewed in Table 5 and Figure 5.

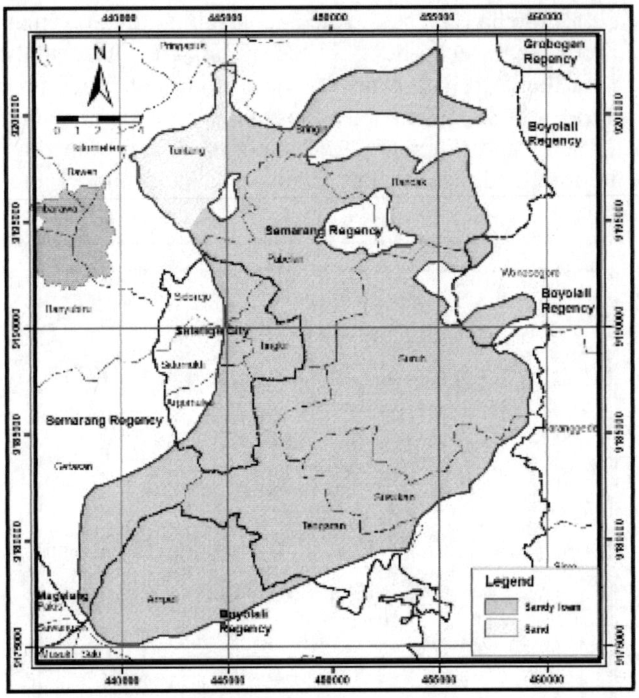

Figure 5. Soil Media.

Table 5. The Calculation of Soil Media Parameter Score.

No	Soil Media	Rating	Weight	Index
1	Sandy Loam	6	2	12
2	Sand	9	2	18

4.1.5. Topography. Steepness of an area which allows low slope for high infiltration of pollutant along with rainwater and steep slopes for low infiltration. The topography of study area was prepared based on the Shuttle Radar Topographical Mission (SRTM) satellite data [16]. Based on the slope maps, the studied areas have varies slope between 0 until > 18%. Slope parameter was weighted 1 point. While the Slope Distribution based on the interval and the score calculation for the slope can be seen on Table 6 and Figure 6.

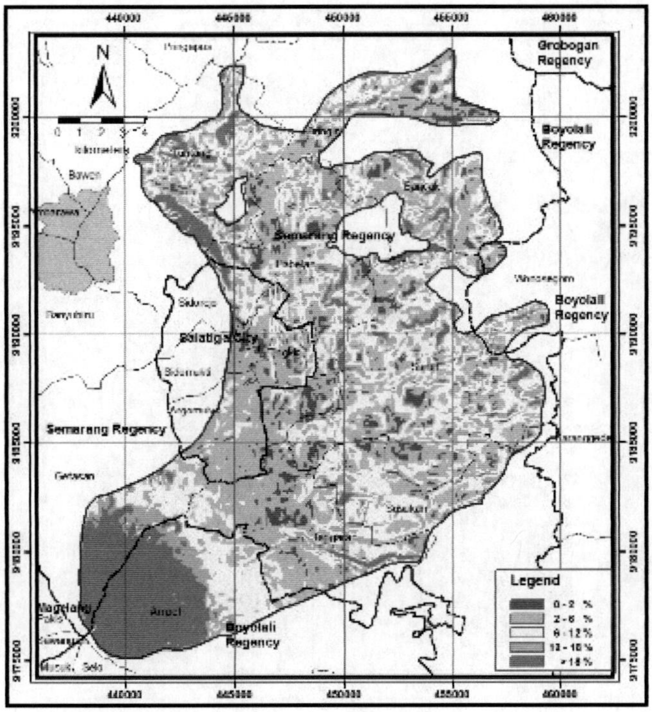

Figure 6. Topography.

Table 6. Slope Value.

No	Slope (%)	Rating	Weight	Index
1	0 - 2	10	1	10
2	2 - 6	9	1	9
3	6 - 12	5	1	5
4	12 - 18	3	1	3
5	> 18	1	1	1

4.1.6. Impact of Vadose Zone. The Vadose (unsaturated) zone is the zone located above the groundwater level. Based on the subsurface data according to the borehole log and geoelectrical data obtained from the Central Java Province Energy and Mineral Resources Department, it can be interpreted that the rocks on the water unsaturated (vadose) zone generally composed by sand and gravel with silt and clay content. Those materials resulted from the weathering of tuffaceous sandstone alternate with volcanic breccia. This parameter was weighted 5 points. The calculation of water unsaturated zone score as well as the distribution maps of that parameter can be viewed in Table 7 and Figure 7.

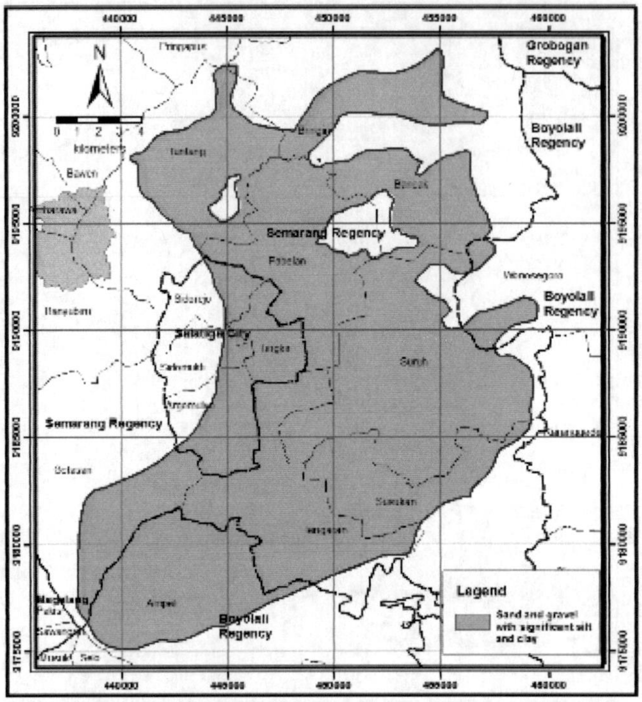

Figure 7. Vadose zone media

Table 7. The Calculation of Vadose Zone Media Parameter

No	Vadose zone media	Rating	Weight	Index
1	Sand and gravel with significant silt and clay	6	5	30

4.1.7. Hydraulic Conductivity. Hydraulic conductivity is the ability of the aquifer to transmit the water and to control the average speed of groundwater flow. Groundwater flow velocity also controls the speed of transport of contaminants in the aquifer. The hydraulic-conductivity data was obtained from the rock permeability tests in the study area. This parameter was weighted 3 points. The calculation of hydraulic conductivity score as well as the distribution maps of that parameter can be viewed in Table 8 and Figure 8.

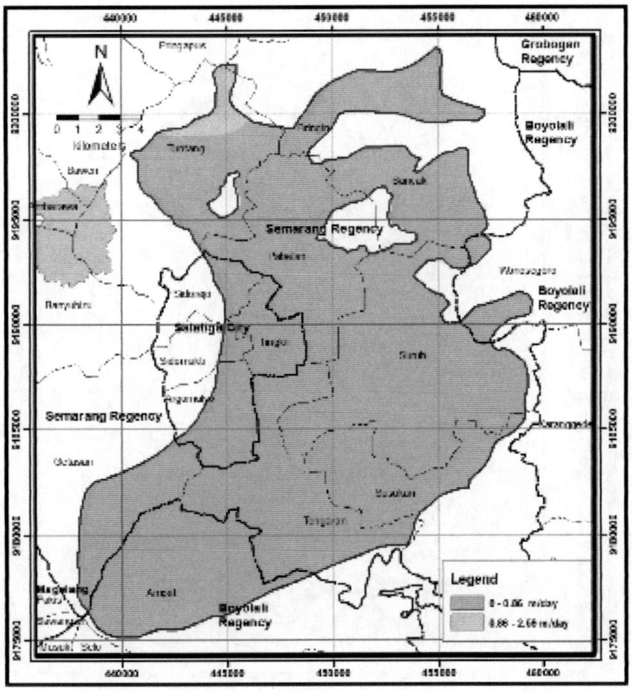

Figure 8. Hydraulic Conductivity

Table 8. The Calculation of Hydraulic Conductivity Parameter

No	Hydraulic Conductivity (m/day)	Rating	Weight	Index
1	0 – 0,86	1	3	3
2	0,86 – 2,59	2	3	6

4.2. DRASTIC Vulnerability Assessment

In this study, the minimum value of DRASTIC Vulnerability Index in Salatiga Groundwater Basin was 85 – 159. The values of DRASTIC Vulnerability Index were classified into five vulnerability levels, which are: Low (85 – 100), Low to moderate (101 – 120), Moderate (121 – 140), Moderate to High (141 – 150), and High (151 – 159). Here is the division of the groundwater vulnerability based on the calculation according to the weight of each parameter and rating applied (Table 9) along with a map of groundwater vulnerability (figure 9).

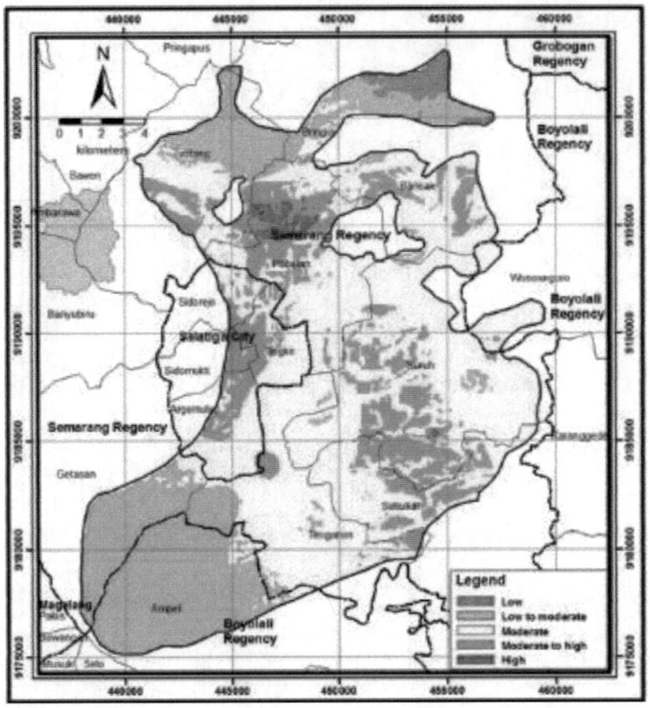

Figure 9. Vulnerability Map.

Table 9. The Vulnerability Class and the Distribution Area.

No	DRASTIC Index	Vulnerability Class	Characteristic [17]	Distribution Area
1	85 – 100	Low	Only vulnerable to conservative pollutants in the long term when continously and widely discharged or leached	Semarang Regency: The southern part of Bringin District, small part of the northern of Bancak District
2	101 – 120	Low to Moderate	Only vulnerable to conservative pollutants in the long term when continously and widely discharged or leached until Vulnerable to some pollutants but only when continously discharged or leached.	Semarang Regency The southern part of Getasan district, small part of the southern of Tengaran district, the northern part of Tuntang district, the southern part of Bringin district, small part of the northern of Bancak district Boyolali Regency : Ampel district
3	121 – 140	Moderate	Vulnerable to some pollutants but only when continously discharged or leached.	Semarang Regency Tuntang, Pabelan, Bancak, Suruh, Tengaran districts and small part of the southern of Bringin district. Boyolali Regency

No	DRASTIC Index	Vulnerability Class	Characteristic [17]	Distribution Area
				The small part of the western of Wonosegoro district <u>Salatiga City</u> Sidorejo, Tingkir, and Argomulyo districts
4	141 – 150	Moderate to high	Vulnerable to some pollutants but only when continously discharged or leached. Until vulnerable to many pollutants (except those strongly absorbed or readily transformed) in many pollutants scenarios.	<u>Semarang Regency</u> Tuntang, Pabelan, Bancak, Suruh, and Tengaran districts <u>Salatiga City</u> Sidorejo, Sidomukti, Tingkir, and Argomulyo districts
5	151 – 159	High	Vulnerable to many pollutants (except those strongly absorbed or readily transformed) in many pollutants scenarios.	<u>Semarang Regency</u> Most of the northern part of the Pabelan district and small part of the southern of Suruh district

Basically, the level of vulnerability in the study area are influenced by all the local hydrogeological factors. However, the most influential factor is the depth of groundwater level and rocks in the unsaturated (vadose) zone. Based on the groundwater vulnerability maps, land use zoning can be determined. In principle, the area with the high vulnerability of aquifers against contamination should be reserved for the low potency of land utilization, whereas in areas with low vulnerability against contamination can earmarked for land utilization with high potency of contamination, such as industrial activities, waste disposal and so on.

5. Conclusion

Based on the DRASTIC method analysis, the levels of groundwater vulnerability against contamination in Salatiga Groundwater Basin were classified into: Low, Low to Moderate, Moderate, Moderate to High, and High level. The high level of vulnerability zone was found in Pabelan District, Semarang Regency. The main factor affecting the high level of vulnerability zone was the very shallow depth to the water table, which was 0.6 – 1.5 meters depth from the local groundwater surface. The shallow position of groundwater table will cause contaminants to easily reach the groundwater, so that the potency of groundwater contamination in the area will be increased.

6. Acknowledgement

The authors wish to acknowledge PUSBINDIKLATREN BAPPENAS for funding this research as well as Zaky and Yusrizal from Geological Engineering Department, Diponegoro University for assisting in the field.

References

[1] Brindha, K., and Elango, L., 2015. Cross comparison of five popular groundwater pollution vulnerability index approaches. *Journal of Hydrology 524 (2015) 597–613*, available at web site www.elsevier.com/locate/jhydrol, accessed at 24 July 2016.

[2] Zuswanto, Musiyam, and Kaeksi, W., 2014. Analisis Spasial Pertumbuhan Ekonomi Kawasan Kedungsapur (Kendal, Demak, Ungaran, Kota Semarang, Kota Salatiga Dan Purwodadi) Provinsi Jawa Tengah Tahun 2008-2012. Available at http://eprints.ums.ac.id/29011/. Accessed at 28 September 2016.

[3] Wijaya, B. and Atmanti, H.D., 2006. Analisis Pengembangan Wilayah Dan Sektor Potensial Guna Mendorong Pembangunan Di Kota Salatiga. Available at http://eprints.undip.ac.id/16953/1/Analisis_Pengembangan_Wilayah_Dan_Sektor....by_Bayu_Wijaya_%26_Hastarini_Dwi_Atmanti_(OK).pdf. Accessed at 9 August 2016.

[4] Tusianto, A.Taryono, and Sigit, A.A., 2015. Evaluasi Kesesuaian Penggunaan Lahan Kota Salatiga Tahun 2010-2014 Terhadap Rencana Tata Ruang Wilayah Kota Salatiga Tahun 2010-2030. Publikasi Karya Ilmiah. Available at http://eprints.ums.ac.id/38672/1/NASKAH%20PUBLIKASI.pdf. Accessed at 28 September 2016.

[5] Al-Hanbali A. and Kondoh A., 2008. Groundwater vulnerability assessment and evaluation of human activity impact (HAI) within the Dead Sea groundwater basin, Jordan. *Hydrogeol J 2008;16:499–510.*

[6] Lathamani, R., Janardhana, M.R., Mahalingam, B., and Suresha, S., 2015. Evaluation of Aquifer Vulnerability Using Drastic Model and GIS: A Case Study of Mysore City, Karnataka, India. *International conference on water resources, coastal and ocean Engineering (ICWRCOE 2015). Aquatic Procedia 4 (2015) 1031–1038.* Available at web site www.sciencedirect.com. Accessed at 22 December 2016.

[7] Todd, D. K., 1980. Groundwater hydrology (2nd Ed.). New York, NY: Wiley.

[8] Vrba, J. and Zoporozec, A., 1994. Guidebook on Mapping Groundwater Vulnerability. International Contributions to Hydrogeology. Volume 16. Hannover.

[9] Rahman A., 2008. A GIS based DRASTIC model for assessing groundwater vulnerability in shallow aquifer in Aligarh, India. *Appl Geogr 2008;28:32–53.* Available at web site www.elsevier.com/locate/apgeog, accessed at 24 July 2016.

[10] Energy and Mineral Resources Department of Central Java Province, 2015. Studi Sistem Jaringan Sumur Pantau Pada CAT Rawapening dan CAT Salatiga. Semarang.

[11] Central Bureau of Statistics, Boyolaly Regency, 2016. Boyolali Regency in Figures. Catalog: 1102001.3309. Boyolali.

[12] Central Bureau of Statistics, Semarang Regency, 2016. Semarang Regency in Figures. Catalog: 1102001.3322. Ungaran.

[13] Central Bureau of Statistics, Salatiga City, 2016. Salatiga City in Figures. Catalog: 1102001.3373. Salatiga.

[14] Aller, L., Bennet, T., Lehr, J. H., Petty, R. J., and Hackett, G., 1987. “DRASTIC: A Standardized System for Evaluating Ground Water Pollution Potential Using Hydrogeologic Settings,” US EPA Report 600/287/035, U.S. Environmental Protection Agency.

IOP Conf. Series: Earth and Environmental Science **70** (2017) 012009 doi:10.1088/1755-1315/70/1/012009

The customer satisfaction towards the service quality of Tawang Alun Malang-Banyuwangi Train

D Permatasari

Department of Urban and Regional Planning, Faculty of Engineering, Universitas Brawijaya, Malang, Indonesia

dea.erta88@gmail.com

Abstract. Service sector which has quiet vital role in supporting people's daily activities is transportation service. Transportation is one of the important and strategic developments in improving economy sector. One of the alternative ways to overcome people's need of transportation is by providing trains. This research was conducted on the weekend that has objectives to analyze the work performance of Indonesian Railway Company towards the service quality that can determine the customers' satisfaction of TawangAlun Malang-Banyuwangi train and to analyze the customers' satisfaction itself towards the service quality of TawangAlun Malang-Banyuwangi train. This research used quantitative descriptive as the research method. There are two kinds of data that were used in this research; the first one is the primary data taken from questionnaire's results and interview meanwhile the second one is the secondary data taken from literature and internet. The sample used in this research is nonprobability sampling using convenience sampling technique. Data analysis used in this research is Importance Performance Analysis (IPA) and Customer Satisfaction index (CSI). The results are the Indonesian Railway Company should make a new innovation to buy the ticket from the ticket machine and add more exhausts in every railway coach.

1. Introduction

Transportation is an important and strategic improving sector in order to expediting people's economy, unite many people and affect all the aspects of life. This condition forces people to be able to work and do many activities efficiently and on time [1]. The importance of transportation reflects to the increasing number of transportation service of mobilization for people and goods since there are growing populations and housings. To fulfil this kind of need, the transportation service needs to be provided more including the numbers and the qualities (safety, comfort, on time schedule and efficiency). One of the alternative ways to overcome people's need of transportation is by providing trains. Trains are kind of transportations with many advantages compare to the other kinds of transportation including the less fuel, less energy used, less pollution, no traffic jam and it is a mass transportation.

As the demand of transportation service has been growing, it also effects to the competitiveness among people who are getting involved in this business. Every company in that kind of business must have effective and efficient strategy to survive and always be number one among the other competitors. Keeping up the customers' satisfaction is one of the good ways to be applied by companies to keep competing and dominating the market. In order to make it happens; Indonesian Railway Company is expected to maintain its good qualities which are the efficiency and effectiveness in providing its best products and services, so it can gain customers' trust and they can be more comfortable using the service.

Indonesian Railway Company is also expected to be more aware of what factors that can raise customers' satisfaction and also to evaluate and fix those factors. Some of those factors are including the fixed departure schedule sometimes has delay, the lack of comfort and the safety either in the rain station or the train, the availability of the staffs to help customers and the lack of information and the other supporting services.

These factors need to get more attention from the company to be solved with the best solution, so in the end, it can have better performance to serve its customers. Indonesian Railway Company also needs to do measurement of the level of customers' satisfaction periodically, by having feedbacks from them to maintain staffs' work performance and the level of the customers' satisfaction itself. [2] This perception is the experience towards objects, tourism or something taken by collecting information and suggestions.

This research was done with objectives: analyzing the work performance of Indonesian Railway Company towards the service quality that can determine the customers' satisfaction of TawangAlun Malang-Banyuwangi Train, as well as analyzing the customers' satisfaction towards the service quality given in TawangAlun Malang-Banyuwangi Train. Service quality can be measured by using theories by [3] expressing five dimensions which are reliability, responsiveness, assurance, empathy and tangible.

2. Research Methodology

This research used descriptive quantitative research method. The sample taken for this research is nonprobability sampling using convenience sampling technique. The data acquisition was done by doing direct observation and interview and spreading questionnaires to get the primary data, meanwhile for the secondary data came from the references. The method to pick the samples used is non-probability sampling method using convenience sampling technique. The number of respondents in this research was determined by using Slovin formula [4] which is:

$$n = \frac{N}{N(e)^2 + 1} \tag{1}$$

Information:

n = the number of samples
N = the measurement of population
e = the percentage of inaccuracy because of sampling error that can be tolerated 10%

The number of people who used Tawang Alun Train is more than 300 people on the weekends, so based on the formula, the respondents are 75 respondents.

The data analysis used in this research is Importance Performance Analysis (IPA) and Customer Satisfaction Index (CSI). The Importance and Performance Analysis is used to get information about the level of customers' satisfaction towards certain service by measuring its importance and implementation level. IPA is an implementation method to measure attribute based on the importance and work performance or satisfaction level which is useful to improve the market strategy for the company to be more effective [5]. In this technique, the respondents was asked to assess the importance and implementation level, and then the average value of importance and implementation level were analyzed with Importance Performance Matrix, in which the x axis represents perception or work performance meanwhile y axis represents hopes or importance. And later, the result of the x and y value (from each variable) in the four quadrant on Kartesius diagram can be drawn. There is also interpretation from squared which is quadrant I (Keep up the Good Work); quadrant II (Possible Overkill); quadrant III (Low Priority) and quadrant IV (Concentrate Here).

The measurement towards customer satisfaction index is needed because the result from the measurement can be used as the reference to determine targets in the next few years. According to [6] the measurement towards the CSI is needed because the result from the measurement can be used as the reference to determine targets to improve the service quality towards the level of overall respondents' satisfaction which can be seen from the criteria of customers' satisfaction level. The steps to count the customer satisfaction index are as follows:

1. Counting *Weighting Factors* (WF)
   The function from the median importance score or median score of the importance level from each variable in a form of % from total median importance score or median score of the importance level from overall attributes that are being tested.
2. Counting *Weighted Score* (WS)
   The function from the median satisfaction score of median score of the importance level from each attribute is multiplied with Weighting Factors (WF) from each variable.
3. Counting *Weighted Median* Total (WMT)
   The total from overall Weighted Score (WS) value.
4. Customer Satisfaction Index
   The counting from Weighted Median Total (WMT) is divided in maximum scale or high test scale which is multiplied by 100%. The level of overall respondents' satisfaction can be seen from criteria of the level of customers' or consumers' satisfaction with the criteria as follows : a. 0,81-1,00 Fair; b. 0,66-0,80 Very good; c. 0,51-0,65 Fair; d. 0,35-0,50 Poor; e. 0,00-0,34 Very poor.

CSI can be taken from (T divided by 5Y) x 100%. 5 (in 5Y) is the maximum value which is used on the measurement scale. The CSI formula which is:

$$CSI = \frac{T}{5Y} x100\% \tag{2}$$

Information:

CSI	:	Customer Satisfaction Index
T	:	Overall numbers of S variable
Y	:	Overall numbers of P variable
S	:	Multiplication from I and P value (each variable)
I	:	The average of satisfaction or Work Performance (each variable)
P	:	The average of the importance (each variable)

The level of overall respondents' satisfaction can be seen from the criteria of the customers' or consumers' satisfaction level, with the criteria as follows:

Table 1. Criteria Consumers' Satisfaction.

No.	CSI Value	Customer Satisfaction Index
1.	0,81 – 1,00	Very Excellent
2.	0,66 – 0,80	Excellent
3.	0,51 – 0,65	Good
4.	0,35 – 0,50	Poor
5.	0,00 – 0,34	Very poor

3. Result

3.1. Importance Performance Analysis (IPA)

The analysis of Important Performance Analysist (IPA) in this research used to analyze the factors related to importance and people's satisfaction in using TawangAlun Malang-Banyuwangi Train. The result will be presented in a form of Kartesius Diagram, in which the diagram has quadrants contains of Keep Up the Good Work (quadrant I), Possible Overkill (quadrant II), Low Priority (quadrant III) and Concentrate Here (quadrant IV). There is also aKartesius Diagram of TawangAlun Malang-Banyuwangi Train as follows:

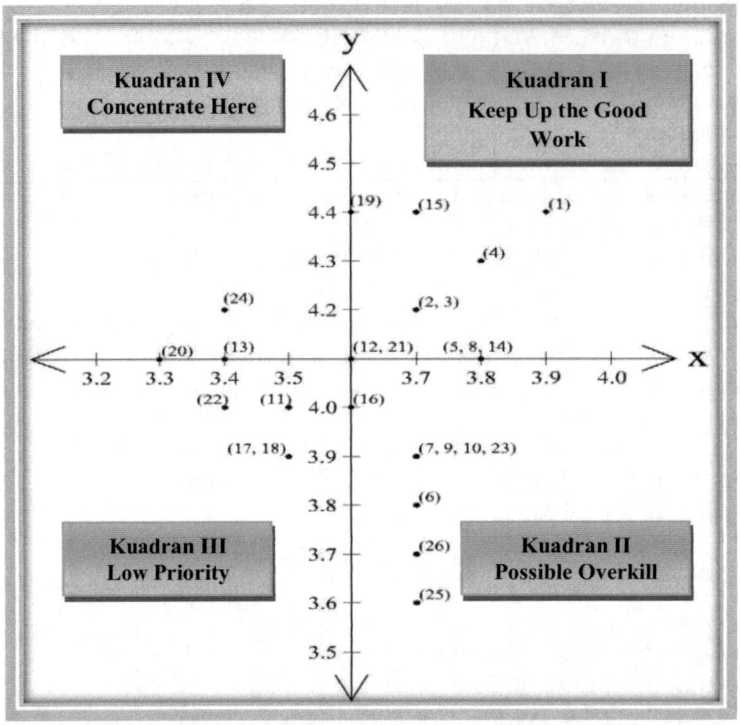

Picture 1. Kartesius Diagram of Tawang Alun Malang-Banyuwangi Train.

In interpreting the quadrant on the Kartesius Diagram as follows:

1) Keep Up the Good Work
In this quadrant, the variable is considered as the important factor or it is expected to be maintained because the consumers feel satisfied. There is also variable that needs to be maintained which is the ease to reach the location of the station, the availability of the information related to the train schedule, the accuracy of train schedule, the ability to give the best service to the customers, the ease to get clear information, the speed to respond emergency either in the train or station, the capability of the staffs to do their job, the comfort when the passengers get off from the train, the availability of insurance or the safety guarantee, the price ticket offered and the cleanliness in the station.
2) Possible Overkill
In this quadrant, the variable is considered as the less important factor, but the implementation in the field is very satisfying. There are variables including the speed and the readiness of the staffs in providing service for customers, the speed and the accuracy in providing information needed by the customers, the speed in responding customers' complaints and problems, staffs always show confidence and the readiness to serve or help customers, the capability of the staffs to give information to the customers that can be received well and easy to understand, the cleanliness of the rest rooms in the station and the numbers of the seats in the train.
3) Low Priority
In this quadrant, the variable is considered as the less important factor or least expected by the consumers. The variables are including the hospitality and good manner of the staffs in helping the customers, the politeness of the staffs to respect and to serve as well as make the customers' need as the priority, the honesty and patience of the staffs in giving service and cleanliness in the train.

4) Concentrate Here
In this quadrant, the variable is considered as the important factor or expected by the consumers but not satisfying yet and there has to be priority to be improved. The variables to be improved are including the safety and comfort either in the station or train, and the 24/7 customer service and the air circulation in the train.

Based on the result of the counting using IPA method, the variables that need to get more attention is the factor in quadrant IV. The result of this analysis shows that the quadrant IV is the factor that is considered important and it's in bad condition so it needs to be noticed and fixed. The analysis of these factors can be seen on the table.

Table 2. Analysis Factors of Customers' Satisfaction towardsTawang Alun Train on Quadrant IV.

No.	Variable	Satisfaction Factors	Analysis
1	ASSURANCE	The safety and comfort either in the station or train	The customers' perception of TawangAlun Train is that the safety and the comfort either they are in the station or train is poor, it is because several stations do not have waiting rooms and do not have enough chairs to sit so many passengers wait for the train uncomfortably.
2	EMPHATY	The 24/7 customer service	Not all passengers of TawangAlun Malang-Bayuwangi Train know and understand how to book ticket online, so the 24/7 customers service needs to be available.
3	TANGIBLES	The air circulation in the train	The service of Indonesian Railway Company nowadays is already good enough and it pays attention to passengers' comfort since in every railway coach has air conditioner. But there is a thing to be noticed including the customers' health whether they have carsick or vomit. This thing can bother the other train passengers because of the smell and since there is an air conditioner in the railway coach, every ventilation or window is closed.

3.2. Customer Satisfaction Index (CSI))
The result of the customer satisfaction index of TawangAlun-Malang Banyuwangi Train can be seen on this table:

Table 3. Customer Satisfaction Index (CSI).

No.	Statement	Importance (I)	Satisfaction (P)	Score (S)
		Scale 1 - 5	Scale 1 - 5	(S) = (I) x (P)
	RELIABILITY			
1	The ease to reach the location of the station	3.89	4.43	17.23
2	The availability of information related to the train schedule	3.67	4.16	15.25
3	The accuracy of the train schedule	3.73	4.24	15.83
4	The capability to give the best service to the customers	3.83	4.29	16.43

No.	Statement	Importance (I)	Satisfaction (P)	Score (S)
		Scale 1 - 5	Scale 1 - 5	(S) = (I) x (P)
5	The ease to get clear information	3.77	4.07	15.34
	RESPONSIVENESS			
6	The speed and readiness of the staffs in giving service to the customers until it's all clear	3.72	3.77	14.04
7	The speed and accuracy in giving information that customers need	3.65	3.92	14.32
8	The speed in responding emergency either in the train or station	3.79	4.08	15.45
9	The speed in responding customers complaints and problems	3.65	3.85	14.08
10	Staffs always show confidence and the readiness in serving or helping the customers	3.65	3.92	14.32
	ASSURANCE			
11	The hospitality and politeness in serving the customers	3.45	3.97	13.72
12	The capability of the staffs in implementing their job	3.63	4.05	14.70
13	The safety and comfort either in the station or train	3.44	4.11	14.13
14	The comfort when the passengers get off from the train	3.76	4.11	15.44
15	The availability of safety insurance or guarantee	3.73	4.36	16.29
	EMPHATY			
16	The capability of the staffs in providing information to the customers that is easy to understand	3.64	4.00	14.56
17	The manner's staffs to respect and serve the customers especially the customers' need	3.53	3.93	13.90
18	The honesty and patience of the staffs in giving service to customers	3.48	3.91	13.60
19	The price of the ticket	3.57	4.43	15.82
20	The 24/7 customer service	3.29	4.13	13.61
	TANGIBLES			
21	The cleanliness in the station	3.57	4.05	14.48
22	The cleanliness in the train	3.44	3.97	13.67
23	The cleanliness of the restroom in the station	3.68	3.95	14.52
24	The air circulation in the train	3.40	4.24	14.42
25	The device availability to inform the routes	3.68	3.64	13.40
26	The numbers of the seats in the train	3.73	3.71	13.84
	TOTAL	**Y = 105,30**		**T = 382.39**

$$CSI = \frac{T}{5Y} x100\% = \frac{382,39}{5 \, x \, 105,30} x100\% = 0,73\% \quad (3)$$

According to the result of the processing data as well as the analysis of Customer Satisfaction Index, they can be concluded that the overall customers' satisfaction of Tawang Alun Malang-Banyuwangi Train is on the satisfaction assessment with the value of 0.73%. There are several aspects that need to be fixed in order to improve the customers' satisfaction to use Tawang Alun Malang-Banyuwangi Train.

4. Conclusion and Suggestion

Based on the analysis of Importance Performance Analysis (IPA), there are several attributes that needs to be maintained which are the safety and comfort when the passengers are in the station or train, the 24/7 customer service and the air circulation in the train. Related to the comfort and safety in the station, it is better for the station and Indonesian Railway Company to be able to improve the passengers' comfort more including providing the waiting room or chairs for the passengers to wait for their train comfortably and safely. Meanwhile to be able to provide the 24/7 customer service, Indonesian Railway Company and the station staff should promote it more so that the passengers can use the service to buy the train ticket online. Besides that, the company should make a new innovation for the passengers to buy the ticket from the ticket machine in the station. And for the air circulation in the train, the Indonesian Railway Company can add more exhausts in every railway coach to prevent unpleasant moments especially to overcome bad smell.

References

[1] Fortuna Zain Hamid, Ernita siambaton, dan mercy kent yanuar. 2012). Pengaruh kualitas pelayanan jasa terhadap keputusan pembelian pada pengguna jasa ojek online go-jek, studi kasus : di Politeknik Negeri Jakarta jurusan administrasi bisnis, Politeknik Negeri Jakarta
[2] Drs.Jalaluddin Rakhmat , M.SC .Psikologi Komunikasi :Penerbit PT Remaja Rosdakarya : Bandung
[3] Prof. Dr. Jasfar, Farida. 2009. Manajemen Jasa. Ciawi, Bogor : Ghalia Indonesia.
[4] Umar, H. 2005. Riset Pemasaran dan Perilaku Konsumen. PT Gramedia Pustaka Utama, Jakarta.
[5] Supranto, J. 2001. Pengukuran Tingkat Kepuasan Pelanggan: Untuk Menaikkan Pangsa Pasar. PT Rineka Cipta, Jakarta.
[6] Irawan, H. (2003). *Indonesian Customer Satisfaction*. Jakarta: PT Gramedia Pustaka Utama.

Attachment :

Table 4. Average Value User Satisfaction and Interests Railway Tawang Alun Malang-Banyuwangi

No.	Statement	Importace		Satisfaction		Integrity X	Integrity Y
		Score	Mean X	Score	Mean Y		
	RELIABILITY						
1	The ease to reach the location of the station	292	3.89	332	4.43	0.04	0.04
2	The availability of information related to the train schedule	275	3.67	312	4.16	0.04	0.04
3	The accuracy of the train schedule	280	3.73	318	4.24	0.04	0.04
4	The capability to give the best service to the customers	287	3.83	322	4.29	0.04	0.04
5	The ease to get clear information	283	3.77	305	4.07	0.04	0.04
	RESPONSIVENESS						
6	The speed and readiness of the staffs in giving service to the customers until it's all clear	279	3.72	283	3.77	0.04	0.04
7	The speed and accuracy in giving information that customers need	274	3.65	294	3.92	0.04	0.04
8	The speed in responding emergency either in the train or station	284	3.79	306	4.08	0.04	0.04
9	The speed in responding customers complaints and problems	274	3.65	289	3.85	0.04	0.04
10	Staffs always show confidence and the readiness in serving or helping the customers	274	3.65	294	3.92	0.04	0.04
	ASSURANCE						
11	The hospitality and politeness in serving the customers	259	3.45	298	3.97	0.04	0.04
12	The capability of the staffs in implementing their job	272	3.63	304	4.05	0.04	0.04
13	The safety and comfort either in the station or train	258	3.44	308	4.11	0.04	0.04
14	The comfort when the passengers get off from the train	282	3.76	308	4.11	0.04	0.04
15	The availability of safety insurance or guarantee	280	3.73	336	4.36	0.04	0.04
	EMPHATY						
16	The capability of the staffs in providing information to the customers that is easy to understand	273	3.64	300	4.00	0.04	0.04
17	The manner's staffs to respect and serve the customers especially the customers' need	265	3.53	295	3.93	0.04	0.04
18	The honesty and patience of the staffs in giving service to customers	261	3.48	293	3.91	0.04	0.04
19	The price of the ticket	268	3.57	332	4.43	0.04	0.04
20	The 24/7 customer service	247	3.29	310	4.13	0.03	0.04
	TANGIBLES						
21	The cleanliness in the station	268	3.57	304	4.05	0.04	0.04
22	The cleanliness in the train	258	3.44	298	3.97	0.04	0.04
23	The cleanliness of the restroom in the station	276	3.68	296	3.95	0.04	0.04

No.	Statement	Importace		Satisfaction		Integrity X	Integrity Y
		Score	Mean X	Score	Mean Y		
24	The air circulation in the train	255	3.40	318	4.24	0.04	0.04
25	The device availability to inform the routes	276	3.68	273	3.64	0.04	0.03
26	The numbers of the seats in the train	280	3.73	278	3.71	0.04	0.04
	Total	**7080**	**94,40**	**7906**	**105,30**	**1.00**	**1.00**
	Mean		**3,63**		**4,05**		

IOP Conf. Series: Earth and Environmental Science **70** (2017) 012010 doi:10.1088/1755-1315/70/1/012010

The difference of level CO2 emissions from the transportation sector between weekdays and weekend days on the City Centre of Pemalang

E Sawitri[1, 4], G Hardiman[2], I Buchori[3]

[1] Master Program of Environmental Science School of Postgraduate Studies Diponegoro University, Semarang, Indonesia
[2] Department of Architecture Engineering Faculty Diponegoro University, Semarang, Indonesia
[3] Department of Urban and Regional Planning Engineering Faculty Diponegoro University, Semarang,Indonesia he author addresses here

eliasawitri@rocketmail.com

Abstract. The high growth of human activity potentially increases the number of vehicles and the use of fossil fuels that contribute the increase of CO2 emissions in atmosphere. Controlling CO2 emission that causes greenhouse effect becomes the main agenda of Indonesian Government. The first step control CO2 emissions is by measuring the level of CO2 emissions, especially CO2 emissions from fossil fuel consumption in the transport sector. This research aims to assess the level of CO2 emissions from transportation sector on the main roads in the city centre of Pemalang both in weekdays and weekend days. The methods applied to calculate CO2 emissions using Intergovernmental Panel on Climate Change (IPCC) 2006 method. For this, a survey on the number of vehicles passing through the main roads using hand tally counter is firstly done. The results, CO2 emissions in working day, i.e. 49,006.95 tons/year compared to weekend i.e. 38,865.50 tons/year.

1. Introduction

The high growth of people in a region is the early emergence of environmental problem due to the increased variety of human needs [1]. Population growth requires new residential developments in the suburbs. The development of the settlement which today is far from community activities to encourage people to find a mode of transportation. Transportation required in support of the activities of the community everyday. Transportation is an important means for the community both in the community and facilitate mobility as well as goods. The transport needs of the community will encourage people in the purchase of the vehicle in particular motor vehicles. This has resulted in an increasing number of motor vehicles and as well as increased emissions of gasses such as CO2 as a b-product of combustion of fossil fuels used by motor vehicles. The level of concentration of CO2 gas emissions that exceeded the normal limit may negatively impact humans and the environment. Despite the use of motor vehicles resulting in increased fuel usage so that gas emissions containing pollutants will also increase. The use of motor vehicles by people on weekdays and weekend will be different to affect the magnitude of the emissions of CO2 are produced and released into the environment.

The growth of the city center of Pemalang became the center of trade, services, education, offices and health services enables an increasing number of vehicles on some streets in the city center of Pemalang and increased energy consumption as a consequence of the rapid growth of the city is the increased mobility of the population, either in the city or between city. A growing number of private two wheel vehicle, three wheels and four wheels are hard to avoid. Procurement and distribution of gasoline in the city of Pemalang have increased more or less by 2.32%, whereas solar procurement experience an increases of 3.65%[2].

Most of the sources of energy used come from fossil fuels. The use of fossil fuels results in the occurrence of air pollution due to some major pollutant derived from fossil fuel use[1], [3]. The largest fossil fuel consumption in the city of Pemalang comes from the transport sector and the sector is contributing the biggest polluters as it does occur in large cities such as occurs in large cities. Research conducted by the Soehodho mention that transportation accounts for 70% of total pollutants that enter the atmosphere in Jakarta[4]. Therefore, the study of inventory of CO2 emissions is done to inventory emissions of CO2 in the transport sector in Pemalang on 1 (one) working day and 1 (one) day off to know how big a difference levels of emissions caused by motor vehicles at the time of the working days and weekend days.

2. Literature Review

2.1. Air Pollution Resources

Sources of air pollution come from natural resources and the result of human activity (anthropogenic) such as fuel combustion in transportation, power plants and use of boilers in the industry[5],[6]. The largest anthropogenic resources that cause air pollution comes from transportation activities.

2.2. Motor Vehicle Emissions

The air in the atmosphere under normal conditions the gas mixture is composed of 78% nitrogen, 20% oxygen, 0.93% argon, 0.03% carbon dioxide and the rest consists of neon, helium, methane, and hydrogen. The condition of the air in the atmosphere can be said to be contaminated when changing the composition of the air. Changes in the composition of the incoming air can occur due to or inclusion of substances, energy and/or other components resulting from an activity that is potential as polluters[7]. The main pollutants produced by motor vehicles which can cause air pollution, among others, dust, carbon monoxide, carbon dioxide, hydrocarbons, nitrogen oxides and sulfur oxides [1], [3]. The resulting carbon emissions from motor vehicles is one of the causes of global warming because most emissions contribute to global warming is carbon dioxide (CO2) besides methane (CH4), dinitro oxide (N2O), perfluorocarbons (PFC), hydrofluorocarbon (HFC), and sulfurhexaflourida (SF6). According to the United Nations Framework Convention On Climate Change in addition to the six types of greenhouse gasses, there are some gasses also enter into the greenhouse gasses include carbon monoxide (CO), nitrogen oxides (NOx), chlorofluorocarbon (CFC) and gas - gas metal organic non-other volatiles. Three main types of gas that most so-called greenhouse gasses are CO2, CH4, N2O, because it is considered as a gas layer which acts as a trap heat wave and the end - the end concentration in the atmosphere continue to rise until doubled [8]. Based on calculations in the last few years, the CO2 contribution towards global warming reached more than 60%. CO2 emissions increase, amounting to 27.0 billion tons of CO2 in 2004 or average increased by 1.6% per year. Contributor to CO2 emissions are spread is the United States (21.9% of total world CO2 emissions), followed by China (17.4%), India (4.1%). While Indonesia accounted for emissions of 1.2%. The main source of CO2 emissions comes from fossil fuel use (74% of total emissions in the world), followed by a land convention (24%) and the cement industry (3%)[9].

IOP Conf. Series: Earth and Environmental Science **70** (2017) 012010 doi:10.1088/1755-1315/70/1/012010

3. Methods

An inventory study of CO2 emissions from the transport sector this is done through a quantitative methodology in 11 (eleven) of the main roads in the city center of Pemalang based on Survey guidelines for counting traffic by manual, Number: Pd, T-19-2004-B using a hand tally counter. Selection of the city center of Pemalang as a location because it's the center of trade, services, education, offices and health service. The kind of vehicles that surveyed was differentiated into 3 main types of motorcycle, light vehicles, and heavy vehicles. Survey on the volume of traffic on the roads done the 12 hours on 1 (one) working day (weekday) and 1 (one) holidays (weekend) which is divided into three periods of time in the morning peak hours (6:00 to 10:00 a.m.), during peak hours (11:00 a.m. to 3:00 p.m.) and afternoon peak hours (4:00 p.m. to 8:00 p.m.).

A method for calculating CO2 emissions is very diverse and constantly evolving, but the calculation of the CO2 emissions that were used by the Government of Indonesia refers to a method of the Intergovernmental Panel on Climate Change. The analytical calculation of the CO2 emissions calculated based on emission factors or the coefficients of fuel types used to then multiply by the amount of fuel consumption in accordance with the method of mathematical calculations the IPCC 2006 Guidelines. The basic model calculations of GHG emissions as follows [10] :

$$GHG\ Emissions = Activity\ Data \times Emission\ Factor \qquad (1)$$

GHG Emissions: Total GHG Emissions
Activity Data: The quantitative magnitudes of human activity or activities that can release and/or absorb GHG
Emissions Factor: A quantity that indicates number of GHG emissions that would be released from a specific activity.

Data activity in equation 1 is the abundance of fuel consumption or the consumption of energy by vehicles passing through a road section, in this case, the energy consumption is converted first into units energy Terra Joule (tj) with the formula :

$$Energy\ Consumption\ (tj) = Energy\ Consumptions\ (ltr/km) \times Number\ of\ vehicles \times Length\ of\ road\ (km) \times Heat\ value\ (tj/ltr) \qquad (2)$$

4. Results

4.1. Data Traffic Volume

Counting the number of vehicles in the CO2 gas emissions calculations done in several main roads in the city center of Pemalang that is Sudirman street, A Yani, Pemuda, Jenderal Gatot Subroto, Slamet Riyadi, Dr. Cipto Mangunkusumo, Perintis Kemerdekaan, Moh. Yamin, Letjend. Suprapto, Mochtar and Urip Sumoharjo as presented in Figure 1.

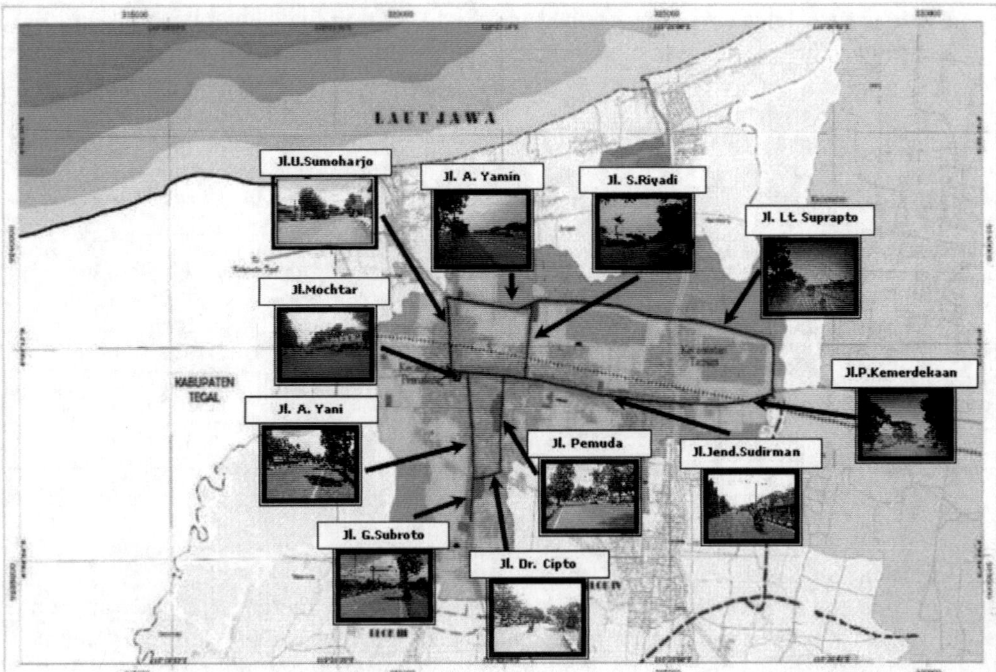

Picture 1. Location research (Source: Bappeda Kab. Pemalang).

Based on the results of the traffic enumeration in 11 (eleven) of roads during the twelve (12) hours, the vehicles dominate on weekdays and weekend days is a motorcycle, followed respectively by car, pick-up/small trucks, freight/small buses, and large/heavy trucks/buses.

Traffic volume retrieved from results of traffic counting is served in units of passenger cars by means of multiplying the value of passenger car equivalent (emp) with the existing traffic volume. The value of the emp to light vehicles is equal to 1, the value of the emp to motorcycle equals 0.25 and emp for heavy vehicles equals 1.2[11]. The value of the emp is not the same as each vehicle types have different characteristics.

From table 1 it can be noted that after converted into units of passenger cars, total volume of vehicles on weekdays is 10,637.27 units of passenger cars /hour while on weekend days is 8,855.86 units of passenger cars/hour. Light vehicles and motorcycle is a vehicle that dominant on most location research. The number of light vehicles on weekdays is 4,971.31 units of passenger cars /hour and weekend days is 4,483 units of passenger cars/hour whereas the volume of motorcycle on weekdays is 4,347.85 units of passenger cars/hour and weekend days is 3,509.96 units of passenger cars/hour.

Table 1 shows that in weekdays, Moch. Yamin street has the highest traffic density with total volume of vehicle is 1,454.84 units of passenger cars/hour, and Perintis Kemerdekaan is the second-highest traffic density after Moch. Yamin street with total volume of vehicle is 1,306,27 units of passenger cars /hour then followed by Jend. Sudirman, Jend. Gatot Subroto, Letjend. Suprapto, Mochtar, Pemuda, Dr. Cipto Mangunkusumo, Urip Sumoharjo, Ahmad Yani and Slamet Riyadi while on the weekend days, Moch. Yamin street has the highest traffic density with total volume of vehicle is 1,155,79 units of passenger cars/hour, and Perintis Kemerdekaan is the second-highest traffic density after Moch. Yamin street with total volume of vehicle is 1,089,74 units of passenger cars/hour then followed by Jend. Sudirman, Jend. Gatot Subroto, Mochtar, Letjend. Suprapto, Pemuda, Dr. Cipto Mangunkusumo, Ahmad Yani, Urip Sumoharjo, and Slamet Riyadi.

Table 1. Data Traffic Volume on the Main Roads in the City Center of Pemalang.

No	Street Name	Long roads [km]	Weekdays [units of passenger cars/hour]				Weekend days [units of passenger cars/hour]			
			MC	LV	HV	Total	MC	LV	HV	Total
1	Jend. Sudirman	3.56	670.29	577.33	3.70	1,251.33	545.06	526.92	2.90	1,074.88
2	Ahmad Yani	1.60	341.13	290.33	2.60	634.06	313.83	264.08	1.30	579.22
3	Pemuda	1.70	489.65	297.83	1.40	788.88	370.13	292.58	1.30	664.01
4	Perintis Kemerdekaan	2.26	636.83	652.83	16.60	1,306.27	519.21	563.33	7.20	1,089.74
5	Urip Sumoharjo	1.19	350.58	284.33	1.90	636.82	274.81	248.25	0.80	523.86
6	Moch. Yamin	1.58	98.35	677.08	679.40	1,454.84	86.54	632.25	437.00	1,155.79
7	Letjend. Suprapto	2.66	77.83	488.25	593.30	1,159.38	60.83	426.67	401.80	889.30
8	Slamet Riyadi	1.20	228.50	179.33	0.60	408.43	174.67	162.25	0.50	337.42
9	Jend. Gatot Subroto	1.36	597.71	615.17	12.70	1,225.58	468.40	552.17	7.00	1,027.56
10	Mochtar	1.00	539.50	541.58	3.30	1,084.38	433.79	476.83	1.30	911.93
11	Dr. Cipto Mangunkusumo	1.30	317.48	367.25	2.60	687.33	262.69	337.67	1.80	602.15
	Total		4,347.85	4,971.31	1,318.1	10,637.3	3,509.96	4,483	862.9	8,855.86

Source: The calculation results.

4.2. CO2 Emissions

Calculation of CO2 emissions in this study refers to a method of the Intergovermental Panel on Climate Change (IPCC) in 2006. Heat value used for calculation of the heat value of using premium fuel and diesel fuel for transportation activities, i.e. amounting to 33 x 10-6 for premium fuel and 36 x 10-6 for diesel fuel[12]. The value of energy consumption for motorcycle of 0.04 liters/km, light vehicle of 0.11 liters/km for premium fuel and 0.17 liters/km for diesel fuel, and heavy vehicle of 0.25 liters/km[13]. While the emission factor for gasoline amounted to 69,300 kg CO2/tj and 74,100 kg CO2/TJ for diesel[14]. Based on the data volume of vehicles, the value of CO2 gas emissions on each road can be calculations. The results of the calculation of CO2 emissions are shown on table 2.

Table 2. The Result of the Calculation of the Emissions of CO2 from Motor Vehicles On the Main Roads in the City Center of Pemalang

No	Street Name	Long roads [km]	Weekdays [tons/yr]				Weekend days [tons/yr]			
			MC	LV	HV	Total	MC	LV	HV	Total
1	Jend. Sudirman	3.56	1,912.16	5,758.23	76.95	7,747.35	1,554.92	5,055.72	60.31	6,670.95
2	Ahmad Yani	1.60	437.37	1,325.33	24.30	1,787.00	402.37	1,166.74	12.15	1,581.27
3	Pemuda	1.70	667.03	1,355.07	13.90	2,036.00	504.21	1,294.30	12.91	1,811.42
4	Perintis Kemerdekaan	2.26	1,153.31	4,263.39	219.17	5,635.87	940.29	3,564.80	66.02	4,571.10
5	Urip Sumoharjo	1.19	334.31	893.50	13.21	1,241.02	262.06	769.58	5.56	1,037.20
6	Moch. Yamin	1.58	124.53	2,994.69	6,271.15	9,390.37	109.57	2,706.30	4,033.70	6,849.57
7	Letjend. Suprapto	2.66	165.90	4,090.48	9,219.79	13,476.17	129.67	3,398.16	6,243.91	9,771.73
8	Slamet Riyadi	1.20	219.72	716.75	4.21	940.68	167.96	627.35	3.51	798.81
9	Jend. Gatot Subroto	1.36	651.39	2,469.53	100.90	3,221.83	510.46	2,148.21	55.62	2,714.29
10	Mochtar	1.00	432.32	1,417.23	19.28	1,868.83	347.61	1,231.80	7.59	1,587.00
11	Dr. Cipto Mangunkusumo	1.30	330.73	1,311.36	19.75	1,661.84	273.65	1,184.84	13.67	1,472.16
	Total		6,428.77	26,595.56	15,982.61	49,006.96	5,202.77	23,147.8	10,514.95	38,865.5

Source: The calculation results.

Based on the results of the calculation of the CO2 emissions at table 2, on weekdays amounted to 5,594.40 kg/hour or 49,006.95 tons/year of CO2 emissions while on weekend is 4,436.70 kg/hour or 38,865.50 tons/year with the largest emissions of contributors is light vehicles followed by heavy vehicles later motorcycles. On weekdays, the percentage of light vehicles accounted for 54.27% of total emissions generated while on weekends, light vehicles accounted for the percentage of 59.56% of the total emissions generated. Roads with the highest emissions of CO2 is the largest Jend. Letjend. Suprapto street where vehicles that dominate on this road are heavy vehicles (trucks and buses) as well as lighter vehicles.

From table 2 it can be noted that in weekdays, Letjend. Suprapto street has the highest emissions of CO2 with total emissions of CO2 is 13,476.17 tons/year, and Moch. Yamin has the second-highest emissions of CO2 after Moch. Yamin street with total emissions of CO2 is 9,390.37 tons/year then followed by Jend. Sudirman, Perintis Kemerdekaan, Jend. Gatot Subroto, Pemuda, Mochtar, Ahmad Yani, Dr. Cipto Mangunkusumo, Urip Sumoharjo, and Slamet Riyadi while on the weekend days, Letjend. Suprapto street has the highest emissions of CO2 with total emissions of CO2 is 9,771.73 tons/year, and Moch. Yamin has the second-highest emissions of CO2 after Moch. Yamin street with total emissions of CO2 is 6,849.57 tons/year then followed by Jend. Sudirman, Perintis Kemerdekaan, Jend. Gatot Subroto, Pemuda, Mochtar, Ahmad Yani, Dr. Cipto Mangunkusumo, Urip Sumoharjo, and Slamet Riyadi.

Based on the results of the survey in table 1 and calculation in table 2, the number of vehicles passing through a road can not be inferred automatically that the CO2 gas emissions will be high because that determine the high or low of CO2 gas emissions are the type of vehicle which passes through the road. For example, on the weekdays data traffic volume of Perintis Kemerdekaan street is 1,306.27 units of passenger cars/hour and Letjend. Suprapto street is 1,159.38 units of passenger cars/hour but total CO2 gas emissions of Perintis Kemerdekaan street lower than Letjend.Suprapto street where the CO2 gas emissions of Perintis Kemerdekaan street is 5,635.87 tons/year and Letjend. Suprapto street is 13,476.17 tons/year. The types of vehicles that pass through Perintis Kemerdekaan street dominated by light vehicles and motorcycles while the types of vehicles that pass through Letjend. Suprapto dominated by light vehicles and heavy vehicles. Heavy vehicles is the type of vehicle that produces the largest of CO2 gas emissios because heavy vehicles have a large engine capacity. The larger engine capacity, the more fuel required therefore CO2 gas emissions produced are high.

5. Conclusion

The levels of CO2 emissions generated by motor vehicles at 11 (eleven) roads of Pemalang City on working days larger than a weekend. Based on the results of the calculation of the CO2 gas emissions, on working days amounted to 5,594.40 kg/hour or 49,006.95 tons/year of CO2 emissions while on weekend 4,436.70 kg/hour or 38,865.50 tons/year.

6. Acknowledgment

I gratefully acknowledge the support of National Development Planning Agency of Indonesian Republic and District Government of Pemalang.

References

[1] J. G. Tyler Miller and S. E. SpoolMan, *Living in the Environment*. 2012.

[2] Badan Pusat Statistik Kabupaten Pemalang, *Pemalang dalam Angka*. Pemalang: Badan Pusat Statistik Kabupaten Pemalang, 2015.

[3] X. Qiao, W. Xiao, D. Jaffe, S. H. Kota, Q. Ying, and Y. Tang, "Atmospheric wet deposition of sulfur and nitrogen in Jiuzhaigou National Nature Reserve, Sichuan Province, China.," *Sci. Total Environ.*, vol. 511C, pp. 28–36, 2014.

[4] S. Soehodho and E. S. Taufick, "Study on correlation between motor vehicle emission and public health," *Proc. ...*, vol. 5, no. 2002, pp. 1841–1856, 2005.

[5] K. Huang, G. Zhuang, C. Xu, Y. Wang, and A. Tang, "The chemistry of the severe acidic precipitation in Shanghai, China," *Atmos. Res.*, vol. 89, no. 1–2, pp. 149–160, 2008.

[6] M. R. Lotfalipour, M. A. Falahi, and M. Ashena, "Economic growth, CO2 emissions, and fossil fuels consumption in Iran," *Energy*, vol. 35, no. 12, pp. 5115–5120, 2010.

[7] Sekretariat Negara Republik Indonesia, *Peraturan Pemerintah Nomor 41 Tahun 1999 Tentang Pengendalian Pencemaran Udara*, no. 41. Sekretariat Negara Republik Indonesia, 1999.

[8] IPCC, Mitigation of climate change: Contribution of working group III to the fourth assessment report of the Intergovernmental Panel on Climate Change. Intergovermental Panel on Climate Change, 2007.

[9] A. Sugiono, "Penanggulangan Pemanasan Global," *J. Sains Teknol. Modif. Cuaca*, vol. 7, pp. 15–19, 2006.

[10] IPCC, *2006 IPCC guidelines for energy and carbon emissions*, vol. 2. Intergovermental Panel on Climate Change, 2006.

[11] Kementerian Pekerjaan Umum Republik Indonesia, *Manual Kapasitas Jalan Indonesia (MKJI) 1997*. Jakarta: Kementerian Pekerjaan Umum Republik Indonesia, 1997.

[12] Kementerian Lingkungan Hidup Republik Indonesia, *Gas Rumah Kaca Nasional Buku I*. Jakarta: Kementerian Lingkungan Hidup Republik Indonesia, 2012.

[13] Kementerian Lingkungan Hidup Republik Indonesia, *Baku Mutu Emisi Gas Buang Kendaraan Bermotor Tipe Baru Kategori L3*. Jakarta: Kementerian Lingkungan Hidup Republik Indonesia, 2012.

[14] IPCC, 2006 IPCC Guidelines for National Greenhouse Gas Inventories, Volume 2 : Energy Chapter 1 : Introduction. Intergovermental Panel on Climate Change, 2006.

The centrifugal and centripetal force influence on spatial competition of agricultural land in Bandung Metropolitan Region

E Sadewo

Department of Regional and Urban Planning, School of Architecture, Planning, and Policy Development (SAPPD), Institute of Technology Bandung, Indonesia

erie@s.itb.ac.id

Abstract. Agricultural activity has suffered a massive land functional shift caused by market mechanism in Bandung metropolitan region (BMR). We argue that the existence of agricultural land in urban spatial structure is the result of interaction between centrifugal and centripetal force on spatial competition. This research aims to explore how several recognized centrifugal and centripetal force influence to the existence of agricultural land in BMR land development. The analysis using multivariate regression indicates that there exists spatial competition between population density and degree of urbanization with agricultural land areas. Its extended spatial regression model suggested that neighboring situation plays an important role to preserve agricultural land areas existences in BMR. Meanwhile, the influence of distance between the location of the city center and employment opportunities is found to be insignificant in the spatial competition. It is opposed to the theory of von Thünen and monocentric model in general. One of the possible explanation of such condition is that the assumption of centrality does not met. In addition, the agricultural land density decay in the southern parts of the area was related to its geographical conditions as protected areas or unfavorable for farming activity. It is suggested that BMR was in the early phase of polycentric development. Hence, better policies that lead redirected development to the southern part of the region is needed as well as population control and regulation of land use.

1. Introduction

Indonesia has experienced rapid economic growth by transforming its national development policy to modern industry. One of the major agglomeration of industrial activity, especially in the textile industry is located in Bandung Metropolitan Region (BMR) [1]. BMR initially consists of two areas: the city of Bandung as core and its surrounded periphery region. It is administratively divided into three regencies and a city, and became the third largest metropolitan region in Indonesia.

Its recent urban development contributes to spatial mismatch of population growth trend and employment growth trends, namely disagreement between the population and employment center locations [2]. To cope with these problems the government has adopted a policy of counter magnet development, aimed at decentralizing the urban economic activities from Bandung City to its surrounding small towns, particularly Soreang, Majalaya, Banjaran, Padalarang and Ujung Berung. However, urban economics decentralizing policy does not work effectively, so that the Bandung city, the core of BMR, has grown faster than the small towns surround [1]. This implies that Bandung city still played a strong role as the center of gravity, whereas their surrounded areas in east and southeastern part of the city have become uncontrolled urban sprawl [3].

Urban land use has become one of the main issues in Indonesia's urban land development. Recent urban development was followed by land use changes in the urban area and conversion of prime agricultural land to residential as well as other urban function in the peripheries [4]. The problems faced by BMR which has suffered massive land functional shift of agricultural activity. Having 340,000 hectares of area and 8.25 million of population, BMR facing serious threat to its agricultural land existence. The data from agricultural census showed that the number of agricultural activities in the region during the decade of 2003-2013 decreased by 22.75 percent, with about 84 percent of whom are smallholders. The conversion has not only taken place in the prime agricultural land of the urban fringe, but has also occurred on land designated as conservation areas, such as Northern Bandung [5]. This result not only could endanger food security, but also have detrimental impacts on the environment.

The integration of urban agriculture into urban economic system and urban ecosystems compete in acquiring land with other urban functions, affected by planning and urban policy [6]. In such case, we argue that the existence of agricultural land in urban spatial structure is the result of interaction between centrifugal and centripetal force on spatial competition. Therefore, the protection of agricultural land is a challenge for spatial planning in the urban area that requires special handling strategy. The policy making needs an accurate information on the extent of linkage characteristic of the spatial structure of urban land use against the competition. Nevertheless, research on spatial competition that is done in this area still has not received much attention. This research aims to derive several centrifugal and centripetal forces which believed to play important role in the spatial competition of BMR. Furthermore, we examine how each factor could explain the existence of agricultural land areas.

Although such literature came from a distinct model of urban structure, but the study of the morphology of the whole area of study has not yet provided so far. Assuming the city of Bandung as center gravity for surround areas [1] [3], we choose monocentric urban models as analysis base of urban spatial structure. It has an eminence of simple assumptions to generate a model of the spatial structure of a city with remarkable predictive power [7].

This paper is organized into six parts. The first part discusses the urban land development and agricultural land issue in BMR. In the second part, we present a review to what extent the existence of agricultural land could be considered as the result of spatial competition. In third parts, we derived several centrifugal and centripetal forces which believed to explain the existence of agricultural land as a result of spatial competition from a literature review. The fourth parts explain the methodology used and the results of the analysis. The fifth parts present the discussion of the model. In the last part, conclusion and some appropriate policy recommendations were proposed for the direction of development, with regard to the sustainability of agricultural land areas.

2. Spatial Competition

Urban spatial structures are shaped by market forces interacting with regulations, primary infrastructure investments and taxes [8]. From the perspective of classical theory, the monopolistic power of a firm against another within an area is determined from its production and transportation cost [9]. While production costs tends to be fixed, firms minimizing its cost by simultaneously choosing the location which could reduce transportation expenses. Since land resource is immobile, while the market located within specific area, the whole economics stakeholder such as household, farm, and firm will always compete for the most strategic spatial environment [10], known as spatial competition. The price elasticity of demand for the agricultural commodity is of crucial importance in determining even the direction of change in rent and the areas in which firms of the industries locate. But there raise questions to the extent of which parcel of land is available or should be taken to gather a spatial equilibrium?. The answers depend on how the urban structure is shaped trough the land use change process. [11] consider that the urban-rural land conversion was a special case of the more general equilibrium location of firms in a von Thünen plain changes in the condition of demand and supply for the commodities they produce.

[12] argue that this theory is appropriate to explaining the loss of agricultural land to non-agricultural uses.

As a human process, land use change is affected by spatial interaction [13]. Gravity model as the most widely used types of spatial interaction model interrelates two basic elements which influencing land use change: scale (mass factors) and distance. The land use change in the urban area shows that it was a dynamic organism constantly in process of evolution, which involve both modification of long-established function and the addition of new functions. Among the forces that determine the urban function, form, and pattern, two groups stand prominently [14]. The first factor is made up centrifugal forces to migrate from the urban center towards, and the second is centripetal force which promotes agglomeration in the center. These two forces usually influence agglomeration trough migration activity. But in broader perspective, [15] argue that its concept is closer to socioeconomic interaction and relate to spatial infrastructure. Empirical result has shows the two forces relation with urban structure dynamics (see [16]–[19] for example). Since the influence of each of these factors would differ depends on the context of the location, therefore it was important to recognize the effect of the two forces on the shape of urban spatial structure.

3. The Driving Forces

The first agricultural location and land rent theories were introduced by Ricardo in 1820 and von Thünen in 1826. Since then, agricultural location theory has played an important role in the development of urban land use theory [13]. Thünen developed a model of agricultural land use and its association with land rent as well as production. It was explained that land use function basically will follow a rule that involves certain spatial configurations. In such case, [20] believe that the Thünen theory not only associated with traditional geography and location theory but also includes the basis of the new geography of economic theory. The theory was later reinterpreted by Alonso and generalized its central concept of big rent curves into an urban context known as monocentric urban model, a critical analysis tool of urban structure.

The monocentric urban model explains a unit of land function as a function of distance from the urban center. It was allocated to the entity which has the highest land bid rent. The land rent declined as the distance from the CBD getting farther. Thus, the morphological pattern of land use close to the central business district (CBD) will be occupied by the economic activity which has the ability of capital to pay the land rent. The location of CBD becomes an attraction for employment seekers. These consequences of worker migration to the urban areas. To reduce housing cost, workers would have to live at a distance around the CBD. As a result, the amount of settlement location decays further from the CBD. The boundary between the urban area and the agricultural periphery is the point at which urban bid rents no longer exceed the amount that farmers will pay for land. Population growth and an improved transportation network cause the urban area to expand at the expense of the agricultural hinterland. From that explanation, we could derive several factors which are considered to have a relationship or influence to the spatial competition of agricultural land in the urban area [21].

3.1. Distance from city center

In monocentric urban models, centrality occurs when distance becoming the main factor in determining spatial structures. It was first considered by Thünen which assume that a plain area will consist a lot of cities within the same distance from each other. Thünen also investigating how the distance can influence the agricultural production and land lease pattern. However, Thünen did not explain how to determine the distance [20]. [16] argues that following the monocentric model, residents assumed to commuting to work towards the city center. Therefore, the distance of the CBD holds a very important role in deciding the location of settlements. The progress of the transport system and technology is an important factor that encourages people to live at a distance from the city center. Thus the function of population is determined jointly by the location of the work and other exogenous factors.

A different perspective of distance importance on land use change process is provided by spatial interaction theory which emphasizes on gravity model [13]. [19] argued that the distance between the residence and the employment location with the market can describe a closeness. The distance function is to hold settlement and business activities agglomerate near to the CBD. But at some tolerable point, living at a distance from their workplace would decrease the profit of commuting. As a result, some workers could change their job location near their settlement, Considering their sub-center in a region of the city, the use of linear distance between the centroid can explain the spatial organization of the urban and the possible impact on the landscape structure on a regional scale [18]. [17] agreed with the idea that the distance is a representation of the location factor. In their analysis, the distance was compared to the cost of transport in order to measure the spatial relationship between cities. Thus the of spatial linkages within the monocentric urban model can affect land use patterns.

3.2. Population density

According to Thünen theory, a city that is within the same distance should have the same size too. However, Thünen didn't explain how to specify the size of the city [20]. In this case the size of the city related to the population density [17]. Based on monocentric urban model, the location of economic activity does not follow a particular dimension. At equilibrium, the particular economic activity will be in the city center. Furthermore, taking into account spatial constraints represented by the cost of transport, the location of settlements can be determined. That is, where the population is an endogenous factor, while the central location of economic activity is an exogenous factor and the density of the population depends on the location of economic activity center.

In monocentric and relatively compact region, urban or peri-urban uses of land have a higher efficiency and lower per-capita consumption of agricultural land [22]. [18] found that an indicator of population density can describe urban expansion and population displacement. The increase in the urban population, originating from the migration process [19]. [16] argues that the measurement location of people and the location where the work is based on the density can be mistaken for causality is hard to describe in a spacious suburban area. But a dense region still considered to indicate a preference in the selection of the location of urbanization. High population density is seen to have negative effects through increased population pressures on scarce resources such as farmland [23]. In densely populated locations, agricultural land spatial competition will be intense. Agricultural land will be reduced productivity due to pollution while increasing production costs due to high rents. As a result, it will be unable to compete and forced to switch its function.

3.3. The level of urbanization

The location of new urban development is guided by a preference over lower density areas, yet in proximity to current urban development [24]. Economic agglomeration generates new growth centers that require manpower. A shortage of available workforce will pull urbanization from rural areas. Monocentric urban model of standard assumption met if residents choose residence that can maximize the function of individual needs with consideration of limited income and location of work [16]. With increasing urbanization, the number of the population will be increased so the need for land becomes larger [12]. [17] argued that population growth is naturally represented by their use of land for housing. As population grows, urban areas in the most central municipalities experience the most significant urban sprawl and loss of farmland. Moreover, the remaining farmland is located in the same areas that already have converted the most. These areas also experience significant pressure for continued land take [25].

Changes in the natural and agricultural land that occurs as a result of urban expansion were influenced by existing forms of spatial organization [18]. The experience in Norway and various countries shows that the main purpose of converting farmland into built-up areas has a strong relation to existing urban settlement areas [25]. [16] argues that employment growth in the city edge will affect the displacement of households out into an area far from the city center. In addition, the consumption of land on monocentric models are more efficient than the polycentric. The issue of land use for urbanization location choice other than to rely on the availability of land will also depend largely on the form of spatial organization. In this case, the type of land available for residential development of the area comes from agricultural land. It could have a diverse impact on agriculture, whether increase agricultural profitability trough increase in demand or decrease of labor wage for the landless household [23]. But the capacity of regions to accommodate urbanization will depend heavily on remaining agricultural land areas.

3.4. Employment opportunity

Von Thünen theory has evolved no just only associated with agricultural issues, but also industrial agglomeration. According to [20] monocentric structure built by Thünen have in common with the core-periphery model which initiated by Krugman in its principle. However, the Theory of Thünen has broader applications that can be used to explain the principles of centripetal and centrifugal in an urban. That urbanization will happen to the location of industrial competitiveness [17]. Due to the growth of regional and industrial agglomeration requires the growth of urban population as a prerequisite that must be met. To that end, mechanical factors such as employment and accessibility should be available. Employment opportunities which offered by a growth pole would attract the influx of people and industry from other areas which led to the raising demand for land. When such case occurs [12] argue that employment opportunities also contribute to the conversion of agricultural land to nonagricultural use.

In the model of monocentric, employment is assumed only in the city center. In other words, the location of residence of residents affected by the location of the workplace. In the agricultural terms, this is confirmed by [19] which states that the existence of farmers was usually found close to its farm location and market. Many experts consider that the city center has lost its economic base so that the new city grew away from the old town center, which called edge cities. As recursive models that concluded that market activity is determined by the location of settlements, [16] found that the greater the distance from downtown edge Cities, then employment will follow the location of the population, creating urban sprawl. Thus, the effect of the existence of employment in the nonagricultural sector is the driving factor that led to diminishing agricultural land.

4. Data and methods

We have identified several factors representing centrifugal and centripetal force of urban evolution, which assumed to have influences to the existence of agricultural land in BMR. These factors include distance and population density [16]–[19], as well as the level of urbanization [16]–[18] and employment opportunities [16], [17], [19]. While distance and employment opportunities were assumed to have a centripetal effect, density and urbanization level is considered to have opposite impact. Due to the data limitation, our object of research is smaller compared to the area of BMR which described in [1], since we exclude the district located in Sumedang Regency.

We construct hypothesis that these four factors have a significant influence in affecting the agricultural land in the metropolitan area of Bandung (Table 1). Each data used here is considered to be a representation, or as an approach to its actual figure. The whole data collected from National Statistics Agency, but since the data on district level available on different census activity, the time span used in this research was also varied. For instance, we use migration data from 2010 population census, agricultural land area data from agriculture census 2013, and employment growth data from 2016 economic census, this could lead to serious bias. Another possible shortage about the data used in this research is the lack of employment information at the district level which could explain employment opportunity. Therefore, we use the growth in the number of non-agricultural firms or business as an approach for determining employment opportunity.

To assess a theoretical model of urban land changes, some research, such as [11] and [12] suggested a test using regression analysis. However, this method has some fundamental assumption that needs to be fulfilled such as the normal distribution of error, appropriated error structure of the variables, independence of variables, and model linearity. Any violation of this assumption would make the forecasts, confidence intervals, and scientific insights yielded by the model inefficient or seriously biased or misleading. A common issue that often occurs in the discussion about location behavior is spatial dependency. [24] argue that spatial dependence is an important factor to take into account when analyzing land use choices and conversion. Therefore, our method would employ spatial auto regression which incorporates location factors as an extension of multivariate linear regression method. As its properties, we evaluate its spatial autocorrelation condition, both global and locally.

Table 1. The description of variable used in linear regression modeling.

Variable Type	Variable Name	Brief description of data used	Notation
Dependent variable	Agricultural land area ratio	The percentage of agricultural land area compared to the entire land area (%)	Y
Independent variable	Distance	Straight centroid distance to the central business district of Bandung (Km)	X_1
	Population Density	Number of inhabitants per Km^2 (person)	X_2
	Urbanization	The percentage of in-migrate residents on years 2005-2010 against the total population in 2010 (%)	X_3
	Employment opportunity	Growth in the number of non-agricultural firms/ businesses during the years 2006-2016 (%)	X_4

5. Result and Discussion

Bandung Metropolitan Region (BMR) initially consist of 80 districts, where the average area of each district is 40.95 km2. Each district has approximately 103.000 people of average population in 2015, so that the density per km2 is about 8,213 inhabitants. The highest density located in Bandung City as the core of BMR, which reached nearly twice of its average. The ratio of agricultural land in the core area is only 11.75 %, half of the average area of agricultural land in the overall BMR. The main agricultural location distributes clustered in the eastern and southern area of BMR core (Figure. 1). The darker color of the map indicates a higher proportion of agricultural land area ratio.

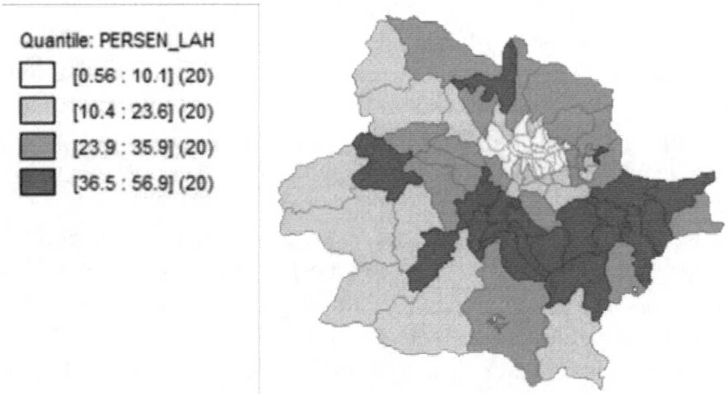

Figure 1. Spatial pattern of district's agricultural land area in BMR, 2013 (percent).

To examine the relationship between the influence of centrifugal and centripetal force to the existence of agricultural land, we made multivariate linear regression model from the data and checks its assumption first. Model (1) indicates strong linear relationship between the agricultural land area and the several centrifugal and centripetal forces, shown by a significant probability value of F statistics ($\alpha = 0{,}05$). But a partial examination on each variables shows that only density and migration variables to have significant influence on the agricultural land area.

$$Y = 40,98 - 0,14X_1 - 0,001X_2 - 1,22X_3 + 0,052X_4 + \varepsilon \quad (1)$$

t-stat: 9,689* -0,697 -5,897* -3,637* 0,943
*) statistically significant at $\alpha = 0,05$

The significance negative value confirms that both variables are a competitor for the existence of agricultural land areas. Population density is found to be less sensitive compared to urbanization. It took an increase of 1,000 people density per district to decrease of 1 % agricultural land area ratio ceteris paribus, compared to urbanization rate which only took 1 % of increase to have a 1,22 % of agricultural land area decrease. The whole assumptions are met, but the variation of the force which could explained is only 48,76 percent of adjusted R^2 value, which seems to be less favorable. This suggests that there might be other factors which could explain the variation of dependent variables.

Table 2. The linear regression's assumption test result.

Indicators	Value
Multicoliearity Index	7,367*
Jarque-Bera probability of normality of error	0,498*
Breusch Pagan probability of heteroskedasticity	0,146*
Koeneker Basset probability of heteroskedasticity	0,120*
Moran's I	0,004*
LM (Lag)	0,001*
Robust LM (lag)	0,004*
LM (error)	0,020*
Robust LM (error)	0,139

*) statistically significant at $\alpha = 0,05$

With regard to the theory of Thünen and monocentric, the finding of distance and employment opportunities influence were contrary to the centrality assumption. One of the possible explanation of such condition is that the assumption of centrally located market or business district does not fulfilled. The agricultural land area density decay in the southern parts of the area affected by its non-plain topography as well as several areas has been protected or being unfavorable for farming and business activity. This support the argument of leap frog sprawl phenomenon of land development [26], which has divide market and employment concentration to several location within. Thus, both variables are excluded in the spatial auto regression model.

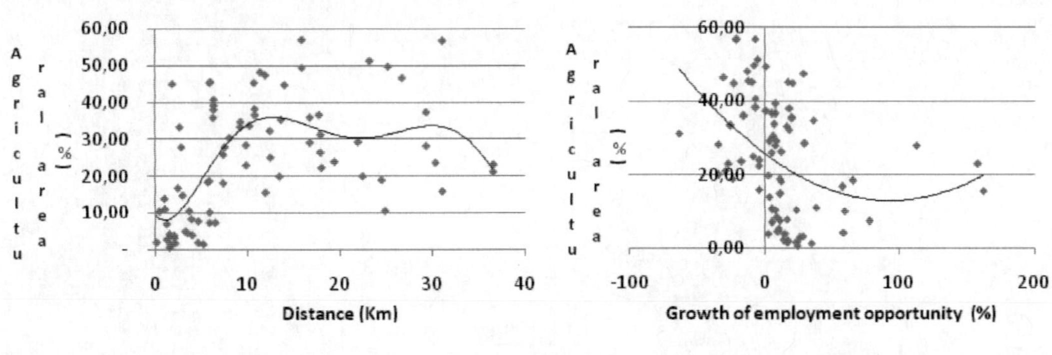

Figure 2. Plot of agricultural area compared to (a) distance to city core, and (b) growth of employment.

On assessing spatial autocorrelation, we use Queen contiguity to determine each district's neighborhood due to the nature of the border between districts. The analysis of spatial effect test shows that Moran's Index value is significant. This indicates that there are significant spatial variations, which could verify of visual interpretation of Figure 1. The Lagrange Multiplier (LM) for a lag effect and the effect of error both showed a significant result. However, examination of Robust (LM) found that a lag effect has statistically significant while the error effect is not. It can be concluded that the spatial model used is the spatial lag as in equation (2).

$$Y = 0,372WY + 24,65 - 0,0007X_2 - 0,69X_3 + \varepsilon \quad (2)$$

t-stat: 3,846* 6,468* -4,443* -2,469*
*) statistically significant at $\alpha = 0,05$

Model (2) has smaller value of AIC and SC so that it can provide better estimation results than the linear regression. The whole assumption regression has also been met so that the model fit for use. With concluded that the amount of agricultural land at a site in BMR could be determined by population density and degree of urbanization, as well as the amount of agricultural land areas in neighboring locations. Higher population density and urbanization would give negative impact to the existence of agricultural land area, while the presence of the area on neighboring district would make an opposite impact. Nevertheless, population density has less elasticity than the degree of urbanization and neighboring effect. This suggests that should be directed on better arrangement between the location for new settlements and existing agricultural land.

Table 3. The goodness of fit model comparison between linear and spatial regression.

Indicators	Multivariate regression	Spatial regression
R^2	51,36	59,38
AIC	620,081	608,02
Schwarz criterion	631,992	617,548

The arrangement of development location should consider the existence of specific clusters in spatial arrangement. To evaluate local association of agricultural land area clusters, we use local indicators of spatial autocorrelation (LISA) and local G statistics [27], whereas the former is indicating a local cluster of similarity value, and the later shows local cluster of high or low value. The result shows that Bandung city as core area has the low dissimilarity and low value of agricultural land area clusters. High similarity clusters are found in the southeast and southwest area of the core, consist of 11 sub districts. Among those sub districts, there were 8 of them, whose exhibit clusters of high agricultural land area density. Such areas should be seen as a greenbelt, in which land use development caused by migration should be controlled.

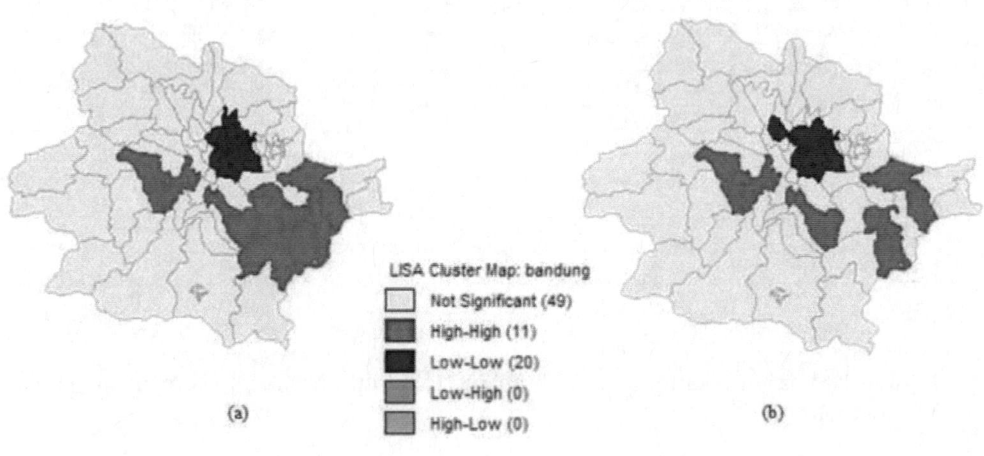

Figure 3. Clusters of significance agricultural land area measured by (a) LISA and (b) Local G Statistics

6. Conclusion

In urban areas, the existence of agricultural land is the result of spatial competition mainly driven by market mechanism between households, farms, and firms. BMR has suffered massive agricultural land use shift to non-agricultural function. In such case we argue that there were several centripetal and centrifugal force characteristics which has contributed to the spatial competition of agricultural land in BMR. The Characteristics assumed to affect the existence of agricultural land area in a location includes distance and population density, as well as the level of urbanization and employment opportunity. We use linear regression method with the addition of spatial effects to verify of how the whole and each factor influence to the area of agricultural land in the region. Assuming the city of Bandung as center gravity for surround areas, we choose monocentric urban models as analysis base.

The result indicates that spatial competition did occur between population density and degree of urbanization with the existence of agricultural land area in BMR showing by its negative relationships. On the opposite, the existence of agricultural land in neighboring locations has a positive influence to maintain its areas. It is implied that when an agricultural land area in a district is being extinct, their neighboring location will be affected too. The current condition shows that it was still exist some greenbelt areas of agricultural land in the southern parts of BMA core is under heavy pressure from a massive urbanization. Contrary to the argument of failing implementation of counter magnet policy, we believe that BMR has actually entering the early phase of polycentric development. However, it must be admitted that spatial mismatch could not be avoided due to the lack of infrastructure which connect between centers within the region. It was suggested a need for policy that lead to population control and provide regulation of land use within the radius of 5 to 15 km from the city center to maintain agricultural areas. The direction of development should deploy towards rural areas behind the agriculture land concentration, for example Ciwidey or Rongga sub districts. Through these policies is expected that the new growth center not only can reduce the socioeconomic impact, but can bring progress to the region bordering on the southwest. At the same time it needs the support of BMR infrastructure plan which mainly supported development in the south area of the metropolitan region. The planned infrastructures should be more compact due to the physical morphology of the Bandung basin area. The development in the north should be limited as well because there is a tendency to form population centers while there is a hill area [2].

However, this research still contains several drawbacks. Since we use discrete areal units for analysis, a certain modeling outcome can result from the underlying aggregation level and the configuration of zones. It means that the result is only a single manifestation of a range of possible results, the problems known as modifiable areal unit problem (MAUP) [28]. Yet, our model could only explain some variation of which agricultural land exists. We consider that there were still many possible factors which could help to explain the existence better. For the next research, it would be desirable to take road development impact on spatial competition of agricultural land into account. This suggestion came from two perspectives. First, the development of road infrastructure is more likely to be built at agricultural location, since its direct rejection of conversion would be minimum. Secondly, the increase of transportation accessibility and mobility due to the development in road project not only has caused raised wages both in agricultural and non-agricultural employment, but also the decline of agricultural employment working time [29]. This would also have an impact to the decrease in farming activity.

References

[1] T. Firman, "Urban development in Bandung metropolitan region: a transformation to a Desa-Kota region," *Third World Plan. Rev.*, vol. 18, no. 1, p. 1, 1996.

[2] F. U. Utami, D. Chikashi, and D. Hudalah, "Characteristics of Metropolitan Urban Growth Trend in Indonesia Evaluated by Klaassens's Spatial Cycle and Its Differences with Spatial Planning (case Study: Jakarta, Bandung, Medan)," *J. Perenc. Wil. Dan Kota B SAPPK*, vol. 2, no. 3.

[3] P. R. Santoso, "Revisited composition: Implementation strategies for mobility based development in Bandung, Indonesia," Master Thesis, Delft University of Technology, Delft, 2016.

[4] T. Firman, "Major issues in Indonesia's urban land development," *Land Use Policy*, vol. 21, no.

4, pp. 347–355, Oct. 2004.

[5] T. Firman, “Rural to urban land conversion in Indonesia during boom and bust periods,” *Land Use Policy*, vol. 17, no. 1, pp. 13–20, Jan. 2000.

[6] D. L. Cahya, “Analysis of Urban Agriculture Sustainability in Metropolitan Jakarta (Case Study: Urban Agriculture in Duri Kosambi),” *Procedia - Soc. Behav. Sci.*, vol. 227, pp. 95–100, Jul. 2016.

[7] L. Salvati and M. Carlucci, “Distance matters: Land consumption and the mono-centric model in two southern European cities,” *Landsc. Urban Plan.*, vol. 127, pp. 41–51, Jul. 2014.

[8] A. Bertaud, “The Spatial Organization of Cities: Deliberate Outcome or Unforeseen Consequence?,” Background Paper 27864, 2003.

[9] H. Hotelling, “Stability in Competition” *Econ. J.*, vol. 39, no. 153, pp. 41–57, 1929.

[10] R. L. Hall, J. H. Dorfman, L. F. Gunter, and others, “Spatial competition and pricing in the agricultural chemical industry,” in *American Agricultural Economics Association Annual Meeting*, 2004.

[11] R. F. Muth, “Economic Change and Rural-Urban Land Conversions,” *Econometrica*, vol. 29, no. 1, p. 1, Jan. 1961.

[12] A. F. Ramsey and F. L. Corty, “Conversion of Prime Agricultural Land to Nonagricultural Uses in One Area of the Sunbelt,” *South. J. Agric. Econ.*, vol. 14, no. 02, pp. 23–29, 1982.

[13] E. Rustiadi and T. Kitamura, “Analysis of Land Use Changes in City Suburbs. A Case Study on Some Subdistricts of the Bekasi Area of West Java, Indonesia.,” *J. Rural Plan. Assoc.*, vol. 17, no. 1, pp. 20–31, 1998.

[14] C. C. Colby, “Centrifugal and Centripetal Forces in Urban Geography,” *Ann. Assoc. Am. Geogr.*, vol. 23, no. 1, pp. 1–20, Mar. 1933.

[15] R. Krzysztofik, “Revisiting the question of centripetal and centrifugal forces in urban systems,” *Geogr. Pol.*, vol. 89, no. 4, pp. 429–442, 2016.

[16] C. Ding and R. D. Bingham, “Beyond Edge Cities: Job Decentralization and Urban Sprawl,” *Urban Aff. Rev.*, vol. 35, no. 6, pp. 837–855, Jul. 2000.

[17] C. Lu, Y. Wu, Q. Shen, and H. Wang, “Driving force of urban growth and regional planning: A case study of China’s Guangdong Province,” *Habitat Int.*, vol. 40, pp. 35–41, Oct. 2013.

[18] L. Salvati and M. Carlucci, “Distance matters: Land consumption and the mono-centric model in two southern European cities,” *Landsc. Urban Plan.*, vol. 127, pp. 41–51, Jul. 2014.

[19] O. J. Taiwo, “Determinants of peri-urban and urban agricultural locational choice behaviour in Lagos, Nigeria,” *Land Use Policy*, vol. 39, pp. 320–330, Jul. 2014.

[20] M. Fujita, “Thünen and the New Economic Geography,” *Reg. Sci. Urban Econ.*, vol. 42, no. 6, pp. 907–912, Nov. 2012.

[21] R. Arnott and D. McMillen, “Urban Land Use: Preface,” in *A companion to urban economics*, R. Arnott and D. McMillen, Eds. Oxford: Blackwell Publishing Ltd, 2006, pp. 77–95.

[22] B. Catalán, D. Saurí, and P. Serra, “Urban sprawl in the Mediterranean?,” *Landsc. Urban Plan.*, vol. 85, no. 3–4, pp. 174–184, Apr. 2008.

[23] Y. Liu and F. Yamauchi, “Population density, migration, and the returns to human capital and land: Insights from Indonesia,” *Food Policy*, vol. 48, pp. 182–193, Oct. 2014.

[24] C. E. Carrión-Flores, A. Flores-Lagunes, L. Guci, and others, “Land use change: a spatial multinomial choice analysis,” in *Agricultural and Applied Economics Association 2009 Annual Meeting*, 2009.

[25] K. L. Skog and M. Steinnes, “How do centrality, population growth and urban sprawl impact farmland conversion in Norway?,” Land Use Policy, vol. 59, pp. 185–196, Dec. 2016.

[26] Y. Budiyantini and V. Pratiwi, “Peri-urban Typology of Bandung Metropolitan Area,” *Procedia - Soc. Behav. Sci.*, vol. 227, pp. 833–837, Jul. 2016.

[27] L. Anselin, “Local indicators of spatial association—LISA,” *Geogr. Anal.*, vol. 27, no. 2, pp. 93–115, 1995.

[28] M. Helbich and M. Leitner, “Spatial Analysis of the Urban-to-Rural Migration Determinants in

the Viennese Metropolitan Area. A Transition from Suburbia to Postsuburbia?," *Appl. Spat. Anal. Policy*, vol. 2, no. 3, pp. 237–260, Oct. 2009.

[29] F. Yamauchi, "Roads, labor markets, and human capital : evidence from rural Indonesia," World Bank, Washington, DC, 2014.

Sustainability study of domestic communal wastewater treatment plant in Surabaya City

E Bahar[1, 3], Sudarno[2], B Zaman[3]

[1] Environmental Science Department, Diponegoro University, Semarang 50241, Indonesia
[2] Environmental Engineering Department, Diponegoro University, Semarang 50275, Indonesia
[3] Author to whom any correspondence should be addressed

erwin.ppekal@gmail.com

Abstract. Sanitation is one of the critical infrastructure sectors in order to improve community health status. The Ministry of Public Works of the Republic of Indonesia to define that word sanitation include: domestic waste water management, solid waste management, rain water management (drainage management) as well as the provision of clean water. Surabaya city as the capital of East Java province and Indonesia's second largest city with a population of 2,853,661 inhabitants in 2014 (the second largest after Jakarta), but the people who have been served by the sanitation infrastructure systems were expected at 176,105 families or about 26.95 % of the population of the city is already using sanitation facilities. In the White Book Sanitation of Surabaya City in 2010, Surabaya City sanitation development mission is to realize the wastewater management of settlements in a sustainable and affordable by the community. This study aims to assess the sustainability of the wastewater treatment plant (WWTP) domestic communal in the city of Surabaya. The method in this research is quantitative method through observation, structured interviews and laboratory testing of the variables analyzed. Analyses were performed using a technique Multidisciplinary rapid appraisal (Rap-fish) to determine the level of sustainability of the management of communal WWTP based on a number of attributes that easy scored. Attributes of each dimension includes the technical, environmental quality, institutional, economic, and social. The results of this study are sustainability index of environmental quality dimension at 84.32 with highly sustainable status, technical dimension at 62.61 with fairly sustainable status, social dimension at 57.98 with fairly sustainable status, economic dimension at 43.24 with less sustainable status, and institutional dimension at 39.67 with less sustainable status.

1. Introduction

Today there are many people in Indonesia who have difficulty accessing clean water and sanitation due to the limitations of the existing infrastructure. In accordance with the Millennium Development Goals (MDGs), the Government has set a target in 2015 that as many as 68.87% of the total Indonesian population should have access to improved drinking water sources, while access to adequate sanitation facilities is as much as 62.41% [1]. Based on data from the Central Statistics Agency (BPS), for the period 2011 to 2014 Q1, the proportion of households to sanitation is 61.04% for basic sanitation in urban and rural areas. But it is merely a quantity not quality. When viewed from the quality is still

51.02% of the families in Indonesia have access to improved sanitation [2]. Report of the Health Research Ministry of Health in 2013 showed that the proportion of households with access to improved sanitation facilities in Indonesia in 2013 amounted to 59.8 %. When compared with the results Riskesdas 2007 and 2010, the proportion of households Indonesia have access to improved sanitation facilities tend to increase (2007: 40.3%; 2010: 51.5%; 2013: 59.8%).

In urban areas easily seen their means of wastewater that flowed through the channels, where wastewater from the household immediately flowed into channels in the area around the settlements to the water bodies nearby creeks and rivers. Besides flowed into the channels that exist, there is an approach to business processing household waste water is to use a communal domestic WWTP. More and more domestic communal WWTP that creates a variety of problems associated with sustainability, because during this time the concept of management of communal WWTP less domestically involving all stakeholders and not be seen from various aspects. Under these conditions certainly needed a study to formulate the concept of effective management to achieve sustainability of domestic WWTP communal [3].

Surabaya city as the capital of East Java province and Indonesia's second largest city, and the city is quite important in eastern Indonesia continues to experience population growth, and to experience growth in population lifestyles. Based on the report Surabaya in Figures 2015 published by BPS Surabaya, Surabaya total population reached 2,853,661 inhabitants in 2014 (the second largest after Jakarta). The population is so big no wonder the Surabaya City development results in more buildings for the centers to attract people outside the city of Surabaya to enter and stay in the city of Surabaya well as seasonal residents and permanent residents. Population growth demanded accretion residences, ranging from the very simple, which is in river side or channel up to luxurious. Urban development is not integrated with the surrounding area will pose complex problems such as environmental health problems, pollution, water supply and disposal of domestic wastewater.

Systems and conditions of the disposal of domestic wastewater in the city of Surabaya, based on information from Surabaya City Health Office, the conditions and the level of sanitation services related to the problem of sewerage in the city of Surabaya is still less qualified technical and health. Some locations have supplied domestic waste disposal facilities, but there are some that can't be used anymore. This is due to lack of awareness of society to live a healthy life as well as the cost factor for the operation and maintenance of sanitation facilities. Based on number examined in each health center can be concluded that, people who have been served by the sanitation infrastructure systems were expected to total 176,105 households, or about 26.95% of the population of the city is already using sanitation facilities. Therefore we need an assessment of the sustainability of wastewater management in particular settlements WWTP domestic communal in Surabaya who can provide fill to related parties.

2. Literature Review

Wastewater resources from domestic activities such as from urine, bathing, washing appliances, washing clothes and other kitchen activities before being discharged into waterways should be processed first in sewage treatment plants. The basic principle is that the waste water released into the environment is not dangerous for the health of the environment. The waste water is not managed properly can have very wide, for example, can poison drinking water, poison pet food, the cause of the imbalance of river ecosystems and soon [4]. Each body of water whose quality has been affected by human activities may be regarded as wastewater. Domestic wastewater can be from the residence, industry or agriculture. Domestic wastewater includes a wide range of contaminants that can be potentially dangerous or concentration, which can lead to degradation of water quality. Potential contaminants include soap and detergent from the bathroom, food scraps and grease from the kitchen and other human activities that involve the use of water. Household waste is waste generated from household activities and activities of man routine sanitation [5].

Environmental pollution in the water body is most often caused by domestic wastewater. Therefore, the handling of domestic waste water needs to be done through wastewater treatment with the aim to lower the concentration and load of pollutants before being discharged into the environment [6].

Domestic waste water management strategies need to consider the environmental, social, economic and cultural. So the technology and systems used in the processing of domestic waste water should be inexpensive and socially acceptable and environmentally friendly [7]. Domestic wastewater management program that is now being developed in urban areas is a centralized processing system with the involvement of the community as empowerment in the form of sewerage piping system that is processed in a communal domestic WWTP. According to the Ministry of Public Works, in the form of communal WWTP piping system that drains waste water from households connected to the piping system and the processing is done centrally. Treatment sludge dewatering pipe network and processed former WWTP, periodically needs to be cleaned and drained, so it needs to be managed by local non-governmental groups [8].

Meanwhile, according to (Balkema et al, 2002), indicators of sustainable development criteria on domestic waste water management includes several things, among others;

1. Technology, one of the functional indicators that affect the wastewater treatment system includes the technical requirements that include the ability of the system to expand the processing capacity, accept fluctuations in the quality of the effluent quality and reliability of the system. Operating system expensive and complicated will affect the sustainability of the system, because it involves the human resource capacity to manage the system.
2. Financing, economic sustainability implies financing undertaken by the community aimed at the interests of the community itself with a value that does not exceed the benefits. Sustainability indicators used domestic waste water treatment systems on the economic aspects could include investment costs, operating and maintenance including the affordability of financing by the people or parties.
3. Institutional, Institutional is an indicator of the sustainability of domestic waste water treatment system of socio-cultural aspects that give an important role. Domestic waste water treatment systems are very diverse in need of regulations and institutions in accordance with the implemented system. Institutional itself generally serves as decision-makers and regulators, development and implementation of operational and service.
4. Community participation, involvement and community involvement in the development both in the planning and implementation be an actualization of the availability of people to contribute in every stage of development. Ignorance, behaviors / habits and perceptions will affect the domestic waste water treatment systems. Different socio-cultural conditions will have a different perception of the concept of waste and sanitation, so the existence of domestic wastewater which is more accepted by society will stimulate increased awareness and social responsibility towards the environment in the vicinity.
5. Environment, one indicator of the sustainability of domestic waste water treatment system is good or bad quality of the waste generated by the system. The quality of wastewater that will either give less environmental pressure against the media receiver wastewater. Some countries set limits wastewater quality with certain standards with several physical parameters, chemical, biological and presence of other living creatures [9].

3. Methods

In analysing the sustainability of one of the methods used are Rapid Appraisal for Fisheries (Rap-fish) is a technique for assessing the sustainability of fisheries using a number of attributes that are multidisciplinary to evaluate the comparative sustainability based on various dimensions revealed to be a number of attributes / indicators to be done scoring [10]. Rap-fish used in calculating the sustainability of communal WWTP, is a modification of Rap-fish developed by the University of British Columbia, Canada and used to assess the sustainability fishing. Sustainability status assessment using software techniques Rap-fish with Multidimensional Scaling (MDS), which has many advantages, including a simple, easily valued, fast and cost required relatively inexpensive. In addition, this technique may explain the relationship of the various aspects of sustainability and also defines a flexible regional development [11].

Indicators of sustainable development criteria of domestic wastewater management includes several things including technology, financing, institutional, community and the environment. Rap-fish modified are used to determine the level of sustainability of the management of communal WWTP based on a number of attributes that easy scored. Attributes or indicators of any dimension or variable (technical, environmental quality, institutional, economic, and social) will be evaluated and may be selected to reflect sustainability, and can be fixed or can be replaced when the latest information was obtained. Attributes in the technical aspects include the physical condition of the WWTP, the suitability of the location of the WWTP with the planning, the production capacity of the WWTP and the removal efficiency of contaminants. Attributes in environmental quality aspects can be seen from the WWTP effluent parameters on the fulfilment of quality standards by East Java Governor Regulation No. 72 Year 2013, namely the parameters pH, BOD5, COD, TSS and Oils and Grease. Attributes in the institutional aspects include local institutions, the role of government and regulations. Attributes in the economic aspects include physical construction costs and operating and maintenance costs. Attributes include participation in social aspects and public perception. An assessment of the attributes of each dimension is done through observation, literature studies, laboratory tests and the results of the questionnaire. Examples of attributes on a technical dimension and weight criteria of assessment can be seen in Table 1.

Table 1. Dimensions and attributes WWTP sustainability assessment communal

Dimension / Attributes	Low	High	Criteria
Technical			
Physical Condition	1	3	(1) not according to plan (2) according to plan but a lot of mistakes (3) in the planning horizon
Conformity Layout	1	3	(1) not according to plan without explanation (2) does not fit no explanation planning (3) appropriate planning
Capacity	1	3	(1) 0-50% of inflow planning (2) 51-99% inflow planning (3) debit entry according to plan
Efficiency	1	2	(1) Not according to plan (2) Based on planning

Whereas in determining the sustainability status communal WWTP use categories as submitted by (Suyitman et al, 2009). Tabulation category status of WWTP sustainability are presented in Table 2[12].

Table 2. Category of WWTP Sustainability

Index Value	Category
0,00-25,00	unsustainable
25,01-50,00	less sustainable
50,01-75,00	fairly sustainable
75,01-100,00	highly sustainable

4. Results

The study on the sustainability of wastewater domestic communal in the city of Surabaya was conducted in Krembangan WWTP where located in the northern city of Surabaya. Based on Rap-fish analysis, sustainability index on each dimensions displayed in a graphic image. For example, sustainability index for technical dimension can be seen in figure 1.

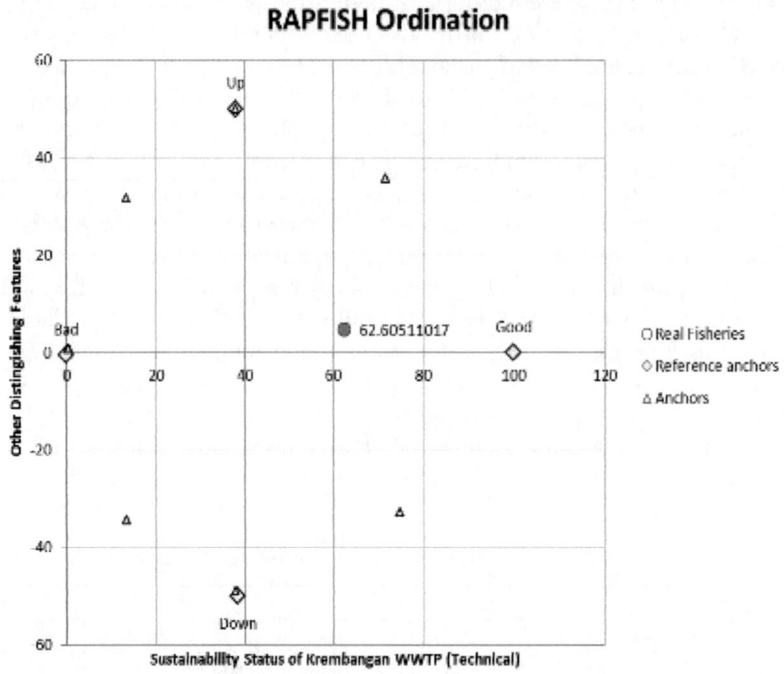

Figure 1. Sustainability Index of Krembangan WWTP (technical)

In addition, that analysis also displayed leverage atrributes and Montecarlo analysis. Leverage is used to see which most sensitive attribute to changes in the value of sustainability dimension. The highest value is the most the most sensitive attribute. Leverage attribute for technical dimension can be seen in figure 2. The Montecarlo analysis was used to compare the sustainability value of the MDS analysis and then see the difference in value to account this study. The value of Montecarlo analysis can be seen in figure 3. Then, detailed results of the analysis on each dimension presented in Table 3.

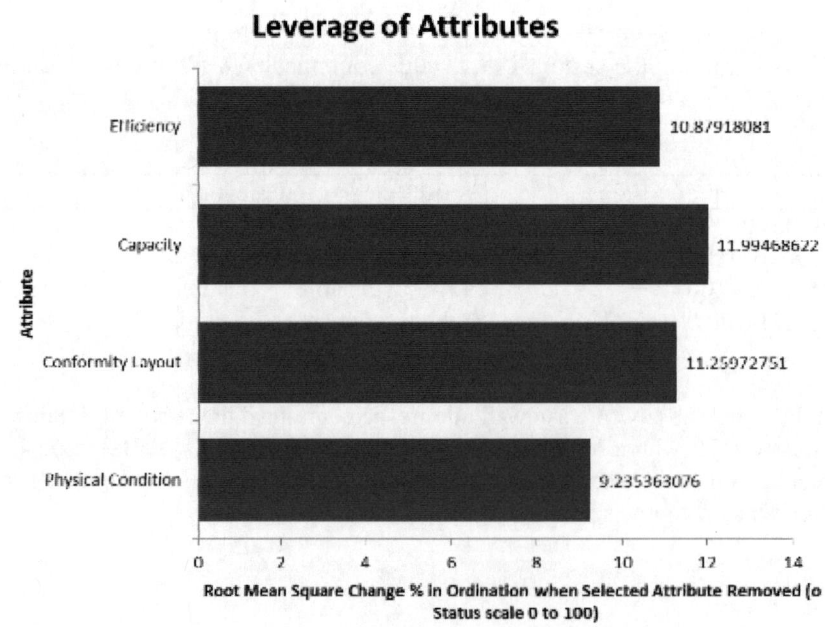

Figure 2. Leverage attributes of Krembangan WWTP (technical)

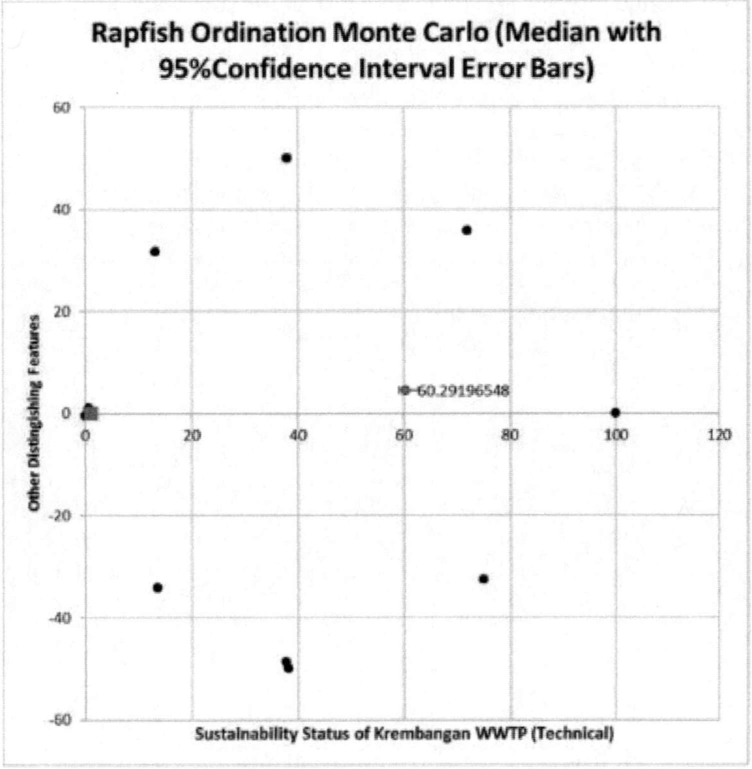

Figure 3. Montecarlo Index of Krembangan WWTP (technical)

Table 3. Sustainability index on each dimension

Location	Dimension	Indeks (%) MDS (Rapfish)	Indeks (%) Monte Carlo	Deviation	R^2	Stress
Krembangan WWTP	Environmental Quality	84.32	81.25	3.07	0.9582	0.1304
	Technical	62.61	60.29	2.32	0.9636	0.1289
	Social	57.98	56.37	1.61	0.9679	0.1291
	Economic	43.24	41.98	1.26	0.9674	0.1307
	Institutional	39.67	39.35	0.32	0.9734	0.1239

The results of the analysis with MDS methods and analysis with Monte Carlo method produces a very small difference as in column 4, not more than 5%. This proves the level of confidence in the total index (Multidimensional), confidence in the value of the index every dimension, and the effect of errors that can affect the whole process of analysis with MDS method is exceeding 95%. From Monte Carlo analysis concluded that (1) the impact of mistakes towards making a score on each attribute is very small; (2) the error caused by because of the understanding, differences of opinion, or judgment of a researcher different from each other, is relatively small; (3) data entry errors or missing data, or the value of "stress" is too high, it is very small; (4) error procedures that could affect the stability of the MDS analysis process is also relatively small. The results of the analysis (Table 2) shows that each dimension and multi-dimensional value "stress" is much smaller than the provision which states that the value of "stress" on the analysis by the method of MDS is sufficient if the obtained value <0.25. Because the smaller the value of "stress" which earned means the better the quality of the analysis performed. In contrast to the coefficient of determination (R2), the quality of the analysis results are better if the coefficient of determination is greater (close to 1). Thus, from the second parameter (the value of "stress" and R2) shows that all the attributes used in the analysis of the sustainability of the shore zone management system is good enough in light of the five dimensions of development are analyzed [13].

The index value of sustainability dimension of environmental quality, economic, social, technical, and institutional can be illustrated in the kite diagram as shown in Figure 3.

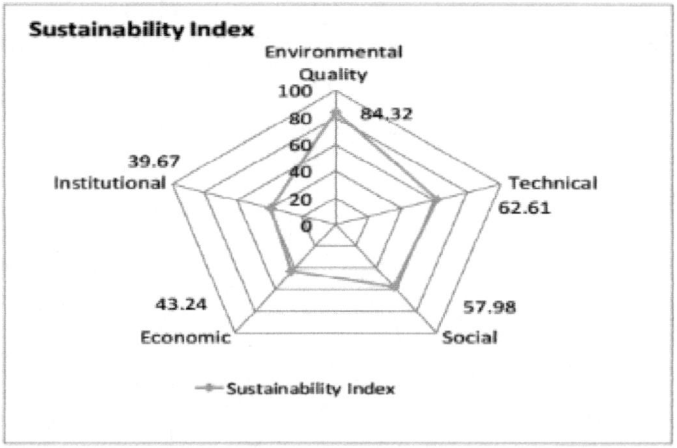

Figure 4. Kite diagram of Krembangan WWTP sustainability index

5. Conclusion

Based on the analysis method Rap-fish, the sustainability status of communal domestic WWTP Krembangan, Surabaya are environmental quality dimension at 84.32 with highly sustainable status, technical dimension at 62.61 with fairly sustainable status, social dimension at 57.98 with fairly sustainable status, economic dimension at 43.24 with less sustainable status, and institutional dimension at 39.67 with less sustainable status.

References

[1] Afriadi, T. & Wahyono, H., 2012. Partisipasi Masyarakat dalam Penyediaan Air Minum dan Sanitasi Berbasis Masyarakat (PAMSIMAS) di Kecamatan Simpur, Kabupaten Hulu Sungai Selatan (Community Participation in Water Supply and Sanitation Community Based (PAMSIMAS) in The Simpur District of Hulu Sungai Selatan). Journal of Urban and Regional Development, 8(4), pp.341– 348.

[2] Wahyuni, S., Setiani, O. & Suharyanto, 2012. Implementasi Kebijakan Pembangunan Dan Penataan Sanitasi Perkotaan Melalui Program Sanitasi Lingkungan Berbasis Masyarakat Di Kabupaten Tulungagung (Implementation of the Development Policy and Planning Urban Sanitation Through Community-Based Environmental Sanitation Program in Tulungagung). Journal of Environmental Science, 10 (Vol 10, No 2 (2012): Oktober 2012), pp.111–122. Available at: http://ejournal.undip.ac.id/index.php/ilmulingkungan/article/view/4534/pdf.

[3] Prisanto, D.E. & Yanuwiadi, B., 2015. Studi Pengelolaan IPAL (Instalasi Pengolahan Air Limbah) Domestik Komunal di Kota Blitar , Jawa Timur (Management Studies WWTP (Wastewater Treatment Plant) Domestic Communal Kota Blitar, East Java). J-PAL, 6(1), pp.74–80.

[4] Soegiharto, 2005. Dasar-Dasar Pengolahan Air Limbah (Basics Wastewater Treatment). Jakarta:

UI Press.
[5] Sari, N. R., Sunarto, & Wiryanto, 2015. Analisis Komparasi Kualitas Air Limbah Domestik Berdasarkan Parameter Biologi, Fisika dan Kimia di IPAL Semanggi Dan IPAL Mojosongo Surakarta (Comparative Analysis of Domestic Waste Water Quality Parameter Based on Biology, Physics and Chemistry at the WWTP Semanggi And WWTP Mojosongo Surakarta). EKOSAINS Journal, 7(5).
[6] Massoud, M. A., Tarhini, A., & Nasr, J. A, 2009. Decentralized Approaches to Wastewater Treatment and Management: Applicability in Developing Countries. Journal of Environmental Management, 90(1), 652–659.
[7] Nurhidayati, A., & Herman, J., 2009. Strategi Pengelolaan Air Limbah Domestik dengan Sistem Sanitasi Skala Lingkungan Berbasis Masyarakat Di Kota Batu Jawa Timur (Domestic Wastewater Management Strategy with System Scale Community-Based Environmental Sanitation In Kota Batu in East Java). National Conference of Technology Management. Surabaya: Sepuluh November Institute of Technology.
[8] Wijaya, A., Purwanto, D. S., & Supriyandani, 2014. Sistem Instalasi Pengolahan Air-Limbah (IPAL) Communal Di RW 11 Kelurahan Kertajaya Kecamatan Gubeng Surabaya (Wastewater Treatment Plant (WWTP) Communal System In RW 11 Kertajaya Gubeng Surabaya District). Journal of Gema Environmental Health, 12(3), 125–130.
[9] Balkema, J. A., Preisig, H. A., Otterpohl, R., F. J. D., 2002. Indicator For Sustainability Assesment Of Wastewater Treatment Systems. Urban Water, 4, 153-161.
[10] Fauzi, A., & Anna, S. , 2005. Pemodelan Sumber Daya Perikanan dan Kelautan (untuk Analisa Kebijakan) (Modeling of Fisheries and Marine Resources (for Policy Analysis). Jakarta: PT Gramedia Pustaka Utama.
[11] Rahayu, A., Bambang, A.N. & Hardiman, G., 2013. Strategi Peningkatan Status Keberlanjutan Kota Batu Sebagai Kawasan Agropolitan (Development Strategy of Batu Sustainability Status As Agropolitan Regional). EKOSAINS Journal, V(1), pp.21–34.
[12] Suyitman et al., 2009. Status Keberlanjutan Wilayah Berbasis Perternakan di Kabupaten Situbondo untuk Pengembangan Kawasan Agropolitan (Status of Regional Sustainability Based Livestock in Situbondo for Development Agropolitan). Agro Economic Journal, 27(2), pp.165–191.
[13] Kavanagh P, & Pitcher, 2004. Implementing Microsoft Excel Software for Rapfish: A Technique for The Rapid Appraisal of Fisheries Status. University of British Columbia. Fisheries Centre Research Reports 12(2):275-450.

IOP Conf. Series: Earth and Environmental Science **70** (2017) 012013 doi:10.1088/1755-1315/70/1/012013

Potential air pollutant emission from private vehicles based on vehicle route

H S Huboyo[1, 3], W Handayani[2], and B P Samadikun[2]

[1] Department of Environmental Engineering, Diponegoro University, Tembalang, Semarang, Indonesia
[2] Urban and Regional Planning Department, Diponegoro University, Tembalang, Semarang, Indonesia

huboyo@undip.ac.id

Abstract. Air emissions related to the transportation sector has been identified as the second largest emitter of ambient air quality in Indonesia. This is due to large numbers of private vehicles commuting within the city as well as inter-city. A questionnaire survey was conducted in Semarang city involving 711 private vehicles consisting of cars and motorcycles. The survey was conducted in random parking lots across the Semarang districts and in vehicle workshops. Based on the parking lot survey, the average distance private cars travelled in kilometers (VKT) was 17,737 km/year. The machine start-up number of cars during weekdays; weekends were on average 5.19 and 3.79 respectively. For motorcycles the average of kilometers travelled was 27,092 km/year. The machine start-up number of motorcycles during weekdays and weekends were on average 5.84 and 3.98, respectively. The vehicle workshop survey showed the average kilometers travelled to be 9,510 km/year for motorcycles, while for private cars the average kilometers travelled was 21,347 km/year. Odometer readings for private cars showed a maximum of 3,046,509 km and a minimum of 700 km. Meanwhile, for motorcycles, odometer readings showed a maximum of 973,164 km and a minimum of roughly 54.24 km. Air pollutant emissions on East-West routes were generally higher than those on South-North routes. Motorcycles contribute significantly to urban air pollution, more so than cars. In this study, traffic congestion and traffic volume contributed much more to air pollution than the impact of fluctuating terrain.

1. Intorduction

Conventional air pollutants and significant levels of greenhouse gases (GHGs) in the atmosphere originate from the transportation sector via vehicle emissions [1]. The major GHGs are carbon dioxide (CO2), methane (CH4) and nitrous oxide (N2O). Existing research shows that a large percentage of urban transport emissions contributions are extremely high.

Factors that affect the magnitude of air pollutants include (but are not limited to) engine combustion technology, the sort-quality of consumed fuel, exhaust device control, type of road surface-terrain and vehicle operation (driver behavior). Appropriate measures for abating air emissions related to the transportation sector have been adopted in Indonesia. Several measures have already been implemented in recent years for managing the transport sector and have co-benefited pollution reduction. These policies include the application of an intelligent transportation system (ITS), implementing traffic impact control, introducing a bus rapid transport (BRT) system, developing non-motorized transport, renewing paratransit public transportation and introducing smart driving training. All of these measures are implemented initially in big cities and claimed for conventional air pollutants mitigation as well as a reduction in GHGs. However, such policies lack supporting data for calculating hypothetical reduction. For example, ITS applications include vehicle speed before and after deploying ITS devices, actual shifting of private vehicles to pedestrian in NMT program, actual shifting mode of private vehicles to BRT system in BRT system program. Many studies have attempted to reveal this data in a bid to better calculate potential emissions reduction. The Ministry of Transportation must sort all data to complete a review of emission reduction efforts.

To date, assessing air quality in a specific area requires a vehicle population estimation [2]. In this case, identifying recent and detailed fleet being investigated is important. A source inventory must be robust and detailed, including statistical data featuring various vehicle categories and their specific operating conditions [3]. The vehicle routes are important for determining real emission according to the vehicles involved. Generally, the more flat a terrain, the higher speed it will produce; however this may not be true if we also consider traffic volume. Thus, in this study, we aim to discover the different air pollutant emissions due to cars/motorcycles travelling different routes.

2. Literature Review

Conventional air pollutants such as nitrogen oxides, sulfur dioxide and carbon monoxide, as well as hydrocarbons, are emitted from internal combustion engine in vehicles. The dominant emission from the transport sector is nitrogen oxide, while for diesel engines, SO2 emission is also a concern, due to the sulfur content in diesel fuel. Emission from the transport sector has diurnal variation, i.e., a variation of vehicle volume during the day. Below is an example of NO2 diurnal variation which has a repetition during the week [4]:

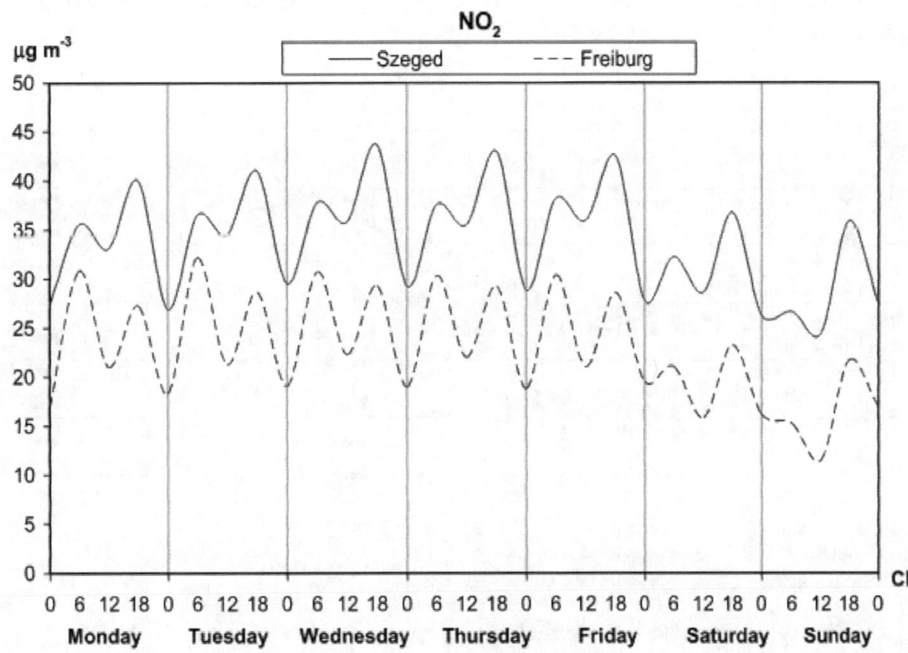

Figure 1. Diurnal variation of NO2 during one week.

Harmful secondary pollutants such as ground level ozone are also triggered by primary pollutants originating from vehicle emissions, e.g., hydrocarbons and nitrogen oxides. Generally, the majority of ground level ozone in urban areas was contributed by vehicle emissions. Additionally, airborne fine particulate can promote the risk of myocardial infarctions, strokes and heart failure. Furthermore, particle deposition in the lungs triggers the release of systemic pro-inflammatory responses.

Air pollution related to traffic is pronounced in urban areas, both in developed and developing countries. Studies on source apportionment have in many areas revealed that air pollution related to traffic is quite high. Traffic air pollution contributes 17% to 49% of urban air pollution in Indian cities [5]. In Bandung city, Indonesia, for example, roughly 60% of fine particles, including secondary aerosols, were attributed to have traffic origins [6].

In India, during a strike period when the traffic volume was considerably lower than normal, a significant reduction of ambient concentrations of BC, PM and CO (48%, 28%, and 20%, respectively) was observed [7]. As a significant contributor to worsening urban air quality, the controlling of air pollution related to traffic (both exhaust control and non-exhaust control) should be intensively promoted. The fuel consumption of vehicles can generally be estimated when we know how far a vehicle has travelled. In general, for vehicles, fuel consumption and air pollutant emissions are related to vehicle type, weight and speed. Several models have been proposed to investigate these factors, as shown in Demir [8].

Road grade was also considered an important factor for vehicle emissions [9] [10]. In Indonesia, driving pattern, particularly those of public transport drivers, are of concern due to their driving styles, which can lead to the emission of high levels of air pollutants. Fonseca [11] found that driving style was an important parameter for vehicle emissions. In general, the method for quantifying emissions from vehicle exhaust comprises two methods, i.e., on-road testing using a portable emission measurement system (PEMS) and using a chassis dynamometer in laboratory tests. On-road emission measurement has an advantage in that it may better represent real world conditions, i.e., it can handle seasonal variations, driving on different types of road materials, as well as road types (flat, ascending, descending). Air pollution emissions during vehicle operation have been researched in several studies such as Wang [12] for NOx and BC emissions, and Fuo [13] for estimating the CO, HC and NOx emissions of light duty gasoline vehicles. In the Indonesian context, the fuel consumption of vehicles related to vehicle velocity speed can be seen in the following graph [14].

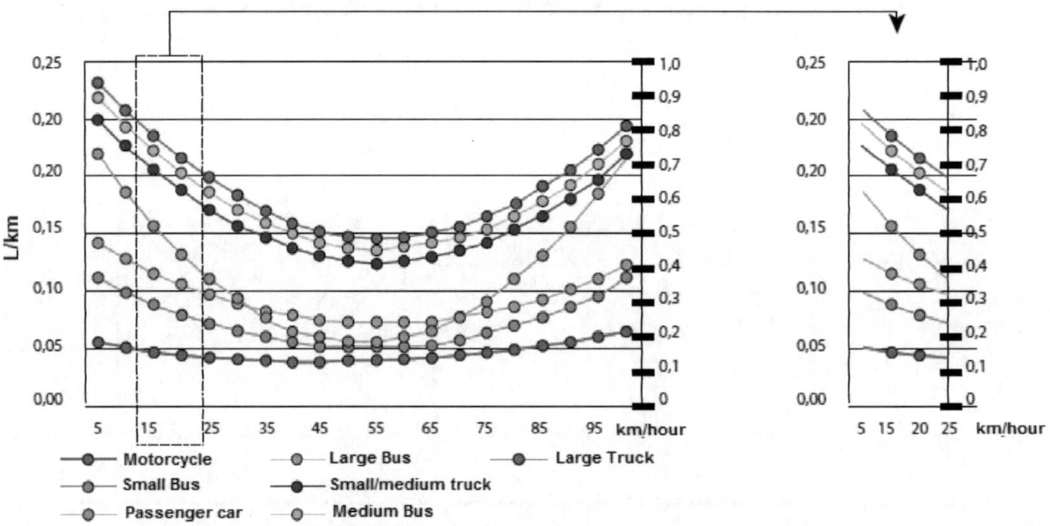

Figure 2. Specific vehicle fuel consumption related to vehicle speed.

Thus, the relationship between specific fuel consumption and vehicle speed can be derived as follow [14]:

Table 1. Specific fuel consumption and vehicle speed.

Vehicle type	Specific fuel consumption formula
Private car	7E-5x2 – 0.0077x + 0.2579
Motorcycle	1E-05x2 – 0.0009x + 0.0601
Small bus	3E-05x2 – 0.0029x + 0.1285
Medium bus	5E-05x2 – 0.0056x + 0.2961
Large bus	3E-05x2 – 0.0029x + 0.1533
Small/medium truck	5E-05x2 – 0.0053x + 0.2771
Small/medium truck	5E-05x2 – 0.006x + 0.3147

Based on Figure 2 and Table 1, the lowest emission for vehicles was achieved during medium speed, i.e., 45 km/hour to 65 km/hour. Even at low speed (around 5 km/hour), emissions were still higher than at 45 km/hour to 65 km/hour. Thus, maintaining vehicle speed while driving at roughly 45 km/hour to 65 km/hour can potentially reduce air pollution emissions. In his research, Fergusson [15], using the ERR transport emissions model estimated that 3.1% of car CO2 emissions can be curtailed if the maximum limit of 50 mph was imposed on ordinary roads.

3. Methods

A questionnaire survey was conducted to get data for technology distribution concerning the vehicle fleet in Semarang. A total of 1150 questionnaires was collected for the survey and included public transport (taxis, buses, paratransit); however, only private vehicles are highlighted in this study. Semarang consists of 16 sub-districts (*kecamatan*).

Semarang is divided into 10 city zones (Bagian Wilayah Kota, BWK). The city center is concentrated in Semarang Utara (north), Semarang Tengah (central), Semarang Barat (west) and Semarang Selatan (south). The number of registered vehicles for each district was used to allocate the number of questionnaires for each type of vehicle.

In addition to the parking lot survey, we conducted a car and motorcycle workshop survey to establish a regular maintenance figure. In this survey, we delivered 94 surveys for motorcycles and 73 surveys for personal cars. The GlobalSat DG-100 GPS Data Logger was employed for monitoring. The DG-100 GPS recorded time, traveling speed, altitude and the location of each monitored vehicle on a second-by-second basis. The output information for GPS monitoring was the driving pattern of the vehicle (including start/stop time and speed) and travel route location. In this study, GPS monitoring was conducted only for private cars.

A combination of emission factors was used to quantify emissions. No single reference was able to provide a complete emissions factor for land transportation. Additionally, a local emissions factor was not identified in this study. A list of emission factors used in this study is shown in the table below.

Table 2: Emission factors applied in this study.

Fuel	Pollutant						
	TSP (g/kg) (a)	NOx (g/kg) (a)	CO (g/kg) (a)	HC (g/mile) (b)	CO2 (g/TJ) (c)	CH4 (g/TJ) (c)	N2O (g/TJ) (c)
Gasoline	0.03	8.73	84.7	0.184	69.3	0.033	0.0032
Diesel	1.1	12.96	3.33	0.29	74.1	0.0039	0.003

Note: (a) CORINAIR (2009) (b) US EPA 1997 © IPCC 2006

IOP Conf. Series: Earth and Environmental Science **70** (2017) 012013 doi:10.1088/1755-1315/70/1/012013

4. Results

The survey was conducted in random parking lots across the Semarang districts. Average kilometers travelled for private cars was 17,737 km/year with number of machine start-up during weekdays and weekends were on average 5.19 km/year and 3.79 km/year, respectively. For motorcycles, the average kilometers travelled was 27,092 km/year with number of machine start-up during weekdays and weekends were on average 5.84 km/year and 3.98 km/year, respectively. The odometer reading for private cars showed a maximum of 3,046,509 km and a minimum of 700 km. For motorcycles, odometer reading showed a maximum of 973,164 km and a minimum of roughly 54.24 km.

Based on the survey conducted in workshops, for motorcycles, the average kilometers travelled was 9.510 km/year, while for private cars the average kilometers travelled was 21.347 km/year. The difference between VKT travelled per vehicle for the parking lot survey and workshop survey may have been due to the distribution of vehicle age, which was quite different between the two. In the workshops, relatively new vehicles (< 5 years) were identified. Semarang city has a complex terrain, where the South-North route is significantly different in terms of altitude (roughly 300 m in height), while the East-West route has relatively flat terrain.

In general, air emissions related to vehicle speed indicated that the East-West route delivered larger levels of air pollutant emissions, both in the morning and at noon, compared to those for the South-North route (see Figures 3 to 6 below). Slower speeds on the East-West route due to higher traffic volume may be the reason of this. Noon generally produced higher emissions than in the morning, which may be due to the use of private cars for more than one daily activity, e.g., lunch, meetings and relaxing. During the afternoon, pollutant emissions were generally indicated to be higher than emissions in the morning.

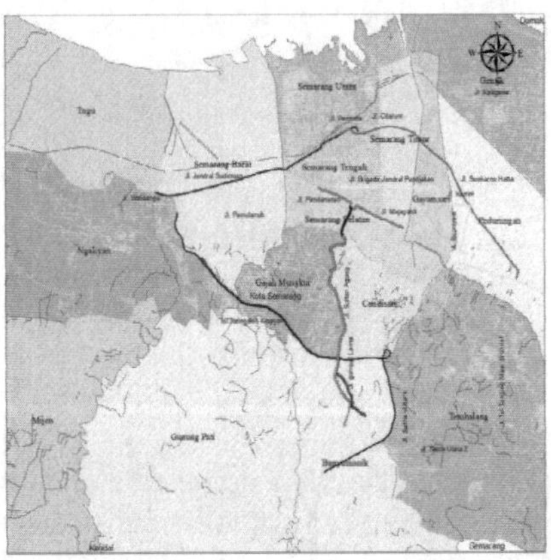

Figure 3. Car weekday routes.

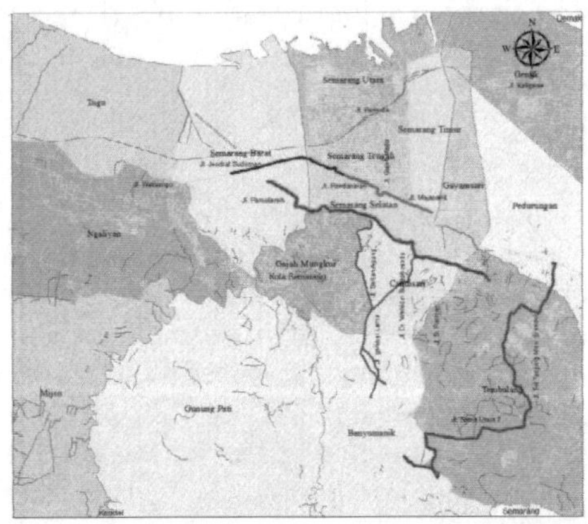

Figure 4. Car weekend routes.

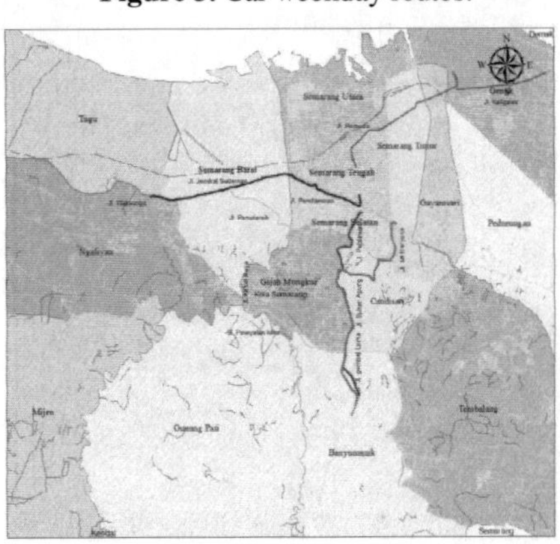

Figure 5. Motorcycle weekday routes

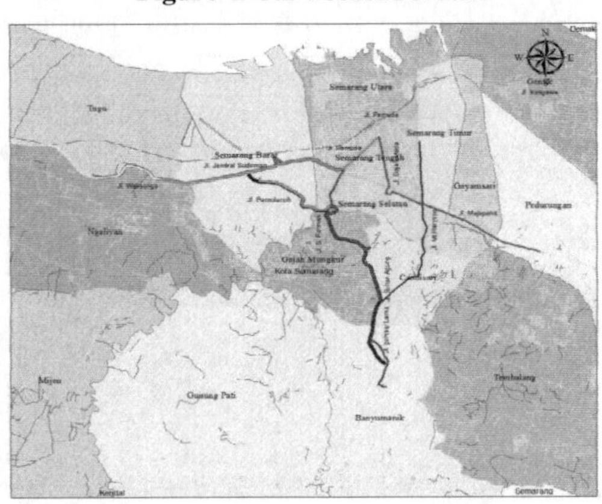

Figure 6. Motorcycle weekend routes.

Vehicle speeds occasionally dropped to almost zero, indicating traffic congestion (this figure represents morning traffic). In general, the speed of vehicles remained under 50 km/h and very rarely exceeded 50 km/h. It was difficult to reach this speed due to severe congestion during rush hour traffic in the mornings and afternoons. This speed could be reached during non-rush periods, particularly by motorcycles. Regarding the fact that the vehicle fleet continues to increase, it is likely that these conditions will worsen in the years to come. Thus, the implementation of a public transport system in Semarang is an important approach for combatting the effects this may have. An example of the relationship between vehicle speed based on routes during the morning periods is presented in the Figure 7 below.

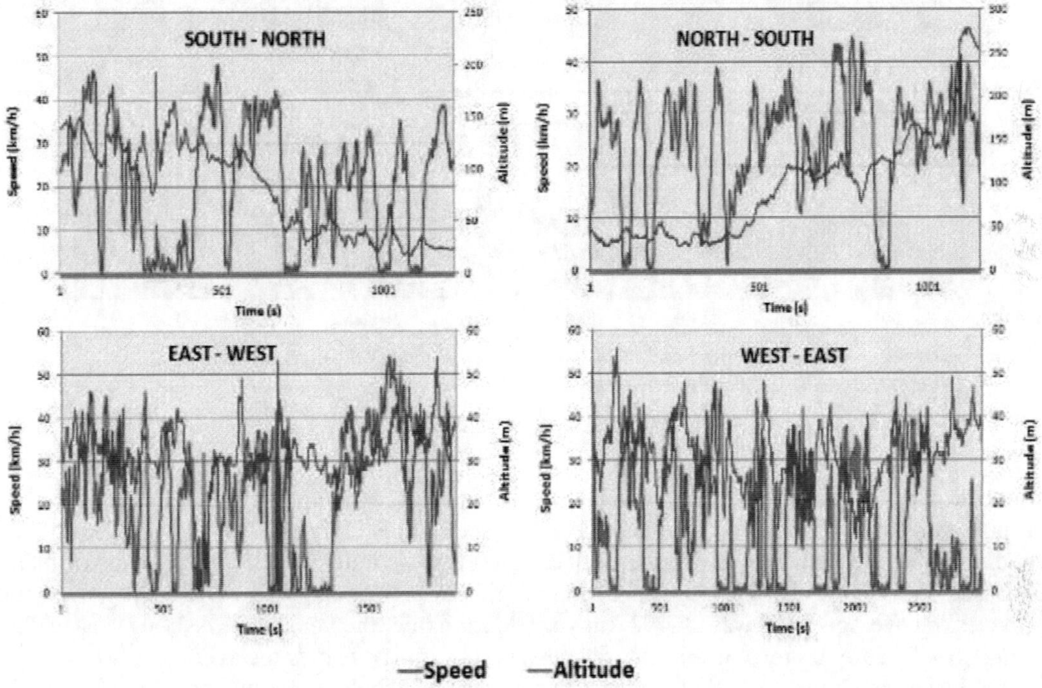

Figure 7. Altitude and time related to vehicle speed.

Based on the emissions results (see Figure 8), the East-West corridor appears to have higher emission than the South-North corridor. This is contrary to our hypothesis that vehicles travelling on the South-North corridor have higher emission rates than those using the East-West corridor. We presumed that by travelling at fluctuating altitudes, the speed of the vehicle would not reach an ideal speed on the road. In fact, traffic congestion and high vehicle volume on the road has a greater influence on emissions than the terrain differences present in Semarang city, as is shown in Figure 8. The East-West corridor connects to the Trans-Java Northern Highway, which may have an impact on this traffic congestion. Moreover, there is essentially no difference in emissions between the measurements periods, i.e., morning, noon and afternoon.

The emissions of private motorcycles is estimated to be higher than private vehicles, as the registered number of motorcycles running on Semarang roads is much higher than the registered number of cars. In 2015, registered motorcycles numbered 886,420 units, while cars were roughly 186,613 units. Based on our calculations using LEAP software (not detailed in this paper), the estimated CO emissions in 2016 for motorcycles and cars was 6.545 Mton and 19.1 Mton, respectively. Traffic arrangement through ATCS may one solution for reducing traffic congestion, which causes high levels of emission air pollution. Shifting from using a private car or motorcycle to a public transport system will have a significant impact on reducing the number of private cars and motorcycles on the road, and will therefore diminish air pollution related to traffic (exhaust and non-exhaust emissions). However, air pollution initially originated from people moving from from one area to another; as such, proper land-use planning, which can minimize people's movement, is also important for a long-term solution for reducing air pollution in urban cities.

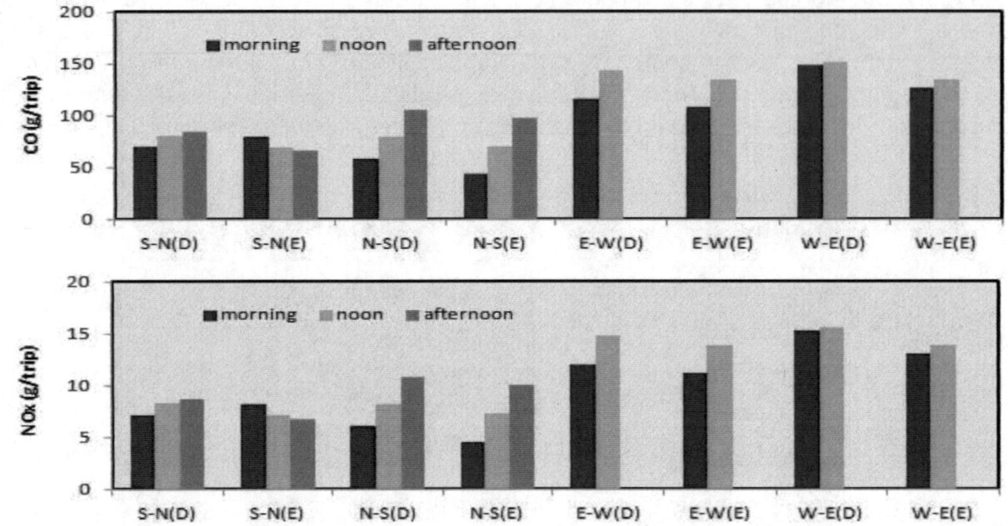

Note: Emissions were derived from recorded speed related to fuel consumption
S: South, N: North, E: East, W: West, (D): Weekday, (E): Weekend

Figure 8. Air pollutant emissions derived from private cars.

5. Conclusion

Air pollutant emissions related to the transportation sector in Semarang city will be higher in the near future, due to the large number of private vehicles used to commute in both the city and for inter-city travel. Based on the parking lot survey, the average VKT for private cars was 17,737 km/year, while for motorcycles the average VKT was 27,092 km/year. Meanwhile, the vehicle workshops survey showed an average VKT of 9,510 km/year for motorcycles, while for private cars the average VKT was 21,347 km/year. During rush hour, the speed of vehicles was generally below 50 km/h and very rarely did the speed exceed 50 km/h. Air pollutant emissions for the East-West routes of Semarang city were generally higher than those on South-North routes, according to private car emissions. Motorcycles contribute significantly more to urban air pollution than cars. In this study, traffic congestion and traffic volume contributed more to air pollution than the impact of fluctuating terrain.

6. Acknowledgement

This study was supported by funds from PNBP DIPA Universitas Diponegoro No: SP DIPA-042.01.2.400898/2016, dated 7 December 2015.

References

[1] PaMorningannaki K, Diakoulaki D. Decomposition analysis of CO2 emissions from passenger cars: the cases of Greece and Denmark. Energy Policy 2009; 37: 3259-67.
[2] Soylu S. Estimation of Turkish road transport emissions. Energy Policy 2007;35: 4088-94.
[3] Southworth, F. VMT Forecasting for National Highway Planning: A Review of Existing Approaches. 1986. Oak Ridge National Laboratory, Oak Ridge, TN 37831.
[4] Makra L, Mayer H, Mika J, Sánta T, Holst J. Variations of traffic related air pollution on different time scales in Szeged, Hungary and Freiburg, Germany. Physics and Chemistry of the Earth 2010 35: 85-94.
[5] Central Pollution Control Board, Air quality monitoring, emission inventory and source

apportionment study for Indian cities, 2011. National Summary Report. Ministry of Environment and Forest India.
[6] Huboyo HS. Source Apportionment of Fine and Coarse Particles at Bandung Airshed Using Chemical Mass Balance (Case Study at Tegalega and Dago Pakar).
[7] Mahalakshmi DV, Sujatha P, Naidu CV, Chowdary VM. Contribution of vehicular emission on urban air quality: Results from public strike in Hyderabad. Indian Journal of Radio & Space Physics 2014, 43: 340-348.
[8] Demir E, Bektas T, Laporte G. A comparative analysis of several vehicle emission models for road freight transportation. Transportation Research D, 2011, 16: 347-357.
[9] Boriboonsomsin, K, Barth M. Impacts of road grade on fuel consumption and carbon dioxide emissions evidenced by use of advanced navigation systems. Trans. Res. Rec.: J. Transport. Res Board (Trans Res Board) 2009,1: 21-30.
[10] Meccariello G, Della Ragione L. Slope effect in emission evaluation to assess real pollutant factors. Int. J. Chem. Mol., Nucl., Mater. Metall. Eng. (ICESE) 2014, 8: 10.
[11] Fonseca, N, Casanova, J and Espinosa, F. Influence of driving style on fuel consumption and Emissions in diesel-powered passenger car. In: "8th International Symposium Transport and Air Pollution, TAP 2010", 18/05/2010 - 19/05/2010, Dübendorf, Suiza. ISBN 978-3-905594-57-7. pp. 1-6.
[12] Wang X, Westerdahl D, Hu J, Wu Y, Yin H, Pan X, Zhang KM. On-road diesel vehicle emission factors for nitrogen oxides and black carbon in two Chinese cities. Atmos. Environ. 2011, 46:45-55.
[13] Fuo H, Yao Z, Zhang Y, Shen X, Zhang Q, Ding Y, He K, 2012. On-board measurements of emissions from light-duty gasoline vehicles in three mega-cities of China. Atmos. Environ. 49:371-377.
[14] JICA. Study of Integrated Transportation Master Plan for Jabodetabek (SITRAM Phase II) 2014.
[15] Fergusson M. The effect of vehicle speeds on emissions. Energy Policy 1994, 22:103-106.

IOP Conf. Series: Earth and Environmental Science **70** (2017) 012014 doi:10.1088/1755-1315/70/1/012014

The modal transfer analysis by adding transport costs. Case study: The use of bus and private vehicle student Institut Teknologi Sumatera

H Muliarto[1], Alhamidi, and M Syahbandi

Department of Urban and Regional Planning, Institut Teknologi Sumatera (ITERA), South Lampung, Indonesia hor addresses here

hendro.muliarto@gmail.com

Abstract. Since the last two years the Institut Teknologi Sumatera to accept new students in large numbers. Increasing the number of students is directly proportional to the increase of private vehicles that enter the campus it is not in line with the policy of the Institut Teknologi Sumatera as Smart, Friendly, and Forest Campus. Institut Teknologi Sumatera as Smart, Friendly, and Forest Campus fact has made some deal with mass public transport, Damri, to provide bus transportation to the Institut Teknologi Sumatera, but the number of bus users is still inferior to the number of private vehicle users. This study was conducted to see the modal transfer of the entire academic community ITERA, from private cars to public transport such as buses, with the addition of transportation costs in the form of parking rates. This study shows the dominant displacement can occur if the respondent charged parking fees of IDR 4.000. Besides the displacement mode of transportation from using private vehicles be using the bus can occur if Trans Lampung fix three systems including bus departure and arrival schedules, facility service providers that support, and the addition of the Bus Trans Lampung.

1. Introduction

Transportation became an important part in human life, because it acts as a transport carriage of goods and human beings in order to facilitate human activities in terms of economic, social and territorial. In accordance with the target goal number 11 on Sustainable Cities and Communities that are discussed in the general assembly of the United Nations (UN) on 25 September 2015 authorizing the Sustainable Development Goals (SDGs) that by 2030, providing access to safe, affordable, accessible and systems sustainable transport for all citizens, improve road safety, particularly by expanding public transport, with special attention to those who need a vulnerable situation such as women, children, the disabled and the elderly [1]. Sustainable public transport system a own positive impact on the sustainability of the economic, social, cultural, and environmental diversity of the community being served [2]. The existence of the transport system is to meet the needs of economic and social relevance, and give people an opportunity to improve mobility [3]. Where the influence of public transport so have a major impact in reducing traffic congestion, reduce air pollution, and reduce costs.

ITERA admit students every year with large numbers, in 2015 ITERA received the students as much as approximately 400 students and in 2016 ITERA received the students as much as approximately 1200 students. This is a new problem that is where the increasing number of students who use private vehicles, with limited parking space on campus ITERA lead to less availability of parking spaces on campus

ITERA. In this study, researchers conducted a survey when the vehicle parking fee imposed on the campus of ITERA, the extent of the student's ability to pay for services it receives to switch modes from using private transport be using public transportation. This study aims to determine how much a student's ability to pay for transportation services ITERA received based on income and is considered ideal (Ability to pay) to the parking area on campus ITERA applied regardless of the location of a student residence.

2. Basis Theory

In this study researchers conducted a study on students ITERA to be able to switch modes from using private vehicles be using the bus. Taking into account the ability and willingness of users to pay the cost of transport, with a view Ability to pay and willingness to pay student. Ability to pay (ATP) is a person's ability to pay for transportation services received based on income and is considered ideal. ATP analysis is viewed allocation of transport costs and travel intensity with a great view of ATP or the ratio of the budget. Ability to pay travel expenses influenced by family income per month, the allocation of transport costs, the intensity and number of family members [4]. Willingness to pay (WTP) is the willingness of users to issue a reward for services obtained. WTP analysis with user perception of the rates of the bus transport service as well as the factors affecting including service quality, quantity of services, utilities, and the user's income. Mode selection model aims to determine the proportion of people who will use each mode. This process is done with the intention to calibrate models of modal choice between a base year to determine independent variables (attributes) that affect the modal choice. Factors that may affect the selection of these modes can be grouped into four (Ben-Akiva and Lerman, 1985), namely: (1) Characteristics of road users; (2) Characteristics of the movement; (3) Characteristics of transportation facilities; and (4) Feature town or zone [5].

Model forms modal choice in European countries is dominated by a model of the distribution of the movement, so that the modal choice models should be used after dispersion modeling stage movement. In this study, there are two types of modes of transportation are: buses and private transport, the analysis in this study using a modal choice model of cost. Costs in modeling modal choice, it is important to distinguish between the estimated cost and the actual cost. Cost estimates are the costs that feel pressure by road users and basic decision-making, while the actual cost is the actual cost incurred after the electoral process modes do.

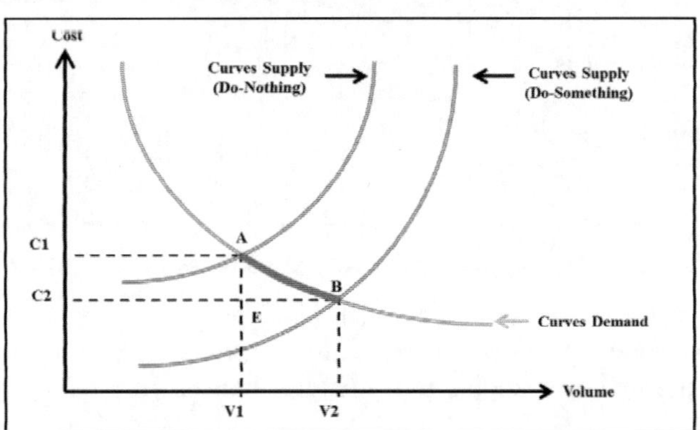

Figure 1. Supply curve Do-Nothing and Do-Something [5]

Demand and supply curves occur when the policies in effect between producers and consumers in achieving equilibrium point. Where the point of equilibrium will shift as a result of their influencing factors demand and supply. The curve above explains that shift the equilibrium point on the supply curve of a do-nothing to do-something. At the moment the number of do-nothing volume v_1 at a cost c_1. Meanwhile, when the do-something with the amount of volume v_2 c_2 costs. Because the supply curve is perfectly elastic at c_1, then if there is additional volume of the number of vehicles to v_2 will benefit (surplus) for producers of area above the line AB.

3. Analysis Method

ITERA to analyze student data, researchers used a descriptive statistical analysis methods. Through methods relating to the collection and presentation of a range of data so as to provide useful information is called Descriptive Statistics [6]. In this study the determination of the number of samples is 100 people with a 90% confidence level using Taro Yamane formula (Cluster Random Sampling) where:

$$n = \frac{N}{\left(N \cdot (d)^2 + (1)\right)} \quad (1)$$

Description :
n = Samples; N = Population; d = Precision

$$n = \frac{1700}{\left(1700 \cdot (0{,}1)^2 + (1)\right)} = 94{,}4\,percent \quad (2)$$

To simplify the calculations the researchers used a sample of 100 of the total population of 1700 students ITERA.

3.1. Data

Researchers conducted a survey directly to all students of the Institut Teknologi Sumateraobtained a sample of 100 students with a variety of diverse opinions on the policy smokeless day. In order to keep transportation costs for the students in a day, the data necessary expenses or income students ITERA. Researchers obtain data on the percentage of students ITERA spending per day in the following diagram.

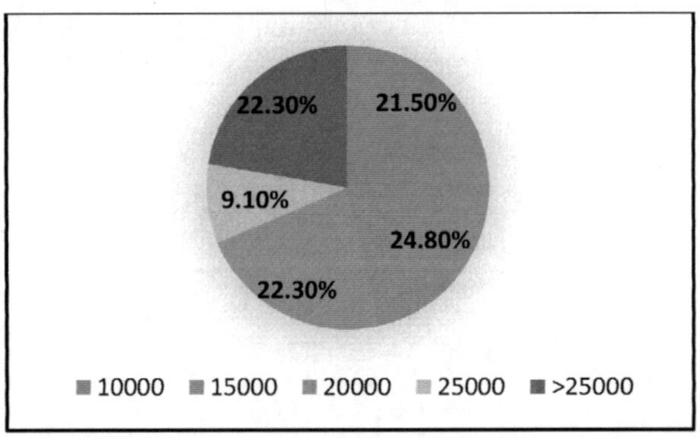

Figure 2. Diagram of a percentage of students ITERA per-day (Source: Survey researchers, processed)

Based on the above data that as many as 24.80% of the students have the IDR 15.000 per day, and while the cost incurred by the person for transport costs in the everyday is 30% of the revenue or expenditure of a person in a day. If the student spending a day fifteen thousand rupiah, the transport costs incurred in a day is as follows.

$$\text{Transport fee} = 30\% \,.\, \text{Spending per day} = \frac{30}{100} \,.\, 15000 = \text{Rp } 4.500 \quad (3)$$

Then the cost of transport students in a day good ITERA is IDR 4.500. From research conducted in the Institut Teknologi Sumatera on the analysis of the transfer modes of transportation from using private vehicles into Bus Trans Lampung with the enforcement of the parking fee using a questionnaire obtained data as follows.

Table 1. Comparison of the percentage of the number of private vehicles and a bus ride will take effect when the parking rates.

Parking fee	The number of users of private vehicles	The number of bus users
IDR 1.000	88%	12%
IDR 2.000	78%	22%
IDR 3.000	59%	41%
IDR 4.000	46%	54%
IDR 5.000	44%	56%

(Source: Survey researchers, processed)

From Table 1 above shows that at the time of parking charges applied to IDR 1,000 the percentage of students who ride personal vehicles 88% and the number of students who ride buses 12% of the total student ITERA. In the parking fee is applied IDR 2.000 the percentage of students who ride personal vehicles 78% and the number of students who ride buses to 22% of the total student ITERA. Then at the parking fees applied IDR 3.000 the percentage of students who ride personal vehicles to 59% and the number of students who ride buses 41%. Furthermore, when the parking fee is applied IDR 4.000 the percentage of students who ride personal vehicles to 46% and the number of students who ride buses 54%. Then when the parking fee was raised to IDR 5.000 the percentage of students who ride personal vehicles dropped to 44% and the number of students who use the bus transport 56% of the total student ITERA. It showed that when the parking fee increase occurred in transportation modal transfer from private cars to bus transportation. So at certain parking charges can hit up all the students ITERA use bus transportation.

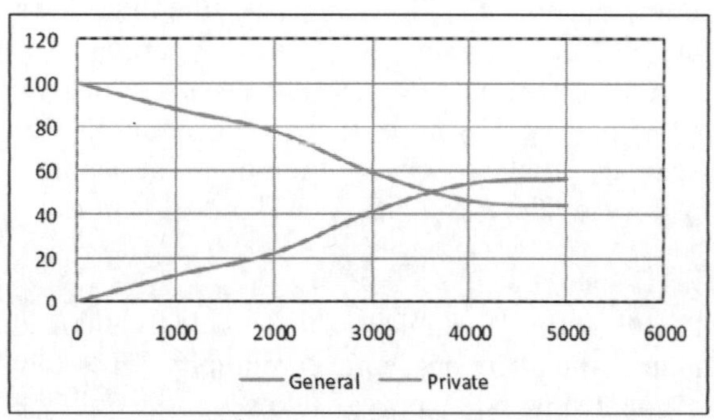

Figure 3. Graph private vehicle transfer to the bus (Source: Survey researchers, processed).

The graph above that at the time of parking fees applied will make the student make the shift mode of transportation to the bus, due to the ATP and WTP owned by ITERA student than he paid parking is more expensive than paying the cost of bus transportation. So if the parking fee to be increased, it will cause all students ITERA use bus transportation.

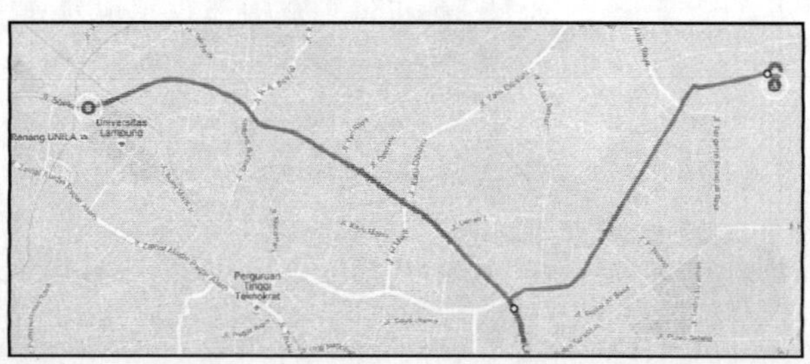

Figure 4. Trans Bus Route Lampung ITERA - UNILA (Source: Google maps)

Trans Lampung Bus is the only bus that has the purpose to the University of Lampung without transit transport to another. The bus route Trans-ITERA Lampung Unila Ryacudu through the canal road, turn left onto the street Hi Pangeran Suhaimi, continue past the primary collector road Mayjend Riyakudu, and turn right into primary arterial road soekarno-hatta to a stop in front of the campus of the University of Lampung. From interviews with the Trans Lampung income earned by Trans Lampung property of Lampung Province which operate on routes ITERA-UNILA calculated based on the average income of the seven Trans Lampung which operates on this route is the lowest incomes per day IDR 16.000 and the largest revenue per day IDR 250.000 with a 1-trip transportation costs for students / student is IDR 2.000. As for public transport costs 1 trip IDR 4.000.

Table 2. Revenue Lampung Trans Bus Route ITERA - UNILA

No	Day	Income (Rupiah)
1	Monday	± 50.000
2	Tuesday	± 200.000
3	Wesnesday	± 30.000
4	Thursday	± 30.000
5	Friday	± 30.000
6	Saturday	± 30.000 ± 100.000
7	Sunday	± 30.000 ± 100.000

(Sources: *Results of Interview, 1 November 2016*)

From Table 2 above illustrates the low income obtained by the Trans Lampung every day. The low revenue showed no effective and efficient operation on the route. Though bus takes money to do maintenance, employee salaries, and so forth. However, from the interviews of the Trans Lampung no bus maintenance costs to date for free bus maintenance costs. After that, the bus departure schedule no one has made irregular bus arrivals. However, the earliest departure time 05:30 pm and the afternoon return schedule at 17.30 pm. With a capacity buses that can accommodate 30 people seated passengers and 40 standing passengers, so the total capacity of the bus is 70 people. With the capacity and the service bus every day can take a trip ITERA - UNILA - ITERA 3-4 times.

4. Discussion

In August 2016 regarding the regulation of smokeless was made by the rector, with the aim to make a transfer mode of transportation to the entire academic community ITERA. With the implementation of the policy of smokeless implemented within one week. However the policy is not entirely well received by the entire academic community ITERA, supported by the lack of public transport leading ITERA. Bus Trans Lampung be the only public transportation that operate to ITERA to stretch Unila-ITERA.

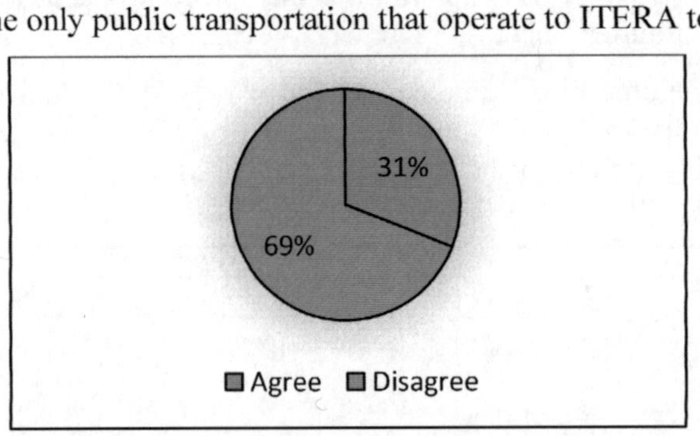

Figure 5. Percentage of student opinion against the policy of smokeless (Source: Survey researchers, processed)

In the diagram above illustrates that not all academic community ITERA agree with this policy. That as many as 69% of students ITERA did not agree with this policy even though there are 31% of students who agree with this policy. This indicates that the policy of smokeless is a program that seems inappropriate to make a transfer mode of the entire academic community of the Institut Teknologi Sumatera.

Table 3. Percentage of ITERA student opinion against the smokeless policy based on the most frequent.

Student opinion	Stake	Percentage
Maintain the environment by reducing air pollution	𝍸𝍸𝍸𝍸𝍷	21%
Assessing the "Day without smoke" is "moving day parking"	𝍸𝍸𝍷	11%
Lack of service providers	𝍸𝍷𝍷𝍷𝍷	9%
The bus route does not pass through a student residence	𝍸𝍸𝍸𝍷	16%
Bus schedule uncertain	𝍸𝍸𝍸	15%
All others	𝍸𝍸𝍸𝍸𝍸𝍷𝍷𝍷	28%

Source: *Data and Analysis, 2016*

In the table above is explained that the percentage of student opinion on policy ITERA's smokeless day. Around 21% of students suggested that the smokeless in the good sense to be implemented because of the need for the movement to care about protecting the environment by reducing air pollution produced by motor vehicles and vehicle private cars. We know that the reaction of organic compounds from the combustion of fossil reaction incomplete combustion in vehicle engines, resulting in residual or air pollution. Especially now that global warming is disturbing the Earth's ozone layer, plus more air pollution from vehicles and fossil fuel exacerbating the situation. The solution is clear policies and systems in tackling this problem, one day smokeless policy in ITERA and afforestation / tree planting. Policies days without a smoke removal methods mode of personal transportation to bus transportation / mass. So from 1700 students and plus employees ITERA everyday use of private vehicles alone were transferred to bus transportation capacity of 30 seated passengers and 40 standing passengers, a significant reduction in air pollution occur and decomposition of traffic jams in the city of Bandar Lampung also be resolved. And also the bus used by the Trans Lampung have to use fuel gas, biogas, or electricity. But as much as 11% of the students considered that "Day without smoke" is "Day moved park" because they lack careful planning, less effective and efficient, because at the time of the day without the smoke was still there academicians who bring personal vehicles and parked his car in front of the gate ITERA. A total of 9% of the students assess violations committed by the academic community because of the lack of provision of service facilities which are not optimal that caused them to have a private vehicle. Furthermore, there are as many as 16% of students who have difficulties in their mobility access from the residence to the campus, is caused by the Bus Trans Lampung is no coverage area of the entire academic campus ITERA. Airport Bus Trans Lampung is not coverage area thoroughly for the entire academic community residence ITERA, this raises new problems that result are often late arriving students into campus. In addition to a range of less than optimal bus service is also caused by scheduled bus departure is uncertain, as many as 15% of the students found Bus Trans Lampung not have a definite timetable and result in delayed students entering college at that hour. And there is also a 28% sample of college students having a different opinion, one of the students found the smokeless needs to be done to make the academic community ITERA switching modes of use of private vehicles into using public transport.

Solving the problem in this case is the need to reform the transportation system Bus Trans Lampung Unila-ITERA is currently being implemented. The system needs to be improved is (1) the scheduled departure and arrival of buses, (2) service providers and support facilities, (3) the addition of Trans Lampung route bus. Schedule bus departures and arrivals are sure to be important in serving the entire academic ITERA in everyday activities. As an example every day bus operates from 05.30 pm with departures every 30 minutes from early departures. If this is applied to the entire academic ITERA will have confidence in terms of time to Bus Trans Lampung, so that the entire academic community ITERA comfortable with the services provided by the Bus Trans Lampung and reduce violations occurring on a day without smoke in ITERA. Lack of service providers that support the activities of Trans Lampung bus, as the bus stop is not yet available, the little bus fleet. Thus making passengers uncomfortable with the situation. Solving the problem is the need to build a bus stop at strategic points that facilitate passengers to reach the bus, not only for the entire academic community ITERA but also for the surrounding community. By building a shelter around the residential area and schools could be me-coverage area several schools, because in the RTRW Bandar Lampung area traversed by bus was one of which the District Sukarame is the area of education. So that it can increase revenue Bus Trans Lampung. Additionally Bus Trans Lampung can also be felt by the local community to access mobility. Bus Trans Lampung route via Jl. Canal next next Ryacudu through Jl. Hi Pangeran-suhaimi primary collector road passing through Jl. Mayjend Ryacudu primary arterial road next to the turnpike middle Sumatran Jl. Soekarno-Hatta and the final destination of Lampung University. This is the route that passed by Bus Trans-ITERA Unila Lampung. The region is an area of education in 2011-2030 RTRW Bandar Lampung. If this is only the route of the bus then causes unscheduled departure and arrival of buses is at least the bus passengers were also due to the service can't be the coverage area of the overall residential district bus passengers, especially the academic community ITERA. The solution for this is to be done adding bus routes and fleet units to be coverage area is mainly residential area that has crowds and high activity.

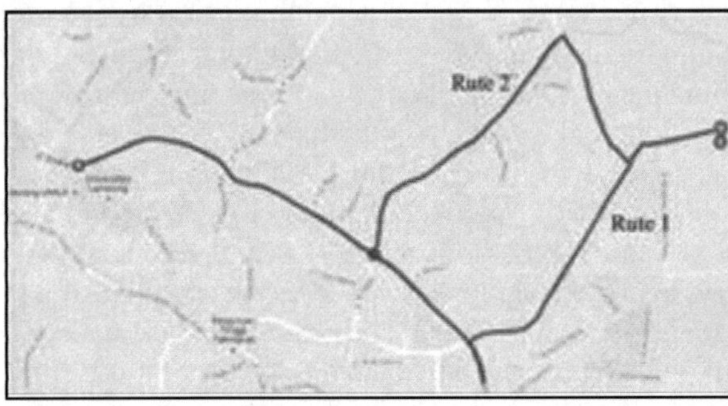

Figure 6. The addition of these buses Trans Lampung (Source: Google Maps)

The addition of these buses could be dilakuan as in figure 7 bus passing through Jl. Terusan Ryacudu then to Jl. Hi Pangeransuhaimi passing through Jl. Raya subsequent flow through Jl. Ratu Dibalau and last through primary arterial road turnpike middle Sumatran Jl. Soekarno-Hatta to a stop in front of the campus of the University of Lampung. Selection of the 2nd because in those areas there are many traditional markets and the way kandis-medium residential housing such as housing griya beautiful promontory, housing the image of the earth scenic, residential scenic headlands highway. So that in the area may ter-coverage area by Bus Trans Lampung. In other responses assess student days without smoke is less decisive because there are still some vehicles such as cars and motorcycles parked in front of the gate ITERA, and it is a thing that deviate towards the smokeless in ITERA. Today smokeless not prove to be a good way to be able to realize the transfer of the mode of transportation to the entire academic community ITERA, considering the vehicle parked in front of the gates of ITERA and violate regulations smokeless day is not without reason. Solving the problem in this case study is to be able to transfer the mode of the entire academic community ITERA, researchers analyzed that need enactment parking rates. In Table 1 produced the data that ITERA students will switch modes when the tariff is parking at a cost of IDR 4.000 per one parking, with this the students will switch from using private vehicles into public transport. In accordance with the calculations performed in the data that the cost of transport a person is 30% of a person's income or income per month or per day, according to the viewpoint taken into account. ITERA that students spend a day transport as many as four thousand five hundred rupiah. In this study the researchers were unable to calculate the amount Ability to pay student ITERA as limitation of this study that the researchers did not see the location of every student residence ITERA, so the absence of data on the total length of the trip respondents per month (T_t).

5. Conclusion

Based on the analysis performed, it can be deduced that to make a transfer mode of personal transportation to the bus transportation should pay attention to three systems (perception of respondents), including: (1) the scheduled departure and arrival of buses, (2) facilities service providers and support, (3) the addition of bus route Trans Lampung. Three systems are related to each other, and the system also greatly affect the level of willingness of people to use the bus transportation is convenient. After that, should the addition of parking fees high enough outstrip capacity transport costs of issuance, which is IDR 4.500. So if a parking fee to be paid beyond that will make the entire academic community ITERA switch to bus transport Trans Lampung with the support by the supporting facilities.
Furthermore the Trans Lampung also have to add bus lines to meet demand. Extra bus route can be done is a bus passing through Jl. Terusan Ryacudu then to Jl. Hi Pangeransuhaimi passing through Jl. Raya subsequent flow through Jl. Ratu Dibalau and last through primary arterial road turnpike middle Sumatran Jl. Soekarno-Hatta to a stop in front of the campus of the University of Lampung. On the route was done to coverage area around the road there are many residential and traditional markets.

References

[1] U. Nations, "United Nations Sustainable Development Goals," United Nations, [Online]. Available: United Nations. (n.d). Sustainable Development Goals. Retrieved from . [Accessed 28 September 2016].

[2] F. Miro, Perencanaan Transportasi Untuk Mahasiswa, Perencana, dan Praktisi, Jakarta: Penerbit Erlangga, 2005.

[3] R. Sulistyorini, Perencanaan dan Pemodelan Transportasi, Yogyakarta: Penerbit Graha Ilmu, 2014.

[4] O. Z. Tamin, Perencanaan, Pemodelan & Rekayasa Transportasi: Teori Contoh Soal, dan Aplikasi, Bandung: Penerbit ITB, 2008.

[5] R. Walpole, Probability and Statistic for Engineers and Scientists, New Jersey: Pearson Education, Inc, 2007.

[6] M. R. Permata, "Analisa Ability to Pay Dan Willingness to Pay Pengguna Jasa Kereta Api Bandara Soekarno Hatta – Manggarai," Thesis UI, Depok, 2012.

The effect of industrial vehicles on the road's level of service of industrial area in Malang City

I W Agustin[1], Y Sumantri[2]

[1] Department of Urban and Regional Planning, Faculty of Engineering, Universitas Brawijaya, Malang, Indonesia
[2] Department of Industrial Engineering, Faculty of Engineering, Universitas Brawijaya, Malang, Indonesia

immasaitama@ub.ac.id

Abstract. Malang was a great city that has the characteristics of diverse industries and it was spread in each district. Industry was one of the activities that have the characteristics of the movement with the goal of shipping and distribution. The vehicles used are large vehicles such as trucks and trailers. Problems related to the road's performance that is the emergence of several points of delay and congestion due to the movement and the volume of vehicles as well as the limited capacity of existing roads. It encourages researchers to do research related to the effect of the industrial vehicle movements on the performance of existing roads in the city of Malang. The main purpose of the research is to analyze the effect of industrial vehicles movement on the road's level of service of industrial area and to find out alternative solutions to improve the road's level of service of industrial area. The research used multiple linear regression and do something analysis. The results showed that the types of industry in the city of Malang are ILMETTA, IATT, Agrokim, Sentra and Large Industries. Determining the location of the main observation based on the distribution of industry, have a major impact directly on the street, and the class of the national road are located in Sunandar Priyo Sudarmo Street. Existing land use in Sunandar Priyo Sudarmo Street are industrial, healthcare, office and it dominated by trade and services. Related to the variables that affect the movement of industrial vehicles are the number of employees (X3) and the delivery frequency (X6). The road's level of service of Sunandar Priyo Sudarmo Street are D and E. While the effect of industrial vehicles movement on the road's level of service which ranged from 13.77 per cent to 22.13 per cent of the total volume of vehicles on the road. Some referrals will be used to handle the problems that the road widening and the selection of alternative routes for the industrial vehicles.

1. Introduction

Good transport planning is the planning of transport in accordance with the conditions of the existing land use. Types of land use covering trade, services, industry, public facilities and other land uses. The linkage between land use and transportation is how much the generation and pull on the land use. The increase of the generation make the greater burden of the existing road around the land use [1]. According to the Spatial Plan in East Java (RTRW) in 2011 to 2031, the establishment of Malang as the Center of National Activities (PKN) encourage the development of economic activities in the services sector, trade and industry which became pattern of economic activity. It makes Malang City has a wide variety of industries that are part of the economic system, both engaged in small-scale (in the city) and large scale (inter-provinces). Industry sectors in Malang City control of 35.61 per cent of economic activity in the city of Malang, with the largest contributor to the local government in the amount of 5.14

Trillyun per year [2]. The existence of the industry can not be separated from the transport groove distribution of raw materials and finished goods because it requires a means of transport to be brought to their destination. The shipping of goods by four-wheeled vehicles accounted for 15 per cent to 25 per cent of the volume of vehicles that overload the street [3]. Of course, if it is not managed properly then it can cause traffic problems are quite severe, such as traffic jams. Transportation problems began in the 1960s and 1970s, some of the problems associated with transport is congestion, air and noise pollution, accidents and delays [4]. The source of the cause of problems that occur in transport comes from several things, one of which is the activity of the Industry. Industry in Malang quite a lot lead to some impacts associated with transportation. One of the effects caused by that activity of loading and unloading of goods or raw materials that occur can lead to delays up to congestion. In addition, the activity of industrial workers in the morning (it comes) and afternoon (return) that uses the vehicle caused delays on roads. Therefore, this study was conducted to identify the impact of the movement of large and medium industrial vehicles on the road's service of level of existing roads in Malang City, which is expected to be used as material for evaluation in the planning of industry in Malang City related to transportation and the completion of the problem.

2. Methods

The research location is in Malang with a focus on Sunandar Priyo Sudarmo Road, it was determined by:

- Class national roads.
- The amount of the distribution of large and medium industries most.
- Industry is located near the highway, and it has a direct impact on the way.

This study consists of several stages for the solution of problems in the study area. Firstly, identify and classify industries in Malang City. Secondly, determine the road with some consideration. Thirdly, calculate and analyze the trip generation / attraction model. Finally, analyze the road's level of service and calculate the effect of industrial movement on the road capacity. The first paragraph after a heading is not indented (Bodytext style).

2.1. The Road's Level of Service Analysis

The road's level of service in this study only calculated traffic volume and road capacity. Based on the Highway Capacity Manual Indonesia in 1997 [5], the road capacity is the amount of vehicle traffic maximum that can be accommodated on the roads during certain conditions (design of geometry, environment, and traffic composition) can be specified in the passenger car unit (pcu per hour). The degree of saturation is defined as the quotient of traffic flow on the road section to capacity of the point of saturation degree can be used as a measure to indicate whether a particular road segment will meet the capacity problem or not. The degree of saturation equation is as follows:

$$DS = \frac{V}{C} \tag{1}$$

Notes:
DS = Degree of Saturation
V = Volume of traffic flow (pcu / h)
C = capacity of the road (pcu / h)
Calculation of the performance conducted in Sunandar Priyo Sudarmo Road which is the location chosen by the consideration that has been done.

2.2. *Multiple Linear Regression Analysis*

Linear regression analysis is a statistical method used to find linkage of more than one independent variable [6]. The first step in this regression analysis phase is the selection of variables that serve as the independent variable or dependent variable. Here is the independent variable of industry are used:
Y1 = Number of Movements
X1 = Building
X2 = Area Parking
X3 = Number of Employees
X4 = Number of Visitors
X3 = Number of Shift
X4 = Frequency Delivery

$$Y = a + b_1 x_1 + b_2 x_2 + ... + b_n x_n \tag{2}$$

Notes:
Y: The dependent variable in this study is the amount of movement
A: Constanta parameters which means that if all independent variables (X1 through X4) remained unchanged or equal to zero, then Y1 would be equal to a constant
b_n: Coefficient parameters are: the value of which will be used to determine the effect of variables on the amount of movement
X_n: Independent variables such as all factors are included in the model and that may affect the amount of movement.

3. Results

3.1. *Characteristics of Industries in Malang City*

Malang has the characteristics of the industry, namely the classification types Agrokim, Metal, Machine Textiles, and Miscellaneous (ILMETTA), Transport Equipment and Telematics (IATT), Sentra and Large Industries. A description of the classification can be seen as follows:

1. ILMETA
   Divided into three classifications, namely basic materials industry of metal, Manufacture of material machinery and agricultural machinery, Textile and Miscellaneous.
2. Transport Equipment and Telematics (IATT).
   The division of the classification of industries Transportation Equipment and Telematics (IATT) is a combination of several types of industries such as industrial land transportation and Electronics and Telematics.
3. Agrochemicals industry
   Agrochemical industry is an industries classification types engaged in food processing, beverages, forest products, up to cigarettes and tobacco.
4. Industrial centers
   Industrial centers is a collection of several small industries or households are clustered in a region with the same characteristics. Industrial centers are generally engaged in the production of small and medium scale.
5. Large industry
   An industry which has a production performance in a large scale are some examples such as the tobacco and timber.

Small industries in Malang has the largest percentage and spread throughout the district. Total industry in the city of Malang is 617 industrial units. The distribution of industries in Malang (Table 1) separated by district and type of industry. Agrochemical industry is a kind of industry that have a degree of dispersion highly. The highest is located in Klojen with 117 units, while for large industries is only located in Blimbing District and Sukun District (Figure 1).

Table 1. The Distribution of Industries in Malang City

No	Classification	Sukun	Klojen	Lowokwaru	Kedung Kandang	Blimbing
1	ILMETA	30	35	12	12	33
2	IATT	12	10	6	9	11
3	Agrochemicals industry	110	117	53	53	92
4	Industrial Center	2	2	7	2	6
5	Large Industry	2	-	-	-	3

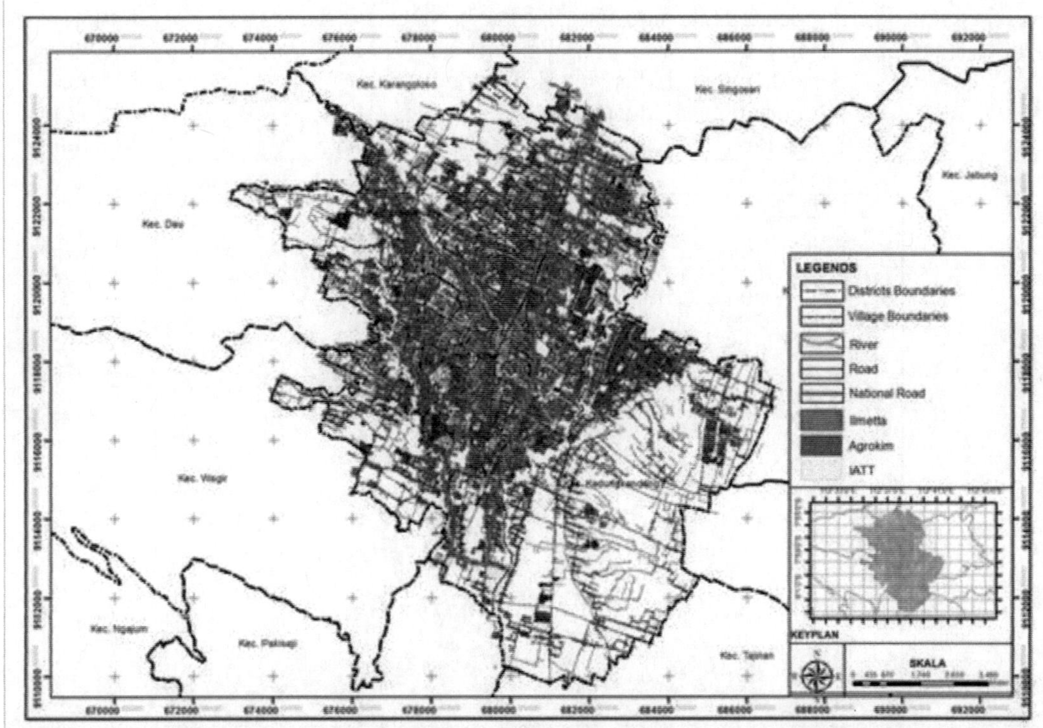

Figure 1. The distribution of industries in Malang City.

Distribution of industries in Blimbing District is still relatively evenly for each district is the largest in Bandungrejosari with the type of small industry. As for Kedungkandang District dispersion more centered in Kotalama Village with Agrokim number is 16 units of the industry.

3.2. Characteristics of Industries in Selected Location

Results of identification which have to be got Blimbing District locations that have the largest industrial distribution and in accordance with predetermined criteria (Table 2). Roads are selected in accordance with the consideration that the focus of observation is Sunandar Priyo Sudarmo Road. The function of land use around Sunandar Priyo Sudarmo Road dominated by commercial area which this road is a potential for such activities (Figure 2).

Table 2. Characteristics of Industries in Selected Location

Location	Type of Industry	Name	Size
Blimbing District	Agrochemicals industry	HM. Sampoerna, Tbk	Large
	Agrochemicals industry	Bronson Prima Industri	Large
	Agrochemicals industry	Nanas	Large
	Agrochemicals industry	Merry May	Medium
	Agrochemicals industry	Naga Laut	Medium
	Agrochemicals industry	Lestari Biscuit Factory	Medium

Location	Type of Industry	Name	Size
	Agrochemicals industry	credo	Large
	ILMETA	panglima alumunium	Medium
	ILMETA	benang merah	Medium
	ILMETA	Mandiri Garmen	Medium
	Agrochemicals industry	CARAGENAN	Medium
	Agrochemicals industry	Heksa Manunggal Jaya	Medium
	Agrochemicals industry	Kharisma Matahari Katulistiwa,PT	Medium
	Agrochemicals industry	PT karya Niaga Bersama	Large
	ILMETA	Palatino	Medium
	Agrochemicals industry	Hollywood system	Medium
	Agrochemicals industry	Gracia	Medium
	ILMETA	Lancar Jaya	Medium

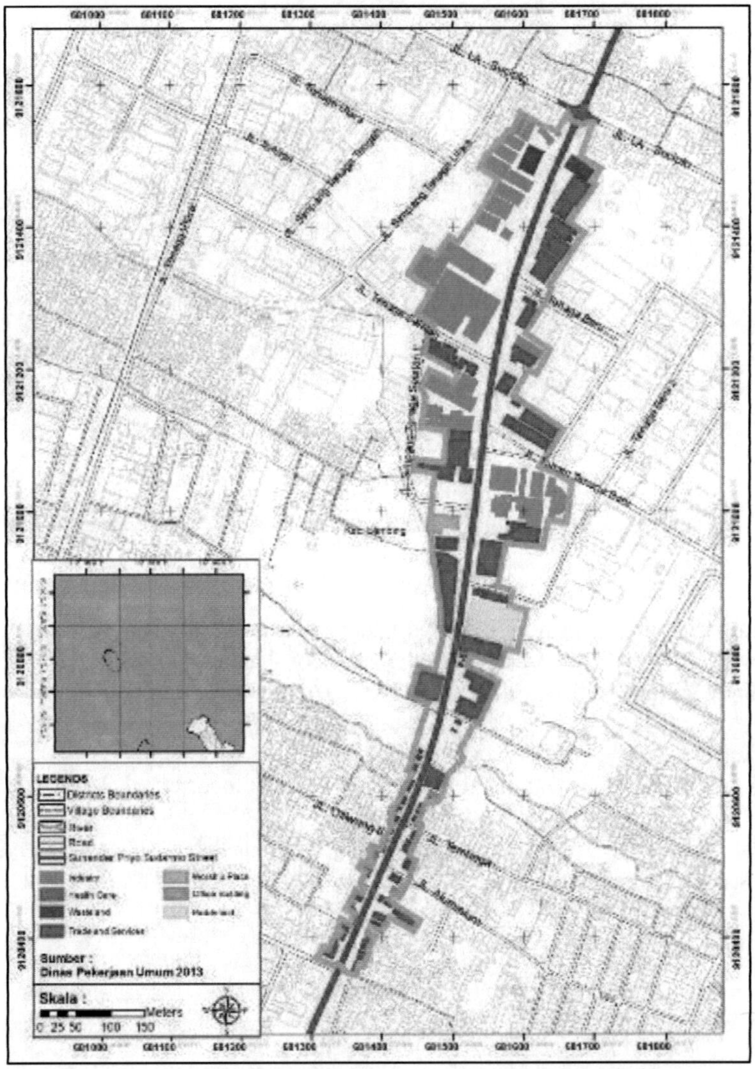

Figure 2. Characteristics of land use in Sunandar Priyo Sudarmo Road.

3.3. Trip Attraction Movement Model

Based on movement analysis, it obtained the classification of land use is industrial, healthcare, offices, and commercials. Thus, in the calculation of the regression analysis included variables that affect the land use. So that the results of the regression analysis found that:

1. Trip Attraction Model of Industry

$$Y_{industri} = 5.104 + 0.736\ (X3) + 1.832\ (X6) \quad (2)$$

Information:
$Y_{industri}$ = dependent variable, the amount of movement
X_3 = independent variable, the number of employees
X_6 = independent variable, frequency of delivery
The amount of R square for the regression model were performed in the amount of 0.941, which means 94.1 per cent of the influence of the independent variables on the dependent variable

2. Trip Attraction Model of Healthcare and Commercial

$$Y_{Health} = 12.545 + 0.277\ (X10) + 0.364\ (X11) \quad (4)$$

$$Y_{Office} = 38.076 + 0.19\ (X12) + 0.301\ (X15) \quad (5)$$

$$Y_{store} = 4.766 + 0.445\ (X16) + 0{,}049\ (X19) \quad (6)$$

$$Y_{Building\ and\ furniture\ store} = 37.958 + 0.463\ (X20) + 0.14\ (X23) \quad (7)$$

$$Y_{Joinery} = 2.074 + 0.252\ (X24) + 0{,}702\ (X27) \quad (8)$$

$$Y_{Restaurants\ and\ cafes} = 1.811 + 1.84\ (x31) + 0.443\ (X32) \quad (9)$$

$$Y_{Electronics\ \&\ electrical\ stores} = 4.359 + 0.206\ (x33) + 0.459\ (x36) \quad (10)$$

$$Y_{Dealership\ /\ showroom} = 13.189 + 0.23\ (X38) + 0.392\ (x42) \quad (11)$$

$$Y_{Other} = 7.678 + 0.662\ (X46) \quad (12)$$

Tabel 3. Total of Movement based on Land Use according to Operational Time

Operational	Total of Industry Movement	Total of Healthcare Movement	Total of Office Movement	Total of Commercial Movement
08.00-09.00	316.68	2.80	19.32	482.36
09.00-10.00	180.96	6.40	33.12	533.38
10.00-11.00	180.96	5.20	27.60	600.61
11.00-12.00	158.34	2.80	30.36	758.91
12.00-13.00	180.96	4.40	41.40	664.95
13.00-14.00	180.96	4.00	38.64	744.5
14.00-15.00	203.58	1.60	30.36	625.01
15.00-16.00	203.58	4.80	24.84	494.87
16.00-17.00	316.68	4.40	19.32	421.79
17.00-18.00	339.30	3.60	11.04	222.56
TOTAL	2,262.00	40.00	276.00	5,548.94

Figure 3. Sunandar Priyo Sudarmo Road.

3.4. Implementation of Trip Attraction Model in Sunandar Priyo Sudarmo Road

The application of the trip attraction model based on land use aimed to find out how much of the movement of incoming and outgoing vehicles for the land. So with these calculations could be more significant of the magnitude of the movement is in each land use related contributions from existing volumes on the road (Table 3). Calculations performed for industry, healthcare, office buildings and commercials.

$$Y_{industri} = 5.104 + 0.736\ (X3) + 1.832\ (X6) \tag{13}$$

$$Y_{industri} = 5.104 + 0.736\ (90) + 1.832\ (56) \tag{14}$$

$$Y_{industri} = 173.936 \approx 174\ pcu/h \tag{15}$$

3.5. The Road's Level of Service in Sunandar Priyo Sudarmo Road

Class of Sunandar Priyo Sudarmo Road is 1st class hierarchy collector-1 and it has a length of 2.05 kilometers. The road status is a national road and other treatments which are carried out by national (Figure 3 and Table 4). Priyo Sunandar Sudarmo Road has 10 alleys where 5 are the alley that connects them to the land of industry, so that the calculation of the road's level of service of traffic should consider to constant volume, the amount of alleys, and the movement of the industry (Figure 4).

$$V_{sunandar} = V_{continuous} + V_{tenaga\ baru} + V_{tenaga\ Selatan} + V_{taman\ Tenaga\ Baru} + V_{Simpang\ Tenaga\ Selatan} + V_{S.P.Sudarmo\ A} \tag{16}$$

$$+ V_{gg.\ Settlement\ A} + V_{gg\ Settlement\ B} + V_{Ciliwung\ II} + V_{Tembaga} + V_{Aluminium} \tag{17}$$

$$V_{sunandar} = 1337.8 + 108.45 + 90.1 + 107.15 + 24.40 + 10.80 + 3.75 - 0.75 - 2.50 + 0.5 - 9.00 \tag{18}$$

$$V_{sunandar} = 1670.7\ pcu/hour \tag{19}$$

So that the total volume of vehicles at 9:00 to 10:00 o'clock in the amount of 1670.7 pcu/hour. Based on the calculation of volume, capacity and the road's level of service, the road's level of service affected by the amount of movement of land use and alleys. The road's level of service at operational hours of land at 8:00am to 18:00pm on the road's level of service is D and E (Table 4).

Table 4. Characteristic of Sunandar Priyo Sudarmo Road

Road Characteristics	Sunandar Priyo Sudarmo
Type	2/2 UD
Current direction	North-South South-North
Lanes	4
Direction system	2 ways
Effective width (m)	10.5
Material	Asphalt
Bandwidth (m)	5.25
Sidewalk (m)	-

IOP Conf. Series: Earth and Environmental Science **70** (2017) 012015 doi:10.1088/1755-1315/70/1/012015

Road Characteristics	Sunandar Priyo Sudarmo
Kerb	-
Roadside	1 m and 1 m
Median (m)	-
Parking	available
Land use	commercials
City size	0,5-1 million population

3.6. The Influence of Industry Movement to Road's Level of Service in Sunandar Priyo Sudarmo Road
The movement of the vehicle industry to the overall volume of vehicles in Sunandar Priyo Sudarmo Road ranged at 13 per cent to 22 per cent. Which at certain hours has a different contribution. The contribution of the vehicles by the industry that occur at 14.00 pm to 15.00 pm in which the vehicle industry has a number of moves by 22.13 per cent. This is because it is during these hours many industrial vehicles coming out of an industrial site. In addition, during the early morning at 09.00 am to 10.00 am contribution vehicle that is equal to 21.52 per cent (Table 5).

Table 5. The Road's Level of Service in Sunandar Priyo Sudarmo

Time	$V_{continuous}$	$V_{industry}$	$V_{non\text{-}industry}$	$\sum V_{gg}$ industry	$\sum V_{gg}$ Settlement	V_{TOTAL}	C	DS	LOS
08.00-09.00	1623.80	316.68	504.48	263.20	8.75	2,716.91	3235.22	0.85	E
09.00-10.00	1337.80	180.96	572.90	340.90	8.00	2,424.56	3235.22	0.75	D
10.00-11.00	1158.35	180.96	633.41	259.00	10.25	2,241.97	3235.22	0.76	D
11.00-12.00	1595.80	158.34	792.07	226.95	25.50	2,798.66	3235.22	0.87	E
12.00-13.00	1451.45	180.96	710.75	323.10	1.00	2,665.26	3235.22	0.82	E
13.00-14.00	1842.95	180.96	787.14	278.35	5.00	3,094.40	3235.22	0.96	E
14.00-15.00	1346.15	203.58	656.97	356.75	7.00	2,581.25	3235.22	0.80	D
15.00-16.00	1511.05	203.58	524.51	238.95	2.25	2,475.84	3235.22	0.77	D
16.00-17.00	1772.55	316.68	445.51	234.75	6.75	2,774.74	3235.22	0.86	E
17.00-18.00	1911.50	339.30	237.20	169.00	26.00	2,683.00	3235.22	0.85	E

IOP Conf. Series: Earth and Environmental Science **70** (2017) 012015 doi:10.1088/1755-1315/70/1/012015

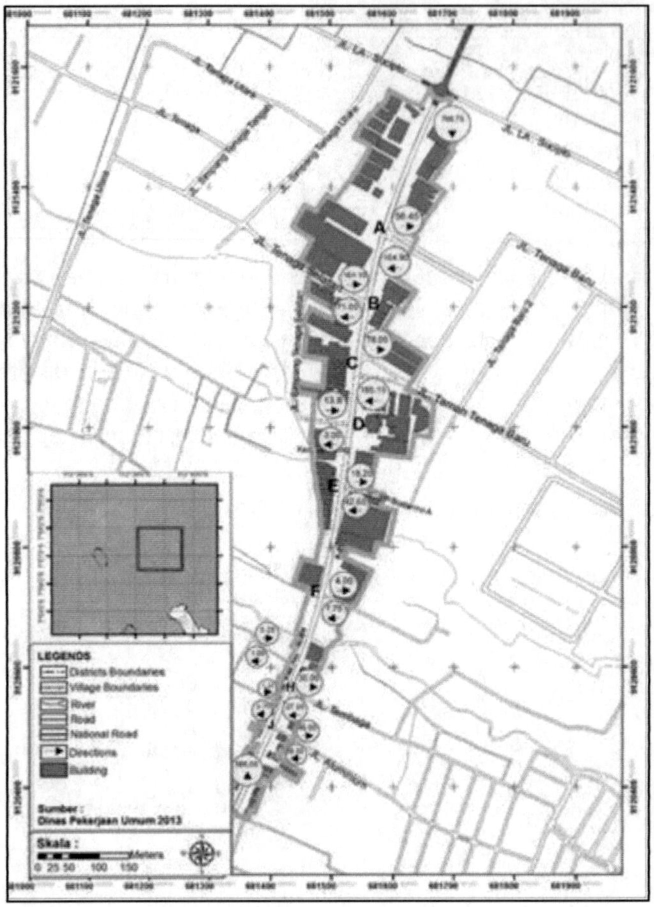

Figure 4. Volume of alley in Sunandar Priyo Sudarmo Road.

Table 6. Influence of Vehicles Movement of Industry to The Road's Level of Service of Sunanadar Priyo Sudarmo

Time	$V_{industry}$	$\sum V_{gg\ industry}$	V_{Total}	Contribution (%)
08.00-09.00	316.68	263.20	2,716.91	21.34
09.00-10.00	180.96	340.90	2,424.56	21.52
10.00-11.00	180.96	259.00	2,241.97	19.62
11.00-12.00	158.34	226.95	2,798.66	13.77
12.00-13.00	180.96	323.10	2,665.26	18.91
13.00-14.00	180.96	278.35	3,094.40	14.84
14.00-15.00	203.58	356.75	2,581.25	22.13
15.00-16.00	203.58	238.95	2,475.84	17.87
16.00-17.00	316.68	234.75	2,774.74	20.31
17.00-18.00	339.30	169.00	2,683.00	18.95

4. Conclusion

The road's level of service in Sunandar Priyo Sudarmo is C, D and E. The road's level of service has a difference every time, and the effect of the volume of industrial vehicles in Sunandar Priyo Sudarmo amounting to 4.57 per cent to 21.51 per cent. Which is where the highest peak is the greatest contribution

IOP Conf. Series: Earth and Environmental Science **70** (2017) 012015 doi:10.1088/1755-1315/70/1/012015

at 14.00 pm to 15.00 pm. The influence of the volume of industrial vehicle to vehicle volume is 13.77 per cent with total lowest at 11.00 am to 2.00 pm and the highest at 14:00 to 15:00 of 22.13 per cent.

5. Aknowledgement

Special thanks to all member of EIIS Laboratory and some students of Urban and Regional Planning (2012): Fauzan, Syavitri, Elita, etc for their cooperation and kindly help to this research. Thank you for your support.

References

[1] Tamin, Ofyar, Z., Frazila, RB., 1997. The application of land use interaction concept and transport systems in the planning of the transport network system. Urban Regional Planning Journal, ITB, Vol. 8, No 3, pp. 34-52. http://digilib.itb.ac.id.
[2] Spatial plan in East Java in 2011-2031
[3] Herzog, Bernhard. 2013. Urban Goods Transport in Cities Developing Countries. Modules Source Book: Sustainable Transportation. BMZ. Germany
[4] Tamin, Z, Ofyar. 2000. Transportation Planning and Modelling. ITB. Bandung
[5] Manual Highway Capacity Indonesia, 1997
[6] Miro, Fidel. 2004. Transportation Planning. Jakarta: Erlangga

IOP Conf. Series: Earth and Environmental Science **70** (2017) 012016 doi:10.1088/1755-1315/70/1/012016

Disaster mitigation action plan: Digital media on improving accountability and community relationships

I Adila[1,2], W W A Dewi[1] D Tamitiadini[1], W R Syauki[1]

[1] Universitas Brawijaya, Malang, Indonesia

adila.isma@gmail.com

Abstract. This study wants to address on how communication science is applied to Disaster Mitigation Plan. Especially, the implementation of Community Media and Development of Communication Technology that synergize to create a Disaster Mitigation Medium, which is appropriate for typology of Indonesia. Various levels of priorities that include disaster mitigation information, namely, increasing chain system of early warning systems, building evacuation, improving alertness and capacity to face a disaster, as well as minimizing disaster risk factor. Through this concept, mitigation actions plan of Tulungagung Coastal areas is expected to be applied in other regions in Indonesia by BNPB (Badan Nasional Penanggulangan Bencana). Having this strategy to be implemented based on region characteristics, it is expected that risk reduction process can be run optimally. As a result, the strategy is known as Community-Based Disaster Risk Reduction (PRBBK), which means as the organized-efforts by society for pra-, during, and post- disaster by using available resources as much as possible to prevent, reduce, avoid, and recover from the impact of disasters. Therefore, this result can be a Pilot Project for BNBP Indonesia, as a government decisive attitude for the next steps in protecting people residing in the region prone to natural disasters all over Indonesia.

1. Introduction

National Medium Term Development Plan (RPJMN) 2010-2014 of Indonesia proposed disaster management program as government priority because Indonesia is prone to disaster. United Nations International Strategy for Disaster Reduction (UNISDR, 2010) shows that Indonesia was ranked 12th out of 35 countries of the world's highest mortality risk threat hazard. In the same vein, National Board for Disaster Management shows that Indonesia was in the first rank for tsunamis and landslides with the highest number of casualty's world. Possible explanations for catastrophic events cannot be separated from geographic, geological, and demographic of Indonesia [1]. Geographically, Indonesia is located on the equator, so the drought and forest fires due to droughts can occur.

Using various form of communication is a common strategy to deal with real catastrophe in Indonesia. The implementation of intense mitigation are occupying mass communication medium, such as television, radio, newspapers, and magazines, to deliver relevant news. Especially for television, its usage is intended to reach people in a rural area due to high penetration of this broadcast media. However, the effort was not without problems such as there are many distorted media coverage, which caused victims. In Sinabung, residents testified that news in media sounds too had said that mountain situation has eased, thus it encourage people to enter the villages in the red zone. As a result, a number of people died as a heat cloud swept the region on February 1, 2014. This is one example of communication failure in emergency situations, which takes many forms. Some other failures are people's rejection to be evacuated, confusing information, news' errors, delays of information delivery, as well as the failure of outreach. Unfortunately, society is the victims of those failures. When communication fails, the fault entirely belongs to communicators and to a certain extent, in the medium, because both are the process of formulation and delivery of messages is determined. Thus, society as the recipient of information cannot be mistaken. Messages that fail to be delivered during an emergency are a reflection of the failure of risk reduction communication in pre disaster. Therefore, in this case, successful human rescue is determined by the success of communication. Communication is not always in the form of verbal communication, but as far as communication activities may enable humans to take necessary action to rescue.

Haddow explains the importance of media involvement both conventional and new media in the process of crisis management [2]. It refers to a great revolution of mass media has been created a new media including computer, telephone and communications network, Internet, and multimedia technologies [3]. Internet became the medium of computer mediated communication (CMC), which interactivity as its superiority.Besides, social media is part of Internet that gives people power to inform their ideas to others both in interpersonal and massive. However, there are problems that need to be solved, such as who should be trusted and how to filter information that is useful for readers [5].

Based on the use of social media that are contradictory in the crisis and the importance of helping government (BNPB and BPBD) as the pivot of disaster information to clarify the flow of information in public communication through social media, this study tries to analyze the communication network of tweets of Twitter's users about hoax or rumor of Kelud eruption on February 13, 2014. In doing so, network analysis has been used widely in order to understand virtual world phenomenon, one of which is Woo-young and Park that identify public preferences toward national issues in Korea within blog network structure [6]. They found a communication network including sizes, density, reciprocity, and centrality.

Disaster Mitigation based on Communication Network and Communication technology, has actually been initiated by "JalinMerapi.org" and "Jangkar Kelud" that are two social movements by local community. They have managed to rescue when disaster was occured while running the principle of accountability in the distribution of disaster aid. Referring to RIK (*Rencana Induk Kebencanaan*) seminar by National BNPB that Indonesia is in need of disaster mitigation, especially Cultural Mitigation, which means that not only focuses on the construction and rehabilitation of infrastructure but also means reducing disaster risk by taking into account cultural aspects. This action is important because disaster mitigation that is culturally-based requires the participation of local community or local community groups who are aware of Natural Disasters.

Dukuh Sidem, a hamlet located in the southern sea area through which tsunami plate. It is located in Tulungagung that has abundant marine potential that makes Dukuh Sidem is potential to develop. In the last 2006, Tsunami has happened in Tulungagung area that includes Popoh Beach, Sine Beach, and Sidem Beach. BNPB itself has done disaster mitigation trainings regularly annually to cope with tsunami disaster as well as to educate society to be always aware of disaster. One of causes of many disaster's victims is an unpredictable disaster. if the early symptoms of disaster can be predicted, it can help people to be on alert and avoidance. Therefore, the attempt to develop the ability of people to recognize threats, reduce vulnerabilities, and the ability to deal with threats. Disaster management is the responsibility of all parties, society as well, because the society is the first who will directly deal with threats and disasters. Therefore, the readiness of society determines the impact size of disasters on society.

Community-Based Disaster Risk Reduction (PRBBK) is the organized-efforts by society for pra-, during, and pasca- disaster by using available resources as much as possible to prevent, reduce, avoid, and recover from the impact of disasters. As the importance of this program, the entire territory of

Indonesia where is prone to disasters must implement a program of community-based disaster risk reduction. Yet, regional autonomy may be a challenge when PRBBK is simply based on administrative villages, where the cost of investment for enabling PRBBK is only project-based procurement, which are not integrated into development planning strategy. Therefore, this is a new problem in the implementation of PRBBK in Indonesia.

2. Literature Review

2.1 Disaster Communications

The term of disaster communication has not become a popular concept in the field of communication studies as well as the field of disaster. Despite, in Indonesia, disaster communication research has been conducted after earthquake and tsunami in 2014.

One important aspect in communication is the concept of reduction of uncertainty. Communication itself emerged from the need to reduce uncertainty; in order to act effectively to protect or strengthen the ego is concerned in interacting individual and group. In disaster management, accurate information is needed by public or private organizations that have concern for the victims. Communication in a disaster is not only needed in emergency conditions, but also important at the time and pre-disaster. As it is said, that communication is the best way for the success of disaster mitigation, preparedness, response, and recovery during a disaster situation. The ability to communicate messages about disasters to public, government, media and opinion leaders can reduce risks, save lives, and reduce the impact of the disaster [17]. They also stated that there are 4 major cornerstones in building effective disaster communication, namely:

a) Audience Focus: this has an objective to understand what kind of information is needed by public of the community and volunteers. Thus, communications mechanism should be build in order to ensure information is delivered with precise and accurate. The information contains messages about handling hazards, the degree of possible disasters as well as how big the scale and the impact of natural disasters.
b) Leader commitment: leadership role in emergency response must be committed to effective communication and be actively involved in the communication process. Leader's commitment may a representative of community organizations, local leaders, or even people who are influential or opinion leaders in the region.
c) Situational awareness: effective communication is based on the collection, analysis, and dissemination of information relevant to disaster control. Principles such as transparency and effective communication are believed to be its trusted key. Regarding that, the principle of effective communication is run by well-targeted information delivery. Effective communication in the community can build trust of information they received. The information submitted must be relevant to what is happening around the region.
d) Media partnership: media such as television, newspapers, radio, and others are very important medium to convey correct information to public. Having a training to be in same page about information to be delivered in both parties is important. In doing so, the media should be managed by local community in the form of socialization region directly.

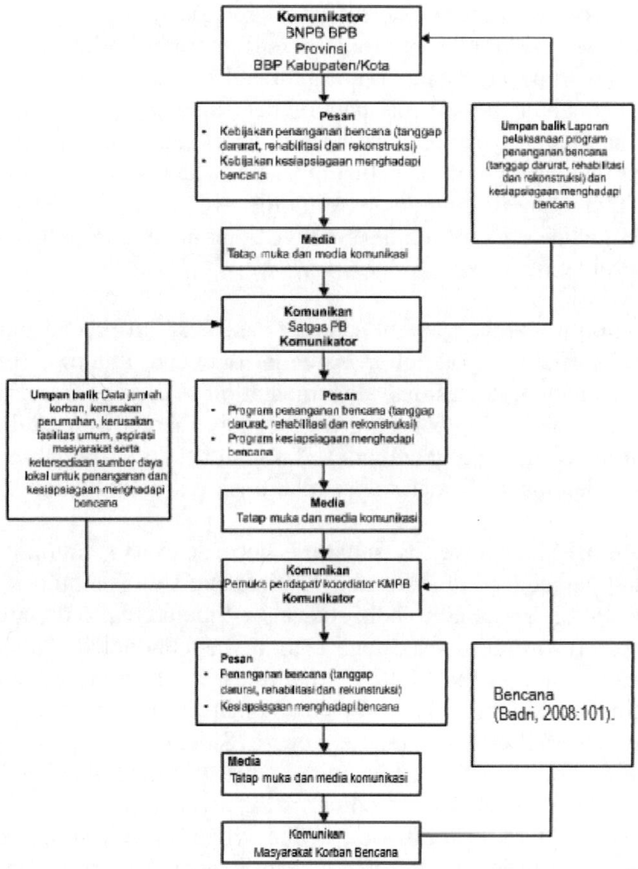

Figure 1. Table disaster communication handling through community media

3. Methods

Moleong [7] states that research focus has two purposes: 1) to restrict the study, and 2) to meet the criteria of new information's flow that is obtained in the field. Thus, through the focus of research, it is known what kind of data to be collected. The focuses of this study are as follows:

1. How the set of process of disaster mitigation in Sidem village
2. Appropriate community media for disaster mitigation, based on communication technology or group of people.

In order to achieve the objectives, Action Research is used in this study. Action Research Type used in this study is Participatory Action Research. According to Gonsalves et al [8], participatory action research is a combination of research and action in a participatory manner in order to improve aspects of society life. Regarding that, integration and participation among researchers, object studied by stakeholders, and other elements of society is an element that cannot be separated.

In its application, Kurt Lewin [9], which is the pioneer of using the term action research suggests cyclic step, consisting of problems identification, action planning, implementation, evaluation, and reflection. The next spiral step of Lewin that is explained by Edi Basuno et al [10] consists of a circle of planning, execution, and facts discovery for final evaluation purpose of the second step; and preparing the premises of third step, it also possible to modify the overall plan.

Participatory Action Research consists of research key principles as follows [10, 11].

1. Society should be involved in the whole research process, starting from problem formulation to discussion on how to seek resolution of problems and interpret findings.
2. The research team should be a combination of all the elements that make changes.
3. The research process should be seen as part of the educational experience that seeks to build community needs and increase public awareness and community commitment.
4. The process should be viewed as a dialectical process, which shows dialogue all the time and not in the form of a static picture of a single point in time.
5. The object must be the liberation of human creative potential and mobilization of human resources for solving social problems.

In the participatory action research, relevant activities are very different from other common action research. This is because participatory action research is conducted in a more practical manner, which is adapted to society's knowledge. Besides, documentation is a very important part, especially as reference material for society regarding commitment they have agreed among themselves. On the other hand, the interaction among community members and researchers is sustainable. Sustainability element is an indicator of a close relationship. Thus, a close relationship is an indicator of a huge opportunity for development [10].

Susceptibility in the area of Sidem towards tsunami is because of its location. It is located in the south area of Tulungagung that passed by tsunami slab. Besides, that area is a home for approximately 1000 people. Therefore, this study is expected to help society in Sidem area to practice a disaster mitigation appropriately. In order to do so, informant categorization was conducted. This categorization is based on informant's proximity to access relevant information. First, a group of people who are agent of change including the chief of Sidem Village, the chief of institution related to disaster mitigation for Sidem area, and the member of disaster awareness group in Sidem. Second, a supported group including people who already live in Sidem for 5 years who are member of mentioned group earlier.

In this study, a researcher also acts as a facilitator. Researchers facilitate Dusun Sidem society in reviewing and implementing disaster mitigation in the village. Facilitation can be seen as part that determines effectiveness and speed of reaching an agreement in various empowerment programs. Basically, facilitators' companion is relevant to how to facilitate society in solving and developing their potential independently [10]. The potential of disaster in the Sidem village is a potential for tsunami, since Sidem Beach is located in the southern sea area through which the plate tsunami. Thus, it requires disaster mitigation for Dusun Sidem. Disaster mitigation can be implemented by using community media of either communications technology or group of people.

4. Expected Results

Disaster is an event or series of events that threaten and disrupt the lives and livelihoods caused by both natural factors and / or non-natural and human factors that cause human fatalities, environmental damage, loss of property, and the psychological impact. Disaster has a nature of uncertain, which means that it may occur or may not occur. If disaster is occurred, severity of impact is also uncertain. Disasters can also give a trivial impact, but it may transform into a morbid, and thus, it may swallow hundreds of thousands of lives. Generally, the cause of disaster is divided into two: the consequences of natural (natural disaster) or manmade (man-made disaster). Therefore, a series of disaster relief efforts include the establishment of development policies that has risk of disasters ocurred, disaster prevention, emergency response, and rehab. These are characteristics' description of a number of disasters that often occur in Indonesia.

a) Flood
b) Landslides
c) Drought
d) Forest fires and land
e) The wind storm

f) The storm surge / tide
g) Earthquakes
h) Tsunami
i) Volcano eruption.

Having aware of the aforementioned dangers, understanding about threat of disaster, including knowledge of the following matters:

a) How does the danger arise
b) The level of possible disasters as well as how big the scale are.
c) The mechanism of physical destruction
d) Its sector and what activities will be greatly affected by disaster
e) The impact of damage, asone of the best efforts is to give knowledge to society about various disasters that might occur and its impact / consequences.

As an archipelago, Indonesia has vulnerabilities potential of the largest natural disasters, tsunami. Tsunami is a series of long waves caused by shifting plates beneath the sea, or in general is the seaquake. The speed of tsunami waves is capable of reaching 170 km / h. However, when it reached the superficial velocity, the speed may decrease but its wave is heighten. Besides, tsunami may ask for vistims, damage economic activities, as well as damage the ecosystem in coastal areas [12]. In the region of South Tulungagung, especially around the area of Sidem Beach, Popoh Beach, and Sine Beach, Tsunami is one threat of natural disasters that is prone to occur. As the impact, tsunami may heavily afect the community's economy because the majority of people work as farmers. Especially, the main impact of natural disaster can damage boats, industry sector, residence, and marine ecosystems.

Claude Gilbert proposes a summary concept of disaster [13] in three paradigms. First, disaster is the result or consequence of an external pressure. Second, as a result of social vulnerability. The last, as a result of uncertainty. This concept is in vein with opinion of Pelanda who interpret disasters as follows. First, disaster is the result of social and environmental conditions that are bad. Second, disaster is the result of collective pressure by community. Third, there was differentiation of capacity to handle damages and its negative impact. Thus, there are things to be observed which is associated with various definitions of disaster that means a perspective used to see a disaster. One of definition of disaster by Webster's New World Dictionary of the American Language cited by Porfiriev [13]. "*..any Happening that causes great harm or damage; serious or sadden misfortune; calamity. Disaster implies great or sudden misfortune that results in loss of life, property, etc. or that in ruinous to an undertaking; calamity suggests a grave misfortune that brings deep distress or sorrow to an individual or to the people at large.* "Dahl's Tolkovi Slovar Zhivogo Velikorusskogo Yazika (Explanatory Dictionary of The Great Russian Living Language):"*Misfortune or bedstviye (disaster) is an incident, accident or other harmful occasion the resulting in losses and sorrow. Disaster is more related to the known cases of crop failure, epidemics, storms, floods, To suffer means being a disastrous disaster or dangerous situation; badly needing something; dying.*"Based on these two sources, it is mentioned that disaster was an event that resulted a damage, death, loss, and other severe effects, including some types of natural events that caused catastrophic. Another opinion by Maloney and Cappola [14] about disaster: "*Disaster are measured in terms of lives lost, injuries, sustained, property damaged or lost, and environment degradation. Reviews These consequences manifest themselves through direct and indirect means and can be tangible or intangible.*"Some social scientists emphasize understanding of disaster as a social construction. Based on these experts, disaster is a result of social process or social consequences that produce hazards, or increase the vulnerability of a social system from the effects of hazard [13]. Furthermore, study concept and pragmatic about disaster by Quarantelli tries to define disaster as follows:

"a state/condition destabilizing the social system that manifest itself in malfunctioning or disruption of connection and communications between its elements or social units (communities, social groups, and individuals), partial or total destruction/demolition, physical and psychological overload suffered by some of these elements; thus, making it necessary to take extraordinary or emergency countermeasures to re-establish stability."

5. Research Plan

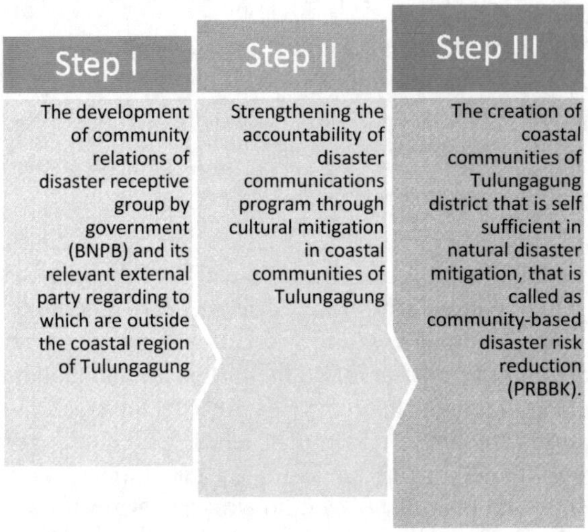

Figure 2. (a) first picture; (b) second picture

The roadmap shows the target of achievement of program that will be generated based on cooperation between universities, BPBDs, as well as community groups. The expected results are in the form of disaster response technology-based application that can be applied by community. The process of disaster mitigation shows the important role of technology, especially things relevant to cultural mitigation. Having understanding of cultural mitigation, community will be able to become an independent people who understand how to address the impact and face the possibilities of natural disasters in the scope of individual and family.

6. Research Plan

The discussion above results following conclusions. Carter states that disastrous split into 5 segments, namely: Prevention, Mitigation, Preparedness, Response and Recovery. In the beginning of the segment, such as prevention, the focus of attention is the importance of prevention, an area that needs attention to do prevention, positive approach through prevention, and what resources that are needed for prevention. In this segment, Mitigation is as an action taken is usually a specific program to reduce the impact of disaster on a nation or community. Moreover, an understanding of wider implications in mitigating the impact of disaster, however, these impacts can be modified or reduced by appropriate action. Some actions or programs related to mitigation, among others: the regulation of the use of area, the regulations on the safety of high-rise buildings, control of hazard material, security transport systems for land, sea and air; development system to protect electrical installations and vital communication tool, the development of infrastructure such as the construction of highways that are not adjacent to disaster-prone areas. The third segment is preparedness. In this phase there are some important things people need as an attempt of early warning, namely: early detection is done impending disaster, early warning process should be done by the party who has the authority, thus avoiding unnecessary information, the chain of transmission of effective information, decision-making after their early warning and early warning information dissemination to the public appropriately. Among many experiences initial reaction to the disaster mitigation, an early warning can save lives and property. Segment recovery, the recovery process can take a long time even could reach between 5-10 years or more. Post-disaster recovery usually involves various aspects such as restoration and reconstruction. In carrying out post-disaster recovery programs are indispensable involvements of various institutions, among others, as the government has the authority in terms of planning and government building of disaster management, including both central and local institutions that provide funding [15].

Disasters can result in short-term and long-term social, economic and environmental consequences with effects often felt country wide. These impacts are often magnified in small atoll island environments and have the potential to set back development gains. Recognizing the multi-faceted nature of the problem, disaster mitigation plan that is implemented in Tulungagung area is expected to be a pilot project related policies. Based on the data obtained in the study, researchers gather information related to the needs of society to mitigate the disaster. In the community in Sidem, they hold the principle of cultural disaster mitigation. People are still very adhering to the tradition that has been done from the first. Public awareness of the importance of the communication media is still very lacking, so the research team provides training for the use of communication media for the dissemination of information. Dissemination of such information requires effective communication. In this case, BPBDs Tulungagung is expected to convey information related to disaster mitigation. BPBD Tulungagung need to improve and strengthened effective communication with local Group (Kepala Desa, Kelompok Nelayan, dan Kelompok Keagamaan). Based on this research, BPBD needs to conduct effective communication is more active in using network-based media (Twitter and Fan pages) to provide important information on disaster mitigation.

References

[1] BNPB 2011

[2] Haddow, G. D, dan Kims. 2008. Disaster Communications, In A Changing Media World. London.Burllington USA.Butterworth-Heinemann.

[3] Mahmoud E.A. dan Auter J. P. 2009. The Interactive Nature of ComputerMediated Communication. Dalam American Communication Journal Vol. 11 No.4. Hal. 19-21.

[5] Winkellmann, Simon (ed). 2012. The Social Media (R)evolution? Asian Perspectives On New Media. Singapore. Konrad-Adenauer-Stiftun

[7] Moleong, Lexy. J. 2004. Metode Penelitian Kualitatif. Remaja Rosdakarya. Bandung.

[8] Iqbal, M; E. Basuno dan GS. Budhi. 2007. Esensi dan urgensi kaji tindak partisipatif dalam pemberdayaan masyarakat perdesaan berbasis sumberdaya pertanian. FAE. Vol 25, No 2, Desember 2007.

[9] Pickard, Alison Jane. 2007. Research Methods in Information. London: Facet Publishing.

[10] Edi Basuno, et al. 2008. Kaji Tindak (Action Research) Pemberdayaan Masyarakat Pertanian Daerah Tertinggal (Tahap II). Bogor: Pusat Analisis Sosial Ekonomi dan Kebijakan Pertanian.

[12] (UNESCO, 2011)

[13] Porifiev, Boris N. (1998). Issues in The Definition And Delineation Of Disaster s and Disasters Areas in Quarantelli. E. L, What Is A Disaster? Perspective on The Question. New York: Routledge.

[14] Maloney,Erin K and Coppola,Damon P. (2009). Communicating Emergency Preparedness: strategic for creating a disaster resilient. USA: Auerbach Publicationa Taylor & Francis Group.

[15] Carter, W. Nick, 2008, Disaster Management: A Disaster Manager's Handbook, Asian Development Bank, Philippines: Mandaluyong

[17] Haddow dan Haddow. 2008. Disaster Communication in a Changing Media World. Burllington

USA. Butterworth-Heinemann (diakses 28 Februari 2014) dari (http://libgen.org/book/index.php?md5=07F87FF80072E2305558FAC6276 FEB01&open=0)

[18] Randolph Kent, 1994. Disaster Preparadness. 2nd Edition. UNDP, Disaster Management Training Programme.

IOP Conf. Series: Earth and Environmental Science **70** (2017) 012017 doi:10.1088/1755-1315/70/1/012017

Infrastructure and social tie: Spatial model approach on understanding poverty in Malang regency, Indonesia

I R D Ari[1], A W Hasyim, B A Pratama, M Helmy, and M N Sheilla

Department of Regional and Urban Planning, Faculty of Engineering, Universitas Brawijaya, Malang 65147, Indonesia

Is2_mu@yahoo.com

Abstract. Poverty is a problem that requires attention from the government especially in developing countries such as Indonesia. This Research takes Place at Kasembon District because it has 53,19% family below poverty line in the region. The purpose of this research is to measure poverty based on 3 poverty indicators published by World Bank and 1 multidimensional poverty index. Furthermore, this research invesitigas the relationship between poverty with social and infrastructure in Kasembon District. This study using social network analysis, hot spots analysis, and regression analysis with ordinary least squares. From the poverty indicators known that Pondokagung Village has the highest poverty rate compared to another region. Results from regression model indicate that social and infrastructure affecting poverty in Kasembon District. Social parameter that affecting poverty is density. Infrastructure parameter that affecting poverty is length of paved road. Coefficient value of density is the largest in the model. Therefore it can be concluded that social factors can give more opportunity to reduce poverty rates in Kasembon District. In the local model of paved road coefficient, it is known that the coefficient for each village has not much different value from the global model.

1. Introduction

Poverty is a problem that needs to get attention because of its strategic issue for developing countries. In addition, according to United Nation's Universal Declaration of Human Rights, it is stated also in the universal declaration on human rights, that every individual has the right to a decent standard of living. To create a decent life, it is very necessary to know the level of poverty and what the main causes of poverty in the community [1].

In general, it has been known that the availability of the infrastructure was the main catalyst for decreasing or increasing poverty. The role of infrastructure is to provide ease of access, access to education; access to health care; trade access; and others. Infrastructure gives advantage to the poor especially in isolated areas [1]

Although, some of the opinions stated, there is a link between social capitals in the form of social bonds with economic development. Social bonds can overcome poverty through: sharing of resources; building support between communities; and initiate community to do collective action [2].

As the district located on the border of Malang and Kediri, Kasembon is passed by the arterial road network with good quality. Nevertheless, with the ease of access, there are 53.19% of households which are below poverty line [3]. Moreover, Kasembon district consist of six villages, nanely Pondokagung, Bayem, Pait, Wonoagung, Kasembon, and Sukosari. In this research, there are two research questions we search to answer. First, how the poverty level of the villages of Kasembon District? Second, how do infrastructure and social condition affecting poverty level in Kasembon District?

2. Literature Review

The main causes of poverty can be seperated into three level characteristics, regional, community, and households. At the regional characteristic, the major cause of poverty are natural disasters, remoteness of the region, and quallity of the local goverment. As for the cause in the community level are the infrastructure (roads, water, electricity), public services (health, education), and proximity to markets [1]. Causes of poverty was examined in this study are the cause of poverty at the community level.

In addition to the physical factors of poverty can also be caused by social conditons that exist in an area. In terms of social condition, we want to focusing on the social tie of the communities in each village of Kasembon District. Social bonds can be a form of social capital in the form of norms and networks that get people to act collectively [4], [5]. It can be said that a society which has a good connection and social network will have advantages in the mobilization towards the development of poor area.

3. Methods

3.1. Data collection and sampling

The method used in collecting data is questionnaire survey to households in each village and also complementary data from local government institution. Institutions' target are BAPPEDA Kabupaten Malang (Regional Planning and Development Agency of Malang Regency), Government of Kasembon District, Health Service of Malang Regency, and Public Works Agency of Malang Regency.

In this research, we use proportionate stratified random sampling. First, we split the sample into two different groups, poor households and non-poor households. Next, we proportionate sample in the two groups to each village. The reason why we stratified the samples is because, we want a clearer portrait of poverty in Kasembon. So, the poor and non-poor household must have the same opportunities to be sampled. To determine the sample size, we use Slovin method. The overall number of samples are 383 households, wherein head of households are the representative respondent for the face-to-face questionnaire survey.

3.2. The poverty indicators

There are four poverty indicators to determine poverty level of each village. The results from poverty indicators used as input in regression analysis. In this research, we use Foster, Greer Thorbecke index (FGT) (1984) and Human Poverty Index (HPI) (UNDP, 1997) as poverty level indicators. The FGT index will measure headcount index (HCI), poverty gap index (PGI), and poverty severity index (PSI). Headcount index (HCI) measures the proportion of the population that is categorized as poor. Formula for headcount index as follows:

$$P_0 = \frac{N_p}{n} \tag{1}$$

Here, N_p is the number of poor an N is the total population (or sample). The households are classified as poor when their expenses are below the poverty line. The advantage from using HCI are it is simple to construct and easy to understand. These are important reason why we use this indicator to get information about poverty level. Poverty Gap Index (PGI) is a moderately popular measure of poverty. The PGI add up the extent to which individuals on average fall below the poverty line by following formula (2).

$$P_1 = \frac{1}{n}\sum_{i=1}^{n}\frac{G_i}{z} \tag{2}$$

In the PGI, G_i is the gap between expenditure and poverty line, while z is the poverty line. This masure is the mean proportionate poverty gap in the population (where the non-poor household have zero

poverty gap). Some people use it to think of this measure as the minimum cost of eliminationg poverty (relative to the poverty line)
To measuring inequality among the poor, we use the poverty severity index (PSI). This is a weighted sum of poverty gaps. In PSI, the poorer households are fall well below the poverty line, puts more weight on observations. Formally,

$$P_2 = \frac{1}{n}\sum_{i=1}^{n}\left[\frac{G_i}{z}\right]^2 \qquad (3)$$

As seen on (3), the PGI and PSI formula almost same. In PSI, we square the poverty gap. So, the increases in measured poverty because of a fall in one's standard of living will be deemed greater the poorer one.

3.3. Hot Spots Analysis
In this research, we also scrutinize expenditure distribution of households using hot spots analysis. So, we can get information where are households with less or more expsense. Hot spots analysis will be using Getis-Ord Gi* statistics, as follows:

$$G_i^* = \sum_{j=1}^{n} w_{i,j}x_j - \bar{X}\sum_{j=1}^{n} w_{i,j}\left(S\left(\frac{\left[n\sum_{j=1}^{n} w_{i,j}^2 - \left(\sum_{j=1}^{n} w_{i,j}\right)^2\right]}{n-1}\right)^{1/2}\right)^{-1} \qquad (4)$$

In hot spots analysis, G_i value is normalized. The data used in this calculation is based on expenditure of each households and coordinates of the respondents. Where $w_{i,j}$ is the spatial weight between i and j respondents. Distance between the respondents is calculated by using euclidean distance. The output from hot spots analysis are cold spots and hot spots. Cold spots is location of households with less expenses, while hot spots is location of the households with more expenses.

3.4. Social Networks Analysis
In this study, we use rate of participation and density approach to analyze the community social tie. To measure rate of participation and density we use affiliation data of the respondents. The affiliation data obtained by looking community groups that followed by the respondents. Formula for rate of participation of the respondents as follows [6].

$$\bar{a}_{i+} = \frac{\sum_{i=1}^{g} x_{ij}^{N}}{g} \qquad (5)$$

Where X_{ij}^{N} is the affiliation matrix of the respondent *i* and respondent *j*. Its data are assumed as binary relation whereby attendance in a certain community group is equal to 1, otherwise equals as zero. The density is used to see how the average number of activities that occur by any pair of actors (Wasserman & Faust, 1994). In density, the value ranged from 0 to 1. To calculate density, we use the same affiliation data used in the rate of participation calculation. The formula for density seen on (6)

$$\Delta(N) = \frac{\sum_{i=1}^{g}\sum_{j=1}^{h} x_{ij}^{N}}{g(-1)} \qquad (6)$$

Here, *g* -1 is the respondents who have not affiliation with other respondents, while *g* is the respondents that have affiliation with other respondents.

3.5. Regression Analysis

Regression analysis is a statistical analysis of the process that aims to find out the relationship between dependent variable and one or more independent variables (or predictor) [7]. The regression model will explain how the phenomenon of poverty occurred in Kasembon District.

Regression techniques were used to find the model of poverty is the ordinary least squares (OLS). General equation of the OLS model is as follows [8].

$$Y = \beta_0 + \beta_1 X_1 + \beta_2 X_2 + \cdots \beta_n X_n + \varepsilon \quad (7)$$

As Y is the dependent variable and X is the predictors (explanatory variables). The ε is the error vector of regression line. And β is the coefficient of the predictors. After the diagnostic results obtained from the OLS method, then the model will need to undergo test of regression assumption. We use multicolinear test, heteroscedacity test, significance test and residual normality test.

4. Results

4.1. Expenditure distribution and poverty level

Expenditure distribution per capita analyzed uses hot spots analysis. Besides expenditure, coordinates of respondent also inputted in hot spot analysis. Then, we will have information where is the community with the high levels of expenditure clustered. Figure 1 portrays levels of expenditure across the district. Red, beige, and blue color represents hot spot-neutral-cold spot. Higher level of expenditure is showed by increasing red color. Beige color means the levels of spending are on average. The bluer, then the level of expenditure lower below the average.

According to the results of hot spot (Figure 1), majority of community expenditure levels are below the average. Cold spot clustering occurs in the region of Pondokagung Village and Pait Village. Hot spots area looks very significant in Bayem Village. In addition, there is also a hot spot in the Sukosari Village toward Malang and Kediri Regency border.

Figure 1 explains that the expenditure is not affected by the existence of the arterial road, the location of border region, as well as center of district location. Even though, there is a hot spot region in Sukosari Village bordering Kediri. However, Kasembon Village as the district center of service and passed by arterial road, thus it has expenditure level below the average. Pait Village, passed by arterial road and located on the border also have expenditure level below the average.

There are 4,986 poor households in Kasembon District. Based on survey results, average households expenditure is Rp 1,151,724 per month. With average family member of 3 persons, therefore average expenditure per capita is Rp 333,199. For reference, Malang Regency poverty line is Rp 254,380 (BPS, 2015). The results of poverty indicators calculation shows Pondokagung Village is the highest poverty level in Kasembon District. HCI value of Pondokagung Village is 42.11%, PGI value is 9.16%, PSI value is 1.22% and HPI value is 26.541%.

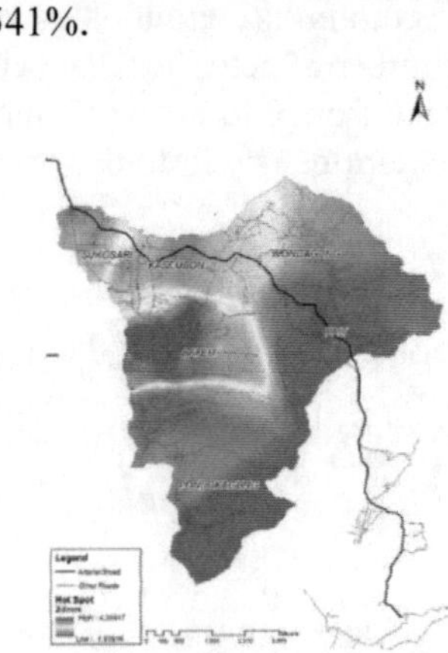

Figure 1. *Spot* and *Cold Spot* for Expenditure per Capita.

Table 1. Poverty Indicator per Village.

Village	HCI (%)	PGI (%)	PSI (%)	HPI (%)
Pondokagung	42,11	9,16	3,41	26,451
Bayem	18,92	3,07	0,87	0,022
Pait	39,66	8,43	3,2	0,009
Wonoagung	35,09	5,48	2	0,405
Kasembon	27,59	5,96	2,07	0,007
Sukosari	28	4,16	1,22	11,957

There are three villages that needs to be noticed, namely Wonoagung, Kasembon, and Sukosari. Table 1 illustrates that Wonoagung and Sukosari villages have figures higher than Kasembon Village on HCI. However, PGI and PSI figure out that Sukosari and Wonoagung Villages are lower. Meaning that people of Sukosari and Wonoagung Village are at risk for falls to more severe poverty.

Table 2. Human Poverty Index Value.

Village	Percentage of Death before age 40	Percentage of Illiteracy	Percentage of malnutrition
Pondokagung	0,03%	28,77%	0,42%
Bayem	0,05%	2,70%	0,45%
Pait	0,02%	2,00%	0,25%
Wonoagung	0,07%	7,14%	0,23%
Kasembon	0,14%	1,85%	0,17%
Sukosari	0,02%	22,08%	0,38%

Visible differences are quite striking on parameter value percentage of illiterate population in Pondokagung and Sukosari Village (Table 2). Difference between smallest and largest illiteracy is 26.82%. Conclusion from these findings are the illiterate population give largest contribution to high value of Pondokagung HPI value.

4.2. Road quality and accessibility

Road quality data is observed by looking at length of asphalt road and non-asphalt road. The data used in this research is sourced from Malang Regency Public Works Service of year 2015. The non-asphalt road pavement consists of cobblestone road and dirt road.

In this research, accessibility observed from distance to Malang Regency center; distance to Kasembon District center; Distance to district market; Distance to education facility such as elementary school, junior high school, and senior high school; also, the distance to health facility.

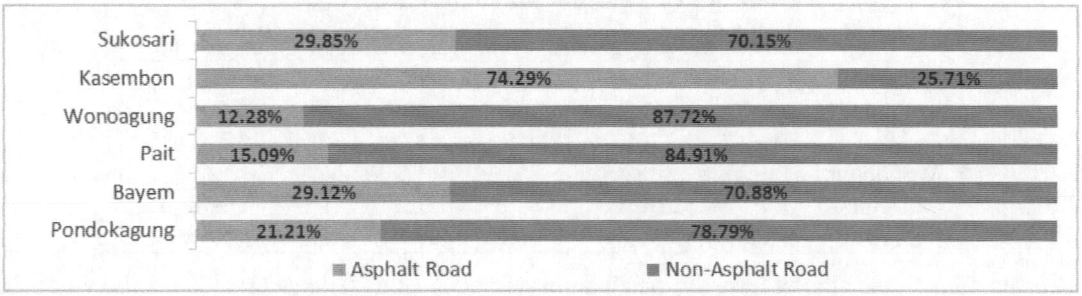

Figure 2. Road Quality Comparison per Village Area.

Length of asphalt road is 24.11km, or 17% of the total length of roads in Kasembon District. Wonoagung and Bayem Villages are the village with the lowest percentage of asphalt road length in Kasembon District. While the village with highest percentage of asphalt road length are Kasembon Village.

In this research, accessibility observed from distance to Malang Regency center; distance to Kasembon District center: Distance to district market; Distance to education facility such as elementary school, junior high school, and senior high school; also, distance to health facility. The distance is the average distance from the neighborhoods towards the facilities Kasembon Village

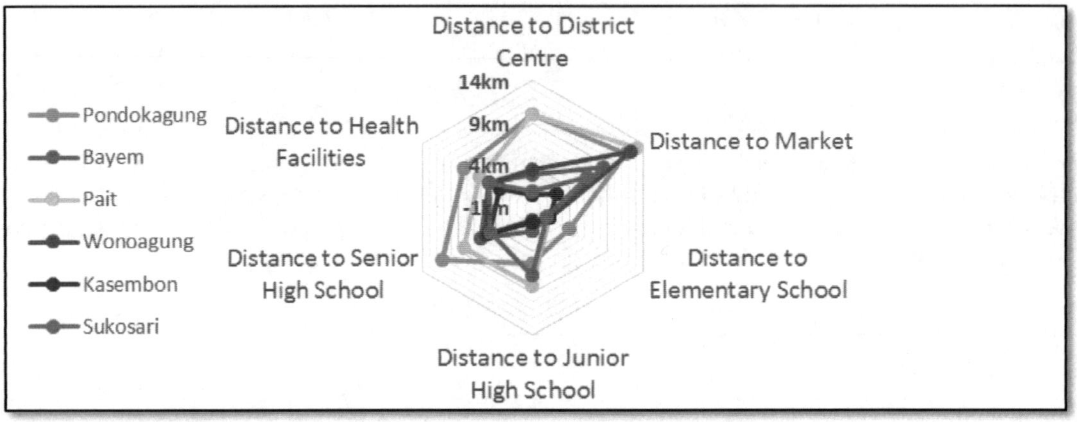

Figure 3. Distance to Facilities from Nearest Settlement for Each Village.

The lowest accessibility to the center of Kasembon District are Pait Village (10 km) and Pondokagung Village (10 km). Pondokagung village is also the farthest distance towards the health facilities (8,3 km), senior high school (11,34 km), and toward elementary school (2,83 km). Pait Village is the farthest distance to junior high school (8,2 km) and district market (6,26 km). Pondokagung village can't meet the elementary school minimal standard of service, because the average distance to elementary school more than 3 km. Pait Village also cannot meet junior high school standard. Junior high school must be ≤6 km from nearest settlements.

4.3. Rate of participation and density

Social tie of Kasembon District community is measured with social network analysis approach using rate of participation and density indices. Rate of participation in the villages are calculated by number of groups that is followed by individual in the villages. The higher the level of participation, the higher awareness of the communities will be. The rate of participation will be classified into three levels – low, medium, and high based on the village value of rate of participation. The rate of participation is calculated through affiliation data between the respondents.

Table 3. Rate of Participation for Each Village.

Village	Rate of Participation	Number of Groups	Classification
Pondokagung	1,5789	24	High
Bayem	1,4189	8	Low
Pait	1,4800	13	Low
Wonoagung	1,3889	12	Low
Kasembon	1,2609	15	Medium
Sukosari	1,5195	10	Low

As seen on Table 3, the highest level of participation is Pondokagung Village (1,5789). This value means the average of Kasembon peoples are joint participation on 2 community groups. Mostly villagers are joint in a kind of religious group, namely Tahlilan. According to the respondents, besides tahlilan people also hold regular social gathering for women known as Arisan. The arisan mostly held by close neighbors. Calculation of density aims to know about density of connections between the respondents in one village. The density value is also a mean for measure kinship value in the village.

Table 4. Density Value and Level of Kasembon District.

Village	Density Value	Classification
Pondokagung	0.404	Low
Bayem	0.722	High
Pait	0.595	Medium
Wonoagung	0.632	Medium
Kasembon	0.493	Medium
Sukosari	0.793	High

Amongst all, the lowest density is Pondokagung Village. It shows also that the Pondokagung villagers tend not to have similarities in the organization participation. Otherwise, it can be said that one organization member don't know with the other member of the organization. In contrast, for Bayem and Sukosari Village, the density is in high level.

4.4. Regression analysis

The regression model analyzed uses ordinary least squares (OLS) technique. In selecting appropriate model we use exploratory regression from ArcGIS. With exploratory regression, we might find the right model combination with the desired parameters. The parameters inputted are, maximum number of explanatory variables is 12 and minimum is 2; minimum adjusted R2 is 0,5 (50%); minimum p-value is $\leq 0{,}1$; minimum VIF (multicolinearity test) is $\leq 7{,}5$; and residual autocorrelation is $\geq 0{,}1$.

Table 5. Variables of Poverty Indices.

Dependent Variable	X1	X2
Head Count Index	(-)Density	(-)Asphalt Road Length
Poverty Gap Index	(-)Density	(-)Asphalt Road Length
Poverty Severity Index	(-)Density	(-)Asphalt Road Length

The three models have the same explanatory variables, i.e. density and the length of asphalt road. Those two variables even have the same inverse relationship with three poverty indices. Since HCI, PGI, and PSI representing poverty, it can be concluded that greater density and asphalt road length, the lower the poverty in Kasembon District.

Regression OLS diagnostics (Table 6) is useful to illustrate the results of calculation that will be used to testing hypothesis, multicolinearity test, and residual normality test. Standard error column represents difference of observation value and the regression line. The coefficient column is the independent variable influence to poverty indices. Intercept is the intersection value between regression lines with y axis

Table 6. OLS Regression Diagnostic.

Dependent Variable	Independent Variable	Coefficient	Std. Error	Probability	VIF	Koenker BP Statistic	Jarque Bera Statistic
HCI	*Intercept*	72,377740	8,613644	0,0000	-	0,7333	0,6196
	Density	- 36.696453	11,551253	0,0472	1,000083		
	Asphalt Road Length	-0,003059	0,000813	0,0208	1,000083		
PGI	*Intercept*	17,425756	2,461122	0,0000	-	0,3148	0,7116
	Density	- 13,138299	3,300467	0,0133	1,000083		
	Asphalt Road Length	-0,000573	0,000232	0,0893	1,000083		

Dependent Variable	Independent Variable	Coefficient	Std. Error	Probability	VIF	Koenker BP Statistic	Jarque Bera Statistic
PSI	*Intercept*	7,127937	0,90227	0,0000	-	0,2714	0,6804
	Density	-5,639812	1,20992	0,0014	1,000083		
	Asphalt Road Length	-0,000265	0,00008	0,0503	1,000083		

The regression equation describes how the relationships between poverty and infrastructure with social condition and its influence. Findings from Table 6 shows that on the three models, the coefficients of density is greater than the asphalt road length coefficient. Then, density known have bigger influence to poverty than asphalt road length. For example, asphalt road length of Kasembon Village is 17 km longer than Sukosari Village. However due to higher level density of Sukosari Village, so the poverty in the Sukosari Village is lower.

In the probability column, it could be illustrated whether the independent variables have influence or not with the poverty indices. P-value significant test in this research using $\leq 0,1$ value. It is different than the commonly used in other research. So, it brings more model options to us.

Density and asphalt road length have meet significance test (p-value $\leq 0,1$) However, density had lower p-value in PGI and PSI model (lower than 0,05). While the p-value of asphalt road length on PGI and PSI model are above 0,05. In this case, density showing to have stronger connection with poverty. Although asphalt road length has significant p-value, its relation to poverty is not as big as the density.

Heteroscedacity test uses to see if the Constanta of variables are stationary or not. Heteroscedacity happens when the Koenker BP value > 0,05. Koenker BP value of the three models (Table 7) are above 0,05. It means, heteroscedacity does not occur in the three models.

Multicolinearity is condition where two or more variables are highly correlated. That is not allowed to occur in a regression, because it is violates the assumption of regression model. In this research, multicolinearity detected by using VIF calculation. If the VIF value is greater than 7.5, then the model have multicolinearity problem. For the three poverty models, the VIF values are below 7.5. So it can be concluded multicolinearity does not occur in the model

Residual normality is the one of linear regression assumptions. In this study, the residual normality is seen through the Jarque-Bera. Residual is normally distributed when the Jarque-Bera value greater than 0,05. For the three poverty model, the residual values have normaly distributed. As seen in last column of Table 7, Jarque-Bera value for the three poverty model above 0,05.

5. Conclusion

The level of Kasembon District community expenditure is not affected by the availability of arterial road, the border location, and the center of activity in Kasembon Village.

According to the 4 indicators of poverty that is used in this study, Pondokagung Village is the worst village in poverty. The largest contribution to the high HPI value is from percentage of illiterate person. Other findings from the poverty level analysis is the Wonoagung Village and Sukosari Village have more risk towards more severe poverty. That is because Wonoagung Village and Sukosari Village HCI are higher than Kasembon Village, but those have lower PGI and PSI level.

Regression analysis results through OLS indicates that social conditions (density) and infrastructure (asphalt road length) have an inverse relationship with poverty. It might assume that density has more influence to poverty compared to asphalt road length. In other words, in order to deal with poverty alleviation in Kasembon District, development of physical infrastructure does not the merely solution, but it is also a must to deal with the development of capacity buidling of the villagers. Existence of social tie between villagers may open more possibility for them to enhance their possibility towards local and extra local resources mobility, especially information resources.

References

[1] B. N. Bhattacharyay, "Asian Development Bank Institute," Tokyo, 138, 2009.

IOP Conf. Series: Earth and Environmental Science **70** (2017) 012017 doi:10.1088/1755-1315/70/1/012017

[2] A. Afridi, "Social networks : their role in addressing poverty," York, 2011.
[3] Centre Bureau of Statistics, "Malang in figures 2016," Malang, 2016.
[4] M. Woolcock and D. Narayan, "Social capital: Implications for development theory, research, and policy," *World Bank Res. Obs.*, vol. 15, no. 2, pp. 225–250, 2010.
[5] R. D. Putnam, "Bowling Alone by," *J. Democr.*, pp. 65–78, 1995.
[6] S. Wesserman and K. Faust, *Social network analysis: Methods and applications*. New York: Cambridge University Press, 1994.
[7] R. E. Walpole, *Introduction to statistics*, Third Edit. London: Collier Macmillan Publishers, 1982.
[8] A. Mitchell, *The ESRI guide to GIS analysis*, Volume 2. Ohio: ESRI Press, 2005.

IOP Conf. Series: Earth and Environmental Science **70** (2017) 012018 doi:10.1088/1755-1315/70/1/012018

Reduction of CO2 emission from transportation activities in the area of Pasar Besar in Malang City

K E Sari, D E Sulistyo and D M Utomo

Department of Urban and Regional Planning, Faculty of Engineering, Universitas Brawijaya, Malang, Indonesia

kartikaeka@ub.ac.id

Abstract. The number of vehicles increases every year. Where vehicles are the largest contributor to air pollution up to 70%-80%, while 20%-30% caused by industrial activities. The increasing number of vehicles which perform movements will result in more emissions of vehicles in the free air of the city. Traffic is also influenced by the presence of land use. One of the types of land use that have considerable influence against the movement of traffic is trade. Along with the development of transport activities in the area of Pasar Besar Malang city (the Biggest Traditional Market in Malang), it will cause problems such as traffic jam and air pollution. Therefore, the need for proper handling of the problem of traffic jam and air pollution in the area of Pasar Besar that is to identify the performance of road traffic around area of Pasar Besar and calculate the quantity of CO2 emissions based on the footprint of transport on the area of Pasar Besar. Where is produced that level of service roads on its way around area of Pasar Besar have an average value between LOS A and B, while the quantity of CO2 emissions is based on the footprint of transport on area of Pasar Besar that is amounting to 4,551.42 tons/year. The magnitude of the emissions have exceeded the standard of composition in the air so that the need for recommendations. Recommendations in this research is in the form of simulated users of private vehicle redirects to public transportation based on the level of willingness to switch by the users of private vehicles. The selected simulation that is the second of four simulations with the output of emissions amounting to 3,952.91 tons/year in which can reduce emissions amounting to 598.51 tons/year or approximately 13.15%.

1. Introduction

Indonesia is a developing country with a high rate of urbanization. The increasing number of population leads to the increasing need of transportation. Transportation is defined as actions or activities of transporting or transferring loads (objects or people) from one place to another or from the original place to the destination [1]. According to Fidel Miro (2005), transportation is defined as efforts at transferring or moving both passenger and goods from one location to another using a certain equipment [2].
The high increase in population results in the high need of transportation. In Indonesia, motor vehicles have the biggest contribution to the air pollution in which the number is 70% - 80%, and the rest 20% - 30% are contributed by industrial activities [3]. The sources of exhaust gas emission come from non-point sources and point (stationary) sources. Non point sources of emission come from the activities of motor vehicles, while point sources comes from activities in one place like industrial activities [4]. One of exhaust gas emission from vehicles is carbon dioxide (CO2). The gas is one of the glass house gases.

The highest contributors of glass house gases (GHG) in increasing the emission are CO2, CH4 and N2O [5]. The highest contributor of GHG is CO2, that is 55%.

Transportation traffic is influenced by the land use. A travel is caused by the difference in the land use. In fulfilling the basic needs, each individual moves from a certain land use to another land use. It creates a very close relationship between the land use and transportation. One type of land use which has a great influence to the traffic movement is trade [6]. Trade facilities which are realized in the form of traditional market and stores is one type of land use that has attracted a significant number of travel. The boost and pull of travel by land use in the form of trade needs the support of transportation facilities performance. Without the support of transportation facilities, the demand and supply in travel will not be balances, and in turn it will create many traffic problems.

Pasar Besar of Malang City (the biggest traditional market in Malang) is one of trade facilities which is located in a strategic aerea in Kota Malang. Pasar Besar of Malang which is located in the center of the city has a potential to attract high movement of people in Malang. In the Urban Planning of Malang 2010 – 2030, Pasar Besar of Malang is planned to be a regional trade center.

The increase of activities in the area of Pasar Besar results in the increase of motor vehicles' movement in that area. In turn, the movement of motor vehicles also increases the number of exhaust gas produced. It is line with the theory that the increasing number of vehicles which perform movements will result in more emissions of vehicles in the free air of the city). The volume of vehicles influences the level of road services. The higher VCR (volume capacity ratio)/service level of a road in certain time, the higher exhaust gas emission from transportation [7].

According to the Regulation of East Java Governor No. 67 Year 2012, East Java targeted the reduction emission as much as 5.22% in the area of energy and transportation. To support the Regulation of East Java Governor, identification of CO2 Emission Reduction in Pasar Besar of Malang is needed. The results of CO2 emission analysis will be used in one of implementation of GHG emission reduction program in East Java. In this study, carbon emission will be measured from transportation activities based on the quantity and types of vehicles in the roads around the study area, Pasar Besar of Malang, using transportation ecological footprint.

2. Research Method

The study was conducted in four streets in the area of Pasar Besar of Malang namely Jalan Pasar Besar, Jalan Sersan Harun, Jalan Kyai Tamin, and Jalan Kapten Usman. The four streets was then divided into 7 (seven) segments namely Jalan Pasar Besar Segment 1, Jalan Pasar Besar Segment 2, Jalan Sersan Harun Segment 1, Jalan Sersan Harun Segment 2, Jalan Kyai Tamin Segment 1, Jalan Kyai Tamin Segment 2, and Jalan Kapten Usman. Methodology used in this research consist of, Road Level of Service Method and Total Amount of CO2 Emission Method.

3. Results

The observation of traffic volume in this study is conducted on weekdays and weekend. The duration of observation is 16 hours/day between 05.00 – 21.00. Survey method used is direct observation in the field. The results of the survey observation of the traffic volume in seven segments on weekdays and weekend. (Figure 1).

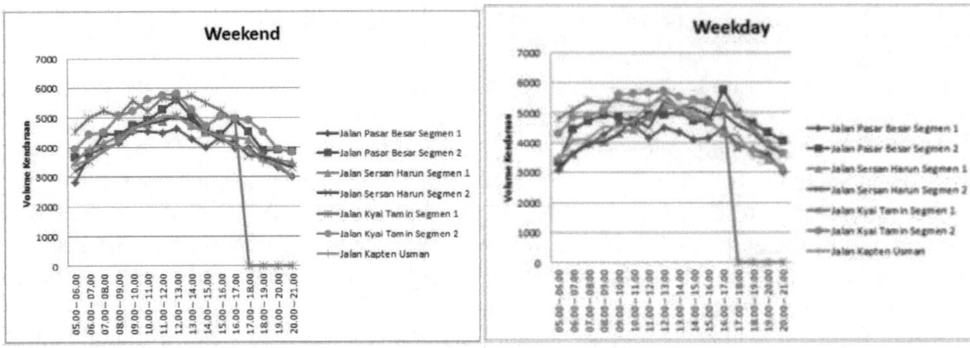

Figure 1. (a) Traffic Volume on Weekdays Recapitulation) and (b) Traffic Volume on Weekend Recapitulation

Based on the two graphics' information above, it can be identified the average peak of rush hour on weekdays and weekend are at 07.00 - 08.00, 12.00-13.00, and 16.00-17.00. It happens because at 07.00-08.00 most stores at the market start to operate, and then at 12.00-13.00 the activities in Pasar Besar Kota Malang are in the peak, and at 16.00-17.00 the stores end their operation/are closed.

3.1. Road Level of Service

The value of Road Service is influenced by traffic volume and road capacity. The lower the traffic volume, the better the road service level. Below are the values of road level of service in the seven segments of the streets on weekdays and weekend. (Table 1).

Table 1. Values of road level of service

Street Names	*Weekdays*		*Weekend*	
	Average V/C Ratio	**Average LOS**	**Average V/C Ratio**	**Average LOS**
Jalan Pasar Besar Segment 1	0.60	A	0.63	B
Jalan Pasar Besar Segment 2	0.66	B	0.64	B
Jalan Sersan Harun Segment 1	0.65	B	0.62	B
Jalan Sersan Harun Segment 2	0.66	B	0.61	B
Jalan Kyai Tamin Segment 1	0.54	A	0.54	A
Jalan Kyai Tamin Segment 2	0.65	B	0.67	B
Jalan Kapten Usman	0.66	B	0.63	B

Based on table 1, it can be identified that the road level of service on the seven segments has the values of A and B, on the average. The highest v/c ratio can be found on Jalan Pasar Besar Segment 2, both on weekdays and weekend. Meanwhile, the lowest v/c ratio can be found on Jalan Kyai Tamin Segment 1, both on weekdays and weekend.

3.2. Total CO2 Emission of Vehicles

In order to identify CO2 emission, we can multiply the total consumption fuels with factor emission of fuels. The factor emission is stated in kilogram/liter. The value of factor emission for gasoline fuel vehicles is 2.31 kg/liter; while, the factor emission for diesel fuel vehicles is 2.68 kg/liter (Andriono, et al, 2013). Below are the results of CO2 emission for each type of vehicles, both private and public transportation (Table 2).

Table 2. Total CO2 emission of private vehicles

Street Names	Types of Vehicles	Total CO_2 Emission (tons/year)
Jalan Pasar Besar Segment 1	2-wheel (Gasoline)	326.61
	Private 4-wheel (Gasoline)	272.92
	Private 4-wheel (Diesel)	135.73
	Truck (Diesel)	4.07
	Bus (Diesel)	0.26
Jalan Pasar Besar Segment 2	2-wheel (Gasoline)	418.70
	Private 4-wheel (Gasoline)	173.74
	Private 4-wheel (Diesel)	161.46
	Truck (Diesel)	2.49
	Bus (Diesel)	0.05
Jalan Sersan Harun Segment 1	2-wheel (Gasoline)	143.40
	Private 4-wheel (Gasoline)	57.95
	Private 4-wheel (Diesel)	56.18
	Truck (Diesel)	0.21
	Bus (Diesel)	0.00
Jalan Sersan Harun Segment 2	2-wheel (Gasoline)	217.39
	Private 4-wheel (Gasoline)	88.16
	Private 4-wheel (Diesel)	75.65
	Truck (Diesel)	0.31
	Bus (Diesel)	0.00
Jalan Kyai Tamin Segment 1	2-wheel (Gasoline)	386.39
	Private 4-wheel (Gasoline)	138.20
	Private 4-wheel (Diesel)	90.41
	Truck (Dicscl)	7.55
	Bus (Diesel)	0.08
Jalan Kyai Tamin Segment 2	2-wheel (Gasoline)	479.96
	Private 4-wheel (Gasoline)	167.69
	Private 4-wheel (Diesel)	100.33
	Truck (Diesel)	27.41
	Bus (Diesel)	0.37
Jalan Kapten Usman	2-wheel (Gasoline)	368.43
	Private 4-wheel (Gasoline)	142.35
	Private 4-wheel (Diesel)	123.16
	Truck (Diesel)	2.99
	Bus (Diesel)	0.00
Total		4,170.60

Table 3. Total CO2 emission of public transportation

Street Names	Types of Vehicles	Total CO_2 Emission (tons/year)
Jalan Pasar Besar Segment 1		36.35
Jalan Pasar Besar Segment 2		26.36
Jalan Sersan Harun Segment 1		11.58
Jalan Sersan Harun Segment 2	4 Wheel-Public (Gasoline)	17.09
Jalan Kyai Tamin Segment 1		21.67
Jalan Kyai Tamin Segment 2		27.42
Jalan Kapten Usman		30.01
Total		140.47

Based on table 3, it can be identified that the total CO2 emission of private vehicles is 4,170.60 tons/year in which the biggest emission is on the seven segments for 2-wheel vehicles. Then, based on table 13, it can be identified that the total CO2 emission of public transportation is 170.48 tons/year. Overall, the total CO2 emission for both private and public transportation is 4,341.08 tons/year.

4. Recommendation

There are two recommendations in this study. The first is a simulation CO2 emission reducation by efforts to migrate the private vehicles users to the public transportation, based on the willingness of the visitors of Pasar Besar Kota Malang who use private vehicles to migrate to the public transportation. The second is a simulation the expected change in road level of existing service and the service at the time of simulation because of the simulation of CO2.emission reduction.

4.1. Simulation of the Reduction Vehicles' CO2 Emission

Before conducting a simulation of CO2 emission of vehicles, we need to identify the precentage of willingness to shift from private vehicles to public transportation from the visitors of Pasar Besar area. Data of the level of willingness to migrate to public transportation by the visitors of Pasar Besar was taken from questionaires given to 400 respondents of Pasar Besar. The result is that the willingness to migrate to the public transportation by the visitors of Pasar is 34% (willing), and 64% (not willing) (Figure 2)

Figure 2. Willingness to use public transportation

Having identified the percentage of willingness to migrate by the visitors, the next step is calculating the simulation of willingness to migrate to public transportation by the visitors. The simulation of CO2 emission reduction consists of four, those are: Simulation I which is call do nothing simulation or a simulation which maintain the existing condition; Simulation II, that is a simulation in migrating the users of private vehicles by reducing 34% 2-wheel private vehicles, and then the result of reduction is added into the public transportation capacity; Simulation III, that is a simulation in migrating the users of private vehicles by reducing 34% 4-wheel private vehicles, and then the result of reduction is added into the public transportation capacity; Simulation IV, that is a simulation in migrating the users of private vehicles by reducing 34% 2-wheel and 4-wheel private vehicles, and then the result of reduction is added into the public transportation capacity. Below is the result of simulation of CO2 emission of vehicles (Table 4).

Table 4. Simulation CO_2 emission of vehicles

No.	Simulation	CO_2 Emission (tons/year)		Total (tons/year)
		Private Vehicles	Public Transportation	
1.	Simulasi I	4,170.60	170.48	4,341.08
2.	Simulasi II	3,374.70	416.32	3,791.02
3.	Simulasi III	3,563.77	222.57	3,786.34
4.	Simulasi IV	3,391.41	399.22	3,790.64

Based on table 4, it can be identified that the lowest CO2 emission from the simulation is simulation III with the number CO2 emission of 3,786.34 tons/year. In simulation III we can reduce CO2 emission as much as 554.74 tons/year or about 12.78% of CO2 emission in the existing condition.

4.2. Road Level of Service Simulation

The simulated calculcation of road level of service is conducted to identify the change in road level of service caused by CO_2 emission simulation which results in the change in the volume of vehicles . The simulated calculation of road level of service can only be done in simulation III. Below is the results of the comparison between the road level of service of the existing condition and the level of service in simulation III, both on *weekdays* and *weekend*. (Table 5 and Table 6).

Table 5. Simulation of the road level of service on weekdays

Street Names	*Weekdays*			
	Existing		Simulation III	
	Average Ratio of V/C	Average LOS	Average Ratio of V/C	Average LOS
Jalan Pasar Besar Segment 1	0.60	A	0.49	A
Jalan Pasar Besar Segment 2	0.66	B	0.56	A
Jalan Sersan Harun Segment 1	0.65	B	0.55	A
Jalan Sersan Harun Segment 2	0.66	B	0.57	A
Jalan Kyai Tamin Segment 1	0.54	A	0.56	A
Jalan Kyai Tamin Segment 2	0.65	B	0.57	A
Jalan Kapten Usman	0.66	B	0.56	A

Table 6. Simulation of the road level of service on weekend

Street Names	*Weekend*			
	Existing		Simulation III	
	Average Ratio of V/C	Average LOS	Average Ratio of V/C	Average LOS
Jalan Pasar Besar Segment 1	0.,63	B	0.50	A
Jalan Pasar Besar Segment 2	0.64	B	0.54	A
Jalan Sersan Harun Segment 1	0.62	B	0.54	A
Jalan Sersan Harun Segmenr 2	0.61	B	0.53	A
Jalan Kyai Tamin Segment 1	0.54	A	0.56	A
Jalan Kyai Tamin Segment 2	0.67	B	0.55	A

Weekend				
	Existing		**Simulation III**	
Street Names	**Average Ratio of V/C**	**Average LOS**	**Average Ratio of V/C**	**Average LOS**
Jalan Kapten Usman	0.63	B	0.52	A

Based on tables 5 and 6, it can be identified that there is a significant change in the road level of service in simulation III compared to simulation I in which the road level of service in simulation III has the total value of LOS A.

5. Conclusion

The road level of service on the seven segments has the values of LOS A and B, on the average. The highest v/c ratio can be found on Jalan Pasar Besar Segment 2, both on weekdays and weekend. Meanwhile, the lowest v/c ratio can be found on Jalan Kyai Tamin Segment 1, both on weekdays and weekend.

The total CO2 emission of vehicles based on the transportation footprint on the seven segments oi Pasar Besar Area of Malang City is 4,341.08 tons/year. The total CO2 emission for private vehicles is 4,170.60 tons/year, and the total CO2 emission for public transportation is 170.48 tons/year.

Based on the four simulation of CO2 emission reduction the lowest CO2 emission is in simulation III with CO2 emission 3,786.34 tons/year. In simulation III we can reduce around 554.74 tons/year or about 12.78% from CO2 resulted from the existing condition.

Based on tables 15 and 16, it can be identified that there is a significant change in the road level of service in simulation III compared to simulation I in which the road level of service in simulation III has the total value of LOS A.

References

[1] S. A. Adisasmita, *Jaringan transportasi: Teori dan Analisis*. Yogyakarta: Graha Ilmu, 2012.
[2] F. Miro, *Perencanaan transportasi: Untuk mahasiswa, perencana, dan praktisi*. Jakarta: Erlangga, 2005.
[3] D. Maryanto, S. A. Mulasari, and D. Suryani, "Penurunan Kadar Emisi Gas Buang Karbon Monoksida (CO) Dengan Penambahan Arang Aktif," *Kesehat. Masy.*, vol. 3, no. 3, pp. 198–204, 2009.
[4] Presiden Republik Indonesia, *Peraturan Pemerintah Nomor 41 Tahun 1999 Tentang Pengendalian Pencemaran Udara*, no. 41. 1999, pp. 1–17.
[5] A. Meiviana, D. R. Sulistiowati, and M. H. Soejachmoen, *Bumi makin panas: Ancaman perubahan iklim di Indonesia*. Jakarta: Kementrian Lingkungan Hidup Republik Indonesia, Yayasan Pelangi Indonesia, 2004.
[6] O. Z. Tamin, *Perencanaan dan Pemodelan Transportasi*. Bandung: Penerbit ITB, 2000.
[7] M. Zhongang and G. Shengan, "Traffic pollution in Xi'an City, P.R. China," no. 5, pp. 2000–2001, 2014.

Urban development control based on transportation carrying capacity

M Miharja, A H Sjafruddin

Regional and City Infrastructure Systems Research Group, School of Architecture, Planning, and Policy Development, Bandung Institute of Technology, Bandung, Indonesia

mimingm@pusat.itb.ac.id

Abstract. Severe transportation problems in Indonesian urban areas are stimulated by one fundamental factor, namely lack of awareness on transportation carrying capacity in these areas development control. Urban land use development towards more physical coverage is typically not related with the capability of transportation system to accommodate additional trips volume. Lack of clear connection between development permit with its implication on the transportation side has led to a phenomenon of exceeding transport demand over supply capacity. This paper discusses the concept of urban land use development control which will be related with transport carrying capacity. The discussion would cover both supply and demand sides of transportation. From supply side, the analysis regarding the capacity of transport system would take both existing as well as potential road network capacity could be developed. From demand side, the analysis would be through the control of a maximum floor area and public transport provision. Allowed maximum floor area for development would be at the level of generating traffic at reasonable volume. Ultimately, the objective of this paper is to introduce model to incorporate transport carrying capacity in Indonesian urban land use development control.

1. Introduction

In Indonesia urban areas, land use development keeps going on to fulfil the residents needs for housing as well as other facilities. However, in most cases, urban land use development does not seem to have a clear reference on transport carrying capacity. Despite long awareness of land use development impact on transport system: land use development causes new trip generation/attraction in urban road network that increases the traffic volume [1], we hardly found a systematic method to take it into land use development control account, especially for Indonesia case. It would be necessary then, to develop a model that systematically incorporate transport carrying capacity as one crucial element in urban land use development control.

In environmental study, carrying capacity is the ability of the environment to support humans and other living beings. Analogically, in transportation study, carrying capacity could be define as the ability of the whole transport system to accommodate trip volume at the level in which a reasonable level of service is maintained. Since we understand that land use development implies additional trip volume, then it should be stopped at the level in which its implication on transport system start to threat the reasonable level of service. (e.g. by using VCR/volume to capacity ratio criteria).

Transport level of service is the reflection of volume to capacity ratio. Solving the problems of traffic that comes from an imbalance proportion between volume and capacity can be solved in three ways [2]:

1. Adding supply capacity by building new roads or widening existing roads. This method may not be carried out continuously since there is certainly limitations of the space for road widening as well as economic-social-cultural problems.
2. Reducing the volume of traffic by reducing the number of vehicles through land use management, public transport provision, etc.
3. Combining the first and the second method through traffic management.

This paper uses those three ways to further discuss the analytical model of using the transport carrying capacity as the basis for urban land use development control that is able to provide guidance for land use development control to maintain reasonable transport level of service. In this study, land use development control is approached by supply and demand management. According to Massachusetts Highway Department [3] in Setiawan [4], traffic management is a process of setting on supply and demand in existing road system for specific purpose without adding new infrastructure, through the reduction and regulation of traffic movement. From that definition, it is clear that traffic management is done by optimizing the supply and controlling the demand [5],.

From demand management point of view, land use development control is done by determining maximum floor area allowed for development, as this will manage trip generation at the required level. Next section will discuss model development to incorporate all the above variables which basically incorporate transport carrying capacity to urban land use development control. This model may be further develop as a practical tools for urban land use planning permit.

2. Methodology of The Research

This section describe the data collection and analysis methods of this urban land use control-transport carrying capacity model.

2.1 Data Collection Method

The type of data in this model include primary data and secondary data. The primary data is obtained from field survey. The secondary data is obtained from institutions, literatures, newspaper, and other sources. This survey is needed to get a description about transport condition and land use in urban area.

2.2 Data Analysis Method

Methods of data analysis in this section is the result of analysis from secondary data. The method developed in this paper is expected to be the initial form of model for future urban land use development control that accommodate transport carrying capacity. The sequential step of the model is based on transport supply and demand management as can be seen in Figure-1.

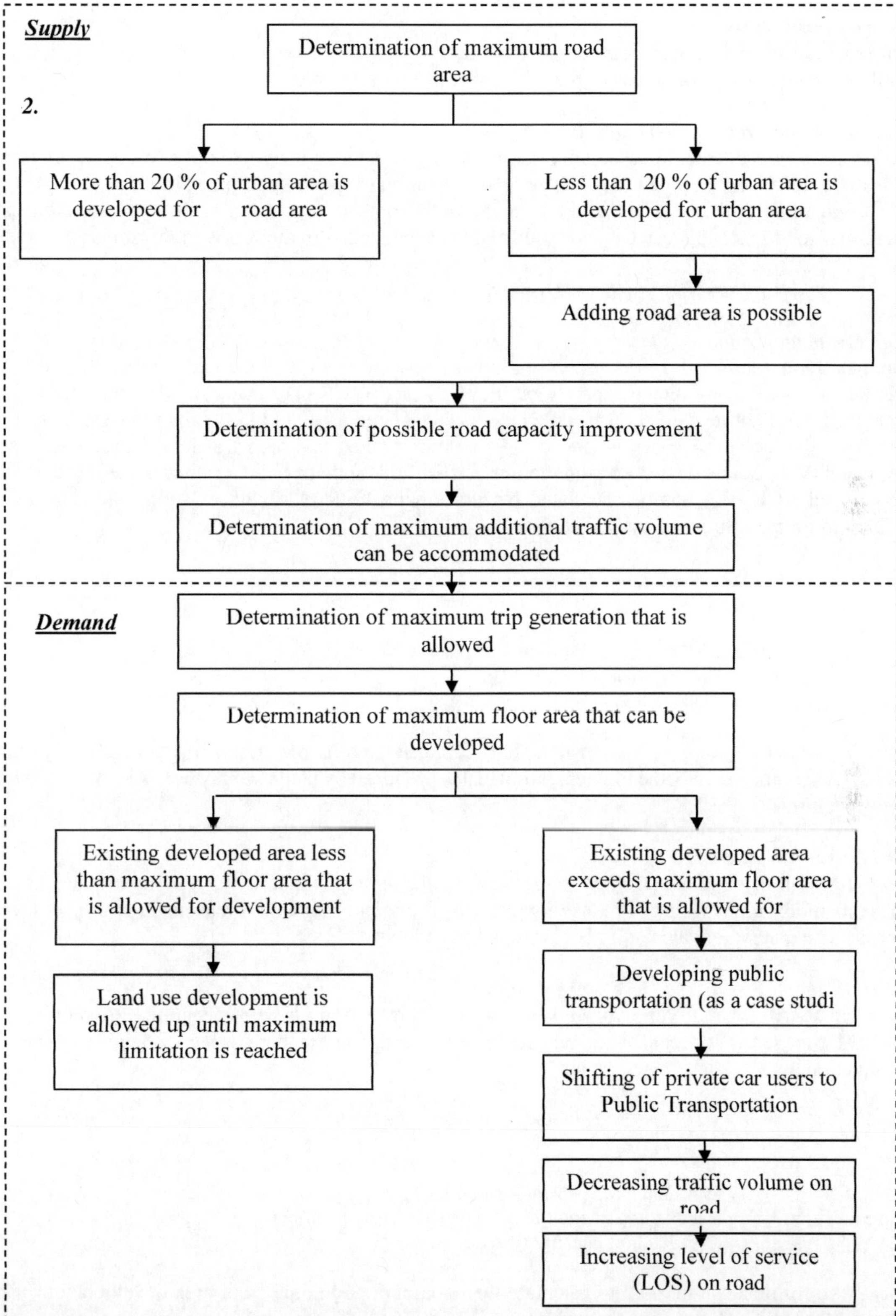

Figure. 1. Flow chart: Analysis of transport carrying capacity for urban development control

2.3 Transport Supply
From supply side, the analysis is proceed by calculating maximum road space that still can be developed in urban area. Here are the stages of the analysis carried out from the supply side.

2.3.1 Maximum Area for Road Development
Construction on the street in urban area in this study refers to Ministry of Public Work Regulation No 05/PRT/M 2008 which guide the provision and utilization of green open space in Urban Area [6]. The proportion of the road network in urban areas based on this regulation is 20% of the urban area. Therefore, road space that can be developed can be calculated with the following formula:

$$Road\ Area_{mx} = 20\%\ x\ Urban\ Area \quad (1)$$

2.3.2 Maximum Traffic Volume.
The maximum volume of traffic on this study is maximum traffic volume that can be allowed to pass road network based on volume capacity ratio (VCR) plan. VCR plan that is used in this study is 0.75 which is the limit of level of service (LOS) that is considered a good LOS for road in Urban Area.

Once the Volume Capacity Ratio (VCR) is known, the maximum traffic volume (V_{max}) can be calculated. V_{max} can be described as maximum traffic volume that can be accommodated by the road at the planned VCR. V_{max} can be calculated by multiplying VCR plan with existing road capacity (C). Maximum traffic volume (V_{max}) is calculated as follow :

$$V_{max} = VCR_{plan}\ x\ C_{existing} \quad (2)$$

where :
V_{max} = Maximum traffic volume allowed (pcu/hour)
VCR_{plan} = Volume Capacity Ratio Plan
$C_{existing}$ = Existing Road Capacity (pcu/hour)

Once V_{max} has been known, V_{max} in this study is considered as maximum trip generation maximum (TG_{max}) that is allowed. So, the trip generation that is produced by land use may not exceed the maximum trip generation.

2.4 Transport Demand
From transport demand management point of view, the approach that is by reducing private car users. These methods are expected to reduce total traffic volume to maintain a reasonable transport level of service. The method of calculation is developed as follows:

2.4.1 Maximum Floor Area Development Allowed
To calculate trip generation produced by one area, trip rates of each type of land use are required. The value of trip generation can be calculated by using trip rate standard for each land use type. The formulation to calculate trip generation is :

$$TG_{max} = \sum Trip\ Rate_{GL}\ x\ A_{GL} \quad (3)$$

where:
TG_{max} = Maksimum trip generation (pcu/hour)
$Trip\ Rate_{GL}$ = standard trip rate for land use type GL (pcu/ hour/m^2)
A_{GL} = Area Floor of land use type GL (m^2)

The above formula can be used to calculate the maximum additional floor area development (of land use type GL) allowed, by using the formula:

$$A_{GL} = \frac{TG_{max}}{\sum \text{Trip Rate}_{GL}} \qquad (4)$$

where:

A_{GL}	= maximum additional floor area development, of land use type GL (m^2)
TG_{max}	= Maximum trip generation (pcu/hour)
Trip Rate $_{GL}$	= standard trip rate for land use type GL (pcu/ hour/m^2)

Therefore, A_{GL} can be referred as the parameter to control land use development in that area. Land use development permit should be subject to criteria that no further permit would be granted if the total area development has reached A m^2.

2.4.2 Percentage of Private Vehicle Users that can be converted to Public Transport

The objective of this section is to estimate the potential traffic volume reduction, by shifting some of the traveller from private to public transport. There are numbers of method could be chosen to proceed the calculation, e.g. based on modal choice questionnaire. Results of the survey is the percentage of private vehicle users who are shifted to public transport. In this section, the percentage of mode shift from private car to public transport is based on types of private vehicles used (this study assesses private car and private motorcycle).

2.4.3 Traffic Volume Reduction by Private to Public Transport Shift

Traffic volume reduction depend on the number private vehicle user shift to public transport. Once the percentage of private vehicle users who are shifted to public transport has been known, we can calculate the traffic volume reduction. There steps are :

1. Calculate the total passenger of private vehicle (P) by multiplying the number of private vehicle (Vp) with its average Load Factor (LF) :

$$P = Vp \ x \ LFv \qquad (5)$$

2. Calculate the total passengers who are shifted from private to public transport (SP) by multiplying the probability of private to public transport shift (%P) with total passenger of private vehicle (P):

$$SP = \%P \ x \ P \qquad (6)$$

3. Calculate the number of public transport (Npu) by dividing the total passengers who are shifted from private to public transport (SP) with the public transport average load factor (LFp)

$$Npu = Sp / LFp \qquad (7)$$

4. Then multiplying the number of public transport (Npu) with standard of pcu (sn) to get the number volume public transport in pcu unit (Vpu) :

$$Vpu = Npu \ x \ sn \qquad (8)$$

5. To calculate the new total traffic volume (Vn), the Vpu is added to the remaining volume of private vehicle (Vpr) whose passengers are not shifted to public transport

$$Vn = Vpu + Vpr \qquad (9)$$

6. Lastly, to know the new traffic performance (VCRn), the new total traffic volume (Vn) is divided by road capacity (C). Whenever the VCRn shows the value of less than 1, it means there would be some additional traffic volume are allowed.

3. Model Application

The objective of this paper is to introduce model to incorporate transport carrying capacity in Indonesian urban land use development control. To do so, this paper examines the model in 2 Indonesian cities: Bandung and Cilegon. Those cities represent big city and small city. The results of the examination are as follow:

3.1 Bandung City

Bandung city is a city in West Java Province. The development of Bandung City is massive, so it would be hard to build new road in Bandung City. Therefore, Tansport Demand Management (TDM) is needed to manage the transport. There are two ways of TDM that can be introduced: controlling maximum floor area of development and promoting public transport.

3.1.1 Maximum Floor Area Development Allowed

Four roads are examined in Bandung City: Jl. AH. Nasution, Jl. Ahmad Yani, Jl. Raya Kopo, and Jl. Wastukencana. The following table show the existing traffic condition in peak hour in the four roads.

Table 1. Existing Condition of Road in Bandung City

No	Road Name	Type of Road	C (pcu/hr)	V (pcu/hr)	VCR	LOS
1	Jl. AH Nasution	Primary arterial road	4236.96	4130.667	0.974913	E
2	Jl. Ahmad Yani	Secondary arterial road	4236.96	3202.254	0.75579	C
3	Jl. Raya Kopo	Primary collector road	4236.96	4158.996	0.981599	E
4	Jl. Wastukencana	Secondary collector road	3643.2	1869.75	0.513216	C

When the exisiting VCR compared with planned VCR, all roads have exceeded the plan. The following table shows a comparison between the existing peak based-VCR with VCR plan. VCR plan is set at 0.75, which is the limit of degree of saturation (MKJI, 1997)

Table 2. Comparison of LOS Existing and Los Plan in Bandung City

No	Road Name	VCR Existing	LOS Existing	VCR Plan	LOS Plan	Land Use Development Permit
1	Jl. AH Nasution	0.974913	E	0.75	C	Not Allowed
2	Jl. Ahmad Yani	0.75579	C	0.75	C	Not Allowed
3	Jl. Raya Kopo	0.981599	E	0.75	C	Not Allowed
4	Jl. Wastukencana	0.513216	C	0.75	C	Allowed, but Limited

Based on above table, the two roads still meet requirement: Jl. Ahmad Yani and Jl. Wastukencana. However, Jl. AH Nasution and Jl. Raya Kopo have exceeded LOS Plan. So, one way to improve LOS is through transport demand management (TDM) by land use development control. Before calculating floor area of land use maximum, we calculate the maximum traffic volume (Vmax) of the four roads. Table-3 shows the results.

Table 3. V max of roads in Bandung City

No	Road Name	VCR Plan	C Existing	V max
1	Jl. AH Nasution	0.75	4236.96	3177.72
2	Jl. Ahmad Yani	0.75	4236.96	3177.72
3	Jl. Raya Kopo	0.75	4236.96	3177.72
4	Jl. Wastukencana	0.75	3643.2	2732.4

Once the Vmax is determined, the maximum land use area can be calculated. The following section shows maximum floor area that can be developed in Bandung city based on Vmax for each road. Land Use proportion can be determined by the government plan. In this paper proportion of land use is determined by sample of land use area exisiting that is next to the road.

Table 4 Maximum Land Use Area

No	Land Use	Vmax	Land Use Proportion	Trip Generate Max	Trip Rate	Maximum Land Use Area (m^2)
			Jl. A.H. Nasution			
1	Commercial	3177.72	29.16%	926.54	0.012*	77212.21
2	Industry	3177.72	40.65%	1291.80	0.118***	10947.46
3	Office	3177.72	14.10%	447.91	0.01*	44791.65
4	Education	3177.72	0.28%	8.77	0.0237**	370.06
5	Residential	3177.72	15.82%	502.68	0.0037*	135861.003
			Jl. Ahmad Yani			
1	Commercial	3177.72	80.24%	2549.73	0.012*	212477.13
2	Office	3177.72	8.52%	270.86	0.01*	27085.55
3	Worship	3177.72	1.32%	41.81	0.0193**	2166.17
4	Education	3177.72	9.92%	315.33	0.0237**	13305.14
			Jl. Raya Kopo			
1	Commercial	1906.632	78.54%	2495.63	0.012*	207969.06
2	Industry	1906.632	19.84%	630.45	0.118***	5342.76
3	Residential	1906.632	0.72%	22.99	0.0037*	6214.29
4	Office	1906.632	0.90%	28.65	0.01*	2865.21
			Jl. Wastukencana			
1	Commercial	2732.4	9.85%	269.25	0.012*	22437.91
2	Residential	2732.4	2.45%	67.02	0.0037*	18112.99
3	Perkantoran	2732.4	78.41%	2142.42	0.01*	214242.51
4	Worship	2732.4	5.68%	155.075	0.0193**	8034.98
5	Education	2732.4	3.61%	98.62	0.0237**	4161.46

Description :
[a] Trip Rate CBD DKI Jakarta,1990
[b] Trip Rate Country Sandiego, 2003
[c]Trip Rate Kota Cirebon and Semarang, 2000

Table 4 shows the results of the estimated maximum land use area, which can be used as a guideline for land use development control in each areas. Given the fact that maybe the existing land use level has exceeded these "normal level", public transport development approach could be introduced as alternative control mechanism.

3.1.2 Traffic Volume Reduction by Private to Public Transport Shift

Firstly, we need to know the composition of vehicles. In this research, the scenario is to shift some of the private vehicles (motorcycles and private car) passengers to public transport (*angkot*). The following is the vehicle composition in four roads in Bandung City.

Table 5. Traffic Volume off Road Existing According to Vehicle Composition

No	Road Name	Traffic Volume (pcu/hour)				Total (pcu/hour)
		Motorcycles	Light Vehicles - Private Car	Light Vehicles - Public transport	Heavy Vehicles	
1	Jl. AH Nasution	2350.80	1094.08	438.27	247.51	4130.68
2	Jl. Ahmad Yani	1530.09	1354.32	170.1	147.74	3202.25
3	Jl. Raya Kopo	1163.61	2266.98	343.23	385.15	4158.99
4	Jl. Wastukencana	1356.75	137.7	351	24.3	1869.75

The second step, calculate the prediction of private vehicle passengers that could be shifted to public transport. In this research the the number of passenger is calculated by multiplying the number of private vehicles with load factor (LF) of each private vehicle. The LF in this research is 2.5 (for private car) and 1 (for motorcycles). It is also assumed that the probability of public transport choice is 40% (Al Faruq, 2015)[7]. The number of passengers who use public transport is divided with load factor of public transport to get public transport needs. In this research the load factor of public transport (*angkot*) is 12. the following table is the prediction of public transport needs in four roads in Bandung City.

Table 6. Public Transport Needs

No	Road Name	Existing Private Vehicle (Vehicles/hour)		Passengers of Existing Private Vehicle (Passenger/hour)		Passengers who shift to public transport (passenger/hour)			Public Transport Needs (vehicles/hour) LF= 12		
		Motorcycles	Private Car	Motor cycles (LF =1)	Private Car (LF = 2.5)	From Motorcycles	From Private Car	Total	From Motor cycles	From Private Car	Total
1	Jl. AH Nasution	9403	1094	9403	2735	3761.2	1094	4855.2	313.43	91.16	404.6
2	Jl. Ahmad Yani	6120	1354	6120	3385	2448	1354	3802	204	112.83	316.83
3	Jl. Raya Kopo	4654	2267	4654	5667.5	1861.6	2267	4128.6	155.13	188.92	344.05
4	Jl. Wastukencana	5427	138	5427	345	2170.8	138	2308.8	180.9	11.5	192.4

Once the number of passengers that shift to public transport is known, the remaining passengers who use private vehicle can be known. The following is the remaining passenger who still uses private vehicles.

Table 7. The Remaining Private Vehicles

No	Road Name	The remaining passengers who are still in private vehicle (passengers/hour)		The Remaining of Private Vehicle (Vehicles/hour)		The Remaining Traffic Volume of private vehicles (pcu/hour)	
		From Motorcycles	From Private Car	Motorcycles	Private Car	Motorcycles	Private Car
1	Jl. AH Nasution	5641.8	1641	5641.8	656.4	1410.45	656.4
2	Jl. Ahmad Yani	3672	2031	3672	812.4	918	812.4
3	Jl. Raya Kopo	2792.4	3400.5	2792.4	1360.2	698.1	1360.2
4	Jl. Wastukencana	3256.2	207	3256.2	82.8	814.05	82.8

Lastly, the number of traffic volume change can be calculated. The number of private vehicle traffic will be decreasing and the number of public transport will be increasing. The calculation of traffic volume when the private vehicle users shift to public transport is as follow.

Table 8. The Changes of Traffic Volume According to Vehicle Composition

No	Road Name	Volume traffic (pcu/hour)				Total (pcu/hour)
		Motorcycles	Light Vehicles – Private Car	Light Vehicles – Public Transport	Heavy Vehicles	
1	Jl. AH Nasution	1410.45	656.4	842.86	247.51	3157.23
2	Jl. Ahmad Yani	918	812.4	486.93	147.74	2365.07
3	Jl. Raya Kopo	698.1	1360.2	687.28	385.15	3130.74
4	Jl. Wastukencana	814.05	82.8	543.4	24.3	1464.55

As a result, the LOS (level of service) of Bandung City roads are now better. Jl. A.H Nasution VCR changes from 0.97 to 0.75. Jl. Ahmad Yani, VCR changes from 0.76 to 0.56. Jl. Raya Kopo VCR changes from 0.98 to 0.74. Meanwhile, Jl. Wastukencana VCR changes from 0.51 to 0.4. The following tables shows the comparison between road existing condition with "after public transport"condition.

Table 9. The comparison of road existing condition and road condition after using public transport

No	Road Name	VCR Existing	LOS Existing	VCR (After Encourage Public Transport Used)	LOS (After Encourage Public Transport Used)	Description
1	Jl. AH Nasution	0.97	E	0.75	C	VCR reduced, LOS increased
2	Jl. Ahmad Yani	0.76	C	0.56	C	VCR reduced
3	Jl. Raya Kopo	0.98	E	0.74	C	VCR reduced, LOS increased
4	Jl. Wastukencana	0.51	C	0.40	B	VCR reduced, LOS increased

3.2 Cilegon City
Similar with Bandung City. The density of Cilegon City is also quite crowded. It would be very difficult to build new roads. Transport demand management would be more sensible.

3.2.1 Maximum Floor Area Development Allowed
Four roads are analysed: Jl. Tirtayasa, Jl. Lingkar Selatan, Jl. Bojonegara, and Jl. K.H Wasyid. The following table shows existing condition of roads in peak hour in Cilegon City. There are two ways of TDM to be analysed: controlling maximum floor area of development and developing public transport.

Table 10. Existing Condition of Road in Bandung City

No	Road Name	Type of Road	C	V (pcu/hour)	VCR	LOS
1	Jl. Tirtayasa	Primary arterial road	5702.4	3699.78	0.65	C
2	Jl. Lingkar Selatan	Secondary arterial road	5702.4	1022.88	0.18	A
3	Jl. Bojonegara	Primary collector road	2453.4	1071.23	0.44	B
4	Jl. K.H. Wasyid	Secondary collector road	2134.5	1393.40	0.65	C

If value of level of service (LOS) is compared with VCR plan, there are no roads that do not meet the requirement of VCR plan. The following table shows a comparison between LOS on existing condition and VCR plan. VCR plan which used in this reasearch is 0.75 which is the limit of degree of saturation (MKJI, 1997)[8]

Table 11. Comparison of LOS Existing and Los Plan in Cilegon City

No	Road Name	VCR Existing	LOS Existing	VCR Plan	LOS Plan	Land Use Development Permit
1	Jl. Tirtayasa	0.65	C	0.75	C	Allowed, but Limited
2	Jl. Lingkar Selatan	0.18	A	0.75	C	Allowed
3	Jl. Bojonegara	0.44	B	0.75	C	Allowed
4	Jl. K.H. Wasyid	0.65	C	0.75	C	Allowed, but Limited

Based on the analyses results, all roads that still meet the requirement of VCR plan. However, the land use development control is needed to manage level of service in Cilegon City in good performance. Giving land use development permit is allowed. However, the limit of land use development is still needed to control trip generation that is provided by land use. So, land use develpment control will be exercised. Firstly, Volume Traffic Max of the four roads are calculated.

Table 12. V max of Roads in Cilegon City

No	Road Name	VCR Plan	C Existing	V max
1	Jl. Tirtayasa	0.75	5702.4	4276.8
2	Jl. Lingkar Selatan	0.75	5702.4	4276.8
3	Jl. Bojonegara	0.75	2453.4	1840.05
4	Jl. K.H. Wasyid	0.75	2134.5	1600.875

Once Vmax is determined, the maximum land use area can be calculate. The following table shows maximum floor area that can be developed in Cilegon city based on Vmax for each Road. Land Use proportion can be determined by the government plan. In this exercise, proportion of land use is based on land use area existing next to the road.

Table 13. Maximum Land Use Area

No	Land Use	Vmax	Land Use Proportion	Trip Generate Max	Trip Rate	Maximum Land Use Area (m^2)
			Tirtayasa Steet			
1	Commercial	4276.8	46.7%	1,997.68	0.012*	166,473.24
2	Public Facilities	4276.8	50.3%	2,153.27	0.0193**	111,568.37
3	Green area	4276.8	0.0%	-	0*	0
4	Office	4276.8	2.7%	115.51	0.01*	11,550.76
5	Residential	4276.8	0.2%	10.34	0.0037*	2,795.67
			Lingkar Selatan Street			
1	Commercial	4276.8	14.8%	634.41	0.012*	52,867.52
2	Industry	4276.8	69.8%	2,986.10	0.118***	25,305.96
3	Residential	4276.8	15.3%	656.29	0.0037*	177,374.71
			Bojonegara Street			
1	Commercial	1840.05	78.7%	1,447.74	0.012*	120,644.87
2	Public Facilities	1840.05	0.6%	10.90	0.0193**	564.64
3	Residential	1840.05	20.7%	381.41	0.0037*	103,084.86
			K.H. Wasyid Street			
1	Commercial	1600.875	100%	1600.87.6	0.012*	133,406.25

Description :
[a] Trip Rate CBD DKI Jakarta,1990
[b]Trip Rato Country Sandiego, 2003
[c]Trip Rate Kota Cirebon and Semarang, 2000

3.2.1. Traffic Volume Reduction by Private to Public Transport Shift

The calculation and scenario are similar to Bandung City case. The following is the vehicle composition in four roads in Cilegon City.

Table 14. Traffic Volume off Road Existing According to Vehicle Composisition

No	Road Name	Traffic Volume (pcu/hour)				Total (pcu/hour)
		Motorcycles	Light Vehicles - Private Car	Light Vehicles - Public	Heavy Vehicles	
1	Jl. Tirtayasa	1155.77	1725.3	717.86	100.84	3699.78
2	Jl. Lingkar Selatan	176.43	411.07	4.05	431.32	1022.87
3	Jl. Bojonegara	620.86	341.21	101.25	7.89	1071.22
4	Jl. K.H. Wasyid	714.42	442.46	184.275	52.24	1393.40

The second step is to calculate the number of passengers shift to public transport by multiplying the number of private vehicles with load factor (LF) of each private vehicle. The LF is set 2.5 (for private car) and 1 (for motorcycles). The probability of public transport choice is set at 40% (Al Faruq, 2015)[7]. And the number of passengers who use public transport is divided with load factor of public transport to get public transport needs. The public transport (*angkot*) load factor is 12. The following table shows the prediction of public transport needs in four roads.

Table 15. Public Transport Needs

No	Road Name	Existing Private Vehicle (Vehicles/hour)		Passengers of Existing Private Vehicle (Passenger/hour)		Public Transport Shift (passenger/hour)			Public Transport Needs (vehicles/hour) Load factor = 12		
		Motorcycles	Private Car	Motorcycles (Load Factor =1)	Private Car (Load Factor= 2.5)	From Motorcycles	From Private Car	Total	From Motorcycles	From Private Car	Total
1	Jl. Tirtayasa	4623	1725	4623	4312.5	1849.2	1725	3574.2	154.1	143.75	297.85
2	Jl. Lingkar Selatan	706	411	706	1027.5	282.4	411	693.4	23.53	34.25	57.783
3	Jl. Bojonegara	2483	341	2483	852.5	993.2	341	1334.2	82.76	28.42	111.18
4	Jl. K.H. Wasyid	2858	442	2858	1105	1143.2	442	1585.2	95.26	36.83	132.1

Once the number of passengers that shift to public transport is known, the remaining passengers who still use private vehicle is also known.

Table 16. The Remaining Private Vehicles

No	Road Name	The remaining passengers who are still in private vehicle (passengers/hour)		The Remaining of Private Vehicle (Vehicles/hour)		The Remaining Traffic Volume of private vehicles (pcu/hour)	
		Motorcycles	Private Car	Motorcycles	Private Car	Motorcycles	Private Car
1	Jl. Tirtayasa	2773.8	2587.5	2773.8	1035	693.45	1035
2	Jl. Lingkar Selatan	423.6	616.5	423.6	246.6	105.9	246.6
3	Jl. Bojonegara	1489.8	511.5	1489.8	204.6	372.45	204.6
4	Jl. K.H. Wasyid	1714.8	663	1714.8	265.2	428.7	265.2

Lastly, the number of traffic volume changing can be known. The following table shows the calculation of traffic volume when the private vehicle users shift to public transport.

Table 17. The Changes of Traffic Volume According to Vehicle Composition

No	Road Name	Traffic Volume (pcu/hour)				Total (pcu/hour)
		Motorcycles	Light Vehicles - Private Car	Light Vehicles - Public	Heavy Vehicles	
1	Jl. Tirtayasa	693.45	1035	1015.71	100.845	2845.01
2	Jl. Lingkar Selatan	105.9	246.6	61.83	431.325	845.66
3	Jl. Bojonegara	372.45	204.6	212.43	7.8975	797.38
4	Jl. K.H. Wasyid	428.7	265.2	316.38	52.245	1062.52

The results shows that the for roads LOS are getting better. In Tirtayasa street, the VCR changed from 0.65 to 0.50. In Lingkar Selatan Street, VCR changed from 0.18 to 0.15. In Bojonegara Street, VCR changed from 0.44 to 0.33. And in K.H. Wasyid street, VCR changed from 0.65 to 0.5. The following tables shows the comparison between road existing condition with "after public transport"condition.

Table 18. The comparison of road existing condition and road condition after using public transport

No	Road Name	VCR Existing	LOS Existing	VCR (After using PT)	LOS (After using PT)	Description
1	Jl. Tirtayasa	0.65	C	0.50	C	VCR Reduced
2	Jl. Lingkar Selatan	0.18	A	0.15	A	VCR Reduced
3	Jl. Bojonegara	0.44	B	0.33	B	VCR Reduced
4	Jl. K.H. Wasyid	0.65	C	0.50	C	VCR Reduced

4. Conclusions

Urban land use development control based on transport carrying capacity is very crucial to assure that urban physical development would not imply a severe transportation problem. While the concept of incorporating transport carrying capacity into urban land use development control has been previously realised, we are still facing lack of clear method to implement the concept in todays urban land use development control. This paper, among others, develop the calculation method of this concept for Indonesia city application.

Whenever road development reaches its maximum level, there is a need to control development both through supply as well as demand management. In this case, transport demand management is needed to maintain the level of service (LOS) on the road. Transport demand management approaches in this paper is land use development control. Land use development control is performed by controlling maximum land use that is allowed in each of every road classifications. Land use area will affect how much trip generation produced. Therefore, land use area should be controlled in the context of maximum capacity of the road. Meanwhile, from supply management side, this paper examines public transport development to reduce total traffic volume.

In Bandung City case, LOS in Jl. A.H. Nasution and Jl. Raya Kopo have exceeded the LOS plan. While LOS in other roads such as Jl. Ahmad Yani and Jl. Wastukencana still in accordance with LOS plan. On the road network which has exceeded LOS plan, there should be an effort to create an alternative transport which can reduce the use of private vehicles. For the road network has not exceeded the limit level of service then the government can make a restriction of land use area that can be developed. The government can adopt the model that is performed in this paper for making development permit.

In Cilegon City, the roads that are analysed have not exceeded the limit of LOS plan. However, the effort to control urban development should be done with control land use area. This way can control trip generation that is produced by land use development in Cilegon City. While the urban development is controlled, other ways should be done to manage road performance in Cilegon City by promoting a better public transport services. Public transport improvement is one of the promising strategy to maintain a good LOS, while allowing further land use development up until a certain level of development.

References

[1] Morlok, E.K. Pengantar Teknik dan Perencanaan Transportasi, Translated by: Johan K. Hainin, Jakarta: Erlangga; 1985.

[2] Warpani, Suwaedjoko P. Pengelolaan Lalu Lintas dan Angkutan Jalan. Bandung : Institut Teknologi Bandung (ITB); 2002.

[3] Massachusetts Highway Department, Chapter 16: Traffic Calming and Traffic Managemen. [online] Available at <http://www.massdot.state.ma.us/portals/8/docs/designguide/ch_16.pdf > [Acessed on 2 October 2016]

[4] Setiwan, Rudi. Simulasi Manajemen Lalu Lintas Untuk Mengurangi Kemacetan di Jalan Jemursari dan Raya Kendangsari. Surabaya. Universitas Kristen Petra; 2009. [pdf] Available at <http://repository.petra.ac.id/16925/1/Publikasi1_01065_1874 .pdf> [Acessed On 2 October 2016]

[5] Putranto LS. Rekayasa Lalu Lintas. Jakarta: Indeks;2008.

[6] Ministry of Public Work. Regulation No 05/PRT/M 2008 about guidance for the provision and utilization of green open space in Urban Area. Jakarta; 2008. [pdf] Available at <http://www.penataanruang.com/pedoman-ruang-terbuka-hijau.html> [Acessed On 5 October 2016]

[7] Al Faruq, Umar. 2015. Peluang Pergeseran Pemilihan Moda Pengguna Kendaraan Pribadi ke Bus Rapid Transit di Koridor 2 Trans Metro Bandung. Tugas Akhir. Bandung: Program Studi Perencanaan Wilayah dan Kota ITB.

[8] Manual Kapasititas Jalan Indonesia (MKJI). 1997. Departemen Pekerjaan Umum, Direktorat Jenderal Bina Marga. Direktorat Bina Marga Jalan Kota. Jakarta

Mangrove canopy density analysis using Sentinel-2A imagery satellite data

M N Wachid, R P Hapsara, R D Cahyo, G N Wahyu, A M Syarif, D A Umarhadi, A N Fitriani, D P Ramadhanningrum, and W Widyatmanti

Geographic Information Science Department, Universitas Gadjah Mada

muchsinnw@gmail.com

Abstract. Teluk Jor has alluvium surface sediment that came from volcanic materials. Sea wave that relatively calm and the closed beach shape support the existence of mangrove forest at Teluk Jor. Sentinel-2A imagery has a good spatial and spectral resolution for mangrove density study. The regression between samples and the NDVI values of Sentinel-2A used to analyze the mangrove canopy density. Mangrove canopy density was identified using field survey with transect method. The regression analysis shows field data and NDVI value has correlation R=0.7739 and coefficient of determination R2=0.5989. The result of the analysis shows area of low density 397,900 m2, moderate density 336,200 m2, the high density has 110,300 m2 and very high density has 500 m2. This research also found that mangrove genus in Teluk Jor consists of Rhizopora, Ceriops, Aegiceras and Sonneratia.

1. Introduction

The balance of coastal ecosystems influenced by the organisms that grow and develop around the coast such as mangrove, estuary, meadow/seagrass, coral reefs, small islands, and the deep sea. The length of Indonesia's shoreline with its physical condition can support mangroves to grow, thus Indonesia has 3.5 million ha of mangrove area [1], had been the largest in the world (18-23 %), after Brazil (1.3 million ha), Nigeria (1.1 million ha), and Australia (0.97 million ha [1].

Mangrove is a coastal vegetation that grow in tidal areas with muddy shore. The mangrove forest is one of the tropical and subtropical forest that grow along the coast or a river that is affected by the tide. The mangrove forest is able to provide considerable benefits for the balance of nature, especially the coastal areas and coastal communities. Mangrove forest is a transition ecosystem between land and sea with the complex interactions between the physical and biological properties. Mangroves are found along the coastal area and can grow in tidal area according to their tolerance to salinities, flooded area, substrates, and coastal morphology [2].

Mangrove forest has some advantages and functions for human life. Mangrove could be used for fuel, building materials, fisheries, textiles, agricultures, papers, home equipment, foods and beverages [1]. Mangrove has an important role in coastal protection of currents, winds, and storms. For fishery, mangrove keeps the life cycle of fishes, shrimps, and molluscs, keeps the fertility of coastal area, and also become organic matter supplier [1]. Forest canopy density is the most useful considered parameter for implementation in planning and environmental rehabilitation [3]. One of the methods to measure canopy density is from canopy cover. Canopy cover is the ground area that covered by vertical projection of vegetation canopy or tree crown. [4]. Measuring canopy cover use remote sensing data nowadays became more often because the remote sensing data has ability to observe the earth surface.

Remote sensing is the science and art of obtaining information about the object, area, or phenomenon by analyzing data obtained using instruments without direct contact with the objects [5]. Sentinel-2A imagery has a high spatial resolution and can be downloaded for free. Sentinel-2A is a program of the Global Monitoring and Environmental Security (GMES) and the initiation of the European Commission (EC) and European Space Agency (ESA). This satellite was launched in June 2016. Studies that use Sentinel-2A imagery is still very rare, especially for mangrove forests study. Sentinel-2A imagery has a great potential in mangrove research because it has a high spatial resolution about 10 meters for visible and near-infrared bands.

Teluk Jor (located in Lombok Timur Regency) has a physical condition that allows mangroves to grow. Teluk Jor has alluvium surface sediment that came from volcanic materials. Sea wave that relatively calm and the closed beach shape support the existence of mangrove forest. Vegetation structure is related to ecological conditions, thus it is important to know the environment conditions. This study is to find out the application of Sentinel-2A in the study of mangrove canopy density at Teluk Jor.

2. Methodology

2.1 Data used and Pre Processing

Mangrove canopy density analysis is carried out with remote sensing data and field survey. Sentinel-2A was launched in June 23rd 2015 and was prepared for providing earth observation for environmental, agriculture and security monitoring. Sentinel-2A satellite data that recorded on 16 January 2016 is used in this study, has various spatial and spectral resolution with 13 bands consists of 4 bands with 10 meters spatial resolution, 6 bands with 20 meters, and 3 bands with 60 meters. This research uses visible and near infrared data (band 2, 3, 4 and 8) for data modelling.

Sentinel-2A data is pre-processed with radiometric, atmospheric correction and finished with image transformation before can be applied for canopy density modelling. Radiometric correction retrieved by calibrating Digital Number (DN) divided by default quantification value that provided in metadata. Atmospheric correction uses Dark of Pixel (DOP) method to decrease the atmospheric effect on reflectance value. Image transformation method has done with Normalized Difference Vegetation Index (NDVI) to shows the information about vegetation density.

2.2 Field Survey Method

Measurement of mangrove canopy density uses two methods, hemispherical photography and subjective density method. Hemispherical photography method usually used for canopy identification with upward photo taking or downward using wide-angle camera [6]. Hemispherical photography method has more advantages compared to other indirect methods, such as faster measurement, cheaper and permanent result. This method gives information about gap fraction distribution that can be used for forest canopy properties, for examples leaf area index, leaf angle distribution, and canopy openness [7]. Subjective density method is used for sample comparison. This method considering field canopy density and measure it based on density card. The sample data are obtained using transect method that directed from the land to the sea.

Technically, canopy cover data obtained by taking photo upward with constant height, 1 meter in every sample location. Photo taken 5 times; 1 photo on the center of plot and others on the surrounding. According to Sentinel's pixel size (10 m) was build sample area 15 x 15 m to anticipate the effects of surrounding object and geometric error. Can Eye software is used to obtain canopy cover values. The process is usually use hemisphere camera but in this study we use pocket camera. We also analyze the horizontal structure of mangrove by observing dominant and non-dominant mangrove genus in every sample location. The leaf, flower, stream, and root of mangroves tree that found in sample location are identified and matched with mangrove genus characteristic table.

2.3 Mangrove Canopy Density Mapping

Mangrove canopy density map is established from two parameters, NDVI values of Sentinel-2A and field survey data of mangrove canopy density. The spectral transformation NDVI image gives optimal result for identify the variation of vegetation in the relation to density [8]. Statistical relationship like regression is the simplest and the most common approach for modelling ecological parameter [9]. Linear regression analysis is used for modelling mangrove forest canopy density at Teluk Jor. The independent variable is field data, and the dependent variable is NDVI values. The linear regression analysis give the equation formula to extrapolating the canopy density values in the defined area. The strong correlation between NDVI and field survey data indicates the model is good enough for estimating the mangrove canopy density at Teluk Jor. Correlation value explains the relationship; strength, direction, and the possibilities to analyze in regression. Whether or not model depends on coefficient of determination value from the regression.

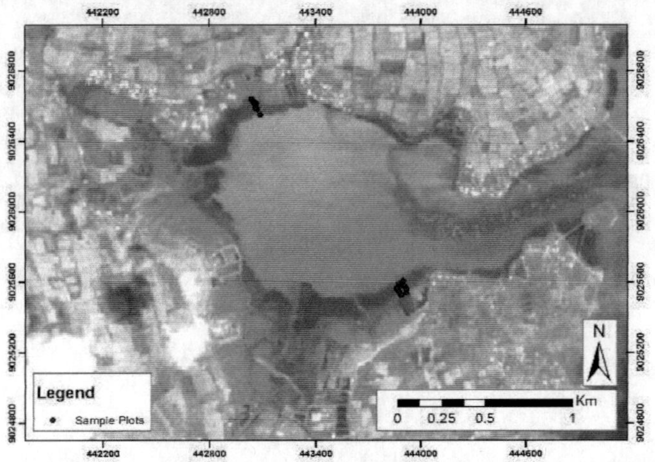

Figure 1. Mangrove Area in Teluk Jor from Sentinel-2A Satellite Data

3. Results and Discussion

3.1 NDVI Result

Canopy density model is established from the regression equation between field data and NDVI value. The results of image processing Sentinel-2A indicates that NDVI imagery can clearly appearance of water, soil and vegetation in the study area. Fig.1. shows the appearance of water or rivers are shown in a low NDVI values (red color), and the appearance of vegetation are shown with high NDVI value (green color). NDVI Transformation has an ability to distinguish objects reliable in identifying vegetation density, assuming higher canopy density is also has higher NDVI value.

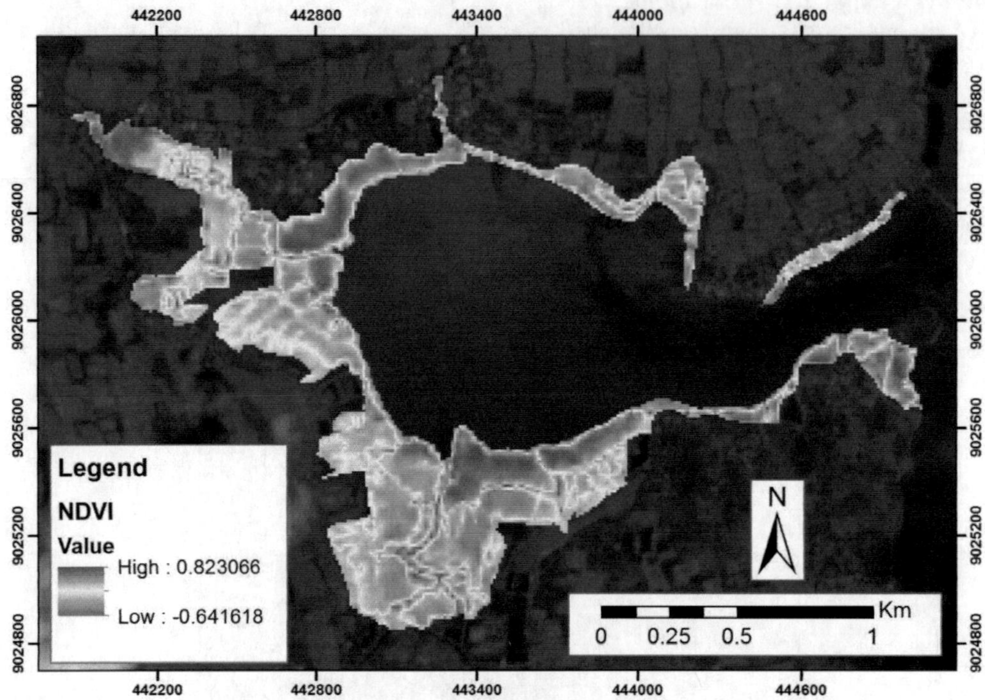

Figure. 2. *NDVI value at Teluk Jor*

Table 1. Field data and canopy cover model

Sample	Coordinate		Genus		Density (%)		
	X	Y	Genus Dominant	Other Genus	Subjective Density	Hemispherical Photography	Canopy Cover Model
1	443036	9026638	Sonneratia	-	20	20.93	31.04
2	443048	9026622	Sonneratia	Rhizophora	40	68.83	57.15
3	443054	9026620	Ceriops	Sonneratia	20	45.59	50.66
4	443061	9026611	Sonneratia	Rhizophora	40	61.18	54.05
5	443060	9026596	Sonneratia	Rhizophora	25	59.62	60.04
6	443069	9026581	Sonneratia	Rhizophora	30	58.78	56.24
7	443087	9026550	Sonneratia	Rhizophora	25	42.67	26.30
8	443061	9026586	Sonneratia	-	80	62.42	56.09
9	443915	9025540	Ceriops	-	25	10.96	16.94
10	443913	9025562	Aegiceras	-	70	57.56	40.25
11	443900	9025579	Rhizophora	-	50	44.41	49.85
12	443898	9025608	Rhizopora	Sonneratia	75	34.50	26.61
13	443872	9025590	Sonneratia	Rhizophora	50	59.23	50.82
14	443869	9025572	Sonneratia	Rhizophora	50	36.15	52.08
15	443859	9025562	Rhizopora	-	65	43.99	52.42
16	443873	9025542	Sonneratia	Rhizophora	20	28.95	38.29
17	443886	9025525	Rhizopora	-	10	8.06	14.32

IOP Conf. Series: Earth and Environmental Science **70** (2017) 012020 doi:10.1088/1755-1315/70/1/012020

3.2 Data Collection

The data was collected using transect method and produced 3 transects by 17 the number of sample plots. Transect method has chosen because of the difficulty of penetrating dense mangrove fields and muddy. Data collection could only be done on 14:00 until 17:00 because at that time, the tide was low thus the mangrove forests can be passed.

Based on field survey, four mangrove genus are identified in Teluk Jor. They are *Rhizophora*, *Sonneratia*, *Ceriops* and *Aegiceras*. The most genus that can be found is *Sonneratia*. But most commonly found near the sea is *Rhizophora*. *Ceriops* and *Aegiceras* can be found in the middle of the mangrove forest, but they are not the dominant genus in Teluk Jor. Mangrove structure in Teluk Jor has irregular pattern. According to field survey, the structure of the mangrove genus does not have a distinct arrangement. For example in mangrove area that dominated by *Sonneratia*, we can also found *Ceriops* genus that grow between them.

Density value is extracted using two methods: subjective methods and hemispherical photography method. Values of subjective method shows the density ranges from 10% to 80% while the hemispherical photography method is processed using Can Eye software, and produces density ranges from 8.06 % to 68.83 %.

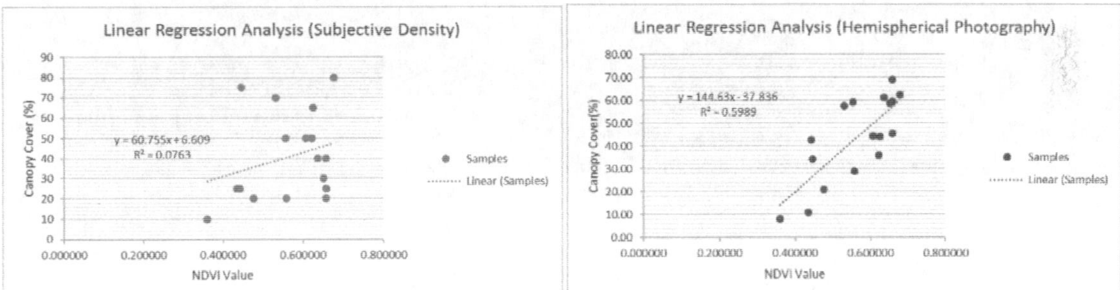

Figure 3. Regression graphic

3.3 Regression Analysis

Vegetation density sample using subjective and hemispherical photography is analyzed using linear regression analysis to show the correlation with NDVI value of Sentinel-2A data. The best correlation between two methods will be used for modelling the mangrove canopy density in Teluk Jor. Based on regression analysis, the subjective density has very low correlation with NDVI value with R=0.2762 and determination coefficient R^2=0.0763. This result shows that the subjective density is not good enough to be modelled with NDVI value. In other hand, hemispherical photography has much better correlation with NDVI value. The result shows canopy density measured with hemispherical photography method has a correlation R=0.7739 and coefficient determination R^2=0.5989. The good correlation between canopy density and NDVI value shows the regression formula is good enough for modelling mangrove canopy density.

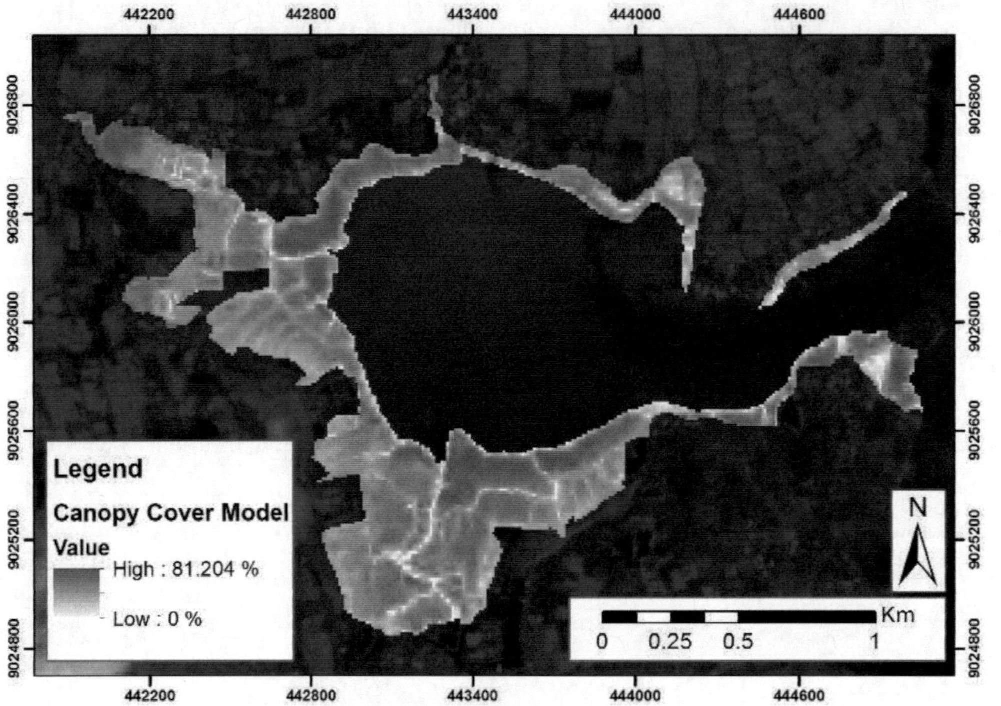

Figure. 4. Canopy cover model

3.4 Mangrove Canopy Density Analysis

Mangrove canopy cover that produced by the mangrove density model has a minimum value of 0 % and a maximum value of 81.204 %. The result of the density map can be seen in Figure 4. The map shows the percent of mangrove canopy cover in result of regression equation. The 0 % value indicates the cover is not vegetation, it could be either water or bare land.

The canopy cover or density of mangrove is classified into 4 classes to simplify the analysis. The result explains mangrove vegetation with low density has an area 97,900 m^2, medium density has 336,200 m^2, the high density 110,300 m^2 and very high density 500 m^2. Qualitatively, it explains there are many high density mangrove near sea, although there is also few area of low density between the high density of mangroves and sea. There are many mangrove medium density in the middle of the forest. Low density mangrove widely located in the edge of medium density mangrove and has the highest number of extents compare to the other classes. Very high densities mangrove are only found at some point between the high density and has the smallest area with only 500 m^2.

This irregularly pattern of mangrove cover density depends on many factors. Some Teluk Jor mangrove sites especially on the near land is the former pond. It can be identified from the pattern looking at Sentinel-2A imagery is made some square shapes. On the former pond, the existence of mangrove is the result of planting programs. Former pond and mangrove forest is divided by mound, therefore restrict the tide.

Mangrove forest at Teluk Jor is directly adjacent with settlement and farming land. The dense settlement gives household waste supply that indirectly affects mangrove growth. Mangrove zonation could not clearly identified in the location because there is no large estuary. The largeness of estuary, and also the aridity climate at the location affects the amount of sediments that become medium of mangrove growth. Meanwhile the medium tidal amplitude also does not really flat coastal morphology at some parts affects the narrowness mangrove at Teluk Jor.

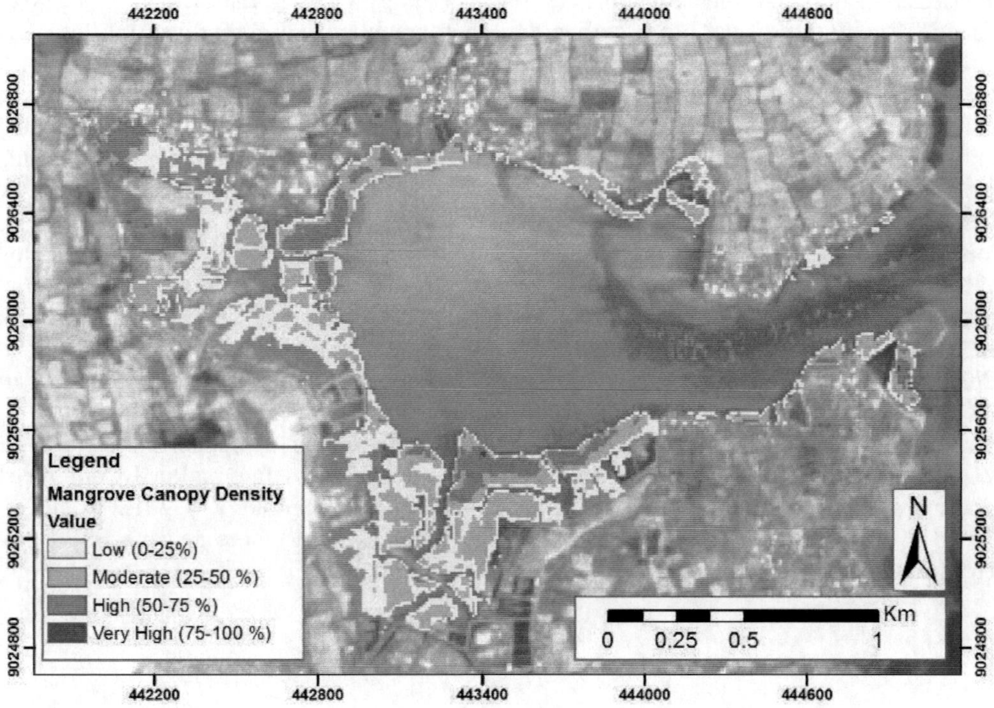

Figure. 5. Mangrove Canopy Density Model simplified in four classes

4. Conclusions

Remote sensing data give a lot of solution to model the environmental condition. Mangrove has a crucial part on environment but has fragile characteristic. Remote sensing has an ability to observe the earth condition, including mangrove forest. This paper shows Sentinel-2A satellite data has an ability to model the mangrove canopy density. The regression analysis between field canopy cover data using hemispherical photography method has a good correlation with NDVI value that computed from Sentinel-2A data. The regression analysis shows field data and NDVI value has correlation R=0.7739, and coefficient determination R^2=0.5989.

The mangrove canopy density model shows Teluk Jor mangrove forest has varies canopy density. The density exists from 0 % to 81.204 %. The largest area is low canopy density, with covering 397,900 m^2 of the mangrove forest, while the smallest area is high mangrove density class with only 500 m^2. The moderate canopy density is covering 336,200 m^2 and the high density covering 110,300 m^2. This research also found the mangrove genus at Teluk Jor consists of *Rhizopora*, *Ceriops*, *Aegiceras* and *Sonneratia*..

Acknowledgement
This Research is the part of Field Work Course III, Cartography and Remote Sensing Study Program, Department of Geographic Information Science, Universitas Gadjah Mada, held in Lombok Timur, Nusa Tenggara Barat, 2016.

References

[1] Noor, Y. R., Khazali, M., and Suryadiputra, I. N. Panduan Pengenalan Mangrove di Indonesia. Bogor: PKA/WI-IP; 2012.
[2] Dahuri, R., Rais, Y., Putra, S. G., and Sitepu, M. J. Pengelolaan Sumber Daya Wilayah Pesisir dan Lautan Secara Terpadu. Bogor: Pradnya Paramita; 2001.

[3] Azizia, Z., Najafia, A., and Sohrabia, H. Forest Canopy Density Estimating, Using Satellite Images. The International Archives of the Photogrammetry, Remote Sensing and Spatial Information Sciences 2008; Part B8, 37 : 1127-1130.

[4] Jennings, S.B., Brown, N.D., and Sheil, D. Assessing forest canopies and understory illumination: canopy closure, canopy cover and other measures. Forestry 1999; 72(1): 59–74.

[5] Lillesand, T. M., and Kiefer, R. W. Penginderaan Jauh dan Interpretasi Citra, translated by Dulbahri, Prapto Suharsono, Hartono, Suharyadi. Yogyakarta: Gadjah Mada University Press; 1990.

[6] Jonckheere, I., Fleck, S., Nackaerts, K., Muys, B., Coppin, P., Weiss, M,. & Baret, F. Review of Methods for In Situ Leaf Area Index Determination: Part I. Theories, Sensors and Hemispherical Photography. Agricultural and Forest Meteorology 2004; 121 (1-2): 19-35. - doi: 10.1016/j.agrformet.2003. 08.027

[7] Chianucci, F and Cutini, A. Digital Hemispherical Photography for Estimating Forest Canopy Properties: Current Controversies and Opportunities. iForest 5 2012; 290-295.

[8] Bhandari, A. K., Kumar, A., and Singh, G. K. Feature Extraction using Normalized Difference Vegetation Index (NDVI): a Case Study of Jabalpur City. 2nd International Conference on Communication, Computing & Security [ICCCS] Procedia Technology, 6, 2012; 612-621.

[9] Green, E. P., Mumby, P. J., Edwards, A. J., and Clark, C. D. Remote Sensing Handbook for Tropical Coastal Management. UNESCO Publishing.Paris; 2000.

IOP Conf. Series: Earth and Environmental Science **70** (2017) 012021 doi:10.1088/1755-1315/70/1/012021

The integration level of public transportation in Makassar City

A M R Kasim[1 2], A D Wicaksono[1] and E B Kurniawan[1]

[1] Department of Urban and Regional Planning, Faculty of Engineering, Universitas Brawijaya, Malang, Indonesia

ridha.kasim@gmail.com

Abstract. Multimodal transportation is transportation modes that can combine one with the other modes properly, efficiently and effectively so that people can move from one type of transportation to another modes with fast, cheap and convenient. The integration of transport services and infrastructure network is not fully realized, among others, can be seen from the public transport service between modes, one with the other modes, which makes people have to travel by other modes and is not served by public transport canal. Furthermore, intramoda displacement can not be done easily and quickly, some things that makes people tend to prefer private rather than public transport vehicles in doing movement. The main objective of this study was to determine the level of integration of modes of land transportation in the city of Makassar. By using analysis of physical alignment, non-physical, and analysis of travel time, which is then summarized into a canvas ratings to get an idea of the level of integration in general. The results showed that the level of integration of modes consist of two criteria with a very low value, 4 criteria with low and medium value, two criteria with a high value, as well as one criterion with very high value. Meanwhile, the variable of integration that influence people's preferences in order from the highest value is the number of the fleet, the availability of routes, the number of passengers compared with the load factor, the location where the stopping points up and down, and the last is the location of the terminal with the place up and down the number of the fleet, availability of service, number of passengers, the location of the terminal to place up and down the passenger, as well as the location where the up and down passengers on the origin and destination. Some of the variables that have a low and very low value, is a variable that should receive greater attention from the municipality, so the concept of the integration of transport modes could be reached, and indirectly will make the public are encouraged to use public transport.

1. Introduction

The problems that were encountered in major cities of the world, including the city of Makassar is the problem of transport. In general, problems arise when the mode is selected communities are private vehicles. The problem is increasing, while the growth of private vehicles is not balanced with increasing length of the road. Policy maker, began switching to suppress the use of private vehicles and improving public transport facilities and infrastructure. Way through the intermodal transport systems and integrated public transport system (multimodal). That is, various modes can were combined properly, efficiently and effectively so that people can move from one type of transportation to another carrier with fast, cheap and convenient.

The integration of transport services and infrastructure network is not fully realized, among others, can be seen from the public transport service between modes, one with the other modes, which causes people to have to travel by other modes and can not be served by transit the canal. In addition, the displacement intramoda can not be done easily and quickly, Likewise, the integration of network infrastructure, such as terminals, ports, airports and setting a schedule that is not satisfactory.

The integration of transport modes can be viewed from two aspects, is the integration of physical and non-physical alignment. The integration of this physical form of the integration of facilities and infrastructure, while the integration of non-physical form of the integration time, information systems, services, and ticketing. (Mulyono, Agus T. 2012). Acceleration handling in Makassar transport problems by providing for public mass transit. Integrated public transport can be a solution, (Irwan, Forum Diskusi Publik Sektor Transportasi "Percepatan Pembangunan Transpotasi Angkutan Umum di Makassar).

Network of intermodal transportation services / multimodal realized through the integration of each route / traffic / transport service road, rail, river and lake crossings, sea and air, with attention to excellence and the concept of combining modes between the main modes, feeder mode and advanced mode. And based on the suitability of the technology and the characteristics of the service area, and cross-level transport which consist Tatranas, Tatrawil and Tatralok.

Some of the problems regarding the use of public transport as well as the integration of the network is:

- The people of Makassar preferred mode of private vehicles instead of public transport in conducting the movement (RTRW Kota Makassar 2005-2025)
- System Development of intermodal transport in Makassar yet well integrated
- The number of public transport passengers who are not proportional to the number and type of public transport available in the city of Makassar

2. Methods

This research uses descriptive analysis method using several analytical tools. The level of integration is obtained based on an assessment using assessment canvas, as the overall conclusions. The assessment criteria used is to use the physical integration of variables such as facilities and infrastructure, as well as the integration of non-physical form of the travel time. The analysis used to determine the level of integration among others are the main modes and modal analysis feeders, transport performance analysis (headway and loadfactor), connectivity index, the scale stops and terminal services, and the travel time index

This study uses a mode of transport as the unit of analysis, which in Makassar City alone there are miraculous mode of transportation. The modes include public transport, namely city transport and BRT, as well as other transport tricycles, taxi, motorcycle and pedestrian paths.

3. Results

The length of roads in the city of Makassar in 2009 throughout 1593.46 kilometers; Compared to 2008, road length has not changed. In 2009, for better road conditions increased 32.82% compared to 2009. Severely damaged down 45.74% from 2008. Public transport is located in Makassar include public transit (usually called pete-pete), Bus Rapid transit (BRT) Trans Mamminasata, as well as motorcycles, tricycles and motor tricycles. City Transport contained in Makassar consists of 15 stretch in the outskirts of town and one stretch. As for the Bus Rapid Transit (BRT) Transmamminasata is a metropolitan area buses that serve Mamminasata. Mamminasata an abbreviation of Makassar, Maros, Sungguminasa, and Takalar. Maros, Sungguminasa, and Takalr a district located around the city of Makassar.

In masterplan, this BRT corridor will have 11 buses with a capacity of 33 people. BRT uses a mix lines, yaiut mixed with other vehicles, but for setting up and down passengers were required off at the stop that has been provided. For existing conditions, BRT Transmamminasata which has been operating as many as 1 corridor, namely corridor II (GTC Mall - Mall Panakkukang) 11 units bis.

Makassar City has two terminals, namely the Regional Power Terminal northern part of the city, as well as Terminal Mallengkeri in the south. Another terminal is located in the Makassar Mall, but the terminal is just a place up and down the ordinary passengers, and no buildings. Regional Power Station is the terminal with type A, as well as terminal mallengkeri who have type B. Regional Power Terminal 3 is traversed by city transportation route (route D, G, W), while for Terminal Mallengkeri traversed by four designated route (route B, B1, F, and F1). Also there are 27 stops, spread over 11 streets..

3.1 The physical integration (infrastructure)

Physical The primary survey results show that the type of Moda contained in Makassar, consists of a Bus Rapid Transit (BRT), Public Transportation, tricycles and motor tricycles, motorcycles, as well as pedestrian paths. The fifth mode will be classified according to the type of use, route determination, the determination of the schedule, determination of tariffs, the area of operation, density service areas, service configuration, the right of priority roads, as well as the travel route. Based on this analysis, it can be concluded that the mode of the main mode is the mode BRT and urban transport, while the rest tricycles / motor tricycles, motorcycles, as well as pedestrian / sidewalk a feeder mode. Analysis of the main modes and feeder modes are described in Table 2.

Shown in Table 1 below, note that 12.59% of people using the 3 modes in one trip, namely public transportation, walking, and pedicabs / rickshaws motors. Meanwhile, sisasnya using two modes of public transportation or a mix between BRT and walk, rickshaw / bentor, motorcycles, as well as private vehicles. People who use public transportation with two different routes that are as many as 20 passengers. Based on the concept, the more selection modes and more modes are used, then the condition is already integrated mode. The average usage mode is as much as 2.38 with the use of at least 1 mode and up to 4 modes.

Table 1. Main and Feeder Mode in Makassar

Feeder	Main	
	City Transport	BRT
Walk/Pedestrian	88	0
Pedicab	18	0
Motorcycle Taxi	7	0
Private Vehicle	4	0
Walk/Pedestrian and Pedicab	17	1
Total	134	1

3.2 Index of Connectivity

Value connectivity or public transportation route network complexity and BRT in Makassar City is 1.36 for the beta value, gamaa value of 0.46, and an alpha value of 0.18 with the average value of the index is 0.67. The greater the value of this index, the network will get better service.

Table 2. Road Connectivity Index in Makassar

Region	Node	Internode	Beta	Gamma	Alpha
Biringkanaya	11	12	1,09	0,44	0,12
Bontoala	21	36	1,71	0,63	0,43
Makassar	10	19	1,90	0,79	0,67
Mamajang	9	13	1,44	0,62	0,38
Manggala	13	16	1,23	0,48	0,19
Mariso	7	7	1,00	0,47	0,11
Panakkukang	23	30	1,30	0,48	0,20
Rappocini	13	19	1,46	0,58	0,33
Tallo	16	22	1,38	0,52	0,26
Tamalannrea	11	12	1,09	0,44	0,12

Region	Node	Internode	Beta	Gamma	Alpha
Tamalate	16	20	1,25	0,48	0,19
Ujung Pandang	25	30	1,20	0,43	0,13
Ujung Tanah	4	4	1,00	0,67	0,33
Wajo	25	37	1,48	0,54	0,29
Total	204	277	1,36	0,46	0,18

3.3 Travel Time Index

The average total time spent by passengers using public transport was 41 minutes, compared with a total time of use of private vehicles is 27 minutes, meaning there is a time difference of about 14 minutes between the use of private vehicles and public. But, when are compared only time in public transport and private vehicles, public transport passenger menghabuskan a time of 26 minutes, whereas for private vehicles menghabuskan 21 minutes. This means that the difference between the total time is more influenced as the use of public transport, people will use a lot of time because there is extra time to walk to the shuttles, waiting, and walk to the destination. Table 3 below describes the condition of an average travel time of passengers

Table 3. Comparison of average travel time of passengers

Public Transport (PT)						Private Vehicle			
To PT	Waiting	Transfer	In Vehicle	To Destination	Total	In Vehicle	Parking	To Destination	Total
4.61	3.30	2.31	26.04	4.89	41.16	21.19	3.49	2.99	27.67

3.4 Level of Integration

The level of integration of land transport modes Makassar City based on 13 criteria (Figure 1), there are two categories that have a very low value, is the number of passengers up and down the terminal and passenger waiting time index (PWI). Moreover, there are four criteria that are at a low value, ie the number of modes, the index alpha route, the number of the route which passes through the terminal, and the ratio OVTT / IVTT. For that is in the medium category, there are also four criteria: the number of modes used, the index alpha, gamma index, as well as the number of passengers up and down at the stop. As for the rest, two criteria are at high criteria, namely the ratio of travel time (RWP) and Running Index, as well as one criterion on the value is as high as the number of passengers.

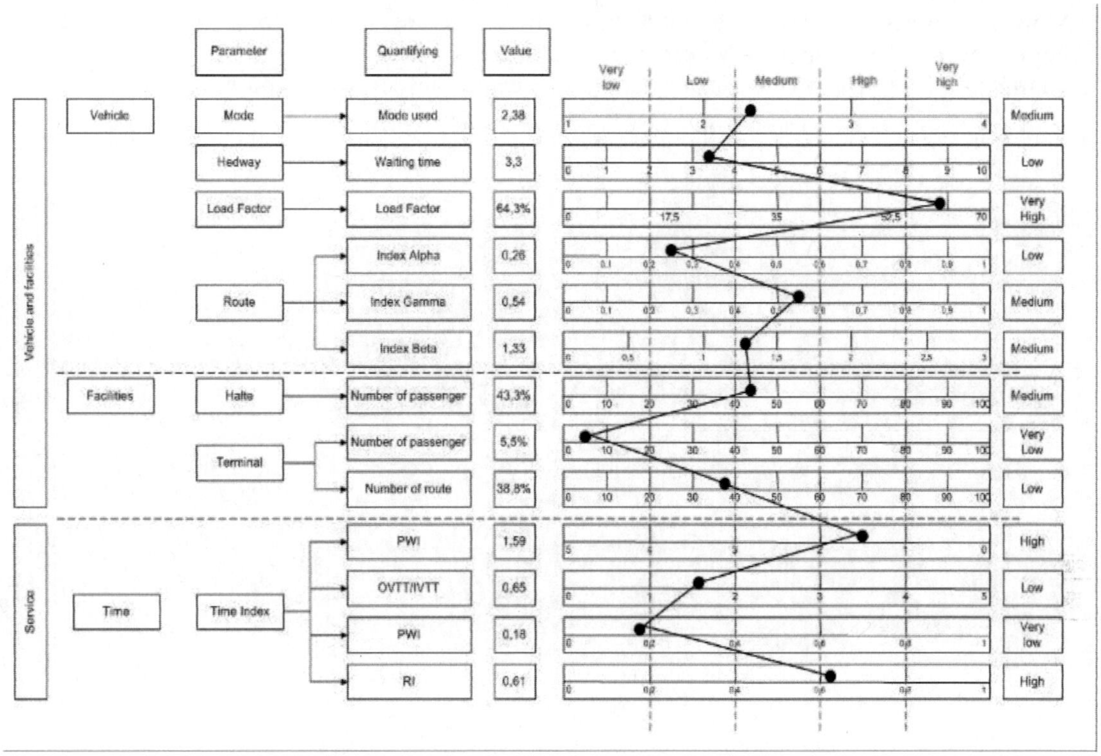

Figure 1. Integration Mode Level in Makassar

4. Conclusions

The level of integration of transport modes Makassar City based on 13 criteria, there are two categories that have a very low value, ie the number of passengers up and down the terminal and passenger waiting time index (PWI). Thus, there are four criteria that are at a low value, ie the number of modes, the index alpha route, the number of the route which passes through the terminal, and the ratio OVTT / IVTT. For that is in the medium category, there are also four criteria: the number of modes used, the index alpha, gamma index, as well as the number of passengers up and down at the stop. As for the rest, the two criteria that are in the high criteria: The ratio of travel time (RWP), and Running Index (RI) and 1 criteria with a very high value on the criterion of the number of passengers. Increased integration of land transport modes can be done by improving variables that have a low coherence value, as well as having a great influence based on public perception.

References

[1] Departemen Perhubungan, Sistem Transportasi Nasional (Sistranas), Jakarta, 2005
[2] Faizal dan Arif. (2009), Estimating Contingency Cost In Construction By Contractors. Department of Civil Engineering, University Technology of PETRONAS. Malaysia
[3] Indarsa, D.P. dan Anak Agung Gde Kartika. Seminar Nasional Aplikasi Teknologi Prasarana Wilayah 2011 ; Studi Penyebab Penurunan Demand Penumpang Angkutan Umum Mikrolet di Kota Surabaya (Studi Kasus Mikrolet Lyn-X) . Surabaya.
[4] Jinca, M Yamin. 2009. Jurnal Transportasi Vol. 9 No. 1 Juni 2009: 1-14. Jakarta
[5] Kanafani,Adib dan Rui Wang. 2010. University of California Transportation Center UCTC-FR-2010-26 : Measuring Multimodal Transport Level of Service. Berkelay..
[6] Kementerian Perhubungan. 2012. Transmedia edisi 8. Jakarta
[7] Miro, Fadel 2002. Buku Perencanaan Transportasi. Jakarta. Graha ilmu.

[8] Mulyono, Agus T. 2012. Seminar Nasional : Tantangan dan Kebutuhan Standarisasi Penyelenggaraan Transportasi Multimoda/ Antarmoda. Padang
[9] Tamin, OZ. 2000. Buku Perencanaan dan Permodelan Transportasi. Bandung.

IOP Conf. Series: Earth and Environmental Science **70** (2017) 012022 doi:10.1088/1755-1315/70/1/012022

Potential to increase active commuting level in university area (Case study: Universitas Gadjah Mada)

M K Devi

Urban and Regional Planning Department, Sekolah Tinggi Teknologi Nasional Yogyakarta, Caturtunggal, Depok, Sleman, Indonesia

mutiasari@sttnas.ac.id

Abstract. In order to alleviate the negative impacts of motorized vehicle use as well as create sustainable environment within campus area, it is pivotal to encourage mode shifting among university students. Active transport modes such as walking, cycling, and using public transport can be considered as alternative modes. This paper tried to identify the potential to increase active commuting in UGM by understanding student's travel behavior. ANOVA test was employed to identify the perceptions between students across residential zones toward motivators and barriers to actively commute. The findings were used to propose strategies for increasing active commuting level in UGM, which are: reducing barriers to actively commute, improving public transport services, improving walking and cycling facilities, and introducing programs to discourage motorized vehicle use.

1. Introduction

Nowadays, universities all over the world has been increasing their attention on encouraging the usage of environmental friendly transport modes as strategy for implementing sustainable transportation in university environment [1][2]. This idea is emerged since university area has attracted trip with a regional scale. Therefore, transportation problem in university will significantly contribute to a problem in the bigger setting. Implementing sustainable transportation policy for university area will not only create a more livable university area but also contribute to the overall sustainability of the city [3].

One of the oldest university in Indonesia which posited in Yogyakarta, Universitas Gadjah Mada (UGM), has high number of students which increase over time. The increasing number of students enrollment are associated with high number of motorized vehicles in campus area as students tend to choose motorized vehicles for their mobility. University students are a group that tends to use various types of travel mode, including large proportion of active transport used. Active transports can be alternative way for bringing sustainable context in the university. Walking, cycling, and using public transport are considered as active forms of transport since these type of modes involve physical activity

Moreover, according to Gatersleben and Appleton [4], student population were the easier group to target for active commuting since they mostly could not afford their own vehicles and do not like to rely on infrequent bus services. Thus, active modes can provide an excellent form of flexible transport for them.

Most people lived very close to their daily destinations and traveled less by motorized vehicles [5]. University students come from various regions, thus forced them to live in temporary residence. In choosing their residences, distance to the campus becomes student's main consideration [6]. Therefore, majority of students who comes from other region will live in campus surrounding. Relatively close to campus however does not make high use of non-motorized vehicle. There are also plenty of trips which made by students causing high traffic in UGM area. Student mobility following with their complex travel behavior, indeed become one of contributor for the traffic in the area. The daily travel of people to and from campus in motorized transports is one of the biggest impacts which an academic institution can put on environment [7].

This paper tried to identify the potential for increasing active commuting in UGM by understanding student travel behavior. Student travel behavior are identified through their travel characteristics as well as their perception towards the barriers and motivators to actively commute. Results of what actually motivate and/or hinder students to actively commute will be used to propose several suitable strategies for increasing the levels of active commuting among students in UGM.

2. Literature Review

2.1 Sustainable Transportation

The University environment tends to have an economical attractiveness, which then resulting on the emergence of various businesses in surrounding area to support university member's activities. The development of university's surrounding area causes mixed activity that triggers complex mobility within university area. In fact, commuting is the single largest impact which a university might have on the environment [8]. Moreover, many university-related activities also contribute significantly on the environment, including transportation [9]. Therefore, implementing sustainable concept in university campuses is needed.

Due to the high number of admitted students each year as well as their number of employee, it is indeed that universities become one of major traffic generators which demand on wide parking areas. Beside of its loss of natural environment and noise disturbance, expropriation of visual environment for parking provision is main environmental impacts of transportation on universities [8]. Based on Velazquez et al. [10], a sustainable university will encourages a less of negative environmental, economic, societal, and health effects in the use of resources for university's activities in order to help community for achieving a sustainable lifestyle. Thus, a university is responsible in managing their university members to commute effectively for creating sustainable campus.

2.2 Active Transports

One way to reduce the automobile use can be done by encouraging people to shift into more sustainable travel modes. Active transport is one of the most possible alternative to replace motorized vehicles. Active transport refers to walking, cycling, and the use of other non-motorized vehicles which then being defined as human power based transports [11]. However, trips using public transport were also considered within active transport due to walking on access or egress roads [12].

Active transport offers various advantages both to individuals and even the broader communities. Beside its low emission consumption, some studies found that people who travelling more by active transports possess better physical and mental health [13]. Moreover, improving active transport also have a possibility to achieve social equity since walking and cycling are often being used by physically, economically, and socially disadvantaged communities [11].

2.3 Motivators and Barriers to Actively Commute

Many previous studies have been identified several factors influencing active commuting, particularly factors which will motivate and/or deter individual to actively commute. Nkurunziza, Zuidgeest, Brussel, & Maarseveen [14] defined factors which may affect cycling behavior in Dar-es-Salaam into three distinct categories, which are individual factors, socio-environmental factors, and physical environmental factors. Individual factors refer to cultural factors, the social norms of the community, and behaviors considered normal or appropriate. While the physical environment is represented the natural (topography, climate, geography, and others) and built physical environments (land use patterns and transport infrastructure). The findings discovered that low bicycle price, quality of bicycle, and cycling training were the major influencing perceived motivators on cycling commuting. In the other hand, factors such as weather, lack of safe parking, lack of cycling paths, water showers, social status, social insecurity, and not being comfortable on a bicycle were found as the most negative influence which impeding people to cycling. Moreover, regarding the interventions factors, exemption of bicycle tax, car congestion charges, and guarding bicycles at public places were the most important policy interventions. Meanwhile, in another study conducted by Nkurunziza, Zuidgeest, & Maarseveen [15] who also investigated cycling commuting in Dar-es-Salaam, the availability of bicycle paths and special bicycle infrastructure gave bigger motivation to people to cycling.

Regarding barriers to actively commute, travel time and distance were identified as the most important barrier in an Australian study [2]. Improvement to health, potential to save money, and avoid the need to find parking were reported as motivating factors among university communities. Meanwhile, an introduction of a U-Pass for using public transport was considered as the most likely intervention to influence travel behavior for active commuting. However, regardless its findings, the study also involved several other motivator and barrier items. Infrequent public transport between residential and university, weather, the need to run errands, cheap parking at the university, lack of continuous cycle path, danger from vehicular traffic, lack of secure bike parking at the university, and involvement of physical effort were another example of barrier items included in the study. Whilst a contribution for reducing air pollution, socialize opportunity, and the difficulty in obtaining parking permit were identified as potential motivators included in the study as well.

Moreover, study on university students which examined the association of environmental, social, and personal factors on cycling commuting was conducted by Titze, Stronegger, & Janschitz [16]. This study included traffic safety, bicycle theft, availability of bike lanes, clothes, and mobility as several factors correlates with cycling. Meanwhile, a similar study further explored factors associated with active commuting to the university in Spain. The findings revealed that access to motorized vehicles, walking and cycling facilities, and psychosocial barriers were significant correlated to active commuting [17]

3. Research Methodology

Data was collected by randomly disseminating questionnaires to active students in UGM through social media, personal messages, and electronic mail. The questionnaire consisted of two sections. In the first section, questions related with socio-demographic information were asked. Then, in the further section, respondents were asked to measure several motivator and barrier items for revealing the importance level of those items in encouraging and/or impeding them to use active transport modes in their trips to university.

The motivator and barrier items were rated by using a Likert scale of 1-5; 1 being 'not at all important' or 'not at all likely' and 5 'very important' or 'very likely'. In order to understand the possible active transport mode based on students' residential distance, the region was divided into three zones (Zone 1: < 1 km away from UGM, Zone 2: 1-5 km away from UGM, and Zone 3: >5 km away from UGM). Residential distance below one kilometer away from campus was chosen to represent a feasible walking distance which also can be reached by bike. While residential distance within one to five kilometers away from campus represents a feasible cycling distance. Finally, distance of more than five kilometers away from campus will have public transport as travel choice to actively commute.

As the purpose of this paper is to identify the potential for increasing active commuting level based on students current travel behavior, data interpretation become important where some trends and facts will be revealed which may useful for further analysis. To begin the data analysis process, descriptive statistics were calculated to summarize and describe the data obtained. The analysis was executed by using one-way ANOVA test which can be used to identify the different perceptions of motivators and barriers toward active commuting behavior among students. This finding will be used to propose strategies for increasing active commuting level in UGM.

4. Results

4.1 Travel Characteristics

Most of students reported motorcycle (73%) as their mode choice for university travel. While car only accounted for about 7% of students' travel mode. Accordingly, there were 80% of the students using motorized vehicles for the university trips. In contrast, there were only few students who reported active transport as travel modes. Walking, cycling, and public transport were reported to be used by 12%, 7%, and 1% of the students respectively.

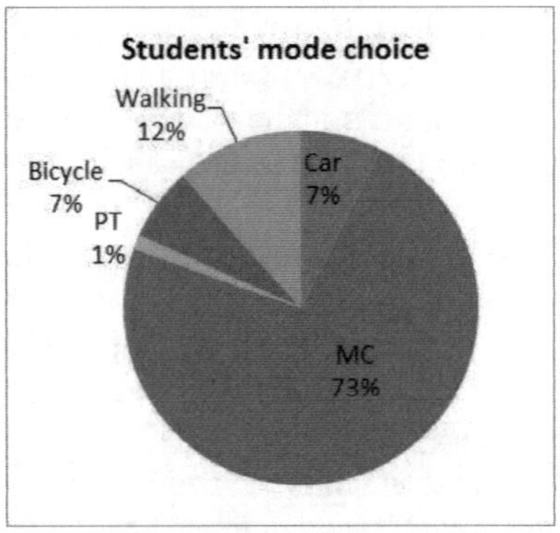

Figure. 1. Proportion of mode choice among students

Regarding their residential location, result showed that walking and cycling were quite high for those living in Zone 1 and Zone 2, most likely because this zone is feasible to be reached by this type of travel modes. Meanwhile, it is no doubt that in Zone 3, only less than 2% of students using public transport and bicycle, and none of them walking to campus. In addition, public transport use was generally not popular among students within all zones. It is only reported less than 4% of students in all zones.

Table 1. Main mode of transport among students based on residential distance zone (%)

Mode Choice	Distance		
	Zone 1 (<1 km) n = 137	Zone 2 (1-5 km) n = 213	Zone 3 (>5 km) n = 120
Car	3.6	7.0	12.5
Motorcycle	55.5	77	85.8
PT	1.5	1.4	0.8
Bicycle	8.8	8.9	0.8
Walking	30.7	5.6	0

Figure 2 illustrate trip frequencies made by students in a day. The figure show that most students generated an average 3-4 trips/day (47.66%) followed by 31.28% of them generated 2 trips/day. However, there is sizeable proportion of students who generated an average 5-6 trips/day. Students with high mobility needs tend to avoid in using active transports, hence they will choose motorized transport as their travel mode.

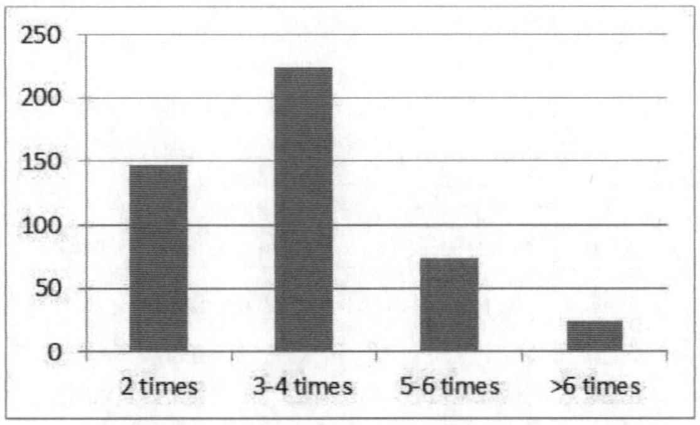

Figure 2. Daily trip frequencies of students

4.2 Motivators ad Barriers for Active Commuting

In order to elaborate students' perceptions towards active commuting behavior according to their travel distance to the university, the motivator and barrier items were analyzed toward residential distance zone categories. There were significant differences for several motivator items across three zones. Significant differences were occurred in motorized free vehicle area ($F(2, 414) = 3.651$, $p < 0.05$); parking restrictions ($F(2, 414) = 3.483$, $p < 0.05$); and protected lanes from weather ($F(2, 414) = 3.241$, $p < 0.05$). Those within closer residential distance from the university rated these three items higher than those living far from university. This finding also reveals that students more motivated with pushed measures rather than pull measures.

Table 2. One-way ANOVA results for perceptions of motivators between residential zones

Motivators	Zone 1 (<1 km)		Zone 2 (1-5 km)		Zone 3 (>5 km)		Total		F	p-value
	Mean	SD	Mean	SD	Mean	SD	Mean	SD		
Lanes are protected from weather	4.73	0.621	4.55	0.725	4.51	0.743	4.59	0.705	3,241*	0.040
Safe parking area	4.59	0.692	4.54	0.732	4.57	0.673	4.57	0.704	0.170	0.844
Environmental awareness	4.54	0.721	4.41	0.837	4.44	0.753	4.46	0.784	1,016	0.363
Health	4.44	0.828	4.39	0.788	4.41	0.786	4.41	0.798	0.176	0.838
Saving expenses	4.43	0.876	4.32	0.860	4.20	0.945	4.32	0.889	1,769	0.172
Presence other active commuters	4.29	0.844	4.14	0.848	4.22	0.835	4.21	0.844	1,168	0.312
Bike shelters and campus bike	4.18	0.967	4.15	0.994	4.06	0.994	4.13	0.985	0.464	0.629
Car free day	4.12	0.980	4.01	1,037	3.93	0.983	4.02	1,007	1,032	0.357
Motorized vehicle free area	3.88	1,185	3.57	1,163	3.49	1,148	3.64	1,173	3,651*	0.027
Parking restrictions	2.79	1,296	2.59	1,271	2.35	1,186	2.59	1,264	3,483*	0.032
Paid parking	2.28	1,353	2.29	1,319	1.98	1,192	2.21	1,301	2,192	0.113

[a]F = ANOVA score, p = Significance level, * ≤ 0.05

For barrier items, there were also significant differences across three zones, including travel distance and travel time. Moreover, the biggest significant differences were also occurred in these two items (travel distance ($F(4, 414) = 14.085$, $p < 0.001$); travel time ($F(4, 414) = 8.939$, $p < 0.001$)). Furthermore, barrier items associated with public transport services, the coverage area and frequent service, were also reported with high importance for impeding students to actively commute with statistically significant differences between those in Zone 1 and 3. This reveals that students in Zone 1 and 2 have slight similarity in perceiving barriers for active commuting. The perception toward barriers are incrementally increased in hindering them to actively commute.

Table 3. One-way ANOVA results for perceptions of barriers between residential zones

Barriers	Zone 1 (<1 km)		Zone 2 (1-5 km)		Zone 3 (>5 km)		Total		F	p-value
	Mean	SD	Mean	SD	Mean	SD	Mean	SD		
High mobilities and activities	4.34	0.82	4.37	0.83	4.39	0.81	4.36	0.82	0.094	0.910
Travel distance	3.88	1.20	4.17	0.97	4.56	0.80	4.19	1.03	14.085[a]**	0.000
Covered area of PT services	3.98	0.98	4.12	0.84	4.30	0.85	4.12	0.89	3,573*	0.029
Longer travel time	3.95	0.92	3.97	1.01	4.41	0.81	4.08	0.96	8,939**	0.000
Infrequent PT services	3.94	0.97	4.14	0.90	4.25	0.87	4.11	0.92	3,394*	0.035
Clothes	3.53	1.08	3.74	1.11	3.72	1.00	3.67	1.08	1,621	0.199
MT vehicle access	3.44	1.08	3.69	1.02	3.83	1.00	3.66	1.04	4,286*	0.014
Social status	3.12	1.31	2.80	1.36	2.35	1.24	2.78	1.34	9,704**	0.000

[a]F = ANOVA score, p = Significance level, * ≤ 0.05, ** ≤ 0.001, [a] = Levene's test for homogeneity of variance has been violated ($p = \leq 0.05$) so the Welch test (adjusted F) has been used instead

4.3 Strategy Implications

It is clear that a variety of different strategies are needed to be adopted in order to increase active commuting among students in UGM. The potential to shift into active transports among students is not dependent solely on making active transport more attractive but also restricting the use of motorized transports. Moreover, the proposed strategies have to be compiled based on which conditions might motivate or hinder students to perform active commuting behavior. Some strategies which will effectively increase the level of active commuting among students in UGM can be suggested as a result. The strategies that could be employed to increase the level of active commuting among students in UGM as well as some supporting evidences will be discussed in below.

4.3.1 Reducing barriers to actively commute

Results revealed that most of students are currently non-active commuters. Generally, non-active commuters perceive more barriers than active commuters. Therefore, reducing barriers to actively commute is proposed as the first intervention in order to target this large proportion of students. Giving more attention to personal barriers can be more important than providing infrastructure. Some approaches reducing the perceived barriers to actively commute are needed.

The biggest barrier to actively commute for students is having high mobility and activities which then forced them to run errands in several destinations during the day. UGM has sprawl characteristic with a large area which then engender its community to commute from one point to another point within campus area in difficult ways. It is needed some approaches in order to reduce this barrier for active commuting. One of the possible strategy can be through developing the university to be more compact area. Specifically, it is important to integrate the clusters in the university as there are distances between clusters. Building integrated bridge or lanes which connecting inter-faculty buildings within campus may give an easy access for university communities to visit one to another points.

Travel distance and time were also perceived as biggest barriers to actively commute among students. Even though a sizeable proportion of students are already living close to campus, distance and time still play an important role in the their travel mode choice which discouraging them to choose active modes as their travel mode to the university. Therefore, community designs that provide networks for bike and pedestrian lanes connecting campus with residential areas that have a high student population, are highly recommended to reduce the perceived barriers of travel distance and time. It is also indicated that improving the built environment along walking and cycling network lanes may help to reduce the perceived barriers of time and travel distance.

Meanwhile, it is reported that students living in Zone 3 also perceived travel distance and time as most important factors to deter them for using active transports to the university. This group only has public transport as suitable option for active commuting activities. TransJogja, as the most reliable public transport service in Yogyakarta, has number of shelters in campus area. However, students still have to walk for long distances in accessing the services, particularly for those in east campus (cluster of Sosio-Humaniora and cluster of Agro) as there is no shelter in this area. Therefore, to decrease the perceived barriers of travel distance and time targeting those living in Zone 3, the need to add number of shelters in the campus area is suggested to be considered by the university in collaborating with regional government as public transport authority for TransJogja services.

Furthermore, it is suggested to introduce campus shuttle bus operating in the university and surrounded area. Bus fleets with environmental friendly technology are highly recommended for the operation. This approach can facilitate students who have high mobility and activities during the day for commuting around campus area. Moreover, barrier of travel distance and time can be also reduced through this approach as students will have an easy access to reach all university and surrounded area.

4.3.2 Improving public transport services

According to its transportation master plan document, it is clear that UGM already realized that it is impossible to prohibit the use of car or motorcycle in the campus area without preparing other alternative modes including its facilities. It is also not feasible to enforce academic communities, especially those who live far from the university, for cycling or using public transport from their residential place since the current urban transportation policy can not guarantee cyclists' safety on the road and public transport services are not reliable in serving urban mobility. Meanwhile, there are a sizeable proportion of students living in Zone 3 who only rely on public transport services.

Therefore, the idea of improving public transport services is needed to be addressed in order to attract more students of UGM to actively commute. Indeed, it is not things which can be done by the university itself. The regional government, as the public transport authority, need to be involved for implementing this strategies. Therefore, it is important to underline that the development of campus transportation can not stand independently since it is always affected by the urban transport policies.

4.3.3 Improving walking and cycling facilities

The results in this study demonstrate that active commuters are found to be more attracted by the motivator for active commuting rather than non-active commuters since they are already performing active commuting behavior. These student groups live in both Zone 1 and 2. Therefore, it is suggested for UGM to consider improving walking and cycling facilities in campus area as well as collaborate with regional authority for improving cycling facilities in Yogyakarta area.

Firstly, the improvement strategies will be focused on providing walking and cycling lanes which protected from weather conditions. Results showed that the availability of protected lanes from weather can motivate students to actively commute to university. Therefore, presenting protected lanes from weather condition for pedestrians and cyclists can promote active commuting activities in the university. For executing, developing pergolas and vegetation along the lanes are suggested.

Providing safe bicycle parking area becomes the second improvement strategies which can be employed in UGM as it was an important motivator for students to encourage them actively commute. Students concerned about the availability of safe bicycle parking area as an important motive for them to actively commute, particularly cycling to the university. Moreover, providing bike campus to be rented by university communities which already implemented by UGM, can give an access to bicycle among those who do not own a bike which then may promote more bicycling in university environment. Therefore, providing more campus bikes and shelters as well as improving regulation for renting bike campus facilities are suggested in order to increase active commuting, particularly cycling, among students.

Finally, it is important for UGM to collaborate with the regional authority of Yogyakarta to implement several regulations in order to ensure the cyclists' safety in the road. It is known that the existing condition of cycle lanes in Yogyakarta are still mixed up with other motorized vehicles.

However, by developing some alternative lanes in which involving residential areas could reduce the potential for cyclist to combat with motorized vehicle users. Moreover, socialization for prioritizing pedestrians and cyclists in the road are also suggested to be done among the citizens of Yogyakarta.

4.3.4 Introducing programs to discourage motorized vehicle-use

Implementing strategies to create an active commuting behavior in campus area through reducing the convenience and cost-effectiveness of driving have been introduced in many universities [18]. It also suggested further to employ a comprehensive approach includes both "carrot and stick" strategies, in which encouraging people to consider issues other than travel time, such as the cost of commuting [2]. At the beginning, the schemes are suggested to be implemented by UGM through introducing car free day program and increasing motorized vehicle free area in the university. These two conditions were found to be considered by students as important motives to actively commute and they also feel to be more encouraged for using active modes if the university employed these interventions.

It is also argued to support the cost effectiveness of improving walking and cycling facility schemes by increasing the cost of driving. Paid parking can be a strategy to be considered thereby the cost of driving would be rising up. Therefore, the further schemes that could be implemented by UGM after active transport facilities have been upgraded are implementing several push measures such as parking restrictions and paid parking schemes. By increasing the cost of driving both economically and efficiently through paid parking and parking restrictions, active commuters are expected to be increased..

5. Conclusions

This study was focused on the potential to increase active commuting among students in UGM. Findings revealed that majority of students are using motorized transport as their travel mode to the university. However, there are a potential to increase the level of active commuting since the majority of students are living within walking and cycling distance. Regarding motivator for active commuting, students within closer residential distance from the university rated the items higher than those living far from university. This finding reveals that students more motivated with pushed measures rather than pull measures. While for barrier items, students in Zone 1 and 2 have slight similarity in perceiving barriers to actively commute. However, the perception toward barrier is incrementally increased in hindering them to actively commute.

In order to increase the level of active commuting among students, several strategies are proposed to be implemented by UGM. The proposed interventions are consists of four main strategies. First, it includes strategies in reducing the barriers to actively commute, specifically by developing the university to be a more compact area, improving supporting facilities in the faculty area, building a network for bike and pedestrian lanes connecting campus with residential areas that have a high student population, improving the built environment for walking and cycling, increasing the number of TransJogja shelters in the university area, and introducing campus shuttle bus. Second, it is needed to improve public transport services, particularly increasing the frequency of the bus services and increasing the coverage area of bus services. Third, it is suggested to consider improving walking and cycling facilities, particularly providing walking and cycling lanes which protected from weather conditions, providing safe bike parking area, providing more campus bike and its shelters, developing alternative routes for cyclist in the city, and carrying out socializations among citizens of Yogyakarta for prioritizing cyclists and pedestrians. Finally, it is necessary to introduce programs for discouraging motorized vehicle-use such as introducing car free day program, increasing motorized vehicle free area in the university, implementing parking restrictions and paid parking schemes, and carrying out socializations among students regarding the disadvantages of motorized vehicle use and the benefits of active commuting.

References

[1] Limanond, T., Butsingkorn, T., & Chermkhunthod, C. *Travel behavior of university students who live on campus: A case study of a rural university in Asia.* Transport Policy; 2011, p. 163-171

[2] Shannon, T., Corti, B. G., Pikora, T., Bulsara, M., Shilton, T., & Bull, F. *Active commuting in a university setting: Assessing commuting habits and potential for modal change*. Transport Policy; 2006, p. 240-253

[3] Rybarczyk, G., & Gallagher, L. *Measuring the potential for bicycling and walking at a metropolitan commuter university.* Journal of Transport Geography; 2014, p. 1-10

[4] Gatersleben, B., & Appleton, K. M. *Contemplating cycling to work: Attitudes and perceptions in different stages of change.* Transportation Research Part A; 2007, p. 302-312

[5] Cervero, R. *Linking urban transport and land use in developing countries.* In: Tran, M.T., Zhang, J., Chikaraishi, M., & Fujiwara, A. *A Joint analysis of residential location work location, and commuting mode choices in Hanoi*, Vietnam. Journal of Transport Geography; 2016, p. 181-193.

[6] Muliana, R., & Kustiwan, I. *Perilaku Pemilihan Lokasi Tempat Tinggal dan Karakteristik Perjalanan Mahasiswa di Kota Bandung.* Jurnal Perencanaan Wilayah dan Kota B; 2014, Vol. 1, p. 37-43.

[7] Toor, W. The road less traveled: *Sustainable transportation for campuses. In: Garvey, A., Gossell, L. & Zhai, D. Sustainable transportation on the university campus: overcoming barriers and developing strategies for success*; 2009.

[8] Tolley, R. *Green campuses: cutting the environmental cost of commuting. Journal of Transport Geography*; 1996, p. 213-217.

[9] Alshuwaikhat, H. M., & Abubakar, I. *An integrated approach to achieving campus sustainability: assessment of the current campus environmental management practices. Journal of Cleaner Production*; 2008, p. 1777-1785.

[10] Velazquez L, Munguia, N, Platt A, Taddei J. *Sustainable university: what can be the matter?. in Alshuwaikhat,* H. M., & Abubakar, I. *An integrated approach to achieving campus sustainability: assessment of the current campus environmental management practices. Journal of Cleaner Production;* 2008, p. 1777-1785.

[11] Litman, T. *Evaluating active transport benefits and costs. Victoria Transport Policy Institute*; 2015.

[12] *National Public Health Partnership. Promoting active transport: an intervention portfolio to increase physical activity as a means of transport*. In: Irawan, Z., & Sumi, T. Promoting Active Transport in Students' Travel Behavior: A Case from Yogyakarta (Indonesia). Journal of Sustainable Development; 2011, p. 45-52.

[13] Frank, L., Engelke, P..*The built environment and human activity patterns: exploring the impacts of urban form on public health.* In: Khan, Kockelman, & Xiong. Models for anticipating non-motorized travel choices of the built environment; 2014, p. 117-126.

[14] Nkurunziza, A., Zuidgeest, M., Brussel, M., & Maarseveen, M. V. *Examining the potential for modal change: motivators and barriers for bicycle commuting in Dar-es-Salaam.* Transport Policy; 2012, p. 249-259.

[15] Nkurunziza, A., Zuidgeest, M., & Maarseveen, M. V. *Identifying potential cycling market segments in Dar-es-Salaam, Tanzania. Habital International*; 2012, p. 78-84.

[16] Titze, S., Stronegger, W. J., & Janschitz, S. O. *Environmental, Social, and Personal Correlates of Cycling for Transportation in a Student Population*. Journal of Physical Activity and Health; 2007, p. 66-79.

[17] Molina-García, J., Castillo, I., & Sallis, J. F. *Psychosocial and environmental correlates of active commuting for university students.* Preventive Medicine; 2010, p. 136-138.

[18] Toor, W., Havlick, S.W. *Transportation and sustainable campus communities: issues, examples, solutions. In: Shannon*, T., Corti, B. G., Pikora, T., Bulsara, M., Shilton, T., & Bull, F. *Active commuting in a university setting: Assessing commuting habits and potential for modal change. Transport Policy*; 2006, p. 240-253.

School bus's level of service in Malang City

S Hariyani

Department of Urban and Regional Planning, Faculty of Engineering, Universitas Brawijaya, Malang, Indonesia

septianahariyani@ub.ac.id

Abstract. School Bus began operated on the 12th of January 2015. Provision of school buses is expected to reduce not only the use of vehicles by students, but it is also to reduce the number of traffic jams. Malang school bus facilities provided by the Department of Transport in cooperation with the Department of Education to serve students in elementary school, junior and senior high schools. After the service running two years, based on the preliminary observation not all students are interested in using the school bus. The research objective was to measure the school bus's level service. The method to measure school bus's level of service was used Importance Performance analysis (IPA). The results showed that through IPA, it can be concluded that school bus's level of service in Malang City have been able to serve students/customers with the mean of degree suitability (Tki) is 111. Meanwhile it must be observed and get more attention to improve by government, attributes which is lies in the first quadrant or concentrate here (attribute Adequate space, Seating capacity, Availability trash can, Passenger facility down in points, The availability of information boards in each bus stop, Availability public telephone in each bus stop, and Availability CCTV in each bus), in order to increase its performance.

1. Introduction

Basically, every human being in need the service, even in the extreme it can be said that the service cannot be separated from human life [1]. One form of public services needed by the community is a transportation service. Transport is an activity of the movement of people and goods from one place (origin) to another (destination) by means of a vehicle [2]. There are five categories of trip destination from home-based, namely: 1. The movement into the workplace, 2. Movement to school or university (the movement of the purposes of education), 3. Movement to shopping, 4. Movement for social interests, and 5. Movement for purpose recreation [3].

To go to the destination required modes of transport. Along with the development of transportation technology, users of motor vehicles is increasing, especially 2-wheel vehicles, this affects the selection and junior high school students to choose a motor vehicle as a means to get to school. This becomes problematic because in Indonesia the provision of motor vehicles has been regulated in Law No. 22 in 2009 regarding Traffic and Road Transportation [4]. Described in Article 288 paragraph 2 that any person driving a motor vehicle on the street must have a driver's license, where to get the driver's license minimum age requirement is 17 years [4].

According to [5] the school transportation services is a transportation for students to smooth the learning process. Students will feel safe and be able to get in or get home from school with the right timing. Organizers of school transport is the school itself or the private sector working together with schools, could also be the government.

One of the efforts undertaken by the government of Malang, namely by providing a new transport service. The transport services include the provision of school bus that is used to help students in both departing and return the mobilization of school.

The school bus is an alternative transportation for students provided by the city government for the students, where a school bus has a goal to give the ease, smoothness and comfort for students. Some students find it helpful because the school bus was comfortable. On the other hand there are things to be taken into consideration in the development of the school bus that is still a lack of number of users of the school bus because students prefer other transportation such as motorcycles and public transport for some reason, and there are also students who feel lucky with their school bus because it does not need to pay to ride [6].

Based on preliminary observations not all students are interested in using the school bus, so that is necessary to evaluate the school bus operation to improve services to students. The research objective was to measure the school bus's level service.

2. Methods

Primary survey was conducted by researching and recording the condition of school bus's level of service. Meanwhile, secondary survey was done by finding literatures related to the issues discussed. Furthermore, Importance Performance Analysis (IPA) was used to determine the factors that influence school bus's level of service in the area of Malang City.

2.1. Sampling

The number of sample was determined using Slovin formula proposed by [7], with the formula:

$$n = \frac{N}{1+N\,(e)^2} \qquad (1)$$

Where:
N = number of population
n = sample
e = margin of error (this study used e = 10%).

The calculation using the above formula result to the number of the sample is 100 respondents. Where the number of population or total student in elementary school, junior high school, dan senior high school is 177,893 , and e=10%. 100 respondents divided into 5 group representing the respondents of the passenger school bus in Malang City as showed in Table 1.

Table 1. Distribution of sample.

No.	District	Sample
1.	Kedungkandang	18
2.	Sukun	14
3.	Klojen	32
4.	Blimbing	11
5.	Lowokwaru	25
	Total	100

This research is conducted in Malang City, East Java, Indonesia. The location is shown in Figure 1 and Figure 2. Figure 1 described the orientation East Java Province from Indonesia Country and Figure 2 described orientation Malang City from East Java Province.

Figure 1. East Java Province orientation from Indonesia Country
Source: Badan Informasi Geospatial and Badan Pertanahan Nasional (BPN) (2010)

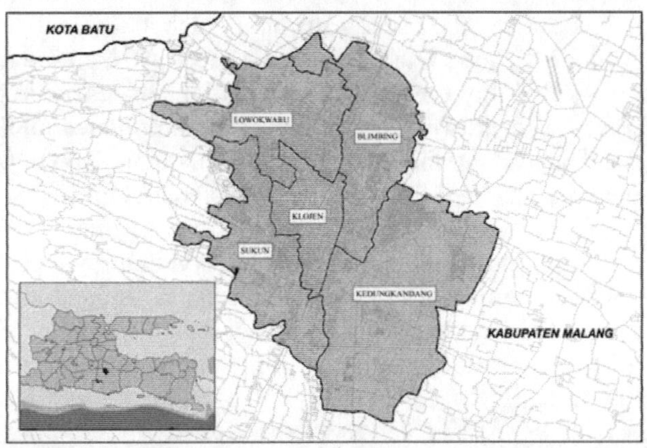

Figure 2. East Java Province orientation from Indonesia Country
Source: Badan Informasi Geospatial and Badan Pertanahan Nasional (BPN) (2010)

2.2. Methods
Importance Performance Analysis (IPA) is used to analyze the importance and performance of school bus's level of service [8,9]. The attributes that were examined were measured using a Likert scale. Likert scale is ordinal measurement scale consisting of five levels and is weighted according to its level. The relationship between the level of interest and the performance perceived by the customer illustrated in the diagram of Importance Performance Analysis as in Figure 3.

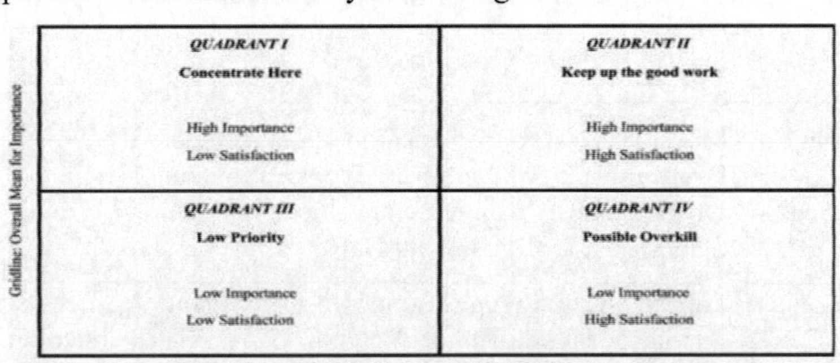

Figure 3. Importance Performance Analysis diagram.
Source: Martilla, 1977; Supranto, 2006

IPA is presented in a diagram consist of four quadrants which are:

1. Quadrant one meaning "Concentrate Here" (high importance and low satisfaction)
   Factors located in this quadrant is considered as very important factors for consumers, but the conditions is not satisfied. Factors located in this quadrant is a priority for improvement.
2. Quadrant two meaning "Keep up the Good Work" (high importance and high satisfaction)
   Factors located in this quadrant are considered as additional factors for customer satisfaction so that the management is obliged to ensure that the performance of the institutions under its management needs to continue to maintain the achievements.
3. Quadrant three, "Low Priority" (low importance and low satisfaction)
   Factors located in this quadrant have a low level of satisfaction and at the same time is not considered too important for the consumer, so that the management does not need to prioritize or paying much attention to these factors.
4. Quadrant Four, "Possible Overkill" (low importance and high satisfaction)
   Factors located in this quadrant are not very important that the management can allocate resources associated with these factors to other factors that have higher priority still need improvement.

To assess the school bus's level of service use 22 attributes such as Friendliness officer, Discipline of diver, Door condition, Altitude School Bus, adequate space, AC/ air conditioning, Comfort seat passenger, seating capacity, Availability trash can, Facilities for self-rescue in an emergency, Facilities for self-rescue in an emergency, security of passengers, The compliance rates for the services given, Punctuality of departure and arrival at the school bus stop point, Down facilities for passengers, Passenger facility down in points, shielded from the sun in each bus stop, The availability of information boards in each bus stop, Availability trash can in each bus stop, Availability public telephone in each bus stop, Availability chair in each bus stop, circulation, and Availability CCTV in each bus.
IPA method requires the usage weighing both for level of satisfaction and level of importance, so that each attributes above was weighted. The weighing consist of 5 values which are 5 = very satisfied / very important; 4 = satisfied / important; 3 = fairly satisfied / quite important; 2 = less satisfied / less important and 1 = not satisfied / not important.

3. Results

Malang government initiative to provide school buses to help students in both departing and return from home to school. School bus provision is for students who live in the city or on the outskirts. The number of buses provided are 7 buses, 6 used to serve five districts in Malang City, while the one buses as backup. School bus provision is expected to reduce the use of vehicles by students of elementary, junior high school or senior high school. It is also expected to reduce the number of traffic jams although there was a polemic with public transport drivers who refused and did not agree with the school bus service. Table 2 shows route of school bus in Malang City.

Table 2. Route of Malang school bus.

No. Bus	Starting Point	Street Route	Halte/Shelter
1.	Gas station Tlogomas or Baiduri Sepah yard	MT Haryono–Soekarno-Hatta–Bundaran Pesawat–Cultural Garden East Java Province – DI Panjaitan-Bogor–Veteran–Bandung– Ijen – Semeru – Kahuripan –City Hall	Dinoyo market/ Griyashanta/ Cultural Garden East Java Province, Ijen, Semeru, Gajayana Stadium
2.	Pisang Candi (Gas station Mergan)	Langsep–Galunggung–Bondowoso–Jombang – Surabaya–Jakarta-Bogor–Veteran–U Turn Senior high school 8– Veteran–Bandung–Ijen–Semeru–Kahuripan-city hall	Ijen, Semeru, Gajayana stadium

No. Bus	Starting Point	Street Route	Halte/Shelter
3.	Sawojajar Madyopuro terminal	Ki Ageng Gribig–Danau Toba–AM Wiyono – Kesatrian–Hamid Rusdi– Junior high school 5–Panglima Sudirman–Dr Cipto– Cokroaminoto–Trunojoyo–Kertanegara–city hall	AM. Wiyono
4.	Sawojajar Madyopuro terminal	Ki Ageng Gribig–Mayjen Sungkono– Hamid Rusdi terminal–Block Office	-
5.	Block Office	Mayjen Sungkono–Gadang Market–PLN Satsuit Tubun–Kolonel Sugiono–RE Martadinata–under Flyover Kotalama–Comboran–Sulawesi–gas station Sawahan–Syarif Al Qodri–Kauman–KH. Hasyim Asy'ari–AR Hakim–Gereja–Mojopahit–city hall	-
6.	Samsat Kacuk	Sudanco Supriyadi–Janti–Halmahera–Tanimbar–Sulawesi–Nusakambangan–Comboran –RE Martadinata–Gatot Subroto – Trunojoyo – Kertanegara – city hall	-

Source: http://mediacenter.malangkota.go.id

The following table 3 shows that level of satisfaction for attribute number 4 is the highest (flow rate during the rainy season) meaning that respondents are satisfied in water consumption demand during rainy season. The reason for this is that probably the supply of water in rainy season is abundant so that the customers can fulfill their needs. Meanwhile, the lowest score for satisfaction is showed by attribute number 3 (flow rate during the dry season) meaning that respondents are not so satisfied compared to other attributes. It may cause by the decreasing quantity of water supply during the dry season.

Level of importance in Table 3 shows that the highest score is attribute number 22 (Availability CCTV in each bus) indicating that it is the most important attribute for the students. Meanwhile, attribute number 12 (The compliance rates for the services given) has the lowest score indicating that the compliance rates for the services given is not so important compared to other attribute for the students.

Table 3. Level of satisfaction.

No	**Attributes**	**Level of Satisfaction**					**Total Score**
		1	**2**	**3**	**4**	**5**	
1	Friendliness officer	0	0	2	37	57	439
2	Discipline of diver	0	0	11	34	51	424
3	Door condition	1	0	13	28	54	422
4	Altitude School Bus	1	1	26	36	32	385
5	adequate space	1	27	13	28	54	339
6	AC/ air conditioning	0	1	4	36	55	433
7	Comfort seat passenger	0	2	17	41	36	399
8	seating capacity	2	55	8	16	15	275
9	Availability trash can	28	13	13	21	21	282
10	Facilities for self-rescue in an emergency	6	8	17	31	34	367
11	security of passengers	4	6	16	28	42	386
12	The compliance rates for the services given	29	32	28	6	1	206
13	Punctuality of departure and arrival at the school bus stop point	3	4	16	20	53	404

No	Attributes	Level of Satisfaction					Total Score
		1	2	3	4	5	
14	Down facilities for passengers	3	5	17	34	37	385
15	Passenger facility down in points	2	34	6	25	29	333
16	shielded from the sun in each bus stop	3	34	13	17	29	323
17	The availability of information boards in each bus stop	5	33	18	23	17	302
18	Availability trash can in each bus stop	6	44	17	15	14	275
19	Availability public telephone in each bus stop	9	29	19	21	18	298
20	Availability chair in each bus stop	6	7	14	20	49	387
21	circulation	1	2	11	27	55	421
22	Availability CCTV in each bus	0	3	2	6	85	461

The IPA getting from the calculation of 22 attributes selected by respondent as shown in Table 3 Level Satisfaction, and Table 4 Level of Importance, afterward it was described in a Cartesian diagram. Mean of degree suitability (Tki) is 111, meanwhile ranged satisfaction level from 62 to 765 means that respondent are satisfied with attributes, it is showed in Table 5.

The IPA is described in a Cartesian diagram as showed in Figure 4. The four-quadrant Importance-Performance matrix was defined by the two axes based on the overall mean of importance (3.7) and satisfaction (3.3) of the 22 attribute (Figure 4). Almost half (ten) of the attribute were located in the "Keep Up of The Good Work" quadrant, two attributes positioned in "Possible Overkill" quadrant, and three attributes are located in "Low Priority" quadrant. Moreover, seven of them appeared in the "Concentrate here" quadrant.

Table 4. Level of importance.

No	Attributes	Level of Importance					Total Score
		1	2	3	4	5	
1	Friendliness officer	1	0	5	34	55	427
2	Discipline of diver	0	1	7	24	64	439
3	Door condition	0	38	9	16	33	332
4	Altitude School Bus	1	37	13	21	24	318
5	adequate space	0	27	0	7	62	392
6	AC/ air conditioning	2	1	12	35	46	410
7	Comfort seat passenger	1	1	11	36	47	415
8	seating capacity	0	9	3	4	80	443
9	Availability trash can	6	35	7	19	29	318
10	Facilities for self-rescue in an emergency	1	2	4	22	65	432
11	security of passengers	0	3	4	19	70	444
12	The compliance rates for the services given	1	18	8	12	57	394
13	Punctuality of departure and arrival at the school bus stop point	4	4	13	26	49	400
14	Down facilities for passengers	1	2	7	25	61	431
15	Passenger facility down in points	0	14	4	8	70	422
16	shielded from the sun in each bus stop	0	2	8	18	68	440
17	The availability of information boards in each bus stop	2	12	6	15	61	409
18	Availability trash can in each bus stop	0	9	4	15	68	430
19	Availability public telephone in each bus stop	2	40	8	18	28	318
20	Availability chair in each bus stop	1	5	7	19	64	428
21	Circulation	0	2	8	31	55	427
22	Availability CCTV in each bus	0	5	5	25	61	430

Table 5. IPA students who ride school bus in Malang City.

No. Attributes	Level of Satisfaction	Level of Importance	Degree of suitability (Tki) (%)	Mean $\overline{X}$	Mean$\overline{Y}$
1.	439	427	103 %	4.6	4.4
2.	424	439	97 %	4.4	4.6
3.	422	332	127 %	4.4	3.5
4.	385	318	121 %	4.0	3.3
5.	339	392	86 %	3.5	4.1
6.	433	410	106 %	4.5	4.3
7.	399	415	96 %	4.2	4.3
8.	275	443	62 %	2.9	4.6
9.	282	318	89 %	2.9	3.3
10.	367	432	85 %	3.8	4.5
11.	386	444	87 %	4.0	4.6
12.	206	57	361 %	2.1	0.6
13.	404	400	101 %	4.2	4.2
14.	385	431	89 %	4.0	4.5
15.	333	422	79 %	3.5	4.4
16.	323	68	475 %	3.4	0.7
17.	302	409	74 %	3.1	4.3
18.	275	68	404 %	2.9	0.7
19.	298	318	94 %	3.1	3.3
20.	387	64	605 %	4.0	0.7
21.	421	55	765 %	4.4	0.6
22.	331	405	82 %	3.4	4.2
Total	7816	7067	4188 %	81.4	73.6
Average	355.27	321.23	111 %	3.7	3.3

Note No. Attribute:

1=Friendliness officer,
2=Discipline of diver,
3=Door condition,
4=Altitude School Bus,
5=Adequate space,
6=AC/air conditioning,
7=Comfort seat passenger,
8=Seating capacity,
9=Availability trash can,
10=Facilities for self-rescue in an emergency
11=Security of passengers,
12= The compliance rates for the services given,
13=Punctuality of departure and arrival at the school bus stop point,
14=Down facilities for passengers,
15=Passenger facility down in points,
16=shielded from the sun in each bus stop,
17=The availability of information boards in each bus stop,
18=Availability trash can in each bus stop,
19=Availability public telephone in each bus stop,
20=Availability chair in each bus stop,
21=circulation,
22=Availability CCTV in each bus

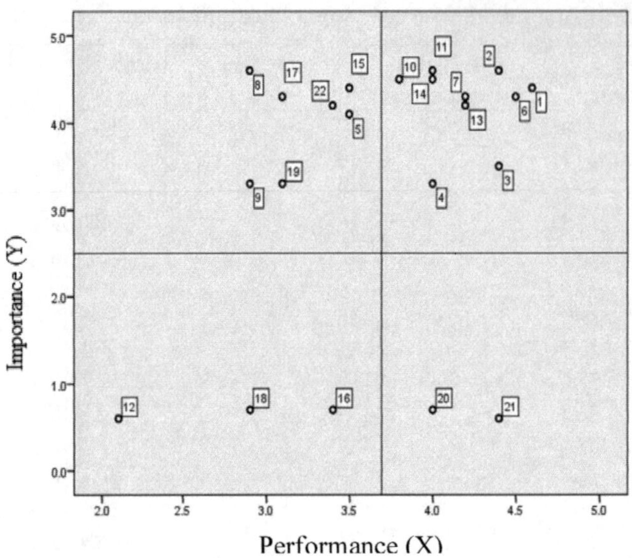

Figure 4. Importance Performance Analysis school bus services in Malang City.

The first quadrant (concentrate here) was considered as high important for students, but in fact these attributes is lower than expected (level of satisfaction is still very low). Therefore, the Adequate space, Seating capacity, Availability trash can, Passenger facility down in points, The availability of information boards in each bus stop, Availability public telephone in each bus stop, and Availability CCTV in each bus should get more attention or be improved so that the performance will increase. In the second quadrant (keep up the good work) attributes are considered to be high important by respondents and high levels of satisfaction. Therefore, the attribute Friendliness officer, Discipline of diver, Door condition, Altitude School Bus, Comfort seat passenger, Facilities for self-rescue in an emergency, Security of passengers, Punctuality of departure and arrival at the school bus stop point, and Down facilities for passengers should be maintain because the respondent think that this attribute is important and has had a good performance.

In the fourth quadrant (possible overkill) attribute number 20, 21 is considered low important by the respondents and high satisfaction. Attributes 20, 21 is maintained as in existing condition.

4. Conclusion

Through IPA, it can be concluded that school bus's level of service in Malang City have been able to serve customers with the mean of degree suitability (Tki) is 111 means that respondent are satisfied with 10 attribute. Meanwhile it must be observed and get more attention to improve by government in these case is the Department of Transport and Department of Education attributes which is lies in the first quadrant or concentrate here (attribute Adequate space, Seating capacity, Availability trash can, Passenger facility down in points, The availability of information boards in each bus stop, Availability public telephone in each bus stop, and Availability CCTV in each bus), in order to increase its performance.

5. Acknowledgments

I would like to express my special thanks to undergraduate students Atika Naggi, Suseno S, Kiki Agustina, Anggrina Merry, Satrio DU, Lusi Hana F, Iresa Lucky, Oase M and Novi Pujiana in Urban and Regional Planning Faculty of Engineering, Universitas Brawijaya (PWK FT UB) for data survey and data compilation in the interest of this paper.

References

[1] Sinambela, Lijan Poltak, dkk. Reformasi Pelayanan Publik: Teori, Kebijakan, dan Implementasi.Jakarta: PT. Bumi Aksara.2014.
[2] Warpani, Suwardjoko P. Pengelolaan lalulintas dan angkutan jalan. Bandung: Penerbit ITB 2002.
[3] Tamin, OZ. Perencanaan dan Pemodelan Transportasi – Edisi Kedua. Bandung: Penerbit ITB. 2000.
[4] Undang-Undang No. 22 Tahun 2009 tentang Lalu Lintas dan Angkutan Jalan.
[5] Kusmintardjo. Pengelolaan Layanan Khusus di Sekolah (Jilid II). Malang: IKIP Malang. 1992.
[6] Nugraha, et al. Kajian Efektivitas Pengoperasian Bus Sekolah di Jakarta. Depok: Universitas Brawijaya. 2013.
[7] Sevilla, C.G., Jesus A.O., Twila G.P., Bella P.R., Gabriel G.U. Research Methods, Rex Printing Co. Inc., Quezon City. 1993.
[8] John A. Martilla and John C. James, "Importance-Performance Analysis" (Journal of Marketing, January, 1977) pp. 77 – 79.
[9] Supranto, J. Pengukuran Tingkat Kepuasan Pelanggan untuk Menaikan Pangsa Pasar. Jakarta.Rineka Cipta. 2006.

IOP Conf. Series: Earth and Environmental Science **70** (2017) 012024 doi:10.1088/1755-1315/70/1/012024

Bus network redesign for inner southeast suburbs of Melbourne, Australia

S T Pandangwati, N A Milyana

Department of Architecture and Planning, Faculty of Engineering, Universitas Gadjah Mada, Jl. Grafika No. 2, Sekip, Yogyakarta 55281, Indonesia

sri.tuntung@ugm.ac.id

Abstract. Public transport is the most effective mode of transport in the era of climate change and oil depletion. It can address climate change issues by reducing urban greenhouse gas emission and oil consumption while at the same time improving mobility. However, many public transport networks are not effective and instead create high operating costs with low frequencies and occupancy. Melbourne is one example of a metropolitan area that faces this problem. Even though the city has well-integrated train and tram networks, Melbourne's bus network still needs to be improved. This study used network planning approach to redesign the bus network in the City of Glen Eira, a Local Government Area (LGA) in the southeastern part of Metropolitan Melbourne. The study area is the area between Gardenvale North and Oakleigh Station, as well as between Caulfield and Patterson Stations. This area needs network improvement mainly because of the meandering bus routes that run within it. This study aims to provide recommendations for improving the performance of bus services by reducing meandering routes, improving transfer point design and implementing coordinated timetables. The recommendations were formulated based on a 'ready-made' concept to increase bus occupancy. This approach can be implemented in other cities with similar problems and characteristics including those in Indonesia.

1. Introduction

Public transport is the most effective mode of transport in the era of climate change and oil depletion. Public transport can address climate change issues by reducing greenhouse gas emission and oil consumption while at the same time improving people's mobility [1]. Public transport is effective because it can carry many passengers with different destinations from anywhere in one trip. Therefore, it is essential that a reliable public transport system or network be developed. Network planning is one example of approaches that enhance the performance of public transport and can be implemented in any urban areas with different densities [2]. In network planning, the design of the network determines how public transport can run effectively. The idea of this 'ready-made' concept is how to produce maximum public transport services by utilizing minimum infrastructure (lines) and resources [2].

Network planning has been implemented in several North American and European cities. It is evident that this kind of multi-destination approach has performed better than the direct route approach [2]. The European Verkerhrsverbund (EVV) is an example of a good implementation of public transport network planning. This model integrates timetables of different transit modes and provides integrated public transport services in every part of the cities especially in the low-density areas where the transit frequency is low [3]. Furthermore, the integration of the grid system of the existing arterial roads network and the development of public transport networks has also been implemented in Toronto and Vancouver, Canada [4, 5].

Nonetheless, network planning has not been implemented in Melbourne. Radial routes are the predominant structure of public transport network in the city [2]. Melbourne's bus routes are tailor-made, which may lead to high operating costs and low frequencies [6, 7]. This study aims to formulate a set of recommendations to improve bus networks in Melbourne's inner southeast suburbs. The grid street pattern of many areas of Melbourne enables the adoption of the multi-destination network from the Squareville concept [2]. In the last part of this paper, lessons learned from Melbourne's context is briefly linked to TransJakarta, a BRT system in Indonesian capital city of Jakarta.

2. Literature Review

The Squareville Model. Squareville is a model that illustrates how the network effect works [2]. Fig. 1 shows that grid network (right) can provide more transit services than radial network (left and middle) in the same area. The picture on the left-hand side shows ten bus routes running north to the south. These bus lines provide less than 10% of the total assumed trips assumed generated in this city. The picture in the middle depicts double frequencies implemented on the ten bus lines.

Based on the traditional transport demand modelling, doubling the frequency of a line will create 50% increase in demand. However, the cost of doubling the frequency will increase more than 50%. In that case, doubling the frequency is not efficient to do. The figure on the right-hand side illustrates the network effect that is generated by 20 bus lines running north-south and east-west in the city.

This model can directly serve roughly 20% of the total trips generated, and even 100% of the total trips if people are willing to transfer. The elasticity of demand for this model will be 550%, which is significantly than the traditional one (50%). This model will be able to cater for more than a five-time increase in demand. This model also suggests that transfers are an important part of this concept as it enables the provision of anywhere-to-anywhere trip.

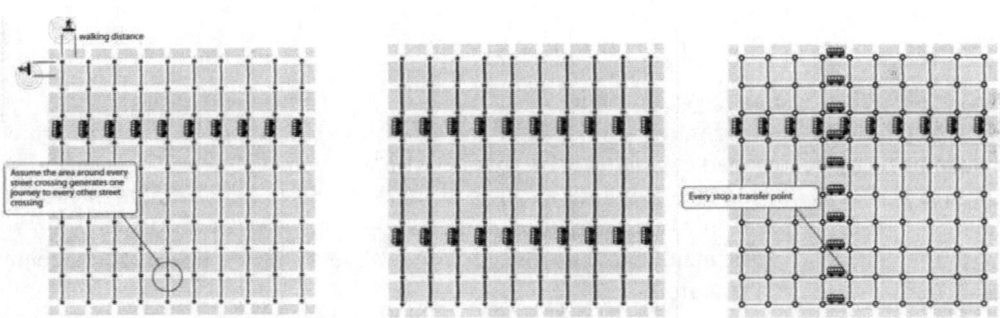

Figure 1. The network effects illustrated by Squareville Model [2]

3. Methods

The study area is located in the City of Glen Eira, an LGA in the inner southeast part of Metropolitan Melbourne, between Gardenvale North and Oakleigh Station, and between Caulfield and Patterson Stations. This area needs network improvement mainly because of the meandering bus routes that run within it (625, 626, 701 and 822). In terms of land use, residential area is predominant in this study area.

The analysis consists of two parts. The first one is the description of existing service supply and the second one is the quantification of service supply within the study area. The first part discusses key trip generators, meandering routes and statistics of bus patronage. This analysis utilises the online database from Public Transport Victoria (PTV) website. The second part is the calculation of the existing service kilometer. This combines the data of route lengths from the Glen Eira local area map and service frequency from the service timetables. The service kilometre was calculated by multiplying the routes length within the study area and the number of services of buses, trams and trains in either directions on the entire week including the weekdays (Monday-Friday) and weekends (Saturday-Sunday).

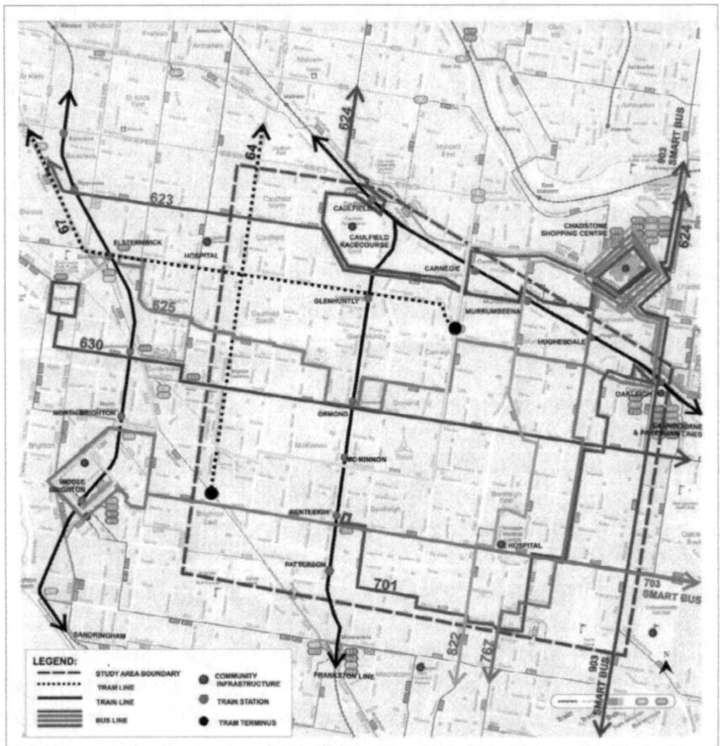

Figure 2. Trip Generators and Existing Public Transport Network.

4. Results

4.1 Description of the Existing Service Supply

The main external trip generator in the study area is the Chadstone Shopping Centre. The majority of bus services in the area is aimed to connect the residential areas with the shopping centre. Another important trip generator outside the study area consists of parks, hospitals, educational institutions and other smaller shopping centres in Middle Brighton and Oakleigh. Inside the study area, train stations and tram terminus points are the major trips generators. Some of these stations have activity centres and retail strips that are also trip generators.

Meandering bus routes and long waiting times are the main issues. There are four meandering bus routes run within the study area, routes 625, 626, 701 and 822. These kinds of tailor-made routes are not effective in terms of operational effectiveness and not attractive for passengers because they do not accommodate directness of travel and high-speed operating time [8]. Moreover, problems of ineffectivity also come from several areas where bus routes are overlapping, as depicted in the map below (Fig. 3).

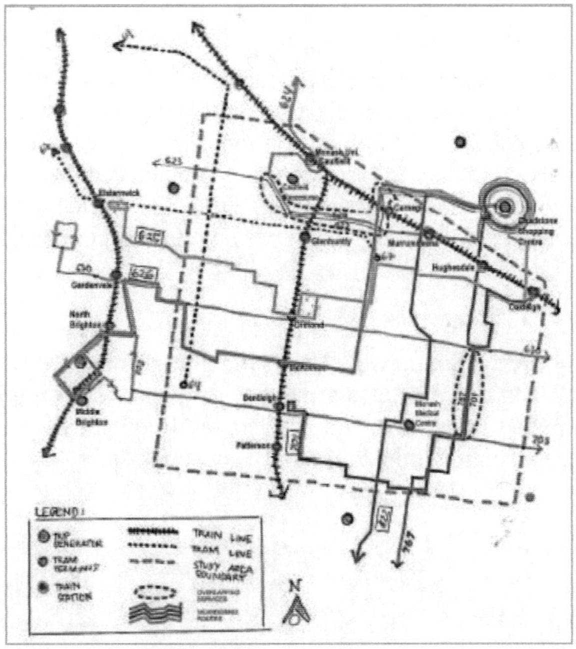

Figure 3. Meandering Bus Routes and Overlapping Services

Furthermore, the long waiting time and transfer time are also a critical problem. Based on the timetable of bus lines running in this study area, the average frequencies are mostly 20 to 30 minutes from Monday to Friday. Only the Smart Bus Route 903 has 10-15-minute frequencies. According to Mees [9], public transport services must have a minimum of 10-minute frequency during weekdays, especially during peak hours. The timetable of existing bus services shows that the frequencies are much reduced on weekends. Buses mostly run every 30 minutes to 1 hour. Meanwhile, even during low demand period, the frequency should be no more than 30 minutes [4, 9].

Lastly, the bus patronage growth data showed an increase in most of the bus routes operating in the study area. Only two routes, 630 and 822, had significant decreases in patronage on both weekdays and weekends. Furthermore, while Route 701 increased significantly by about 20% in terms of annual and average weekday patronage, its patronage in the weekend decreased slightly. Additionally, the patronage of Smart Bus routes, 703 and 903, grew slightly by no more than 10%, and decreased slightly in terms of average weekend use. This data depicts the need of public transport network improvement in fostering the patronage growth as well as addressing the patronage decrease in some bus routes.

Table 1. Bus Patronage Growth 2010 – 2012 (PTV, 2015)

Bus Lines	Patronage Growth (%)			
	Annual	Average Weekday	Average Saturday	Average Sunday
623	22.1	15.0	19.1	4.3
624	15.8	8.6	1.5	15.4
625	N/A	56.3	25.2	32.0
626	N/A	48.7	37.1	50.6
630	-13.1	-19.2	-12.6	-15.9
701	28.0	22.5	-4.6	-8.5
703	6.2	0	-6.9	-9.0
767	18.0	11.6	0.1	2.4
822	-1.6	-6.2	-20.0	-24.3
903	9.6	3.8	-3.0	-4.4

4.2 Quantification of existing public transport service kilometre.

The calculation of the existing service provision is based on the service kilometer in a week (Monday to Friday, Saturday and Sunday). The focus of this calculation is the service kilometer of the bus service. The proposed bus network discussed in the next section should not exceed the existing service kilometer to ensure the minimum cost. The service kilometer of the existing bus service in the study area is roughly 25,663 service kilometers (s km) per week, 4,355 s km per day on weekdays, 2,109 s km on Saturdays and 1,776 s km on Sundays.

4.3 Recommendation.

This study proposes a new public transport network that is integrated and effective in Melbourne's Inner Southeast. This aim can be broken down into two main objectives, i.e. to implement network effect concepts and to create coordinated timetables and services as frequent as needed. These objectives intend to address the main issues mentioned in community consultation.

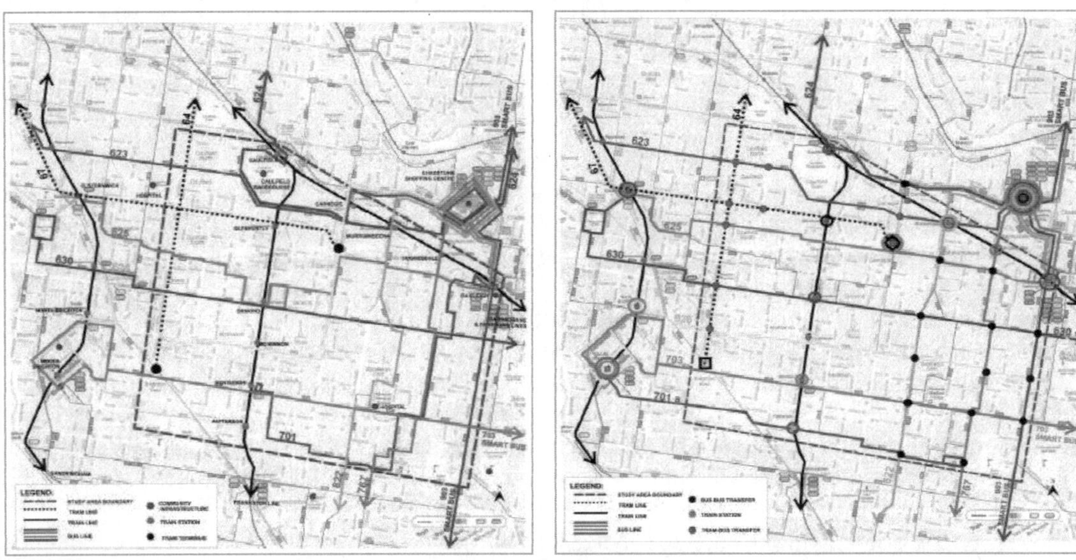

Figure 4. Existing Public Transport Network (left) and Proposed New Bus Lines and Transfer Points (right)

The meandering routes can be replaced by a more effective network so that network effect can be created. The proposed bus network is formed according to the existing arterial road network within the study area (Fig. 4). One siginificant change is to cut Route 701 into 701 A by removing the route duplication with Route 767. This grid network will provide faster journeys and create more transfer points. For example, the length of the new bus route 625 is 1.42 km less than the existing one as it removes the meandering routes around Ormond station. Therefore, passengers from Chadstone Shopping Centre who wish to travel to Elsternwick Station will only spend roughly two minutes (if the bus runs 40 km/hour) in the new bus network. Moreover, people living in the southern part of Glen Huntly Station such as around Oakleigh Road and Mara Road, which originally have no bus services can also benefit from this new 625 route.

As mentioned in the previous section, the existing bus service kilometer is 25,663 service kilometers per week. The proposed new bus network generates roughly similar (25,569) service kilometer as the existing one, with additional frequencies for particular routes. The proposed grid network generates a service kilometer gap and this proposal uses this gap to increase the service frequency for route 624, 626 and the new route, 701A. This will support the objective of this proposal, which is to reduce the waiting time or transfer time for buses.

In addition, integrating rail-bus and bus-bus transfer time can address the long waiting time issue. The tram and train services in this study area are relatively frequent especially during weekdays and peak hours. However, their connections with bus services, especially from rail to bus, need to be improved. These different modes can be integrated effectively by creating time-transfer to connect [3]. Fig. 5 shows the proposed time-transfer to better connect bus and train services.

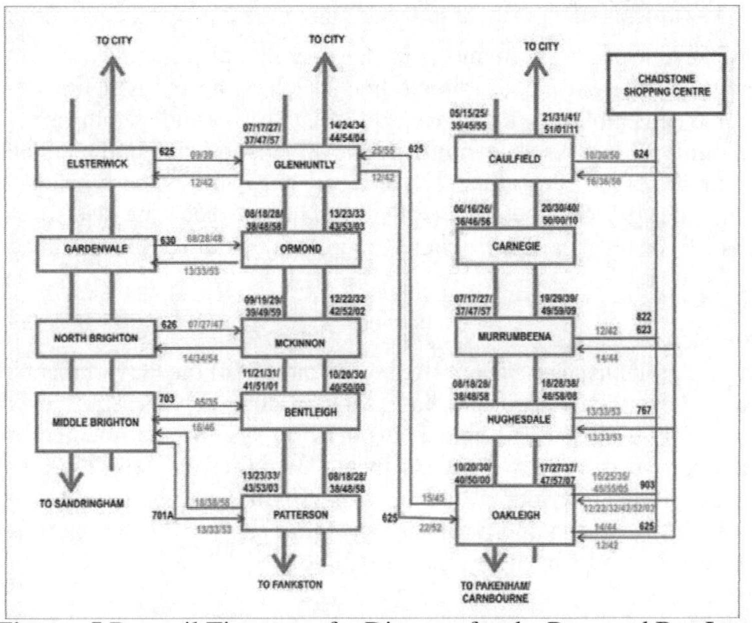

Figure. 5 Bus-rail Time-transfer Diagram for the Proposed Bus Lanes

Instead of increasing the service kilometer of all frequencies that will increase the cost, it will be better to use the existing resources without creating additional cost. As mentioned in the Squareville model, if people are willing to transfer, the grid network will generate a significant increase in bus patronage. In theory, people are willing to transfer if they do not need to wait too long to transfer. Therefore, integrated timetables and time-transfer are important to minimize the transfer time [2]. Good timetabling and integration among different modes of transport contributes in the increase of public transport patronage [2, 3, 6, 9].

Figure. 6 Examples of Convenient Transfer Point within a Transit Stop Highlighting The Minimum Walking Distance and Clock-face Scheduling for Transferring to Other Public Transport Modes [10] [11]

The improvement of transit stop design can also support the transfer between rail-bus and bus-bus services. Coordinated timetables together with a convenient station design that accommodate short walking distance for transfer and provide different kind of amenities will create interesting transit trips environment for the passengers [8]. Public Transport in Zurich has some good examples of a good station design. Fig. 6 shows that the schedules of these different public transport modes are integrated. Cable car, bus, tram and train come at the similar time, and they stop closely to each other. The design of the transfer points in the transit stops is ultimately critical in supporting the proposed integrated and clock-face departure timetables.

Fundamentally, these aspects are included in the operational principles of network planning. It consists of integration of different modes, simple line structure, accommodation of direct and multi-destination routes, and timetabling management [2, 8]. Brown and Thompson [14] suggest that integration between train and bus services improved transit performance in the US. The objectives can be measured according to several indicators such as a rail-bus coordinated timetables and a minimum of 10 minutes time transfer [9]. Grid network is therefore highly recommended for developing a more effective network, as it reduces meandering routes, increases operating speed and achieves maximum operating efficiency [8, p. 186].

4.4 Transjakarta Bus Rapid Transit (BRT) System

The BRT system known as Transjakarta is operated by an agency of the Government of Jakarta with the same name, although the infrastructure and buses are procured by its Transportation Department. Jakarta's 12-corridor, 200 km-long Bus Rapid Transit (BRT) system was initiated by the Governor of Jakarta in 2003, to improve travel for Jakarta's then 8.3 million citizens representing 40% of the metropolitan population. BRT corridor 1, the first in South East Asia, was completed from the commercial hub of Blok M in the southern part of the central city to Kota, the northern area of the city in January 2004. Corridors 2 and 3 opened in January 2006 [13].

4.4.1 Patronage Trends.

Trends in daily patronage from 2004 to 2014 are shown in Fig. 6, along with cumulative BRT length and an index measuring real fares (that have remained at 3,500 rupiahs in nominal terms to date). When the project was finalized in 2012, the BRT system was 180 km in length with the patronage of 370,000 pax/day (2,060 pax/km). As a comparison, in 2006, the patronage was 105,000 pax/day (or 2,500 pax/km) in 2006 [14]. During the project, patronage had increased by 240% although productivity measured in terms of pax/km declined by 18%.

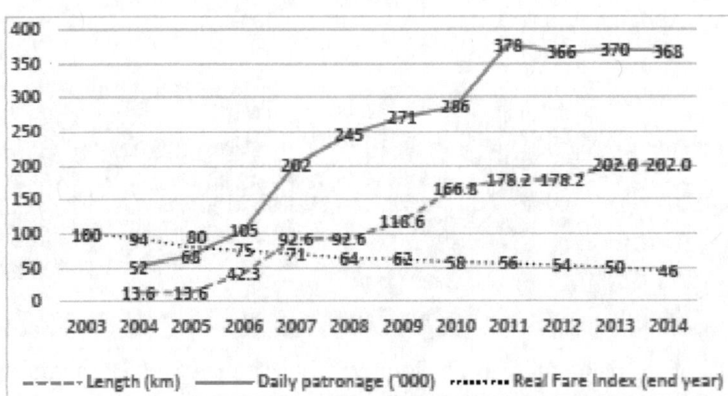

Figure. 7 Trends in patronage, BRT length and real fares 2003-2014 [13]

The actual increase in patronage from 2006 to 2012 was less than half of the planned increase of 600,000 pax/day. By 2012, stagnate declining in patronage growth was revealed to be 3%, then stabilising over 2013 to 2014 due to poor performance, despite the halving of real value of fares from 2004 to 2013 and the addition of a new corridor in 2013.

4.4.2 Passenger Perception

At the close of the project, Transjakarta held a survey to learn passenger perception toward the BRT operational aspects. The survey found that passengers rated the following attributes of BRT as 'generally poor': drivers (poor); officers (very poor); infrastructure (very poor); buses (very poor); operations (good); and ticketing (poor). The UNEP-GEF post-evaluation team's surveys undertaken in January 2014 found that a key attribute that BRT passengers preferred as a whole about the service was its low fare. However, 48% considered waiting times to be 'very long' or 'long' indicating problems with BRT service regularity and reliability. Twenty nine percent of BRT users rated the service as 'fast', 57% rated the service as of 'moderate' speed, and 14% rated it 'slow'. Overall, 43% of users said they were 'satisfied' with the service, 39% said it was 'adequate' and 17% said they were 'dissatisfied' [11, pp. 85-89]. The passenger's overall perception of 'generally poor' to the BRT services they receive underlines the significant challenges facing Transjakarta in its future efforts to improve its performance and increase patronage.

4.4.3 Improvement Recommendation Plan Based on the Melbourne Case.

Given the array of problems currently facing Transjakarta, optimisation of existing corridors, particularly by assessing life-cycle costs and travel time impacts by patronage market segment should be considered. Globally, fares cannot be defined on nodes of future busways because of the free transfer system. Fares should be included on access links of the system and/or transfer links on the actual network, defining a specific mode with a time equivalence of fare that can vary automatically changing attributes in macros. In order to improve the quality of Transjakarta service delivery, it is recommended that further improvement on existing BRT, or other rapid transit projects, focus on collecting travel data on rapid transit users and their households shortly after a project opens; distinguish drivers and passengers when considering shifts from private modes; and identify their household vehicle availability..

5. Conclusions

The meandering routes and uncoordinated timetables are the main public transport issues in Melbourne's Inner Southeastern Suburbs. This study addresses these issues by formulating a set of recommendations, such as implementing coordinated timetables, establishing a good public transport network and improving transit stop design. These recommendations are based on utilizing the existing resources (service kilometer) to gain more benefit with no extra cost. The methodology used in this study can be used for further studies in formulating strategic plans for improving public transport network in other cities including Jakarta, Indonesia.

References

[1] Tripathi, N., Where the Road is without Congestion and Environemnt is Pollution free: Trip to Susutainable Urban Mobility. OIDA International Journal of Sustainable Development, 2013. 6(8): p. 27.

[2] Mees, P., et al., Public Transport Network Planning: a guide to best practice in New Zealand cities. 2010, New Zealand Transport Agency: Wellington.

[3] Pucher, J. and S. Kurth, Verkehrsverbund: the success of regional public transport in Germany, Austria and Switzerland. Transport policy, 1996. 2(4): p. 279-291.

[4] Toronto Transit Commission, Service Standards Manual. 1990.

IOP Conf. Series: Earth and Environmental Science **70** (2017) 012024 doi:10.1088/1755-1315/70/1/012024

[5] Translink, Transit Service Guidelines Technical Report. 2005.

[6] Lazanas, K. and J. Stone, An Investigation of Institutional Arrangements for Design and Delivery of Multimodal Public Transport in Melbourne. 2010, 33rd Australasian Transport Research Forum: Canberra.

[7] Nielsen, G. and T. Lange, HiTrans Best Practice Guide No. 2. Public Transport: Planning the Networks (Stavanger, Norway: HiTrans, c/o Rogaland County Council), 2005.

[8] Vuchic, V.R., Urban transit: operations, planning, and economics. 2005.

[9] Mees, P., Transport for suburbia: beyond the automobile age. 2010: Earthscan.

[10] Creative Urban Projects (2015) Polybahn's integration with city and public transit.

[11] SBB Zurich (2015) The Z-Pass connects up networks., At: Zurich.

[12] Brown, J.R. and G.L. Thompson, Service Orientation, Bus—Rail Service Integration, and Transit Performance: Examination of 45 US Metropolitan Areas. Transportation Research Record: Journal of the Transportation Research Board, 2008. 2042(1): p. 82-89.

[13] Sayeg, P. and Al-Rasyid, L.H., Terminal Evaluation of the UNEP/ GEF Project "Bus Rapid and Pedestrian Improvements Project in Jakarta. Prepared for the UNEP Evaluation Office, May 2014. http://www.unep.org/eou/Portals/52/Reports/2954_2014_TE_UNEP_REGIONAL_CC_FSP_Jakarta_BRT_May_2014.pdf.

[14] Davies, S. and Gleave. Jakarta Feeder Bus Study: base year transport model report. February 2011.

IOP Conf. Series: Earth and Environmental Science **70** (2017) 012025 doi:10.1088/1755-1315/70/1/012025

Development of A Solar-Powered Electric Bicycle In Bikesharing Transportation System

S Adhisuwignjo[1 2], I Siradjuddin[1], M Rifa'i[1], R I Putri[1]

[1]Electrical Engineering, State Politechnic of Malang, Malang, Indonesia

supriatna_s@yahoo.com

Abstract. The increasing mobility has directly led to deteriorating traffic conditions, extra fuel consumption, increasing automobile exhaust emissions, air pollution and lowering quality of life. Apart from being clean, cheap and equitable mode of transport for short-distance journeys, cycling can potentially offer solutions to the problem of urban mobility. Many cities have tried promoting cycling particularly through the implementation of bike-sharing. Apparently the fourth generation bikesharing system has been promoted utilizing electric bicycles which considered as a clean technology implementation. Utilization of solar power is probably the development keys in the fourth generation bikesharing system and will become the standard in bikesharing system in the future. Electric bikes use batteries as a source of energy, thus they require a battery charger system which powered from the solar cells energy. This research aims to design and implement electric bicycle battery charging system with solar energy sources using fuzzy logic algorithm. It is necessary to develop an electric bicycle battery charging system with solar energy sources using fuzzy logic algorithm. The study was conducted by means of experimental method which includes the design, manufacture and testing controller systems. The designed fuzzy algorithm have been planted in EEPROM microcontroller ATmega8535. The charging current was set at 1.2 Amperes and the full charged battery voltage was observed to be 40 Volts. The results showed a fuzzy logic controller was able to maintain the charging current of 1.2 Ampere with an error rate of less than 5% around the set point. The process of charging electric bike lead acid batteries from empty to fully charged was 5 hours. In conclusion, the development of solar-powered electric bicycle controlled using fuzzy logic controller can keep the battery charging current in solar-powered electric bicycle to remain stable. This shows that the fuzzy algorithm can be used as a controller in the process of charging for a solar electric bicycle.

1. Introduction

The world population has been progressively concentrating in the cities [1]. Cities around the world are undergoing rapid urbanization. Mobility in urban areas is one of the challenge that must be addressed and improved to get a better quality of life for the community [2]. The increasing mobility has strong correlation to the traffic conditions, extra fuel consumption, automobile exhaust emissions, air pollution, and quality of life. Urban mobility is a prevalent problem in many cities around the world. Issues on the urban mobility affects the quality of life and environmental sustainability are gaining importance in the world.

In recent years there have been over energy use and carbon emissions in the transportation sector, which then leads to an attempt to reduce the impact of transportation activities. In 2011 the transportation sector uses about 33% of the energy consumption in Europe [3]. So that the promotion of non-motorized modes of transportation received significant attention within the framework of sustainable urbanism and vision of eco-mobility [4].

Apart from being a clean, cheap and equitable mode of transport for short-distance journeys, cycling can potentially reduce traffic congestion. Cycling can be considered as one of the solution for urban mobility problems especially for short distance trips. Cycling offers many benefits to the problem of urban mobility. Cycling offers a fast, cheap and efficient transportation option for end-to-end short-distance trips, with smaller carbon than driving a car. Cycling can also encourage a modal shift from private car to public transport. To get these benefits, many cities have tried promoting cycling using different policies, particularly through the implementation of bike-sharing projects. [5].

Bike sharing scheme as an alternative transportation system has been growing rapidly throughout the world. Cities around the world have invested on bike sharing systems to improve their mobility and reduce the car dependency [6]. In recent years, bike-sharing has become popular in the world [7]. A bike-sharing system is a short-term rental scheme allowing bicycles to be collected and returned at any of self-serve stations. It represents a transportation alternative by complementing the urban public transportation. It enables commuters to flexibly use bicycles without incurring the cost and the trouble of owning and maintaining them. There has been rapid growth in bikesharing systems, and apparent success in attracting riders. In recent years, bikesharing systems have spread throughout the World. As of June 2014, public bikesharing programs existed on five continents, including 712 cities, operating approximately 806,200 bicycles at 37,500 stations [8]

Recently, many cities have implemented the third generation of public bikesharing systems. These new bikesharing systems overcome many of the operational problems associated with older bikesharing programs that operated with free or coin-deposit schemes. The third generation of bikesharing systems have docking stations to rent and return bicycles also use key smartcards for access to a bicycle. The technology improvements have enabled their success of bike sharing programs. In Europe, the deployment and expansion of the third generation bikesharing schemes in recent years have been implemented rapidly. The third generation is now approaching market saturation. At present, the fourth generation bikesharing system has been introduced which utilizes electric bicycles and clean technology implementations. Utilization of solar power is probably the key developments in the fourth generation bikesharing system and will become the standard in bikesharing system in the future [9]. The fourth generation bikesharing program uses a solar-powered electric bicycle as the main vehicle. Electric bikes use batteries as a source of energy and thus require a battery charging system with a source of energy from solar cells. Therefore, it is necessary to develop an efficient battery charging system. This work presents a novel method of a battery charging system using fuzzy algorithm.

2. Literature Review

2.1 Bikesharing.

Bikesharing program is a bike user community network which is distributed around town at low cost. Bicycles can be borrowed at any time at the nearest bike station and can be returned to a bike station to another so bikesharing program ideal for transport from point A to point B. To use a bicycle, a person must register as a member and use the card members, users simply swipe the card members to make the process bicycles [10]. Bike-sharing is a convenient and 'green' transportation mode, therefore, plays an important role as a complementary within comprehensive transportation system [11]. Bikesharing has the potential to play an important role in bridging some of the gaps in existing transportation networks, as well as encouraging individuals to use multiple transportation modes. [8]

At this time the fourth generation has began to appear that integrates with other modes of public transport and clean technology. Utilization of clean technologies including solar power will likely become the standard in the future bikesharing system. [9]. The author in [12] has suggested that the bike sharing system based on electric bicycles provide a high level compared with the conventional bike sharing system, while maintaining low environmental impact. An electric bicycle reduces the trip effort required and the travel time [13], though at a higher cost due to the additional requirement of electricity. E-bike sharing system are expected to contribute, among others [14] include:

- reduction of single occupancy journeys with car, and thus ease of traffic congestion,

- reduction of CO2 emissions from motorized traffic, and thus improvement of air quality,
- increased of physical activity levels and improvement of public health,
- improvement of accessibility and flexibility of mobility,
- improvement of road safety (particularly for cyclists), and
- enhancement of the profile and conditions of living in modern cities, supporting also local economies and tourism.

2.2 Fuzzy Logic.

A basic configuration of a fuzzy logic system consists of four main components, namely a fuzzyfication device, knowledge base, decision making and defuzzyfication logic devices. This fuzzy controller consists of two input variables and one output variable. The input variable consists of an error (Error, E) and the change of error (CE). The concept of fuzzy logic is used for the handling and processing of linguistic information. Since 1980 Fuzzy Logic Controllers (FLCs) has been successfully applied with success in many industries [15], for instance: the process control, image processing, motor control, the automatic train operation, comcorder outfocusing, servo motors control, aircraft control, setting spacecraft position and others.

Major factors in FLC are:

- The value of the inputs and outputs as well as the universe of discourse (universe discourse) input and output,
- Scaling factor input-output variables,
- Fuzzy membership function used in the preparation of fuzzy values for each variable input-output, and
- Rule base.

In this study, FLC planted in memory of the microcontroller ATmega8535 and membership functions to be stated in the charging current error in the microcontroller ATmega8535 expressed in Figure 1. Membership function delta error in the charging current ATmega8535 microcontroller expressed in Figure 2. PWM output membership function is expressed in Figure 3.

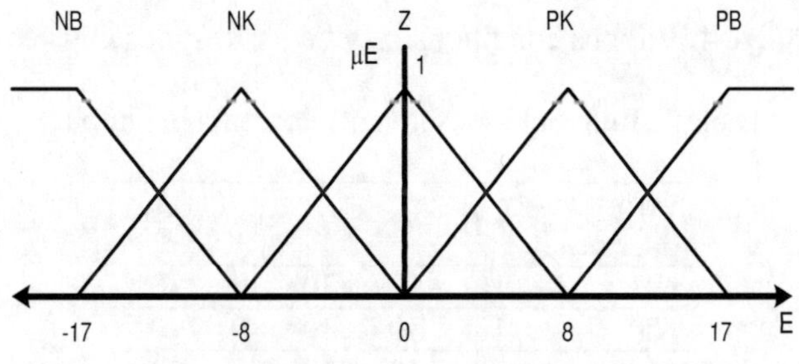

Figure 1. Membership Function error (E) Charging Current on the ATmega8535

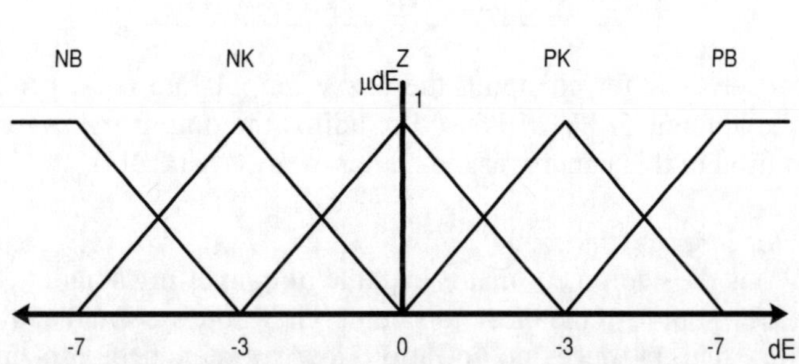

Figure 2. Membership function Delta Error (dE) Charging Current on the ATmega8535

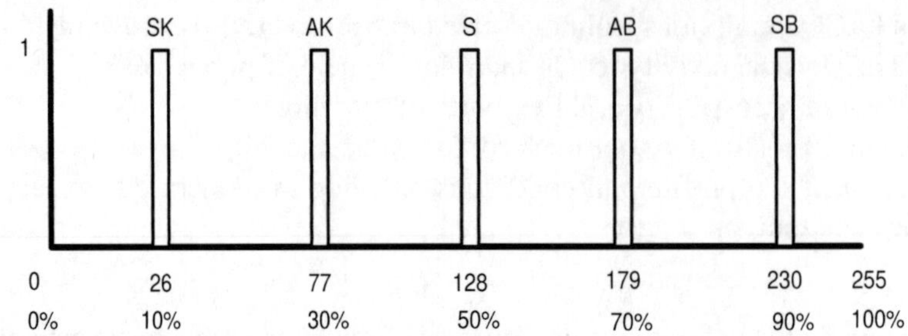

Figure 3. Membership Function PWM Output ATmega8535

A phase of fuzzy rules set needs to consider the form of the system response that may occur as shown in Figure 4. From this figure there are 25 possibilities for error and delta error combinations. The y axis represents the current set point of 1.2 A with an error of ± 1 A while delta-error taken ± 0.4 A. The designed fuzzy rule is shown in Table 1. Decision methods (inference) used by the fuzzy controller fuzzy logic is MAX-MIN method.

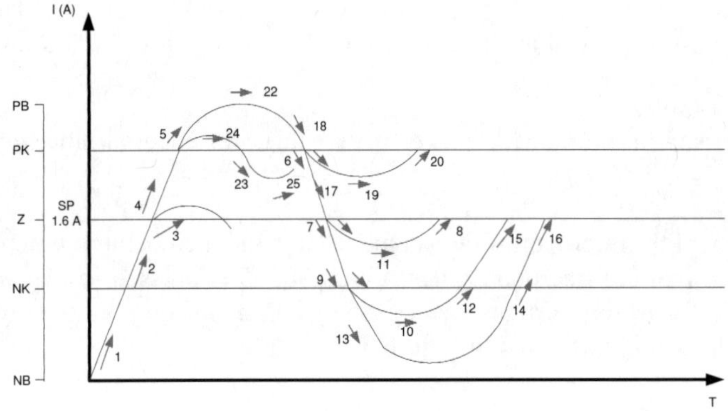

Figure 4. Various conditions may be on Response System

Table 1. Rule Set (consequent) Percentage Output

Error / d-Error	NB	NK	Z	PK	PB
NB	10	10	30	30	50
NK	10	30	30	80	90
Z	10	30	50	50	90
PK	50	50	80	50	50
PB	80	80	80	80	90

Defuzzification is the process for changing the fuzzy output into a crisp output. Defuzzification results that determine the amount of PWM is used to adjust the output voltage of the buck converter. Defuzzification method used in this paper was the center of gravity (COG).

2.3 Photovoltaic and Charging System.

Photovoltaic system (PV) is the equipment that is capable of converting solar energy into DC electrical energy. Solar cells become a source of the most important renewable system that offers many advantages such as it does not require fuel, produces no pollution, low maintenance costs and produces no noise. The solar cells system is a promising renewable energy source in many tropical countries, where the sun energy irradiation does not fluctuate too high throughout the year and radiates about 12 hours a day.

Solar energy received in one day (solar insolation or solar irradiation) can vary from 0:55 kWh / m (2 MJ/m) in a cool area until 5:55 kWh/m (20 MJ/m) in the tropics [16]. I-V characteristics of a solar cell is shown in Figure 5. From the characteristic curve can be seen that the solar cell is a current source, which generates a constant current for a variable output voltage [5].

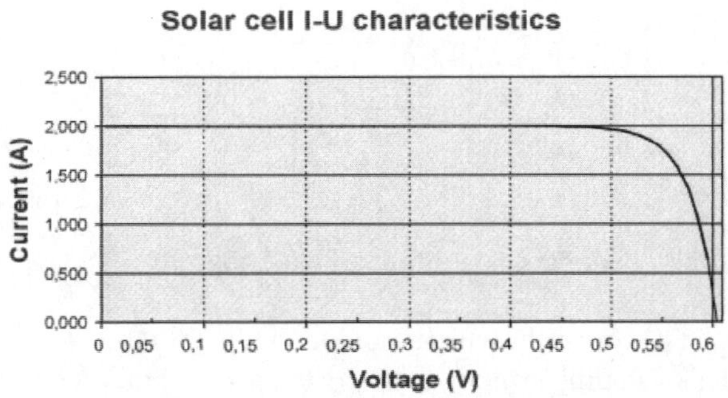

Figure 5. I-V Characteristic of Photovoltaic

The basic principles of Sustainable Tourism is optimally utilize environmental resources while maintaining the ecology and conservation, respect for cultural authenticity and communities and ensure the long term sustainability [6]. Environmental Carrying Capacity of natural attractions is the ability of natural attractions in the area and a certain time unit to accommodate the number of tourists [7]. Tourism carrying capacity has an important role in the management of the area because it was considered as a systematic, strategic policy tool in the planning process [8].

Management of protected areas as a tourist spot is one effort to build a sustainable management. Development of tourism in the conservation area occurred from mass tourism to an individual or small group tourism. The development of this travel pattern focuses on the experience with the motivation to get the expansion of life [9].

Physical carrying capacity is the maximum number of travellers who are physically fulfilled by the space provided at a certain time. The real carrying capacity was the number of visitors allowed to visit a tourist attraction with a correction factor derived from the characteristics of the object that is applied to the Physical carrying capacity [10]. The effective carrying capacity was the maximum number of visitors received at a tourist attraction, was obtained from the value of the real carrying capacity that related to the capacity menagemen attractions manager [8].

3. Methods

The electric bicycle batteries charging system proposed in this paper, as shown in Figure 6, consists of photovoltaic, boost converter, buck converter, microcontroller, current and voltage sensors, and battery. Photovoltaic converts solar energy into DC voltage and the buck converter generates an output voltage that is lower than the input voltage by the PWM pulse that is given to the IGBT switches. While the microcontroller has a function as PWM generator for generating a PWM signal by adjusting the duty cycle. The output voltage is generated depending on the duty cycle of the PWM signal given to the IGBT. Current and voltage sensors will detect current and voltage that flow to the power storage. Fuzzy Logic Controller (FLC) role controls the value of the PWM signal. FLC inputs are error voltage, delta error voltage, error current, and the delta error current. These inputs are used to determine the amount of PWM value. If the battery is full, microcontroller will disconnect the power storage from buck converter. On the other hand, there is a boost converter that serves as a charger circuit for electric bicycles. Boost converter circuit produces a higher voltage than the input voltage from the power storage so that current can flow from the power storage to the electric bicycle batteries.

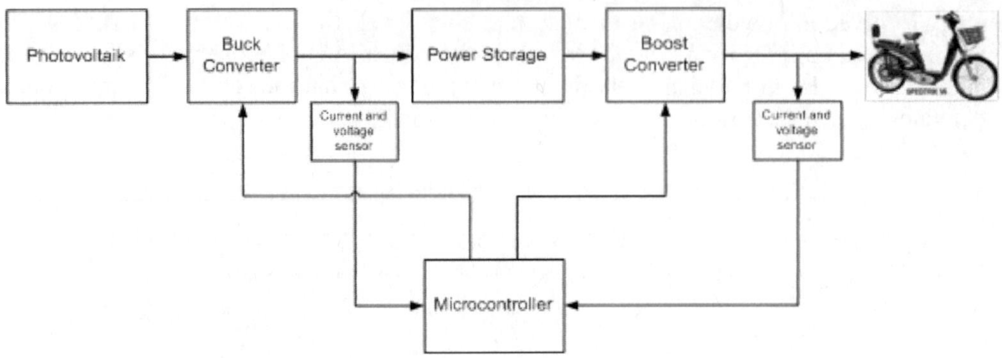

Figure 6. System Block Diagram

Buck converter is a DC to DC converter which provides a lower output voltage than the input voltage. Buck converter circuit on a bike sharing system connects the photovoltaic to the power storage, as shown in Figure 2. The buck converter consists of inductor (L), IGBT as a switch, diode and capacitor as filter. The working principle buck converter is as follows, during the interval when the IGBT is on, the diode in Figure 7 become reverse biased and the photovoltaic provides energy to the power storage as well as to the inductor. During the interval when the IGBT is off, the inductor current flows through the diode, transferring some of its energy to the power storage.

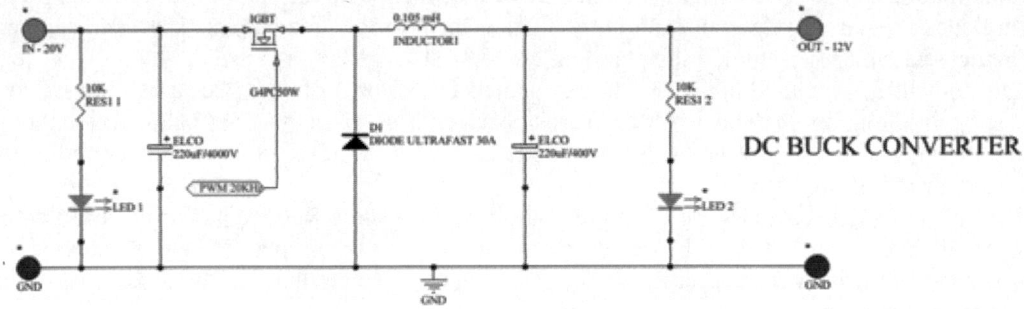

Figure. 7 Buck Converter for Bike Sharing

Boost converter is a DC to DC converter which provides a higher output voltage than the input voltage. Boost converter circuit on a bike sharing system is connecting the photovoltaic and battery, as shown in Figure 8 that consists of inductor (L), IGBT as a switch, diode and capacitor as filter. The workings of boost converter to raise the voltage is divided into two modes, the first mode when charging the inductor current (IGBT on) and the second mode is re-charging inductor (IGBT off).

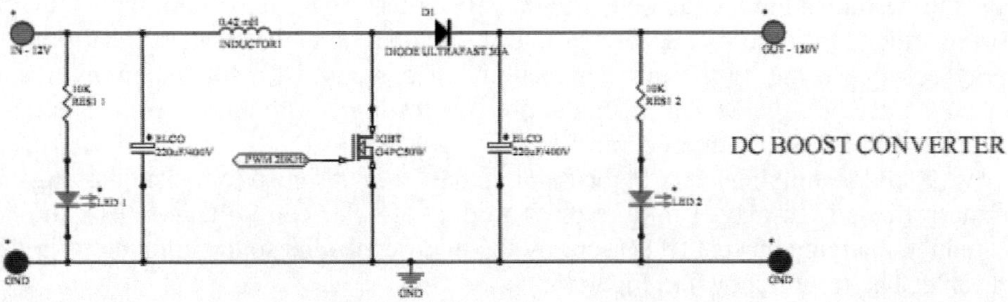

Figure. 8 Buck Converter for Bike Sharing

In the first mode, the inductor will be short circuited with the source. Because all of the current flowing through the switch to a negative source, the current flowing to the load can be ignored. Load voltage (Vb) will be zero when the charging process so that the inductor voltage (VL) will be equal to the voltage source (Vs) which can be expressed by

$$Vs = L\frac{dl}{dt} \tag{1}$$

Since switch is on and duty cycle is on, dt on eq (1) can be changed by ton and di = ΔI so that eq (1) can be expressed by

$$V_s = L\frac{\Delta I}{t_{on}} \tag{2}$$

When the switch is in the open position, the voltage source is coupled to the series inductor and load that it causes inductor in discharge state. In this condition, the inductor acts as a current source to supply the load. In this mode, the load is supplied from two sources, namely from the input source and the inductor discharge, the condition which causes the output voltage is always greater than the input voltage. The time span switch on and switch off will affect output voltage produced. If the phase switch on time then the longer the charging process, so that when the switch off voltage adders of the inductor becomes larger. The relationship between the input voltage and the output of the boost converter can be expressed.

$$V_R = Vs\frac{1}{(1-D)} \tag{3}$$

Boost converter is used to raise the voltage of the photovoltaic for battery charging. In this paper, bike sharing system using photovoltaic with 20Volts maximum output voltage range, input voltage 12 V 60Ah battery and a switching frequency of 20kHz. Inductor value used in this circuit depends on switching frequency (f), duty cycle (D) dan load resistance (R) that can be expressed by

$$L_{min} = \frac{D(1-D)^2R}{2f} \tag{4}$$

While capacitor value is determined by ripple (ΔV) that can be expressed by

$$C = \frac{D}{R(\%\Delta V)f} \tag{5}$$

.

4. Results

This research aims to design and implement electric bicycle battery charging system with solar energy sources using fuzzy logic algorithm. It is necessary to develop electric bicycle battery charging system with solar energy sources using fuzzy logic algorithm. The study was conducted by means of experimental method including the design, manufacture and testing controller systems. The designed fuzzy algorithm planted in EEPROM microcontroller ATmega8535.

The circuit was tested by applying fuzzy algorithm that has been designed to make charging electric bike. The charging process is displayed in a graph. The data of solar cell voltage (Vpv), the battery voltage (Vbat), and the charging current (Ibat) sent by the microcontroller to the computer via RS-232 communication cable. The results shown in Figure 8.

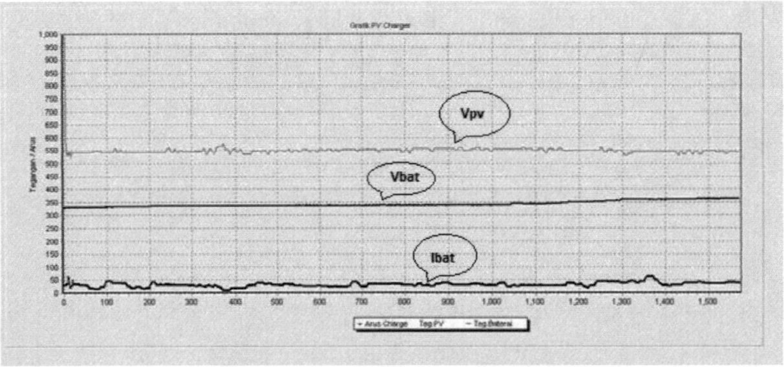

Figure. 9 Fuzzy Logic Controller Test

Based on the graph shown in Figure 9 the charging current is still unstable. This shows that the planning has been done is still not in accordance with the characteristics of the input and output, so it is still necessary to set the optimization rules and membership functions of input and output. After changing the triangular to the singleton output membership, membership input error, error delta, and set the rules obtained regulatory process flow chart as shown in Figure 9.

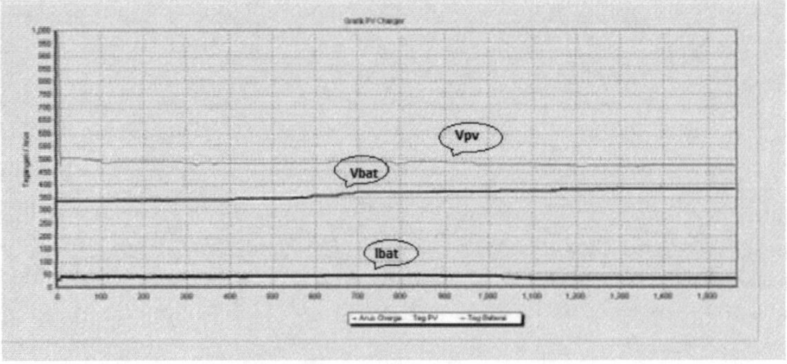

Figure 10. Optimization Results of Fuzzy Logic Controller Test

Figure 10 shows that the charging current is more stable when compared to the previous process. This shows that the fuzzy algorithm optimization results that have been made are in accordance with the characteristics of the system.

The charging current is set at 1.2 Amperes and when the full battery voltage of 40 Volts. The results showed a fuzzy logic controller is able to maintain the charging current of 1.2 Amperes with an error rate of less than 5% around set point. The process of charging electric bike lead acid batteries from empty to full takes times about 5 hours.

5. Conclusions

From the discussed results, it can be concluded that the development of solar-powered electric bicycle controlled using fuzzy logic controller can keep the battery charging current in solar-powered electric bicycle to remain stable. This shows that the fuzzy algorithm can be used as a controller in the process of charging in solar electric bikecycle. So, the solar-powered electric bicycle system developed can be applied to the program bikesharing to reduce traffic congestion which is a prevalent problem in many cities around the world.

References

[1] Christos S. Ioakimidis, Pawel Rycerski, Sesil Koutra, Konstantinos N. Genikomsakis. A University E-Bike Sharing System used as a Real-Time Monitoring emissions tool under a smart city concept. EVS29 Symposium Montréal, Québec, Canada, June 19-22, 2016.

[2] Keskin, D. Analysis of public use bicycle systems from a Product-Service System perspective, Dissertation (Master of Science in Industrial Design). Middle East Technical University, Ancara. 2006

[3] EUROSTAT. Environment and energy, EUROPA eurostat – data navigation tree. [Online]. Available: http://ec.europa.eu/eurostat. 2013. Accessing date October 31, 2016.

[4] S. L. Handy, Y. Xing, and T. J. Buehler, "Factors associated with bicycle ownership and use: A study of six small U.S. cities," Transportation, vol. (37), pp. 967-985, Nov. 2010.

[5] Kumar, Ashwani., Kwong Meng Teo, Amedeo R. Odoni. A Systems Perspective of Cycling and Bike-sharing Systems in Urban Mobility. http://www.systemdynamics.org/conferences/2012/proceed/papers/P1306.pdf. Accessing date October 31, 2016.

[6] P. Baptista, A. Pina, G. Duarte, C. Rolim, G. Pereira, C. Silva, and T. Farias, "From on-road trial evaluation of electric and conventional bicycles to comparison with other urban transport modes: Case study in the city of Lisbon, Portugal," Energy Conversion and Management, vol. (92), pp. 10-18, Mar. 2015.

[7] Tran, T. D., Ovtracht, N., and d'Arcier, B. F. "Modeling Bike Sharing System using Built Environment Factors", Procedia CIRP, Vol. (30), pp. 293-298. 2015

[8] Susan A. Shaheen, Elliot W. Martin, Nelson D. Chan, Adam P. Cohen, Mike Pogodzinski. Public Bikesharing in North America During a Period of Rapid Expansion:Understanding BusinessModels, Industry Trends and User Impacts. REPORT 12-29 A publication of Mineta Transportation Institute, College of Business, San José State University. 2014

[9] Susan A. Shaheen and Stacey Guzman. Worldwide Bikesharing. Access Nomor 39, Fall 2011. . p 22-27. 2011

[10] NYC Dept. City Planning. Bike-Share Opportunities In New York City. Report U.S. Department of Transportation, Federal Highway Administration. 2009.

[11] Zhang, L., Zhang, J., Duan, Z.-y.,and Bryde, D. "Sustainable bike-sharing systems: characteristics and commonalities across cases in urban China", Journal of Cleaner Production, Vol.(97), No.124-133. 2015

[12] J. Dill and G. Rose, "E-bikes and transportation policy: Insights from early adopters," in

Transportation Research Board 91st Annual Meeting, Washington DC, 2012.

[13] Duarte G, Mendes M, Baptista P. Impact on biker effort of electric bicycle utilization: results from on-road monitoring in Lisbon, Portugal. Accepted for publication in LNCS – lecture notes in computer science. Springer; 2014.

[14] M. Ricci, "Bike sharing: A review of evidence on impacts and processes of implementation and operation," Research in Transportation Business and Management, vol. (15), pp. 28-38, June 2015.

[15] Yan, Juan. Michael Ryan, James Power. Using fuzzy logic: towards intelligent systems, Volume 1. Prentice Hall. 1994.

[16] Kenna, J., and Bill Gillet. Solar Water Pumping. A Handbook. Intermediate Technology Publications. Shouthampton Row, london WCI 4HH. 1985

Pedestrian visual recommendation in Kertanegara – Semeru corridor in Malang City

V B Cosalia

Graha Laksana Tidar F4, Malang, 65151, Indonesia

cosaliaverenita@gmail.com

Abstract. Streetscape could be the first impression to see an urban area. One of the streerscape that should be attended to it is corridor of Jl. Kertanegara – Semeru since at that corridor is the road corridor having the strong caracter also as the one of the main axes in Malang city. This research is aim knowing the visual quality also the exact structuring rcommendation for Jl. Kertanegara – Semeru based on pedestrian's visual. The methode used to this research is Scenic Beauty Estimation (SBE) and used historic study. There is several variables used , they are scale space, visual flexibility, beauty, emphasis, balance and dominant. Based on those variable the pedestrians as a respondent doing the assessment. Based on the result of SBE have been done, it is showed that the visual quality in Corridor Kertanegara Semeru is well enough since the result showed that there are 10 photos in low visual quality in Jl. Semeru and 14 photos in high visual quality in Jl. Kertanegara, Jl. Tugu dan Jl. Kahuripan. By the historic study and based on high visual quality reference doing the structuring recommendation in part of landscape having the low visual quality.

1. Introduction

Visual quality is one of the factor which giving the satisfaction to the user or observer directly. Visual quality in difficult enough to be declared objectively but measurable by the human responses. Landscape visual quality cannot be mesured with measuring, but it could be a quality result from the human perspective, the measuring should be involed by human being [2]. Visual perception is the type of external perception, the perception that coming from stimulation from the outside of observer or individu. Perception extitence can persuade somebody to make them realize with the environment arround them. The thing need to do by somebody in order to enjoy their visual environment by preparing the environtment. Society's perpection based on the visual environment can strengthen the meaning and knowing the environment caracteristic itself. The assessment of visual envionment is needed to be done in order to make the result of design of public space integrated in imaging the city.

The development of Malang city from time to time influenced the visual quality. It is not only occurring at the road coridor Jl. Kertanegara to the Jl. Semeru where both of that places having a differentation of visual quality. Meanwhile, if it is confirmed by RTRW Malang city, road coridor Jl. Kertanegara – Semeru have the great potential to strengthen the caracteristic of Malang city. Those things become the basis that need for arranging the structure road corridor. Based on RTRW Malang city in 2010-2030 has been explained that the vision of Malang city is to build a Malang as a Best educated city, Healthy and Sustainable City, Cultural Tourism City, toward the society advancement and independent. The meaning of Cultural Tourism City is based on the mission to build and enlarge the cultural tourism also maintaining and envolving the environment and the heritage building to historical necessary, science, culture, and tourism. Predetermined plan in tourism area are, Gajayana Stadium, Public Library and Archive of Malang city, Tugu Plaza, Railway station of Malang city. In addition, as strategic social culture area in Tugu Plaza, Corridor of Jl. Semeru, Ijen Boulevard and Gajayana Stadium compleks.

Structuring all along the road, espcially based on pedestrians' perception as the subject that feeling the visual caracter Malang city from the corridor that able to strengthen the visual caracter Malang city also support how the planning has been there on RTRW Malang city. The historical and valuable buildings hope that they are able to impress everyone who see it, therefore the impression could be persuaded by the manufactuing and structuring the other element near it.

The assessment of visual quality in one location is influenced by the several landscape elements and the factor could be influenced by the somebody's feeling while on it. The element also the substance mentioned are scale, the visual comfort will cover the spaciousness and aesthetic, the unity which cover the balance and emphasis, and also domination [3]. Visual content refers to the structure and the space of one area giving an infomation to the each individu such a perception in used as purpose of developing the environment [7]. The structural of physical objects and the human activity construing the environment and the relation of the elements that could be the huge character in the shaping the area [8].

The several identifications matters in study area are:

1. Rapidly development causing the changing of image of Malang city, especially in the main road as a corridor of Jl. Kertanegara – Semeru. Those things are able to influence the visual quality reduction in corridor Jl. Kertanegara – Semeru.
2. The identity of th city in the beginning since the development of Malang city in corridor Jl. Kertanegara – Semeru could not be caught the whole part by the pedestrians because the increasing of the activiy in that corridor and there is a development the area.
3. There are several spots visual scenery in corridor Jl. Kertanegara – Semeru was blocked by the another objects such as trees and billboard so that the visual comfort of pedestrian in view unobstructed. It will persuade the visual quality from the coridor Jl. Kertanegara – Semeru.

2. Methods

This research is done through the direct observation, and admission the quisionaire by the pedestrians in the corridor of Jl. Kertanegara – Semeru to get the assessment visual quality in each corridor based on society perception. The variable used in this research and they are scale, circulation, aesthetics, emphasis, balance and dominant [3]. Scale variable which is intended in this research are the high and space between the exist building, in other way the circulation variable and the aesthetic include in sub variable from the unity and it occurs in the emphasis and balance variable which are included in sub variable from the unity. The next variable is dominant (soft and strong element).

Scenic Beauty Estimation method (SBE) is quantitative system for indext the landscape aesthetics quality. The purpose of using this method is to bestow the valid prediction using the observer perception. The aesthetic quality assessment are showed by the scale intervale according to the aesthtic scenery based on perception and human assessment. The observer which intended could be as representative as the society , or the choosen society in the particular population. Scenic Beauty Estimation is the aesthetic notion in using of SBE method consist of three main steps and they are landscape imaging, slide presentation and data analysis [4]. In this research there is changing in the step of determining the value of SBE and there are:

1. landscape photograph based on vantage point is done by applying the human perspective principle based on perspective point, degree, or a space.
2. The photos election. After applying the primary survey with take a photo by observing point then those photos are choosen by the prsenting to the respondens. The choosen respondens in this sudy are the society in pedestrians' area.
3. Assessment by respondents. The using of rating scale 1 – 10 for valuation the landsape quality based on each variable in used by this research, where the point 1 shows the lowwest assessment

toward landscape visual quality and the rating scale of point 10 shows the highest assessment. The assessment by respondents are only taken once, it is not allowed repeatedly

The using of SBE method in this study through the several steps, first is landscape photographing. Landscape photographing is done by the human being point of view, besides that it is doing based on taking the landscape photo. Taking photo need 20 m of space started from the front part of public library Malang city until Jl. Kertengara and the exact place of take a photo in Trunojoyo park to the Jl. Semeru. Whereas the south part of th photos is taken from Jl. Semeru until Kertanegara. The aim of taking a landscape photos in certain space and direction is in order to make the taken photos could represent the whole of coridor. The result of the photos are 323 landscape and after that doing the eliminate. The eliminate of the photo is done based on the similar characteristic in every photos. The result after doing the elimination shows that there are only 53 photos will be valued by respondens.
Calculation of the number of respondens that were used in this study that is using time linear function. The following function of Time Linear Function:

$$N = \frac{(T\text{-}to)}{t1} \qquad (1)$$

Description:
n = The choosen sample
T = The available time (10 days x 24 hours = 240 hours)
to = The timing survey (9hours/day x 10 days = 90hours)
t1 = Time survey used in every unit sampling (0,25hours x 10days = 2.5hours)

$$N = \frac{(T\text{-}to)}{t1} = \frac{(240\text{-}90)}{2{,}5}$$

$$= \frac{150}{2{,}5} = 60 \text{ Respondens}$$

Spreading the quisionaire to the each respondens is using the taking sample method accidental sampling. Accidental sampling is the method which the process of taking enough sample by taking whoever meets by observeer in the field suit with study necessary [5]. The boundaries and survey map of this research could be seen in (Fig 1).
The next step is calculating SBE value. The calculating visual value with the method of SBE started with data tabulation, frequency calculation in every score (f), calculting cumulative frequency (cf) and cumulative probabilities (cp). And after that deteermining the z value in every cp value. The average value of z in every photo enter the SBE rule. And the SBE rule are as follow:

$$SBEx = (Zx - Zo) \times 100 \qquad (2)$$

Description:
SBEx = SBE landscape value toward x
Zx = Average of Z landscape toward x
Zo = The average of Z in standart landscape (average value of Z in accumulation toward zero point)

IOP Conf. Series: Earth and Environmental Science **70** (2017) 012026 doi:10.1088/1755-1315/70/1/012026

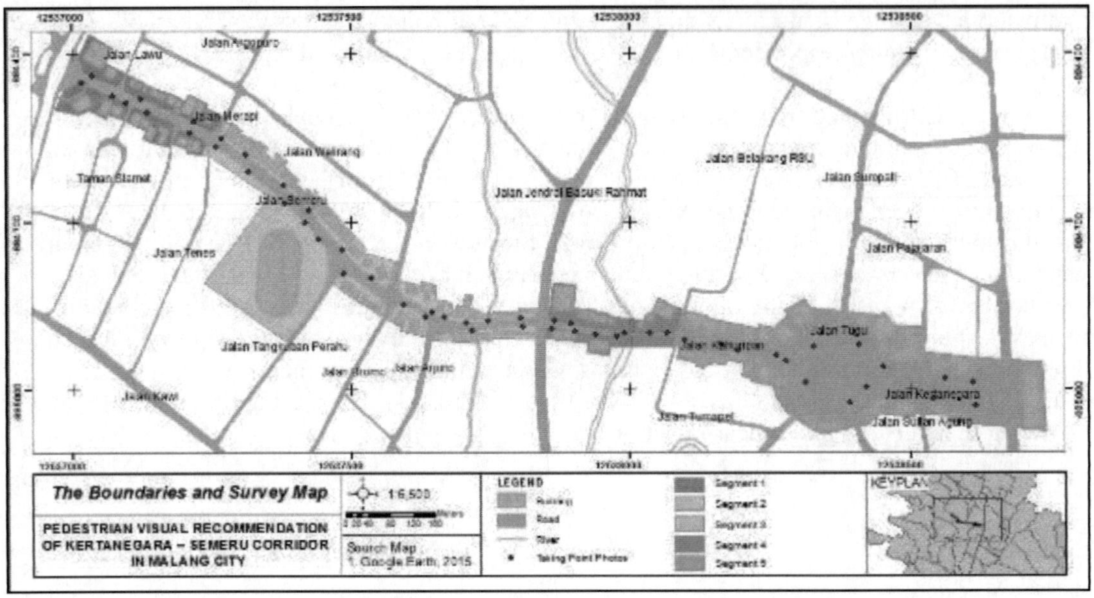

Figure 1. The boundaries and survey map

With the result of SBE value from each photos, we will enter to the low, medium and high category. From those result, there are the resut of visual quality in corridor of Jl. Kertanegara – Semeru. The step is done after the calculating the road visual landscape by doing the visual quality analysis. Visual quality analysis is one after knowing the SBE value in each photo. The purpose of visual quality analysis to knowing how the visual quality of corridor Jl. Kertanegara – Semeru based on pedestrians and how's the visual condition in the past.

Except using SBE analysis, this research also using the serial vision [1]. This method is used to recognize changing of Jl. Kertanegara – Semeru from the first developmental and the condition in present. Serial Vision is the visual imaging which is caught by th observer in occured while they were walking from one to another place in the certain location. Visual record by the observer become the photos college in every phase and it is shaping the unity visual record to the observer. The orientation become one important factor in understanding a city. The characteristic of a city is based on every area could be seen or understood as serial vision. The thing should be emphasized in this case is the mentoring every motion. Those process is called optic wich consist of two main group, the exixting view and the emerging view. Existing view concern in one area only while the emerging view concern in relation between one area and another [1].

3. Results and Explaination

The assessment of corridor visual of Kertanegara – Semeru based on pedestrians in every landscape photos could be calculated using the SBE rule. High visual quality (high SBE value) shows the landspace having the highest assessment as the beutiful landscape, in other way the low visual quality (low SBE value) shows that the bad or poor landscape (Table 1).

The result from the SBE calculating in every photo, then it is divided into three classes , the low, medium, and high class based on the class interval (Table 2) with use of simplified rating by the rules as follow:

$$I = \frac{\text{(the highest value-the lowest value)}}{\text{total class}} \quad (3)$$

$$= \frac{(108.013 - (-73.033))}{3}$$

$$= 60,348$$

Table 1. Calculation of SBE

Picture 1U 01					Picture 1S 01					Picture 5S 06				
Skor	f	cf	cp	z	Skor	f	cf	cp	z	Skor	f	cf	cp	z
1	0	12	1	-	1	0	12	1	-	1	0	12	1	-
2	0	12	1	2.326	2	0	12	1	2.326	2	0	12	1	2.326
3	2	12	1	2.326	3	1	12	1	2.326	3	1	12	1	2.326
4	2	10	0.833	0.967	4	1	11	0.917	1.383	4	0	11	0.917	1.383
5	1	8	0.667	0.431	5	2	10	0.833	0.967	5	1	11	0.917	1.383
6	5	7	0.583	0.210	6	4	8	0.667	0.431	6	2	10	0.833	0.967
7	2	2	0.167	-0.967	7	4	4	0.333	-0.431	7	4	8	0.667	0.431
8	0	0	0.000	-2.326	8	0	0	0.000	-2.326	8	3	4	0.333	-0.431
9	0	0	0.000	-2.326	9	0	0	0.000	-2.326	9	1	1	0.083	-1.383
10	0	0	0.000	-2.326	10	0	0	0.000	-2.326	10	0	0	0.000	-2.326
The total number of Z				-1.685	The total number of Z				0.024	The total number of Z				4.677
Z average				-0.187	Z average				0.003	Z average				0.520
SBE = (-0,187 – 0,003) x 100 = -18,992					SBE = (0,003– 0,003) x 100 = 0,00					SBE = (0,520– 0,003) x 100 = 51,697				

Table 2. Class Interval

Category	Value of SBE
Low	-73.033 - (-12.685)
Medium	-12.684 - 47.664
High	47.665 – 108.013

Based on the class it could be determined in 10 photos in a low visual quality, 29 photos include the medium visual quality and the 14 photos contained as the high visual. The low visual quality mostly in Jl. Semeru, the high visual quality majority in Jl. Tugu and Kertanegara. The low visual quality landscape are 1U 01, 1U 02, 2U 04, 3S 01, 3S 03, 3U 01, 3U 02, 3U 03, 3U 04, 3U 05. From the 10 pictues having the low visual aesthetic so that it would not create the beutiful scenery to the pedestrians. The less of vegetation impact the rigid impression in the corridor since the dominance of strong element. The variety in using the vegetation also could give the monotonous and boring impression.

Globality, there is no harmonism balance in every element will influence the landscape element in visual quality.the soft and strong element should be balance one and other so it will not evoke the rigid or over shady on the corridor cause it is dominaced by one element. The soft element such a tree which is blocked the scenery from the pedestrians also influence the visual quality in that corridor. The average of the vegetation in pedestrian path was blocked them from the scenery. In other case, the space in each building is not the same so it becomes the problem and it is the blocking scenery from the pedestrians.

The landscape in the low visaul quality will be given the repair recommenation based on the landscape in a high visual quality. The landscape in a high visual quality in corridor of Jl. Kertanegara – Semeru amount 14 photos. The high visual quality is dominanced by the photo in Jl. Tugu and Kertanegara, besides it only several corridor in Jl. Kahuripan and Semeru. After got the visual quality from each photo, then it will be analized descriptively in relation with the condition in study field in the past in every variables. The result of the visual quality analysis produced the differentioation and similarity from the condition in the past and in present. After the analysis in every photo, then it is recommended as a strucuring the corridor. The photo with a high visual quality has a several elements it is included in the high visual class, in other way th photo with the low visual quality has a several factor which is valuate as a bad quality by the pedestrians. The next step is done by giving structuring recommendation in photo which is having the low visual quality. The reccomendation given to these photo in low visual quality refers to the pgoto which is having the high visual quality with considering the similarity in it and the variable in photo with the low visual quality. Except that, the recmmendation refers to the differentation and similairy condition between the past and present also it refers to th quality of the photo based on pedestrians. (Fig 2) (Fig 3).

Figure. 2. The Recommendation of Corridor Kertanegara – Semeru

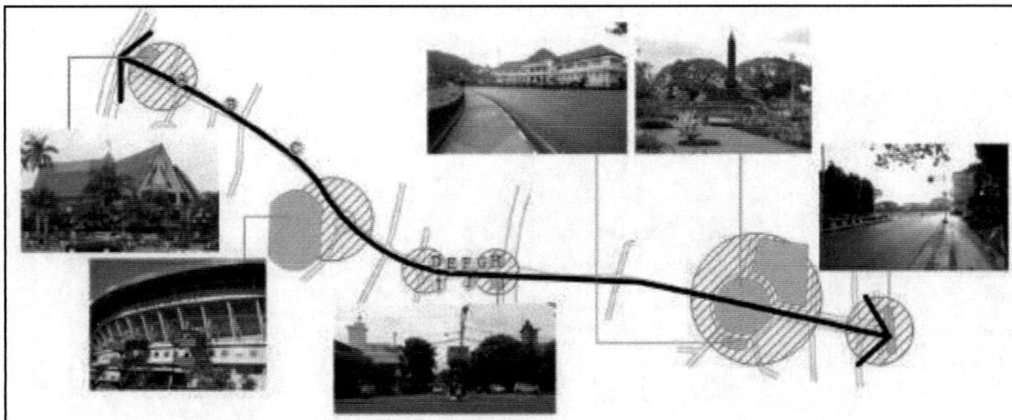

Location | **Recomendation**

Corridor of Jalan Kertanegara – Semeru as main axes

- Strengthen the start and end of the axis or corridor road with the Stasiun Kota Baru and Monument Suropati
- Fixing visual flexibility toward the both end point
- Highlighting some visual transition point

Visual Transition of Pedestrian

- Fixing blocked visual toward landmark so that created **visual flexibility**
- Making existing landmark as a **focal point** that keeps the attention of pedestrians to move from one point to another

Building as focal point

Photos with Low Visual Quality

- **Scale space** :
  - ✓ Selection of forms and types of vegetation, laying billboards, etc. *D E F G H I J
  - ✓ Determination of the maximum building height is 2 floors and setback 1 meter from pedestrian
  - ✓ Guardrail or barrier of shrubs between the building and the pedestrian does not exceed the height of a human chest
  - ✓ Reverse some of the building or building canopies that are too close to the pedestrian *D E F G H I J
- **Visual flexibility** :
  - ✓ At some point there is a billboard that blocks pedestrian visual to focal point or visual transition. So could do removal of the location of the billboard, the billboard with the provisions instead of private property * A B C D
  - ✓ At the point of vegetation that blocked visual pedestrians can be replaced with vegetation that has a crown shape is not too wide. This can be done on trees that haven't a diameter of more than 15 cm. *D E F G H I J
  - ✓ Selection of the vegetation on the narrow road corridor based on the assessment of respondents with high visual quality *D E F G H I J
  - ✓ Widening pedestrian to ± 3 meters and widening at the point interrupted or cut off by another object
- **Beauty** :
  - ✓ Strengthen existing focal points
  - ✓ Adding street furniture as an object that directs pedestrians to keep going up and Jalan Ijen – Jalan Kertanegara (Stasiun Kota Baru) *D J
  - ✓ Repair of damaged pedestrian
  - ✓ Replacement colors on pedestrian so that more attention and create contrast
  - ✓ The color changes from several points of interest to make it more visible monumental
  - ✓ Application of electricity and telephone networks underground
- **Balance** :
  - ✓ Establish a balance coverging towards the focal point and a good visual transition from vegetation and street furniture * A C H
- **Emphasis** :
  - ✓ Strengthen existing focal points
  - ✓ Emphasis on soft elements so that corridor has the impression of a shady, comfortable and relaxing *D E F G H I J
  - ✓ Using street furniture that shows the impression heritage in order to form a contrast to the surrounding buildings and provide a contrast to the corridor
- **Dominant** :
  - ✓ At its has domination of hard elements and soft elements, the addition of spacing between trees 10 meters *D E F G H I J
  - ✓ The use of soft elements are slightly varied both in the size or color. This is done to create contrast on the corridor

Figure. 3. The example of Recommendation of Kertanegara – Semeru

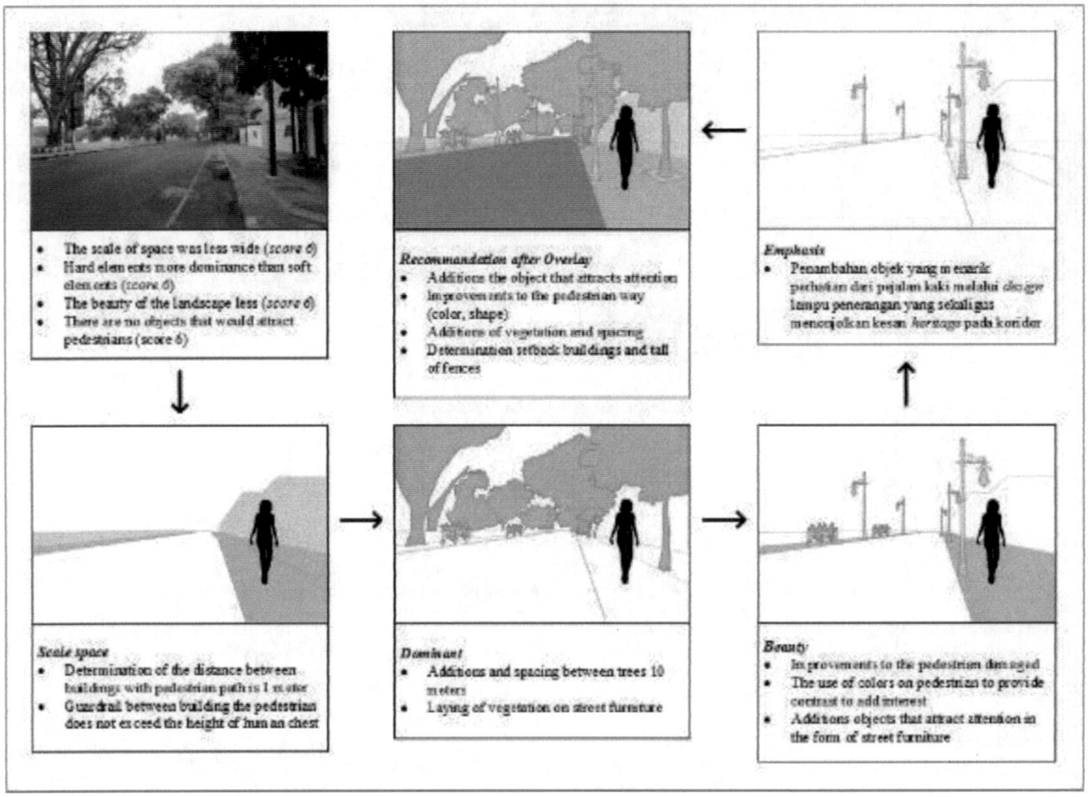

4. Conclusions

The road corridor Kertanegara – Semeru consist of several corridor and thery are Jl. Semeru, Jl. Kahuripan, Jl. Tugu and Jl. Kertanegara. Totally, the coridor of Jl. Kertanegara – Semeru has the good visual quality caused there are 14 and 29 photos from 53 landscape photos with the high visual quality and medium visual quality. The grouping result based on SBE value in each photo with produced of photo with the high visual quality are the photo with the SBE value about 47.665 – 108.013, and the medium visual quality are the photo with the SBE value about -12.684 – 47.664 and the photo in low visual quality are the photo with SBE value about 47.665 – 108.013.

Photo in high visual quality is the one which is having vegtation dominant, having the harmonism between the soft and strong element , having thee point of interest, having well visual comfort, also having the space between the high and large of the building. Those reasons become the basic recommendation is done toward th photo with the low visual quality. Besides, condition in present to the study field are the total objects was blocked visual of pedestrians such as tree and billboard with a huge size. The stong element gives nore recommendation than the soft element, those things evoke the impression of rigid and uncomfortable corridor.

The structuring recommendation is got from th result of overlay between the peedestrian perception with the historic conition in corridor Jl. Kertanegara – Semeru. Besides the recommendation to improve the visual quality such an additional of vegetation. According to the photos in high quality, this recommendation is done by the additional soft element to the landscape which mostly dominate with strong element. The selection of veegetation considering the space scaling. There is several trees was blocked the pedestrians visual could do the changing of vegetation form but with a certain requirements. Several lanscape consist of billboard with a huge size and it was blocke the aesthetic of landscape. In this condition, the billboard has the huge size and was blocked having an impact where it reduces the aesthetic of the landscape itself, so the billboard is moved and changed the size. Besides to th billboard eith the huge size butit won't really block the pedestrians visual, so that doing the changing of the size of billboard become smaller size. The unseen object or blocked in this existing condition such as twin building, Adipura monumen, the building from Bukopin Bank in Jl. Semeru also the flower monumen in Ijen Boulevard. Those objects having the potential to be a point of interest toward the landscape because it has a different shape from the other building so it will give more attention to pedestrians.

References

[1] Cullen, G. 1961. The Concise Townscape. London: The Architectural press.

[2] Lotihan, A. Landscape Quality Assessment of South Australia. Dissertation for Doctorate Philosophy; 2000

[3] Hakim, R. Komponen Perancangan Arsitektur Lansekap: Prinsip - Unsur dan Aplikasi Desain. Jakarta: PT. Bumi Aksara; 2012

[4] Daniel, T. C., & Boster, R. S. Measuring Esthetic Landscape (The Scenic Beauty Estimation Method). USA; 1976

[5] Prijana. Metode Sampling Terapan. Bandung: Humaniora; 2005

[6] Rencana Tata Ruang Wilayah Kota Malang Tahun 2010 – 2030

[7] Smardon, R. C. Methodology For Visual Analysis. United State; 1986

[8] Gosling, David & Barry Maitlasnd. Concept of Urban Design. New York Academy Edition; 1984

IOP Conf. Series: Earth and Environmental Science **70** (2017) 012027 doi:10.1088/1755-1315/70/1/012027

Control mechanisms in the third-generation planning. Case study: Control to realize sustainable cities

A D Wicaksono

Department of Regional and Urban Planning, Faculty of Engineering, Universitas Brawijaya, Malang 65147, Indonesia

agus2wicaksono@yahoo.co.id

Abstract. Since the last few years, Indonesia has experienced important events that bring significant changes to the social, political and economic life. The changes directly or indirectly impact the field of planning. With the challenging condition which grows fast and is more complex ahead, and the greater demands on the role of planning, it is required that planning should have higher quality. This paper seeks to answer some questions as follows: (i) How are changes in paradigm and also the development of planning model for the current transition era?, (ii) What is the best way to improve the quality of planning control on the last generation planning model to realize sustainable city?. Analysis steps that will be used to achieve the paper objectives are: (i) Review of planning and sustainable cities theory, (ii) Pattern recognition, (iii) Identifying control mechanisms and sustainable urban forms, (iv) conceptualization. Based on discussion about sustainable cities and control mechanism, some conclusions can be generated as follows: (i) The third generation planning model is based on the theory of expanded system, emphasizing on the constraint of capacity and the ability of planners within the context of larger environment, (ii) There are various theoretical studies that recommend prescriptive model or solution for sustainable urban form and structure. The concepts of Sustainable Cities can be grouped in Neotraditional Development, Urban Containment, Compact City and The Eco-City. The four models above have criteria, namely (i) high density; (ii) a high level of diversity; (iii) mixed land use; (iv) compactness; (5) sustainable transport; (6) passive solar design; (7) Greening Ecological Design. The three main activities in control mechanisms are: Monitoring and Recommendation; a comparative review of the facts (conditions that exist or are developing) with the purpose (expected conditions, set out in urban planning) and recommendations; Evaluation, a review on the intended purposes and can be followed up with revised purposes; Intervention/Actions toward existing conditions.

1. Introduction

Since the last few years, Indonesia has experienced important events that bring significant changes to the social, political and economic life. Changes that occur among others are in social and political field (from centralized power into decentralization, from Authoritarian into Democratic), in economics (from Centralized Financial Management into local-regional management, from local markets into global markets), in institutional field (the occurrence of globalization) as well as physical sector (the occurrence of Domestication). [1]

The changes directly or indirectly impact the field of planning. The demand for quality planning becomes a necessity to anticipate the changes that occur. It is not a secret that in Indonesia a lot of existing plans merely serve as a 'paper tiger'. They are conceptually and visually interesting and beautiful, but have very little influence on people's daily lives or can be said to be very weak in implementation. In reality, a lot of planning products which simply decorate the cabinets of governmental agencies can be found. Utilization and planning control becomes one of the critical points in the overall planning cycle. [2]

With the challenging condition which grows fast and is more complex ahead, and the greater demands on the role of planning, it is required that planning should have higher quality. Many factors affect the quality of planning, not to mention the dynamic factor of planning which always follows the existing paradigm. Whatever the paradigm is, the source of all paradigms is planning syntagm, as an essential process of planning. The planning model which is still widely used in practical and academic world in schools of planning in Indonesia is comprehensive rational planning model introduced around the 1940s. There have been a lot of criticisms related to the comprehensive rational planning model. As the first-generation planning model, comprehensive rational planning is considered too ideal, not focus, and the planning process requires large resources. Criticisms on the first generation planning model raise an antithesis which results in second generation planning. Since the 1960s there have been many models that attempt to refine and even remodel comprehensive rational planning model, such as advocacy planning model, incremental planning, medium-term planning, strategic planning, radical planning, etc. However, their practical application is still not widely used, especially in spatial or spatial planning.

One of the stages in comprehensive rational planning process that often serves as the critical point is utilization and planning control, a transitional phase of a plan at the conceptual level into an action to realize the plan at operational level [3]. On that basis, this paper seeks to answer some questions as follows: (i) How are changes in paradigm and also the development of planning model for the current transition era?, (ii) What is the best way to improve the quality of planning control on the last generation planning model to realize sustainable city?

Analysis steps that will be used to achieve the paper objectives are: (1) Review of planning theory, and other multidisciplinary literature that is related planning model. The aim is to deconstruct ("take apart") a multidisciplinary text related to control mechanism in planning. The outcomes of this process are numerous themes, "planning model" in this case, that are related to control mechanism, (2) Pattern recognition. The aim is to note major patterns and concepts within the results of the first step. This second step looks for similarities or patterns within the sample and codes the results by concepts, (3) Identifying control mechanisms and sustainable urban forms —to recognize specific and distinctive control mechanism in sustainable cities, (4) Conceptualization—to find theoretical relationships among the identified concepts and control mechanisms.

2. Development of Planning Model

The characteristics of "first generation" planning, which are widely studied in schools of planning and are also practiced until the 1970s, are characterized as a rational process analogized as a scientific process in a research. Although various authors and planners sort first generation planning stages in more detail or more macro, or with different terms, but implicitly the first generation planning is based on the same assumptions and principles, as follows: (1)Formulation of the problem and the solution are two discrete stages independent of each other, (2) Planning approach must be rational and objective. The process is not only supported by one discipline alone, but involves many disciplines (multidisciplinary), (3) Solution in planning is optimal, which means that all relevant aspects are maximized to obtain a single value. This assumption and viewpoint are the basis for rational planning model, (4) Most decisions made by planners are based on the primary consideration of rationality, namely to develop alternative solutions and choose a solution or an alternative with rational criteria. [3]
The characteristics of second-generation planning are contrary with the previous generation planning described above. If the first generation planning process is dominated by a *"tame problem"* or a simple problem, then the second-generation planning is related to complex problem *(wicked problem)*. 'Tame problem' is analogized as a problem in the game of chess or settlement of math problems. It is conceivable that every problem in the game of chess or math equation completion has a clear and definite

solution to resolve. Furthermore, if a problem has been solved, then the solution is used as a reference or even becomes the rule for those who face similar problems. [3]

It is different from the characteristics of complex problems (wicked problem) contained in second generation planning. The characteristics of 'complex problem' are as follows: (1) Every complex problem is essentially unique, (2) There is no fixed limit in complex problems. Each description of the problem is only temporary and is considered as a symptom of other problems, (3) Description of the complex problem can be viewed from various angles. The choices of viewpoint will determine the problem-solving method, (4) The exact number of potential solutions for the 'complex problem' can not be determined, and it does not have the quantitative benchmark that can be used in the planning process, (5) The benchmark of successful solutions does not use the value of 'right or wrong', but 'better or worse', (6) A planner can not do experimental activities, such as those carried out by scientists in the laboratory, so he is expected not to make a fatal mistake in the planning process, (7) Each implementation of a solution to a problem is a final and linear process which does not allow 'trial and error', (8) Considering that the description and understanding of a problem is temporary, there is no fixed rule associated with problem solving.

In this section a more advanced planning model will be described, namely the third generation planning. The basis of the third generation planning is system theory, which is developed by Heidemann [5]. With the intention that the very complex substance of 'planning' can be structured, system theory which serves as the basis for the planning model must be discussed once more.

This approach is based on the assumption that the theory of system isolates not only the relationship between the core or the element, but also its relationship with the environment. Such considerations structurally place elements of a system that is always related with the environment. The new paradigm of system theory, often referred to as a paradigm of system-environment, considers system as the formation of system-core fusing into a larger or bigger environment.

A planning model should consider all barriers and above qualifications. A search for a model is answered by a theoretical approach which was first revealed by a biologist Jacob von Uexkuell [5]. The approach of von Uexkuell theory confirms the theory of system-core/components and explicitly underlines not only limited perceptional ability of humans on the surrounding environment, but also limited cognitive capacities of humans and limited human ability to perform action. According to Uexkuell, every organism including humans, can absorb only some of the stimuli produced by surrounding environment (the external world), and respond to it in a specific way. The response specifically affects the external environment and also re-provides stimuli to the organism. The closed-cycle by Uexkuell is called a functional cycle.

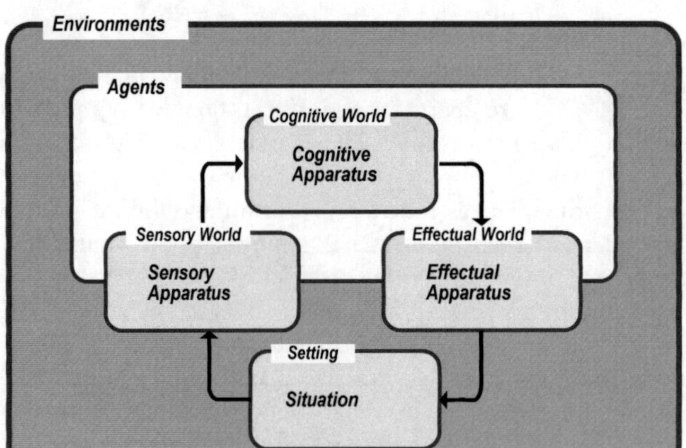

Figure 1. Functional cycle (Heidemann, 1992 adapted from Uexküll)

Functional cycle can be articulated as follows (see figure 1 above): the external stimuli that affect each organism will be captured/perceived by Agents with the help of 'Sensory Apparatus' in 'Sensory

World'. In humans, 'Sensory apparatus' is known as sensory organs that have limitation in capturing all stimuli coming from the environment. If the stimuli from the environment cannot be detected by senses, these stimuli cannot be interpreted afterward. Cognitive Apparatus of an organism in the Cognitive World is a device or organ that has the capability and limitation to affect the desired action. The influence of an organism to its environment will use the 'Effectual Apparatus' in 'Effectual World' which is also limited according to the ability and the extent of 'Effectual World' of each organism.

3. Third Generation Planning Model

The above Uexkuell model provides insight into various possibilities of the role of a planner, who perceives, thinks and acts (and proceeds in an organization), who has background levels of knowledge, and who lives in a neighborhood (which would also limit set of actions owned by a planner).
The basic idea of the third generation planning model can be described briefly as follows (See figure 2)

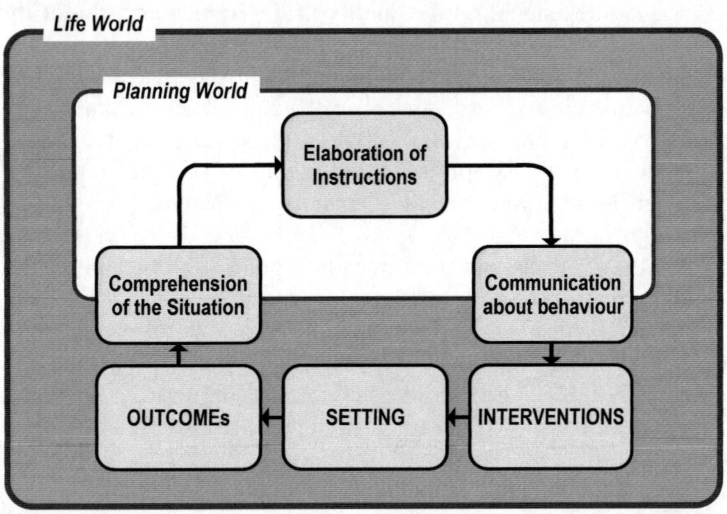

Figure 2. Basic scheme of planning (adapted from Heidemann, 1992)

'Sensory World' in basic scheme of planning is defined as 'Understanding the situation'. This refers to the ability and capacity of a planner to perceive and interpret his environment. The stage 'understanding the situation' is a transition phase between the 'Life World' with the 'Planning World'.

Furthermore, 'cognitive world' in Uexküll model is translated as arranging/formulating instruction, where a planner with his capacity and ability formulating measures to determine or control appropriate actions. Instructions (in the form of plans, descriptions or directions) are measures for the third party to perform certain actions (e.g.: building public facilities, constructing roads, building houses or revitalizing the area).

In the adaptation of Uexküll model for basic planning model, the concept of 'effectual world' is translated into two parts, i.e. 'communication about behavior' and intervention. This is because planning process is dealing with complex conditions. In a more concrete term: planning should be understood as an activity in public domain, which cannot implement all instructions directly and quickly. Modification of the instruction that has been prepared in the 'planning world' is often necessary. The second part, a translation of 'effectual world' is an intervention because the source of planning is a change in the real world. As well as in the 'effectual world', 'setting' is also interpreted into two parts. This is done to distinguish between the situation before and after the intervention of planners.

In brief, it can be concluded that in a planning, minimally there are two (2) critical points prone to errors that can lead to declining quality of plans and planning. The first critical point is in the stage 'Comprehension the Situation', where there is a transformation of information from the factual level (Life World) to the conceptual level (Planning world). This stage is simply known as survey phase to understand the planning area. The point where there is a transformation of information in the form of plan at the conceptual level to the Factual level is the second critical point, known as 'Communication about behaviour' and 'Intervention' stages. Actually, there are still some critical points other than the two points of transition and analysis phase, but at the two critical points of transition, the possibility of information bias will be greater than at other points.

Understanding the planning area is indeed the first critical point. Based on existing experience there are some possibilities that result in bias information between the factual manifestation and conceptual articulation, causing low quality of data or information, and eventually low quality of plan. Events at the factual level can be transformed into data as a quantitative/qualitative description at the conceptual level, if the incident impressively leaves traces. Based on the characteristics, traces can be permanent, generally in the form of memory, or non-permanent, having the shape of a perception perceived by the senses. Representation of factual conditions or transformation of events – Traces – Data will involve three phases of activity, namely: Scanning - Recording - Measurement. The quality of flowing transformation will be highly dependent on the quality of the implementation of the three steps above or observation [5].

Scanning stage includes interpretation cycle, and so it is also known as the exploration-interpretation cycle. Interpretation mechanism is a reverse process of Events–Traces–Data into Data–Traces–Events as a representation of events on factual level. Interpretation process involves three stages of activity, namely: Decoding – Assignment – Ascription. Good quality of this mechanism (understanding) will become the foundation for the interpretation of events on the factual level with cognitive theoretical support.

If the first critical point is the transformation transition point from factual level to conceptual level, the second critical point is the transition point from conceptual level to factual level. The process that occurs is transformation from Plan in the form of directions or guidelines at conceptual level transformed into an event or activity at conceptual level. This mechanism includes three (3) features, namely: INTENTS-INTERVENTION-OUTCOME with development and testing cycle, known as the control cycle. Basic control cycle is composed of the combination of two cycles, they are: Monitoring-Recommendation-Evaluation-Purpose which is at the level of 'Planning World ' and Monitoring-Recommendation-Intervention-Fact cycle which is at the level of' Life World' [5].

4. Green City (Eco City) in the Context of Sustainable City

Since *sustainable development* was conceived as a new paradigm of development in the 1970s, "sustainability" has become a magical terminology that is widespread in planning documents at local, regional, national and even global level. Since then, the concept of sustainability becomes one of the themes that is often debated and used as a concept in urban and regional science as well as transportation. Sustainable becomes the most commonly used concept in urban planning in Indonesia and around the world. The shift of paradigm in the field of urban and regional planning and urban design has led to the need for a more holistic, inclusive approach to meet sustainability criteria.

Riddell in 2004 articulated the relationship between sustainability and urban planning through a statement: "Sustainability requires ways to satisfy the lives of all without exceeding the ecological capacity of the biosphere. Future-friendly towns and cities are key players in building a sustainable future. Since it is at the local level where infrastructure related design and policy shapes peoples' lives and resource consumption, towns and cities can be particularly effective in pioneering sustainability" [6].

There have been many theoretical studies in journals and textbooks that discuss and recommend a variety of models, theories and prescriptive solutions about the form and the structure of a sustainable city, along with criteria and the positive and negative aspects. The models and the concepts can be classified into four typologies of form and structure of urban space which are hypothetically considered as environmentally friendly [7]. Based on the description of typology above and also review of existing theory, it can be concluded that sustainable cities have the following criteria:

1. Density
   The influence of density (along with the type of building) to *sustainability*, occurs through differences in energy consumption, goods and land for housing, roads and infrastructure of a city. High density and integrated land use not only reduce the use of resources, but also encourage social interaction through compaction. Kenworthy concluded some policies that can reduce energy consumption by increasing the density of the city, improving the quality of city center, optimizing land within city, providing good public transport alternatives, as well as limiting the infrastructure for private vehicles. Some alternative arrangements of buildings and KDB in the same level of density are described as follows [8].
2. Diversity
   The diversity of activities is the essence of sustainable cities. Jane Jacobs popularized the concept of diversity which was later adopted widely by many models of urban planning, such as new *urbanism, smart growth* and *sustainable development*. Jacobs also wrote: "In dense, diversified city areas, people still walk, an activity that is impractical in the suburbs and in most grey areas. The more intensely various and close-grained the diversity in an area, the more walking. Even people who come into a lively, diverse area from outside, whether by car or by public transportation, walk when they get there" [9].
3. Mixed Land Use
   Mixed land use or heterogeneous zoning allows for suitable land use which needs to get close to each other so as to reduce the distance between activities. Mixed land use is characterized by the diversity of land use functions in an area, such as housings, trades, services, offices, industries etc. The next impact is a reduction in the use of private vehicles for shopping, and going to work, school and recreation areas. The use of the concept of mixed land use at least ensures that a lot of city service facilities are located within a radius which can be reached by walking and cycling. Besides, the use of mixed land can revitalize parts of the city areas with the addition of new functions.
4. Compactness
   Compactness of built environment is a widely accepted strategy to achieve a more sustainable urban structure. The concept of compaction is applied to direct the development of the city on the existing urban structure, so as to minimize the energy for transportation needs, water, goods and people. With intensification, which is the main strategy to achieve compaction, building density and activity can increase, so it ultimately leads to a more efficient use of urban land. Some studies (Acioly Jr. and Claudio C. 2000) consider that compaction is the most important typology to achieve sustainable city. For example, it is said that a sustainable city must be compact, solid, varied and powerfully integrated. The formed urban structure will be able to encourage people to walk with ease and reduce the desire to use private vehicles.
5. Sustainable Transportation
   There is no doubt that transportation is the biggest problem in debates about environment relating to the form and structure of urban space. The interesting thing is that *sustainability* is defined as the minimization of mobility and length of a trip [10]. According to Handy, the form of a sustainable city must have the appropriate structure to carry out movements by walking and cycling and have efficient public transport. Besides, it must have a compact form and encourage social interaction.
6. Passive Solar Design
   Passive Solar Design is a central point in an attempt to realize the form and the structure of sustainable urban space. The general concept is to reduce the energy requirements and provide passive energy usage through specific design. The design that affects built environment, in general is through building orientation and building density. The assumption is that the design, position, orientation, layout and landscape can encourage optimum use of lighting and microclimate conditions to minimize the need for heating or cooling buildings by conventional energy sources.

7. Greening
   'Greening' of the city or *green Urbanism* also becomes an important design concept to achieve a sustainable urban structure. Green open space has the ability to contribute positively in the development of sustainable cities. Through greening it is expected that natural integration between nature and city will occur and the presence of nature in the life of urban society will be maintained.

5. Control Mechanisms in Creating Sustainable Cities

As noted in section 3 above, control mechanisms have different characteristics from 'understanding the situation' due to its divergent transformation with a high level of difficulty, namely: How to transform a plan in conceptual level so that it is capable of acting as directions or guidelines for desired actions in the factual level. What is the guarantee that the entire community or *Planning Addressee* conducts activities as planned? Monitoring and recommendation that continuously follow and attach to control mechanism are required. Therefore, in planning the term control is actually more appropriate to use than/to replace the term Realization [5].

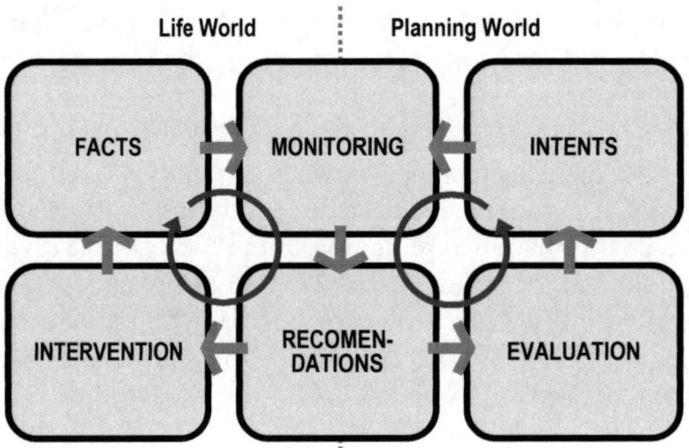

Figure 3. Control mechanism (Heidemann 1992)

A brief explanation about the main activities contained in control mechanism is described as follows:

5.1. Monitoring and Recommendations

Monitoring and recommendations are a comparative review of the **fact** (conditions that exist or are developing) with **the purpose** (expected conditions, set out in urban planning). Monitoring is the first step in the overall mechanism of area controlling to keep the appropriateness in space utilization with the assigned space function. This monitoring stage produces conclusion that serves as recommendation for the next stage. If the result of monitoring concludes that the plan for space utilization can direct and accelerate the development process and can be realized, the conclusion will recommend interventions or actions for achieving the desired state. Likewise, if a predetermined plan is not able to accelerate the development process or even slow it down, it is possible to revise or improve the existing plan. Thus, monitoring activities must be carried out periodically and in a sufficient period of time to be able to quickly find out deviation that occur in the field, and so intervention or required action can be done immediately. At this stage the role of planning information system becomes important.

5.2. Intervention /Treatment of existing conditions.

Transformation of guidelines contained in the plan into action is done through intervention. In this case intervention is a program of action to realize predetermined plan. All this time interventions which are often applied and commonly used are regulation or legislation, namely provisions which must not be violated and their sanctions. However, it is also important to take advantage of other interventional devices with a better planning and this may make up a unity with mutual support. Interventional devices in question are:

1. Motivation
   Motivation aims to influence urban society to participate in using their resources and engage in the planning and realization. Motivation is associated with changes in attitudes or feelings. In a situation where the role of society is claimed to be more dominant in urban development and there are potentials that people have, motivational device is becoming increasingly important. Potentials in question are, for example, mutual cooperation that is still embedded in the community or the willingness to donate some properties or land for public interest. Some examples of instruments of motivation in creating a green city, among others: (1) Organizing events correlated with the aims of green or sustainable city, (2) Giving reward/awards for the achievement or support green city, (3) Training related to green city, (4) Seminars related to sustainable city.
2. Information
   Informational instrument relates to changes in the public's knowledge and views to support established purposes. The purposes can be achieved by introducing knowledge about current condition and facts, as well as procedures to be agreed. This information can make use of the lines of communication that already exist and thrive in urban areas, such as: oral communications through counseling, newspapers (especially KMD), radio, television, performing arts (performance for public).
3. Organization
   This instrument is disincentive, specifically associated with procedures which are permitted and should not be violated, along with the sanctions. Some examples of organizational instruments, among others: PERDA/PERBUP/PERWALI about Green Open Space, *Corporate Social Responsibility (CSR)* for Green Open Space, PERDA/PERBUP/PERWALI about tree planting requirement for arranging ID card, PERDA/PERBUP/PERWALI about tax dispensation for green open space, etc.
4. Installation
   Installation is the development of facilities and infrastructure in accordance with the provisions of urban planning to support desired activities. If regulation devices is categorized as disincentive device, the development of facilities and infrastructure is incentive device, namely to provide an added value in the form of ease, smoothness and benefits for people who abide by the rules or support the creation of expected condition.

6. Conclusion

Based on above discussion about sustainable cities and control mechanism, some conclusions can be generated as follows

1. Development of planning from the first generation to the third generation has brought a fundamental change in any generated planning model. Each model has its own characteristics according to the challenges it must face. The third generation planning model is based on the theory of expanded system, emphasizing on the constraint of capacity and the ability of planners within the context of larger environment.
2. There are various theoretical studies that recommend prescriptive model or solution for sustainable urban form and structure. Those concepts can be grouped in Neotraditional Development model (including the New Urbanism, Transit Oriented Development, Transit Village, Transit Supportive Development, Transit Friendly Design), Urban Containment (including Urban Greenbelt, Urban Growth Boundaries - UGBs, and Urban Limit Line - ULL), Compact City and The Eco-City.
3. The four models above have similar criteria, namely (1) high density; (2) a high level of diversity; (3) mixed land use; (4) the form of a compact city; (5) sustainable transport; (6) passive solar design; (7) Greening Ecological Design. Studies about sustainable urban structure

and its form are followed with a search for city form that is able to (1) reduce dependence on motor vehicles, thus resulting in lower emissions and reducing energy consumption; (2) improve public transport services to be better; (3) increase accessibility; (4) reuse infrastructure and land that have been built; (5) improve the regeneration of urban areas; (6) improve the quality of life; and (7) protect green open space.

4. The three main activities in control mechanisms are: **Monitoring and Recommendation;** a comparative review of the **facts** (conditions that exist or are developing) with **the purpose** (expected conditions, set out in urban planning) and recommendations; **Evaluation,** a review on the intended purposes and can be followed up with revised purposes; **Intervention/Actions toward existing conditions,**
5. Control instrument is an intervention device that consists of Information, Motivation, Organization and installation, which are interrelated and support one another.

References

[1] Wicaksono, Agus Dwi. Model Implementasi Rencana Tata Ruang dalam Perencanaan Generasi Ketiga. Proseding Seminar Nasional ASPI 2015. Denpasar

[2] Boyne, George A; Goud-William, Julian S; Law, Jennifer-Walker, Richard M. Problem of Rational Planning in Public Organizations: An Empirical Assessment of Conventional Wisdom. Adminsitration & Society; Juli 2004; 36, 3.

[3] Schönwandt, Walter L.. 2008. Planning in crisis? : theoretical orientations for architecture and planning. Ashgate Publishing Limited

[4] Rittel, H. and Webber, M. 1973, 'Dilemmas in a General Theory of Planning', dalam Policy Sciences 4(2): 155-169.

[5] Heidemann, Claus. 1992. Planning Theory. Institute for Regional Planning/Science, University of Karlsruhe.

[6] Riddell, Robert. 2004. Sustainable Urban Planning: Tipping the Balance. Blackwell Publishing Ltd

[7] Jabareen, Yosef Rafeq. 2006. Sustainable Urban Form: Their Typologies, Model, and Concepts. Journal of Planning Education and Research. Vol. 26, 38-52.

[8] Kentworthy, Jeffrey R. 2006. The Eco-city: Ten Key Transport and Planning Dimensions for Sustainable Development. Environment and Urbanization. Vol. 18, No.1, hal. 67-85

[9] Jacobs, Jane. 2002. "The Uses of Sidewalks: Safety" from the Death and Life of Graet American Cites dalam The City Reader (editor Ricahrd T. LeGates dan Frederic Stout) . Routledge, Taylor & Francis Group. London and New York

[10] Acioly Jr , Claudio C. 2000. Can Urban Management Deliver the Sustainable City? Guided Densification in Brazil versus Informal Compactness in Egypt, dalam Compact Cities: Sustainable Urban Forms for Developing Countries (editor Mike Jenks and Rod Burgess) .E & FN Spon, London.

Planning of dairy farm and dairy plant based ecotourism

A S Sarnyoto[1], I P Tama, and C F M Tantrika

Department of Industrial Engineering, Faculty of Engineering, Universitas Brawijaya, MT. Haryono Street 167, Malang 65141, Indonesia

amaliasharfina3@gmail.com

Abstract. One of a dairy production company producing pasteurized milk and yoghurt drink, whose brand has widely known in East Java, has a factory plant in Batu City, one of tourism destinations in Indonesia. Behind the factory plant, there is a vacant land with an estimated total area of 2.3 ha and a vacant cowshed which had not been used for cattle ranching. Because of that, the company planned to develop the vacant land as a dairy farm and plant based ecotourism. In addition, dairy farm and dairy plant based tourism attractions are still rarely found in Batu. Thus, the first aim of this study was to analyse the potencies of the company that related to future plans of ecotourism built. The second aim was to set up the strategies that can be done in order to actualize the ecotourism project. The next aim was to plan the ecotourism, especially the facilities planning and the facilities arrangement on the vacant land. Strategic management approach was used to analyse the potencies and to determine the strategies. To select the proper facilities, tourists were asked to give appraisal by using questionnaire. Appraisal result was mapped onto four quadrants spatial map to see advantages and shortcomings of each facility along with choosing the right facilities to be built. Those facilities and tourist activities were compared with ecotourism criteria to make sure that the facilities were appropriate to provide not only entertainment but also ecotourism function. To arrange the chosen facilities, the step in Systematic Layout Planning were conducted to generate a propose layout of facilities arrangement. Based on potencies analysis, in Internal-External matrix, the company current position was on quadrant 2 (grow and build), with the most appropriate strategy was intensive or integrative. The proposed strategies were to build the new infrastructure, to renovate cowshed, and to add new tourism facilities on the land. There were 11 selected facilities based on MDS. Moreover, based on SLP, a facility layout had been designed and been proposed to be built on the vacant land behind the factory as the new ecotourism destination.

1. Introduction

Tourism is a recreational activity outside the residence to get away from routine activities or to look for other atmosphere[1]. One kind of tourist attraction is natural resource that should always be preserved along with tourism attraction development. Travelers' awareness to preserve the nature is needed while traveling. Raising travelers' awareness can be done through provision of ecotourism. One of tourism destination in Indonesia is Batu City. There are some featured products from this city, including pasteurized milk. One of dairy production company had a vacant land behind the factory.

During this time, the company only ran production process of pasteurized milk and yoghurt. Production process cannot be seen by public. Whereas, production process can be a tourist attraction if managed properly and can be one promising ecotourism destination. However, the development of ecotourism was still constrained by a variety of things so that development cannot be run optimally.

In addition of the factors mentioned before, dairy farm and dairy plant based tourism attractions were still rarely found in Batu. Because of that, the company planned to develop the vacant land as a dairy farm and dairy plant based ecotourism destination. Therefore, the company needs to know its internal strength and weaknesses, along with its external opportunities and threat. This information can be used for analyzing the potencies and determining the applicable strategies in order to actualize the ecotourism project.

This paper aimed to evaluate internal strength and weakness along with its external opportunities and threats faced by the company in order to develop its factory to become an ecotourism destination. Based on the evaluation, strategies will be determined. Likewise, in order implement the strategies and to maximize the use of the land, additional tourism facilities will be determined. Furthermore, to increase tourists' interest, the facilities should be arranged as attractive as possible.

2. Methodology

Strategic management is the science concerning in formulation, implementation and evaluation of cross-functional decisions that enable the organization to achieve its objectives[2]. In this research, strategic management approach was used to identify strengths, weaknesses, opportunities, and threats faced by the company. It was also used to generate appropriate strategies based on the company's potencies. There are 3 main stages in determining strategies i.e. input stage, matching stage, and decision stage. But, only 2 stages were done during this research.

1. Input Stage
   In this stage, the information about internal strengths and weaknesses, along with external opportunities and threats were obtained. Identifying information was presented in a matrix of External Factor Evaluation (EFE) and Internal Factor Evaluation (IFE).
2. Matching Stage
   In matching stage, the internal and external factors were analyzed in quantitative models for formulating alternatives strategies. The analysis tool models were SWOT (Strength, Weakness, Opportunities, and Threats) matrix and IE (Internal-External) matrix.
   a. SWOT matrix is an important matching tool that helps researchers develop four types of strategies, namely SO strategies (strength-opportunity), WO strategies (weaknesses-opportunities), ST strategy (strength-threat), and WT strategies (weaknesses threats)[2].
   b. IE (Internal-External) matrix is used for positioning a business or company into a matrix consisting of 9 cells. IE consists of two axis i.e. total score of IFE on the X axis and the total score of EFE on the Y axis[2].

After determining the strategies, the next step was to make ecotourism facilities development plan. Travelers' requirement should be considered. To determine the type of ecotourism facilities, identifying and analyzing the needs of travelers are necessary to be done. The method used in this study was Multidimensional Scaling (MDS). MDS is a method of analysis related to the placement of objects on a multidimensional map which is determined based on the similarity obtained from the respondents' perception based on attributes [3]. Therefore, this method was used to determine the appropriate type of facility to be developed in milk processing plant as ecotourism based on traveler assessment.

In order to determine the appropriate facilities, preliminary research to other similar tourist attraction was conducted, called it as benchmarking. It aimed to compare the existing tourist attraction in order to find new ideas about the new attraction that will be built in Batu. The opinions of company's management were also noteworthy because they would operate the facilities. Hence, they were interviewed to know which facilities and what kind of tourist attraction they want to build.

The next step was to collect tourists' requirement about the facilities that should be added by distributing questionnaire. Facilities obtained from preliminary research were also included in questionnaire to provide some overview to tourists (as respondents) in filling it. Facilities obtained then reassessed based on some criteria in order to select the best and appropriate facilities to be built.

Facilities assessment was conducted by measuring tourists' perception based on some criteria through questionnaire. Perception data then was processed by Multidimensional Scaling (MDS). In

MDS, the criteria were applied as variables. If the assessment criteria are more than two, then the grouping variables are needed to be simplified into two groups [3]. Hence, the mapping and interpreting the mapping result would be easier. The MDS result was used to determine facilities needed to be built. The chosen facilities then organized by applying one of industrial engineering science that was facilities layout planning. Facility layout planning is a way to organize factory facilities to create the most economical, safest, and most comfortable working areas. Therefore, it can raise morale and performance of operators [4]. One method of designing facility layout is Systematic Layout Planning (SLP) [5]. SLP was used because of its systematic approach to design the layout. In designing facility layout, activities were used as input. It was also important to determine the equipment needed to do the activities and to determine the space requirement. Besides that, the closeness relationship among facilities would be determined and be considerate by using Activity Relationship Chart (ARC). ARC was more suitable to determine the facility layout because the objects processed in ecotourism facility was the people whose number is uncertain, unlike the production floor that uses the exact amount of material flow as consideration.

Table 1. Internal factors weighting.

No	Internal Factors Strength	Weight	Rating	Weighted Score
1	Pasteurized milk and yoghurt drink products dominate the East Java market.	0.11	4	0.11
2	Milk can be processed into other kinds of products such as ice cream and cheese. That process can be another attraction for tourists.	0.09	3	0.27
3	Milk that produced is healthier because containing high antioxidant, hence can be categorized as another attraction in educative and health way.	0.09	3	0.27
4	Some process might be exposed to the tourists, not only milk and yoghurt production process, but also waste treatment into compost or biogas.	0.09	3	0.27
5	A 2.3 ha vacant behind the plant as the space for eco-tourism development.	0.11	4	0.44
6	Plant strategic location, which is located in the main road of Batu.	0.11	4	0.44
7	Assured product quality because of hygienic production standard.	0.11	4	0.44
No	**Weaknesses**	**Weight**	**Rating**	**Weighted Score**
8	Production process has not allowed to be seen by tourists or public yet.	0.11	1	0.11
9	Need substantial capital financial.	0.11	1	0.11
10	Lack of adequate infrastructure.	0.06	2	0.12
	Total	1		2.94

3. Results

3.1. Formulating strategies

There were three stages in establish the strategies by using strategies management approach, such as: Input stage, Matching Stage, and Decision Stage. The information was obtained by conducting interview to manager of KUD "BATU" and observation around KUD "BATU". Internal and external factors that had identified further be weighted in the IFE and EFE matrix. IFE matrix can be seen on Table 1 and EFE matrix can be seen on Table 2.

As mentioned before, there would be some strategies that established based on consideration about each pair of internal and external factor. Based on strength and opportunity, there were four strategies that can be arranged i.e. (1) make ecotourism by rearrange facilities inside the plant so that all processes can be exposed to tourists, (2) add the outlets that sell the company's products in some tourism attraction located in Batu or other cities and promote the ecotourism in those outlets, (3) sell products and promote ecotourism in each outlet owned by the parties in cooperation with the company, and (4) maximize the use of vacant land for developing the ecotourism. There were also four strategies established based on weakness and opportunities i.e. (1) more intensive promotion, (2) make ecotourism by rearrange facilities inside the plant so that all processes can be exposed to tourists, (3) build the infrastructure that supports ecotourism with regard to the preservation of nature, and (4) sell products and promoting ecotourism in each outlet owned by the parties in cooperation with the company. The strategy based on strength and threat was held the ecotourism promotion that carries the brand of milk already known, so that the tourists would be more interested. Lastly, the strategy based on weakness and threat was improve infrastructure so that local ecotourism in Batu would be able to contend in tourism attractions in other countries, especially in South East Asia region.

Based on weighting score calculation of internal and external factors, 2.94 was obtained as a total weighted score for internal factors and 3.53 as a total weighted score for external factors. The value were mapped onto IE matrix and resulting the company's position. IE Matrix can be seen on Fig 1.

From the Fig 1, the company was laid on quadrant II, which means the company was on grow and build state. Thus, the most appropriate strategy type was intensive (market penetration, market development and product development) or integrated (forward integration, backward integration and horizontal integration).

Based on strategies that have been formulated, some strategy that fits with the type of appropriate strategies to be applied by the company were to open more outlet in famous tourism object, promote the ecotourism in the outlets, intensive promotion, participate in infrastructure improvement, make a detail about ecotourism development future plan, and renovate the cowsheds and vacant land by adding and arranging more tourism facilities.

Table 2. External factors weighting.

No	External Factors Opportunities	Weight	Rating	Weighted Score
1	Developing products is planned to be penetrate to entire region of Indonesia because of its well known in East Java. Hence, the name of the company will also be known by more people.	0.14	3	0.42
2	Cooperation with big franchise companies that has widely known all over Indonesia, thus the cooperation can be continued for ecotourism development.	0.18	4	0.72
3	Dairy farm and dairy plant tourism object is still rarely found in Batu.	0.18	4	0.72
4	Beautiful scenery of Batu.	0.18	4	0.72
No	Threats	Weight	Rating	Weighted Score
5	Many famous tourism object in Batu and Malang	0.09	2	0.18
6	Dairy products that quickly expired	0.05	1	0.05
7	ASEAN Economic Community allows the company to has competitors from othe countries in ASEAN	0.18	4	0.72
	Total	1		3.53

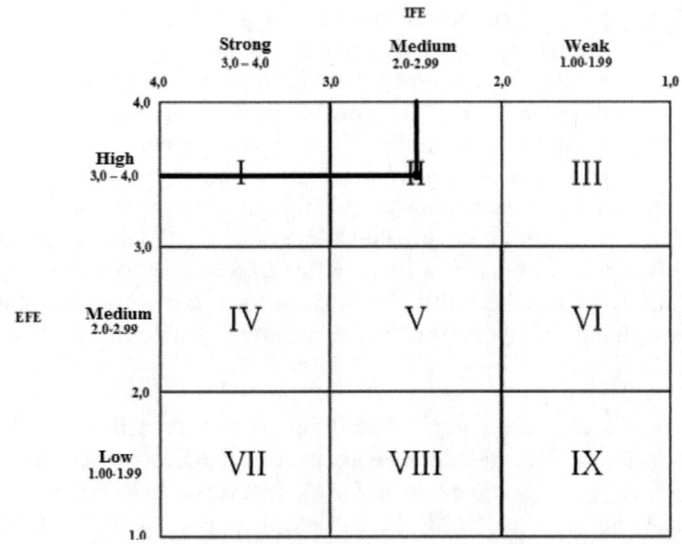

Figure 1. KUD "BATU" position on IE Matrix.

3.2. Determining the facilities

As the continuation in implementing the strategies, the determination of facilities was needed. To determine the facilities, firstly, some choices of facilities were identified by benchmarking to similar tourism attraction located in Batu and Bogor. The company's manager was also asked about which facilities they want to add. The known facilities in other similar tourism attractions are mini farm for cows and calves, biogas installation, simple milk processing station, catfish pond, rest area, cow museum, playground area, souvenir shop, restaurant, and milk ingredient foods store. Moreover, based on interview to company's manager, they intend to add cowshed or mini farm, biogas installation, compost pile, and milk processing station in form of mini factory.

The facilities obtained from benchmark and interview then included as choices in a questionnaire given to tourists. The facilities were included in questionnaire to provide overview to respondents thus their answer would not be too broad. But, the choices were still open if they want to add other facilities outside the facilities given. Respondents' answer was needed in order to determine kind of facilities they want along with the activities they want to do. Because when the facilities are already built, the tourist would be the main target to enjoy the attraction. Therefore, tourist's choices were essentials in determining facilities. These first questionnaires were distributed to 30 respondents.

Based on questionnaires, there were 15 activities that tourists want to do, such as making yoghurt, making ice cream, making cheese, milking, taking photos, viewing scenery, making pasteurized milk, playing games, outbound games, walking around the farm, feeding cows and calves, souvenir shopping, learning how to produce biogas, learning how to produce compost, and buying various milk ingredient products. As a means to perform the desired activities, there were 13 facilities which were wanted by respondents, such as cowshed, biogas installation, compost pile, outbound area, ice cream making station, yoghurt making station, cheese making station, milk processing station, viewing post, souvenir store, milk ingredient products store, biology museum, and playground for children.

3.2.1. Criteria of assessment. Facilities acquired from respondents' answers would be reassessed to determine which facilities meet the criteria as a tourist attraction, but still consider the elements of ecotourism. The tourist attractions criteria were taken based on Sundoro[7]. Those criteria are 'something to do', 'something to see', and 'something to buy'. Assessment based on 'something to do criteria' aimed to gain respondent's level of perception to enjoy a tourist attraction based on the activities that can be done. Assessment based on 'something to see' criteria aimed to understand respondent's level of perception to enjoy a tourist attraction based on the things that can be seen. Assessment based on 'something to buy' criteria aimed to perceive respondent's level of perception to enjoy the attractions based on the availability of facilities for shopping. Besides the tourist attraction criteria, there were also ecotourism elements criteria which used to assess the facilities. The criteria were ecological preservation, education, and economic community [7][8]. Ecological preservation criteria were used to assess the perception of respondents' in enjoying the attractions by a desire that will arise from the travelers themselves to preserve the environment after enjoying the attractions. Education criteria were used to assess respondents' perception to enjoy the attractions based on education that can be obtained. Economic community is a principle that can be rated by respondents to see how much of ecotourism development will be intended for the public, which the facilities/attractions can involve the local community to work in running the operations so that it provides income for the communities around ecotourism objects.

3.2.2. Facilities assessment. The assessment of facilities was conducted by distributing questionnaire to respondents. The questionnaire contained several statements that represent the criteria that must be met by each facility. Assessment used a Likert scale in intervals of 1-5 which states of agreement or disagreement on the statements in assessing the facility. The assessment were later be processed by using Multidimensional Scaling. The number of samples should be taken for data to be processed by using multivariate analysis was minimal 10 times the number of variables. In this research, 6 criteria were used as variables. That means, the samples of 60 was needed for each facilities based on each criteria. Hence, there would be 4680 total data collected.

3.2.3. Data mapping and facilities selection. After validity and reliability test, average data of each facility based on each criterion were calculated. Those values were used as input to calculate Euclidean distance between each facility. Euclidean distance then used to determine each facilities coordinate on spatial maps. Average data can be seen on Table 3. Meanwhile, Multidimensional Scaling spatial map can be seen on Figure 2. SPSS 20 were also used in processing data for Multidimensional Scaling to map the facilities into four quadrants. Before interpret the result of Multidimensional Scaling, variable grouping by using Principal Component Analysis was needed because the spatial map only consist of 2 axes, whereas there were 6 variables. The result of variables grouping can be seen on Table 3. Based on Table 3, 'something to do', 'something to see', and 'something to buy' were grouped as one into Component 1 because the value of their Component 1 were greater than the value of their Component 2. Otherwise, education, ecological preservation, and economic community were grouped into Component 2 because their Component 2 were greater than the value of the Component 1. After grouping the variables, interpretation of MDS spatial map can be completed.

Table 3. Average data of respondents' perception.

Variables	Facility*												
	1	2	3	4	5	6	7	8	9	10	11	12	13
Something to do	4.10	3.47	3.57	4.50	4.38	4.42	4.36	4.67	4.13	4.47	4.13	4.00	4.43
Something to see	4.05	3.98	4.20	3.42	4.51	4.47	4.50	4.56	3.67	3.45	3.80	4.05	3.65
Something to buy	3.88	3.43	3.55	3.60	4.16	4.31	3.88	4.31	3.60	3.90	4.02	3.83	3.77
Education	4.13	4.17	4.35	4.17	4.22	4.53	4.13	4.53	3.58	3.13	3.07	4.23	3.43
Ecological Preservation	4.03	4.18	4.03	4.18	3.60	3.69	3.62	3.69	3.92	3.28	3.32	3.80	3.60
Economic Community	4.18	4.10	4.18	4.10	4.02	4.16	4.18	4.16	3.97	4.33	4.32	3.47	3.68

***1** = Cowshed; **2** = Biogas Installation; **3** = Compost Pile; **4** = Outbound Area; **5** = Ice Cream Making Station; **6** = Yoghurt Making Station; **7** = Cheese Making Station; **8** = Milk Processing Station; **9** = Viewing Post; **10** = Souvenir Shop; **11** = Milk Products Store; **12** = Biology Museum; **13** = Playground.

IOP Conf. Series: Earth and Environmental Science **70** (2017) 012028 doi:10.1088/1755-1315/70/1/012028

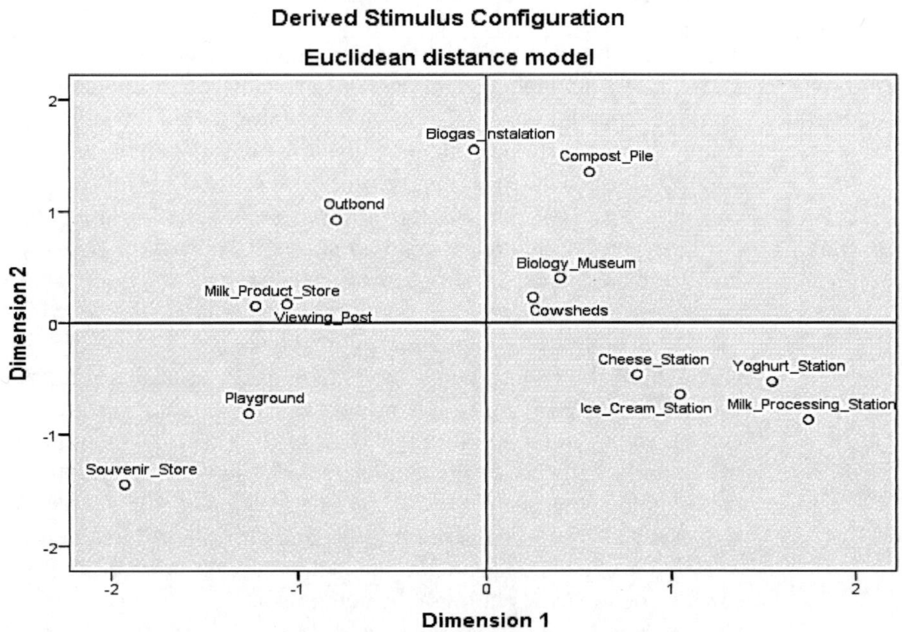

Figure 2. Coordinate of each object on MDS spatial map.

Table 4. Variables grouping result.

Component Matrix[a]

	Component	
	1	2
Something_To_Do	,706	-,423
Something_To_See	,668	,657
Something_To_Buy	,918	-,033
Edukasi	,332	,884
Pelestarian_Ekologi	-,676	,625
Ekonomi_Masyarakat	,108	,181

Table 5. Description of facilities.

No	Facilities	Description
1	Cowshed	Facility to placed cow and cattle. The cowsheds will be design like an open field that the cows can freely move. Tourists can get information on how to care for cattle. Visitors can also perform the activity of cow bathing, feeding, and milking cows.
2	Biogas Installation	Facility where tourists can see the making process of biogas from waste of cattle. Tourists can also learn by direct practice in biogas production.
3	Compost Pile	Facility where tourists can see the making process of compost from waste of cattle and waste from biogas production. Tourists can also learn by direct practice in compost production.
4	Outbound Area	Facility where tourists can perform outbound activities such as playing games, flying fox, pole climbing, and swimming. The help of safety guard is needed in this facility to check the completeness of visitor safety equipment when playing games and ensuring visitor safety procedures already performing well
5	Ice Cream Making Station	Facility where tourist can make ice cream with a simple way without waiting the ice cream to be frozen in freezer.

No	Facilities	Description
6	Yoghurt Making Station	Tourist can learn about how to make yoghurt in simple way. Because of long time producing and short time expiring of yoghurt outside the chiller machine, company can only provide the information through video. But, the tourists will still be given yoghurt drink before exit the station.
7	Cheese Making Station	Facility where company run the production of cheese in low production capacity. Tourists can see the production process through glass wall bordering production room and tourist room. Tourist can also learn about how to make cheese in simple way that they can try at home. Same as yoghurt making station, information about making cheese with simple method is submitted to tourist through video
8	Milk Processing Station	Contains of mini factory and tourists room. In mini factory, production process of pasteurizing milk runs. Like cheese making station, tourists can see milk production through glass wall to keep maintaining hygiene of the milk. After that, they can learn about how to pasteurizing milk by using simple equipment which used daily, such as stove and pan. Tourists will also be given pasteurized milk that already packed in a bottle.
9	Viewing Post	Viewing post can be used to see the natural scenery around ecotourism. It can also be used for photography activities.
10	Milk Products Store	Milk product store will provide a wide range of products made from milk, such as pasteurized milk, that can be purchased by the tourists. Clerk and cashier are needed in this facility to run operational activities and help tourist in shopping.
11	Biology Museum	Facility that has biological objects and other objects related to the environment. Information related is devoted to animal agriculture and dairy cattle, such as how the cows can produce milk, the types of dairy cattle, bovine growth process and so on

From Figure 2, it is known that cowshed, biology museum, and compost pile were located on Quadrant I. It implied that those three facilities had advantages in all six variables. Meanwhile, biogas installation, milk product store, outbound, and viewing post were located on Quadrant II, which implied that those 4 facilities were had advantages in education, ecology preservation, and local community economy variable but having shortcomings in 'something to do', 'something to see', and 'something to buy' variable because the negative value of the Dimension 1. Facilities located in quadrant III had all shortcomings in all six variables. Hence, it is decided not to select facilities in quadrant III. In addition to the main facility as an attraction for the tourist, some supporting facilities were also needed to support tourists' activities and comfort when travelling. Referring to the Indonesian Government Regulation No. 50 Year 2011 concerning the National Tourism Development Master Plan Year 2010-2025 Article 25 Clause a, supporting facilities to be added were security post, ATM, deposit box, medical room, wheelchair and baby trolley rental, pedestrian, parking area, praying room, food court/restaurant, information center, information board or signpost that will be placed in some corner on pedestrian walkway, landscaping (scenery, flower garden, and orchard).

3.3. Designing Facilities

To design the facilities, activities in each facility should be determined first. Activities at each facility are determined by considering respondents' answer in the questionnaire, the existing activities in similar tourist attractions, and a company's desire about the development. In each facility, not only tourists who are going to do some activities but also tour guide and operators in running this ecotourism operational. Description of each facility can be seen in Table 5.

Based on activities determination, space requirement of all facilities can be calculated. To calculate the space requirement, some standards from literatures were used. Standard from Littlefield were used to calculate space requirement of all facilities and especially space allowance for tourists and table set for tourists and employees, and also aisle in each and cross-facilities[6]. Besides that, to calculate space requirement for ice cream, yoghurt, and cheese making station, also milk processing station, the size and capacity of existing machine were used. Machines needed were freezer, chiller, eco pasteurizer, press machine, sealing machine, container, homogenizer, and separator. To calculate space requirement for parking area, standard from Department of Transportation of Indonesian Republic to determine the aisle and space needed for each type of vehicle were used. The result of space requirements calculation can be seen on Figure 3(a).

After determining space requirement, the next step was to determine closeness relationship between each facility by using Activity Relationship Chart (ARC). Determining closeness relationship were based some reason obtained by management consideration. There are 6 levels of relationship, i.e. Absolutely Necessary (A), Very Important (E), Important (I), Ordinary (O), Unimportant (U), and Undesirable (X). Part of Activity Relationship Chart can be seen on Figure 3(b).

Main Facilities		**Supporting Facilities**	
Facilities	**Space Requirement (m^2)**	**Facilities**	**Space Requirement (m^2)**
Cowshed	781.09	*Food court*	427.64
Biogas Installation	85.12	Parking Area	2,313.63
Compost Pile	58.03	Safety Post	7.00
Outbound Area	3504.45	Information Centre	16.21
Ice Cream Making Station	141.87	Praying Room	102.71
Yoghurt Making Station	109.94	Toilet	53.48
Cheese Making Station	178.80	Medical Room	14.00
Milk Processing Station	164.35	Ticketing locket	164.74
Viewing Post	39.97	ATM	10.00
Milk Products Store	46.12	Orchard	2,140.00
Biology Museum	119.40	Generator set	8.00
Total	**5354.16**	**Total**	**5257.41**

(a)

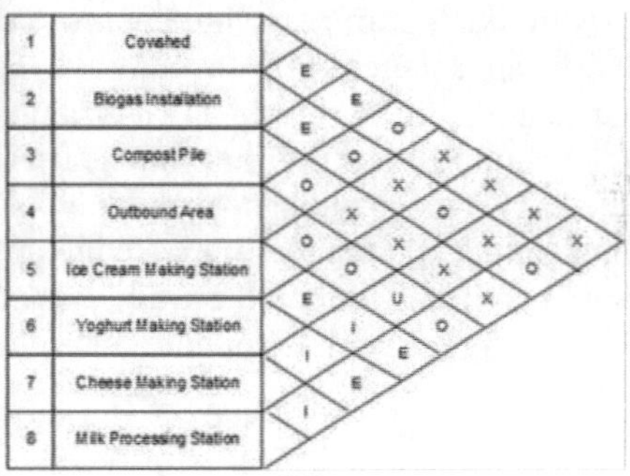

(b)

Figure 3. (a) Space requirement; (b) Part of activity relationship chart.

ARC resulted in 3 "A" relationships, 8 "E" relationships, 32 "I" relationships, 48 "O" relationships, 71 "U" relationships, and 63 "X" relationships. To draw the layout design, some line and color symbols were used as supporting tools to visualize the position of each facility. "A" relationship is symbolized by red line, "E" by orange line, "I" by green line, "O" by blue line, "X" by brown line, and "U" without any line. Each facility was represented by space block according to space requirement when placed on layout design.

After drawing the relationship diagram, some modifications were considered in designing facility layout to adjust the design with actual condition. It is commonly called as modifying consideration [5]. Layout was designed by adjusting the shape of land. Among the facilities, it was also provided pedestrian lane with 3.6 m of width [6]. To add to the beauty of the scenery, in the rest of the land where facilities and pedestrian were not placed, flower garden was set.

There were also two limitations in designing the layout. On the land, the cowshed was already exist but had never been used before. Thus, the location of the cowshed was fixed and should not be changed. Moreover, the land was bordered by rice field in north, west, and east side. The only side which was bordered directly with main highway was in the south side. Therefore, the entrance and exit of vehicles should be located in south side of the land. The facility layout design can be seen on Figure 4.

Based on Figure 4, it can be seen that food court was directly accessible to travelers when passing the entrance, milk processing area that arranged circularly with the orchard as its center which made the scenery look more beautiful and the tourists can also immediately see all areas of milk processing, and boundaries between types of vehicles in the parking area easy to determined. Beside the advantages, there are some shortcomings of this layout, which are not so many alternative route can be pass, the location of outbound area is on the edge and likely to be felt sealed because it is directly adjacent to the border wall, and the south edge of outbound area is close to the parking area, hence the activities may disturbed by the vehicle sounds and pollutions or activities outside of ecotourism because the outbound area better be directly connected with the pristine nature.

4. Conclusions

The appropriate type of strategy based on the company's position on IE matrix was intensive or integrative strategy, such as make a detail about ecotourism development future plan and also renovate the cowsheds and vacant land by adding and arranging more tourism facilities. Based on MDS, the appropriate facilities that selected were cowshed, biogas installation, compost pile, outbound area, ice cream making station, yoghurt making station, cheese making station, milk processing station, viewing post, dairy products store, and biology museum. The arrangement of those facilities was designed into a layout. The advantages of the layout were food court that directly accessible to travelers when passing the entrance, milk processing area that arranged circularly with the orchard as its center which made the scenery look more beautiful, and boundaries between types of vehicles in the parking area easy to determined. The shortcomings of this layout were not so many alternative route could be pass, the location of outbound area was on the edge and likely to be felt sealed because it was directly adjacent to the border wall, and the south edge of outbound area was close to the parking area, hence the activities may disturbed by the vehicle sounds and pollutions from the outside of ecotourism.

References

[1] Damanik J, Weber HF. Perencanaan Ekowisata Dari Teori Ke Aplikasi. Yogyakarta: Pusat Studi Pariwisata (PUSPAR) UGM dan ANDI; 2006.
[2] David FR. Strategic Management. 13th ed. New Jersey: Prentice Hall; 2011.
[3] Masuku TJ, Marline SP, Yohanes ARL. Persepsi Konsumen Terhadap Produk Sepatu Olahraga Di Sport Station Megamall Dengan Menggunakan Analisis Multidimensional Scaling. Jurnal Ilmiah Sains 2014; XIV (2): 68 – 72.
[4] Wignjosoebroto S. Tata Letak Pabrik dan Pemindahan Bahan. Surabaya: Guna Widya; 2009.
[5] Heragu SS. Facilities Design. 3rd ed. Florida: CRC Press; 2008.
[6] Littlefield D. Metric Handbook Planning and Design Data. Oxford: Elsevier Ltd; 2008.
[7] Sundoro DA. Arahan Pengembangan Objek Wisata Alam Cangar Kota Batu Berbasis Konsep Ekowisata. Unpublish Thesis. Malang: Department of Urban and Regional Planning, Universitas Brawijaya; 2011.
[8] Departemen Kebudayaan dan Pariwisata dan WWF Indonesia. Prinsip dan Kriteria Ekowisata Berbasis Masyarakat. Jakarta: Departemen Kebudayaan dan Pariwisata dan WWF Indonesia; 2009.

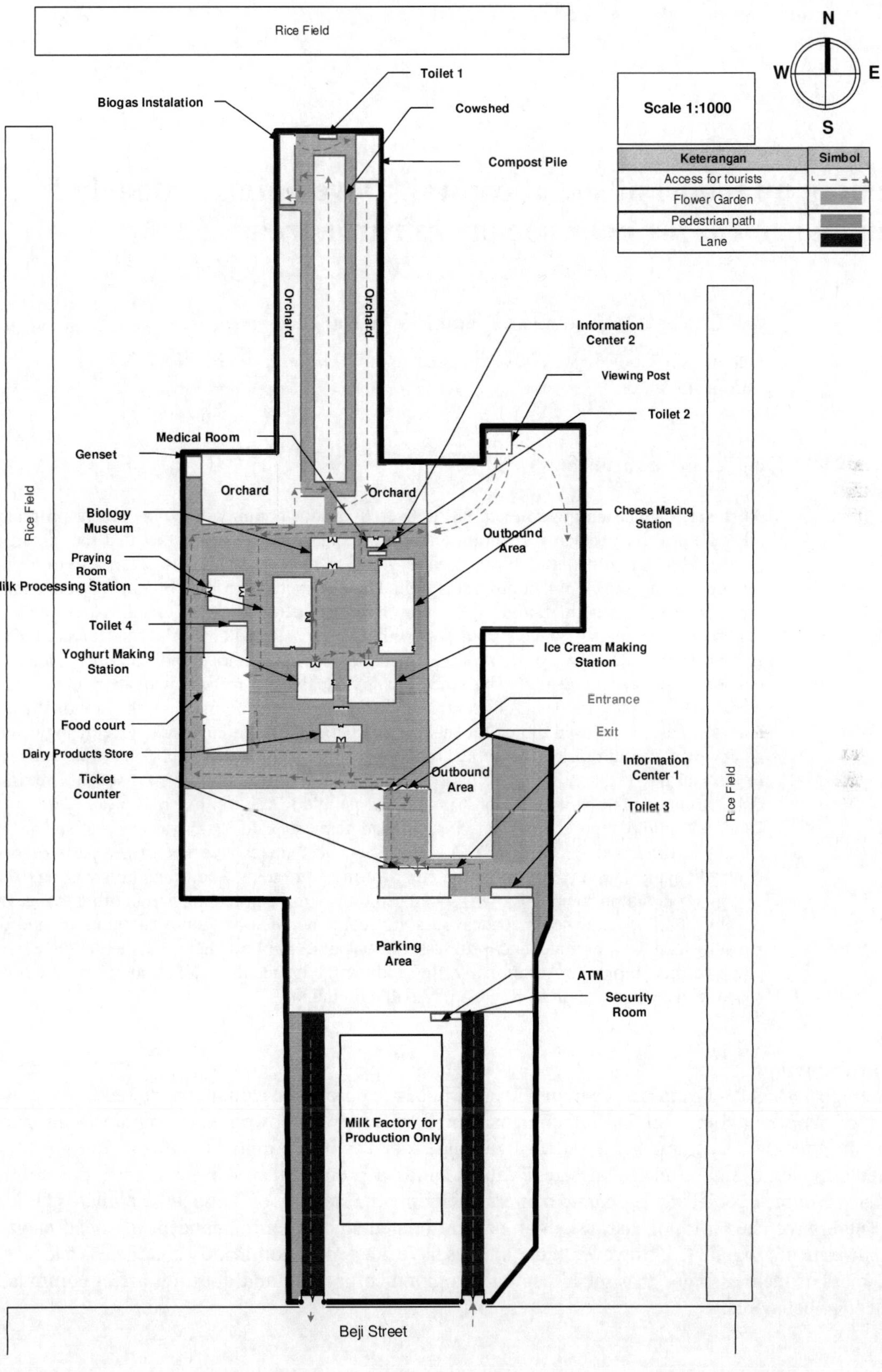
Rice Field
N
W
E
S
Scale 1:1000
Keterangan
Simbol
Access for tourists
Flower Garden
Pedestrian path
Lane
Toilet 1
Biogas Instalation
Cowshed
Compost Pile
Orchard
Orchard
Information Center 2
Viewing Post
Toilet 2
Medical Room
Genset
Orchard
Orchard
Rice Field
Biology Museum
Cheese Making Station
Outbound Area
Praying Room
Milk Processing Station
Toilet 4
Yoghurt Making Station
Ice Cream Making Station
Entrance
Exit
Food court
Dairy Products Store
Information Center 1
Ticket Counter
Outbound Area
Toilet 3
Rice Field
Parking Area
ATM
Security Room
Milk Factory for Production Only
Beji Street

Scaling up model of social capital: Developing a model of empowerment for poor society in rural areas

A F Chawa[1], A Kusumastuti, and I W W Harjo

Department Sociology, Universitas Brawijaya, Veteran Street No 1, Malang, Indonesia

anif.chawa@gmail.com

Abstract. Indonesian government has delivered various community development programs to alleviate poverty problems of rural communities. This is despite the fact that the numbers of people who live in poverty in this area is going to increase. This has given rise to a crucial question in how development programs should be distributed to rural community members. This article present empirical findings of research depicts a model or mechanism by which rural community members who have been involved in Posdaya (Pos Pemberdayaan Keluarga/ Family Empowerment Post) program success in achieving economic empowerment objective by employing their social capital. This study employs qualitative method with an approach of case study and multiple case design. It conducts the logic of 'replica', meaning that it would lead the analysed case to be used either for making predictions of similar results (literal logic) or for achieving different results (theoretical replica). The fundamental assumption of the study refers to the lack involvement or participation of the empowered community or social groups. The domination of top-down approach has allowed people to rely on external party. On the other hand, bottom-up approach has not been fully implemented due to the absence of strong social capital in the empowered society. As a consequence, there have been only parts of people controlling and enjoying the presented empowerment program. Social capital is not an existing or a given capital in society. Rather, social capital is a mere potential, and in order to shape and develop it some strategies or endeavours are required. Thus, the study attempts to find out a suitable model to arrange and progress the social capital in the implementation of social empowerment program. Additionally, the study will advance the scaling up model into a wider space with an expectation to create a powerful community.

1. Introduction

The empowerment program is a program with a purpose to improve the welfare and to reduce the poverty level of people in both rural and urban areas. The community empowerment program is inseparable from the attempt of a community to establish or shape social capital. Along with the previous-mentioned statements, Fukuyama argued that social capital holds a prominent role in economic development, human resources, social and economic matters, including the stability of democratic climate [1].

There have been a lot of discussions in several related studies on the concept of social capital in empowerment program [2]. However, those studies have been more dominantly conducted in local scale and most of them assumed that social capital is a given component and does not affect community's collective behaviors.

The present study, accordingly, intends to reveal that social capital is a potential which requires efforts to build and maintain the existing social capital. In addition, the study attempts to uncover a model of social capital development that is applicable for national scale so it is able to support the community empowerment program initiated by the government.

The empowerment program in this case study is POSDAYA (Pos Pemberdayaan Keluarga – Family Empowerment's Post) program in Ngroto and Pandansari villages, Pujon District, Malang Regency. The selection of these two locations is used as a comparative strategy and as an embryo of family empowerment. POSDAYA of Ngroto village is chosen to be the subject of the study due to its well achievement in regency, province, and national levels.

Meanwhile, POSDAYA of Pandansari and Madiredo village are used as the subject of the study is under a consideration that it has a close relation with Ngroto area, so it will help the researchers to build an integrated empowerment model. Besides, the establishment of empowerment group in Pandansari village is different from that in Ngroto village that is shaped based on the same interests. The empowerment group in Pandansari village is formed by the same Rukun Tetangga (RT – neighborhood association).

According to the above-mentioned discussion, the problems of the study are as follows: 1) What is the attempt to develop social capital in community empowerment program through POSDAYA in Pujon District, Malang Regency? (2) How can the model of developing social capital be scaled up at the wider level in order to support the POSDAYA program in the scale of district?

2. Literature Review

2.1. Empowerment

Empowerment is giving power or authority to act. Prijono and Soenyono said that empowerment is an effort to push social change and possibilities for marginal people to have power and give bigger influences at social and political arenas [3]. Soenyono explained empowerment is collective action that has changing power relations between individual, group and social institution [4]. The basic asumption is every individual has been born with the power of ownnership but the power levels between one indivual and another are different. The individual power is influenced by various factor such as knowledge, capacity, status, wealth, gender, etc. Empowerment initiatives needs support from local government to enhance rural development [5]

The purposes of empowerment program are human development. Object of empowerment program should increase their capacity and skill due to be responsible for their self-development [6]. This principle underlines human as the main source and the final goal of empowerment program. Other principles of empowerment is participation. Participation is members community involvement to has capacity for self-decision making related with implementation of community empowerment program. Ife, 2013; Swanepoel & De Beer, 2006 described the main principle of participation is inclusion. Inclusion related with giving opportunity of every members to reveal their problems and to use their own khowledge to find solution [7].

2.2. Social capital

Social capital offeres alternative community concept. This perspective rejects every form of structural system of authority include the role of government. Social capital emphasizes the role of community for self-help through voluntary principles, cooperation, and mutual help among members [8].

Putnam explained social capital as social relationship's role such as trust, norm and social network that has possibilities to produce community [9]. Furthermore, community can be built through voluntary action and cooperation between members. Fukuyama emphasized the important of norm that has informal characteristic to arrange social relation in community [10]. Coleman described community social capital can be formed by obligation, espectacy, trustworthiness, information channel, norma and effective sanctions [11].

Social capital sees community as entity that can maintain countinuity of life through social network, trust and solidarity as basic norm of social relation. The analysis of social capital can utilize to see how rural poor group empower their self with develop scoial capital through em[powerment program in POSDAYA (Pos Pemberdayaan Keluarga/Family Empowerment Post).

Some theoritical framework states social capital is given and possesed by community. Conversely, social capital should be built by community with their strength and capacity for collective and individual pueposes. Bourdieu stated social network is not given and has to be constructed throught investation strategy that has beneficial for community members [12]. Coleman indicated that social capital has same characteristic with economic capital. They has productive characteristic [13]. Without develop social capital, everyone cannot obtain material profit and reach maximal success.

Social capital can be assumed as economic capital that can be built and accumulated with particular strategy to reach individual or group purposes. Community or group should develop social capital based on their capacity to implement empowerment program.

The process of social capital development should pay attention on social capital dimension. Thus dimension has relation with social capital typology, how the different of interaction pattern in community produces different typology of social capital namely bonding social capital and bridging social capital. Bonding capital has exclusive characteristic. The idea contexs, relation, network and trust which developed are inward looking rather than outward looking. The ties of inward looking formed based on same of value, interaction and homogeneous daily knowledge. According to Putnam, the type of bonding social capital can be seen at sacred society which has particular dogma dominated and tend to maintain social structure [14].

Bridging social capital was generally emphasizes into outward looking that utilize social network of community members. It can give huge contribution for development, progress and strenght of commuity. Idea exchange, experience, knowledge, norm, value from other group create inovation as group problem solving.

Coleman, in Social Capital in teh Creation of Human Capital explains that bridging social capital, the creation of other network from the group, offers dimension to strive together

To maintain group solidarity [15]. In other part, there are other social capital namely linking. Linking social capital connects community with actor who has political and economic influence [16].

3. Methods

The study utilizes qualitative method with an approach of case study and multiple case design to comprehend the model of building and developing the social capital of a group of poor family joining POSDAYA program in Ngroto, Madiredo and Pandesari villages, Pujon District, Malang Regency. Yin stated that multiple case design followed the logic of 'replica', meaning that it would lead the analyzed case to be used either for making predictions of similar results (literal logic) or for achieving different results (theoretical replica) [17]. Along with the context of POSDAYA in Ngroto and Pandesari villages, the logic of replica is utilized to identify the similarities and differences of developing social capital in the implementation of community empowerment model through POSDAYA.

The present study employs an analysis technique which is divided into three stages namely; the researchers analyze the data during the stage of collecting data in the field, form a category or theme to interpret meaning from the collected data. The categories or themes in this study of POSDAYA program are: 1) the activities performed by the group of poor people joining POSDAYA program to establish and develop social capital, and (2) Scaling up model to build and improve social capital in the level of district. The last stage is to organize the pattern to determine the discussion structure, analyze the data according to the concepts of social capital and community empowerment.

4. Results

4.1. Utilization of Social Capital in Ngroto Village

POSDAYA in Ngroto village was established on July 18, 2012 according to an attempt of the women in the neighborhood to increase the income level of the community. There have been several POSDAYA groups exist in the area, they are: Makaryo Bersama group with a specific business of vegetable nurseries, Gemah Ripah group that concentrates on agriculture, Griya Rajut group with its knitwork products, Maju Jaya group for grocery shops, Srikandi Collection group of sewing or convection, and Sarinah Creative group that runs food processing.

The whole process of building the POSDAYA groups in Ngroto village is inseparable from the efforts of society to develop social capital. The social capital itself is, according to Putnam based on trust, norms, and social networks [18]. These essential aspects have been performed by the POSDAYA groups in Ngroto village. On the basis of trust, each member of the group keeps offering hands for those who are in need. One example is the borrowing seeds from the Makaryo group. While in knitting group, the aspect of trust is built through the group sales system.

Another component of social capital is networks. Bourdieu mentioned that social networks are not given but rather constructed [19]. In forming the social networks the members of POSDAYA groups hold a regular meeting in every 1.5 months and improve the frequency of communication through social media or WhatsApp. The trust and networks carried out by the members of these groups will finally create a regulation that aims at strengthening the bond among the members of POSDAYA groups.

In addition to the existence of POSDAYA in Ngroto village, the members do not only create a social bond but also a social bridge. For instance, the POSDAYA group of Makaryo does not only fulfill the need of seeds of the in-group region but also the out-group space. Also, the members maintain social link with the cooperatives service.

4.2. Utilization of Social Capital in Pandesari Village

The existence of empowerment program cannot be separated from the principle of empowerment that is human oriented and participation [20]. It goes well with the establishment of POSDAYA in Pandesari village that is based on the human oriented principle as represented by the women as the subject of empowerment.

Nevertheless, the formation of empowerment program in the village does not involve society in a way it determines the types of empowerment they are about to perform. This situation is due to the principle supporting the empowerment program in Pandesari village that tends to follow the code of practicality. As a consequence, the existence of POSDAYA program had just been realized by the surrounding right after the forming of administrator in village level that is the decree of the head of the village.

This condition later distinguishes the formation process of POSDAYA between Ngroto village and Pandesari village that embrace the bottom-up development principle. POSDAYA in Pandesari village is established on the basis of similarity of neighborhood association (RT – Rukun Tetangga). It was performed to allow the coordination among the empowered members through a direct targeting on the activities carried out by Pembinaan Kesejahteraan Keluarga (PKK – a social organization empowering women to actively contribute to national development) in each of the neighborhood administration. To add, there are 48 RTs in Pandesari village with a distant gap of area distribution from one to another.

One example of the empowerment programs in Pandesari village is to make the village a strawberry tourist village. However, the idea was originally derived from the empowerment agent that was carried by the community service of University of Brawijaya Malang. The community service also provided strawberry seeds and polybags for free to stimulate the community around to have an intention to plant strawberries. Consequently, since the program was mostly performed by the community service of the university, as their service period was over, the planting activity faced a failure.

One background of the failure in the empowerment program of strawberry tourist village is due to the lack of network in Pandesari group that depends much on the pioneer, that is the community service of University of Brawijaya Malang. The limited intensity of interaction on the neighborhood level and does not cover the POSDAYA group in RT level triggers the decrease of community's level of trust. Additionally, the formation of POSDAYA that is based on the role of society to perform self-governing communities and the principle of practicality or administrative district has caused the absence of the binding rules on planting the strawberries and another obligation of producing strawberries in particular amounts applied to each member of the groups.

The formation of social capital in Pandesari village that puts greater emphasis on the role of people to perform self-governing communities has caused a decline of bonding social capital and bridging social capital. As previously mentioned above, a weak bonding social capital may lead to the failure of the empowerment program. This is all because the formation of POSDAYA program in Pandesari village is not built on the basis of common interest, trust, norms, and networks. Thus, it consequently affects the reducing sense of solidarity among people in the society and they tend to prefer to prioritize personal interest instead.

Furthermore, the decline of social bonding has affected the weakness of bridging social capital since the network that the POSDAYA group of Pandesari village has made with the outside party only refers to the community service of University of Brawijaya Malang.

4.3. Utilization of Social Capital in Madiredo Village

The POSDAYA program built in Madiredo village has put a basis on economic potential of the area that is represented into processing apple for snacks. The apple used in the POSDAYA program is the plants that the people around the village grow in their own yard. It is due to the apple they plant is intended to be consumed or sold whenever the harvest time comes with the best apple products. While the apple grown in the field is oriented to a mass production that is located outside of the Pujon District or Malang Regency.

The POSDAYA program in Madiredo village that exploits apple as their empowerment program attempts to create a business group of Carangmas. This group is a business group that processes apple into carangmas (a snack product made from sweet potato). The members of the POSDAYA consist of five villages in which every village will have two representatives to take care of the business, so there will be 10 in total.

In terms of management, the Carangmas business group refers to the system of work division. However, the division does not restrict the presence of any cooperation among the members of the group. In addition, there is also a regular contribution as much as Rp. 10,000.00 to fulfill the needs of rice. Analyzed from the principle of social capital by Putnam encompassing trust, norms, and networks, the trust and norms components in Carangmas group have been displayed through the collective cooperation and a working system that are built according to their interest on the expertise they are confident with. However, this sytem does not guarantee the long existence of the group.

Another aspect supporting the failure of POSDAYA in Madiredo village is the problem of the low network. For instance in terms of marketing, the group is only able to sell the products to the surrounding stores or food stalls so that they cannot optimize the income as compared to the production cost.

Internal factors have also been an element causing the failure of the POSDAYA program in the village. The internal factors are the production site that is centered and the social gap among the members in terms of working hours. Eventhough at the early implementation of the working hours each member is allowed to select the hours based on their interest of expertise, the marketing department has still experienced a double responsibility of work. It is because the marketing department does not only market the products but also participate in the production process.

The final cause of the failure is the lack of initiative from the chief of the POSDAYA in Madiredo in which he does not claim his purpose to protect the members' welfare but rather personal profit. It is against the standard of community development which believes that the chief of a group should play an important role in building and managing programs.

Basically, the POSDAYA program formed in Madiredo with the Carangmas business group has performed the principle of human oriented and involved people's participation in the process of determining the types of empowerment that they are going to carry out. However in the real practice, there has been an absence of the involvement of members in the process of making decisions. The decision is made personally by the chief of Carangmas group himself.

Referring to Cooke and Kothari, participation may trigger the presence of 'new tyranny' in the society with an aim to obtain illegal or unfair issues [21]. This context has been applied in Carangmas business group in which the decisions have always been made personally and not collectively by the chief of the group. One clear example is when the chief decided to organize the cooking utensils only by some particular groups of people. Indeed there is not any distribution of business treasures to the whole members.

4.4. The Development Model to Establish Social Capital for Community Empowerment

The design of POSDAYA proposed in the present study is a bottom-up POSDAYA. This type of POSDAYA development program will create a domination for the society to decide on their own programs. Despite the role of government as a ruler of policies, the apllied design of POSDAYA is adopted from the existing POSDAYA in Ngroto village.

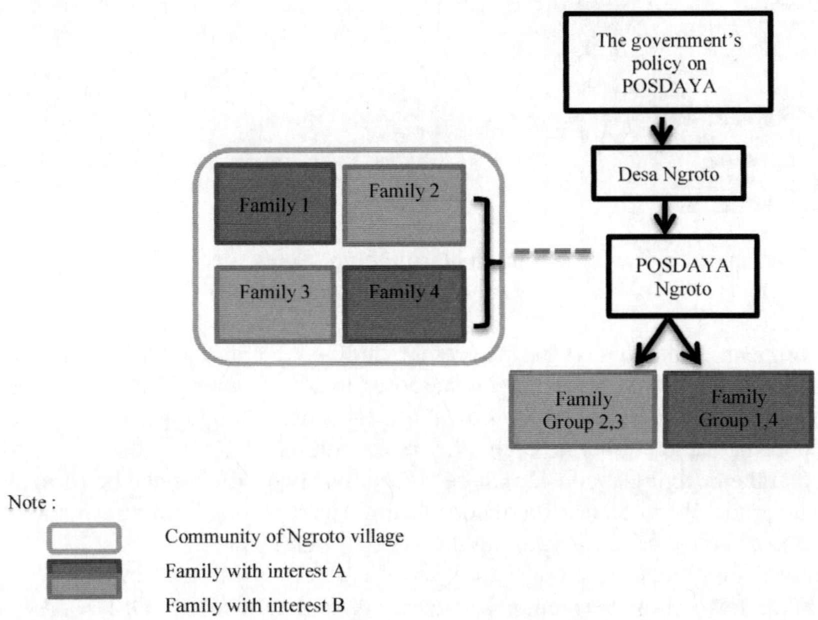

Figure 1. Model design of POSDAYA.

The program has been early started with the existing rule by the government about POSDAYA that is later implemented by the society. For example, if there are two families of A and B with different interest, they will be able to choose the interest group that is similar to their own interest.

The design of POSDAYA is expected to strengthen the society's participation and teh solidarity among the members of the group. So, it might influence the social capital namely trust, network, and norms to create bonding and bridging social capital.

4.5. The Development and Implementation of Empowerment Model (Scaling Up) in a Wider Scale in Pujon District

The empirical result found in the field comparing the two models of developing POSDAYA in Ngroto and Pandesari villages has been a standard to develop and implement the empowerment program with scaling up model to a wider coverage that is Pujon district. The success of POSDAYA development with scaling up model in Ngroto village will be adopted to advance another village, that is Pandesari by performing a set of integrated coordination and cooperation between one village to another.

One example of developing the empowerment model in Pandesari village is to change the basis of similarity between the administration of RT and POSDAYA into the basis of similar interest. To realize the idea, the social connection between Ngroto and Pandesari village to create a cooperative relation is highly required. The relation is expected to create a similar stimulus. In the end, the people of the two villages will be able to organize the resources collectively.
As soon as the social collective capital is established, the model of empowerment will be expanded into wider coverage of POSDAYA with the basis of sharing similar interest among the people in the society.

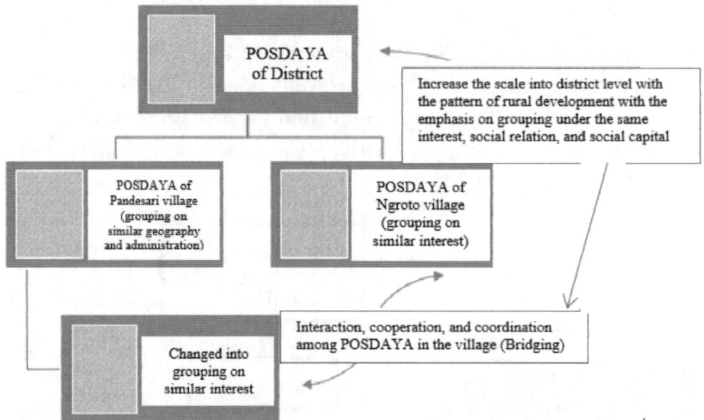

Figure 2. Implementation of empowerment model (scaling up) in a wider scale.

5. Conclusion

The scaling up model of development in the context of this present study is applied by performing the empowerment modeled by Ngroto village. Pandesari village is supposed to adopt the way Ngroto village carry out the POSDAYA by grouping people based on their own interest. Then, the villages may create a relation that will be able to coordinate and cooperate each other in the future.

By doing so, it is expected that the bridging social capital between those two villages can be formed and strengthen. There will be automatically a connection of social capital as represented in information, network, norms, expectations, and effective sanctions among the existing POSDAYA.

The empowerment of POSDAYA requires some considerations of collective potential, interest, and purpose. Besides, the scaling up model of empowerment needs to synergize the model of POSDAYA in the rural level. When the empowerment model in the village has not referred to the similarities that might tighten the bond of all members collectively, then the model will tend to be relatively top-down, and not bottom-up.

References

[1] Fukuyama, F. (1999). *Social Capital and Civil Society.* Institute of Public Policy. George Mason University.
[2] Astuti, A. K. (2015). Modal Sosial dan Mekanisme Adaptasi Masyarakat Pedesaan dalam Pengelolaan dan Pembangunan Infrastruktur. *Masyarakat: Jurnal Sosiologi, 20*(1), 81-97.
[3] Thobias, E., Tungka, & Rogahang. (2013). Pengaruh Modal Sosial Terhadap Perilaku Kewirausahaan : Suatu Studi pada Pelaku Usaha Mikro Kecil Menengah di Kec. Kabaruan Kab. Kepulauan Talaud. *ACTA DIURNA*.
[4] Soenyono. (2012). *Sosiologi Pemberdayaan Masyarakat.* Surabaya: Jenggala Pustaka Utama.
[5] Soenyono. (2012). *Sosiologi Pemberdayaan Masyarakat.* Surabaya: Jenggala Pustaka Utama.
[6] Ahmad, M. S., & Talib, N. B. A. (2015). Empowering local communities: decentralization, empowerment and community driven development. *Quality & Quantity*, *49*(2), 827-838.
[7] Bhattacharyya, J. (2004). Theorizing Community Development. *Journal of the Community*

Development Society, 34(2).
[8] Kenny, S. (2006). *Developing Communities* . Melbourne: Thomson.
[9] Swanepoel, H., & De Beer, F. (2006). *Community Development: Breaking the Cycle of Poverty.* South Africa: Juta.
[10] Ife, J. (2013). *Community Development in An Uncertain World.* New York: Cambridge University Press.
[11] Campfens, H. (1997). *Community Development Around The World: Practice, Theory, Research, Training.* Toronto: University of Toronto Press.
[12] Putnam, R. (1993). The Prosperous Community: Social Capital and Public Life. *American Prospect*, 35-42.
[13] Fukuyama, F. (1999). *Social Capital and Civil Society.* Institute of Public Policy. George Mason University.
[14] Coleman, J. S. (2012). Social Capital in the Cration of Human Capital. *American Journal of Sociology*, 95-120.
[15] Bourdieu, P. (1977). *Outline of Theory if Practice.* Politi Press: Cambridge.
[16] Coleman, J. S. (2012). Social Capital in the Cration of Human Capital. *American Journal of Sociology*, 95-120.
[17] Putnam, R. (1993). The Prosperous Community: Social Capital and Public Life. *American Prospect*, 35-42.
[18] Coleman, J. S. (2012). Social Capital in the Cration of Human Capital. *American Journal of Sociology*, 95-120.
[19] Woolcock, M. (2001). The Place of Social Capital in Understanding Social and Economic Outcomes. *ISUMA, 2*(1), 11-17.
[20] Yin, R. K. (2009). *Case Study Research: Design and Methods.* California: Thousand Oaks.
[21] Putnam, R. (1993). The Prosperous Community: Social Capital and Public Life. *American Prospect*, 35-42
[22] Bourdieu, P. (1977). *Outline of Theory if Practice.* Politi Press: Cambridge.
[23] Swanepoel, H., & De Beer, F. (2006). *Community Development: Breaking the Cycle of Poverty.* South Africa: Juta.
[24] Cooke, B., & Kothari, U. (2001). *The case for participation as tyranny* (pp. 1-15). Zed Books.

IOP Conf. Series: Earth and Environmental Science **70** (2017) 012030 doi:10.1088/1755-1315/70/1/012030

Assessment of community satisfaction index of population and civil registration office in Malang municipal

A N Hakim

Department of Urban and Regional Planning, Faculty of Engineering, Universitas Brawijaya, Malang, Indonesia

annisanurulhakim@gmail.com

Abstract. The demands and needs of the community will be the quality of public services in Malang increased, so that the Government of Malang as organizers and executors of public services must meet these demands. The Municipal Government in its efforts to improve public service performance, forming the regional One Stop Operator (PPTSP) with the goal will be to simplify and improve efficiency in administrative proceedings. But the existence of the one stop service is still not optimal because of the persistence of public complaints about the performance of the one stop service. This study will discuss the performance of services in service counter of Population and Civil Registration through community satisfaction index to measure the service level and the perception of satisfaction and interest to determine which variables are less optimal and need to be improved using IPA method. The results showed that the level of service at the service counter of Population and Civil Registration quite good. Meanwhile, according the results of IPA analysis, there are two important variables for the community on the performance / quality was lacking, namely the service procedure and time.

1. Introduction

The demands and needs of public services in Malang will increase so that the Municipal of Malang as organizers and executors of public services must accommodate it. According the Law on Public Service No. 25 year 2009 and Regulation of the Minister of Administrative Reform No. 16 year 2014 about the Implementation of Public Service, the organization of public services must involve the community from the preparation of service standards, evaluation and award, as well as an active role in arranging public service's policy.

Malang City Government in its efforts to improve public service performance constructed one stop service located in integrated office block one of them is Population and Civil Registration Office. The existence of one stop service was expected to serve society throughout the administration of population and civil registration in one location. On the contrary, the existence of one stop service is not optimal due there are many complaints of citizens in Malang Municipal website against the public service, such as long service time and costs that are not in harmony with regulation [1].

In previous research, such as Frederik (2008) entitled Analysis of Public Satisfaction Index of the Public Service Health Center Ngersep Semarang, showed the quality of public service increase significantly after doing a survey of community satisfaction index [2]. As on Agustina research the Public Satisfaction Index Service Unit of Government Agencies in Tulungagung District Office, said that to improve the quality of public services required a survey or study on community satisfaction index

[3], since government agencies are institutions that serve the needs of society. Therefore, the quality of services and satisfaction of the people should be main focus in government agency's performance. Thus, to fix the public service performance, there should be a study of public perceptions of agency performance. To measure the level of public services by calculating community satisfaction index and continued with IPA method to determine which variables are less optimal and needs to be developed.

2. Methods

2.1. Sampling Technique

The variables used in this research to assess the performance of public services based on the Minister of Administrative Reform No. 16 year 2014 on Guidelines Survey Satisfaction Index Societies Operation of Public Services, are requirements, procedures, time of service, fees / tariffs, product specification services, officers competence, officers behaviour, edicts services, as well as handling, complaints, suggestions and feedback.

The sampling technique used simple random sampling, because respondents who selected are represent of a population [4][5]. The basis to determine the number of samples is from the population who is receiving the service counter on Civil Registration Office with an estimated error of 10% [4]. Slovin formula:

$$n = \frac{N}{1 + Ne2} \tag{1}$$

Population of this research is about 717.916 people, drawn from data in year 2015 based on interview survey with the officer. To determine samples used Slovin formula, hence obtained 99,97 sample which is rounded up to 100 people.

2.2. Community Satisfaction Index

To calculate the value of community satisfaction index using the reference calculation of the decree of Minister Administrative Reform No. KEP/25/M.PAN/2/2004 [6] by counting the weighted average for each element first, and then be able to calculate the index's value.

$$Weightedaveragevalue = \frac{Totalweight}{Totalelements} \tag{2}$$

After find out the weighted average value, so it can calculate the value of community satisfaction index using the formula:

$$Index = \frac{Totalperceptionvalueperelement}{Totalelements} xscales \tag{3}$$

To make easier in interpretation of the value of community satisfaction index ratings, then made classification class into four classes with range between 25-100. Then, the results of the assessment of community satisfaction index will be multiplied by 25 in order to get into classification classes.

$$Classification = community\ satisfaction\ index \times 25 \tag{4}$$

Later, classification classes of community satisfaction index was created which have been converted as follows based on decree of Minister Administrative Reform No. KEP/25/M.PAN/2/2004. It is divided into four classes which has also been determined the ranking from the lowest score until the highest score, also its description.

Table 1. Classification class of Community Satisfaction Index (CSI).

Perception Value	Range Interval	Service Quality	Description
1	25 - 43,75	D	Poor
2	43,76 – 62,50	C	Less
3	62,51 – 81,25	B	Good
4	81,26 - 100	A	Very Good

Source: Decree of Minister Administrative Reform No. KEP/25/M.PAN/2/2004

2.3. Importance Performance Analysis (IPA)

Importance Performance Analysis used to determine the level of interest and satisfaction or perceptions and expectations of the community can use the analysis of visitor. This method combines measurement of factors of importance and satisfaction levels in two-dimensional graph that ease data explanation.

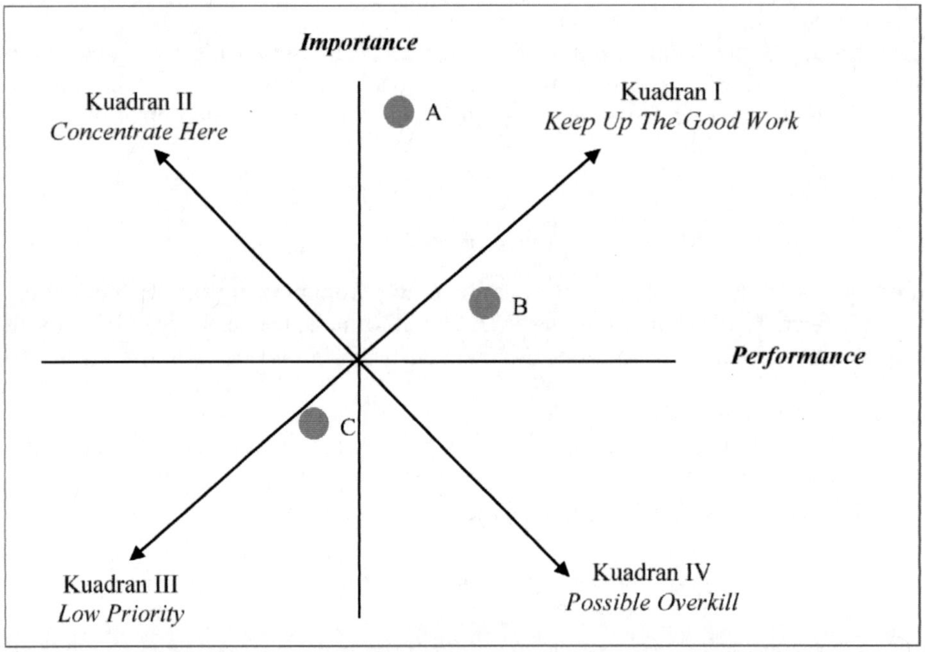

Figure 1. Importance Performance Analysis (IPA) quadrant.

The four quadrants in IPA analysis will described as follows:

2.3.1. Quadrant I (Keep Up The Good Work). The attributes of the performance of the service unit, considered important by the community as the basis for decisions by the performance and quality of service is very good.

2.3.2. Quadrant II (Concentrate Here). The attributes of the performance of the service unit is very important in the decision, but did not have a good quality service.

2.3.3. Quadrant III (Low Priority). Some attribute the performance of the service unit has decreased, due to both the importance and the quality of service is lower than the average value.

2.3.4. Quadrant IV (Possible Overkill). The attributes of the performance of the service unit is less important for the community, but has a good quality service.

3. Results

3.1. Public Service Characteristics

Analysis of the characteristics of public service describes the performance's condition in Population and Civil Registration office elaborated on each variables of public service.

3.1.1. Requirements. In general, public service requirements on service counter of Population and Civil Registration has been suitable as the type of service. This fact also supported by the results of primary survey shows that 72% of respondents assess the requirements in accordance with the type of service. Although respondents felt that the requirement is very accordance with the type of service as much as 28%.

3.1.2. Procedures. Service procedures at the service counter of Population and Civil Registration is shown in Figure 2:

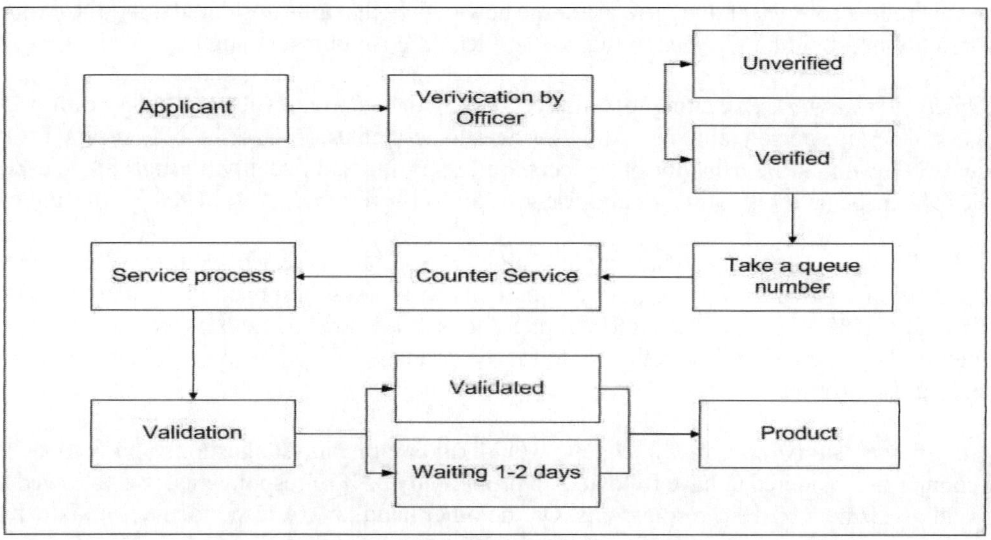

Figure 2. Procedures in services counters of population and civil registration.

Based on a primary survey respondents' opinions about the public service counters of Population and Civil Registration procedures consists of four votes, as many as 73% respondents rate the service procedure is easy, 7% rate it is very easy, 19% assess less easy and 1% judging difficult. The respondent' opinion who rate the service procedure is less easy and difficult due to the complicated procedure of having to adjust to the type of service counter and have to move to another counters.

3.1.3. Time Service. Based on a primary survey have known the time service in public service counters of Population and Civil Registration judged not fast by 11% respondents, less fast 38% respondents, 50% respondents said fast while very fast as much as 1% respondents. The existence of respondents who considered a public service counters is not fast enough and not easily influenced by the length of time waiting queue registration and the number of people who take care of administrative purposes, although the number of counters provided quite a lot and the counters are able to handle multiple types of services, but it is still deemed less.

3.1.4. Fees. The fees charged for getting public services on the of Population and Civil Registration considered reasonable by 60% respondents. While respondents were offered free charge approximately 60%. The fees charged generally to reimburse the cost of printing the ID card and certificate of civil registration. Respondents who do not charged fees such as to arrange family card update or data information or take care of other administrative services without having to print products.

3.1.5. Product Satisfaction Service. Product service specification is the result of a public service of Population and Civil Registration. Based on the survey results in mind as much as 88% respondents considered that the product specification of the service was accordance with the provisions. As many as 10% respondents rate the product specifications of the service is very accordance with the terms and 2% of respondents rate the product specification is less pursuant. Some respondents believed their product specifications are less suitable for general service due errors in the information / product type on the service.

3.1.6. Officers Competence. Officers competence includes knowledge, skills, expertise and experience of executive officers. Respondents assess officers competence consists of three votes, are less competent as much as 2%, 79% competent and very competent as much as 16%. Certain respondents believed their competence are less competent due there are some new officer that are considered not quite experienced and less readily in providing services so that the service time is not fast enough.

3.1.7. Officers Behavior. Executing of officers behavior can be evaluated from politeness and friendliness of service personnel in providing services to service users. Based on the survey results, 5% respondents believe that the behavior of officers are less polite and friendly, though 84% respondents found the officers behavior is courteous and friendly, and 11% respondents said very polite and friendly.

3.1.8. Edict Services. The respondents opinions about the service at service counter of Population and Civil Registration divided in 3 judges, are respondents who assess responsible as many as 87%, very responsible are 11%. Whereas respondents rate the service less liable services as much as 2%. Respondents who are consider less responsible for the counters officers who are joked with colleagues while serving the citizens.

3.1.9. Complaints, Suggestion, and Feedback. Handling complaints, suggestions and feedback on the service counter is considered to have followed up properly by 82% of respondents, and followed up with a very good assessment by 10% respondents. On the other hand, 8% of respondents found the handling of complaints, suggestions and feedback are actionable but less well.

3.2. Community Satisfaction Index

Based on the survey data obtained of public opinion to the variables / public service element. The result of the calculation, the index of satisfaction can be seen in Table 2 below.

Table 2. Variables values of public service.

No.	Service Elements	Average Values	Descriptions
1	Requirements	82	Very Good
2	Procedures	71.5	Good
3	Time	60.5	Less Good
4	The cost / fees	84	Very Good
5	Product specification services	77.5	Good
6	Officers Competence	77.75	Good
7	Officers Behavior	76.5	Good
8	Edict services	77.25	Good
9	Handling of complaints, suggestions and feedback	75.5	Good
Community Satisfaction Index Service Unit		75.83	Good

Based on the calculation, the value of community satisfaction index toward public service counters of Population and Civil Registration is 75.83, meaning that it has a good level of public services. Generally all variables public service has a good value, but there are some variables that need to improve their performance, namely the service procedure and service time. Service procedures are still considered complex and convoluted, such as many types of counters available were confuse the public. In addition, in taking care of population administration, sometimes not only in the same counters, but had to move to another booth. It is considered less effective and efficient. Time service variable has the lowest average value among other services, because the queue waiting time is relative long and the number of people who take care of administrative tasks were so many. In addition, the malfunction of the queue number machine, so that people who take care of each become disorderly.

3.3. Importance Performance Analysis
To determine which service variables are still less optimal and needs to be developed used IPA (Importance Performance Analysis). The survey showed that the results of the public perception of the service counter of Population and Civil Registration Office. Variables that are considered good are requirements, procedures, time of service, fees / tariffs, product specification services, executive competence, executive behavior, edicts services, as well as the handling of complaints, suggestions, and feedback. On the contrary, variables that assessed need to improve are service procedures and time services.

Table 3. Variables values of Importance Performance Analysis (IPA)

No.	Service Elements	Performance (X)	Importance (Y)
1	Requirements	3.46	3.63
2	Procedures	2.68	3.65
3	Time	2.34	3.62
4	The cost / fees	3.51	3.67
5	Product specification services	3.35	3.71
6	Officers Competence	3.31	3.71
7	Officers Behavior	3.39	3.64
8	Edict services	3.46	3.42
9	Handling of complaints, suggestions and feedback	3.41	3.43
Average value		3.21	3.60

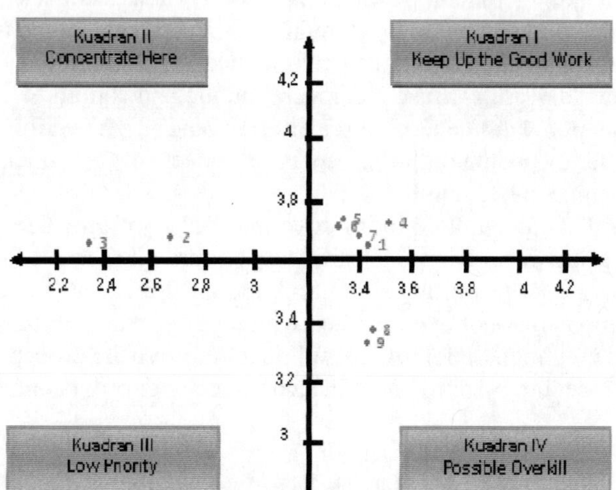

Figure 3. IPA quadrant of citizen perception in population and civil registration office in Malang municipal.

Based on calculations using the technique IPA perception, public perception of the known results of the service counters' performance of Population and Civil Registration in Municipal Malang.

3.3.1. Quadrant I, (keep up the good work / maintain performance). The quadrant that includes factors that are importance to service users and the level of perceived performance has been good. The variables that include this quadrant should be maintained, among other requirements, costs, product specification services, officer competence, officer behavior, edicts services, as well as the handling of complaints, suggestions, and feedback. Variables that are in Quadrant I assessed compliance as existing condition, in which the variables do not get many complaints from respondents.

3.3.2. Quadrant II, (concentrate here / enhance performance). The quadrants that load important factors to service users but the level of performance were considered unfavorable. The variables that in quadrant II must be improved among other variable, were service procedure and time service. Local people have complained about service procedures, related with the number of counters available and must be moved to other counters to taking care of the administration of population and civil registration. There was also complaint from citizens about time service. Citizens assessed that there was a long queue happened to take care the administration. The main problem is the number of people who manage the administration was so many. In addition, the malfunction of the queue numbers machine makes the atmosphere less orderly.

3.3.3. Quadrant IV, (possible overkill). Attributes on service performance are less important for service users but have a good service quality value, which includes edict service variable and handling complaints, suggestion, and feedback.

4. Conclusion

Based on the analysis of community satisfaction index and IPA method can be concluded that the public service at the service counter of Population and Civil Registration quite good. It is shown from the value of the satisfaction index and IPA method that seven services variables are in first quadrant. Those are requirements, costs, product specification services, officer competence, officer behaviour, edicts services, as well as the handling of complaints, suggestions, and feedback. Meanwhile, two other variables, namely the service procedure and time are in quadrant IV, where the two variables are crucial variable for public service users counters but the performance is not good. Lack of good values of both variables due to (1) counters diverse confuse the general public and should be moved to other booth to take care of administration, population and civil registration, and (2) the waiting time is relatively long because of the large number of people with the paperwork necessary and queues less orderly because the malfunction of the machine queue number.

Some strategic actions that can appllied to improve the quality of time services are accelerate time service by improving the quality of officers and technology, improving service facilities especially repair the queue machine number, enlarge waiting room and adding some seat. In addition, to shorten time service is by online system, so that not every citizen who take care the administrative purposes should goes to the population and civil registration office. While to improve the procedures quality can be done by developing an easier procedure system, but still accordance with provisions.

References

[1] Puspitasari, Dian., Sumartono., Mindarti L.I., Administrative Reform of Public Services On Integrated Services Office (Study on Disperindag and Dispendukcapil Malang). Journal of Public Administration (JAP), 2 (2), pp. 338-343.
[2] Titin. 2012. Analysis of Public Satisfaction Index of the Services Rendered by the Clerk Village In Village Tumenggungan Lamongan. Ekbis Journal, 2(1), pp. 1-5.
[3] Agustina, Holy. 2013. Community Satisfaction Index Services Unit Against Government

Institutions (Study on Implementation of Decisions of the Minister of State for Administrative Reform No. KEP / 25 / M.PAN / 2/2004) in Tulungagung District Office, Tulungagung. Journal: UB Faculty of Law.

[4] Uditrinurcahyo, Mohammad. 2005. Assessment of Public Perception of the General Spatial Plan Kendal. Unpublished thesis. Semarang: Diponegoro University.

IOP Conf. Series: Earth and Environmental Science **70** (2017) 012031 doi:10.1088/1755-1315/70/1/012031

Creative industry in supporting economy growth in Indonesia: Perspective of regional innovation system

AR R T Hidayat[1, 3] A Y Asmara[2]

[1] Department of Urban and Regional Planning, Faculty of Engineering, Universitas Brawijaya, Malang, Indonesia
[2] Centre for Science and Technology Development Studies, Indonesian Institute of Sciences, JL. Gatot Subroto, Jakarta, 12710, Indonesia

taufiqplano06@gmail.com

Abstract. Creative Industry is one of the most influential economy sources in the world in era 2000 years. It was introduced by John Howkins [1] in which economy growth is dependent on new ideas. This concept answers concerning to industrial-based economy and has shifted from industrial economy (manufacture) to creative economy (intellectual as main asset). As developing countries, Government of Indonesia has seriously paid attention on creative industry sectors since 2009 through President Instruction Number 6 Year 2009 about Development of Creative Economy in Indonesia [23]. Since Joko Widodo has been President of Republic of Indonesia, creative economy is more developed by forming creative economy agency (Bekraf). Now, economy creative is one of new economy sources which is promoted by Government of Indonesia. Many creative sectors are pushed to complete national economy in Indonesia. In this term, perspective of regional innovation system is also important to understand what is creative industry expected by Government of Indonesia. Innovation and creative economy is two terms which is not separated each other. This paper uses case study in Indonesia as research methodology, also perspective of regional innovation system is to be main perspective in this study. The result is that creative industry and innovation are mutual relation each other in conceptual level. Practically, both are aimed to support national economy growth in Indonesia

1. Introduction

Creative becomes popular in this decade, even since John Howkins has launched the first book of "How People Makes Money from The Ideas" in 2001 [1]. The term of creative word is debateable by many scholars and practitioners at economy field. Firstly, creative term is one of new approaches to develop economy sector in developed countries like European Union countries, or even in Australia by which creative economy was exposured in 1997 when cultural industries were as commercial activities [2]. This sector as a answer related to industrial-based economy (post-industrial economy). Scott in Lorentzen [3] said that post-industrial economy as a cognitive cultural economy, characterized by a high level computerization and with high levels human capital.

What Scott say is linear with human capital revealed by Florida [4] that creative class is formed and reinforced by human capital that generate and run new economy sources. In Howkins' idea [4], human capital is a main asset, it is embedded at everyone. Principally, creativity depends on the individual's capacity to dream, wander, think, challenge, disagree, and invent. Creativity express diversity, which is the source of culture. Without diversity, there is no a culture. All kind of creativity generate the same buzz of excitement, whether by artist, scientiest, or business executive.It is noted that creative economy is strikingly different to industrial economy sector. Notably, creative economy is boosted through idea (intangible asset) as main capital, whereas industrial economy is boosted from tangible asset.

Creative economy concept has emerged as means of focusing attention on the role of creativity as a force in contemporary economic life, stating that economic and cultural development are not separate but can be a part of a larger process of development [2]. Not only in advanced countries, this concept can be currently applied in developing countries. This policy is a part of economy policies offering and providing new economy opportunities in those countries to boost and to accelerate econmy growth into emerging high-growth areas as by which it is the same as economy growth in advanced countries [6][7].

In many dicourses and studies, term of creativity and inovation is often interchangeably [8]. Notwithstanding creativity and innovation have distinct meaning. Basically creativity is the ability to make or otherwise bring into existence something new whether a new solution to a problem, a new method or device, or a new artistic object of form. While innovation is defined adding something new to an existing product or process. Creativity is starting point to innovation or innovation is the implantation of creative inspiration [9]. Creativity, in other words, is the process through which new ideas are produced, while innovation is the process through which they are implemented [8].

For purpose of this study analysis, creativity and innovation is interchangable each other We call creativity-innovation in a unity as a new from people's to ideas (conceptual level) in many sectors including its implementation (practical level). According to OECD [10], innovation is a core growth driver in knowledge-based economies, and a subject for a wide-ranging policy efforts. While, creative minds are necessary in order to succeed as scientists, inventors and entrepreneurs. In innovation driven societies, education and training systems must empower people to innovate and quickly respond to the need for new skills generated by innovations in the workplace [11].

Creativity and innovation concept are widely dispersed in many countries, creativity and innovation are adopted and adapted to economy programs initiated by central and local governments around the world. Creativity and innovation is interesting topic and issue at regional development studies or geography studies like regional planning [8] [12] [4] [13]. At common, regional studies focus on spatial of entity of the region [14]. The concept of regional is subordinated to the concept of space. In 1950s, regional studies were dominated by regional science, a discipline with combined economic, geographic, and planning approaches and focused on theoritical and quantitive analysis ef economy regional issue. After long term debate, in 1970s, regional studies concerns on social and economic structure. In1990s, regional studies comprises not only social, but also cultural category. It is the system of belies, valueas, and meanings of a given community that shapes both institutional and social relations [15].

Definion of regional studies has been widely dispersed and adopted by scholars in many newly contextual concept. Doloreux and Parto [16] asserts that regional development ensues as competitiveness occur in places where localized capabilities such as institutional endowment, built structures, knowledge, and skills exist. OECD [10], across the Organisation for Economic Co-operation and Development (OECD) countries, regional development increasingly focus on innovation-driven growth. With this shift, regional development policies have included more integrated policies portfolios to promote the complementarity of policies in a given places (place-based approaches) and to leverage regional assets. Place matter for innovation, as it is regions and cities that compete to be hubs in global networks. One of innovation-driven growth models is creative city. Conceptually, creativity or creative city is closed to proximity concept or geography which both are discussion material on regional planning study.

While Florida [4] says that the human capital theory establishes creativity people, it is also the driving force in regional economic growth. From that perspective, economic growth will occur in place that have highly educated people. Again he reinforces that new geography and creativity shift to powerful migratory trends and emerging new economic geography. Also, Markusen [17] argues that discourses of creative city around the world has attracted may scholars from urban planning, economist, and public policy analyst. In this case, economic development planning is increasingly emphasizing the human capital side of the growth equation, stressing occupation as well as industry in analysis and policy.

Practically, creative city is innovative cities on many fields which both are approached by special studies like regional planning, geography studies, public policy, economy development, and also innovation studies. Within public policy context, Audretsch [18] reinforces that if entrepreneurial and creative people need positively social-economy environment, a government policy is important to grow economy considering culture, institution, ethics values in a region. The important thing is whatever those elements can influence robust entrepreneurial ecosystem.

Even, many scholars and studies are concerning to creativity-innovation and regional planning, and we usually call two key concepts as "creative city". At practice, creative city is not absolutely separated from role of government including the role of city planner [18]. Creative cities are also fusion of culture industries and creative class [19]. It means that creative city is not single actor, but it comprises many actors involving from government, academician, bussiness, and community. Therein, application of creative city is materialized into innovation. Many innovation scholars study creative city as part of their innovation-driven economy researches.

The literatuire regional innovation system (RIS) has provided substantial description and analysis of relationship between innovation, learning, and economic performance of particular regions. For example, Lim [20] studies concept of regional studies and innovation as regional innovation system as a new model for regional development and it could be a new means for a balanced economic development in nation. According to OECD [21], RIS is appropriate mechanisms for promoting innovation at the regional level when they succeed in identifying the specific competitive advantages of the region, as well as innovation systems bot tlenecks, and if the strategies are detailed and robust enough to tackle these opportunities and challenges.

In Indonesia, concept of regional innovation system (RIS) is translated to term of sistem inovasi daerah (SIDA) initiated by Taufik [22]. Conceptually and practically, RIS and SIDA are interchangebly used to avoid bias of concept of regional innovation perspective.

In this study, regional innovation system is concerned to creative and innovative activities created and conducted by human, promoted or intervened by government, supported by related communities and academician. Easily understood, regional innovation system perspective is focused on creative industry which is depended on readiness of region to receive and to open access of new economy sources. Easily, creative industries are located in creative cities. Case of Indonesia, there are several creative cities like Malang, Bandung, Yogyakarta, Denpasar, and Makassar. These cities are cities which generate creative products as new economy income for local people. Besides, the cities are tourism destination attracting many local and internatinal tourists to spend their time and money in those cities and its surrounding area.

Emergence and sustainbility of creative industries in creative cities engender three questions? Why do local people produce creative products and services? Who are specific targets of creative products and services? and How does local government support creative industries in region? This study will analyze and review three questions from national and local policy in bringing new economy programs to improve economy growth at local and national context. Surely, each region has special or similiar strategy to grow new economy sources in region.

2. Methodology

This research uses two methods namely direct observation in field and literatures review. Those methods are linear to how data was collected to complete this study. Direct observation was data which were derived from interview and data collection in field, as while literatures review was data which were derived from researches/studies results, books, scientific journals and articles, acts, and supporting documents. This research was conducted along 2016 year in two cities in Indonesia, namely Malang city at East Java Province (Java Island) and Denpasar City at Bali Province (Bali Island). Both cities are choosen as case study because they represent creative cities in Indonesia as well as popular tourism destinations in Indonesia.

To easily understand this study, we propose analytical framework which is combined from several concepts of creativity, innovation, regional innovation system, economy and policy (Figure 1).

Practically in Indonesia, creative sector is closed to innovation sector by which both support national economy in the different way. Although, both provide new strategy to increase people income, to widen job opportunity, to open access, and to increase added value. In the end, creative economy and innovation in particular region aimed to increase human well-being, not only to satisfy interest for academic field merely. Surely, regional innovation system is inevitably tied to colour creative economy sector initiated by local people and local government.

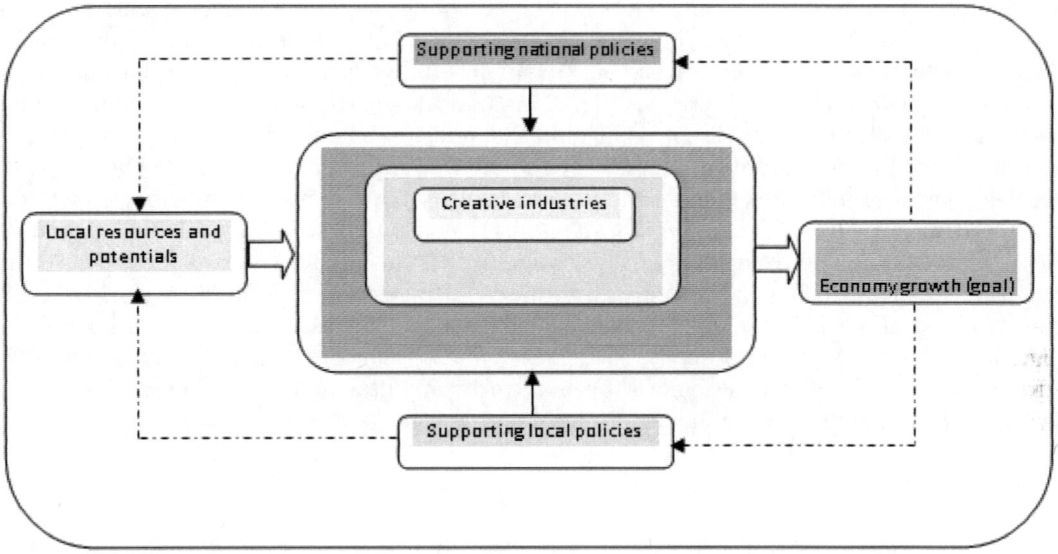

Figure 1. Analytical framework of creative industries in regional innovation system perspective.

3. Highlight of Research Results

3.1. Growth of Creative Industries to National Economy in Indonesia

Government of Indonesia has seriously developed creative industries since 2009 year within Presidential Instruction Number 6 Year 2009 about Development of Creative Economy in Indonesia [23]. This regulation was enacted and promulgated when President Susilo Bambang Yudhoyono (SBY) leaded Indonesia. Two years later, in 2011, Government of Indonesia increased role of Ministry of Tourism (Kemenpar) to be Ministry of Tourism and Creative Economy (Kenmenparekraf) based on Indonesian Regulation of President Number 92 Year 2011 about Nomenclature Change of Menteri Pariwisata (Kemenpar) to Menteri Pariwisata dan Ekonomi Kreatif (Kemenparekraf)[24].

Based on Indonesian Statistic Board (BPS) in 2013 year [25], creative economy sector has grown as much as 5,76% of total national economy growth by 5,74%. Concerning to gross domestic product (GDP), this sector has contributed to gross domestic product (GDP) by 7% or $US 46,930 billion and has employed or opened new jobs about 11,8 million people to work in this creative economy sector. This number is significant to prove that creative economy sector really contributes to national economy growth, even more in uncertainity of economy situation in Indonesia. Craft is one entrepreneurship-based creative sectors which has achievement to national economy growth by 5,76%, while average economy growth at national level is only 5,74%.

Linear with data which were derived BPS [24], creative sector reported 11,8 million people who worked in creative industry jobs in 2013 year. It comprised 3,1 million people who worked at creaft industry, 3,8 million people who worked at fashion industry, 167.000 people who worked at design industry, and 43.000 people who worked at architecture industry. This fact shows that significance of creative economygrowth is dominated from low technology sectors or products related to recent tourism and lifestyle fields. For example at culinary sector, this sector has ranked first level of other creative sectors, but this sector is still questionable due to inapropriate count and research method. Culinary is counted from all food-beverage units which is dispersed at many levels like food street, small-medium-large restaurants, bakery shops, cafe, caterers, confectionery and nut stores, and many others.

The monumental policy issued by Government of Indonesia is when President Joko Widodo (Jokowi) who substituted President SBY formed and legitimated Creative Economy Agency (Bekraf) in 2015 year. Bekraf Bekraf deals with creative economy replacing Mministry of Tourism and Creative Economy (Kemenparekraf) formed in 2011. Currently, Ministry of Tourism is specific in dealing with tourism affairs in Indonesia. Bekraf plays key role in promoting and boosting creative entrepreneurs in many regions in Indonesia. President Joko Widodo expects that Bekraf as leading sector to live national economy growth based new creative sector. Bekraf has 16 creative industry sectors namely: 1) application and game developer, 2) architecture, 3) interior design, 4) visual communication design, 5) product design, 6) fashion, 7) movies, 8) animation and video, 9) photography. 10) craft, 11) culinary, 12) music, 13) publishing, 14) advertisement, 15) art performance, 16) television and radio.

To accelerate creative industries growth, Ministry of Industry targets creative industries growth by 7-7,5% along 2015-2019. It is noted that regional domestic growth product of creative industries accounted by 5-6% yearly. Again, improvement of creative workers of creative industries is targeted by 10,5-11% of total national employment [27]. In vein with Ministry of Industry's goal, In 2016 year, Bekraf had six key fields to grow and to fertilize creative industries in Indonesia, namely: 1) film/movie; 2) animation and design; 3) music; 4) craft; 5) fashioon, and 6) culinary. Those sectors are considered because sixth sectors are high contributor for national economy growth to gross domestic product, job opportunity, and also improvement income for local people in local region.

3.2. *Creative Industries in Malang City*

3.2.1. Profile of Malang City. Malang city is one biggest city in the East Java province. It is located approximately 80km southern of Surabaya city, the capital city of East Java Province. This city serves as growth pole. Its development is being expected to drive development of the surrounding areas. Malang city is home for 851 thousand inhabitants. They live in the 5 districts. In the Blimbing district, there is a center of home industry for tempeh chips. In the Lowokwaru and Klojen District, there are prominent universities.

3.2.2. Resources and Potentials of Malang City. Malang city is one of famous tourism destination. It was well-known as "Kota Bunga" (flower city). With Malang regency, including Batu City, tourists enjoy various attractions. It offers such as eco-tourism, natural tourism, and historical tourism. Tourists who visit to Malang area always visit Malang city first because it serves as interchange. Malang city provides accommodation for tourists who want explore Malang Area. Currently, Malang City has rebranded its image. Malang city is a creative city. However, tourism activities still flourish. It also supports tourism activities of Batu city. Batu city was a part of Malang regency. Batu City separated from Malang Regency in 2002. This city declares itself as tourism city. Malang city provides accommodation and souvenirs. Malang city and Batu city are close. It is approximately 25Km apart. There are more than 30 Higher education institutions in the Malang city. Five of them are famous and big institution. Those are University of Brawijaya, Sunan Kalijaga Islamic Institute, University of Malang, Ma Chung University, and University of Muhammadiyah Malang. Their students are coming from across the nation and accept foreign student. Therefore, except holiday, total population of Malang city reaches 1 million inhabitants. These university provide wide range major that able to support city development as well as creative industry, for example tourism studies and food industry. Tourism and education activities are potential for creative industry in term of economic creative. Tourism activities stimulate culinary industry. Tourists need to buy souvenirs. Local people create unique culinary that resemble this city. They produce fruit chips, dairy products, and tempeh chips. Raw materials of those delicacies are mainly produced in this area. Previously, farmers sell their yield in the market. Nowadays, home industries buy their yield and produce various products. University also act important role. It provides a place to increase capacity by offering lecture that support creative industry, for example food processing and tourism studies. University has support creative industry trough research and community services. University also offers high quality of work force. Base on BPS, Malang city has high labor force of its total population [28]. Approximately, 60% of total population is labor force. It can support the development of creative industry in Malang city. Food is a basic human needs [29]. For tourist, food is part of tourism attraction, something to buy [30]. Malang city is a market. The number of tourist and huge population are potency as well as local demands. Mostly, local farmers of Malang city and Malang regency provide raw materials. It establishes mutual benefits among them. Therefore, culinary industry is one of prominent sector.

3.2.3. Creative Products of Malang City. Government of Malang city declared 4 sector as seed sector, namely 1) game and application; 2) film, video and animation; 3) culinary; and 4) graphic design. However the most prominent sector is culinary. It has 40% of total producer is food producers. Industry culinary sector produces various products. They produce not only for food souvenirs but also for daily consumption. Unique and new products are produced in the last decades. Fruits chips and tempeh chips are famous for souvenirs. Tempeh chips industries are located in Sanan area, Blimbing district. However, this industry has spread out in the surrounding area. Local Food service is arousing. For example "Ayam Goreng Nelongso" and "Mie Jogging". This food services offers unique and new variety of culinary. Their consumers are mainly local residents. They provide food for daily consumption. Fruit chips mainly produced in the Malang regency. However, main market of this industry is in Malang city. They buy in high quantity, repack, and sell in the Malang city. It gives mutual benefit to each actor. Raw material of these products come from Malang regency and other areas. This industry is highly dependent to other region. It creates linkage system among Malang city and surrounding regions. Culinary industry is facing challenges. Currently, market demands food certifications such as halal certification and food safety certification. Most of culinary product of Malang city based on halal material. However, it is demanding halal certificate on every pack of the product. Producers need to certify their products to get the label. Food safety certification is also essential. This certification is needed to convince consumer that products are safe. This certification allows producers sell it abroad. Main market of culinary sector is Malang city. But their consumers of this industry are from other region, mainly are tourists. Local demand from local resident is high as tourist demands. Just few producers are able to sell their products to other regions.

3.2.4. Supporting Policies (Municipality of Malang and Central Policy). Government of Malang city declared Malang city as a creative city. In order to support it, government established creative economic committee [31]. The main task of committee is elaborating creative economic actors and provide supports any creative industry. This committee consists of government representatives, academician, creative industry communities, and producers. Government of Malang also promoted creative industry to establish creative economic industry community, Malang Creative Fusion. This community is a place for creative industry actors to learn and growing their business. Main producers of creative industry are small and medium enterprise. Through industry and commerce department, Malang government supports SMEs. This department established strategy to develop SME This strategy involves private sectors to help SMEs' actors develop their capacity under partnership agreement. Government encourages SME to establish SME Holding. It creates collaboration environment rather than competition environment.Government also set SME center. This policy allows government and producers create good environment for production. In addition, it helps government to control and manage spatial development and support infrastructure for creative industry.

3.2.5. Economy Contribution to Malang City. Manufacturing contributes highest gross regional domestic product (26.5%) [28]. Food and beverages are part of manufacturing sector. It shares second highest percentage in the manufacturing sector after tobacco industry. Accommodation and food services contributes 4.37% to gross regional domestic products. Food service has high share (59%) of accommodation and food services. Culinary industry contributes approximately more than 3.5 trillion rupiah annually. This figure is growing year by year. However, this industry is not the biggest contributor to gross regional domestic products. Creative industry also contributes job opportunity. It allows labor force to create new business as well as job opportunity. According to BPS, labor force of Malang city reach 450.000 people. They are mainly work at manufacturing sector. Only 7.28% of labor force are unemployed. This figure is low than other regions. It gives benefit to creative industry to get employee.

3.3. Creative Industries in Denpasar City

3.3.1. Profile of Denpasar City. Denpasar is the biggest city and as province capital city in Bali Province. Denpasar is in Bali Island where its position in east of Java Island (East Java Province) and in west of Lombok Island (West Nusa Tenggara Province). The width of Denpasar city is 127,78 km2 or 21,8% of Bali Province area. Denpasar is the centre for business and government activities in Bali Province. In this city many public facilities to fulfill needs of local people in Bali. Denpasar is central for Balinese people mobility including education centre in this province. Regarding to spasial location, Denpasar city is also tourism and culture centre in Bali Province. Denpasar is surrounded cultural and tourist areas like Kuta and Ubud. Therefore, Denpasar is also called SARBAGITA area that it means Denpasar, Badung, Gianyar, and Tabanan. Recently, Government of Indonesia is preparing Denpasar City as one of three new metropolis cities beside of Medan and Makassar City. This policy is based on government plan for space and region according to President Law No. 45/2011 [32].

3.3.2. Resources and Potentials of Denpasar City. Local natural resources is limited in Denpasar city compared to its surrounding areas like Bangli, Tabanan, and Gianyar where these cities have much local natural resources to be developed as new economy sources. As the largest city in Bali, Denpasar is a hub of local and international people to do activities in many sectors like business, education, tourism, government, and others. Surely, Denpasar city has human capital whom they mutully interact each other and have similiar interest. New ideas are daily generated by them including new business scheme. Denpasar city has several prominent universities to support many life sectors like Udayana university (state-owned university), Indonesian Institute of Denpasar Art or ISI Denpasar (state-owned university), Marwadewa University (Bali government-owned university), and many private universities. In creative sector, Institute of Denpasar Art is education institution that largely contributes to creative economy development in Bali and Indonesia. Local and international students study in that university and part of them also study Indonesian Language in Udayana University. Although people is a key of local resources and potentials in Denpasar city, there is also tourism destinations like Sanur beach and cultural art performance performed by Denpasar community.

3.3.3. Creative Products of Denpasar City. Denpasar city is different to its surrounding cities in generating creative products. In this city, creative products are materialized in form of animation, modern game, advertisement, music, architecture, movie, information technology-based products, fashion, culinary, and art performance. These products are created from creative and wild ideas generated by local people in Denpasar. Uniqueness of Denpasar creative products is combination between local values (Hinduism) and advanced and simple technology. Market of Denpasar creative products ranges from Bali Province, Indonesia, and to other countries aroun the world. In 2014, municipality of Denpasar proposed to join UNESCO creative city network as a Creative City of Crafts and Folk Art [33].

3.3.4 Supporting Policies (Municipality of Denpasar, Government of Bali Province and Central Government). Municipality of Denpasar supported Government of Bali Province has same goal to grow Bali economy and to improve local people income by activating local natural resources as tourism source and also empowering local people as creativity source. Municipality of Denpasar has regulation at Major of Denpasar level coined into vision of Denpasar city that Denpasar is creative city. In central policy, Ministry of Tourism (Kemenpar) and Creative Economy Agency (Bekraf) support creative economy in Denpasar City through creative programs and also regulations that facilite creative product development including creative people. Even Ministry of Industry in collaboration with PT Telkomsel (information technology corporation) provide public space for local people especially youth to study as well as work (internship) in producing information technology-based creative products. Until now, this program have passed many creative actors whom they have opened business by themselves at information technology sector.

3.3.5 Economy Contribution to Denpasar City. Economy development of Denpasar city has significantly increased through creation of added value. In 2007, economy contribution of Denpasar city to gross Regional Domestic Product [34][35] was accounted by Rp. 795 trillion and it increased in 2008 which was accounted by Rp. 940 trillion. At average, real economy growth of Denpasar City was 6,60% in 2007 and 6,83% in 2008. This economy growth is prmarily bolstered by trade, hotel and culinary (restaurants) sectors (35,56% in 2007 and 35, 72% in 2008). Other sectors like rental, communication, transportation, and industries also contribute to economy growth of Denpasar City [35].

3.4. Malang and Denpasar: Is Creative City also a Tourism City and an Education City?

A tourism city is highly closed to creative city because prominence of each tourism city is able to attract many people (local and foreigner tourists) into those areas. In this case, tourism city is a city where has proximity to new creative economy sources such as culinary, craft, fashion, photography, etc. For example at culinary sector, scholars [37][38] found that culinary development is closely tied into tourism destinations where many people visit and spend their time and money to enjoy tourism object in a region. The increasing globalisation of innovation is also forcing regions to think beyond their borders, but those borders do remain a barrier, even for neighbouring regions [10]. At implementation stage, creative economy faces many challenges that are from internal and external factors. Internal factors are like lack of capable human resource as creative actors and internal bureaucracy system/management. While external factors are regulations, market, social problems, and political support. Those factors are to be homework in each of local government in Indonesia, included for central government. This implies that local people and Government of Indonesia proactively in supporting creative economy in Indonesia concurrent with implementation of innovation programs in many regions.

Malang and Denpasar city show similarity in creative industry. Their industries are driven by tourism activity. But it shows different products that they offer. Malang city offers culinary, Denpasar city offers art products and culinary. However, tourism development deals great it industry development.

They also show their similarity in local consumption market for culinary industry in Malang city is not only for tourist but also for local demands. Number of food services that offers daily demand of local product contributes more than Rp. 1 trillion to gross regional domestic products. Denpasar shows the same way. Local residents also buy their art productions.

Education institution also contributes to creative industry. Both cities are home for hundreds higher education institution. It offers education facility for local resident and attract outsider to the city. The output is this institution offers high quality of labour forces.

3.5. Conclusion

Creative economy has begun to do as leading component of economic growth, employment, and innovation in advanced countries. At conceptual stage, creativity and innovation are absolutely closed to improve people income through new creative economy strategy in a region. Recently, this concept is emulated in developing countries through creative city policy offering new economy source

opportunities to leapfrog into emerging high-growth areas of world economy. It is worthwhile to show that regional innovation system is appropriate approach to answer how creative cities run in several region. In Indonesia, many local government initiates creative cities using regional innovation system to run creative economy as sustainable program recently.
Case of Malang city and Denpasar city shows that creative industry is likely affected by activity within city a vice-versa. Tourism activity provides market for creative products. However, local demands also important market. Education in both cities contributes to supply quality labor forces. There are more than 40 higher education institutions in this city and offer education in the related field. In the end, creative industry in these cities significantly contributes to gross regional domestic products. Therefore, it demands government support to sustain and develop the creative industry. Government of Malang city creates special committee to elaborate creative industry actors and boost their activity. In addition, government of Malang city encourages actor of creative economy to create community as a place for discussion. This activity supported by department of commerce and industry of Malang city. Government of Denpasar city has collaborated with private sector to encourage artisan to open their own business. Moreover, central government also backed up government of Denpasar city to develop creative industry that supports tourism activity where Bali, in general, is main tourism destination of Indonesia.

References

[1] Howkins, John. 2013. The Creative Economy: How People Make Money From Ideas. England-UK: Penguin Books.
[2] The Canadian Policy Research Group. 2013. The Creative Economy: Key Concepts and Literature Review Highlights, 1-48.
[3] Lorentzen, Anne. 2011. Postindustrial Growth: Experiences, Culture, or Creative Economy. Working Paper. Havecity Universitat Hamburg.
[4] Florida, Richard. 2003. *Cities and Creative Class.* City and Community. American Sociological Association. 2 (1): 1-19.
[5] Howkins, John, 2001. *The Creative Economy*: How People Make Money from Ideas. Penguin. London.
[6] UNDP & UNCTAD. 2008. *The Challenge of Assessing the Creative Economy*: Towards Informed Policy--Making Creative Economy Report.
[7] UNDP. 2013. *Creative Economy Report 2013 Special Edition*. Widening Local Development Pathways New York: UNDP and UNESCO.
[8] Landry, Charles and Bianchini, Franco. 1995. *The Creative City*. London-UK: Demos.
[9] Okpara, Friday O. 2007. *The Value of Creativity and Innovation in Entrepreneurship*. Journal of Asia and Entrepreneurship And Sustainbility, Vol 3 (2): 1-14.
[10] OECD. 2013. Regions and Innovation: Collaborating Across Borders (OECD Review of Regional Innovation Series). OECD Publishing.
[11] OECD-CCE. 2013. *Educating for Innovation in Asia: Theory, Evidence and Practice*.Singapore International Workshop Singapore, 15-16 January 2013.
[12] Morgan, Kevin. 1997. The Learning Region: Institutions, Innovation, and Regional Renewal. Regional Studies 31 (5): 491-503
[13] Pratt[a], Andy C. 2008. *Innovation and creativity.* In: Hall, Tim, Hubbard, Phil and Short, John Rennie, (eds.) The Sage Companion to the City. SAGE Publications, London, UK, pp. 138-153.
[14] Markey, Sean. 2011. A Primer on Regionalism. Canadian Regional Development. A Critical Review of Theory, Practice, and Potentials.
[15] Sagan. 2003. Contemporary Regional Studies: - Theory-Methodological-And Practices . Geografia regionalna a przemiany metodologii badan´ społeczno-ekonomicznych. (in:) Sagan, I., Czepczyn´ ski, M., (eds.), Wymiar i wspo´ łczesne interpretacje regionu. Uniwersytet Gdan´ ski, Katedra Geografii Ekonomicznej, Gdan´sk.Poznan´ , 2003, pp. 43.54.
[16] Doloreux, David and Parto, Seed. 2004. *Regional Innovation System: A Critical Synthesis.*

Discussion Paper Series #2004-17, 1-38.
[17] Markusen, Ann. 2006. *Cultural Planning and The Creative City*. Paper presented at the Annual American Collegiate Schools of Planning Meetings, Ft. Worth, Texas, November 12, 2006.
[18] Audretsch, David B. 2005. *The Emergence of Entrepreneurship Policy in Local Heroes*, pages: 21-43 in *The Global Village: Globalization and New Entreprenurship Policies*, edited by David B. Audretsch, Heike Grimm, Charles W. Wessner. New York-USA: Springer.
[19] Pratt[b], Andy C. 2008. *Creativie Cities: The Cultural Industries and The Creative Class*. Geografiska Annaler: Series B Human Geography, 90 (2): 107-117.
[20] Lim, Jung Duk. 2006. Regional Innovation System and Regional development: Survey and a Korean Case. Working Paper Series Volume 2006-05, 1-21.
[21] OECD. 2010. *Regional Innovation Strategies*. Retrieved from www.oecd.org/innovation/policyplatform.
[22] Taufik, Akhmad Tatang. (2005). *Pengembangan Sistem Inovasi Daerah: Perspektif Kebijakan* [Development of Regional Innovation System: Policy Perspective] . Bidang Pengembangan Sistem Iptek Nasional – KNRT. dan Peningkatan Kapasitas Masyarakat BPPT Bekerjasama dengan Deputi Jakarta: Pusat Pengkajian Kebijakan Teknologi Pengembangan Unggulan Daerah. *(In Indonesian).*
[23] Presidential instruction no 6 tahun 2009 about Development of Creative Economy in Indonesia
[24] Regulation of President Number 92 Year 2011 about Nomenclature Change of Menteri Pariwisata (Kemenpar) to Menteri Pariwisata dan Ekonomi Kreatif (Kemenparekraf)
[25] Statistic Indonesia 2013. Retrieved from http://bps.go.id
[26] Statistic Indonesia 2013. Retrieved from http://bps.go.id
[27] Pusat Komunikasi Kemenperin. 2015. Menperin: Kontribusi PDB Ekonomi Kreatif Ditargetkan 7,5% [Ministry of Industry: Contributor of Bruto Domestic Product to Creative Economy is Targeted by 7,5%]. Accesed from *http://www.kemenperin.go.id/artikel/ 13182/Menperin:-Kontribusi-PDB-Ekonomi-Kreatif-Ditargetkan-7,5*.
[28] Malang city Statistic 2016. Retrieved from http://Malangkota.bps.go.id
[29] Maslow, A. H. 1943. A Theory of Human Motivation. *Psychological Review, 50(4)*, 370-96.
[30] Karyono, A. Hari Karyono. 1997. Kepariwisataan.Gramedia Widia Sarana Indonesia. Jakarta.
[31] Peraturan walikota No. 188.45/90/35.73.112/2016
[32] President Law No. 45/2011 about Plan for Cities Space of Denpasar, Badung, Gianyar, and Tabanan.
[33] Febianti. Memperkuat Citra Pariwisata Budaya: Denpasar Sebagai Nominasi Jaringan Kota Kreatif UNESCO [Strenghtening Image of Culture Tourism: Denpasar is as Network Nomination of UNESCO Creative City. 2015. JUMPA 1 [2] : 57 – 66.
[34] Denpasar city Statistic 2007. Retrieved from http://Denpasarkota.bps.go.id
[35] Denpasar city Statistic 2008. Retrieved from http://Denpasarkota.bps.go.id
[36] Rencana Pembangunan Jangka Menengah Daerah (RPJMD) Kota Denpasar Tahun 2010-2015 [Planning for Medium-term Development of Denpasar City in 2010-2015 Period].
[37] Horng, J.S., and Tsai, C.Y. 2010. Government websites for promoting East Asian culinary tourism: A cross national analysis. Tourism Management, 31 (2010) 74-85
[38] Lee, A. H., Wall, G., &Kovacs, J. F. 2015. Creative food clusters and rural development through place branding: Culinary tourism initiatives in Stratford and Muskoka, Ontario, Canada. Journal of Rural Studies. 39 (2015) 133-144.

IOP Conf. Series: Earth and Environmental Science **70** (2017) 012032 doi:10.1088/1755-1315/70/1/012032

Development of Indonesia-Papua New Guinea border, Muara Tami District, Jayapura City through agropolitan concept

A Subagiyo, F Dwiproborini and N Sari

Department of Urban and Regional Planning, Faculty of Engineering, Universitas Brawijaya, Malang, Indonesia

arissubagiyo@ub.ac.id

Abstract. The border of RI-PNG Muara Tami district is located on the eastern part of Jayapura city, which has agricultural potential. The past paradigm put the border as the backyard caused underdevelopment in border RI-PNG Muara Tami district, so that needed acceleration development through agropolitan concept. The purpose of the research is to define the aspect of physical, social, economic and border security to support agropolitan concept in border RI-PNG Muara Tami district. The analytical research method are border interactionan analysis, border security analysis, land capability analysis, land availability analysis, schallogram analysis, institutional analysis, leading comodity analysis (LQ and Growth Share), agribusiness linkage system analysis, accessibility analysis and A'WOT analysis. The result shown that mobilization from PNG to Muara Tami district could increase the economic opportunities with agricultural based. Border security of RI-PNG Muara Tami district is vulnerable, yet still condusive to mobilization. There is 12.977,94 Ha potensial land for agricultural (20,93%). There are six leading commodities to developed are rice, watermelon, banana, coconut, areca nut and cocoa. The border of RI-PNG Muara Tami district is ready enough to support agropolitan concept, but still have problems in social and economy aspect.

1. Introduction

The paradigm of the past put the border area as a backyard implications Muara Tami district border are under development in Jayapura city. Under development in Muara Tami district border as a result of the development of the region in Jayapura leads to the central city can be seen from the limited facilities and infrastructure services and low quality of human resources (RTRW Jayapura 2008-2027). Based on preliminary studies, the quality of human resources in Muara Tami district border is measured by the Human Development Index (HDI). HDI Muara Tami District border about 66.3 is still lower than the achievement of HDI Jayapura City. Muara Tami District border is located on the eastern part of Jayapura city with farmland, 68.38% of the population work in the agricultural sector and as a food supplier at Jayapura, but agriculture has not followed the development of agribusiness.

The Development of border areas currently is done by changing the direction of development policy in the past that is inward looking as the defense area being more outward looking approach by taking into account the community welfare, so that in addition to be the defense region is also to increase economic activities [1].

Border development model that can be developed in accordance with the conditions and potentials can be a fast-growing area, agropolitan area, transit area and tourist areas [2]. One of the concepts can be applied in Indonesia-PNG border Muara Tami District Jayapura City in accordance with the conditions and potential in the agricultural sector is agropolitan concept. Based on Nugroho and Rokhim (2012: 270), Agropolitan is a concept and development methodology that planned and integrated to a particular area, which is based on agriculture sector in terms of on-farm and off-farm and all supporting with the aim of enhancing region economic growth, increasing revenue, improve and maintain the quality of natural resources and the environment, and improve the function and effectiveness of government institutions and social in the region.

Development in Indonesia-PNG border Muara Tami District through agropolitan concept that should be encouraged to improve the economy and increase social welfare without compromising the security aspect as a crucial issue at the border. Therefore, this study aims to determine the support of physical, social, economic and border security in favor agropolitan concept.

2. Methods

Research location included 8 villages/towns on the border between Indonesia and PNG Muara Tami District in Jayapura city are Skow Mabo village, Skow Sae village, Skow Yambe village, East Koya townMiddle Koya town, West Koya town, Hotekamp village, and Mosso village. The collecting data used primary survey and secondary survey. Primary survey is conducted by direct observation, semi-structured interview, and questionnaire techniques. Besides, secondary survey is conducted by literature studies and doing surveys to some institutes to obtain data. The analyzing data used in this research namely.

1. Descriptive analysis consisted of border interaction analysis, border security analysis, and potential analysis (identification of the strengths, opportunities) and problems (identification of weaknesses, threats).
2. Evaluative analysis consisted of land capability analysis, land availability analysis, central analysis and services sub central (schallogram), institutional analysis (Venn diagram), superior commodity determination analysis (LQ and Growth Share), agribusiness system linkage analysis, and accessibility analysis.
3. Analysis of Development is using A'WOT analysis.

3. Methods

3.1. The Supporting of Pyshical, Social, Economic Aspects and Border Security Encouraging the Agropolitan Concept

The ability of land between Indonesia and PNG Muara border Muara Tami District refers to the standard decision letter of Minister of Agriculture No. 837 / KPTS / MU / 1980 based on the data of topography, soil type, and rainfall intensity. The broad area functions in border area between Indonesia and PNG Muara Tami District Jayapura City based on the land ability that is owned are the season cultivation / settlements zones about 23277.36 hectares (37.32%), annual year cultivation zones covering of 12110.7 hectares (19.42 %), buffer zones covering of 14050.74 hectares (22.53%) and protected zones covering of 12923.17 hectares (20.73).

The availability of land aims to know the broad area of agricultural land that can be developed for the application of the agropolitan concept at RI-PNG border Muara Tami District Jayapura city. The existing of land use is dominated by forest area of 37421.76 ha so we need a review of Government Regulation No. 10, 2010 about Procedures of allocation changing and the functions of forest area. Based on the regulation, forest can be changed the allocation by the convertion of production forest, while production forests and production forest which is limited is still maintained as forest area.

Agricultural land that can be developed based on the existing of land use and land capability that is an area of 12 977 ha, or 20.81% of the total area. Agricultural land in Indonesia-PNG border Muara Tami District need more attention to the land governance status on customary rights (hak ulayat) in order to avoid a conflict ownership in land utilization. Kinship between RI-PNG populations that still in one tribe cause the customary rights of PNG population is claimed in Tami River Skow Sae village Muara Tami District. The customary land ownership of PNG population does not affect the availability of land for agricultural development because it based on the existing of land use are in production forests used for farming subsystem.

3.2. Border Interaction
Internal mobility which is done by the residents in Muara Tami district border sell agricultural products, education and health. External Mobility inter-district that conducted by resident in RI-PNG border Muara Tami District is to sell agricultural products to Youtefa market in Abepura district, while External Mobility between countries that did by the residents at Muara Tami District border is to visit member relatives in PNG country as a form of social networking. The mobility of PNG Population at border is much compared with the border of Muara Tami District. The mobility of PNG population going to the Muara Tami District border is also to visiting members of relatives in Muara Tami District border also did to demand the economies need at market of Muara Tami district border, in addition, so it can be said the relationship between residents of both countries is not balanced that residents of PNG border relative tend to be oriented to the Muara Tami district.

3.3. Border Security
Border region potentially cause a conflict that will have an impact on security stability. Criminality behavior that occurred from 2009-2013 in Indonesia-PNG border Muara Tami District is carried out by the group of security disturber movement, but the current condition of peacefulness in Indonesia-PNG border is conducive to do cross-border activities.
Security system / supervision on RI-PNG border Muara Tami District is characterized by the existence of Cross Border Inspection Post (PPLB) official, border control posts and the construction of the border monument. The problems that exist now is the role of Skow PPLB as the place for go in and out of people and goods to and from the border of Muara Tami District that is not optimal because there is no supporting facilities and infrastructure completely and there is illegal trade at Lhoncin market that break the provisions of buffer area 5 km.

3.4. Superior Commodities
The determination of superior commodities in RI-PNG border Muara Tami District used LQ analysis method and Growth Share. Based on the LQ calculation, a basis commodity (LQ> 1), namely commodity paddy rice, string beans and bandeng fish. Based on the result of growth share calculation, the superior commodity is paddy rice and rambutan. After the comparation between LQ and share growth calculation, then it can be found the valuable commodity LQ> 1 and a valuable growth (+) is paddy rice.
In this research, the researchers considered the potential commodities on the result of growth share calculation becomes the superior commodity based on the criteria of government support, the selling price, the endurance of products and the connectedness other sectors in order to get the superior commodity in RI-PNG border Muara Tami District, namely rice, watermelon, banana, coconut, cocoa and areca nuts.

3.5. Development of Agropolitan Concept
The border region of RI-PNG on Muara Tami District Jayapura City has full support on physical aspect for the application of agropolitan concept namely the available of agricultural land of 12977.94 hectares for agricultural development with the area function is season cultivation zones covering an area of 23277.36 hectares (37.32%), annual year cultivation zones covering an area of 12110.7 hectares (19.42%), a buffer zones covering an area of 14050.74 hectares (22.53%) and protected zones covering an area of 12923.17 hectares (20.72%). The border of RI-PNG on Muara Tami District Jayapura city has a support and an obstacle on the social aspects for the implementation of agropolitan concept. The support of social aspect is communitys'activities that is dominated by agricultural (68.38% of the population work as farmers). The obstacle on social aspects for the implementation of agropolitan concept is the low quality of human resources in order to develop the agricultural activities on

agribusiness and farmers have a low ability to support the agropolitan development. The RI-PNG border on Muara Tami District Jayapura city have a support and an obstacle on the economic aspects for the implementation of agropolitan concept. The support of economic aspects has six commodities that can be as a superior. However, the obstacle on economic aspects for the implementation of agropolitan concept, namely trade in agricultural products is still in a local scale and have not been supported by adequate infrastructure to support the agropolitan development. The RI-PNG border on Muara Tami District Jayapura city has enough support in security aspects, namely the security between Indonesia and PNG border Muara Tami District that relatively stable to conduct cross-border activities.
The application of agropolitan concept on RI-PNG border Muara Tami District Jayapura city to eclipsed the lagging of development can be conclued quite ready to be developed but the current situation still has a constraint on the supporting of social and economic aspects. Agropolitan on the border of RI-PNG Muara Tami District that the development need to be encouraged based on the conditions and the potential of superior commodities that is owned can be the characteristics for the distrcit in order to improve the economic growth in borderland with competitiveness, so it can realize the development of border as the front porch of the state concerning the aspect of well-being in addition to security aspects that become a crucial issue in that region.

3.6. Recommendations of Border Develpoment throught Agropolitan Concept
The development on RI-PNG border Muara Tami District Jayapura city through the agropolitan concept on the result of space matrix analysis are in quadrant II room C that is developing strategies actively and aggressive. Therefore, the strategies that used in SWOT matrix using WO (Weakness-Opportunities) strategy as the main strategy. The focus on the main strategy targets in this study are:

1. Support the border cooperation opportunities between Indonesia and PNG in agriculture so it can increase the reach on agricultural marketing with exports scale that can encourage the economic development Muara Tami district boder.
2. Develop the industry of agricultural product processing in Muara Tami district border to capture the market opportunities that the PNG population orientation is more attracted to Muara Tami District border.
3. Utilize the support of government policy through efforts to increase the development of supporting infrastructure such as roads, irrigation, storage of infrastructure for agricultural production and agriculture products in Muara Tami district border and supply the modal for farmers to be able manage the farm.

Recommendations of the development on Agropolitan space pattern divided into cultivated zones and protected zones, those are:

1. Cultivation zones of agricultural allocation
   Zoning superior commodity in agriculture cultivation zones is based on the physical condition of the land, namely:
   a. Slope 0-8% prioritized for the development of rice plants, watermelon and bananas, while for the existing land use that are perennial crops, will be defended.
   b. Slope 8-15% prioritized for the development of coconut, cocoa and areca nuts, whereas for the existing land use which are annuals cultivation that padiakan is retained by terracing management.
   c. Slope 25-45% prioritized for the development of coconut, cocoa and areca nuts that already exist on the existing land use as a buffer between the cultivated and protected zones.
2. The cultivation zones of production forests and limited production forest will be defended as forest area. Those zones need the intercropping pattern between cocoa, coconut and areca nuts in the limited production forest zones with agroforestry system in Mosso village so that farmers in Mosso village still have an income without destroying the production forest.
3. Protected zones such as protected forests, sempadan river area and beach are not allowed doing agriculture activities to prevent land degradation and land resources intrusive.

4. Settlement zones for farmers in Holtekamp village, Middle Koya, Skow Mabo, Skow Yambe, Skow Sae and Mosso.

3.7. Recommendations on the Development of Superior Commodities

The development of superior commodities can improve the economy of the citizenry by integrating the agribusiness subsystems

1. Sub system on upstream agribusiness
   It needs for supply the superior local seed and certified to ensure the quality and the yield by planting the excellent seeds in the central and sub-central agropolitan.
2. Sub system on farming agribusiness
   Farmers have limited modal to develop their agricultural business, so the government needs to provide the modal assistance to farmers with educative principle. The direct aid to community and the aid for village empowerment strategic plan that is given should be selective based on the needs and right target.
3. Sub system on downstream agribusiness
   Recommendations on the development of sub systems downstream agribusiness is procurement and increment of processing industries except rice. However, it will be retained for national food defense.

3.8. Recommendations on the Development of Infrastructure Net

The adequate of facilities and infrastructure is necessary to support the agropolitan development on RI-PNG border Muara Tami District Jayapura city. Hence it needs for recommendations on infrastructure development such as:

1. The need to increase the easier access to the market on the road network which have low accessibility through road widening, pavement improvement and the provision of public transportation which is equipped with the procurement of terminals to support the distribution and marketing the agricultural products to the market.
2. Operationalize agribusiness sub-terminal and Sehito markets that does not work, water supply, maintenance of technical irrigation network by the related institution and the development of electricity networks to Skow Mabo village, Skow Yambe village and Mosso village.

3.9. Recommendations on Institution Improvement

The employee of Agricultural extension need to make the counseling system more effective throught the approachment to local farmers in village of Muara Tami District border with learning system that more easier accepted by farmers on cultivation techniques, product marketing and agribusiness.

3.10. Recommendations on Human Resources Development

Recommendations for human resources development are to improve the ability and independence to do with farmers' empowerment individualy although groups through formal and informal education, apprenticeship and training to change the mindset of local farmers particularly

3.11. Security Recommendations in RI-PNG Border Muara Tami District

Security approach needs to be noticed in agropolitan development in Indonesia-PNG border Muara Tami District Jayapura city to support economic activities over there. Based on Jayapura City RTRW security development is by establishing the command belt of state border along the ± 5km as defense space and state safety that supported by the addition of control posts and the provision of facilities and infrastructure to support the Skow PPLB.

4. Conclusion

1. The border of RI-PNG Muara Tami District has a border characteristic that is not much different from other land border in Indonesia. Special characteristic possessed by Muara Tami District

border is the orientation potential of PNG population that is more attracted to Muara Tami district border to fulfill the economy needs.

2. Agropolitan concept is already to be developed in RI-PNG border Muara Tami District by full support on physical aspect with the available of agriculture land covering area of 0.977,94 Hectares (20,93%) and have land capability of cultivation functions covering area of 35388.06 (56.74%). RI-PNG border Muara Tami district has full support and obstacle on social aspect that is the citizenry mostly work as farmer (68.38%), but the ability of farmers and farmer groups are still low. The border of RI-PNG Muara Tami district also have support and obstacle on economic aspect that can be found six superior commodities but the marketing of yields is still in local scale and have not be supported by an adequate facilities and infrastructures. This district also has full support on security aspect that is the safety in borderland is relatively stabil to conduct cross-border activities.
3. Agropolitan in Indonesia-PNG border Muara Tami District need to be encouraged the development based on the condition and the potential superior commodity that be possessed can be their own characteristics to increase the border economic growth with competitiveness, so it can realize the border development as the front porch of the state concerning the aspect of welfare in addition to the security aspect that becomes a crucial issue in that region.

References

[1] B. Omar, "Arahan dan Strategi Pengembangan Kawasan Perdesaan dengan Pendekatan Agropolitan di Kabupaten Blitar". Dissertation. IPB, Bogor, 2013.
[2] H. Suprayoga. (2009, Mei-June). "Program Pembangunan Kawasan Perbatasan". *Buletin Penataan Ruang.*
[3] Ditjen Penataan Ruang, *Pedoman Pengelolaan Ruang Kawasan Sentra Produksi Pangan Nasional dan Daerah (Agropolitan).* Jakarta, Indonesia: Ditjen Penataan Ruang, 2002.
[4] "Undang Undang Republik Indonesia Nomor 26 Tahun 2007 tentang Penataan Ruang", 2007.
[5] "Undang-Undang Republik Indonesia Nomor 43 Tahun 2008 Tentang Wilayah Negara", 2008.

IOP Conf. Series: Earth and Environmental Science **70** (2017) 012033 doi:10.1088/1755-1315/70/1/012033

Building a transformative adaptation: Comparing municipal government' and community's initiatives on minimizing the risk of coastal inundation in Pekalongan

Artiningsih[1, 3], S Worosuprojo[2], R Rijanta[2], S R Hardoyo[2], M H S Pratama[1], and N C Putri[1]

[1] Urban and Regional Planning Department, Diponegoro University, Semarang-50271, Indonesia

[2] Faculty of Geography, Gadjah Mada University, Yogyakarta-55281, Indonesia

artiningsih@undip.ac.id

Abstract. In 2006, coastal inundation was firstly reported to start entering community's agriculture land in Northern part of Pekalongan, Central Java, Indonesia. The exposure covered most of paddy fields and fishponds. The disturbance has become bigger, when the exposure of coastal inundation has started to cover some part of settlement areas in 2010. Increasing salinity has prevented farmers to grow paddy cultivation and has forced them to find a new way for living. Pekalongan Municipal Government prepared Pekalongan City's Resilience Strategy (PCRS) in 2010 in cooperation with PAKLIM-GIZ, and it involved significant and meaningful local stakeholders' participation process. One of the concerning strategies was to minimize the risk of coastal inundation. In terms of PCRS implementation, observed local community has different planning interpretation in comparison to the municipal government. This paper aims to evaluate the implementation of PCRS by comparing the municipal government' and community's initiatives - either in the stage of planning or/and implementation - from a transformative adaptation perspective. Data for this study was collected through interviews with several key informants, who were selected by purposive sampling method. These key informants consist of Pekalongan Municipal Government Agencies' members and local community figures in Northern Pekalongan Sub-district. This research reveals a double development standard from the side of municipal government, when it comes to prioritize both the economy and the environment. However, the local community prefers to choose a new livelihood, which provides them not just economic security, but also social and ecological benefits.

1. Introduction

Transformative adaptation, defined as adaptation that is recognized for its potential address root causes of poverty and failures in sustainable development, including the need for rapid progress on mitigation. Thus, municipal government has to prepare a framework that support relevant investment, behavior among household, communities and enterprise [1]. Case study on several cities and regencies in Central Java conveyed that there are 4 types of transformative adaptation scenario in terms of economic development namely 1) a profitable economic scale, 2) fostering fisheries and marine resource on livelihood shifting and diversification, 3) vertical (from agriculture as primary sector to industry, commercial and trade, and end up on tourism) and horizontal (from food-crop to fisheries/livestock or from an-organic to organic farming) transformation and 4) modernization [2].

Sustainable Development Goals' (SDGs) implementation on urban areas has many challenges. Achieving sustainable nations need encouragement on creating resilience cascade, start from community, cities, regions and end up on national level [3]. In order to make cities and human settlements inclusive, safe, resilient and sustainable (Goal no 11), many attempt related to underpinning combating climate changes (Goal no 13), end of poverty (Goal no 1) and hunger (Goal no 2) are needed [4].

Many coastal areas are vulnerable to increasing inundation as an impact of Sea Level Rise (SLR). Recently, many coastal areas in Central Java Indonesia are already subject to coastal hazard due to tidal inundation and flood. The impact of the inundation is predicted to be even more severe with the prediction of sea level rise in the next 20-100 years [5]. Extended coastal inundation has become endangered coastal community livelihood.

In 2006 coastal inundation was firstly reported start to entering community's agriculture land in Northern part of Pekalongan, Central Java, Indonesia. The exposure covered most of paddy fields and fishponds. The disturbance has become bigger, when the exposure of coastal inundation has started to cover some part of settlement areas in 2010. Increasing salinity has prevented farmers to grow paddy cultivation and has forced them to find a new way for living. Loss and damages of community assets and public facilities has been occured. This why, human and biophysical vulnerability are needed to consider accordingly. Community readiness is a key of success on adaptation process. This readiness include social learning and flexible networking which has become an important practical tools for finding sustainable livelihoods which enhance community persistence to live at risk [6].

People come up with some choices as alternative for livelihood shifting/diversification and adaptation. They tried to learn new thing, such as milk-fish, tilapia-fish, vaname-shrimp or tiger-shrimp and sea-weed cultivation. Considering the difficulties and opportunities in doing those new experience and regardless their existing skill capacity, technique and knowledge, they force to decide what the best option for living. It is *...crucial to support the ability of urban communities to negotiate their need and rights in order to increase their the flexibility and inclusiveness.....* [7]. Nevertheless, there are lack of information about how people in urban area build a transformative adaptation through their community's initiatives.

This paper will evaluate what choices have made by coastal farmers on Northern Pekalongan Sub-district. These choices would represent as community's initiatives on build a transformative adaptation.

Regarding on inclusive development, Pekalongan Municipal Government has prepared Pekalongan City's Resilience Strategy (PCRS) in 2010 in cooperation with PAKLIM-GIZ (Donors from Germany). It involved significant and meaningful local stakeholders' participation process [8]. One of the concerning strategies was to minimize the risk of coastal inundation. In terms of PCRS implementation, observed local community has different planning interpretation in comparison to the municipal government. Therefore, this paper aims to evaluate the implementation of PCRS by comparing the municipal government' and community's initiatives - either in the stage of planning or/and implementation - from a transformative adaptation perspective.

2. Research Method

Using qualitative approach, case study research was occupied to answer the question of 'why and how' [9]. Especially for answering some question such as:

1. Why did the community live in an inundated settlement? How did they tackle their environment changes?
2. Why did they choose to livelihood shifting or diversification? What kind of livelihood experience did they have?
3. How did the community build transformative adaptation to reduce the risk of coastal inundation and floods? What kind of initiatives have they done?

4. How was the local government tackling the coastal hazard? What kind of initiatives have they done?

Deep knowledge of an individual could give much useful information instead of hundred respondents who do not know anything or have limited knowledge [10]. Data for this study was collected through interviews with several key informants, who were selected by purposive sampling method. These key informants consist of Pekalongan Municipal Government Agencies' members and local community figures in Northern Pekalongan Sub-district. However, the local community prefers to choose a new livelihood, which provides them not just economic security, but also social and ecological benefits. Some triangulations have been done to gain data validation and verification. These triangulations involve related donors and NGO.

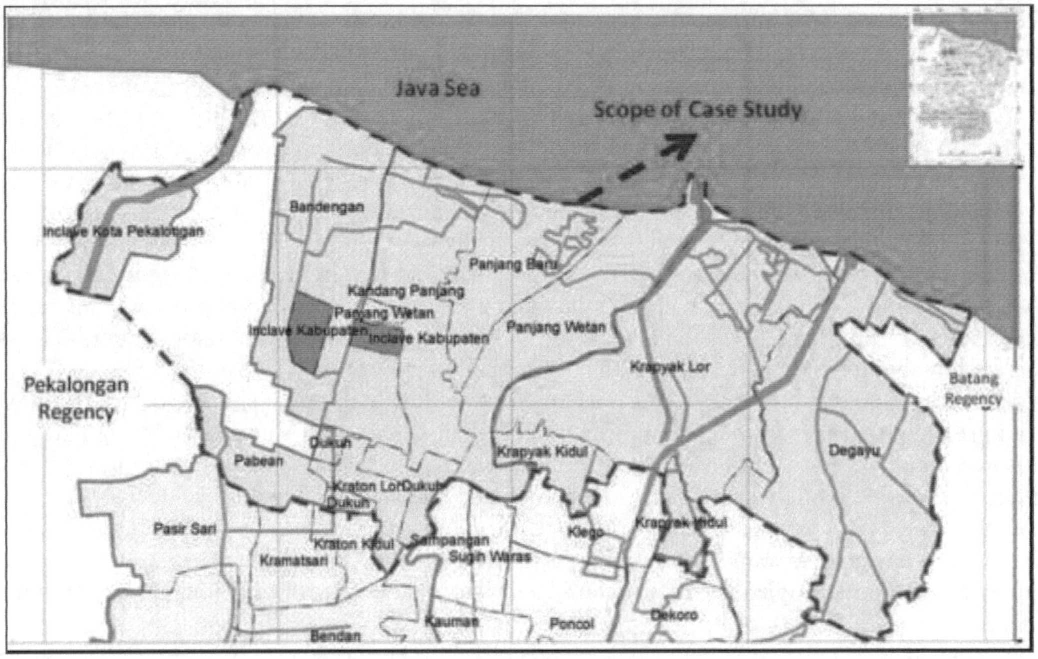

Figure 1. Scope of case study [11].

Scope of case study area consists of 6 sub-districts on North Pekalongan District, namely Bandengan, Padukuhan Kraton, Kandang Panjang, Krapyak, Panjang Baru and Degayu. See Figure 1 which ilustrated the location of scope of case study has taken. Those areas selected based on the most recently inundated area, which have triggered communities to elevate their capacity from vulnerable to cope with, adapt and short term resilience. Considering increasing community's capacity as a process, there was various achievement on each process that followed on transformative path ways respectively [6]. Important critical point on how the community decide to move on further process has proven both of transformation adaptation and resilience thinking. Some of them has atained either environment recovery or higher capacity in reducing the risk of coastal inundation. Those transformative adaptation was identified by community experienced, novel knowledge on reducing their vulnerability by minimized risk of coastal inundation, especially on livelihood adjustment either alteration or diversification.

Evaluation analysis has been done based on PCRS implementation performance by assessing development outputs/outcomes on social, economical, physical and ecological aspects. Elaboration from some important findings which were supported to related SDGs achievement were needed in order to compare Pekalongan Municipal Government and communities' initiatives.

3. Result and Findings

Community initiatives on economical and ecological development aspect were determined based on their transformative adaptation which was indicated by livelihood choices. Table 1 provides information of coastal farmer's consideration on livelihood choices. Their choices mostly was determined based on how does their livelihood alternatives give them sustainable benefit with low of production cost, what kind of potential risk and failure related each choices respectively and what kind of consequences in terms of livelihood sustainability. When coastal inundation was entering their paddy fields in 2006, land salinity has been increasing and prevented them to grow paddy and jasmine any longer. Thus the community tried to find a new way of life by livelihood shifting or diversification.

In line with Handayani et al, 2015 [2], which reveal with vertical and horizontal livelihood transformation in Central Java Province. Vertical livelihood transformation in North Pekalongan District slightly revealed as outcome of aqua-cullture livelihood, indicated by emerging local tourism activities. Mostly transformative adaptation illustrated by horizontally shifting livelihood transformation from former community livelihood on mangrove cultivation and agriculture (paddy and jasmine cultivation) to aqua-culture (fishery on fishpond) and combination of it. Litle part of community conveyed livelihood diversivication as simultant activity to gain multiple income from agriculture and informal sector employment such on household enterprise or batik industry.

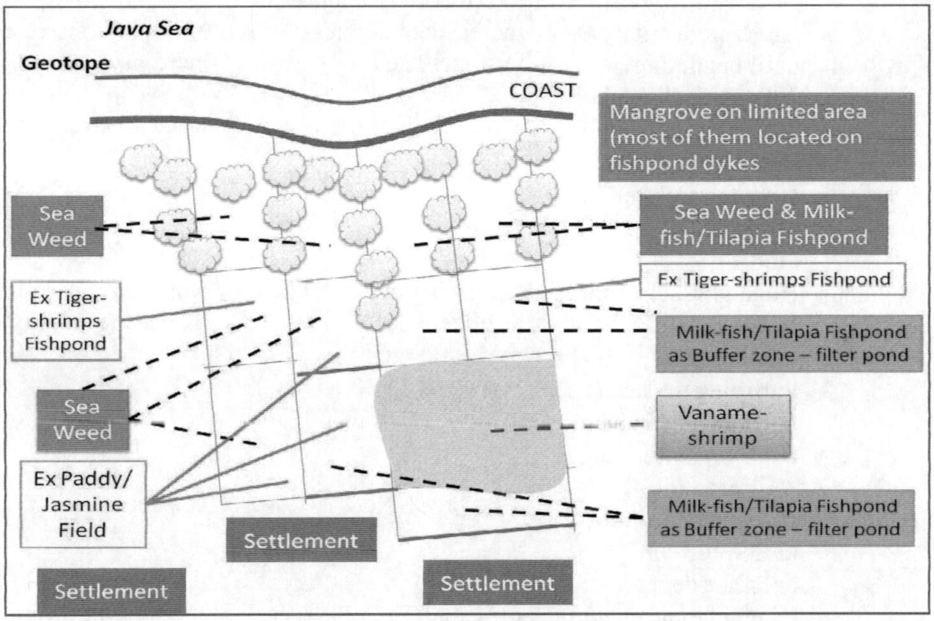

Figure 2. Imagery of adjustment space as livelihood shifting consequences.

Livelihood shifting triggered by exploitation of mangrove forest which convert to fishpond, due to high-demand export of tiger-shrimp in early 1990s. After a decade of growing mangrove conversion, livelihood shifting continued with paddy and jasmine cultivation into milk-fish/tilapia-fish cultivation at the beginning of transformative adaptation phase and continued with vaname-shrimps or sea-weed cultivation on the advance process of it. Those livelihood shifting illustrated on Figure 2. Parcel of land has facing different occupancy from time to time. This phenomena represented on adjustment space as livelihood shifting consequences.

Table 1. Coastal farmers consideration on livelihood choices as community initiatives.

Livelihood Alternatives	Possible Benefit	Possible Cost	Supporting to SDGs Achievement
Mangrove Cultivation	• Protection from abrasion • Bio-diversification (mangrove, prawn, shrimp, & wild fish) & economic benefit	Reducing mangrove intensity due to lack of maintenance; community's illegal cutting	High support on conservation attempt which give important effect for providing resources supply recovery
Tiger Shrimp Cultivation on Fishpond	• High Economic Value • Fulfilling Export Demand • Tiger-shrimps cultivation give bigger income for farmers	• Total production was decreasing after 2 years due to chemical feeds and drugs • Increasing threat of Mangrove extinction due to altering land cover to fishpond	Contra-productive with Mangrove Conservation. So give low support to achieve sustainable agriculture.
Paddy Crops on Paddy field	• Huge productivity until beginning of 2004 & decline until 2006 • High economic value • Give farmers ability to pay the cost of pilgrim	Increasing chemical fertilizer uses	None, as consequence of increasing salinity
Jasmine Plantation	• Huge production until 2010 as paddy intercropping • Fulfilling the need of Jasmine Tea Industry • High economic value • Ability to fulfill children education tuition fee until bachelor degree	Increasing chemical fertilizer uses	None, as consequence of increasing salinity
Milk-fish Cultivation	• High tolerance with pollutions • Small capital and low cost of maintenance. • Simple technology • Fishing as attraction activity on local tourism	soil-smelling milk-fish due to low water supply in dry season, push down the selling price	• Medium sustainability • Farmers did Milk-fish cultivation as side job not primary livelihood
Tilapia-fish Cultivation	• High tolerance with pollutions • Small capital and low cost of maintenance. • Simple technology	Risk of drift out fishpond due to flooding	• Medium sustainability • Farmers did Tilapia-fish cultivation as side job not primary livelihood

Livelihood Alternatives	Possible Benefit	Possible Cost	Supporting to SDGs Achievement
Vaname Shrimps Cultivation	• Fishing as attraction activity on local tourism • High price and demand • Milk-fish and Tilapia fish on filtered pond as positive externalitas of vaname cultivation	• Big capital • High cost of maintenance • Vulnerable with some desease • High risk of crop failure • Total production was decreasing after 2 years due to chemical feeds and drugs	• Low sustainability, • need government support in cooperation with private sector
Sea-weed cultivation	• Small capital with low cost maintenance • High tolerance with pollution • The role of Milk-fish and Tilapia fish as pest-plant eater	• Need dyke-fishpond support • Threat of chemical utilization as farmer choices on anorganic farming due to push up the production volume	• High sustainability (if kept on organic farming) • Potential to give either economic, social or ecological benefit as a production package

Compare to community initiatives, Pekalongan Municipal Government Agencies performed selected programs implementation from PCRS related to several aspects of developments. Those implementation have been experiencing some difficulties, since it have not been reinforced and supported by any special Cities Act. However, part of them has similarity with Municipal Agency's Programs and Projects.

Considering goals of Pekalongan City's resilience, many of them mostly intend to reduce city's emissions through related development's sectors as mitigation attempts. Therefore, transformative adaptation program related to community capacity and resiliency enhancement attempts become less priority.

Table 2 shows Government initiatives. The development orientation was considered more on physical aspect such as public infrastructures, which has become not just local or regional but also national authorities. Social economical aspect, especially on building socio-ecological resilience has become the second layer of development. Controlling flood and coastal inundation strategy due to increasing rain intensity have been targeted as risk reduction on 2015. Pekalongan City's drainage plan and program improvements done by DPUPR (Pekalongan Public Works and Settlement Agency). Neither selected city road nor paved road elevation on settlements area were consider on Pekalongan City's Resilience Action Plan. But surprisingly, regarding community satisfaction, it gave good performance effectively on North Pekalongan District risk reduction attempts.

Table 2. Selected development sectors as government initiatives.

PCRS	Pekalongan Municipal Government Initiatives	Supported Planning & Development policy
Mitigation: Infrastructure improvement on inundated area for vulnerable public facility/ utility and settlements in particular	• Improve access (road elevation) to Fish Market and Port on coastal core area (2015) • Encourage Local Tourism Attraction on coastal area (boat-ride, fishing, fish market, culinary) • Fishery & Marine Techno-park upgrading (2016) • Build 'Geotope' as sea wall along the coast to reduce abrassion and minimize coastal inundation (2014-2015) • Housing upgrading for the poor • Cleaning beach environment regularly	• Developing *Minapolitan* (Fish-city) on North Pekalongan District • Drainage Master Plan (2014)
Mangrove Conservation	• Mangrove plantation on limited coastal area (10.000 mangrove was planted, but less than 50% were survive)	Developing Mangrove Information Centre
Community Empowering	• Sending selected farmers for training on vaname-shrimps cultivation • Manage Public Private Partnerships on vaname-shrimps cultivation	

Drainage Masterplan has been made by BAPPEDA (Pekalongan's Development and Planning Board) in 2013. Therefore, there were some difficulties on financing polder system due to limited support from national budget (which need more than 250 billion rupiahs). This polder system were planned to reduce inundation on east part of Pekalongan. However, some significant progress of drainage system improvement reveal, since a big portion of inundated area in Slamaran, located on northern-east part of Krapyak Sub-district, has been succeeded to break free from inundation. This succeed was supported by prior river diversion management on Old Banger River.

Meanwhile, other sub-districts such as Kandang Panjang and Panjang Wetan have reduced the risk of inundation as an impact of city main road and local pave road elevation, from 30 to 60 cm high. However, there were some emerging negative externalities which has not been considered before, since road elevation at some area caused some new inundated area on surrounding settlements which have lower elevation.

Ecological sustainability has become less important since mangrove conservation experiencing difficulties due to limited area for plantation, disturbance from coastal flood/inundation and irresponsible visitors. On the contrary, local government encourage on developing park and city forest on the city centre instead on the coastal areas. Greening program on the coastal area mostly were intend to develop green pathways along the beach with shade's tree planting. However, there was Mangrove Information Centre (MIC) which give new experience for people to learn how mangrove conservation were managed. Information include plantation-demo on at least 2 mangrove species, namely Red Mangrove (Rhizopora Mongle sp.) and Black Mangrove (Avicenia Germinans sp), from seedling, planting, and keep them growing in limited area

Table 3. Building a transformative adaptation: comparing government's and community's initiatives.

Government's Initiatives	Community's Initiaves
• Mitigation Oriented • Provide Basic Need Infrastructure and maintain its services • Get international/national/ regional financing to enhance infrastructure services (case of Geotope, road elevation, Old Banger River diversion) • Supporting on Public Private Partnership (Case of Vaname Shrimp Cultivation) • Encourage more on Vaname-shrimp cultivation rather than Mangrove Cultivation • Mangrove Information Centre	• Adaptation Oriented • Altering Paddy/Jasmine to Milk-fish/Sea-weed Cultivation as sustainable livelihood (instead of Vaname-shrimp) • Protect fishponds from abrassion and flood by mangrove plantation on fishpond dykes • Minimize cost/crop failure, grab potential social & economic benefit • Enhancing Capacity through social networking, partnership and gain novel knowledge • Build social learning by establishing self-help group

In spite of community sustainable livelihood become a key of creating communities' resilience, there were limited related initiaves from municipal agencies. Some exception programs were done in cooperation with donors, regional NGOs and higher education institution, particularly on integrating the mitigation and adaptation on environmental improvement attempts. BINTARI as local NGO has incorporated an integration of mitigation and adaptation in 2015 through serial training and practising mangrove plantation on Bandengan Sub-District. This initiave was undertaken in cooperation with Mercy Corp Indonesia.

Pekalongan Farming, Fisheries and Marine Agency (DPPK) prefer to encourage vaname-shrimp cultivation which are high cost and contra-productive with mangrove conservation. Some Mangrove leaves floated on surrounding fishpond could attract any bacteria and viruses which caused vaname's decease. Furthermore, vaname-shrimps cultivation also threatened the environment capacity due to utilization of chemical disinfectant for preventing vulnerable vaname-shrimps from any potential decease.

4. Conclusion

The main conclusion comes from evaluation of Pekalongan Municipal Government initiatives on both of managing public private partnerships on vaname-shrimps cultivation for encouraging economic benefit and mangrove conservation in limited coastal area. Since both initiatives are contra productive, the more vaname-shrimps production the less mangrove conservation. These initiatives indicated a double development standard from the side of municipal government, when it comes to prioritize both the economy and the environment. Regarding on resilience thinking, build a transformative adaptation means taking any initiatives which reboost environment recovery as driving factor on saving resources supplies for further benefit on achieving sustainable development. This finding neglected Goal No 12 responsible consumption and production. However, the local community prefers to choose a new livelihood, which provides them not just economic security, but also social and ecological benefits. It means the community succeeded in proven their capability on building a transformative adaptation.

5. Acknowledgement

This paper was written based on 2 research findings. First, dedicated as one of publication series from on going doctoral research in Environmental Studies Gadjah Mada University, under title of '*Pola Kerentanan Wilayah Akibat Rob dan Banjir: Studi Kasus Kognisi Spatial Masyarakat di Kecamatan Pekalongan Utara*' (Spatial Pattern of Region Vulnerability as the Impact of Coastal Inundation and Flood: a case of Community Spatial Cognition on North Pekalongan District.' Second, as part of fundamental research funded by Urban and Regional Planning Departement through DIPA budgeting under Faculty of Engineering, Diponegoro University in 2016, under title of '*Adaptasi Transformatif Penghidupan Masyarakat Akibat Rob Di Pesisir Kota Pekalongan*' (Transformative Adaptation on Community Livelihood As Coastal Inundation Impact on Coastal City of Pekalongan). Draft of this paper was presented on March 6-7 in Malang, at paralel discussion on The 3rd International Conferrence on Planning in The Era of Uncertainty: Sustainable Nation which held by Urban and Regional Planning Department, Brawijaya University.

References

[1] Satterhwaite, D. (2014). Editorial: Getting local governments, residents and enterprise to respond to the new IPCC assessment. *Environment and Urbanization,* Vol 26 No 1. Pp. 3-10.

[2] Handayani, W, A Artiningsih, I Rudiarto, J A Syahbana, Mussadun, I Kusumastuti, A F Zain. (2015). *Membangun Kapasitas Transformatif Masyarakat dalam Upaya Pemanfaatan Bonus Demografi: Kajian Sektor Primer di Jawa Tengah.* Direktorat Analisis Dampak Kependudukan (Ditdamduk) BKKBN, Jakarta.

[3] United Nations. The Sustainable Development Goals Report 2016. http://unstats.un.org/sdgs/report/2016/The%20Sustainable%20Development%20Goals%20Report%202016.pdf accessed on Feb 14th 2017

[4] Osborn, Derek, Amy Cutter and Farooq Ullah. (2015). *Universal Sustainable Development Goals: Understanding the Transformational Challenge for Developed Countries*. Report Of A Study By Stakeholder Forum.

[5] Diposaptono, Subandono. (2008). Teknologi Adaptasi Kenaikan Paras Muka Air Laut Di Wilayah Pesisir Dan Pulau-Pulau Kecil. Direktorat Pengelolaan Pesisir dan Lautan terpadu. http://pirba.hrdp-network.com/e5781/e5795/e6331/e10038/eventReport10187/06TEKNOLOGIADAPTASI-Subandono-.pdf

[6] Artiningsih, A, S Worosuprojo, R Rijanta, S R Hardoyo. (2016). Enhancing social-ecological resilience in Indonesia: A case of North Pekalongan District, Central Java. Jurnal Wilayah dan Lingkungan, 4(3), 187-198. doi:10.14710/jwl.4.3.187-198.

[7] Wamsler,C and E Brink.(2014). Moving beyond short-term coping snd sdsptstion. *Environment and Urbanization.*Vol 26 No 1. April 2014. pp. 86-111.

[8] GIZ PAKLIM (ed). 2011. *Strategi Ketahanan Kota Pekalongan*. Badan Lingkungan Hidup Kota Pekalongan.

[9] Yin, Robert K. (1989). *Case Study Research: Desain and Methods*. Applied Sosial Reasearch Methods Series, Vol 5. Sage Publication, London.

[10] Gering, John. (2007). *Case Study Research: Principles and Practices.* Cambridge, University Press. New York.

[11] Bappeda Kota Pekalongan. *Peta Administrasi Kota Pekalongan* in Artiningsih, A, MHS Pratama, N C Putri. (2016). *Adaptasi Transformatif Penghidupan Masyarakat Akibat Rob Di Pesisir Kota Pekalongan.* Laporan Akhir Penelitian Dasar (Unpublished Fundamental Research Report). Hibah Bersaing Dana DIPA Fakultas Teknik Universitas Diponegoro, Semarang.

IOP Conf. Series: Earth and Environmental Science **70** (2017) 012034 doi:10.1088/1755-1315/70/1/012034

Analysis of perception and community participation in forest management at KPHP model unit VII-Hulu Sarolangun, Jambi Province

B Purnomo[1, 3], S Anggoro[2], M Izzati[2]

[1] Master Program in Environmental Science, Graduate School, Diponegoro University, Semarang 50241, Indonesia
[2] Graduate School, Diponegoro University, Semarang 50241, Indonesia

obay230586@gmail.com

Abstract. The concept of forest management at the site level in the form of forest management units (KPH) implemented by the government in an effort to improve forest governance in Indonesia. Forest management must ensure fairness for all stakeholders, especially indigenous and local communities that have been the most marginalized groups. Local communities have become an important part in the efforts to achieve sustainable forest management. Public perception as one of the stakeholders in forest management need to be analyzed to determine their perspectives on the forest. This study aimed to analyze the perception and the level of community participation in forest management activities in KPHP Model Unit VII- Hulu Sarolangun, as well as examine the relationship between these two variables. Perception variables are divided into three categories: good, moderate and bad, while the participation variable is also divided into three categories: high, medium, and low. Data was obtained through semi-structured interviews with the key informants and questionnaires to randomly selected respondents. Statistical analysis was conducted to determine whether there are differences of perception and participation between the two villages and the relationship between perceptions of participation or not. The results showed 90,16 % of people have a good perception and the remaining 9,84% have a moderate perception. In general, community participation is at a low level that is as much as 76,17 % and only 1,55% had a high participation rate. The analysis showed differences in levels of participation between the two villages and there is no relationship between the perception and the level of community participation in forest management. The results of this study can be taken into consideration for KPHP and other stakeholders in forest management policy in the region KPHP.

1. Introduction

Deforestation in Indonesia is a reflection of the failure of the management of forest resources. The rate of deforestation in Indonesia in the period 2009-2013 is estimated to reach approximately 1.13 million hectares per year [1]. Jambi province alone in the period 2012-2013 has recorded a deforestation rate of 91248.3 ha / year. This figure is the second highest after the province of West Kalimantan with a deforestation rate of 273 355 ha /year [2]. The high pressure by activities outside the forest sector such as agriculture and mining, poor forest management performance and weak supervision are several factors that cause more damage to forests. Improving forest governance through the concept of forest

IOP Conf. Series: Earth and Environmental Science **70** (2017) 012034 doi:10.1088/1755-1315/70/1/012034

management at the site level in the form of forest management units (KPHP) is one way to overcome this problem.

KPHP Model Unit VII- Hulu Sarolangun is one of 17 KPH region in the province of Jambi. The decline in the condition of forest resources and poverty in forest communities encourage the strengthening of forest management approaches that involve the community. As a forest management organization at the site level, KPHP required to accommodate the interests of local communities. It is in line with the increasing recognition of the rights of local communities and their ability to manage their environment. Authority devolution to the local people in the management of natural resources and the environment is a response to the poor performance of centralized management system and communities are given the authority to be involved in decision making [3]. Emphasis participatory approaches in natural resource management aims to align the conservation efforts with the achievement of sustainable development goals. Community based natural resources management is a popular approach in environmental management in order to achieving towards the objectives of economic, social and environmental [4]. Some studies indicate that the involvement of local communities in forest management encourage them to utilize their resources sustainably [5], gives significant impact on the improvement of forest conditions biologically and household income levels [6].

The public perception as one of the stakeholders in forest management need to be analysed to determine their perspective on the forest area. Public perception of the environment affects their attitudes and behaviours. There is a positive correlation between public perceptions with their level of participation [7]. The success of a project or program is determined by the support of local communities highly affected by perception and their opinion on the perceived impact of these activities [8]. There was a tendency where the community choose to defer involved in a project until they can see favourable results from it [9]. Increased public awareness and active participation is required to support the achievement of sustainable forest management. This study aims to determine the perception and public participation in the management of forests in the region KPHP Model Unit VII- Hulu Sarolangun and to analyse the relationship between perception and level of community participation.

2. Literature Review

2.1. Perception

Perception is the way a person sees and assessing an object or event. Someone will act in his perception, so the perception has a very important role in affecting a person's behavior [10]. Someone who has a perception always through a certain process that begins upon receipt of the stimulus through the receiver, and then transmitted to the brain. Psychological processes that occur in the brain cause a person aware about what happened. During the process of perceiving an object, individual is influenced by internal factors and external factors. Internal factors are factors that exist within the individual, such as experience, feeling, thinking abilities, and motivation. External factors such as the stimulus itself and environmental factors in which the perception takes place [11]. Differences in perception can also influenced by the level of intelligence and expectations about the perceived object [12], age and livelihood typologies [13].

2.2. Participation

Perception is the way a person sees and assessing an object or event. Someone will act in his perception, so the perception has a very important role in affecting a person's behavior [10]. Someone who has a perception always through a certain process that begins upon receipt of the stimulus through the receiver, and then transmitted to the brain. Psychological processes that occur in the brain cause a person aware about what happened. During the process of perceiving an object, individual is influenced by internal factors and external factors. Internal factors are factors that exist within the individual, such as experience, feeling, thinking abilities, and motivation. External factors such as the stimulus itself and environmental factors in which the perception takes place [11]. Differences in perception can also influenced by the level of intelligence and expectations about the perceived object [12], age and livelihood typologies [13].

3. Methods

KPHP Limau Unit VII-Hulu Sarolangun has been established as KPHP Model in 2011 with an area of 121 102 ha. This research was conducted at two the village in around the KPHP region. Both villages (Sungai Bemban and Temalang) are part of some villages that were targeted implementation of collaborative forest management program drafted by the KPHP and both have different distances to the forest area. Temalang directly adjacent to the forest area, while the Sungai Bemban within 3-4 km. Data was collected through semi-structured interviews and surveys. Interviews were conducted with key informants who are considered to have understanding of and influence in the decision-making from respective stakeholders include village apparatus, provincial forestry agencies, the head of KPHP, NGOs, and academics. The interviews were conducted to determine their role in management, perception of the existence of local communities in forest management and aid to understand the factors that affect the community participation. While the method of the survey done by giving questionnaires to 193 respondents who are patriach and randomly selected. Against the questionnaires were distributed to test the validation and reliability to generate the valid and reliable research instruments. Each question from questionnaire of the variable perception and participation using 3 points of measurement scale.

Table 1. Criteria of community perception.

Variable	Criteria	Parameter
Perception	Good	Have a high level of dependency on forest resources and desire to preserve it
	Moderate	Have a high dependency on forest resources, but do not realize that it needs to be preserved
	Bad	The forest does not need to be preserved

Public perception variables analyzed and grouped into three categories: good, moderate and bad (Table 1). The public has a good perception if they realize that they depend on forest resources and desires that these resources are managed sustainably; moderate perception if the community realizes his dependence on forest resources, but do not understand if the resource should be managed sustainably; and included in the category of bad perception if people have other interests that make them inclined to think that preserve the forest is unnecessary [18]. To assess the level of community participation, the analysis conducted on the frequency of community engagement at every stage of management from planning, implementation, benefit sharing, monitoring and evaluation. Variable levels of community participation are grouped into three categories: high, medium, and low. If they more involved in every stage of these activities, then they will have a high participation rate and vice versa. Analysis using the Mann-Whitney U-test was conducted to determine whether there are differences in perception and public participation between the two villages or not where the research was conducted. Correlation between perception and participation were analyzed by using Chi-square test. The entire statistical analyzes were performed using IBM SPSS Statistics software Ver. 22.

4. Results

Analysis of the perception showed that 90.16% of respondent have a good perception and no one has bad perception. Meanwhile 76.17 % of respondent have a low participation rate and only 1.55 % have a high participation rate (Figure 1). The community perception is affected by the experience, livelihood typologies, and expectations about the perceived benefits from the forest resources. Forests have a role in the hydrological cycle to ensure the availability of water in the dry season and prevent flooding in the rainy season. People who work as farmers desperately need water to irrigate their fields and communities in Sungai Bemban feel the impact of the destruction of the forest. During the dry season arrived they started to feel the shortage of water for irrigation due to decreasing river flow. Many forests are converted to plantation region rural communities around the forest. Meanwhile in Temalang village there is an increased frequency of flooding in their area. Formerly big floods occurred in 3-5 years, but now a year has occurred 3 times. It happened because of the destruction in the upstream area due to rampant illegal gold mining activities. Figure 2 shows that 89.12% respondents are farmers which some of them have agricultural land in the forest area. Besides from the farming, their needs largely derived from the utilization of forest areas such as wood, rattan, bamboo, wooden boards, honey, gum, etc. Most of them depend on the use of forest resources for a long time and they hope still can get the benefits in the future. Some of the conditions that cause the majority of them have a good perception about the existence of the forest. They assume that forests need to be preserved for the sake of their lives.

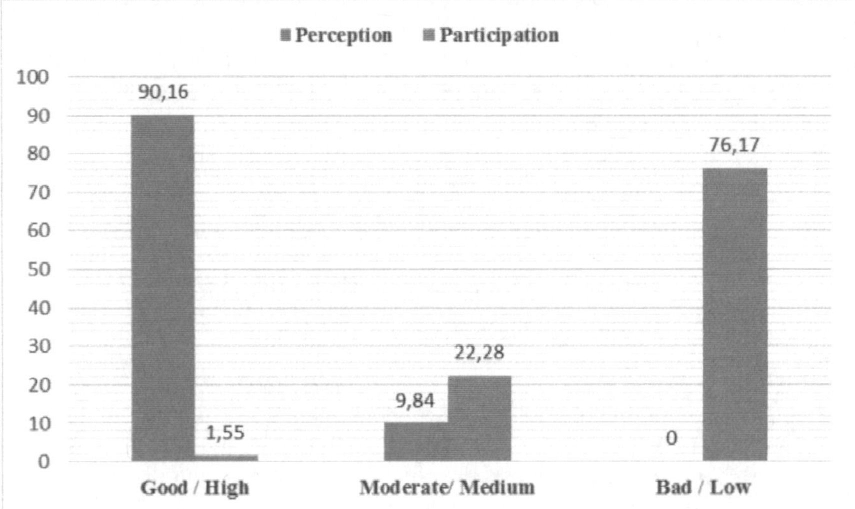

Figure 1. Percentage of perception and participation.

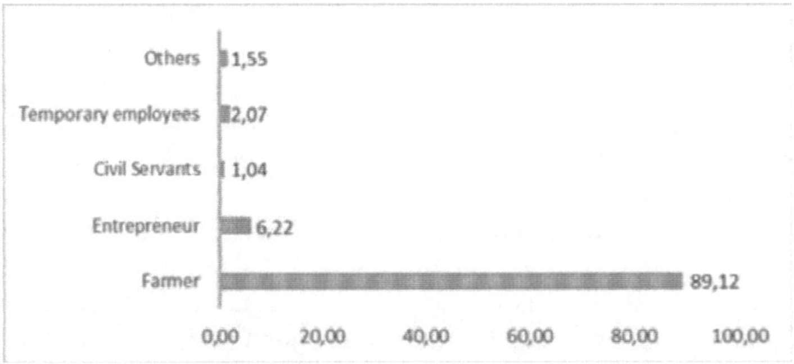

Figure 2. Percentage of livelihood typologies.

Based on Table 2 can be seen that the community participation in each phase of activity is at low level. The community who have a good level of participation is the village officials, administrators of farmer groups, and community leaders. Community as participants not directly involved to determine or to influence in the decision making process. They were just execute the program that predetermined by KPHP. Based on these conditions, the participation is a tool that is used by KPHP to achieve their goals. The low level of public participation is caused by ineffectiveness of the communication that occurred between KPHP with the community, low capacity of human resources in the community, and the benefits that have not been felt by them. The lack of interaction that was built by KPH and communities can be marked from the number of people who are not know the existence of KPHP itself. There are still many members of the community who do not know about the existence of KPHP including its tasks and functions. Good program but is not clearly understood by the community or the program that has been unable to fulfill needs of the community give effect to their interest in the activity.

Limitations of the funds and human resources in KPHP be one reason for the lack of interaction undertaken by KPHP with the community. 77.72% of community have low level of education and this condition affects to their confidence to be further involved in the program. KPHP operational are relatively new so the benefits of the program implemented yet significantly perceived by the public. It cause the low interest of the community to get involved in it.

Table 2. Percentage of community participation in each phase of activity.

Level of Participation	Phases of managements activities			
	Planning	Implementation	Monev	Benefit Sharing
Low	75,65	73,06	81,35	80,83
Medium	11,92	21,76	18,13	14,51
High	12,44	9,84	0,52	4,66

The analysis showed there is no difference in community perceptions both villages but differ in the level of participation. The level of community participation in Temalang is still better than the communities in Sungai Bemban., as shown in Table 3. Temalang villager have higher activity in the forest area compared to Sungai Bemban because most of them have livelihood that derive from it. The existence of customary institutions in Temalang also affects the level of their participation where the levels of adherence to customary rules is still high enough. Generally, community participation at the local level is still quite good compared to their overall participation at the participatory forest management.

Table 3. Percentage of community participation

Village	Level of participation		
	Low	Medium	High
Sungai Bemban	86,67	12,67	0,67
Temalang	51,16	44,19	4,65

The analysis showed that the value of Pearson Chi-Square is smaller (0,905) than the value of the table (5,991). It means that there is no correlation between perceptions of the community with the level of their participation in forest management activities. This is different with the results of research conducted by Ayunita and Hapsari [8] that show a positive relationship between these two variables. This is because most of the people on their research have benefited from the activities of management and these conditions encourage them to be actively involved in it. Meanwhile in forest management activities in the area of KPHP the community have not felt a significant benefit.

5. Conclusions

Based on the above analysis it can be concluded that the community has a good perception of the forest area in the region KPHP, but has a low level of participation in forest management activities. Community has a good perception because they are aware of the importance of forests for their lives, so they felt it necessary to maintain its sustainability. Several factors indicated to provide influence in management activities, among others, economic factors and communication. The principle of equality, trust, and learning should be emphasized in every process followed in this activity.Efforts should be made to provide insight to community about their important role in forest management. Participatory approach through NGOs, local extension workers, and community leaders could become a solution in order to bridge communication between KPHP and society. Limited human resources owned by KPHP can be solved by optimizing the existing role of forestry extension and to note is the skill of the facilitator also needs to be improved in accordance with the dynamics that occur in the community. Transparency in every stage of activities needed to build a relationship based on mutual trust. Government support in terms of funding and human resources required to assist KPHP in program implementation.

Assisting and training programs have been conducted by both NGOs and government agencies to improve community skills. The program is done in order to reduce community dependency upon forest resources in fulfilling their needs, but the program has not had a significant impact for improving the welfare community. The program is generally implemented on the project-oriented and sometimes it can only benefit a handful of parties. The conditions made possible a reluctance of people to participate in management activities, so that future needs to do a paradigm shift in programming for the communities where the program is offered in accordance with what is required by community.

References

[1] Forest Watch Indonesia, 2014. *Potret Keadaan Hutan Indonesia Periode 2009 - 2014*. Bogor: Forest Watch Indonesia

[2] Kementerian Lingkungan Hidup dan Kehutanan, 2015. *Statistik Kementerian Lingkungan Hidup dan Kehutanan 2014*. Jakarta : Kementerian Lingkungan Hidup dan Kehutanan

[3] Brugere, C., 2006. Can Integrated Coastal Management Solve Agriculture-FisheriesAquaculture Conflicts at The Land-Water Interface? A Perspective from New Institutional Economics. In: Hoanh, C.T., Tuong, T.P., Gowing, J.W., Hardy, B. (Eds.), *Environment and Livelihoods in Tropical Coastal Zones*. CAB International Oxon, UK, pp. 258–273.

[4] Dyer, J. et al , 2014. Assessing Participatory Practices in Community-Based Natural Resources Managemen : Experiences in Community Engagement from Southern Africa. *Journal of Environmental Management* 137 : 137-145

[5] Persha, L., Agrawal, A., Chhatre, A., 2011. Social and Ecological Synergy: Local Rulemaking, Forest Livelihoods, and Biodiversity Conservation. *Science* 331, 1606–1608.

[6] Ameha, A., Nielsen, O.J., Larsen, O.H., 2014. Impact of Access and Benefit Sharing on Livelihoods and Forest : Case of Participatory Forest Management in Ethiopia. *Ecological Economics* 97 : 162-171.

[7] Ayunita, D., dan Hapsari, T.D., 2012. Analisis Persepsi dan Partisipasi Masyarakat Pesisir pada Pengelolaan KKLD Ujungnegoro Kabupaten Batang. *SEPA* 9 : 117-124.

[8] Bennett, N.J., and Dearden, P., 2014. Why Local People Do Not Support Conservation : Community Perceptions of Marine Protected Area Livelihood Impacts, Governance and Management in Thailand. *Marine Policy* 44 : 107-116.

[9] Hedge, R. and Bull, G.Q., 2011. Performance of an Agro-forestry Based Payments-for Environmental-Services Project in Mozambique : a Household Level Analysis. *Ecological Economis* 71 : 122-130.

[10] Chartrand TL, Bargh JA. 1999. The Chameleon Effect: The Perception-Behavior Link and Social Interaction. *Journal of Personality and Social Psychology*, 76(6) : 893-910

[11] Swanky OH. 2006. The Self-Perception Theory Versus A Dynamic Learning Model. *Tinbergen Institute Discussion Paper* TI 2006-092/1. Erasmus University Rotterdam and Tinbergen Institute

[12] Pujiastuti, E,. 2011. Persepsi dan Partisipasi Masyarakat dalam Kegiatan Hutan Tanaman Rakyat di Kabupaten Sarolangun, Jambi [tesis]. Bogor: Program Pascasarjana, Institut Pertanian Bogor

[13] Twongyirwe, R. et al, T.D., 2017. Do Livelihood Typologies Influence Local Perceptions of Forest Cover Change? Evidence from A Tropical Forested and Non-Forested Rural Landscape in Western Uganda. *Journal of Rural Studies* 50 : 12-29.

[14] Samah, A.A., and Aref, F., 2011. The Theoretical and Conceptual Framework and Aplication of Community Empowerment and Participation in Processes of Community Development in Malaysia. *Journal of American Science*, 7 (2) : 186-195 *Bayu Purnomo, Sutrisno Anggoro, Muniffatul Izzati*

[15] Suwarno, A. et al, 2009. Participatory Modelling to Improve Partnership Schemes for Future Community-Based Forest Management in Sumbawa District, Indonesia. *Environmental Modelling & Software* 24 : 1402-1410

[16] Loft, L. et al, T.D., 2017. Whose Equity Matters? National to Local Equity Perceptions in Vietnam's Payments for Forest Ecosystem Services Scheme. *Ecological Economics* 135 :

164-175.

[17] Reed, M.S., 2008. Stakeholder Participation for Environmental Management : A Literatur Review. *Biological Conservation* 141 : 2417-2431.

[18] Ngakan, P.O dkk, 2006. *Ketergantungan, Persepsi, dan Partisipasi Masyarakat terhadap Sumberdaya Hayati Hutan (Studi Kasus di Dusun Pampli Kabupaten Luwu Utara, Sulawesi Selatan).* Bogor : Center for International Forestry Research (CIFOR)

IOP Conf. Series: Earth and Environmental Science **70** (2017) 012035 doi:10.1088/1755-1315/70/1/012035

Sustainability and vulnerability: Understanding the anomaly from disaster perspectives. Case study: Glagaharjo Village in Mount Merapi

C D A Depari

Universitas Atma Jaya Yogyakarta, Kampus II Fakultas Teknik Jalan Babarsari No.44, Yogyakarta 55281, Indonesia

catharina2bfree@yahoo.com

Abstract. In its concern for human safety, the Government's policy to relocate those living in disaster prone areas is twofold: it is perceptibly profound yet socially cataclysmic. This anomaly, created by the contradiction between the need for sustainability and the fact of vulnerability, could be found in the case of Mount Merapi. Communities living in the hazardous zone of Merapi, particularly those in Glagaharjo Village, are acknowledged for their persistent rejection of relocation programs despite their high exposure to the catastrophic impacts of eruptions. To mitigate the impacts, a safe and conducive dwelling place which considers the characteristics of these affected communities must be encouraged. This research adopts a consensus method towards responses obtained through a Likert scale-questionnaire and measured with a statistical program. Prior this process, theoretical reviews toward the concept of place attachment, place dependence and place identity was established in order to determine the research variables for the questionnaire.

1. Introduction

The geographical position of Indonesia, which is located above several tectonic plates and chained to the Pacific Ring of Fire, has caused its islands to experience some of the most powerful eruptions and earthquakes ever recorded in the history of disasters. Geologically, Indonesia contains over 130 hazardous volcanoes, mainly concentrated on the island of Java and stretched out along the Sunda Strait [1]. Among these volcanoes is Mount Merapi, a strato-type volcano in Java which is reputed to be the world's most active volcano [2]. Some researches show that its changing eruption characteristics and body deformation during the twentieth century will increase this volcano's catastrophic impacts, particularly due to the density of population around the volcano [3, 4]. Based on the Map of the Merapi Region, areas situated less than ten kilometers from Merapi's caldera and those along the volcano's southern slope are classified as Hazard Zone Level III of Merapi. Of nine sub-villages, three are even found to be extremely vulnerable because of their proximity to the caldera, which is only five kilometers away. Administratively under the jurisdiction of Glagaharjo Village, these areas are now identified as the National Park of Merapi as legalized through a National Government Act in 2011. All kinds of development within this zone are strictly forbidden, and as a result social polemic between these communities and the local authorities has risen. Despite this fact, these communities have long acknowledged their vulnerability to the catastrophic impacts of Merapi's eruptions, which include lahar, pyroclastic flows and other hazards.

IOP Conf. Series: Earth and Environmental Science **70** (2017) 012035 doi:10.1088/1755-1315/70/1/012035

The idea of sustainability usually arises when the existence of particular beings is exposed to certain events, such as wars, starvation, political turmoil and natural hazards. But when it comes to realization, sustainability typically encounters complex circumstances. One instance is the case of Yogyakarta and Central Java after the eruption of Mount Merapi in 2010. Responding to this, the National Government has authorized the revision of Merapi's land use zonation and initiated relocation programs. Relocation is typically regarded as a complex process, because it relates not only to infrastructure reconstruction, but also the social restructuring and the people's political and economic relationships [5]. Therefore, it requires long-term planning [6], and support from the affected communities who must already acknowledge the consequences of relocation [7]. After Merapi's eruption in 2010, almost 2,290 units out of 2,613 temporary shelters have already been occupied [8], and in Yogyakarta alone, almost 2,040 long-term shelters have been established [9]. Although publicly acclaimed as best practices for their quick construction, their effectiveness in terms of occupancy shows a contradictive result. Of the targeted 3,612 households, almost 1,059 households, particularly from Glagaharjo Village, refused to relocate [10]. This refusal is further instigated by the disagreeable experiences of those living in the temporary shelters. According to them, the relocation plan fails to advocate supports which are needed for long-term living. Unable to prevent Glagaharjo communities from reoccupying their disaster-prone neighborhood, the regional authority of Sleman initiated an agreement stressing these communities' commitment to being voluntarily relocated in times of danger. This agreement was initiated by the Government on the concept of living in harmony with disaster risk. Action plans have been generated from it, like encouraging a proper approach to the subject communities, optimizing the roles of a key figure to channel the locals' aspirations and incorporating evacuation routes into the neighborhoods [11]. Given the significance of integrating the aspirations of Glagaharjo communities with the Government's plans, this paper will examine the factors influencing these communities' attachment to their former homes as the essential keys to promoting not only a safe dwelling place, but also one conducive to their cultural sustainability.

2. Literature Reviews

The Concept of Sustainability and Vulnerability

Sustainability defined as a community's ability to recover by utilizing their own resources [12]. It also means recognizing and making the best use of the interconnection between social, economic and environmental goals to reduce hazard risks [13]. Regarding vulnerability, it is the internal factors encouraging the transformation of hazards into disasters [14], or conditions determined by some factors or processes which increase the susceptibility of a community to the impacts of hazards [15]. Consequently, it needs active involvement from each stakeholder in the host region to promote effective disaster management, the chief goal of which is the production of sustainable livelihood along with its protection and recovery during disasters and emergencies. If the goal is achieved, people will have greater capacity to deal with disasters, and their recovery becomes more rapid and abiding [16]. As a part of disaster management, temporary shelters should also be prepared prior to the occurrence of disasters.

Based on the WHO report (2014), emergency shelter involves the provision of temporary shelter during humanitarian crises including the provision of tents, and materials for the construction and repair or modification of public buildings to temporarily accommodate those affected. Actions for semi-permanent shelter relate to the provision of shelter as a long-term solution, including support for hosting arrangements, and the provision of materials and tools for construction or repair [17]. Relocation is aimed at providing a safe and secure existence for those subjected to the sudden onset or eruption of hazards, consequent upon which, returning to the former home for short or long-term inhabitation becomes perilous. Relocation must be considered as the last resort, and for this reason, *in situ* alternatives or solutions should be primarily evaluated, unless the subjected communities already have particular preferences of an ideal home or neighborhood [17].

Place Attachment

Through the Action Plan of Rehabilitation and Reconstruction Post the Merapi Eruption in 2010, the National Government initiated projects to help the disaster prone communities regain their livelihoods. Assistance and immediate aid from disaster relief and humanitarian agencies arrived to support the Government's plan and promote an effective recovery. One of its manifestations was the establishment of temporary shelters typically located in close proximity to the former homes. It relies on the idea that maintaining this proximity will help a transitional adaptation process preceding mobilization to long term shelters to be effective. To support this plan, the Government even incorporated an incentives policy into the program, but it still failed to appeal to the subject communities, particularly those in Glagaharjo Village. Persistent and continuing rejections coming from these communities have exhibited their strong preference for their former homes, despite their exposure to eruption risks. This event reflects an emotional bond phenomenon usually explained from a psychological standpoint through the theory of place attachment [18]. Theories of attachment were originally developed by a British psychologist John Bowbly, through his seminal works of the 1960s. The aims of attachment are twofold. Biologically, it is attuned to one's survival, while from a psychological perspective its aim is to fulfill one's needs of safety, security and protection [19]. Attachment is typically expressed through a set of behaviors towards particular objects, and its intensity is determined by the quality and timing of the bond. An intense desire for fusion or closeness towards the object is usually reflected through extreme behaviors like fear of separation, efforts to maintain proximity, or minimize distance from the object. If this object is able to fulfill one's needs of safety or survival, mechanisms to maintain proximity will occur [20].

The psychological connectedness to a particular place is not determined only by the place and its physical features, but even merely by the meaning of the place and one's experiences of it [21]. In this regard, place could serve as a container or repository of memories, and will carry some meanings for each spectator. Attachment to a place occurs when particular individuals develop a positive sense towards it, which evolves after having prolonged experiences [22]. Therefore, the intensity of attachment depends on personal experience, the meaning and the function of the object or place, and the perceptions of its roles, which enhance the significance of that object or that place [23]. Integrated to place attachment, place identity is conceptualized as a cognitive connection between one's self and an environment [24]. Place identity is also defined as those dimensions of one's self that develop a personal identity regarding a specific environment through a complex pattern of conscious and unconscious ideals, beliefs, preferences, feelings, values, goals, behavioral tendencies, and skills as the means of this process [24]. Environmental setting, in this context, offers an opportunity to express and affirm one's identity. Place dependence, on the other hand, is a concept established from one's positive evaluation of a particular place because of its capability to fulfill needs and to motivate a person to achieve some goals [25]. Place identity, based on emotions, differs from place dependence which is based on a functional relationship. This bond is characterized by the existence of particular facilities and conditions which are required to achieve specific goals and to perform various activities [26]. This bond will be developed if the frequency of travel is increased and interactions with the place's features occur intensively [27]. Therefore, the attributes characterizing the place attachment phenomenon include affection, cognition and behavior, orientation to a place, time and its pattern, social needs, interpersonal relationships, and those related to the place and its community [21]. Socially, each individual tends to share the same emotions with others towards a place, and thus a place will reflect the experiences and the characteristics of its inhabitants. Place is also perceived as satisfying if it can encourage control, creativity, power, privacy, self-esteem, personal display, and security.

3. Methodology

This paper stems from two disciplines, psychology and architecture, and focuses on solving a human problem, which leads this study to select phenomenology as the research paradigm. Phenomenology attempts to eliminate everything that represents a presupposition, and encourages the researcher to look at things openly with intuition and self-reflection [28]. Besides qualitative methods, the research also employs quantitative methods. All data collected from questionnaires are transferred into tabular forms and then calculated by using a statistical analysis program. The result provides the basis for interpreting specific factors that underline the respondents' place attachment based on their experiences and perceptions.

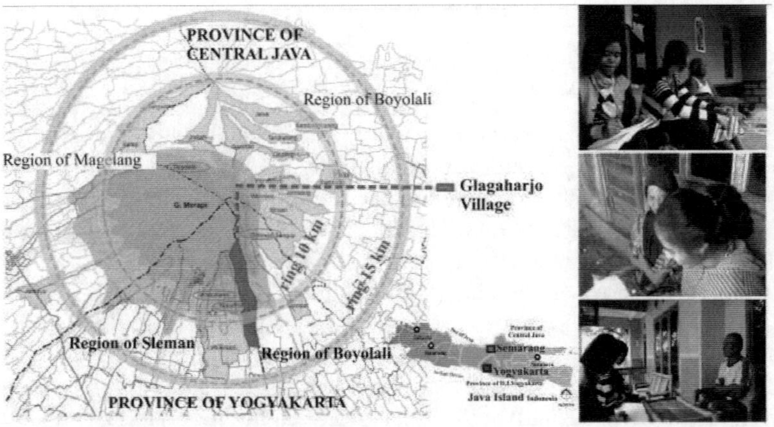

Figure 1. The location of Glagaharjo Village and the field survey with respondents

Considering the limited time given, the sample unit is determined 100 households from 656 households in Glagaharjo Village. The sampling unit is purposive-type towards which criteria of selection must be primarily determined [29]. Related to this, respondents must have settled in the understudy areas for at least ten years. It was based on assumption that at least within those years, their emotional bond to the dwelling place had been strongly developed. Involving ten surveyors, the questionnaire considerably needed 30 minutes of time for response and each response must be filled in on-site. Figure 1 shows the location of the village which is within 15 kilometres from the volcano's caldera and the interview process with the respondents.

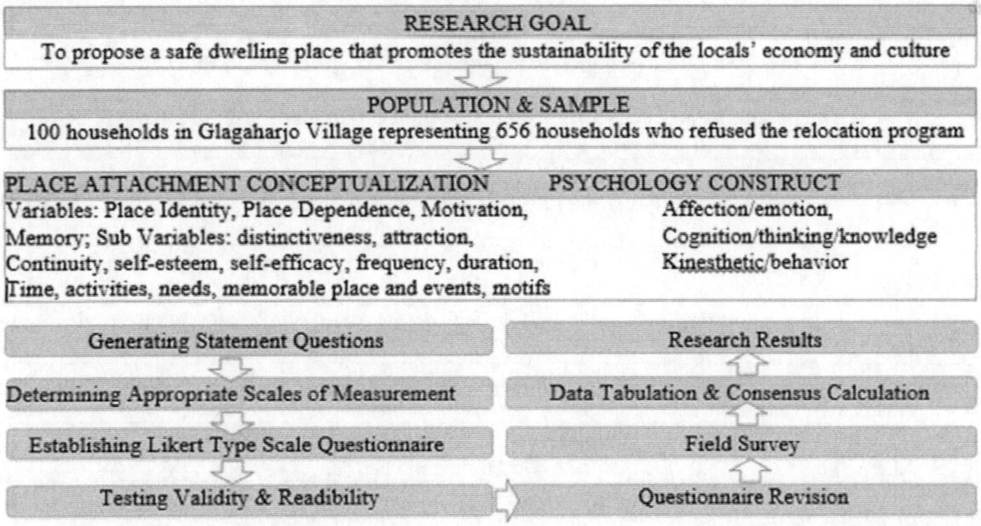

Figure 2. The research framework based on questionnaire survey method

Figure 2 exhibits the systematical process of the research based on field and literature study. The research in this regard, adopts a Likert type questionnaire which comprises a range of response from strongly agree to strongly disagree along with undecided option. After all the responses were entered in the Excel's tabular formats, they are statistically calculated using consensus method and the Average Percent of Majority Opinions (APMO) Cut-Off Rate method [30]. Considering there are 5 scales of response in the questionnaire, the formula was modified by adding the value of majority strongly agreements and disagreements. To reach a consensus, a percentage of strongly agree or strongly disagree must be higher than the APMO cut-off rate to reach [30, 31].

$$APMO = \frac{\text{Majority Agreements} + \text{Majority Disagreements}}{\text{S Opinions Expressed}} \quad (1)$$

As the consequence of adopting a consensus method, the calculation result will reflect the respondents' general attitudes toward each aspect of place attachment. The grade of consensus could also be specified into categories with accordance to the majority level of agreements. Strong consensus shows agreement from more than 95% of total respondents, consensus shows agreement from more than 75% to 95% of total respondents, majority agreement comes from 50% to 75% of total respondents while no consensus category shows agreement from less than 50% of total respondents [32].

4. Research Results

Table 1 presents the general attitude of Glagaharjo's respondents related to place identity. In terms of distinctiveness, most respondents strongly agreed that their neighborhood has no capability of promoting uniqueness and safety compared to other neighborhoods. Regarding safety, this response actually expressed a consciousness of their exposure to eruption risks. Most respondents also strongly stated that their neighborhood's physical and cultural potentials are not attractive and advantageous while in terms of social bond and cultural richness, these factors are not viewed as those that shape the uniqueness of their neighborhood. Table 1 also shows that according to most respondents, their neighborhood's potentials do not contribute to their life improvement. The consensus in this aspect is achieved as 95.7% of respondents agreed to be relocated if a safe, unique and attractive dwellings at the relocation site could be established.

Table 1. Consensus based measurement on the aspect of place identity (1)

APMO	SAA	Consensus Based Measurement of Glagaharjo Respondents						
		SA	A	N	DA	SDA	UD	Missing
		Distinctiveness						
		The neighborhood is capable to encourage safety compared to other areas.						
96.8	3.1	1	2	2	21	69	1	4
		The neighborhood's environmental and cultural potentials are unique.						
87.9	1.10	0	1	2	31	48	9	9
		Attraction						
		The current home has some physical potential which are very attractive and beneficial.						
96.8	1.04	1	0	3	23	69	0	4
		The current home has some cultural potential which are very attractive and beneficial.						
97.9	3.09	3	0	2	22	70	0	3
		Community's strong social bond defines the uniqueness of the neighborhood.						
93.7	1.04	1	0	5	11	78	1	4
		Cultural richness and local wisdom define the uniqueness of the neighborhood.						
73.9	7.61	6	1	9	20	41	15	8
		The neighborhood's potentials contribute to improve my life quality.						
92.4	1.08	1	0	7	21	64	0	7
		If particular attractions are provided at the relocation site, I surely want to be relocated.						
97.8	95.7	88	3	2	0	2	0	5

Table 1 also demonstrates that according to most respondents, their neighborhood's physical and cultural uniqueness are not well-maintained. Contradictorily, they stated that attempts to maintain that uniqueness are not necessary because of its lack of positive impacts to the respondents' lives. As shown from table 1, majority agreement (64.9%) is achieved towards opinion that globalization is the main factor causing physical changes on their neighborhood and more agreement (77.4%) towards its influences on the neighborhoods' cultural changes. Similar to the response towards the importance of maintaining the neighborhood's uniqueness, attempts to maintain the uniqueness of their former homes are also viewed as unnecessary. Despite these negative responses, the Government still needs to provide a unique environment which has the capability of sustaining all potentials that are appealing and contributive, in order to encourage the respondents to relocate. This suggestion is also supported by a strong consensus (94.6%) towards their neighborhood's sustainability. In terms of self-efficacy, most respondents stated that their neighborhood did not enable them to control the environment, perform specific activities and acknowledge the symptoms of eruption. A consensus is achieved as 91.4% of total respondents strongly agreed to be relocated if a neighborhood enabling them to have a control on their environment and perform various activities is encouraged at the relocation site.

Table 2. Consensus based measurement on the aspect of place identity (2)

APMO	SAA	Consensus Based Measurement of Glagaharjo Respondents						
		SA	A	N	DA	SDA	UD	Missing
		Continuity						
		The neighborhood's physical and cultural uniqueness are well-maintained.						
85.2	12.6	4	8	12	22	47	2	5
		Some attempts must be made to maintain the neighborhood's uniqueness.						
91.5	5.2	2	3	5	31	51	3	5
		The current home and neighborhood could support the sustainability of my activities.						
91.4	0.0	0	0	7	19	67	1	6
		Globalization influences have changed the physical aspect of the neighborhood.						
86.1	64.9	54	7	6	16	4	7	6
		Globalization influences have changed the cultural aspect of the neighborhood.						
84.9	77.4	67	5	5	5	2	9	7
		Some attempts must be made to maintain the uniqueness of my home.						
90.1	2.2	1	1	5	19	61	4	9
		If the continuity of uniqueness could be promoted at relocation site, I want to be relocated.						
94.6	94.6	87	2	3	0	0	2	6
		Self-Efficacy						
		The current home enables me to observe my surrounding along with all its changes.						
82.4	3.0	1	2	9	23	54	8	3
		The current home enables me to perform various activities.						
87.1	1.0	1	0	6	18	62	6	7
		The current home enables me to acknowledge the eruption sympthomps.						
86.3	6.3	1	5	6	22	54	7	5
		If freedom, control and various activities could be encouraged, I want to be relocated.						
94.6	91.4	84	2	4	3	0	1	6
		Self Esteem						
		The current home provides a privacy to perform daily activities.						
88.1	3.2	1	2	4	15	64	7	7
		The current home gives supports for specific activities such as self-contemplation.						
82.0	5.6	4	1	5	20	48	11	11
		The current home improves my potentials and life skills.						
76.7	8.2	3	3	7	25	25	10	23
		The current home encourages safety and security from eruption hazards.						
92.7	19.7	18	1	7	17	53	0	4
		If supports for various activities, skills development and safety are encouraged at the relocation site, I want to be relocated.						
93.6	90.4	83	2	5	0	3	1	6

As shown in table 3, most of the respondents strongly stated that their main activities are not only performed inside but also outside the house and generally relate to the fields outside the economy. Areas where economic activities occur are relatively located in a close proximity to the respondents' former houses and include the volcano's forest, agricultural fields and riverside which could be reached within

100 meters to 3 kilometers from the house or within 5 minutes to 2 hours, mainly by walking. Household activities are indicated as the routine that most respondents engage in, particularly because more than 50% of the respondents are housewives. Most of them strongly stated that only a little amount of time is needed to perform their economy activity, which is between 3 to 7 hours per day, and 56.8% of them strongly agreed that a considerable time is always allocated to observing the Merapi's volcanic activities. From the aspect of time, the main activity starts at noon and is finished in the afternoon. A few number of respondents stated that they allocate a considerable time to participating in social activities including in safety control activities and evacuation drills. While regarding the eruption event in 2010, some respondents agreed that they returned to their homes only one month after the event.

Table 3. Consensus based measurement on the aspect of place dependence (1)

AP	SAA	Consensus Based Measurement of Glagaharjo Respondents						
MO		SA	A	N	DA	SDA	UD	Missing
		Frequency						
		Frequently perform various activities at the current home and neighborhood.						
91.4	4.2	3	1	8	16	66	0	6
		Activities frequently perform relates with economy.						
94.7	7.3	6	1	5	10	73	0	5
		Duration						
		A large amount of time is needed to perform the main activity in current neighborhood.						
81.8	6.8	3	3	16	18	48	0	12
		Always provide a considerable time to observe the Merapi's volcanic activities.						
78.9	56.8	45	9	13	10	11	7	5
		Time						
		Starting the main activity in the morning at the current home.						
93.2	2.25	2	0	6	8	73	0	11
		Finishing the main activity in the afternoon at the current home.						
75.5	34.4	29	2	22	10	27	0	10
		Always allocate a considerable time to participate in the neighborhood's social activities.						
93.7	10.4	10	0	6	16	64	0	4
		Always allocate a considerable time to participate in safety control activities.						
92.3	37.3	30	4	6	15	35	1	9
		Always provide a considerable time to involve in evacuation simulations or drills.						
82.6	29.3	24	3	6	16	33	10	8
		After eruption in 2010, I evacueed to relocation center and returned less than one month.						
86.3	47.3	37	8	12	11	26	1	5
		Activities						
		The type of activity at the current home mostly relates with household activities.						
78.1	3.1	2	1	2	12	60	19	4
		The type of activity at the current home mostly relates with economy.						
88.1	48.3	43	2	4	12	25	7	7
		There are various types of activities occur at the current home.						
75.2	62.3	53	5	15	4	8	8	7
		Actively involved both in social activities and disaster mitigation programs.						
90.0	64.4	55	3	5	8	15	4	10
		If socioeconomic activities could be encouraged at the site, I want to be relocated.						
98.9	96.7	88	2	1	1	1	0	7

Table 4 shows agreement on the opinion that their main activity relates with economy, and majority of repondents agreed (62.3%) that various activities usually occur at their homes. Related to active involvement in disaster mitigation and social activities, majority of respondents (64.4%) also agreed with that statement. A strong consensus (96.7%) is achieved regarding their willingness to be relocated if a conducive neighborhood could be established at the relocation site. Regarding the need for particular infrastructure, most respondents stated that their neighborhood fails to provide those needs. The unavailability of good access from house to workplace was asserted as the evidence. Related to the need for safety as a support for activities, majority of respondents (92.1%) agreed to be relocated if the Government could provide infrastructures that support their activities at the relocation site.

Table 4. Consensus based measurement on the aspect of place dependence (2)

AP MO	SAA	Consensus Based Measurement of Glagaharjo Respondents						
		SA	A	N	DA	SDA	UD	Missing
		Activities Needs						
		The current neighborhood could give supports for various types of activities.						
97.8	0.0	0	0	2	17	74	0	7
		The current neighborhood provides my specific needs to perform economy activities.						
77.4	2.1	2	0	5	13	57	16	7
		The current neighborhood is believed to be capable of fulfilling the need on safety.						
86.0	33.3	28	3	7	16	33	6	7
		If all the needs for living could be provided at the relocation site, I want to be relocated.						
95.5	92.1	81	1	3	1	2	1	11

Table 5 shows the motifs influencing the respondents' decision to refuse the relocation program. Their preference for living at the current house is not influenced by perceptions like the house has helped them obtain happiness and sucess, house is a family legacy or a personal asset, or caused by family's suggestions and the availability of neighborhoods' potentials.

Table 5. Consensus based measurement on the aspect of motivations and memory

AP MO	SAA	Consensus Based Measurement of Glagaharjo Respondents						
		SA	A	N	DA	SDA	UD	Missing
		Motivations						
		Living at the current house because it allows me to obtain a happiness and success.						
87.7	0.0	0	0	6	11	68	5	10
		Living at the current house because it is a family's legacy.						
94.1	11.6	6	4	5	15	56	0	14
		Living at the current house because of its status as a personal asset.						
96.3	6.0	3	2	2	13	62	1	17
		Living at the current house because of the family's suggestions.						
98.7	0.0	0	0	1	14	62	0	23
		Living at the current house because of the environment's natural benefits.						
96.8	0.0	0	0	2	17	75	1	5
		Living at the current house because of its proximity to economy sources.						
93.6	0.0	0	0	6	23	66	0	5
		Living at the current house because of its proximity to public and social facilities.						
79.1	13.8	6	4	14	24	23	1	28
		Living at the current house because of its proximity to guard posts and evacuation center.						
77.1	11.4	8	0	15	23	23	1	30
		Living at the current house because of the ancestral messages.						
84.2	31.9	13	10	10	9	29	1	28
		Living at the current house because of a superstitious belief.						
81.9	61.1	38	6	3	5	10	10	28
		If proximity is provided from home to workplace/public facility, I want to be relocated.						
93.6	91.5	84	3	5	2	0	1	5
		Memory						
		The current home or neighborhood reminds me of particular memorable places.						
81.8	32.9	24	5	3	4	39	13	12
		The current home or neighborhood reminds me of particular memorable events.						
80.2	33.3	21	6	4	5	33	12	19

AP MO	SAA	Consensus Based Measurement of Glagaharjo Respondents						
		SA	A	N	DA	SDA	UD	Missing
		The current home or neighborhood reminds me of particular moments of childhood.						
60.0	43.7	29	6	5	3	10	27	20

Table 5 also shows that majority of agreements (61.1%) is achieved towards a statement that mystical beliefs or ancestral messages is what influencing them to return to their former homes. Majority of respondents (91.5%) also asserted their willingness to be relocated if a close proximity from home to workplace and public facilities could be encouraged at the relocation site. While in terms of memory, only a few number of respondents agreed that their homes remind them of any memorable places or events including their childhood.

5. Conclusions

The argument posited in this paper focuses on the revelation of all factors that develop the emotional bond of Glagaharjo disaster prone communities to their homes. The research found that there is no strong consensus or consensus is achieved in this regard. However, a majority of agreement (more than 50% of total respondents) shows a correlation between the people-place bond phenomenon and functional factors. Based on the theory of place dependence, this type of emotional bond is characterized by an individual's intense reliance on particular place with all its features in order to support his or her activities. Although a strong consensus towards the neighborhood's capability to support various activities is not achieved, majority of agreement (more than 50%) shows that this capability along with the locals' strong social bond, are factors influencing them to return to their former neighborhood. To further explain, this bond could be maintained through various cultural ceremonies and social activities like a regular community gathering locally termed as *arisan*. Besides that, the need to develop a solid collusion among the community members of Glagaharjo Village regarding eruption risks, could also strengthen their social relationship and then their emotional bond to their former homes.

Based on the research study, the dwellings planning and design for the subjected communities in Glagaharjo Village must consider the following criteria. First, the house must be designed in a close proximity to the work place and public facilities. Criteria for access connecting those places should be set up and also accompanied by other alternatives for economic activity which avoid Merapi's hazardous areas. Second, the site of the house must be integrated with adequate open spaces which are potential for agriculture or farming. Third, the new neighborhood should encourage the local's social relationship and therefore, their mobilization as one social unit to relocation site must be encouraged. Fourth, the neighborhood should provide a good access from home to each public, social and health facility including to the work place.

References

[1] MacLean, D., 2014. *Reducing the Volcano Risk in Indonesia*. [Online] Available at: HYPERLINK "http://www.irinnews.org" http://www.irinnews.org [Accessed 1 February 2017].
[2] U.S. Department of the Interior, 2011. *USGS Science for Changing World*. [Online] Available at: HYPERLINK "https://www.usgs.gov" https://www.usgs.gov [Accessed 1 February 2017].
[3] Voight, B., Constantine, E.K., Siswowidjoyo, S. & Torley, R., 2000. Historical eruptions of Merapi Volcano, Central Java, Indonesia, 1768-1998. *Journal of Volcanology and Geothermal Research*, pp.69-138.
[4] Kementerian ESDM, 2010. *Peta Kawasan Rawan Bencana Gunung Merapi Jawa Tengah dan DIY*. Direktorat Vulkanologi dan Mitigasi Bencana Geolog.
[5] Gaillard, J.C., 2008. Differentiated Adjustment to the 1991 Mt Pinatubo Resettlement Program among Lowland Ethnic Groups of the Philippines. *The Australian Journal of Emergency Management*, XXIII(2).
[6] Boen, T. & Jigyasu, R., 2005. Cultural Considerations for Post Disaster Reconstruction Post-Tsunami Challenges. In UNDP Conference., 2005.

[7] Martanto, F. & Sagala, S.A.H., 2015. Faktor-Faktor yang Mempengaruhi Persoalan Relokasi Pasca Bencana Lahar Dingin di Kali Putih (Studi Kasus Dusun Gempol, Desa Jumoyo, Kecamatan Salam, Kabupaten Magelang). *Jurnal Perencanaan Wilayah dan Kota B SAPPK*, III(1), pp.69-82.
[8] Pranyoto, V.S., 2011. *Berita*. [Online] Antara Available at: HYPERLINK "www.antaranews.com" www.antaranews.com [Accessed 5 February 2017].
[9] Syaifullah, M., 2014. *News*. [Online] Tempo Available at: HYPERLINK "www.tempo.co.id" www.tempo.co.id [Accessed 5 February 2017].
[10] Badan Nasional Penanggulangan Bencana, 2013. [Online] Pusat Studi Bencana IPB Available at: HYPERLINK "www.http://psb.ipb.ac.id" www.http://psb.ipb.ac.id [Accessed 5 February 2017].
[11] Suryandari, R., Haryono, E. & Sumrahadi, A., 2013. Merapi Pasca Letusan 2010: Polisi Penempatan Semula Penduduk. *Malaysian Journal of Society and Space*, pp.138-49.
[12] Rose, A.Z., 2011. Resilience and Sustainability in the Face of Disasters. *Environmental Innovation and Societal Transitions*, I, pp.96-100.
[13] Rajeev, M.M., 2014. Sustainability and Community Empowerment in Disaster Management. *International Journal of Social Work and Human Services Practice*, II(6), pp.207-12.
[14] Kapucu, N. & Ozerdem, A., 2013. *Managing Emergencies and Crises*. Burlington: Jones and Bartlett Learning.
[15] Kelman, I., 2009. Understanding Vulnerability to Understand Disasters. pp.1-14.
[16] Haigh, 2008. Disaster Management Lifecycle. In *International Conference on Building Education and Research.*, 2008. School of the Built Environment, University of Salford.
[17] Weerasinghe, S., 2014. *Planned Relocation, Disasters and Climate Change: Consolidating Good Practices and Preparing for the Future*. Sanremo: UNHCR: United Nations Agency for Refugee Georgetown University.
[18] Qingjiu, S. & Maliki, N.Z., 2013. Place Attachment and Place Identity: Undergraduate Students'. (91), pp.632-39.
[19] Schaffer, R., 2007. *Introducing Child Psychology*. Oxford: Blackwell.
[20] Dewitte, M., Houwer, J.D., Buysse, A. & Koster, H.W., 2008. Proximity Seeking in Adult Attachment: Examining the Role of Automatic Approach–Avoidance Tendencies. *British Journal of Social Psychology*, pp.557-73.
[21] Altman, I. & Low, S., 1992. *Place Attachment*. New York: Plenum Press.
[22] Hashem, H., Abbas, Y.S., Heidari, A.A. & Nazgol, B., 2013. Between Sense and Attachment: Comparing the Concepts of Place in Architectural Studies. *Malaysia Journal of Society and Space*, pp.96-104.
[23] Falahat, M.S., 1385. The Concept of Place Sense and Its Composing Factors. In *Fine Art*. pp.57-66.
[24] Proshansky, H.M., 1978. The City and Self-Identity. *Environment and Behavior*, X, pp.147-69.
[25] Shumaker, S.A. & Taylor, R.B., 1983. Toward A Clarification of People-Place Relationships: A Model of Attachment to Place. In Freimer, N.R. & Geller, E.S. *Environmental Pschology: Directions and Perspectives*. New York: Praeger.
[26] William, D.R., Roggenbuck, J.W. & Watson, A.E., 1989. *Measuring Place Attachment: Some Preliiminary Results*. San Antonio, Texas. Paper presented at the NRPA Symposium of Leisure Research.
[27] Moghisi, , Mokhtari, S. & Heidari, A.A., 2015. Place Attachment in University Students: Shiraz University. *Procedia: Social and Behavioral Sciences 170*, pp.187-96.
[28] Moustakas, C.E., 1994. *Phenomenologica Research Methods*. Thousand Oaks, California: Sage.
[29] Siregar, S., 2013. *Metode Penelitian Kuantitatif*. Jakarta: Kencana.
[30] Von der Gracht, H. A. (2012). Consensus measurement in Delphi studies Review and implications for future quality assurance. *Technological Forecasting & Social Change*(79), 1525-1536.
[31] Cottam, H.-R., Roe, M., & Challacombe, J. (2004). Outsourcing of Trucking Activities by Relief

Organisations. *Journal of Humanitarian Assistance*, 1-26.

[32] Schaefert, R., Hausteiner-Wiehle, C., Häuser, W., Ronel, J., Herrmann, M., & Henningsen, P. (2012). Non-Specific, Functional, and Somatoform Bodily Complaints. *Deutsches Ärzteblatt International , XLVII*(109), 803-813.

[33] Priyanti, A. & Ilham, N., 2011. Dampak Erupsi Gunung Merapi terhadap Kerugian Ekonomi pada Usaha Peternakan. *Wartazoa*, XXI(4), pp.153-60.

IOP Conf. Series: Earth and Environmental Science **70** (2017) 012036 doi:10.1088/1755-1315/70/1/012036

The readiness of farmer communities in biogas development (A case study: Wiyurejo Village, Malang Regency Indonesia)

D Dinanti, D F Erlina and C Meidiana

Department of Urban and Regional Planning, Faculty of Engineering, Universitas Brawijaya, Malang, Indonesia

dede_planoinc@yahoo.com

Abstract. Wiyurejo Village has the potential for biogas development with ±75% of breeders wanting to build biogas but do not have enough vacant land. Biogas is a renewable energy source which requires the involvement of the community and public awareness in its implementation. The purpose of this research is to know the readiness of breeders for the development of biogas with limited land availability for the development of biodigester in Wiyurejo Village, Malang Regency, Indonesia. Based on the analysis, the value of the stage of readiness of farmers community in Wiyurejo Village is 2.20, which means that from nine stages of community readiness, the stage of community readiness of Wiyurejo Village is in stage three, namely vague awareness. Vague awareness means most people feel that there is a local concern, but there is no motivation to do anything about it (Plested, Edwards, & Jumper-Thurman, 2006). The value of the dimension that gives the lowest contribution and is below the average value of the community readiness is community knowledge on the issue.

1. Introduction

The Government of Indonesia has a target to increase the share of renewable energy in which 10.81% of final energy consumption comes from biomass, including its derivatives (biogas) by 2020. Biogas is an alternative energy that can replace fuel oil or natural gas. Biogas energy can be obtained from household waste, livestock waste, organic waste from market and industry.

Therefore, some regulations concerning renewable energy development have been approved to achieve the target, especially in rural areas. Utilization of biogas tends to increase in Indonesia because it offers many social, economic and environmental benefits [1].

Renewable energy development is promoted as a promising method to solve the problem of rural energy supply and to improve the lives of rural households as it can reach remote areas far from the city gas network. The Government of Indonesia realizes that many benefits can be achieved from the utilization of alternative energy. Reducing greenhouse gas emissions, reducing fossil fuel dependence, and strengthening energy security are the advantages of alternative energy [2]–[5]. Specifically, the objectives of biogas development are to increase public awareness about green energy, reduce greenhouse gas (GHG) emissions, promote sustainable village development, accelerate environmentally friendly agriculture, increase rural household savings, and improve rural energy equity and rural life quality [Lidija, 2012]

One of the potentials for biogas development in Malang Regency is located in Wiyurejo Village, Pujon Subdistrict. Currently, there are 1,077 farmers in Wiyurejo Village, according to its 2013 Village

Profile, but there are only 20 farmers who have built biogas reactors from a total of 1,077 active farmers in Wiyurejo Village. This means that the potential for biogas development in Wiyurejo Village is very big.

According to the Head of SAE Cooperative, there are several obstacles in the development of biogas in Pujon Subdistrict, including in Wiyurejo Village, namely limited land availability for biogas development. Even according to the head of breeder group of Kalangan Hamlet, currently almost 75% of breeders want to build biogas but do not have enough vacant land. Thien Thu et al (2012) outlines some reason why farmers do not have biogas, and one of them is caused by land insufficiency [6]. As a result of this land problem, there are only a small number of people who volunteers to develop biogas. In addition, there are many breeders who do not know about biogas development technology, so they are not ready to find the solution on land problems.

Various theories state that the awareness and participation of the villagers are the key to the success of village development. Community involvement is an important element to consider when discussing biogas development [7]. Meanwhile, people mostly rely on village head to raise awareness among villagers on the importance of development efforts as a way to improve social conditions and increase the participation of villagers in development [8]. Thus, readiness is the basic foundation for a community or government in making a follow up on the activities to be done in the future.

Therefore, if one wants to develop biogas, then one must know the condition of the society and the readiness of the society in running a business and also in developing a region, including biogas development. Local community readiness is important in a development effort as it will make changes that will occur from the development of biogas. If the community is ready, there is a potential that the region will have a positive impact so that later biogas can prosper the community by encouraging the empowerment of the community to be able to participate actively to get the maximum benefit by managing resources.

2. Literature Review

2.1. Stage of Readiness

Readiness means the whole condition of a person or an individual that makes him ready to respond in some way to a situation and condition. Readiness is the overall condition of an individual to respond to or practice an activity which contains mentality, skills and attitude that must be possessed and prepared during a particular activity. In starting any kind of job, readiness is very important because by having readiness, any job will be resolved and can be done smoothly with good results.

2.2. Biogas

Biogas is a gas produced by anaerobic activity or fermentation of organic materials including human and animal waste, domestic waste (household), biodegradable waste or any biodegradable organic waste under anaerobic conditions. The main ingredients in biogas are methane and carbon dioxide. Biogas is a gas mixture of anaerobic fermentation process from cattle dung (cow). Cow dung can meet the need for waste used for biogas. Biogas is a process of producing bio gas from organic material with the help of bacteria. Biogas energy serves as replacement energy for fossil fuels, thereby reducing greenhouse gases in the atmosphere and other emissions [9].

2.3. Community

Community is part of society which consists of various characteristics. Community has a more limited capacity for self-sufficiency than society, but this limitation makes community has closer relationships and greater sympathy. There is often a misunderstanding between society and community. The two terms are often interpreted in the same way, but they are very different. Society is different from community. The definition of society is more general and broader, while the definition of local community is more limited and is also limited by the area of the region and the number of citizens. However, in terms of the activity of their relationship, community is tighter than society, and its unity is also greater. In the aspect of society, this research is aimed at the villagers of Wiyurejo as a whole, while in the aspect of the community, it is aimed at breeders in Wiyurejo Village. This study aims to determine the stage of readiness in the development of biogas, so the target in this study is more specified at breeder community in Wiyurejo Village who are directly involved in the development of biogas.

Table 1. Stage of community readiness.

No.	Stage	Explanation
1	*No awareness*	There is no self-readiness in the community or leader to respond to certain thing based on understanding and reason and moral considerations which accompanied by freedom so that it can be consciously accounted for, such as an issue that is not generally recognized by the community or leader as a problem.
2	*Denial/resistance*	Member of community can minimally recognize a concern, but there is little confession that this will take place locally.
3	*Vague awareness*	Most of them feel that there is a local concern, but there is no motivation to do something on existing efforts/programs.
4	*Preplanning*	There is a clear recognition that something must be done, but the efforts are not focused or done in detail.
5	*Preparation*	A leader begins to actively plan, and the community supports the existing efforts or programs.
6	*Initiation*	There is adequate Information for justification (decision/ reason/ consideration) on the efforts/programs made.
7	*Stabilization*	There are activities/efforts made by the community supported by government or community decision maker for people interest.
8	*Confirmation/expansion*	The members of community feel comfortable with the efforts/programs, and they support expansion, namely an activity to increase/broaden the effort/program marked by creating new effort/innovation, expanding the facility, and improving economic activity.
9	*High level of community ownership*	Members of the community have detail knowledge on the cause, result, and consequence of an effort/program.

Source: Plested, Edwards, & Jumper-Thurman, 2006

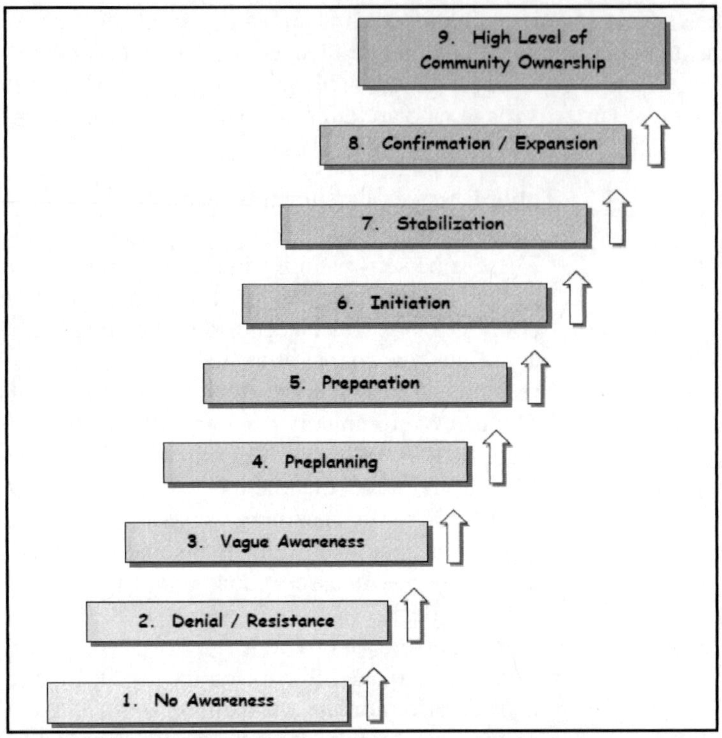

Figure 1. Stage of Community Readiness

3. Methods

3.1. Area of the Study
Wiyurejo village is about 28 km west of Malang City Indonesia with an area of 249.06 Ha which consists of 3 smaller areas called Dusun, namely Dusun Wiyu, Dusun Kalangan and Dusun Bagean. The number of breeders in Wiyurejo village is 1,077 farmers.

3.2. Data collection
Data were collected through questionnaires by collecting information from key persons who are deemed most aware of the condition of the community and existing issues regarding biogas development in Wiyurejo Village

3.3. Key respondents
There are 6 respondents in Wiyurejo Village who are given a questionnaire, namely those who are considered most aware of the community climate and issues about the development of biogas in Wiyurejo village. Key respondents provide community readiness assessment by answering the questionnaire that has been provided. In the Community Readiness Handbook, 4 to 6 speakers are neeeded as key respondents who are able to provide information in biogas development efforts [10].

Table 2. Key respondens in Wiyurejo Village

No.	Respondents
1	Mr. Mufid Farid, Head of Wiyurejo Village
2	Mr. Sugeng, Head of farmers group in Dusun Kalangan
3	Mr. Suwarno, Head of farmers group in Dusun Wiyu

IOP Conf. Series: Earth and Environmental Science **70** (2017) 012036 doi:10.1088/1755-1315/70/1/012036

No.	Respondents
4	Mr. Nur Rohmat, Head of farmers group in Dusun Bagean
5	Mr. Kayin, Head of Biogas Development of SAE Milk Cooperative in Pujon Subdistrict
6	Mr. Yoko, Secretary of Biogas Development of SAE Milk Cooperative in Pujon Subdistrict

3.4. Community Readiness Model

Stage of community readiness can be identified through a community readiness assessment based on six readiness dimensions or variables [10]. Readiness is used to determine the extent to which the community is ready to take action on an issue or a problem. To measure community readiness, there is a dimension of readiness which becomes the key factor to influence community readiness to take action on issue/problem. The readiness dimension consists of six dimensions, including: community efforts, community knowledge of efforts, leadership, community climate, community knowledge about issue, and resources related to the issue. The explanation of the readiness dimension is as follows:

1. Community Effort
   Community efforts are used to find out to what extent the business, program or policy has been implemented to address the issue.
2. Community Knowledge about Effort
   Knowledge is used to see to what extent the community knows about the local effort being undertaken, and whether the efforts are known by the whole community.
3. Leadership
   Leadership is used to find out to what extent the leader has an influence on the community in supporting the issue.
4. Community Climate
   Community climate can be seen from the attitude of the community to the effort as well as community contribution to the effort.
5. Community Knowledge about Issue
   Community knowledge about issue is used to find out to what extent the community knows about the issue.
6. Resources Related To the Issue
   Related resources are used to know to what extent local resources, people, time, money, space and others are available to support the business.

Scoring is a process or steps to assess community readiness based on the six dimensions or variables used [10]. Scores of dimensional measurements are as follows:

Dimension of Community Effort	: 2.84
Dimension of Community Knowledge About Effort	: 3.76
Dimension of Leadership	: 4.67
Dimension of Community Climate	: 3.32
Dimension of Community Knowledge About Issue	: 2.27
Dimension of Resource Related to the Issue	: 3.83
Amount	**: 20.7**

20.7: 6 (number of dimensions) = 3.4

After acquiring readiness value from the community, the ladder of community readiness is then determined. Based on the ladder of readiness, the community climate can be known so that appropriate handling can be given based on the stage of readiness. In addition, in the Community Handbook Manual there is also a strategy based on the ladder of community readiness to improve the readiness of Wiyurejo Village community in developing biogas. The assessment of readiness is based on key respondents using scores one to five. To simplify the scoring assessment, parameters based on the variables and the score are made. The use of likert scale in readiness scoring assessment is because likert scale is generally used to measure the attitude or response of a person to an object. Disclosure of attitude using Likert scale is very popular among researchers. This is because in addition to being practical, a well-designed likert scale generally has a satisfactory reliability [11]. Likert scales use several questions to measure individual behavior by responding to 5 choices on each question: strongly agree, agree, does not answer, disagree, and strongly disagree [12]. However, in this study the measurement of value is adjusted to the variables and questions needed to determine the stage of readiness of breeders in Wiyurejo village in the development of biogas.

4. Results

4.1. Issue related Land Availability for Biogas development

According to Enny Ariani, 2011 one of the success of biogas development is the land availability. Land availability becomes one of the important conditions in developing biogas. According to Thien Thu, et al which discusses some reasons farmers do not have biogas reactor caused by the insufficiency of land owned [6]. Land problems faced by farmers in the village Wiyurejo. Farmers who want to have biogas must undoing their intentions because they realize they do not have enough land to build a digester reactor.

Biogas development in Wiyurejo Village is still considered less, because form 807 farmers in Wiyurejo Village, only 20 biogas units were build. According to the Head of SAE Cooperative there are several obstacles in the development of biogas in Pujon Sub-district, including in Wiyurejo Village, which is limited land available for biogas development. From the survey results it is known that the land owned by farmers are still not sufficient to build bio digester installations that fit the standards, especially in Dusun Kalangan which has the most cows, all farmers do not have enough vacant land

The average vacant land owned by farmers in Wiyurejo Village is classified into 5 classes, which is 0-5 m2, 6-10 m2, 11-15 m2, 16-20 m2, 21-25 m2. The Farmers mostly have 6-10 m2 whereas the land needed to build the size of the reactor, the smallest, ie 4 m3 requires 13.75 m2 of land (See Table 1).

As much as 59% of farmers still do not know that the land availability is one of the conditions that at the time of submission of biogas development proposal will also be seen the feasibility of the land owned. If the width of their land is still less than then the biogas cannot be built. This proves that the farmers in the village of Wiyurejo is not much to realize that the limitations of vacant land is also a problem

Table 3. Required Land for Biogas Installation

Reactor size (m3)	Required Land (m2)
4	13.75
6	18
8	26.25
10	36
12	49.5

Source: Biogas Household, 2015

4.2. Stage of Readiness

The calculation of community readiness in Wiyurejo Village is calculated from the sum of each variable value used, among others, community effort, community knowledge of effort, leadership, community climate, community knowledge about issue, and resources related to the issue. The stage of community readiness can be seen from the total value of the assessment of each variable. It is taken from the total average value that has been divided by the number of respondents in each variable. Then, the total sum is acquired and divided by the number of variables used, namely those six variables. From the average result, the value of the stages of readiness of farmers in Wiyurejo village in the development biogas is related to limited land availability

4.2.1. Community Effort. Community effort scores in Wiyurejo Village are at 1-3 [see table 4]. It's indicate that the effort of Wiyurejo farmers to finding solutions for limited land in the development of biogas is still low. The low level of awareness and understanding of farmers due to the lack of community in the village Wiyurejo. So far, the community is only a group of livestock and cooperative communities. Communities in the village of Wiyurejo so far are still very lacking in conducting joint activities so that cause a lack of means to exchange opinions and inputs for biogas development and business level because there has been no socialization and discussion together to find solutions to the land availability problem that occurred in Wiyurejo Village.

Tabel 4. Score of community effort.

	Community awareness level of village efforts		Community understanding level		The duration of village community efforts / programs that have been running		Score Average
Responden	Score	condition	Score	condition	Score	condition	
Head of Wiyurejo Village	3	Neutral	2	Do not understand	2	>1-2 year	2.3
Head of farmers group in Dusun Kalangan	2	Do not care	2	Do not understand	1	0-1 year	1.7
Head of farmers group in Dusun Wiyu	3	Neutral	2	Do not understand	4	>3-4 year	3.0
Head of farmers group in Dusun Bagean	2	Do not care	1	Very Unfamiliar	5	>4 year	2.7
Head of Biogas Development of SAE Milk Cooperative in Pujon Subdistrict	3	Neutral	1	Very Unfamiliar	1	0-1 year	1.7
Secretary of Biogas Development of SAE Milk Cooperative in Pujon Subdistrict	3	Neutral	1	Very Unfamiliar	1	0-1 year	1.7
					Total Score		13.0
					Average (Total Score/total interviewer)		2.17

4.2.2. Community Knowledge of Effort. The knowledge level of farmers score are 1-2 which means low [see table 5]. The farmers still do not understand the issue because the lack of information. Currently there is an effort by building biodigester under the house. But very few farmers are willing to do so, there are still many farmers who do not want to build biodigester under the house due to leakage. The lack of knowledge about alternatives in building biogas installations made farmers couldn't solved the problem.

Table 5. Score of community knowledge effort.

	Community Knowledge		Community Knowledge Level		Advantages / Weaknesses of Village efforts		Score Average
Responden	Score	condition	Score	condition	Score	condition	
Head of Wiyurejo Village	2	Do not Understand	3	Neutral	3	Neutral	2.7
Head of farmers group in Dusun Kalangan	2	Do not Understand	3	Neutral	3	Neutral	2.7
Head of farmers group in Dusun Wiyu	2	Do not Understand	4	Knowing	4	Helpful	3.3
Head of farmers group in Dusun Bagean	2	Do not Understand	3	Neutral	1	Very Unhelpfull	2.0
Head of Biogas Development of SAE Milk Cooperative in Pujon Subdistrict	1	Very Uninformed	1	Do not Know	1	Very Unhelpfull	1.0
Secretary of Biogas Development of SAE Milk Cooperative in Pujon Subdistrict	2	Do not Understand	1	Do not Know	2	Unhelpfull	1.7
					Total Score		13.3
					Average (Total Score/total interviewer)		2.2

4.2.3. Leadership. Leaders in Wiyurejo Village are still considered less active in assisting biogas development. This is indicated by an assessment of the influence of the leader, the level of involvement and support is 1-3 [see Table 6]. Currently some leaders of Wiyurejo Village have been aware of problems in the biogas development. However, until now the leaders and government are still lacking in helping farmers to settle the problem. The interview with the head of the livestock group stated that the head of the livestock group had tried to offer the farmers, but when the farmer rejected the offer from the head of the livestock group there was no follow-up to help the farmers to build biogas. There has been no initiative from leaders to try to invited the farmers for a discussion together to find a solution for the problems.

Table 6. Score of community leader effort.

	The role of Community Leader		Involvement Level of Community Leader		Support of Community Leader		Score Average
Responden	Score	condition	Score	condition	Score	condition	
Head of Wiyurejo Village	2	No effect	3	Neutral	2	Doesn't support	2.3
Head of farmers group in Dusun Kalangan	2	No effect	2	Not involve	3	Neutral	2.3
Head of farmers group in Dusun Wiyu	3	Neutral	3	Neutral	2	Doesn't support	2.7

	The role of Community Leader		Involvement Level of Community Leader		Support of Community Leader		Score Average
Responden	Score	condition	Score	condition	Score	condition	
Head of farmers group in Dusun Bagean	2	No effect	2	Not involve	2	Doesn't support	2.0
Head of Biogas Development of SAE Milk Cooperative in Pujon Subdistrict	2	No effect	3	Neutral	2	Doesn't support	2.3
Secretary of Biogas Development of SAE Milk Cooperative in Pujon Subdistrict	2	No effect	1	Never involve	2	Doesn't support	1.7
					Total Score		13.3
					Average (Total Score/total interviewer)		2.2

4.2.4. Community Climate. Community climate in the development of biogas is considered high [see table 7] eventhouh the level of constraints is quite high. Community attitude not support for biogas development efforts related to land limitations. There has been no initiative from the community to help or motivate farmers to come together for solutions. Currently there are alternative solutions for building biodigester reactors under the house, but there are less experiences barriers make farmers unwilling to build biodigester reactors. Farmers find it difficult to care for livestock because they do not know where to move their animals temporarily and construction leak

Table 7. Score of community climate.

	Community Contribution		Community Attitude		Level of Obstacles		Score Average
Responden	Score	condition	Score	condition	Score	condition	
Head of Wiyurejo Village	3	Neutral	4	Good	1	Very High	2,7
Head of farmers group in Dusun Kalangan	3	Neutral	3	Neutral	2	High	2,7
Head of farmers group in Dusun Wiyu	3	Neutral	4	Good	2	High	3,0
Head of farmers group in Dusun Bagean	3	Neutral	3	Neutral	2	High	2,7
Head of Biogas Development of SAE Milk Cooperative in Pujon Subdistrict	3	Neutral	4	Good	1	Very High	2,7
Secretary of Biogas Development of SAE Milk Cooperative in Pujon Subdistrict	1	Very unsupported	1	Very Bad	2	High	1,3
					Total Score		15.0
					Average (Total Score/total interviewer)		2.5

IOP Conf. Series: Earth and Environmental Science **70** (2017) 012036 doi:10.1088/1755-1315/70/1/012036

4.2.5. Community Knowledge about Issue. Community Knowledge about issue in the development of biogas is considered low [see table 8]. Limited land has not become the community concentration in the development of biogas. According to respondents there is still no information availability or data collection for the development of biogas related to land limitations. Limitations of such data and information can make it difficult to assist land-ridden farmers to build biogas.

Table 8. Score of community knowledge about issues.

	Information availibility		Database availibility		Level of community Knowledge regarding biogas development programs		Score Average
Responden	Score	condition	Score	condition	Score	condition	
Head of Wiyurejo Village	1	None	1	Not Available	3	Neutral	1.7
Head of farmers group in Dusun Kalangan	1	None	1	Not Available	3	Neutral	1.7
Head of farmers group in Dusun Wiyu	3	Neutral	1	Not Available	2	Do not Know	2.0
Head of farmers group in Dusun Bagean	3	Neutral	1	Not Available	2	Do not Know	2.0
Head of Biogas Development of SAE Milk Cooperative in Pujon Subdistrict	3	Neutral	1	Not Available	2	Do not Know	2.0
Secretary of Biogas Development of SAE Milk Cooperative in Pujon Subdistrict	2	Not available	1	Not Available	1	Do not Know	1.3
					Total Score		10.7
					Average (Total Score/total interviewer)		1.78

4.2.6. Resources Related to Issue. Resources related to issue in biogas development is considered low [see table 9]. There is less socialization and training related to biogas development, especially for farmers who didn't have land. Farmers didn't get more information about issues related to the development of biogas and only information was about biogas development subsidy development for normal construction. If farmers want to build a reactor under the house, it will cost more when compared with normal construction. It will also make it difficult for farmers to spent more due to economic condition. Assistance in addition to cost biogas aside subsidy until now still not exist in the Village Wiyurejo.

Table 9. Score of resources related to issue.

	Level of need and intensity of training procurement (training) to the community		Community support to provide help		Intensity of external assistance		Score Average
Responden	Score	condition	Score	condition	Score	condition	
Head of Wiyurejo Village	1	Not support at all	3	Neutral	3	Neutral	2.3
Head of farmers group in Dusun Kalangan	3	Neutral	3	Neutral	3	Neutral	3.0
Head of farmers group in Dusun Wiyu	3	Neutral	3	Neutral	3	Neutral	3.0
Head of farmers group in Dusun Bagean	2	Not suport	2	Not support	2	Not Satisfied	2.0
Head of Biogas Development of SAE Milk Cooperative in Pujon Subdistrict	2	Not support	2	Not support	2	Not Satisfied	2.0
Secretary of Biogas Development of SAE Milk Cooperative in Pujon Subdistrict	2	Not suport	1	Not support at all	2	Not Satisfied	1.7
					Total Score		14.0
					Average (Total Score/total interviewer)		2.3

The calculation results interpret the stage of readiness of Wiyurejo Village community. The total value of all variables is 13.22, if divided by the number of dimensions which is six, then the result is 2.20 [see table 10]. This value shows that, from nine existing stages, the stage of readiness of farmers in Wiyurejo village is in stage three (vague awareness) [see table 11]. Vague awareness means most people feel that there is a local concern, but there is no motivation to do anything about it [Plested, Edwards, & Jumper-Thurman, 2006]. The value of dimension that gives the lowest contribution and is below the average value of the community readiness is community knowledge about the issue. This could be the main cause why community readiness is still in the third stage.

Table 10. The calculation of combined variables in Wiyurejo Village.

No	Variables	Total
1	Community Efforts	2.17
2	Community Knowledge of Effort	2.20
3	Leadership	2.22
4	Community Climate	2.50
5	Community Knowledge about Issue	1.78
6	Resources Related to the Issue	2.30
Total		13.22
Average		2.20

Table 11. The value of the stages of community readiness.

No	Stage	Range
1	No awareness	1.00-1.44

No	Stage	Range
2	Denial/resistance	1.45-1.88
3	Vague awareness	1.89-2.33
4	Preplanning	2.34-2.77
5	Preparation	2.78-3.22
6	Initiation	3.22-3.66
7	Stabilization	3.67-4.11
8	Confirmation/expansion	4.12-4.55
9	High level of community ownership	4.56-5.00

Source: Plested, Edwards, & Jumper-Thurman, 2006

Community climate in Wiyurejo village shows the same thing with the assessment of the stage of readiness. 36% of farmers have known that the constraint to build biogas is lack of vacant land. It shows that there is local concern from the farmers. However, until now there is still no motivation from farmers to try to find a solution. Meanwhile, 87% of farmers have vacant land, but the width of their land is still less than the need for biogas development, so biogas development cannot be done. According to Mr. Sugeng, the head of farmer group in Dusun Kalangan, many farmers are actually interested in developing biogas, but they feel that their land is not wide enough, and there has not been any follow-up to help the farmers so that they can still build biogas.

At the stage of vague awareness, the goal is to increase community awareness about the problems of land limitations that can inhibit the development of biogas and explain the benefits of biogas for the life of farmers. When their awareness has increased, the next stage can be done, that is helping farmers to look for a solution to overcome the problem of land limitation.

Because 59% of farmers still do not know that the land is one of the conditions. This proves that the farmers in the village of Wiyurejo is not much to realize that the limitations of land owned is also a problem. Previously known that as many as 87% of farmers whose land is still not able to be built biogas due to empty land area that is not in accordance with needs. However, only 36% of farmers are aware that the problem is the lack of vacant land. There are still many Farmers who do not know that at the time of filing the development of biogas will also be seen the feasibility of the land owned, if still less then the biogas cannot be built. Lack of information in the development of biogas in the village Wiyurejo can be one reason there are still many farmers who do not know that land compliance to be a requirement for the construction of biogas reactor.

References

[1] X. Yuan and J. Zuo, "Pricing and affordability of renewable energy in China – A case study of Shandong Province," 2011.

[2] R. Chen, "Livestock-biogas-fruit systems in South China," *Ecol. Eng.*, vol. 8, no. 1, pp. 19–29, Mar. 1997.

[3] Y. Chen, G. Yang, S. Sweeney, and Y. Feng, "Household biogas use in rural China: A study of opportunities and constraints," *Renew. Sustain. Energy Rev.*, vol. 14, no. 1, pp. 545–549, 2010.

[4] S. Cheng, Z. Li, J. Shih, X. Du, and J. Xing, "A field study on acceptability of 4-in-1 biogas systems in Liaoning Province, China," *Energy Procedia*, vol. 5, no. September 2016, pp. 1382–1387, 2011.

[5] L. X. Zhang, C. B. Wang, and B. Song, "Carbon emission reduction potential of a typical household biogas system in rural China," *J. Clean. Prod.*, vol. 47, no. November, pp. 415–421, 2013.

[6] C. T. Thien Thu *et al.*, "Manure management practices on biogas and non-biogas pig farms in developing countries – using livestock farms in Vietnam as an example," *J. Clean. Prod.*, vol. 27, pp. 64–71, 2012.

[7] W. Liu, C. Wang, and A. P. J. Mol, "Rural public acceptance of renewable energy deployment: The case of Shandong in China," *Appl. Energy*, vol. 102, pp. 1187–1196, 2013.

IOP Conf. Series: Earth and Environmental Science **70** (2017) 012036 doi:10.1088/1755-1315/70/1/012036

[8] I. Ulumiyah, A. J. G. Gani, and L. I. Mindarti, “Peran Pemerintah Desa dalam Memberdayakan Masyarakat Desa (Studi pada Desa Sumberpasir Kecamatan Pakis Kabupaten Malang),” *J. Adm. Publik*, vol. 1, no. 5, pp. 890–899, 2013.
[9] W. Febriyanita, “Pengembangan biogas dalam rangka pemanfaatan energi terbarukan di Desa Jetak Kecamatan Getasan Kabupaten Semarang,” Universitas Negeri Semarang, 2015.
[10] B. A. Plested, P. Jumper-thurman, and R. W. Edwards, *Community Readiness Manual*. Fort Collins, CO: Tri-Ethnic Center for Prevention Research, 2006.
[11] Risnita, “Pengembangan Skala Model Likert,” *Edu-Bio*, vol. Volume 3, pp. 86–99, 2012.
[12] W. Budiaji, “The Measurement Scale and The Number of Responses in Likert Scale,” *J. Ilmu Pertan. dan Perikan. Desember 2013*, vol. 2, no. 2, pp. 125–131, 2013.

IOP Conf. Series: Earth and Environmental Science **70** (2017) 012037 doi:10.1088/1755-1315/70/1/012037

Social capital and temporal migrant characteristic in rural Indonesia – A case of Malang Regency

G Prayitno

Department of Urban and Regional Planning, Faculty of Engineering, Universitas Brawijaya, Malang, Indonesia

gunawan_p@ub.ac.id

Abstract. Public participation in the planning system is often implemented in order to encourage participatory planning. The result is the planning will be implemented softly and the society can managed and maintained continuity the project by themselves. Social capital as the concept of collective action, could increase the possibility the people solve their problem together. In this research we tried to implement the concept of social capital from the migration respondent in rural community activities. In rural area, migration is commonly used by rural inhabitants to ensure the survival of their families or to pursue economic mobility to supplement dwindling household resources. Households are generally selected and invest in a family member who is viewed to have the greatest potential for generating migrant earnings and sending remittances. The increase in the number of migrants has an impact on not only household members but also activities in communities. The labor movement might affect social capital in communities. In this paper, the relation between characteristics of migrants and the level of social capital is analyzed. Characteristics of migrants and households who send them are investigated thorough questioner survey data, which were conducted in Malang Regency, Indonesia.

1. Introduction

Indonesian Ministry of Manpower and Transmigration reported that in the previous 10 years, around 400,000 migrant workers have been sent abroad annually. A quick calculation, therefore, tells us that there are at least four million people who have, at some point, been overseas workers in the past 10 years.

According to The Agency for the Placement and Protection of Indonesian Migrant Workers (BNP2TKI) [1], the number of migrant workers from Indonesia has been decreasing over the years, from 748,825 in 2008, to 512,168 in 2013. In addition, the remittances sent by these workers according to Indonesian Central Bank [2], amounted to the USD 10.9 billion in 2013, increasing from USD 6.9 billion in 2012. It became an important factor in the country's renewed economic growth in the past few years.

Although the remittance has been increasing over years, the impacts on rural development are not significant at the macro levels; remittances contribute relatively small, and out of migration usually do not radically transform poor areas. However, at micro level, the studies show that migration reduces the uncertainty of family income, provides investment fund, and contributes to livelihoods for those with small plots.

However, before chooses the decision to migrate, the household or potential migrant will look for opportunities of jobs in destination countries. The potential migrant or head household will find out their relatives or friends who worked abroad or out of the village to reduce the risk. These relations know as migrant network or social ties in migration literature, in line with the research from Lin, Cook and Burt 2001; and Taylor 2006 [3]. They found that the candidate of migrants may obtain direct or indirect benefit which can reduce the risk and cost at the time of relocation and job seeking from the relation with former or active migrant. This also applies particularly with the ties between migrant and family/relative/kinship members [4] and [5].

With the presence of social ties, an individual may mobilize the available social capital in him/her to obtain valuable information, moral and material supports, which will reduce the cost and risk during migration. As the result, the individual that owns social capital with other migrants will be more likely to do migration than that without social capital. The study attempts to find the relation between social capital and decision choice made by the family or individual to migrate or stay.

This study examines the impacts of migration and social capital on rural areas. The concept of public participation is used to design the concept of regional development, especially for rural development. Specifically the following research questions will be addressed: (1) what are the characteristic of migrant workers (local migrant, active and former international migrant); (2) how is the correlation between inhabittans and village attribution of migrant and sense of community; and (3) how is the correlation between personal and village attribution of migrant and community activities.

2. Literature Review

2.1. Migration

The migration literature, unsurprisingly, indicates that relationships between migration and rural development are complex and context-specific. Macro level studies suggest there is some consensus that migrants tend to help to increase welfare: migrants often contribute much to economy of the host society, and have high rates of labor force participation [6]

Migration is thus bound up with contests and negotiation over one's place in society. For some, decisions of when to migrate and for which kinds of work reflect ascribed identities – not every act is an act of resistance. Moreover, migrants make use of social resources and networks – who they know – to access or avoid particular types of work and/or migration destinations. Collective identities contribute to the meanings of migration for individual migrants – as Rogaly et al. suggest, they may be deployed positively to decrease the loss of dignity in hiring out for wage work in "foreign" place.

The meaning of migration to different individuals within a household varies according to age as well as sex [6]. The migratory effects of social capital vary by gender within and across countries [7]. In addition to varying by gender, social capital effects are not uniform across communities [8]. Individual social capital generally has stronger effects on out-migration in rural than urban areas, and the mechanisms of cumulative causation likewise operate more powerfully in rural than in urban areas. Cumulative causation also operates more effectively when social capital is equally distributed among members of a community and when the migration in question is unauthorized rather than legal [7].

2.2. Social Capital

Social capital, which is interpreted with the term of trust, norms, networks and identity as governing human behavior, is significant to motive and coordinate collective action towards collaboration. Social capital such as trust, norms, and networks that can improve the efficiency of society by facilitating coordination action ([9]; [10]). Social capital improve participants' monitoring, reduces free riders, thus mutual bonds of trust.

Communities with high levels of social capital are more effective at exercising social control over deviant and uncivil behaviors [11]. Similarly, weaker social ties directly increase vulnerability to crime by decreasing the likelihood of receiving deterrence, help or information from neighbors and limiting connections to and thus help from police and other institutions although this effect may vary by type of community or, in the present study, cross-culturally between countries [12].

At the individual level, Perkins and Long [13] proposed a two-by-two social capital framework: one dimension distinguishes intrapsychic (cognitive/affective) from behavioral responses; the other contrasts informal/spontaneous versus formally organized responses. This yields four components of psycho-behavioral social capital: the informal, affective component is social bonding (or sense of community); the informal behavior is neighboring; the organized, cognitive component is collective efficacy (or empowerment); the organized behavior is citizen participation.

People gain access to social capital through membership in interpersonal networks and social institutions and then convert it into other forms of capital to improve or maintain their position in society [14]. Migrant networks are sets of interpersonal ties that connect migrants, former migrants, and non-migrants to one another through relations of kinship, friendship, and shared community origin. Network connections increase the likelihood of international migration because they lower the costs and risks of movement and increase the expected net returns to migration. Having a tie to someone who has migrated yields social capital that people can draw upon to gain access to an important kind of financial capital, that is, high foreign wages, which offer the possibility of accumulating savings abroad and sending remittances home [15]

2.3. Neighboring

Neighboring is the help we informally provide, and receive from, neighbors. Unger and Wandersman [16] identified three components of neighboring behavior: (1) social support and network ties (community-level bonding SC); (2) cognitive mapping of the physical environment and symbolic communication (captured in our model by place definition, interdependence, and identity); (3) affective attachment to neighbors and to place (which we see as sense of community and place bonding). Neighboring also operates at both the individual and community levels, and is closely linked to place attachment [17], and the strongest, most consistent predictor of citizen participation [18]. Those people and communities who are so busy participating in civic organizations are the same ones also informally helping their neighbors.

2.4. Identity and Community Activism

In the related researches, Akerlof and Kranton ([19], [20]) argued that a strong in-group identity (i.e., collective identity) is effective in facilitating cooperation among group members. Stronger in-group identification leads to perceived "group entitativity" – that is, the perception that a group is like entity, a cohesive whole and "a sense of symbolic immortality" – that is known to raise group members' self-esteem [21]. It is important issue to explore community activism for sustain and manage regional identity [22]

Tarn said that identity had grounded in local community. But after the Legendary Journeys of Alexander of the Great (330BC) people became increasingly "detachable" from their local societies, and their sense of identity began to shift from being centered in their own individuality [22]. Other researchers discussed about the relationship between place and identity as "moral capital", "regional consciousness", "power of region", and "psychological sense of community" derived from community network, norm, and cultural activities in the place. The identity rooted in place mobilizes pride and attachment to the place and foster community activism.

3. The Field Survey Design

Field survey was conducted in one phase whereby the main goal is to investigate the impacts of migration and social capital in rural areas. In addition, we have done survey by staying in the village location for 7 days to get deep information from the respondents. All respondents were interviewed individually with help from some students as interviewers.

Face to face interviews effectively have been done within 7 days by 10 surveyors. Interviews were scheduled from 07:00 a.m. to 09:00 p.m. depending on the respondent's readiness. Through systematic sampling, 250 migrant households living at Arjowilangun village, Kalipare district, Malang regency, East Java Province Indonesia, are selected as the respondents for the study. Five hamlets are selected covering Pangganglele, Lodalem, Lotekol, Duren and Barisan.

4. The Households Characteristic

The respondent of this research is the household that have one or more migrant workers. If the migrant worker is still active, the question will ask to the member of family (husband, wife, grandfather, grandmother or children from migrant families). As we mentioned before the numbers of respondents are 250 respondents.

The types of households among 250 respondents are 222 households with complete parents (father and mother), 6 single parents (widower), and 20 single parents (widow). The majority of families have one migrant worker (97.4% or 243 families), after that the families with two migrant and one family consists of three migrant workers.

Referring Figure 1, The most common size of family is families with 3 inhabitants live in the same one unit single house consist of parents (father and mother) plus one member (34.8% or 87 households). Another six types of family size are (i) family with 4 member (73 household or 29.2% of total households), (ii) family with 5 member (33 household or 12.8 % of total households), (iii) family with 2 member (31 household or 12.4 % of total households), (iv) family with 1 member (15 household or 3.6 % of total households), (v) family with 6 member (9 household or 3.6 % of total households), and (vi) family with 7 members (3 household or 1.2 % of total households). For the six and seven members, the family consists of parents, two children or one child and grandfather and grandmother.

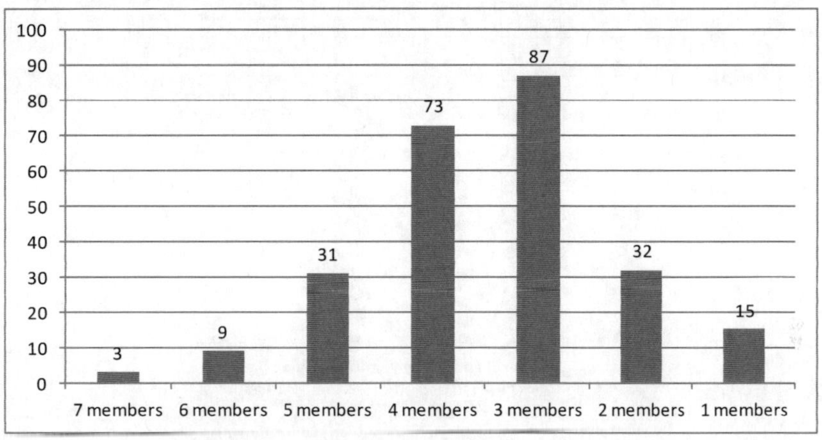

Figure 1. The Family Members of Households

5. The Correlation Analysis

In the household questionnaire survey, we investigated 7 demographic attributes for each household consist of household income, education, age, gender, family members, occupation, and duration of work. Table 2 depicts the best result of cross tabulation test for the relation of demographic attributes and 12 questions about values and belief to village and neighbors. The 12 question consist of : Place attachment to your village as your hometown (Y1); Nature and landscape of your village is nice (Y2); Foodstuff of your village is nice (Y3); Important to involve in community events activities (Y4); Important to consult people who in trouble (Y5); Important to keep daily communication with neighbors (Y6); Important to respect ancestors and manage community grave (Y7); Important to communicate with relatives living in the village (Y8); Neighbors are very important for me (Y9); Neighbors will take care of my children and my parent when I going abroad (Y10); Neighbors will help me and my family when we have some economic trouble (Y11); and Want to continue living in this village (Y12).

The result is, for the sample size of 250 respondents only type of migration and income has strong relation with almost answer of the questions (the value of chi-square test lest than *p values* <0.05). It means that type of migration and income could influence the decision of respondents to answer the questions. Another attributes such as family members and occupation has significant value only with activities Y4, Y5, Y6, Y11, and Y2, Y3, Y12 respectively.

Table 1. Chi square test (Pearson value) between demographic attributes and 12 questions related to values and belief to village and neighbors

Attributes	Y1	Y2	Y3	Y4	Y5	Y6	Y7	Y8	Y9	Y10	Y11	Y12
Education	0.407	0.712	0.419	0.708	0.102	0.944	0.845	0.577	0.95	0.451	0.998	0.899
Income	0.047	0.003	0.033	0.024	0.337	0.365	0.217	0.125	0.348	0.204	0.027	0.18
Type of Migration	0.017	0.151	0.012	0.026	0.029	0.015	0.011	0.023	0.086	0.277	0.277	0.283
Family members	0.095	0.255	0.456	0.001	0.035	0.042	0.538	0.132	0.299	0.452	0.02	0.069
Occupation	0.337	0.01	0.01	0.706	0.836	0.693	0.972	0.084	0.398	0.212	0.319	0.02
Gender	0.139	0.417	0.44	0.506	0.45	0.124	0.405	0.073	0.136	0.392	0.5	0.071

Figure 2. shows the relation between education and important to consult people who are in trouble or Y5 (the value of Pearson chi square 0,102, closest value to $p<0.05$). In this figure we can see that for elementary school or lower and junior high school the higher answer is no comment, and in the high school and higher level the answer of yes and extremely yes is the higher. We can conclude that there are some relation between education level and the respondent answer to the question.

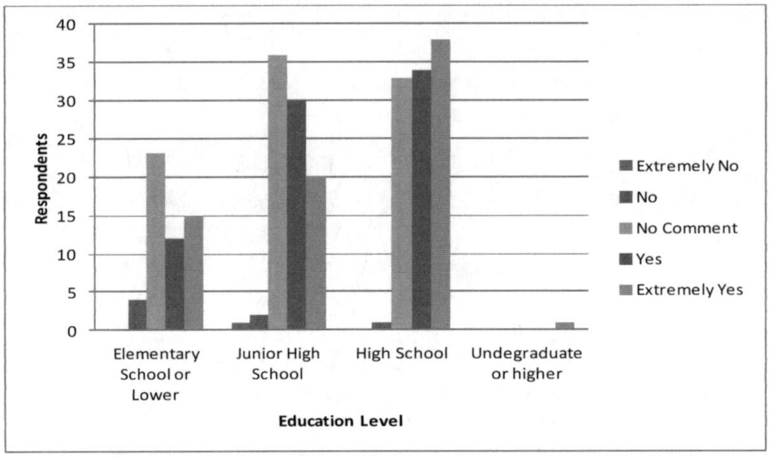

Figure 2. The relationship between education and Y5

Figure 3. below depicts the relation between income and respondent feeling about their neighbors (question no 11 or Y11, Neighbors will help me and my family when we have some economic trouble). The value of Pearson chi square is 0,027 less than the p value ($p<0.05$), it mean that the income has relation with the feeling of their neighbors. Even though the most answer is 'no comment', from the result of chi square test shown the significant value. So, we can conclude that the income has relation with their answer about their neighbor who will help them when they face some difficulties. It is common in rural area that their will help each other.

The next, Figure 4. shows the relationship between type of migration and place attachment (Y1) to their village. The value of Pearson chi square in the cross tabulation analysis is 0,017 less than the p value ($p<0.05$), it mean that the type of migration has relation with place attachment to their village. International migrant workers has more ties to their village than local migrant, majority of respondent answer yes and extremely yes more that no and no comment.

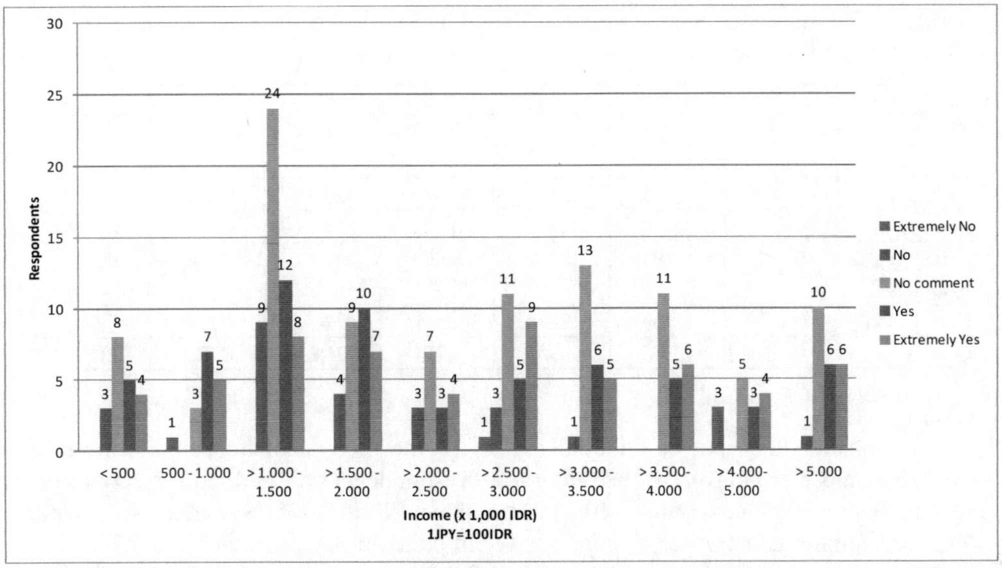

Figure 3. The relation between income and Y11

The relation between demographic attributes and activities in the community are shown by the calculation of cross tabulation (Table 5.2.). We ask the respondents about their activities in the community, and divided as: the activities no 1 (ACT1) is village, hamlet or RT/RW meeting; ACT 2, village cooperative meeting; ACT3, religious activities (praying in mosque, muludan, Ramadhan, sedekah and etc.); ACT4, cultural festival; ACT 5, working together to clean street, pavement local road and etc.; ACT6, sport even at independent day (17 August each year); and ACT7, social gathering (Kenduri, Arisan).

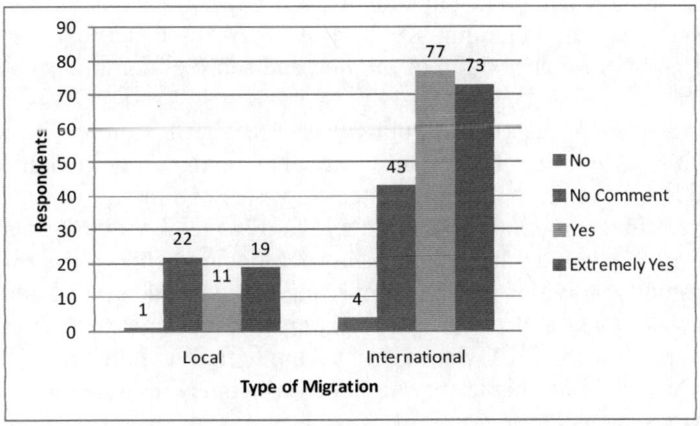

Figure 4. The relation between type of migration and Y1

The result of the calculation using cross tabulation data (Table 3.), the chi-square test shown the value of Pearson's Asymp.Sig (2-sided) less than 0,05 are in type of migration with ACT4, and ACT5, family members with ACT 3 and ACT 4, and gender with ACT 6 and ACT7. For demographic attribute: income, education and family members have no significant values and from this result, we may conclude that relation between demographic attributes and activities of the respondents in community activities are not significant or independent.

Table 3. Chi square test (value Pearson chi square) between Attributes Data and Activities in the community (activities 1-7)

Attributes	ACT1	ACT2	ACT3	ACT4	ACT5	ACT6	ACT7
Type of Migrant	0.131	0.131	0.959	0.006	0.048	0.508	0.008
Income	0.284	0.333	0.567	0.449	0.301	0.241	0.847
Education	0.038	0.84	0.414	0.323	0.649	0.188	0.428
Family members	0.493	0.293	0.079	0.094	0.316	0.449	0.926
Occupation	0.177	0.169	0.298	0.63	0.457	0.376	0.678
Gender	0.364	0.278	0.367	0.277	0.025	0.012	0.01

6. Conclusion

From the results of the study can be concluded that: (i) the income of households increase during migration of the member of family, (ii) the majority of respondents satisfied with the conditions of the village and neighbors and want to continue live in the village, (iii) most of respondent want to participate on the village or community activities, except involve in political party or union labor.

Based on statistical analysis, the result shows that for the sample size of 250 respondents only type of migration and income has strong correlation with almost answer of the questions (the value of chi-square test lest than p values<0.05). It means that type of migration and income could influence the decision of respondents to answer the questions.

Another attributes such as family members and occupation has significant value only with activities Y4, Y5, Y6, Y11, and Y2, Y3, Y12 respectively. The result of the calculation using cross tabulation data, the chi-square test shown the value of Pearson's Asymp.Sig (2-sided) less than 0,05 are in type of migration with ACT4 (Cultural festival), and ACT5 (Working together to clean street, pavement local road and etc.), family members with ACT 3 (Religious activities (Muludan, Ramadhan, sedekah and etc.), and ACT 4 (Cultural festival), and gender with ACT 6 (Sport even at independent day (17 August each year)), and ACT7 (Social Gathering (Kenduri, Arisan)). Its mean that there are any correlation between inhabittans and village attribution of migrant and sense of community that represent with the questionaire survey.

The result shows that event respondent going abroad they still want to involve in the activities in rural area (ACT 4: cultural festival). The cultural festival as annual program of the village attracts more respondents involve to the event. The minimum activities are send money for donation.

For demographic attribute: income, education and family members have no significant values and from this result, we may conclude that relation between demographic attributes and activities of the respondents in community activities are not significant or independent. This result shows there are a relation between social capital and public participation in community activities for migrant people. Even after migration process their still want to contribute to their community. These results show that social capital in this village is still higher because even when they migrate they still want to contribute to their village activities.

7. Recommendations

It goes without saying that this paper has some limitations and constrains. The focus of this study is to investigate the characteristic of the respondent (migrant workers) and the relation between social capital and demographic attribute from rural area in Indonesia. The other aspects such as political effect are not within the scope of this study.

This research is an individual based empirical research of social capital and migration characteristic in rural Indonesia. The formation of social capital is explained firstly by the questions on the questionnaire survey related to the feeling of respondents about the values and believe to their neighbors and living environment secondly by the questions in questionnaire survey related to respondent's participation in the community activities. This construct could be expanded by the designed data collection related to the closeness relation of the houses, group memberships and policy simulation shock to the theoretical model.

It looks promising to develop new analysis based on the data and theoretical model for the level of social capital among individuals' interaction in rural area and their decision choice. Therefore, it is necessary to scrutinize into more details about the relation of social capital and migration decision choice, which we may assume that there is a significant influence from the key person in the village or from government policy
Some text.

References

[1] Placement and Protection of Indonesian Migrant Workers (BNP2TKI), 2014. Accessed from http://www.bnp2tki.go.id/, in November 15, 2016
[2] http://m.koran-jakarta.com/?2921-optimalkan-pangsa-pasar-tki, Accessed March 15, 2017
[3] Lin N, Cook K, Burt RS, (editors). 2001. Social Capital: Theory and Research. Hawthorn, NY: Aldine de Gruyter.
[4] Massey, Douglas. S., Alarcón, R., Durand, Jorge, González H. 1987. Return to Aztlán: The Social Process of International Migration from Western Mexico, Berkeley: University of California Press.
[5] Davis B, Stecklov G, and Winters P., 2002. "Domestic and International Migration From Rural Mexico: Disaggregating the Effects of Network Structure and Composition" Population Studies.
[6] de Haan, Arjan and Rogaly, Ben. (2002). Labour Mobility and Rural Societies. Great Britain, Frank Cass Publisher.
[7] Massey, D. S., and M. Aysa-Lastra. (2011). Social Capital and InternationalMigration from Latin America. International Journal of Population Research, Volume 2011, Article ID 834145.
[8] Garip, F. (2012). Discovering Diverse Mechanisms of Migration: The Mexico-U.S. Stream from 1970 to 2000, Department of Sociology, Harvard University.
[9] Bourdieu, Pierre, and Wacquant, Loic J. D. 1992. An Invitation to Reflexive Sociology, Chicago: University of Chicago Press.
[10] Kobayashi, Kiyoshi, Hans Westlund and Hayeong Jeong (Edt). (2012). Social Capital Formation and Reformation for Rural Lives and Disaster Relief, MARG Group. Kyoto-Japan.
[11] Subramanian, S., Kim, D., & Kawachi, I. (2002). Social trust and Selfrated health in US communities: A multilevel analysis. Journal of Urban Health, 79, 21–34.
[12] Mihaylov, N., & Perkins, D.D. (in press). Community Place Attachment and its Role in Social Capital Development in Response to Environmental Disruption. In L. Manzo & P. Devine-Wright (Eds.), Place Attachment: Advances in Theory, Methods and Research. Routledge.
[13] Perkins, D.D., & Long, D.A. (2002). Neighborhood sense of community and social capital: A multi-level analysis. In A. Fisher, C. Sonn, & B. Bishop (Eds.), Psychological sense of community: Research, applications, and implications (pp. 291-318). New York: Plenum.
[14] Coleman, James S. (1988). 'Norms as Social Capital', in Gerard Radnitzky and Peter Bernholz (eds), Economic Imperialism: The Economic Approach Applied Outside the Field of Economic, New York: Paragon House Publisher, 133-55
[15] Palloni, Alberto., Douglas S. Massey, Miguel Ceballos, Kristin Espinosa, and Michael Spittel. (2001). Social Capital and International Migration: A Test Using Information on Family Networks. American Journal of Sociology, Vol. 106, No. 5 (March 2001)
[16] Unger, Donald G. and Abraham Wandersman. (1985). The Importance of Neighbors: The Social, Cognitive, and Affective Components of Neighboring. American Journal of Community Psychology, Vol. 13, No. 2.
[17] Brown, B. B., Perkins, D. D., & Brown, G. (2003). Place attachment in a revitalizing neighborhood: Individual and block levels of analysis. Journal of Environmental Psychology,

23, 259–271.

[18] Perkins, D. G., & Taylor, R. B. (1996). Ecological assessments of community disorder: Their relationship to fear of crime and theoretical implications. American Journal of Community Psychology, 24, 63–107.

[19] Akerlof, G.A. and Kranton, R.E. (2000). Economic and Identity, The Querterly Journal of Economic, 115 (3): 715-753

[20] Akerlof, G.A. and Kranton, R.E. (2000). Identity Economic: How our Identities Shape our Work, Wages, and Well Being, Princeton, NJ. Princeton University Press.

[21] Sani, F. Bowe, M. Herrera, M., Manna, C., Cossa, T., Miao, X., and Zhou , Y. (2007). Perceived Collective Continuity: Seeing Groups as Entities That Move Through Time, European Journal of Social Psychology, 37: 1118-1134.

[22] Jeong, Hayeong., Kakuya Matsuhima, Kiyoshi Kobayashi. (2011). The Role of Identity in Rural Development: Place Attachment and Community Activity. Kyoto University.

IOP Conf. Series: Earth and Environmental Science **70** (2017) 012038 doi:10.1088/1755-1315/70/1/012038

Problem in application carrying capacity approach for land allocation assessment in Indonesian municipal spatial planning: A case of Kutai Kartanegara Regency

I N S Wijaya, B Rahadi, N Lusiana and I Maulidina
Universitas Brawijaya, Malang, Indonesia

wijaya.suluh2@gmail.com

Abstract. Urbanization in many countries, such as Indonesia, is commonly appeared as a dynamic population of developed areas. It is followed with reducing rural uses of land for improving urban land uses such as housing, industry, infrastructure, etc. in response to the growth of population. One may not be sufficiently considered by the urban planners and the decision makers, urbanization also means escalation of natural resources consumption that should be supported by the natural capacity of the area. In this situation, balancing approach as carrying capacity calculation in spatial planning is needed for sustainability. Indonesian Spatial Planning Law 26/2007 has already expressed about the balance approach in the system. Moreover, it strictly regulates the assessment and the permission system in controlling land development, especially for the conversion. However, the reductions over the rural uses of land, especially agriculture, are continuously occurred. Concerning the planning approach, this paper aims to disclose common insufficiency of carrying capacity considerations in Indonesian spatial planning practice. This paper describes common calculation weaknesses in projecting area for the urban development by recalculating the actual gap between supply and demand of agriculture land areas. Here, municipal spatial plan of Kutai Kartanegara Regency is utilized as single sample case to discuss. As the result, the recalculation shows that: 1) there are serious deficit status of agriculture land areas in order to fulfil the demanded agriculture production for the existed population, 2) some calculation of agriculture production may be miss-interpreted because of insufficient explanation toward the productivity of each agriculture commodity.

1. Introduction

In term of the sustainability, dynamic population of urban area escalates forces for expanding agricultural land use to accommodate basic need of the population [1], [2], [3], [4]. Dynamic population means escalation of natural resources consumption that needs to be supported by the area. Meanwhile, for many developing countries, urbanization is still interpreted as a circumstance where rural use of land must be converted into housing, industry, infrastructure, public facilities, etc. It is followed by insufficient consideration toward balance system between the population and the capacity of the developed areas [5], [6]. By the above circumstance, the balance system is a crucial idea to be considered by the urban planners and the decision makers.

In case of Indonesia, rapid urbanization has brought consequences as shrinking agricultural areas in many municipalities. In fact, each year, 6% agriculture areas are converted into non-agricultural areas [7], especially in Java Island. Desolately, this situation is continuously happened by the designed land use of spatial planning. Beside strictly regulates land the assessment and the permission of land utilization, Indonesian Spatial Planning Law 26/2007, conceptually, has already expressed the balancing consideration in the system. It is defined as carrying capacity (CC) [8]. For this concern, Spatial Planning Law 26/2007 is not the only law. There is Environmental Protection and Management 32/2009 (Law 32/2009) that obligates spatial plan assessment toward CC of the area [9].

However, the system implementation becomes practically uncertain in municipal level. It has been a fact that decentralization of spatial planning in Indonesia escalated the problems [10]. It brings skepticism toward the practiced planning process in municipalities. Concerning to the land use plan, many argue that the uncertainty is caused by uncertainty of the system and the implementation [11], [12], [13], [14]. Among them, one interesting argumentation is the improper calculation in allocating land use corresponds to supply and demand balance between the natural capacity and the population [7].

Regarding the scepticism, this paper concerns about the problem in defining land utilization in Municipal Spatial Plan. This paper aim is disclose common insufficient consideration toward spatial CC in Indonesian spatial planning practice, particularly the supporting capacity. CC in Indonesian spatial planning and land use allocation has been discussed by some disciplines especially toward the importance and the administration, such as [1], [15], [16]. For the resulted gap between the practical calculation and the plan is still rarely done. Therefore, this paper takes this issue as the focus of discussion.

Since Law 32/2009 doesn't define specific method of balance land use allocation, this paper utilizes equations of agriculture production supply and demand defined by Ministerial Regulation of Environment 17/2009 (Permen LH 17/2009) for projecting the required areas of agriculture land [17]. As individual sample case, one of Municipal General Spatial Plan (RTRW) is evaluated. Here, Municipal General Spatial Plan of Kutai Kartanegara Regency 2016-2036 (RTRW Kutai Kartanegara) is the best sample case to study. Comparing to other Indonesian municipalities, Kutai Kartanegara is one of richest municipalities by the natural resources with level of poverty is relatively high. Moreover, it is one of Indonesian municipalities that shows large area of land conversion.

2. Literature Review

According to Miller, CC is maximum number of population can be supported by the natural resources and environment in a region, to support well-being without a significantly depleting or degrading it resources [18]. In accordance, ecologically, CC is interpreted as capacity of environment with the natural resources to support the inhabitant without any threat over the sustainability [19]. It is also about balance among communities and the capacity of the environment [20]. By Law 32/2009 CC is divided into supportive capacity (SC) and assimila¬tive carrying capacity (AC) [9]. SC means the capacity of the environment to support the lives of the people and other living organisms. AC is the capacity of the environment to absorb matters, energy and/or other components that come into the environment, ei¬ther by them or through human intervention. Concerning to the CC concept, the objective of spatial planning is to enhance the area´s compat¬ibility, harmony and balance with its sub-system [21]. By Permen LH 17/2009 one approaching indicator for practical supporting capacity may be built based upon supply and demand calculation of agricultural land in response to the dynamic consumption of the population [17].

3. Methods

This paper evaluates area of land for agriculture that is defined by RTRW Kutai Kartanegara 2016-2036. The evaluation is conducted as recalculation of the existed supply and demand of agricultural land area in response to the existed population. Here, the equation of supply and demand defined by Permen LH 17/2009 is utilized. By the calculation, the gap between supply and demand is identified. Here, the gap reflects the amount areas of agriculture land must be provided for the development. Then, it may compare to the existed land area defined by the plan. Principally, the calculation uses basic assumption that rice production reflects the supporting capacity of the planned area. Since the agriculture is not merely rice, the calculation required conversion of all agriculture products into rice product, that are all Agriculture Product stated by Municipal Statistic Bureau in 2015 [22]. In the calculation, money value of each agriculture products is converted into rice money value. This conversion is conducted in order to draw the same value for further converted into amount area of land.

In order to build the discussion, this paper is built into some sections. First is introduction, it describes the current situation that leads to the necessity of research's discussion. Second part gives some previous theoretical discussion that supports the perspective toward land use conversion, carrying capacity, and the basic calculation. The second part also contains with result of calculation and the discussion related to the defined land use plan of RTRW Kutai Kartanegara. Finally, this paper builds full conclusion based upon the discussion. It also points some findings that may give opportunity to be observed further and developed as a specific discussion

4. Result and Discussion

4.1. Overview of Municipal Development

Kutai Kartanegara Municipality is located in Eastern part of Kalimantan (Borneo) Island. It is one municipality under East Kalimantan (Borneo) Province administration. Kutai Kartanegara Municipality has land area of 27.263.10 km2 and water area of 4,097 km2. This municipality is divided into 18 Kecamatan (district). Similar to other Indonesian municipalities, most of population are reside in each district centre, especially district of Tenggarong. Between 2010-2014, the population increase by rate of 4,64% per year. It is categorized mid-high comparing to other municipalities in East Kalimantan Province. Four main dominant sectors by their contribution for local income are mining (83,84%), agriculture, farming, forest, and fisheries (6,34%), building (3,21%), trading, hotels & restaurant (2,86%) (KLHS Kutai 2016). However, the biggest contributor is still on Oil and Gas. By Indonesia decentralization, Kutai Kartanegara faces the boost economy development and becomes one Indonesian municipality with biggest contribution to national economy development.

In 2016, the municipal government revised the RTRW by some new development agendas of new plan of harbour development, growing trend of palm oil plantation [23]. Some local characteristics are influencing the new plan. In social-economy statistic, Kutai Kartanegara faces large number of population growth, jobless population and improper people settlements. Regarding the geological circumstance, turf and swamps dominate the area. It is one constraint in allocating land for development. In environmental issue, this municipality faces problem of water and soil pollution caused by mining and domestic activities [24]. Moreover, current housing development is spread over the municipal areas [23]. It becomes a constraint for urban infrastructure development.

4.2. Demonstrated CC Calculation for the New RTRW

Some basic data are utilized in re-calculating the SC of the planned area. One is population data. Second is land allocation. The third data consists with various agricultural production data that are taken from the Municipal Statistic Bureau in 2015 [22]. This data is useful in determining the area of agriculture. In line with the examined data in the process of RTRW revision, population and land allocation data for this research are taken from municipal basic data by year of 2014.

Commonly, the population are concentrated in centre area of development. For Kutai Kertanegara, the concentration is existed in Tenggarong Disctrict. Table 1 shows the distribution of municipal's population by year of 2014. From the table, the less concentrated area of population is Muara Wis.

Table 1. Population Growth of Kutai Kartanegara Municipality

No	Name of District	2010	2011	2012	2013	2014
1	Samboja	54.515	56.619	58.347	60.107	61.387
2	Muara Jawa	33.923	35.562	36.988	38.455	39.932
3	Sanga-Sanga	17.588	18.276	18.834	19.404	19.965
4	Loa Janan	56.071	57.785	59.087	60.394	61.647
5	Loa Kulu	39.938	41.136	42.040	42.945	43.183
6	Muara Muntai	17.315	17.515	17.570	17.623	17.645

No	Name of District	2010	2011	2012	2013	2014
7	Muara Wis	8.557	8.702	8.772	8.841	8.894
8	Kota Bangun	31.292	31.913	32.286	32.653	32.978
9	Tenggarong	96.209	100.323	103.801	107.363	110.090
10	Sebulu	36.420	37.070	37.433	37.786	38.090
11	Tenggarong Seberang	61.441	63.762	65.656	67.585	69.477
12	Anggana	32.688	34.432	35.979	37.586	39.210
13	Muara Badak	39.834	41.271	42.424	43.595	44.734
14	Marang Kayu	23.394	23.697	23.807	23.913	23.984
15	Muara Kaman	33.909	34.615	35.055	35.489	35.876
16	Kenohan	9.861	9.905	9.950	9.994	10.038
17	Kembang Janggut	23.817	25.634	27.371	29.217	31.145
18	Tabang	9.908	9.998	10.089	10.541	10.274
	Jumlah	626.680	648.215	665.489	683.131	700.439

Source: Kabupaten Kutai Kartanegara Dalam Angka, 2015

Table 2. Land Use Allocation

No	Land Use	Area (Ha)				
		2010	2011	2012	2013	2014
A.	Conservation Area					
1	Conservation forest	212.614	212.614	212.614	212.614	210.097
2	Natural Preservation	20.556	20.556	20.556	104.177	104.177
3	National Forest	50.726	50.726	50.726	50.726	40.974
4	National Park	67.766	61.776	61.776	61.776	61.776
	Total	351.662	345.672	345.672	429.293	412.024
B	Cultivation Area					
1	Forestry:					
	Permanent Production (HP) (incld. mangrove)	787.675	787.675	787.675	787.675	760.799
	Limited Production (HPT)	504.294	504.294	504.294	504.294	542.413
	Total	1.291.969	1.291.969	1.291.969	1.291.969	1.303.212
2	Non-Forestry					
	Agriculture	46.286	38.300	31.182	38.069	32.563
	Plantation	44.747	51.179	54.797	54.832	56.686
	Mining	378.655	378.655	533,74	533,74	533,74
	Settlement	19.591	20.808	21.470	22.105	22.252
	Industry					
	Fisheries	67.776	61.776	61.776	61.776	61.776

Source: Kabupaten Kutai Kartanegara Dalam Angka, 2015

For land allocation, Table 2 shows the existed situation in 2014. It shows land use Cultivation Area of Forestry dominates the allocation. Meanwhile the non-Forestry that refers to urban activities and settlement is the lowest. However, this situation doesn't reflect the real capacity that is potentially developed further. Practically, the situation faces wide area may not cultivated by the geological, geographical, and other physical factors of the land. Here, Table 3 shows the land allocation based upon the geological, geographical, and other physical factors of the land. It is developed based upon the GIS areas calculation upon the registered 2014 existing land use map of Fig.1. The resulted area is quite different to Table 2 because the basic method of calculation is different. From this point forward, this paper uses Table 3 as the actual land condition in order to compare with the result of supply-demand calculation based upon the population and agriculture productions. Particularly, the utilized data of Table 3 are: Plantation, Wet and Dry Agriculture Area.

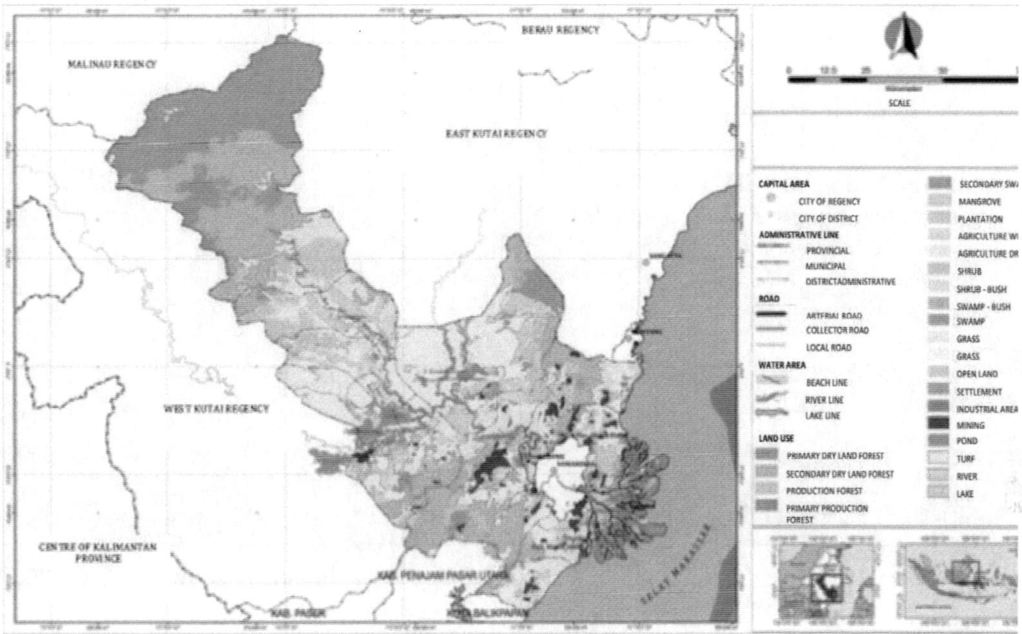

Figure 1. Land use map
Source: RTRW Kutai Kartanegara 2016

Table 3. Land Allocation in 2014

No	Land Allocation (based upon spatial data 2014)	Area (ha)
1	Sailing	139.448
?	Grove	183104.899
3	Mangrove	39568.844
4	Dry Forest	1065887.720
5	Swamp (Forest)	92049.722
6	Production Forest	104663.164
7	Industrial Area	934.449
8	Plantation	181502.022
9	Settlement	19922.664
10	Open Land	25051.184
11	Wet Land Agriculture	42984.6228
12	Dry Land Agriculture	17401.88617
13	Swamp	47847.220
14	Bush	418640.889
15	Grass-land	99459.289

Source: GIS calculation over RTRW 2016 map

By conducting supply and demand approach for agriculture production, the required area for agriculture may be predicted in response to the population growth. Specifically, the calculation is done by the following formula. Then, table 4 shows the result of calculation for each district.

IOP Conf. Series: Earth and Environmental Science **70** (2017) 012038 doi:10.1088/1755-1315/70/1/012038

$$SL = \frac{\sum(P1 \ x \ H1)}{Hb} X \frac{1}{Ptvb} \quad (1)$$

$$Ptvb = \frac{Pb}{Lb}$$

Where:

SL : Agriculture Land Supply (Ha)
P1 : Production of each agricultural commodity
H1 : Money value of each commodity (rupiah)
Hb : Rice money value
Ptvb : Rice productivity
Pb : Rice Production
Lb : Padi field area

Table 4. Calculated Supply Area for Agriculture

No	District	∑P1 x H1	Hb	Pb (Kg)	Lb (Ha)	Ptvb (Kg/Ha)	SL (Ha)
1	Samboja	Rp. 725.212.824.951	Rp. 9.375	14.333.000	2.888	4.963	15.586,70
2	Muara Jawa	Rp. 618.158.556.000	Rp. 9.375	1.636.000	432	3.787	17.411,21
3	Sanga-Sanga	Rp. 19.124.478.376	Rp. 9.375	289.000	74	3.905	522,34
4	Loa Janan	Rp.308.723.042.714	Rp. 9.375	4.332.000	887	4.884	6.742,69
5	Loa Kulu	Rp. 387.101.039.650	Rp. 9.375	33.162.000	6.507	5.096	8.102,02
6	Muara Muntai	Rp. 118.145.312.800	Rp. 9.375	3.064.000	733	4.180	3.014,81
7	Muara Wis	Rp. 19.025.393.389	Rp. 9.375	1.031.000	239	4.314	470,44
8	Kota Bangun	Rp. 282.619.182.742	Rp. 9.375	12.232.000	2.407	5.082	5.932,11
9	Tenggarong	Rp. 225.714.307.545	Rp. 9.375	17.963.000	3.278	5.480	4.393,57
10	Sebulu	Rp.234.491.164.233	Rp. 9.375	15.044.000	3.141	4.790	5.222,28
11	Tenggarong Seberang	Rp. 446.703.892.074	Rp. 9.375	40.009.000	7.601	5.264	9.052,35
12	Anggana	Rp. 123.705.143.252	Rp. 9.375	8.288.000	1.861	4.454	2.962,87
13	Muara Badak	Rp. 110.393.460.104	Rp. 9.375	1.687.000	378	4.463	2.638,45
14	Marang Kayu	Rp. 498.723.990.885	Rp. 9.375	14.937.000	2.949	5.065	10.502,69
15	Muara Kanan	Rp. 277.266.327.374	Rp. 9.375	17.603.000	3.725	4.726	6.258,43

Table 4. Calculated Supply Area for Agriculture

No	District	∑P1 x H1	Hb	Pb (Kg)	Lb (Ha)	Ptvb (Kg/Ha)	SL (Ha)
16	Kenohan	Rp. 60.926.332.604	Rp. 9.375	1.182.000	284	4.162	1.561,47
17	Kembang Janggut	Rp. 87.405.067.320	Rp. 9.375	2.010.000	588	3.418	2.727,39
18	Tabang	Rp. 93.719.918.021	Rp. 9.375	5.700.000	1.707	3.339	2.993,78
Total		Rp. 5.180.100.524.735	Rp. 9.375	194.502.000	39.679	4.902	106.095,60

Using similar assumption, demand on agriculture land reflects the supporting capacity should be provided as the consequences of dynamic population. The utilized equation in defining demand of agriculture land is depicted as the following. Table 5 shows the result of the calculation.

$$\mathrm{DL} = Nx\ KHLL \qquad (2)$$

$$\mathrm{KHLL} = \frac{1\ Ton}{Ptvb}$$

DL : Agriculture Land Demand (Ha)
N : Population
Ptvb : Rice productivity
KHLL : Each person's demanded area for adequate living condition (converted into rice consumption rate per person)

Table 5. Calculated Demand of Agriculture Land

No	District	Population	Ptvb (Kg/Ha)	KHLL (Ha)	DL (Ha)
1	Samboja	61.837	4.963	0,201	12.459,73
2	Muara Jawa	39.932	3.787	0,264	10.544,39
3	Sanga-Sanga	19.965	3.905	0,256	5.112,15
4	Loa Janan	61.647	4.884	0,205	12.622,55
5	Loa Kulu	43.813	5.096	0,196	8.596,92
6	Muara Muntai	17.645	4.180	0,239	4.221,21
7	Muara Wis	8.894	4.314	0,232	2.061,75
8	Kota Bangun	32.978	5.082	0,197	6.489,38
9	Tenggarong	110.900	5.480	0,182	20.237,72
10	Sebulu	38.090	4.790	0,209	7.952,72
11	Tenggarong Seberang	69.477	5.264	0,190	13.199,40
12	Anggana	39.210	4.454	0,225	8.804,27
13	Muara Badak	44.734	4.463	0,224	10.023,39
14	Marang Kayu	23.984	5.065	0,197	4.735,14
15	Muara Kanan	35.876	4.726	0,212	7.591,78
16	Kenohan	10.038	4.162	0,240	2.411,84
17	Kembang Janggut	31.145	3.418	0,293	9.111,07
18	Tabang	10.274	3.339	0,299	3.076,79
Total		**700.439**	**4.902**	**0,204**	**149.252,20**

Further, comparison between the calculations results the surplus or deficit status of the area. In this case, it is articulated as each district's status. Table 6 shows the distribution of status. It shows most of district in Kutai Kartanegara are in deficit status of agriculture land, especially in Sanga Sanga that the demanded agriculture area over than 10 times from the available supply area.

Table 6. Allocation Status After Supply-Demand Comparison

No	District	SL (Ha)	DL (Ha)	Comparison	Status
1	Samboja	15.586,70	12.459,73	SL > DL	Surplus
2	Muara Jawa	17.411,21	10.544,39	SL > DL	Surplus
3	Sanga-Sanga	522,34	5.112,15	SL < DL	Deficit
4	Loa Janan	6.742,69	12.622,55	SL < DL	Deficit
5	Loa Kulu	8.102,02	8.596,92	SL < DL	Deficit
6	Muara Muntai	3.014,81	4.2221,21	SL < DL	Deficit
7	Muara Wis	470,44	2.061.75	SL < DL	Deficit
8	Kota Bangun	5.932,11	6.489,38	SL < DL	Deficit
9	Tenggarong	4.393,57	20.237,72	SL < DL	Deficit
10	Sebulu	5.222,28	7.952,72	SL < DL	Deficit
11	Tenggarong Seberang	9.052,35	13.199,40	SL < DL	Deficit
12	Anggana	2.962,87	8.804,27	SL < DL	Deficit
13	Muara Badak	2.638,45	10.023,39	SL < DL	Deficit
14	Marang Kayu	10.502,69	4.735,14	SL > DL	Surplus
15	Muara Kanan	6.258,43	7.591,78	SL < DL	Deficit
16	Kenohan	1.561,47	2.411,84	SL < DL	Deficit
17	Kembang Janggut	2.727,39	9.111,07	SL < DL	Deficit
18	Tabang	2.993,78	3.076,79	SL < DL	Deficit
Jumlah		**106.095,60**	**149.252,20**	**SL < DL**	Deficit

4.3. Evaluation over the New Plan

Table 4 shows different result of agriculture supply land areas to the data of agriculture land allocation showed by Table 3. This situation is possible occurred by the different utilization basic data. Table 3 is calculated based upon the physical characteristic of land meanwhile the Table 4 is calculated based upon the conversion of agriculture production data into the land area. However, the identified gap may insignificant to be considered further instead of the comparison with the resulted deficit areas.

By comparing total area of agriculture land of Table 3 to the resulted deficit area of Table 6, there are wide gap of land supply indicated. Table 3 represents the current area of agriculture land and the other maximum potential area may be converted to agriculture land. Wide gap between these two tables indicates serious problem of land allocation defined by new spatial plan. In this sense, the land use plan is designed under high insufficient SC of the planned area. In other words, the new spatial plan has already suffered overcapacity in responds to the dynamic population.

Concerning the planning process under law 26/2007, common practiced CC consideration is conducted merely by projecting the population growth in order to be accommodated into housing land allocation. In this case, the housing land use is allocated based upon land classification analysis. This analysis doesn't have any calculation toward the capacity of the land in providing the natural resources for the population. Theoretically, it determines the land capacity based upon the geographical and geological characteristic. Unfortunately, this analysis doesn't have restriction in order to conserve the potential land for agriculture. Then, the housing (urbanization) land uses are continuously provided without further consideration toward the CC.

IOP Conf. Series: Earth and Environmental Science **70** (2017) 012038 doi:10.1088/1755-1315/70/1/012038

5. Conclusion

Based upon the discussion, some concluding remarks are identified that spatial planning process of RTRW Kutai Kartanegara is suffering overcapacity of the natural resources. The unpleasant new plan situation is caused by insufficient consideration about CC, especially the supporting capacity. Particularly, the above discussion leads to some important points that are:

1. Insufficient consideration results uncertain land use allocation corresponds to the agricultural production capability of the area in supporting the dynamic population.
2. Insufficient consideration is indicated by unavailable certain data of agriculture development that are integrated to the spatial analysis for land use allocation.
3. Gap between the calculated land supply for agriculture and the existed data used in practical planning process indicates less compatibility of agriculture land supply-demand approach to the spatial analysis for land use allocation.
4. The incompatibility may caused by money valuation and unidentified harvesting frequency of each agriculture product in order to be converted into area of rice production land. Further, these factors may be clarified and the equation may be improved

References

[1] Stavins, R.N., Wagner, A.F., Wagner, G. Interpreting Sustainability in Economic Terms: Dynamic Efficiency plus Intergenerational Equity. Econ. Lett. 2003. 79, 339–343.

[2] Balmford, A., Green, R.E., Scharlemann, J.P.W. Sparing land for nature: Exploring the Potential Impact of Changes in Agricultural Yield on the Area Needed for Crop Production. Glob. Chang. Biol. 2005. 11, 1594–1605.

[3] Pirard, R., Belna, K. Agriculture and deforestation: Is REDD+ Rooted in Evidence? For. Policy Econ. 2012. 21, 62–70

[4] Laurance, W.F., Sayer, J., Cassman, K.G. Agricultural Expansion and its Impacts on Tropical Nature. Trends Ecol. 2014. Evol. 29, 107–116

[5] Rahadi, B., Lusiana N., Nurlaelih E. E. The Assessment of Carrying Capacity As Endorsing Human Activities in Batu, Indonesia. American Journal of Environmental Sciences. 2014

[6] Wei Y., Huang C., Li J., Xie L. An Evaluation Model for Urban Carrying Capacity: A Case Study of China Mega-Cities. Habitat International. 2016. 53, 87-96

[7] Permatasari P, A., Fatikhunnada A., Liyantono, Setiawan Y., Syartinilia, Nurdiana A. Procedia Environmental Sciences 2016. 33, 27 – 35

[8] Government of Indonesia. Indonesian Law of Spatial Planning 26/2007. 2007.

[9] Government of Indonesia. Indonesian Law of Environmental Protection and Management 32/2009 (Law 32/2009). 2009

[10] Burgess, R., Hansen, M., Olken, B., Potapov, P., Sieber, S., 2011. The Political Economy of Deforestation in the Tropics. NBER Working Paper Series taken from: http://dx.doi.org/10.1017/ CBO9781107415324.004.

[11] Brockhaus, M., Obidzinski, K., Dermawan, A., Laumonier, Y., Luttrell, C. An Overview of Forest and Land Allocation Policies in Indonesia: Is the Current Framework Sufficient to Meet the Needs of REDD+? For. Policy Econ. 2012. 18, 30–37

[12] Sahide, M.A.K., Giessen, L. The Fragmented Land Use Administration in Indonesia—Analysing Bureaucratic Responsibilities Influencing Tropical Rainforest Transformation Systems. Land Use Policy 2015. 43, 96–110

[13] Srinivas, S., Bell, K.C., Toha, K., Zaenal, A., Collier, W., 2015. A Review of Indonesian Land-based Sectors with Particular Reference to Land Governance and Political Economy. Washington DC. Retrieved from http://www.oicrf.org/document. asp?ID=14940

[14] Toha, K., Collier, W.L. Land Control, Governance and Agrarian Conflict in Indonesia. 2015.

[15] Astuti L. T., Koestoer R. H. Green Well-being Through the Cities Carrying Capacity: A Special Reference to Bekasi, West Java, Indonesia. Procedia - Social and Behavioral Sciences 2016. 227, 738 – 746

IOP Conf. Series: Earth and Environmental Science **70** (2017) 012038 doi:10.1088/1755-1315/70/1/012038

[16] Santoso E, B., Martha Erli K. D. H., Aulia B. U., Ghozali A. Concept of Carrying Capacity: Challenges in Spatial Planning (case Study of East Java Province, Indonesia). Procedia - Social and Behavioral Sciences. 2014. 135,130 – 135
[17] Ministry of Environment of Republic of Indonesia. Ministerial Regulation of Environment of Guidelines of Carrying Capacity for Spatial Planning 17/2009. 2009
[18] Miller, Jr., G. Living in The Environment, Principles, Connection, and Solution. Canada: Brooks/Cole. Thomson Learning. 2004.
[19] Kormondy. Concept of Ecology. Prentice Hall. US. 1996
[20] Ress, E. William. Ecological Footprint and Appropriated Carrying Capacity: What Urban Economics Leaves Out. Environment and Urbanization 1992. 4 (2)
[21] Henning Shroll, et al. Carrying Capacity : An Approach to Local Spatial Planning in Indonesia. The journal of Transdisciplinary Environmental Studies. 2012. Vol. 11 No.1 pp 27-39
[22] Municipal Statistic Bureau. Kutai Kartanegara Regency in Numbers (Kabupaten Dalam Angka). 2015
[23] Kutai Kartanegara Municipal Development Planning Board. Municipal General Spatial Plan of Kutai Kartanegara Regency 2016-2036. 2016
[24] Kutai Kartanegara Department of Environment. Strategically Environmental Assessment: Municipal Spatial Plan 2016/2036 Assessment. 2016

IOP Conf. Series: Earth and Environmental Science **70** (2017) 012039 doi:10.1088/1755-1315/70/1/012039

Implications of Pearl, Gold, Silver (PGS) Craft Industrial Cluster towards Settlements Regionin Karang Pule Village, Sekarbela District of Mataram City

I R Sushanti[1], and I S Fitri

Department of Urban and Regional Planning, Universitas Muhammadiyah Mataram, Indonesia

Imarahmawati77@gmail.com

Abstract. The existence of industry clusters in Mataram City gave effect to the surrounding residential areas [1]. In accordance Spatial Plan of Mataram City in 2011-2031 PGS industry cluster in the village of Karang Pule, Sekarbela district established as shopping tourism area. Distribution of industrial locations were in four of seven environments in Karang Pule. Distribution of PGS industry is divided into three (3) groups: craftsmen, craftsmen who is also a seller, and the seller (merchant). The location of the craftsmen, craftsmen who are also businessmen and entrepreneurs are also used as a dwelling house or workshop and store. So most of the people living in settlements around clusters of industry that there is a link between industry cluster and settlements. This study aims to determine the implications or the impact of the presence of PGS industry clusters the surrounding residential areas. The method used in this research is descriptive qualitative with the collection of primary data through direct observation and questionnaires. Based on direct observation on the shopping tourism area there are inequality between the conditions of industry clusters and settlements area by the presence of slums. The results showed that the PGS industry cluster impact on social, economic and environment near settlements area. Impacts that occur are: 1) the social aspect, there is a significant change in the level of education, social welfare and social disparities but less significant to the formation of institutions, particularly in community participation , 2) the economic aspect of the change status of community work, and increased revenue and 3) the environmental aspects of the change to the condition of the building, the quality of public housing, the presence of slums, changes in infrastructure and the environmental pollution.

1. Introduction

Mataram city is a city that has a lot of potentions, one of those is the potential industry. The existence of this industry has a very important role in economic development and growth of a region because it can increase revenue and provide jobs for the local community. The existence of the industry itself can provide a wide range of effects both positive and negative impacts.

Karang Pule Village, Sekarbela district of Mataram City established as a small industry cluster featured to craft Pearl, Gold and Silver (PGS). [2] Sekarbela district included in the development of settlements area while Karang Pule village established as shopping and tourism area of small business development.

Beside those potentions, the presence of issues such as environmental, economic and social. Settlements region can be used as industry activities lead to the presence of slums. The slums can cause the inequality between environmental, economic and social aspects that exist in the PGS industry cluster of Karang Pule village. This research was conducted in order to determine the implications of PGS industrial clusters on the settlements region.

According to Law Number 1 of 2011 on Housing and Settlement region, referred to the settlements which consisted of more than a unit of housing that has the infrastructure, facilities, public utilities, as well as having to support activities other functions in urban or rural areas [3].

According to the Law of Industry Number 3 of 2014, the industry is all forms of economic activity that processes raw materials and or take advantage of industry resources to produce goods that have additional value or higher benefits, including industry services [4]. From the geographical perspective, the industry as a system, sub-system is a mix of physical and human sub-system [5]. Ministry of Industry of Indonesia's national industry grouping into 3 major groups: 1) Basic Industry, 2) Miscellaneous Industry, and 3) Small Industries.

Agglomeration by modern location theory is one of the factors that influence economic activities, agglomeration also be one of the factors in addition to compare advantage and economies of scale that explains why the industry raised areas and cities [6]. Some things that can cause agglomeration [7] are: 1) labor provided many of which have the ability and expertise better than the outside area, 2) a company becomes an attraction for other companies, 3) development of a companies from small to large, causing other companies to support the enlarged company, and 4) transfer of production activity from one place to some other place. Other companies approach the source material for the production activity generated by existing companies to support each other. So agglomeration is the concentration of economic activity spatially arise because of the savings gained due to nearby locations.

The effect by each industrial area varies according to the characteristics and potential sector at different certain sectors. Thus, if an industry or industry area built in a location and then the industry is growing then it would be predicted that other economic units tend to pick a location connected directly which is ultimately going to give mutual benefits.

The existence of PGS industry clusters in the village of Karang Pule have an impact on social, economic and environmental. Social impact, with the presence of PGS industry clusters raises the number of people that affect the building density and increased economic activity. It is also impact on the environment, such as pollution, and the formation of slums around the industry area due to industry conditions that are in settlements [1].

2. Research Methods

The location of this research is located in the village of Karang Pule, Sekarbela district of Mataram. The approach used in this study is derived from the rationalistic approach theory and empirical truth. Kind research used a qualitative descriptive that systematically describe the situation, the situation, the specific program [8]. Data collection methods used were primary data collection in the form of direct observation and questionnaires.

3. Results and Discussion

3.1. Region overview

Karang Pule village has an area of 106.7 square kilometres. Karang Pule village have 544 households with a population of 10423 people divided into seven environments. Karang Pule village is divided into seven environments, they are: 1) Karang Pule, 2) Pande Besi, 3) West Pande Mas, 4) East Pande Mas, 5) Karang Seme, 6) Mas Mutiara and 7) Kekalik BTN.

3.2. Distribution
PGS for the distribution industry Karang Pule Village is divided into three groups: craftsmen, craftsmen who is also the seller, the seller (merchant). There are four environments is a core area of industry PGS Karang Pule village is in the Environment of Pande Besi, Pande Mas West, East Pande Mas, and Mas Mutiara. For the spread of the location of PGS the industry cluster in the village of Karang Pule, can be seen in the table 1.

Table 1. Number of Handicraft Industry Location PGS divided Per Type and Environment

No	Environment	Craftsmen	Craftsmen and Sellers	Seller
1	Pande Besi	1	3	4
2	West Pande Mas	13	0	9
3	East Pande Mas	62	1	14
4	Mas Mutiara	0	4	7
	Jumlah	**76**	**8**	**34**

Source: Observations and Interpretations Map

For the distribution of craftsmen more dominant location in the East Pande Mas environment. The location of the craftsmen is housing that is also used as a workshop while the spread of craftsmen who is also the seller is housing that is also used as a workshop or shop. Location spread of sellers is largely a shop and a small portion of which is used as homes and shops. For the spread of sellers (traders) were in four environments, which are scattered along major roads namely Sultan Kaharudin street which is indeed the most strategic location to sell products.

3.3. Relationship Between Industrial Zone PGS with Settlement Region
Cluster) is the geographic concentration of companies and institutions that are interconnected in a particular area [9]. In this case of the PGS industry cluster is not only related to the industry sector alone but also to the surrounding residential areas associated with the location of industry craftsmen PGS who also serve as homes and workshops and stores. It shows that the agglomeration occurs between industry clusters with surrounding residential areas. Thus indirectly, the industry activity in the region of course affect the surrounding residential areas, both in terms of social, economic, and environmental.
Judging from the distribution of industry location map can be seen that the location PGS craft industry are located in settlements areas. This can be seen in the following figure.

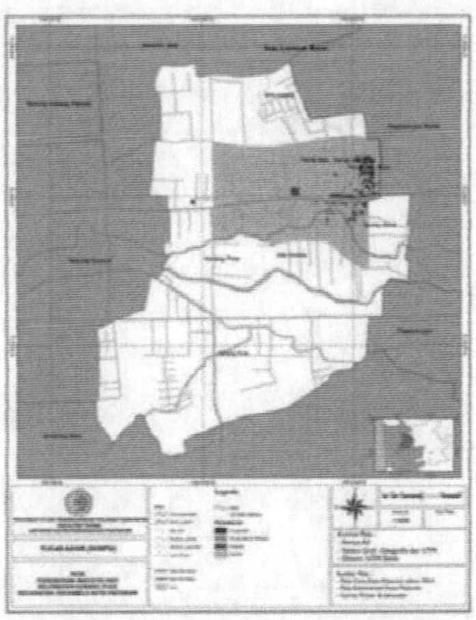

Figure 1. Distribution Map

From mapping the distribution of the industry, it is obtained a regional division based on its potential in the development of PGS industry cluster area, here is a map of the distribution of potential area for the PGS industry cluster.

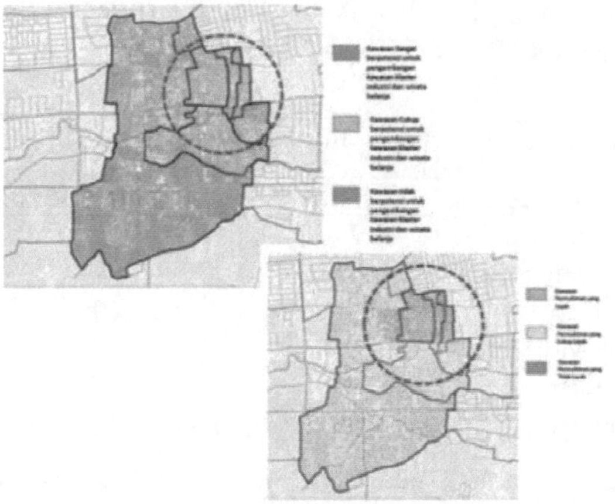

Figure 2. Linkage Clustre Industry.

Based on the distribution map of potential area PGS industry can be seen that the most potensial area in the development of industry cluster is settlements that is unfeasible. This suggests that industry activity PGS's in the Village Karang Pule affect the surrounding residential areas. The slum area located at Pande Besi, West Pande Mas and East Pande Mas environment with an area of 13.50 ha, or approximately 10.23% from the total area of Karang Pule village. As for the quite feasible settlement contained in Mas Mutiara environment with an area of 18.65 ha, or approximately 14.12% and had a feasible settlement are outside the region, namely in Kekalik BTN and Karang Pule environment of 95.20 Ha or about 72.10%. For Karang Seme environment which are outside the PGS industry cluster included in the category of settlements region is quite feasible with an area of 4.69 ha, or approximately 3.55%. For more details can be seen in the following table 2.

Table 2. Feasibility of Settlement Region On PGS Industry Cluster and PGS Non Industrial Cluster

No	Environment	Land Area (Ha)	Persentase (%)	Eligibility Settlement Region	Cluster type
1	Karang Pule	92,40	69,98	Feasible	Non cluster
2	Kekalik BTN	2,80	2,12	Feasible	Non cluster
3	Mas Mutiara	18,65	14,12	Quite Feasible	PGS Cluster
4	Karang Seme	4,69	3,55	Quite Feasible	Non Cluster
5	Pande Mas Timur	2,80	2,12	Unfeasible	PGS Cluster
6	Pande Mas Barat	2,87	2,18	Unfeasible	PGS Cluster
7	Pande Besi	7,83	5,93	Unfeasible	PGS Cluster
Total		**132,04**	**100,00**		

3.4. Implications of PGS Industrial Cluster

The implication is a change caused by an activity, a business investment in development activities have the potential ability to create an impact. So the implications are the PGS industry cluster effect of the emergence of human activity in the construction on the environment, including human PGS industry group.

3.5. Social Implications
The impact of the presence of PGS industry clusters towards the surrounding settlements in the social aspects include: level of education, public welfare, institutional and social inequality. The social implications of the presence of PGS industry clusters can be seen in Figure 3 below:

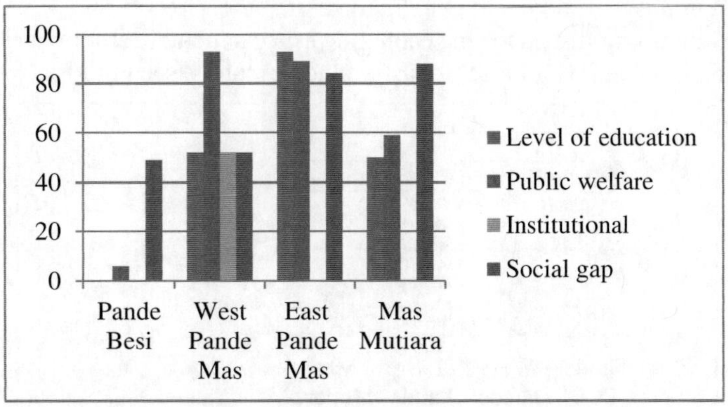

Figure 3. Social Aspects Diagram

The social implications of the presence of PGS industry clusters in the village of Karang Pule are :

1. There were changes in levels of public education in four environments and the highest in Mas Mutiara environment by 90%.
2. There were changes in the welfare of society in four environments and the highest in the neighborhood West Pande Mas by 90%.
3. The establishment of the institutional environment West Pande Mas at 50%, is still not evenly distributed in the other surroundings.
4. There was social inequality in four environments, and the biggest is in Mutiara Mas environment up to 90%.

So the social implications of PGS industry clusters to settlements region is the presence of a significant change in the level of education, social welfare and social disparities, but less significant to the formation of institutions, particularly in community participation and or community.

3.6. Social Expectations
Based on social implications, people who lived in those environments have some expectations to build that region more better than before. People's expectations are:

1. With the change in the level of education in 4 (four) PGS industry cluster environment, especially in environment Mas Mutiara with percentage of 90%, it shows that the presence of industrial clusters PGS affects levels of public education so that the presence of so great influence society also wants their non-formal education in the form of training or courses related to the manufacture and craft PGS's creativity.
2. In terms of public welfare to the existence of the PGS industry cluster in 4 (four) the environment, especially in the environment Pande Mas west with a percentage of 90% which indicates that the presence of industrial clusters PGS affects the level of welfare so that with these changes the public wants PGS's handicraft industry can be an attempt hereditary and make this industry become more organized and orderly.
3. With a percentage of 50% of the institutional, it shows that the presence of PGS industrial clusters does not affect the existing institutional development while craftsmen PGS industry itself wants the institution related to shade and protect the PGS's handicraft business.

4. Social inequality that occurred in 4 (four) industrial cluster environment PGS with a percentage of 90% indicates that the presence of industrial clusters PGS affect the level of social inequalities between communities. This can occur because people just hire their own family so that people outside do not have access to have a job. Though the presence of this industrial cluster in the community look forward to working with both among the people in the surrounding areas so that there is social inequality.

3.7. Economic Implications

The existence of PGS industry clusters for the economic aspects include: employment status and income growth. The economic implications of PGS industry clusters can be seen in Figure 4 below:

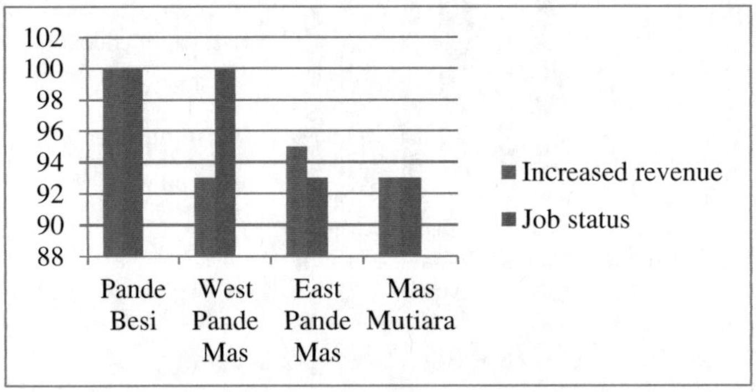

Figure 4. Economic Aspects Diagram

1. There were changes in the people's income fromfour environments and the highest in Environmental West Pande Mas and Pande Besi amounted to 100%.
2. There were a change in the status of community work fromfour environments and the highest in Environmental Pande Besi and East Pande Mas of 100%.

So the economical implications of the PGS industry cluster settlement area is a change in the status of community work, and significant increases in revenue in four the environment, especially in the Environment Pande Besi and West Pande Mas.

3.8. Economic Expectations

Based on economic implications, people who lived in those environments have some expectations to build that region more better than before. People's expectations are:

1. A change in people's income with a percentage of 100% indicates that the presence of PGS industrial clusters obviously affects the income of the people in the surrounding areas and communities want this business can continue to grow and energized.
2. A change in the status of people with a percentage of 100%, which indicates that the presence of PGS industrial clusters can create jobs for people who do not have jobs.

3.9. Environmental Implications

The existence of the industry on the one hand the changes that have an impact on the socio-economic, but on the other hand also brought changes that have a negative impact, the negative impacts include pollution of the environment surrounding the industry. Environmental aspects with the PGS industry cluster include: the condition of the building, the quality of buildings, slums, infrastructure and environmental pollution. Environmental implications of the presence of industry clusters PGS can be seen in Figure 5 below:

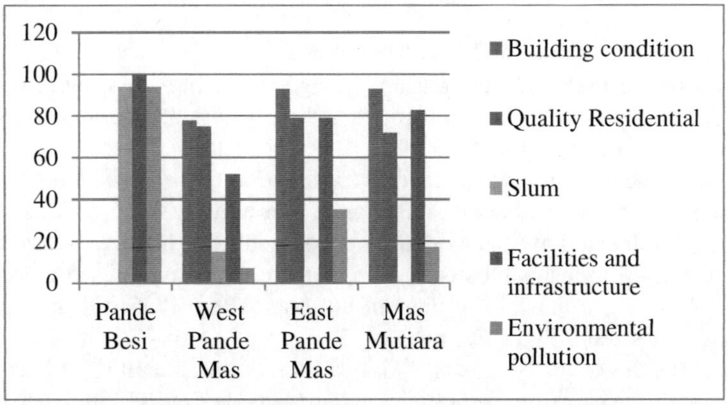

Figure 5. Environmental Aspects Diagram

1. The existence of the PGS industry affect the condition of the building in four environments and the highest is in the East Pande Mas environment about 85%.
2. The existence of the PGS industry affect the quality of residential communities in four environment and the highest is in the East Pande Mas environment amounted about 75%.
3. PGS industry gave rise to slums in four environments and the highest is in the Pande Besi environment amounted about 80%.
4. The presence of PGS industry cause changes in the number of facilities and infrastructure in four environments and the highest is in the Pande Besi environment amounted about 85% .
5. PGS industries cause environmental pollution that occurred in four environments, and the highest is in the Pande Besi environment about 90%.

Thus, the environmental implications of PGS industry clusters towards settlements region to the environmental aspect is the change of the condition of the building, the quality of public housing, the presence of slums, infrastructures changes and the environmental pollution.

3.10. Environment Expectations

Based on environment implications, people who lived in those environments have some expectations to build that region more better than before. People's expectations are:

1. The presence of PGS industrial clusters affect the quality of the building with a percentage of 85% so that the necessary arrangement of the layout of the building so as to create an environment with the condition of the buildings that deserve to be occupied.
2. The existence of the PGS industry that affects the quality of residential with percentage of 75% especially in the neighborhood East Pande Mas show that the conditions that exist today must do system management of garbage and waste disposal so as to reduce the impact on the quality of residential communities that surround PGS industry clusters.
3. The presence of PGS industrial clusters led to the emergence of slums with a percentage of 80%, which means the current circumstances that influence the MEP industry cluster in the surrounding area that causes slums so that the necessary arrangements for management and good systems in various aspects in order to minimize changes that occur as a result of this PGS industrial cluster.
4. The existence of the existence of industrial clusters led to the addition amount of infrastructure by 85%, which means that this situation shows the people want their supporting facilities are inadequate in supporting PGS craft business.
5. The existence of the PGS industry presence led to the emergence of environmental pollution is very high, reaching 90%, so the conditions are thus expected their processing systems and good management, and friendly environment will not cause pollution to the surrounding area.

IOP Conf. Series: Earth and Environmental Science **70** (2017) 012039 doi:10.1088/1755-1315/70/1/012039

4. Conclusion

From the research that has been done can be concluded as follows :

1. The existence of the PGS industry cluster linkages to settlements regions as the presence of slums thus indirectly impact or implications for the joints of people's lives both in terms of social, economic and environmental.
2. There were settlements region is unfeasible or slums on three environment located in the PGS industry clusters, namely Pande Besi, East Pande Mas and West Pande Mas while quite feasible settlements region located at Mas Mutiara environment. For Karang Seme environment which are outside the PGS industry cluster included in the category of settlements region is quite feasible too. Also had a feasible settlement are outside the PGS industrial cluster, namely in Kekalik BTN and Karang Pule environment.
3. The presence of PGS industry clusters implications on the feasibility of settlements region in the presence with the slums while settlements are outside the PGS industry cluster included in the category of feasible and quite feasible.
4. The social implications of PGS industry clusters to settlements region is the presence of significant changes in the level of education, social welfare and social disparities, but less significant to the formation of institutions, particularly in community participation.
5. The social expectations of people towards social implications that they want non-formal education in the form of training or courses related to the manufacture and craft MEP's creativity also PGS's handicraft industry can be an attempt hereditary and make this industry become more organized and orderly.
6. The economic implications of PGS industry clusters towards the settlements region of the economic aspect is the change in employment status of the community and significant increases in revenue from four the environment, especially in the Pande Besi and West Pande Mas.
7. The economic expectations of people towards economic implications that they want this business can continue to grow and energized also can create jobs for people who do not have jobs.
8. The environmental implications of PGS industry clusters on the settlements region are the changes of the building conditions, the quality of public housing, the presence of slums, infrastructure changes and the environmental pollution.
9. The environmental expectations of people towards environmental implications that they want that the necessary arrangement of the layout of the building so as to create an environment with the condition of the buildings that deserve to be occupied, must do system management of garbage and waste disposal so as to reduce the impact on the quality of residential communities that surround PGS industry clusters, arrangements for management and good systems in various aspects in order to minimize changes that occur as a result of this PGS industrial cluster, people want their supporting facilities are inadequate in supporting PGS craft business also create processing systems and good management, and friendly environment will not cause pollution to the surrounding area.

References

[1] Sushanti, Ima Rahmawati. 2015. Impact of Industry of the Housing and Settlement Region in Mataram. Journal of Social Economics Humanities. University of Mataram, Vol.1 No.1.
[2] Regional Regulation No. 12 of 2011 on Spatial Planning of Mataram City in 2011-2031.
[3] Anonymous, Law - Law No. 1 of 2011 on Housing and Settlement Region.
[4] Anonymous, Law - Law No. 5 of 1984 regarding Industry
[5] Sumaatmadja, Nursid. 1981. An Approach Study of Geography and Spatial Analysis. Bandung: Alumni.
[6] Soepono, Prasetyo. 1999. "Theory Area: Representations of Micro Basis for Regional Development Theory", Journal of Economics and Business, Vol. 14, No. 4: FE UGM, Yogyakarta.

[7] Adriand, Indra Jaya. 2008. Review of Literature Theory and Pattern Space Area.
[8] Singarimbun, et al. 2008. Survey Research Methods. Reader LP3ES Indonesia, Jakarta.
[9] Porter, Michael E. 1998. Clusters and the New Economics of Competition. Harvard Business Review, November-December.

IOP Conf. Series: Earth and Environmental Science **70** (2017) 012040 doi:10.1088/1755-1315/70/1/012040

Disaster mitigation at drainage basin of Kuranji Padang City

L Utama, M Yamin

Civil Enginering Department, Faculty of Civil Engineering and Planning, University of Bung Hatta Padang, West Sumatera, Indonesia

lusi_utamaindo115@yahoo.co.id

Abstract. Floods is flooding of effect of exit water groove river because big river debit sudden its accomodation energy, happened swiftly knock over areas which is debasement, in river basin and hollow. Flow debris or which is recognized with galodo have knock over river of Kuranji year 2012 in Padang city. Area is floods disaster are: 19 Sub-District in 7 district, and hard that is district of Pauh and district of Nanggalo. Governmental claim tired loss of Rp 263,9 Billion while Government of Provinsi West Sumatera appraise loss estimated by Fourty Billion Rupiah (Padang Ekspress 28 July 2012), with detail of damage house counted 878 unit, damage religious service house 15 unit, damage irrigation 12 unit, damage bridge 6 unit, damage school 2 unit, damage health post 1 unit. Result of calculation, by using rainfall of year 2003 until year 2015 with method Gumbel, Hasper and Wedwen, got high rainfall plan is 310,00 mm, and method Melchior and Hasper floods is 1125,86 m³ / second. From result of study analyse at Citra map of correlation and image to parameters cause of floods, and use software Watershed Modelling System (WMS) this region have two class that is middle susceptance and low susceptance. Middle susceptance area is there are in middle river and downstream river, with inclination level off. Low susceptance area there is middle river. Area which have potency result the happening of floods is headwaters, because having keen ramp storey level (45 - 55%) and is hilly. For the mitigasi of floods disaster determined by three area evacuate that are: Sub-District Of Kelurahan Limau Manis District Of Pauh, Sub-District Of Surau Gadang District Of Nanggalo, and Sub-District Of Lambung Bukik District of Pauh, in the form of map.

1. Introduction

Forest Flood due to water runoff that flow out of the river because the river discharge was suddenly enlarged beyond capacity, it happens quickly struck areas of humility and in the valley of the rivers. With the increase of population that is increasingly rapidly brings impact to the increased needs of land and demand fulfilment services and urban infrastructure that can impact declining environmental quality such as environmental degradation and natural disasters. Floods of Bandang or which recognized with galodo have knock over Bar of Kuranji, according to Zahrul Umar on Tuesday the 24 July 2012 and Wednesday of[is 12 September 2012 at 16.30 WIB by leaving over sorrowful and uliginous circumstantial distress which very which mixed mud have house and his equipments. Government claim loss of floods effect of bandang Rp 263,9 Billion while Government of Provinsi West Sumatra appraise loss whereas estimated Fourty Billion Rupiah (Newspaper Padang Ekspress, 28 July 2012), with detail of damage house 878 unit, damage religious service house 15 unit, damage irrigation 12 unit, damage bridge 6 unit, damage school 2 unit, damage health post 1 unit. In general there are some factor causing the happening of floods. According to Sutopo and Van of Zuldam the factors is the condition of nature (regional geographical position, condition of toporafi, river geometry and sedimentation), natural event (rainfall and the duration rain, tide, current return from especial river, river stream barrage effect of sliding, and sedimentation), and human being activity (floods plain area conducting, allotment of planology inappropriate floods plain, there is no management pattern him and development of floods plain, setlement river, drainage system which is not adequate, the limited action of mitigasi floods, lack of awareness of society alongside groove river, depilatory of forest headwaters, the limited conservancy effort. From field evaluation, damage of DAS river pate, upstream of Kuranji, with the happening of change of farm function, collapse natural barrage him and aqueduct. Though headwaters is functioning conservation area arrange river stream to middle and river downstream. From some research which have been by former researcher Nur Yuwono, Onrizal and of Raharjo, for the determination of floods susceptance area, parameter the happening of floods is : lack of drainage (diffusion irrigate based on ground type), level of inclination of bevel, changing of it farm function, rain intensity height, height settlement nearness and place woke up from river. By using map of image and correlation with parameter cause of floods and also use program of Argis X.1, will be made mitigasi in the form of band evacuate to society alongside Bar of Kuranji.

2. Literature Review

2.1 Strengthening Regulatory Legislation
In April 2007, two laws have been passed by the legislature, namely: ACT on disaster relief (Act No. 2/2007) and the ACT on Spatial (Act No. 2/2007) which is a revision of the previous ACT No. 24/1992. Laws and regulations are made given the vacuum of legal framework and laws and regulations pertaining to the handling of disaster risk must be paid dearly with the fall of casualties and enormous material losses

2.2 Public participation
Public participation is a process provided an opportunity for technical and broader authority to the community, so that the community is able to solve problems together. The Division of authority was conducted based on the level of community participation in such activities. Community participation (local wisdom) aims to find solutions to problems in a community, by opening up more opportunities for the public to contribute so that the implementation of activities running more effective, efficient, and sustainable.

2.3 Flood mitigation Stakeholders
Stakeholder response to flooding is generally classified into three, namely: (a) a society gets the impact directly or indirectly due to flooding; (b) community groups or individuals who may give consideration or facilitation in flood mitigation, such as: consultants, experts, NGOs, and professionals in the field of flood; (c) decision/policy makers/decision makers, institutions/institutional authorities make decisions and legal basis, such as government agencies and water resources Council.

2.4 Solving
Solving the flood with the approach of non flood disaster mitigation in non phisik/non structural can be: spatial arrangement (land use arrangement) due to the occurrence of conversion of land into urban areas, intensive and continuous outreach involving community, NGOs. Making flood plain zoning, zoning-zoning arrangements, namely the utilization of space is differentiated according to the threat level because of the flood. So the flood plain zoning is effective instruments to avoid the occurrence in areas not yet awakened to the brakes on the occurrence of a change of function of the land according to law No. 25/2007 on Spatial and Act No. 7/2004 about Water Resources

2.5 Soil type
Alluvial soil is formed from the mud of the river that settles in the lowlands, it has fertile soil and is suitable for agricultural land. The soil is alluvial young who came from the results of the deposition.

Alluvial soil at Kuranji has widely 5,445.90 ha b. an area of 12,921.16 ha. Because suitable as agricultural land, then this soil type will cause the occurrence of floods.

2.6 Environmental change

With the ever increasing human population has led to increasingly sensitive environmental conditions. Change the function of the land very large influence on environmental changes that result in a decline in land cover. The longer the amount of vegetation on the wane, particularly in urban areas. From the results of the inverstigasi BPN Padang city, that of the years 2009 – 2012 land use changes such as table in the following:

Table 1. Land use change

No	Land use	Area Land (Ha)		Land Use Change (Ha)
		2009	2012	
1	Residential	176,84	327,55	150,71
2	Area Rain Waters	399,55	289,74	109,81
3	Gardens Mixed	972,06	1005,29	33,23
4	Moor Forest	2,50	3,65	1,15
5	Shrubs	75,52	58,27	17,25
6	Similar	7,62	6,87	0,75
7	The Others	116,78	59,5	57,28
	Amount	1.750,87	1.750,87	378,18

Are land cover vegetation that grows above the Earth's surface will cause the increasing surface flow (run off). Surface flow occurs when rainfall has exceeded infiltration rate of the soil. According to research in Onrizal (2005) DAS Ciwulan, deforestation led to the increase in surface flow of 624 mm/th. It's the new calculation performed on the area of forest is cut down where there are still lucrative land who could infiltrasi. Research of Lusi Utama, 2014, with the use of a map Image and the correlation with the parameter causes flooding, gained the flash flooding potential at Kuranji WATERSHED. Based on the results of the analysis of the parameters of the flood areas there are Kuranji Rods obtained two (2) classes of vulnerabilities and vulnerabilities middle are vulnerabilities that is low. There is a flood-prone area in the middle and downstream areas, with the slope of the slopes. Are areas that could potentially lead to the occurrence of floods is upstream, because it had a sharp level and hilly. The vast pool of skoring value based and the level of insecurity area turns Kuranji flood is an area in the vulnerabilities middle. Of the 15 villages there are three wards that low levels of vulnerabilitieswhich means it is safe from the dangers of flooding, Lambung Bukik, Surau Gadang and Limau Manis, but easily the occurrence of landslides, since the slope a sharp slope. 12 the village is an area of vulnerability middle, that means the sloped area is relatively flat, sharp, and altitude as a function of the land.

3. Methods

Using a map image and map study of vulnerability to flooding along the Kuranji, program ARGIS x. 1 and Watershed Modelling System (WMS) , is expected to come by the evacuation line and place the temporary shelter. A temporary shelter was planned in areas closer to the location of the floods, but have height, so that people can be evacuated. Besides conducting cooperation between communities and Government in reducing the risk of flooding. These activities are based on the parameters of the cause of the flooding is like making a well infiltrasi, normalization of river rock material, giving the building permission by the rules. In addition need to flood control planning buildings such as making buildings and building control sediments. Besides made containing booklet pocket and band evacuate and also orders how society make provision against floods According to Sosrodarsono count dimension of river used formula of Manning, that is :

$$Q = V \, . \, A \tag{1}$$

$$V = velocity \ and \ A = area \ river \tag{2}$$

According to Van of Te Chow, factors influencing coefficient value of Manning, are: crudity of surface, channel Trase and irregularity of channel. With the theory Gumbel, Hasper and Wedwen can calculated rainfall and flood debit used the Hasper and Melchior theory.

4. Results

4.1 Map of Routes for Evacuation and shelters

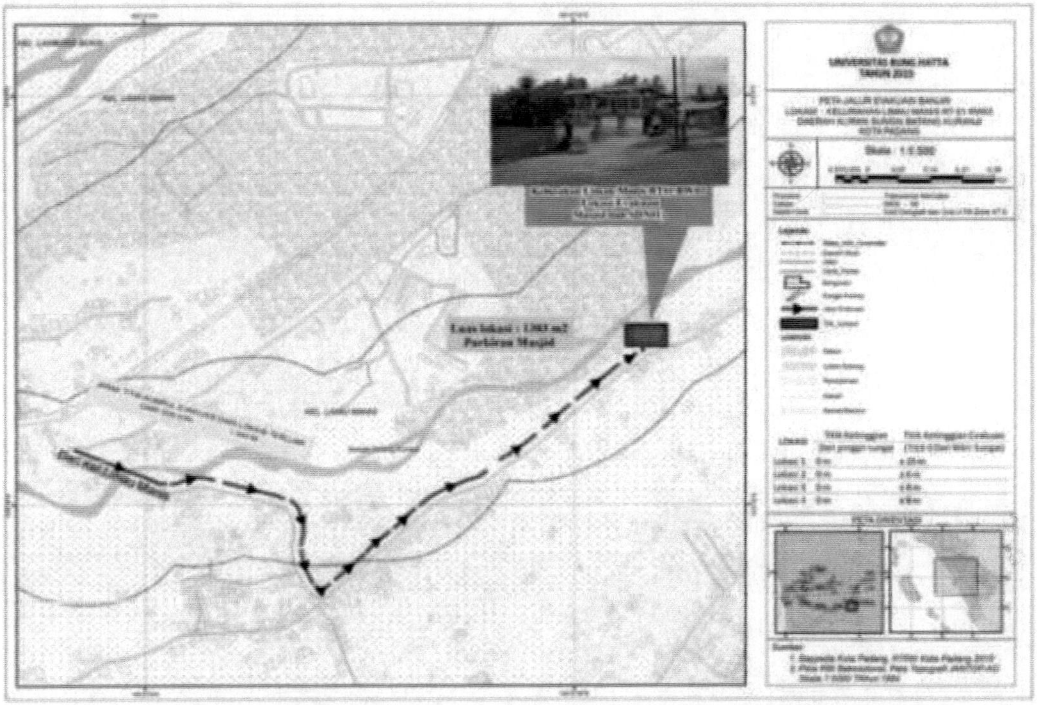

Figure 1. Map of Routes for Evacuationand Shelter Limau Manis

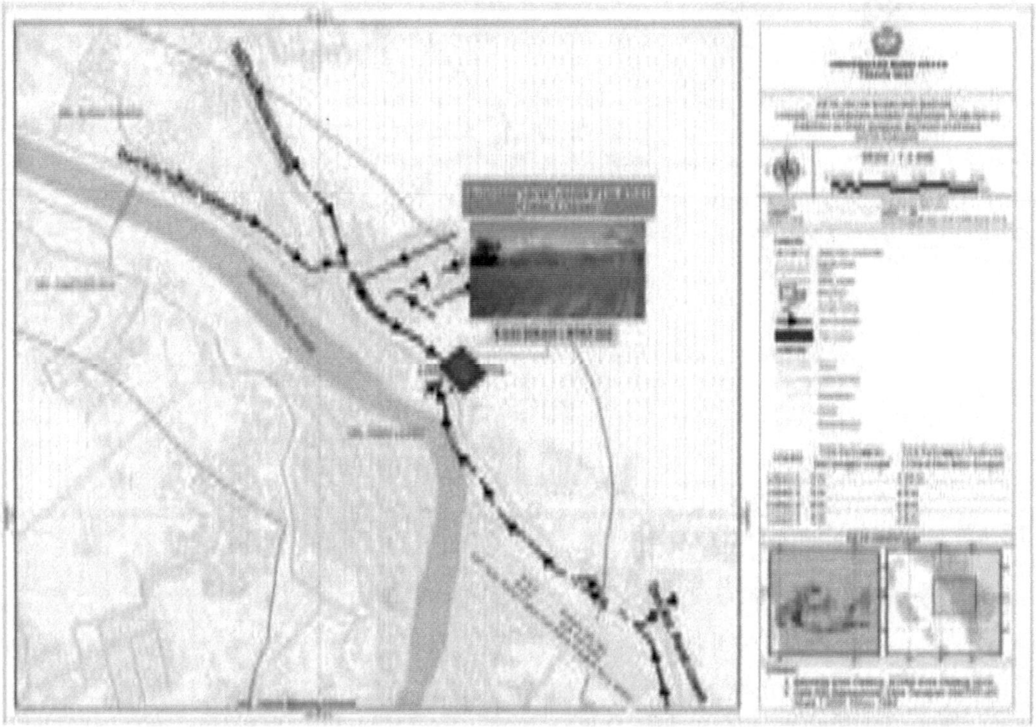

Figure 2. Map of routes for evacuation and Shelter Surau Gadang

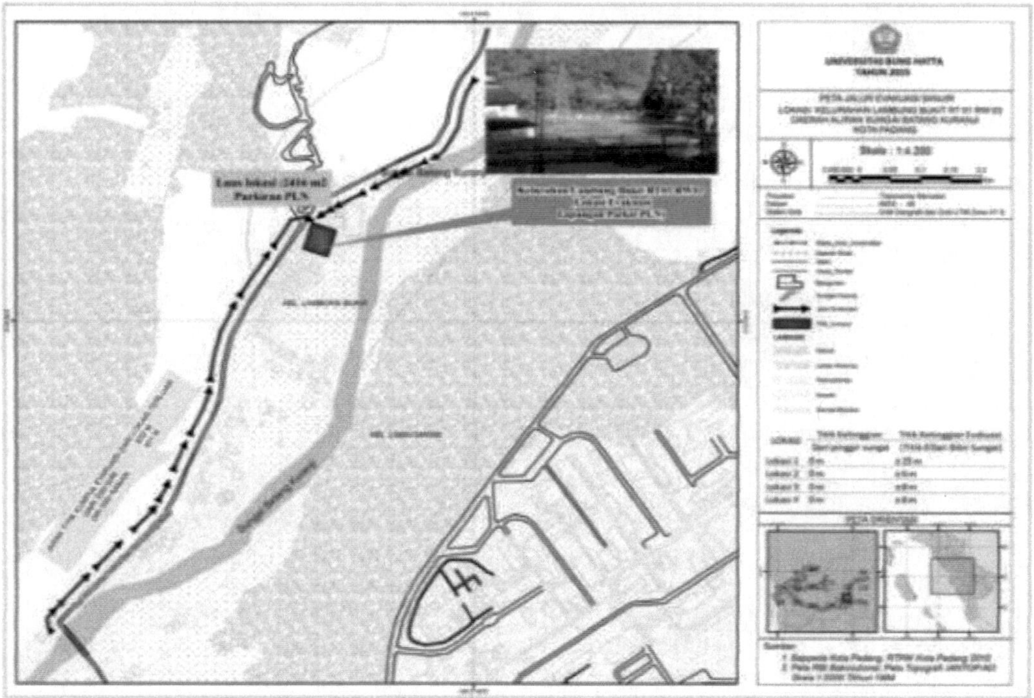

Figure 3. Map of routes for evacuationand shelter Lambung Bukik

4.2 Pocket book

Figure 4. Pocket book: Ready to accept floods

4.3 From 3 station record-keeping of rainfall

Station Gunung Nago, Station Gunung Sarik and Batu Busuk Station , with daily rainfall data of year 2003 until year 2015, got maximum monthly rainfall 310 mm, floods debit 1125.86 m³ / second

References

[1] C. M. Pickering and W. Hill, "Impacts of recreation and tourism on plant biodiversity and vegetation in protected areas in Australia," *J. Environ. Manage.*, vol. 85, pp. 791–800, 2007.

[1] Keputusan Presiden Republik Indonesia No. 114 tahun 1997, Tentang Penataan Ruang Kawasan Bogor-Puncak-Cianjur dan Penanganan Pengungsi, 1997

[2] Nur Yuwono. Penyebab banjir. Pengurangan luas Lahan. sebuah penelitian. 2005

[3] Onrizal dalam DAS Ciwulan. Mengapa sering terjadi banjir? tulisan sebuah pemikiran, 2005

[4] Peraturan Menteri PU N0. 63/PRT/1993, Tentang Garis Sempadan Sungai, Daerah Manfaat Sungai, Daerah Penguasaan Sungai dan Bekas Sungai, 1993

[5] Peraturan dan Perundang-undangan Undang-Undang Republik Indonesia No. 11 tahun 1974, Tentang Pengairan

[6] Undang-Undang Republik Indonesia No. 24 tahun 1992, Tentang Penataan Ruang Undang-Undang Republik Indonesia No. 25 tahun 2000, Tentang PROPENAS 2000-2004

[7] Peraturan Pemerintah Republik Indonesia No. 22 tahun 1982, Tentang Tata Pengaturan Air Peraturan Pemerintah Republik Indonesia No. 27 tahun 1991, Tentang Rawa Peraturan Pemerintah Republik Indonesia No. 35 tahun 1991, Tentang Sungai Peraturan Pemerintah Republik Indonesia No. 47 tahun 1997, Tentang Rencana Tata Ruang Wilayah

[8] Peraturan Menteri PU N0. 39/PRT/1989, Tentang Pembagian Wilayah Sungai Peraturan Menteri PU N0. 48/PRT/1990, Tentang Pengelolaan Atas Air dan Atau Sumber Air Pada Wilayah Sungai atau Sumber Air

[9] Peraturan Menteri PU N0. 63/PRT/1993, Tentang Garis Sempadan Sungai, Daerah Manfaat Sungai, Daerah Penguasaan Sungai dan Bekas Sungai

[10] PSDA Propinsi Sumatera Barat, DED Batang Maransi dan Batang Luruih Kota Padang pada

IOP Conf. Series: Earth and Environmental Science **70** (2017) 012040 doi:10.1088/1755-1315/70/1/012040

Batang Kuranji, Padang,2009
[11] Raharjo, P.D. Pemetaan Potensi Rawan Banjir Berdasarkan Kondisi Fisik Lahan Secara Umum Pulau Jawa. (Online), (http://www.puguhdraharjo.wordpress.com, diakses 22 Desember 2012), 2009
[12] Sosrodarsono, S dan Takeda, K. Hidrologi Untuk Pengairan.Penerbit PT. Pradnya Paramita. Jakarta, 2003
[13] Sutopo, P. N. Analisis Curah Hujan dan Sistem Pengendalian Banjirdi Pantai Utara Jawa Barat. Jurnal Sains dan TeknologiIndonesia,Vol.4, No.5, 2002
[14] Utama Lusi. 2014 Kajian Kerentanan Kawasan Berpotensi Banjir di Batang Kuranji , Penelitian Dikti
[15] Undang-Undang Republik Indonesia No. 24 tahun 1992, Tentang Penataan Ruang
[16] Van Zuidam, R. A. Aerial Photo-Interpretation In Terrain Analysisand Geomorphologic Mapping. International Institute for AerospaceSurveys and Earth Sciences (ITC). Smith Publishers. Netherland, 1985
[17] Van Te Chow, Hidrolika Saluran Terbuka, Jakarta, Erlangga, 2003
[18] Zahrul Umar, Tinjauan Penyebab Banjir Bandang Batang Kuranji menggunakan data Curah Hujan, Penginderaan Jarak Jauh dan SIG, 2012

IOP Conf. Series: Earth and Environmental Science **70** (2017) 012041 doi:10.1088/1755-1315/70/1/012041

Assessment of Land Allotment Support Power Industry In Grati, Pasuruan Regency

M R A Muzaqqi [1]

[1] Department of Urban and Regional Planning, Faculty of Engineering, Universitas Brawijaya, Malang, Indonesia

rizaalmuzaki.ram@gmail.com

Abstract. The industrial sector is always in need of land for factory as well as other supporting facilities, on the other side of the ability of the environment (support) the uneven terrain of every area in favor of intensive activities such as industry. Land uses that are not adapted to the support power, will cause pollution, damage, disaster and loss that generally uses the environment. The purpose of this research was to assess the resources support neighborhood Grati district associated with the existence of a plan to build an industrial area in accordance with the direction of Grati utilization of space in the spatial plan of the Pasuruan Regency area. In this study of land carrying capacity power comparison capability and land use. The Analysis technique used is the technique of overlay with analysis tools namely software using the *software* Arcgis 10.1. The parameters of the ability of land-adapted to the characteristics of the land for industry, namely the slope the slope ranges 0-25% on the slope of 25-45% can be developed with industry improvement area contours, and on a slope above 45% not allocated as an industrial area, the type of soil that is not easy slopes, the intensity of the rain of less than 3000 mm, potential landslide and flood-prone lowlands. Each parameter will be provided scoring between 1-5. Score of 1 was given to the condition of land the most harm, and a score of 5 is given for the condition of the land which supports most of the location industry. The result scoring is divided in 5 clases those are bad (5-9), is bad (9.1-13), medium (13.1), good (17,1-9) and good (21.1-25). The need for industrial land, calculated from the vast land of existing industries. Based on research results, obtained the ability to land on the area of research has 3 classes of 5 classes, i.e. good, moderate and bad. The results of the comparison between the broad capabilities and the needs of the farm industry, it can be concluded that the power of the land to support the industry in Grati still has not been exceeded.

1. Introduction

In the regulatory area of the Pasuruan Ragensi number 12 in 2010 about Plan Room System Regency Pasuruan 2009-2029 Pasuruan Regency Government wanted to expand the industrial park in the Eastern District, this is done to reduce the disparity of development between the eastern region and Western region. Seen from accessibility, Grati will pass by Surabaya-Pasuruan toll road. Accessibility is important because of the large and medium industry trend in big cities is no longer oriented on the market and raw material but on transportation infrastructure sector (roads) and labor (Farid, 2004). The toll road development plan currently in the work by the Ministry of public works to support the traffic of heavy payloads.

IOP Conf. Series: Earth and Environmental Science **70** (2017) 012041 doi:10.1088/1755-1315/70/1/012041

The industry desperately needs land to set up factories and its supporting facilities. The land becomes an important parameter to consider his ability especially for large industry events because the land has a very limited with the ability that vary and tend to experience a decrease in serving the demands of development (Hadi, 2005). To find out how the ability to do analys the ability to land by using some parameters. Land that could be used for industry have in common parameters with land for settlements (Lutfi, 2012) i.e, slope, rainfall, soil type, and the potential for disaster. See the potential and that problem, then research has the objective to analyze how the ability to land on the area of research and identify how the land use activities of big industry. Land use is important to know because it can show where the big industry is located. Location of big industry mandatory are in industrial areas. This is contained in Government Regulation No. 2 of 2009 about industrial area. In addition the identification of land use can provide information land area used large industry activities. The land area required for the land needs to see of great industry. A comparison between the requirements of the land and the ability to land this is called power carrying capacity of the land. Power support's assessment of land important to do so that development remains sustainable. Any activity that will be carried out, power support into a mandatory note because of land use which does not comply with supporting power will result in decreasing a quality environment such as pollution, damage, disaster and other losses.

The purpose of this research was to assess the resources support neighborhood Grati district associated with the existence of a plan to build an industrial area in Grati district. This is in accordance with the direction of the utilization of space in the spatial plan of the Pasuruan Regency area.

2. Literature Review

2.1 The ability and suistability of the land

Land capability is land quality that judged as a whole, while the land suitability is the quality of the land with respect to the request that is balanced with the demand and supply in a scope of special interests. Land suitability is determined by comparing the results of measurements of parameters in the standard field or criteria that apply (Senoaji, 2009).

2.2 Geogerafi information system (GIS)

Geographic Information System (GIS) is a component consisting of software, hardware, data, geography and human resources working together effectively to catch, keep refine, reform, manage, manipulate, integrate, analysis and show the data based on geographic (Budianto, 2002). Additionally the method of interpretation of the map used to help analyze environmental impact so that when applied to identify the suitability of power support industry of the land is suitable.

2.3 Overlay

Overlay is the main of a geographic information system operation that seems to define geographic information system (GIS) itself. Overlay process requires precision and in common locations. Overlays a graphic is to merge data between two or more graphic data to gain new graphics owned a combined mapping of some of the graphics data (fedra, 1996).

3. Methods

Basis of preparation of the study, starting from the dotted issues arising, namely the question of industrial and needs industrial area. The methods that are used are evaluative analisis uses overlay.

IOP Conf. Series: Earth and Environmental Science **70** (2017) 012041 doi:10.1088/1755-1315/70/1/012041

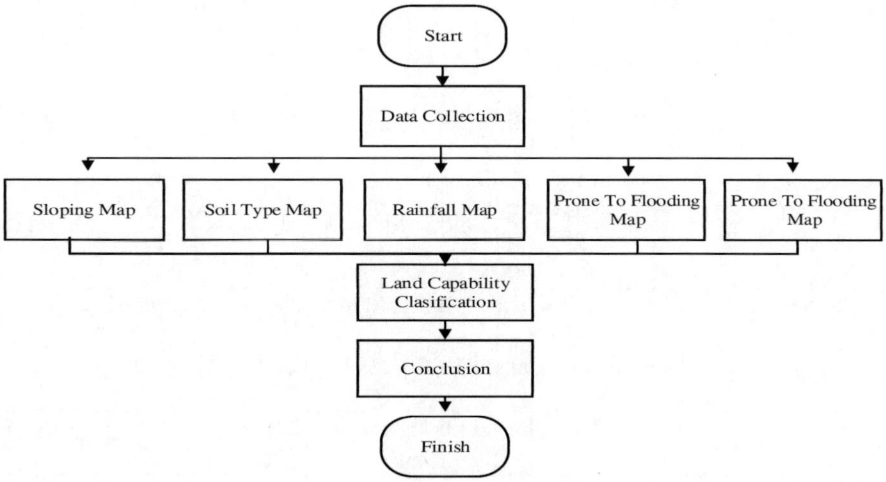

Figure 1. Research flow chart

4. Results and Discussion

To know the ability of land suitable for industrial activities done with information systems approach using geographical analysis of overlay, and to know the power support industry land conducted with interpretation using the software Arcgis 10.1.

Engineering overlay done with using 4 parameters the unit ability (SKL), i.e. the units the ability of land slope slopes of 0-25%, or 45%, unit ability rain intensity less than 3000 mm/th, unit ability of soil types and Disaster-Prone Parameter. Each parameter of a unit's ability to landwill be mapped, and provided scoring between 1 to 5. Method of weighting (scoring factor) is technical type in analyzing data by creating a value against the circumstances that exist, and are arranged according to the ranking. A score of 1 was given to the condition of land which is most detrimental to the industry, location and score 5 given for the condition of the land that supports most industrial locations. The results of the scoring are divided into 5 classes, namely bad (5-9), it is bad (9.1-13), medium (13.1), rather good (17,1-9) and good (21 1-25).

Table 1. The slope line parameter

No.	The slope of the	Description	Score	Interpretation
1	0-8	Flat	5	Good
2	8-15	Ramps	4	Good
3	15-25	A Bit Steep	3	Is being
4	25-45	Steep	2	Bad
5	> 45	Very Steep	1	Bad

Source: Agricultural Minister DECREE No. 683/KPTS/UM/8/1981

Table 2. The Rainfall Parameters

No.	Rainfall Intensity (Mm/Yy)	Score	Interpretation
1	0 – 1500	5	Good
2	1500 – 2000	4	Good
3	2000-2500	3	Medium
4	2500-3000	2	Bad
5	> 3000	1	Bad

Source: DECREE of the Minister of agriculture no. 683/KPTS/UM/8/1981

Table 3. The Parameter Types Of The Land

No.	Soil Type	The Level Of Sensitivity Of Soil Erosion	Score	Interpretation
1	Alluvial, Gley, Palnosol, Hidromorf Grey	Insensitive	5	Good
2	Latosol	Rather Sensitive	4	Good
3	Timberland Bergamping No Chocolate, Mediteran	Less Sensitive	3	Medium
4	Andosol, Laterite, Grumosol, Padosol, Padsolik	Sensitive	2	Bad
5	Regosol, Litosol, Organosol, Renzina	Very Sensitive	1	Bad

Source: DECREE of the Minister of agriculture no. 683/KPTS/ UM/8/1981

Table 4. Disaster-Prone Parameter

No.	Potential Landslide-Prone	The Amount Of Surface	Score	Interpretation
1	Low	< 0.9	1	Good
2	Medium	0.9 – 0.18	3	Is being
3	High	> 1.1	5	Bad

Source: Muta'ali 2012 BPBD 2012

To find out the impact of the development of the industry to changes in land use (Fandelli; 2000.177) then must power level value is known to support industrial land is divided into 5 classes i.e. good, rather good, rather bad, and bad. Power analysis results support against the conditions of land in Pasuruan Regency Grati in more details can be seen in the following table:

Table 5. Land Capability Classification

No.	Soil Type	Total Score
5	Good	13.1 -25
4	Good	17.1 -21
3	Medium	13.1 -17
2	Bad	9-13
1	Bad	5-9

Source: Fandeli,2000

The Data is the slope of the slopes, rainfall parameters and the soil type of the map Spatial Plan area of East Java province in 2009-2029.

From the data unit of the ability of land above, then overlayed to get Grati ability and usage data existing to assess land resources carrying capacity of the land Grati with software Arcgis 10.1. Based on the regulation of the Minister of the environment number 17 in 2009 regarding the determination of the power guidance support environment, then an assessment of the status of land support resources are formulated as follows:

$$DDLi = \frac{SL}{DL}$$

Description:
DDLi = Power Support Industrial Land
SL = Availability Of Land
DL = Land Needs

The formula above indicates if:

- SL more than DL or where land availability is greater than their needs then the power support land declared surplus has not been exceeded
- The SL less than DL or where the availability of land smaller than on its needs, then the power support its deficits have been exceeded.

The Grati district is located in the town of Pasuruan Regency and has 14 villages and 1 village. The geographical conditions of the Grati district is dominated by rice fields reached 1,013.70 hectares, then settlement 595.90 hectares, bordering with 2,448.90 acres, 595.90 acres of State forest and other 348.90 hectares. In the current study produced four map IE sloping map, soil type, rainfall maps and map the disaster-prone therefore each can be seen in Figure 2.a and 2.b , 3.a, 3.b. In the area of research, the results of the analysis a map of the layout of the Jawa Timur planning Spatial in 2009-2029 it is known that 2811 hectares having average slope between 0-2%, 1267 hectare land has average slope 2-5%, 408 hectares of land has slope an average of between 5-15% and the rest is about 6 acres have average slope between 15-40%.

The results of the analysis of the map of soil types, note that on the area of research have 2 soil types that is Alluvial and non-cal (Tanah Hutan Coklat) . The soil is Alluvial soil that is relatively young, because normally this type of soil is derived from sea water or sediment deposits of the river water and usually the ground situated around the coast or located around upstream of the River, the land of this type has properties that are labile, thus having high sensitivity towards erosion, this type of soil is not suitable for buildings with heavy loads, because the soil is prone to move are right soil type non cal or commonly called the land literit have less soil is sensitive to erosion, so that still allows for the establishment of the buildings. At Pasuruan Regency especially in Grati district there is land of about

2,816.12 Ha Alluvial type and soil type non cal (literit) covering an area of 1,675.59 Ha, while scoring soil type has a score of 3-5. Parameters for rainfall, spatial analysis of interpolation, known in the area of research has an average rainfall of between 1500-2000 mm/year. By using that data, then scoring for this parameter is 4. For the parameters of flood prone, according to BPBD, in the area of research has a region prone to floods covering an area of 222.98 Ha with an average of flooding height reaches 1 meter so that flood prone area for scoring is 4, while the avalanche-prone parameter to data BPBD sub Regency Pasuruan Grati has an area of 44.68 Ha had the threat of avalanche, landslide-prone area for scoring is 3. So the scoring disaster-prone areas for floods and landslides is 4.

From the results of these maps overlay noted that in the area of research has only 3 the class of 5 classes, namely the classes either with an area of approximately 1,675.59 Ha, class good with an area of 2,816.12 Ha and poorly with a broad 267.66 Ha. This is due to the total scores range between 9-25. With those results, then the land area which can be used for industrial activity is a good class and being with a total area of 326.65 Ha. For the needs of the land area of the industry in East Java around 200 Ha located on the ability of the land well and being. With that data, then comparisons between the ability of land to the use of its land is as follows :

$$DDLi = \frac{326{,}65}{200} = 126{,}65$$

From those results indicated that the availability of land (SL) is still larger than in land use (DL) or can be said to support the region's power industry in Grati district surplus or has not exceeded. The following is slope map, map of soil types, rainfall maps and map the disaster-prone Grati, Pasuruan Regency, Subdistrict:

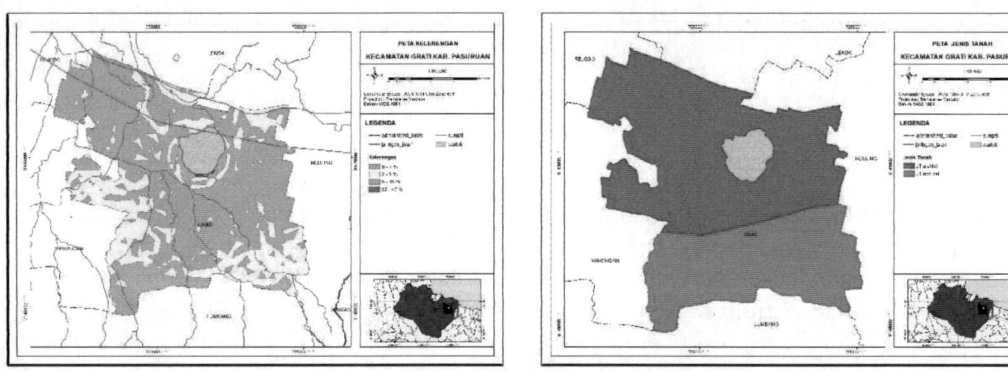

Figure. 2. (a) Sloping Map; (b) Soil Type Map

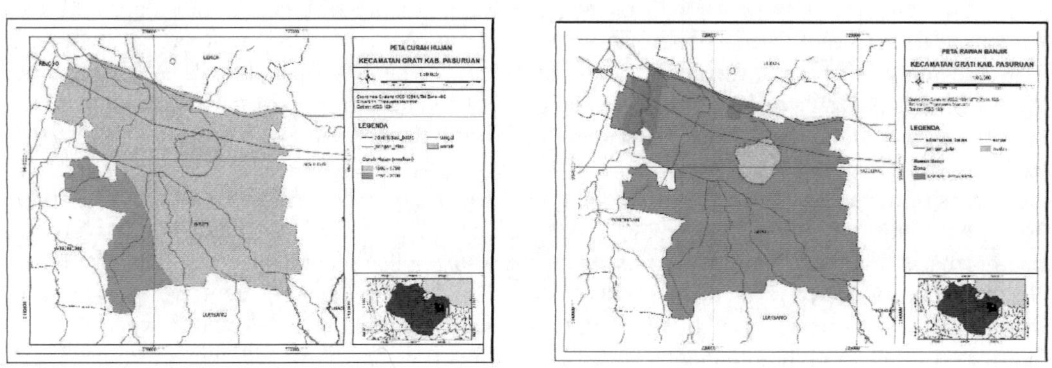

Figure. 3. (a) Prone To Flooding Map; (b) Prone Erosion Map

IOP Conf. Series: Earth and Environmental Science **70** (2017) 012041 doi:10.1088/1755-1315/70/1/012041

5. Conclusions

Based on the results obtained in the research conclusions as follows:

a) Ability to land in Pasuruan Regency especially in Grati district is divided into three classes, namely class good 1,675.59 Ha, good 2,816.12 Ha and bad 267.66 Ha. Land area which can be used for the area of industry is 267.66 Ha.
b) Comparison of Results obtained by power support environment Grati surplus or has not been exceeded. This means that the physical condition of land in Grati district capable to support industrial activities are directed in this district.

References

[1] Fandcli, C, Analisis Mengenai Dampak Lingkungan Prinsip Dasar dan Pemapanannya Dalam Pembangunan. Liberty Offset Yogyakarta, 2000.

[2] G1S Konsorsium, Modul Pclatihan AreGis Tingkat Dasar. Modul. GIS Kosorsium. Aceh, 2007.

[3] Hadi, S. P. Dimensi Lingkungan Pcrencanaan Pembangunan. Gadjah Mada University Press Yogyakarta, 2005.

[4] Muta'ali. L, Daya Dukung Lingkungan Untuk Pcrencanaan Pengembangan Wilayah. Badan Penerbit Fakultas Geografi (BPFG) Universitas Gadjah Mada Yogyakarta, 2012.

[5] Rencana Tata Ruang Kabupaten Pasuruan Nomor 12 Tahun 2010 Tentang Rencana Tata Ruang Wilayah Kabupaten Pasuruan Tahun 2009- 2029

[6] Peraturan Menteri Lingkungan Hidup nomor 17 tahun 2009 mengenai Pedoman Penentuan Daya Dukung Lingkungan.

[7] Irianta, Gunarsa. 2008. Kajian Dampak Perkembangan Industri Terhadap Kondisi Lahan Di Kawasan Bawen Kabupaten Semarang. Tugas Akhir. Semarang. Universitas Diponegoro.

[8] Oktarina, Rosa. 2014. Evaluasi Daya Dukung Lahan Untuk Industry Besar Di Kecamatan Ungaran Barat Dan Timur.Prosiding Seminar Nasional Dan Sains Dan Pendidikan Sains IX, Fakultas Sains Dan Matematika, UKSW Salatiga.

Strategic Planning of small and medium industries. Case study: Hulu Sungai Selatan Regency, South Borneo Province

N Elya, F Shoimah, A P Kartika and A B Sukanto

Department of Urban and Regional Planning, Faculty of Engineering, Universitas Brawijaya, Malang, Indonesia

nidyaulelya@gmail.com

Abstract. Hulu Sungai Selatan Regency has a potential of SMI (Small and Medium Industries) sectors can be developed as economic development. Based on RTRW of Hulu Sungai Selatan Regency, the region has 14 SMI are a propeller, pottery, blacksmith, dried fish, purun webbing, pastries, dodol, crackers, imitation jewelry, woven water hyacinth, bamboo, syrup, brown sugar, and saber. There are several issues related to SMI development such as low quality and quantity of human resources, local raw material, limited capital, low competitiveness, conventional production equipment, and lack of media for marketing the product. The purpose of this study is to develop the leading sectors of SMI and improve the economy and quality of the resident. The research method is descriptive qualitative, leading sectors analysis and force field analysis. Data were obtained from primary and secondary survey of relevant institutions and interview to the community. Based on leading sectors analysis, there is six leading sector is a propeller, blacksmith, dodol, dried fish, pottery, and crackers. Based on force field analysis, determined the strategy for using operational excellence's concept, so that we can develop the industrial sector by minimizing productions cost so SMI's product can compete by the price and efficient production process.

1. Introduction

1.1 Background

Forest Industrial sector development is very important to face stiff competition, both in the domestic market and the international market in the era of globalization and liberalization of world trade. Despite industrial sector as the main support of national development in Indonesia, this sector still has some issues which have a negative effect because it is not obedient to public policy. The problem must be solved by the government for continuity of national development, especially in the industrial sector. Industrial sector as a non-farming activity has a huge potential to expand and open job vacancy [1]. South Kalimantan Province also have Industrial Area Industrial Area which are Batulicin, Tanah Bumbu and Jorong Industrial Zone, Tanah Laut, and Hulu Sungai Selatan [2]

Hulu Sungai Selatan is one of regency in South Kalimantan, which have 14 SMI. They are the propeller, pottery, blacksmith, dried fish, purun webbing, pastries, dodol, crackers, imitation jewelry, woven water hyacinth, bamboo, syrup, brown sugar, and saber. The SMI are located in different districts such as Simpur, Kalumpang, Sungai Raya, Kandangan, Telaga Langsat, and Daha Utara. This condition indicates that Hulu Sungai Selatan Regency has a big chance to develop SMI.

This development has something to do with ASEAN Economic Community (AEC) as well. AEC is agreement among the members of Southeast Asia Countries which agree to make an economic integration. AEC give benefits such as to facilitate the movement of goods, services, investment, capital, and skill, increase trade and investment among member states, promote and expand regional production sharing and network [3]. Indonesia is one of the countries which signed the agreement. Aside from the benefits Indonesia will get, it also means that Indonesia is facing international economic rivalry. One of the challenges facing the national industry of Indonesia is low competitiveness in the international market. Factors that caused the lack of competitiveness among others is the increase in energy costs, the high economic costs, as well as inadequate bureaucratic service.

Attempts the government to do in the era of the AEC is to help the community to survive in the face of free trade between ASEAN countries to support the SMI in Hulu Sungai Selatan through the empowerment of local communities, training, counseling latest information, provision of funds and technology to help the production process in order to develop any existing SMI [4]. AEC enables many sectors are opening jobs vacancy for the local and international community [5]. So that the employee is an important component to survive in the competition in the era of the AEC. We also hope the more the quality of Indonesian workers that are capable of working and to produce goods or services with the best quality.

1.2 Problem Formulation

Based on the identification problems above, the formulation of the problem for SMI in Hulu Sungai Selatan are as follows:

1. Which is the leading SMI can be developed in Hulu Sungai Selatan?
2. What kind of the strategic planning which relevant for SMI in Hulu Sungai Selatan?

1.3 Purpose of the Research

There are so many things could be research on this problem, but this paper would explain the main purpose of this research as follows:

1. Analyze leading SMI which could be developed in Hulu Sungai Selatan.
2. Determine public policy which relevant for the SMI in development plan based on the strategic planning and innovative planning in Hulu Sungai Selatan.

2. Literature Review

2.1 Small and Medium Industries

SMIs are productive business opportunity belonging to individuals or entities that meet the criteria of individual businesses micro enterprises as stipulated by law [6] [11]. Small businesses are the chances of productive economic activities that stand alone, carried out by an individual or business entity that is not a subsidiary or not a branch of the company owned, controlled, or be a part either directly or indirectly from medium or large businesses that meet the small business criteria as defined in the legislation. Criteria of SMIs, micro business opportunities have a maximum asset Rp 50 million, with a maximum turnover of USD 300 million / year. A small business opportunity has assets> Rp 50 million-Rp 500 million with a turnover> Rp 300 million-Rp 2.5 M / year. Opportunities medium-sized businesses have assets> Rp 500 million-Rp 10 M with a turnover> USD 2.5 M - USD 50 M / year.

A form of micro, SMI in the form of a sole proprietorship, a partnership, such as firms and CV as well as a limited liability company. From the perspective of the world recognized that SMI plays a very vital role in the development and economic growth, not only in developed countries (NSB) but also in advanced countries (NM). In developed countries the SMIs are very important not only for the business groups absorb the most labor than large businesses, as well as in developing countries but also in many countries, its contribution to the formation or growth of gross domestic product (GDP) compared to the contribution of large companies.

2.2 Strategic Planning

Strategic planning is a long-term planning that is comprehensive and provides a related formulation of the area will be directed and allocating resources to achieve the goals for a certain period in a range of possible circumstances. Strategic Planning has another sense, namely the process of selecting the organization's goals, strategy determination, wisdom, strategy programs necessary for these purposes. There are three reasons that show the importance of strategic planning, as follows:

1. Strategic planning provides a basic framework for all forms of planning which could be taken.
2. An understanding of strategic planning will facilitate an understanding of other forms of planning.
3. An understanding of strategic planning will facilitate an understanding of other forms of planning.

Strategic planning has the benefit that the direction of development of the area become clear that it is easier to formulate targeted and effective. so it can be said that strategic planning can determine the success of an organization or company. This is because:

1. Strategic planning is the most important type of planning.
2. Perform strategic planning means setting clear organizational mission.
3. Strategic planning allows managers prepare for the possibility of changes to the environmental organization.

Strategic planning is not familiar with standards and processes have unlimited variation. Each implementation of the strategic need to design according to the needs, circumstances and local conditions.

3. Methodology of The Research

In order to collect the data, we do the primary and secondary survey. The primary survey we use the questionnaire to interview the owners of the potential SMI in Hulu Sungai Selatan Regency and for the secondary survey is being used to collect data from the institutions related to the SMI. The questionnaire consists of several questions that are prepared by the researchers related to the industrial activities. The research itself uses two main analysis, which is leading SMI analysis and force field analysis (FFA) to develop the strategy of the SMI in Hulu Sungai Selatan Regency.

3.1 Leading Small and Medium Industries

Leading SMI analysis can be decided by these variables which are scoring the LQ result criteria, product durability, availability of raw material, market location, labor, and the related regulation that applied to the SMI.

3.1.1 LQ Result Criteria

LQ technique is one of the most common approaches that used in the basic economic model as a first step to understanding the activity sector that triggers the growth [10]. LQ result criteria (Table 1) is done based on the LQ formula that is applied to every SMI industries in Hulu Sungai Selatan Regency. If the $LQ > 1$, that means the SMI has a high production and has a bigger chance to become a basis sector. However, if the $LQ < 1$, it means the SMI has a low production and does not have a chance to become a basis sector.

Table 1. Scoring LQ Result Criteria

No.	LQ Result	Score
1	LQ Negative	10
2	LQ <1	20
3	LQ >1	30

3.1.2 Durability Product Criteria

Durability product criteria (Table 2) is done based on how long the product can be used without any problems. The higher durability product score, the farther products can be marketed and vice versa.

Table 2. Scoring Durability Product Criteria

No.	Durability Product Criteria	Score
1	Low	10
2	Medium	20
3	High	30

3.1.3 Availability of Raw Material Criteria

The availability of raw material is seen based on its location (Table 3). The closer raw material can be collected, the cheaper it transportation cost. Conversely, further location can increase the transportation cost.

Table 3. Scoring Availability of Raw Material Criteria

No.	Availability of Raw Material Criteria	Score
1	Located outside Borneo Island	10
2	Located on South Borneo Province and other provinces in Borneo Island	20
3	Located on Hulu Sungai Selatan Regency	30

3.1.4 Market Location Criteria

Market location criteria are seen based on distance and accessibility to the location (Table 4). Market location depends on the type of industrial products that marketed. It is because the market location is affected by durability product as well. For instance, industrial products such as food have to be made near the market location so that the food will not break easily.

Table 4. Scoring Market Location Criteria

No.	Market Location Criteria	Score
1	Located only in Hulu Sungai Selatan Regency	10
2	Located on Borneo Island	20
3	Located outside the Borneo Island	30

3.1.5 Labor Criteria

Labor criteria (Table 5) is seen based on the amount of labor that can work in an SMI. The growth of the SMI is expected to reduce the number of the unemployment in Hulu Sungai Selatan Regency.

Table 5. Scoring Labor Criteria

No.	Labor Criteria	Score
1	Low level of employment	10
2	Medium level of employment	20
3	High level of employment	30

3.1.6 Related Regulation Criteria

Related Regulation criteria (Table 6) can be seen based on regulations that are related to developing industrial sector in Hulu Sungai Selatan Regency. The more regulations that have programs for the SMI, the higher score will be given.

Table 6. Scoring Related Regulation Criteria

No.	Related Regulation Criteria	Score
1	There is no program for the SMI in the regulations	10
2	There are programs for the SMI in the regulations but not a leading SMI	20
3	There are programs for the leading SMI	30

3.2 Force Field Analysis

Force field analysis is a tool to make a strategy in developing industrial sector in Hulu Sungai Selatan Regency. There are several steps to do in order to enhance the industrial sector with this analysis;

1. Identify the driving and resisting factors in industrial activity
2. Give score to each factor
3. Determination of the key factors

4. Research Results

4.1 Leading Small and Medium Industries

Activity in the SMI Hulu Sungai Selatan Regency consists of SMI food and beverages, handicrafts and IKM metals, machinery and equipment [7]. From the criteria and variables in Table 1, Table 2, Table 3, Table 4, Table 5 and Table 6, here are the scores in every SMI (Table 7). Six SMI that get the highest score is the leading SMI in Hulu Sungai Selatan Regency which are propeller, blacksmith, pottery, dodol, dried fish, and crackers (Fig. 1 & Table 8).

Table 7. Scoring Leading SMI in Hulu Sungai Selatan Regency

No.	SMI	Score	No.	SMI	Score
1	Propeller	180	8	Pottery	170
2	Blacksmith	170	9	Crackers	150
3	Dodol	160	10	Woven water hyacint	140
4	Pastries	120	11	h	120
5	Purun webbing	140	12	Bamboo	140
6	Dried fish	150	13	Syrup	130
7	Imitation jewelry	80	14	Brown sugar	130
				Saber	

Table 8. The Six Leading Industries

No.	SMI	Photos	No.	SMI	Photos
1	Propeller		4	Dried Fish	
2	Blacksmith		5	Pottery	
3	Dodol		6	Crackers	

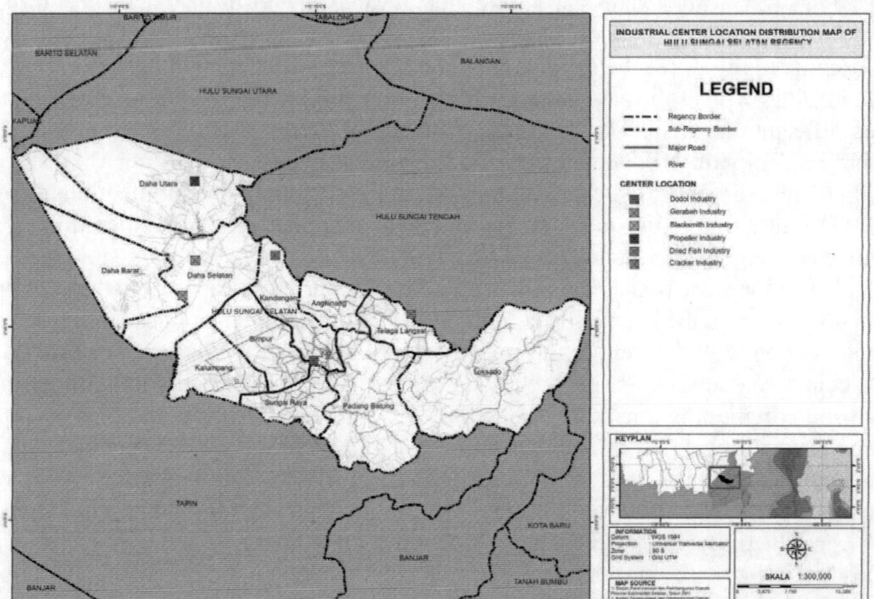

Figure. 1. Leading SMI Sector Distribution Map of Hulu Sungai Selatan Regency

4.2 Strategic Planning of Leading Small and Medium Industries

Force Field Analysis [8] [9] is based on six SMI in Hulu Sungai Selatan Regency is obtained each of the factors of the driving factors and resisting factors (Table 9). Based on those factors, we can get the urgency score and factor percentage of industrial activity in Hulu Sungai Selatan Regency. Key success factors have the highest score among all factors in industrial activity. Here is a table of key success factors from industrial sector (Table 10).

Table 9. Driving and Resisting Factors in Hulu Sungai Selatan Regency

Driving Factors	Resisting Factors
D-1 Raw material from Hulu Sungai Selata	H-1 Lack of capital
n	H-2 Still using traditional equipment
D-2 Local labor from Hulu Sungai Selatan	H-3 Does not have brand
D-3 Marketing outside Hulu Sungai Selatan	H-4 Marketing through the collectors
D-4 Waste recycling	H-5 Lack of skills
D-5 Good quality product	H-6 No regeneration in labors
D-6 Government assistance in form of tools	H-7 Lack of marketing media

Table 10. Driving and Resisting Key Factors in Hulu Sungai Selatan Regency

No	Driving Key Factors	Resisting Key Factors
1	Raw material from Hulu Sungai Selatan	Lack of capital
2	Local labors from Hulu Sungai Selatan	Lack of skills
3	Good quality product	Still using traditional equipment
4	Marketing outside Hulu Sungai Selatan	Does not have brand

After deciding the driving and resisting factors, the next step is to make strategy in developing the industrial sector. The concept of the development of the industrial sector in Hulu Sungai Selatan Regency based on force field analysis, so we get a value proposition that is the direction of the development of operational excellence. Operational Excellence means products and services based on competitive pricing, to minimize the price through benchmarks of efficiency, organization and industry are multifunctional, as well as the advancement of the industry and organizations through the prices that the process is efficient and easy. Thus, the industrial sector products Hulu Sungai Selatan Regency is expected to have a competitive edge in terms of product quality and price.

Operational embodiment excellence did through linkage system development plan, the plan sector linkages, SMI product marketing plans to the management plan of SMI. Operationalization of the concept is carried out with strategies in accordance with their respective development plans. The strategy linkage system development plan includes an increase in the working system from input to output process with good management and structured. Strategy development plans linkage sectors include improving cooperation with the sector Agropolitan and Minapolitan in the supply of raw materials, the development center for new industrial processing sector Agropolitan and Minapolitan, utilization of local tourism as a marketing medium, the active promotion of SMI marketing facilitated by the government or the private sector, an increase in media campaign to support network expansion of the market, the use of e-commerce to support the expansion of the network market. Strategy development plan processing SMI include increased productivity products through the development and utilization of technology, productivity improvement products through training in processing and packaging of industrial products, increase productivity through the capital increase with the cooperation of government and private sector, improving the competitiveness of industrial products, industrial waste processing with the 3R concept (Reuse, Reduce, Recycle), increased use of local resources effectively efficient professional. Then supported with advice and infrastructure plan include the marketing of products closer to the node movement (docks, airports, and seaports), optimization of transport facilities and infrastructure to support the distribution of industrial activities.

5. Conclusions

Based on the leading sector analysis, the SMI sector that has to be developed is a propeller, blacksmith, dodol, dried fish, pottery, and crackers. Therefore, to develop those leading sector of SMI we can use operational excellence concept for strategic planning.

References

[1] Rencana Tata Ruang Wilayah South Kalimantan Province Year 2015-2035
[2] RPJMD South Kalimantan Province Year 2011
[3] Gatot. (2015). Masyarakat Ekonomi Asean MEA dan Perekonomian Indonesia. Jakarta: Kementerian Keuangan
[4] Manuhutu, Y. (2014). Kesiapan Pemerintah Menghadapi MEA 2015. Ambon: Universitas Pattimura
[5] BBC. (2014). Apa Yang Harus Anda Ketahui Tentang Masyarakat Ekonomi Asean.
[6] Rencana Tata Ruang Wilayah Hulu Sungai Selatan Regency Year 2015-2035
[7] Undang-Undang Number 20 Year 2008 About Small and Medium Idustries
[8] Seth Heinert, Sebastian Galindo, Gonzalez. Making Decisions Using Force Field Analysis. Florida: University of Florida; 2014, p. 1–4
[9] Ajimal, K. S. (1985). Force field analysis: A framework for strategic thinking. Long Range Planning. 18(5): 55–60
[10] Fahmana, Arief. (2012). Strategi Pengembangan Usaha Kecil Menengah Sektor Industri Pengolahan.. Jurnal Teknik Industri. XIII (1): 14-21. Universitas Widyatama
[11] Rachmad H,M etc. (2009). Pengembangan Tata Kelola Industri Kecil Menengah di Madura. Jurnal Teknik Industri. XI (1): 61-71. Madura: Universitas Trunojoyo

IOP Conf. Series: Earth and Environmental Science **70** (2017) 012043 doi:10.1088/1755-1315/70/1/012043

Minapolitan Region Development Analysis at Penajam Paser Utara using Blue Economy Concept

P M Mawarsari[1,2], A N Dewanti[1], and F Nurrahman[1,]

[1] Institut Teknologi Kalimantan, Balikpapan, 76127 Indonesia

putrimmsari@gmail.com

Abstract. Penajam Paser Utara (PPU) Region in East Kalimantan was assigned as one of Minapolitan area in Indonesia focusing on aquaculture based on Indonesia Ministry of Maritime Affairs and Fisheries Decision Number 35 in 2013. Unfortunately, it is shown from PPU region statistics data that decreasing fisheries production in PPU region was occur especially for land fisheries from 2013 until 2015. The aim of this study is to formulate development strategies of Minapolitan in PPU Region by the Blue Economy concept. Several variables such as fisheries and aquaculture aspects, processing and marketing, incomes and workers, also sustainability and economic paradigm is evaluated in this study. Firstly it needs to identify the potential of character and problem by reviewing Minapolitan area in PPU Region with descriptive comparative analysis. Secondly, the fruitfulness of Minapolitan activity in PPU Region impact factors was identified by using internal and external factor (IFAS and EFAS) analysis. Then some of strategic considerations was formulated by using the successful development indicator in Minapolitan area that associate to the Blue Economy concept with SWOT analysis. The result of IFAS analysis show the highest score of strength that is production and commodity productivity. Contrarily, the income level factor is become the highest score at weakness. Then, the EFAS analysis declare that marketing system is the highest factor in opportunity. The position of SWOT quadrant indicate that the Minapolitan area of PPU region is in quadrant I /or first quadrant which means progressive. Therefore, the Minapolitan in PPU have a great chance to increase the cultivation optimally. The results of SWOT matrix is to increase its product and productivity by upgrading the quality and quantity production facilities, raising the control of production activites and also increasing the supervision of processing business with sustainable management business principle.

1. Introduction

Minapolitan region is originated from Agropolitan, which is a region containing of one or more centers in rural areas as fisheries production. It have a system and management of natural resources represented by an association of functional and spatial unit hierarchy settlement systems and business-mina system. Friedman and Douglas (1975) offering this concept to face against the failure of industrial system development in Asia. This failure caused by the hyper urbanization phenomena, which emerge a highly amount of unemployment, poverty, lack of food resources and decrease of welfare. Similar to Agropolitan, Minapolitan concept as well, creatre by Friedman and Douglas (1985) as development activities are concentrated in rural areas inhabited between 50,000-150,000 inhabitants.

Penajam Paser Utara (PPU) Region in East Kalimantan was assigned as one of Minapolitan area in Indonesia focusing on aquaculture based on Indonesia Ministry of Maritime Affairs and Fisheries

Decision Number 35 in 2013. Unfortunately, it is shown from PPU region statistics data that decreasing fisheries production in PPU region was occur especially for land fisheries from 2013 until 2015. The aim of this study is to formulate development strategies of Minapolitan in PPU Region by the Blue Economy concept.

2. Literature Review

Based on the Indonesia Ministry of Maritime Affairs and Fisheries (2013), the concept of Minapolitan can be realized through several approaches, such as (1) alternative models / strategies with superior commodities with guaranteed quality and quantity as well as having a market that can accommodate the product, (2) the priority of regional development depends on the characteristics of each region (3) development model that focuses fishery resources as economic potential, (4) integrated economic development clusters from upstream to downstream, (5) locus-based development that integrates resources, finance, information, technology and human resources. While in evaluating Minapolitan performance, seen from (1) aquaculture fisheries, (2) capture fisheries, (3) processing and marketing, (4) revenue and labor.

Pauli (2010) argues that the concept of Blue Economy developed to answer the challenge, so that the world economic system tends to exploitative and environmentally destructive. Beside the effect of wastes, the nature also damaged due to exploitation beyond its capacity or carrying capacity. This concept aims to prosper the community and reduce the potential for natural damage due to exploitation. It can be derived from the essence that blue economic elements can be viewed from (1) sustainability aspects consisting of natural efficiency, waste free, social awareness and innovation and creativity, and (2) an economic paradigm consisting of ecosystems and business method.

3. Methods

By reviewed the Minapolitan and Blue Economy theories, several variables such as fisheries and aquaculture aspects, processing and marketing, incomes and workers, also sustainability and economic paradigm is evaluated in this study. Then further identification of potential character and its problem is needed by reviewing Minapolitan area in PPU Region with descriptive comparative analysis. This descriptive comparative analysis is required to determine those variables.

Secondly, the success of Minapolitan activity in PPU Region impact factors was identified by using internal and external factor (IFAS and EFAS) analysis. Then, the result of IFAS and EFAS were assessed by the stakeholder (representative of Department of Marine and Fisheries of PPU region) using scoring analysis. The score resulted from this assessment were determined by the importance level of each factor.

Then formulate some of strategic considerations using the successful development indicator in Minapolitan area that associate to the Blue Economy concept with SWOT analysis.

4. Results

Several data that used to analyze and identifying characteristics of each factor are production realization data of sea and land fishery commodities in 2015, development and requirement data for supporting Minapolitan infrastructure, expansion support data for production systems and fishery farmers in PPU district, the data of procurement assistance for the community in 2015, the amount of marine fisheries and inland farmers data in PPU district in 2015, the diversity data of inland fisheries farmers in PPU district, data of poor families in PPU district in 2012, data about fishing gear and ownership of boats quantity in PPU district 2015, primary survey through direct observation in supporting infrastructure such as the Port Fish facilities (TPI) as well as an interview with Mr. Roni Tuska (Department of Marine and Fisheries) about infrastructure, accessibility, harvesting system, abrasion disaster potential and irrigation systems.

Through a comparative descriptive analysis by comparing the existing condition of the field either by primary or secondary data, the identity of each factor obtained in this study are:

Table 1. Influential Factors in Minapolitan Success with Blue Economy Principles in the District PPU

Aspect	Factor	Identity	Remarks
Fisheries and aquaculture	Production And Productivity of Superior Commodity	S1	High fisheries production in PPU district is a strength point for activity in PPU Minapolitan
	Multiplier Effect	O1	Minapolitan activities resulted in a positive impact for the economy in PPU district
	Quality of production facilities	W1	The condition of supporting facilities and accessibility is still less than optimal in its development
	Development of Cultivation Systems	S4	Have planned an integration between upstream development system in Babulu District, with downstream system in Penajam, Waru and Sepaku District
	Management Quality	W2	The need for oversight and management activities that deal with the conditions of service in the port of fish (TPI) as the unit of management-related outcomes, inputs and processes.
	Port Development	W3	Analyzing the condition of port infrastructure and fishery activities which are still not optimal
Processing and marketing	Production, productivity, and result value	O4	PPU district have a variety of community activities unit such as home industries that process fishery products and aquaculture
	Quality and diversity of UMKM products	S2	fish processing results are quite varied and the quality is good enough
	Amount of Processing Business	W4	the number of processing business in PPU District still not increased especially UMKM process for fishery products
	Marketing System	O5	able to penetrate the regional market to Banjarmasin, Samarinda and Java.
	Business Development System	O6	there is a commitment between local government and fishery farmer, in the form of financial contributions, personnel, and facilities management and development in the Minapolitan region
Income and employment	Business Opportunities And Employment	O7	high business opportunities and employment in PPU District
	Public Welfare	W5	The level of welfare is still low, especially in fulfilling health and education facilities
	Income Level	W6	The level of income is still inadequate, judging from the large number of poor people in PPU District
Sustainability	Natural Efficiency	S3	How to harvest and tools that used are still traditional and harmless to the environment
	Pollution Free	T1	Lack of community commitment to waste-free fishery production
Economic paradigm	Ecosystem	O8	Fishermen pay attention to cycles of natural ecosystems
	Business Method	T2	Product innovation has not been developed with optimal production systems and management

Then the assessment by the representative of Department of Marine and Fisheries of PPU region using scoring analysis generate the score based on their weighting of importance level on each factor from IFAS and EFAS. The IFAS analysis declare that Production and productivity of superior commodity in strength and the amount of processing business are the highest factor in weakness.

Table 2. Internal Factor Analysis Summary (IFAS)

Identity	Factor	Weight	Score	Total Score
Strength				
S1	Production and productivity of superior commodity	0,15	4	0,6
S2	Quality and diversity of UMKM products	0,15	3	0,45
S3	Natural efficiency	0,1	4	0,4
S4	Development of cultivation system	0,1	3	0,3
	Sum	0,5		1,75
Weakness				
W1	Management quality	0,05	-2	-0,1
W2	Development of harbour	0,05	-2	-0,1
W3	Amount of processing business	0,15	-4	-0,6
W4	People welfare	0,1	-3	-0,3
W5	Income level	0,15	-4	-0,6
	Sum	0,5		-1,7
	Total	1		0,05

Then, the EFAS analysis declare that marketing system is the highest factor in opportunity also the free pollution and business method have the same importance level in threat.

Table 3. External Factor Analysis Summary (EFAS)

Identity	Factor	Weight	Score	Total Score
Opportunity				
O1	Multiplier effect	0,1	3	0,3
O2	Quality of production facilities	0,1	4	0,4
O3	Production, productivity, and result value	0,15	4	0,6
O4	Marketing system	0,15	4	0,6
O5	Business development system	0,05	3	0,15
O6	Business opportunity and employment	0,1	3	0,3
O7	Ecosystem	0,1	4	0,4
	Sum	0,85		2,75
Threat				
T1	Free pollution	0,075	-3	-0,225
T2	Business method	0,075	-3	-0,225
	Sum	0,15		-0,45
	Total	1		2,3

The position of SWOT quadrant indicate that the Minapolitan area of PPU region is in quadrant I /or first quadrant which means progressive. Therefore, the Minapolitan in PPU have a great chance to increase the cultivation optimally.

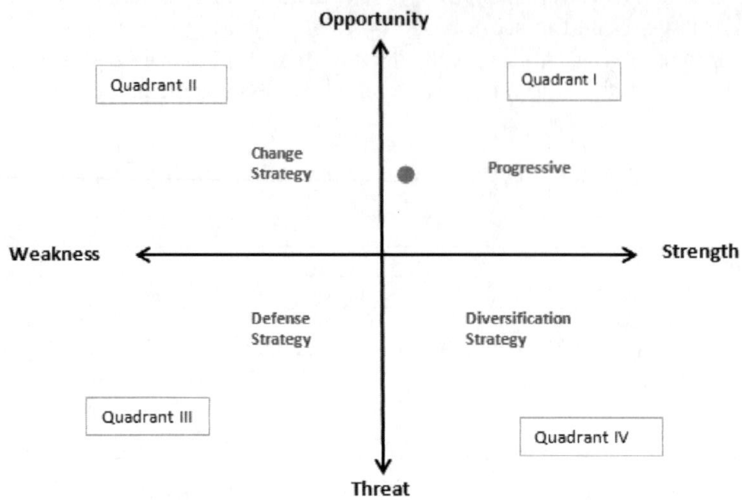

Figure. 1. SWOT Quadrant

Then the SWOT matrix analysis results are :

Table 4. SWOT Matrix Analysis

IFAS / EFAS	**Strength (S)** 1. Production and productivity of superior commodity 2. Quality and diversity of UMKM products 3. Natural efficiency 4. Development of cultivation system	**Weakness (W)** 1. Management quality 2. Development of harbor 3. Amount of processing business 4. People welfare 5. Income level
Opportunity (O) 1. Multiplier effect 2. Quality of production facilities 3. Production, productivity, and result value 4. Marketing system 5. Business development system 6. Business opportunity and employment 7. Ecosystem	1. increase its product and productivity by upgrading the quality and quantity production facilities 2. product diversification for marketing system development 3. preserving environmentally friendly harvest system 4. increase the system cultivation to enhance business opportunity	1. raising the control of production activities 2. infrastructure development as effort to increasing the multiplier effect 3. increasing the unit process with raising supervision by the government 4. Raising income level and wealth by increase the opportunity business and employment absorption in community
Threat (T) 1. Free Pollution 2. Business Method	1. increasing the supervision of processing business with sustainable management business principle	1. increasing the amount of production facilities with sustainable business management principle should be applied

IOP Conf. Series: Earth and Environmental Science **70** (2017) 012043 doi:10.1088/1755-1315/70/1/012043

5. Conclusions

Increasing Minapolitan development strategies in Penajam Paser Utara (PPU) Region using Blue Economy concept could be realized by increasing its product and productivity by upgrading the quality and quantity production facilities, product diversification for marketing system development, preserving environmentally friendly harvest system, increase the system cultivation to enhance business opportunity. Other strategies that resulted from this study is raising the control of production activities, infrastructure development as effort to increasing the multiplier effect, increasing the unit process with raising supervision by the government, raising income level and wealth by increase the opportunity business and employment absorption in community. And finally, increasing the supervision of processing business and the amount of production facilities with sustainable management business principle should be applied.

References

[1] J. Zuzana and L. Zuzana, "Monitoring System of Sustainable Development in Cultural and [1] Anonim. 2012. Sumber Agropolitan & Minapolitan : Konsep Kawasan Menuju Keharmonisan. Jakarta : Kementrian Pekerjaan Umum Direktorat Jenderal Cipta Karya.

[2] Anonim. 2013. Pengembangan Kawasan Minapolitan. Jakarta : Kementrian Kelautan dan Perikanan.

[3] Friedman, J. 1975. Regional Policy: Reading in Theory and Applications. Cambridge, Massachusetts : the M.I.T press.

[4] Friedman, Paul G dan Yarbrough, Elaine A. 1985. Training Strategist. New Jersey : Prentice-Hall, Inc.

[5] Pauli, Gunter. 2010. The Blue Economy: 10 Years, 100 Innovations, 100 Million Jobs. United States.

IOP Conf. Series: Earth and Environmental Science **70** (2017) 012044 doi:10.1088/1755-1315/70/1/012044

Community-based livelihood management in relations to natural disaster – A study on Teknaf (coastal) area of Bangladesh

R Khanam

MPhil Researcher, Bangladesh University of Professionals, Mirpur Cantonment, Dhaka, Bangladesh

rozina1981@gmail.com

Abstract. Teknaf is an Upazila under Cox's Bazar District of Bangladesh, it's a coastal area with strong influenced by the Naaf river estuary of the Bay of Bengal. The study outlines the major livelihood groups or community in the area. It was observed that the livelihoods are severely affected by climatic and non-climatic changes. For example, the increased salinity of both soil and water has seriously affected all livelihood resources, in particular agriculture, fishery, livestock and forestry. The increase in frequency and intensity of natural disasters - floods and cyclones, has made it difficult for the local people to secure their livelihood. In addition to natural factors, several anthropogenic factors remain the major form of vulnerability for the farmers, fishers and other livelihood sections of the society. This study was an exploratory research with questionnaire survey by random sampling, focus group discussion, and review secondary data. The study observed that the local people have evolved many local adaptive practices to deal with the difficult climatic conditions. Outcome of the study is capacity building of the community with in their available resource; combined crop and fish culture need to encourage; control excessive collection of Natural resources like marine fish, forest tree, alternative income generating activities for farmers & fisherman at lean season and disaster situation need to start.

1. Introduction

Geographically Teknaf is located at 20.8667° N 92.3000° E 20.8667; 92.3000. It has total area 288.68 km2. Teknaf port is the main attraction in southern or south – eastern coastal under the Cox's Bazar district of Bangladesh. Because of tropical weather of Bangladesh most tourist usually go to visit this place during winter or in between winter and autumn (October to March). Maximum of household heads directly depends on artisanal fishery activities such as fisher (57%), fish traders (4%), net menders (3%), boat makers (1.5%), wild fry collection (0.5%) in the study area [4]. Major livelihoods in the coastal zone are agriculture, fishery, salt farming, shrimp culture, industrial and agricultural labour, and extraction of forest resources etc. Livelihoods activities of the people's dependent on natural resources are degrading the coastal ecosystems and invading spaces of biodiversity. The major environmental issues faced by the country includes cyclones and storm surge, land erosion, flood, drainage congestion, salinity intrusion, drought, earthquake, shortage of drinking water & arsenic contamination, ecosystem degradation, pollution and climate change [11].

The study reported herein is based on a survey conducted by Rozina on vulnerability and local practices of the study area as well evaluate secondary data and institutional involvements in various aspect. In this report tried to identify the scopes and prospects of improving practices of lives and livelihood of study area. Part 1 introduces the paper and Part 2 discusses the objective of the study, Part 3 literature review of related papers, the methodology of the study reported herein Part 4, part 5 discusses the results of the study and lastly Part 6 concludes the paper.

2. Objective of The Study

Climate have major physical impacts on agriculture, industry, infrastructure, disaster, health and energy and consequently on people's livelihood in terms of employment, income and consumption. Various groups in society has experience the impacts in various degrees dependent upon their initial economic conditions (poor or non-poor), location (coastal or non-coastal, rural or urban) and gender. Low economic strength, inadequate infrastructure, low level of social development, lack of institutional capacity, and a higher dependency on the natural resource, make the country more vulnerable to climate stimuli (including both variability as well as extreme events). Addressing present and future problems related to the climatic effect is appeared to be a complex issue for Bangladesh. Considering the above circumstances, it was necessary to conduct current research on "Community-based Livelihood Management in relations to Natural Disaster– A Study on Teknaf (coastal) area of Bangladesh".

2.1 Specific Objective of The Study

Specific objective of the study intended to

- Review livelihood related management policies and practices of the disaster-prone coastal zone communities of Bangladesh;
- Identify climate change issues affecting the livelihood of the area people;
- Identify/Document livelihood systems of the communities in Teknaf upazila of Cox's Bazar district;
- Identify the major problems of the community, ranking and solutions perceived by the community;
- Recommend appropriate livelihood management strategies for Teknaf area.

3. Literature Review

Different level of research has been conducted on the coastal management, livelihood management and related issues in home & abroad. From the existing literature review gives an way forward to know about coastal livelihoods an introductory analysis, living in the coast people and livelihoods, living in the Coast Problem, Opportunities and Challenges, living in the coast measuring quality of life, living in the coast urbanization, resource use by indigenous community in coastal zone of Bangladesh, coastal livelihoods situation and context, generating sustainable employment in the coastal zone Bangladesh - present situation and future potentials etc. All the existing research are divers and covering vast area of coastal livelihood of the country. All the prior studies/ research has objective that are related to this study and not similar. From review all these related documents this study has got a path to go forward.

4. Methods and Materials

This paper is fully based on primary and secondary information collected from different sources. Secondary data collected from existing literature review of various book, journals, reports, website etc. Primary data has collected from the field by visiting physically as well interviewing people form the study area. The integral part was to identify and collect data; they were classified, analyzed, interpreted and presented in a systematic manner to find the vital points. Primary data collected by interviewed of local people with different occupation. Respondent selected randomly. Along with this Key respondent interview, focus group discussion and various stakeholder like government official, NGO officials etc. people interviewed to get a clear scenario of the study area. The assessment activities were carried out in four major steps given below figure:

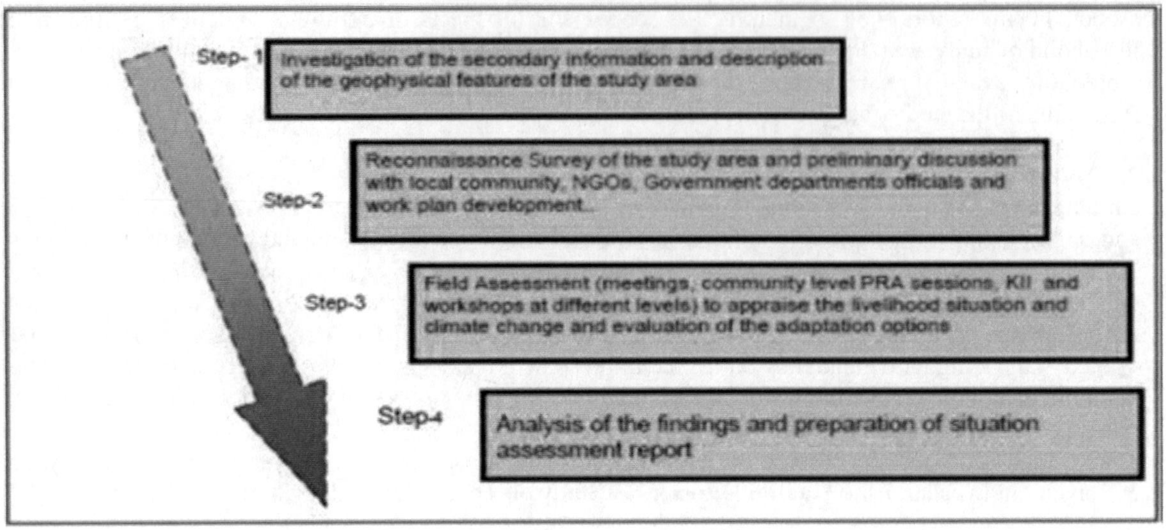

Figure 1. Steps of assessment activities.

5. Analysis and Findings

Different livelihood groups and patterns, including their respective vulnerability, were identified. Assessed local livelihood options and their seasonal dimensions and changing trends. A livelihoods classification was done through a participatory discussion with the community people and based on national data in order to understand the status of the livelihood groups of the study area in below table.

Table 1. Union wise livelihood classification of the community in the study area

Livelihood group	% per union			
	Teknaf Sadar	Nhila	Subrang	Baharchara
Farmer (Salt)	3	7	4	3
Wage laborer (Fishing & Salt production)	19	8	15	21
Fishermen	66	63	71	66
Businessmen	6	14.5	6	5
Service holder	5	4.5	3	4
Others	1	3	1	1
Total	100	100	100	100

The local community in the study area Unions consists of various livelihood groups as mentioned above. Almost all livelihood groups are affected by different types of natural and man-made risks (like natural disaster - cyclone, storm surge, inundation with saline water etc.; climate change – sea level rise, rise of temperature, increase of soil & water salinity etc.) However, the vulnerability of these livelihood groups to different risks depends on various physical and socio-economic factors:

a. Physical assets – cultivable land, irrigation facilities, agriculture/ fishing equipment, livestock, housing conditions, orchard/ homestead garden etc.
b. Human resources- literacy, education, knowledge and skills, availability of health facilities, overall health of the family etc.
c. Socio-economic conditions - Overall social status, access to local decision making groups, cash savings credit and markets, cash valued assets)

As the same risks, may have different impacts on different sectors and members of society, the local communities have been divided into various livelihood groups, i.e. small/ marginal farmers, large farmers, rural wage laborers, fishers, large business men and petty traders. Special attention was given to the specific vulnerability of women.

IOP Conf. Series: Earth and Environmental Science **70** (2017) 012044 doi:10.1088/1755-1315/70/1/012044

The degrees of vulnerability (very high, high and medium) of farmers in the four unions of the study area to various risk factors are shown in below table, which is based on focus group discussion and interaction with different organization.

Table 2. Union wise levels of vulnerability of small & large farmers

Major risk factors	Nature	Teknaf Sadar	Nhila	Subrang	Baharchara
Salinity	Climatic	VH*	M*	H*	VH
Tidal water intrusion	Climatic	VH	M	VH	H
Tidal Surge	Climatic	VH	M	VH	H
Drought	Climatic	H	H	H	M
Heavy rainfall	Climatic	H	M	M	M
Water logging	Non-Climatic	M	H	M	M
Flood / Flash Flood	Climatic	M	M	VH	M
Kalboishakhi	Climatic	M	M	M	M
Depression	Climatic	M	M	M	M
Hail Strom	Climatic	M	M	M	M
Thunderstorm	Climatic	H	M	M	M
Cyclone	Climatic	VH	VH	VH	VH
River Bank Erosion	Climatic	VH	M	VH	NA
Insect- pest infestation	Non- Climatic	VH	H	M	VH
Fog	Climatic	NA*	NA	NA	NA
High price of agricultural inputs	Non-Climatic	M	M	M	M
Unavailability of agricultural inputs	Non-Climatic	M	M	M	M
Crop yield reduction	Non-Climatic	H	M	M	H
High Temperature	Climatic	M	M	M	M
Change in land use pattern (from agriculture to shrimp/ Salt cultivation)	Non-Climatic	VH	M	NA	

*Legend: VH- Very high; H- High; M- Medium, NA – Not Applicable

A list of the major local climatic and environmental hazards and the frequency of their occurrence was generated through documents from a website. Concerning vulnerability and the risks, a perceptions assessment was done by taking opinion, about the occurrence, frequency, intensity of the hazards and the impact on livelihoods of the local people. The risk from coastal hazards is characterized by the frequency of occurrence and severity of the hazards. People had been asked to rank the natural disasters by assigning points to them. The most common hazards considered by the respondents are cyclone, storm surges, sea erosion. Below figure shows that in Teknaf Area, 82.7% people, irrespective of their background consider cyclone be most risky. Another 13.9% people consider cyclone to be risky. Around one fourth of coastal people rank tidal surge to be most risky event. Around 67.6% people categorize tidal surge to be risky. Around 85% -95% of the coastal population have rated Earthquake and Draught as either "least risky" or "moderately risky", whether livelihood groups deviate from this general perception is explored then. It is found that people from all the livelihood groups are consistent in rating riskiness of natural hazards. These findings help to deduce that cyclone and tidal surge appear to be major threats for all the livelihood groups.

A possible future risk scenario including its impact on the livelihood was drawn by the community people, using a risk identification matrix as shown in below table. The scenario was based on the peoples' past experiences and perceptions, and current trends of increasing risks due to changing climatic conditions and accelerating intensity and frequency of natural hazards. The people perception is that if the climatic conditions keep on changing at the current rate, the impact of the main natural hazard on agriculture and allied sectors will be even more adverse in the next ten to fifteen years.

Table 3. Future risk scenario (based on local community perceptions and linking it with the climate change/ impact data)

Hazard	Vulnerable sector	Climate change impact/risk in coming future
Salinity	Agriculture/crop	Major crop losses per year, lower growth rate of plants, leaf injury and yield losses
	Livestock	Approx. 20% less production
	Fisheries	Loss of domestic fish varieties, approx. 70% yield losses
	Forestry	Plants become infested with different diseases, plants death rate increases, less production of fruits, enormous loss of betel nut, mango, jack fruit and litchi tree
Cyclone	Agriculture/crop	Huge crop damage, especially of Rabi crop
	Livestock	Increased death rate of livestock
	Fisheries	Approx. 50% of fish washed out from water bodies, increased fish death rate
	Forestry	Approx. half of the total tree stock damage
Storm surge	Agriculture/crop	Huge crop losses
	Livestock	Approx. 30% livestock losses
	Fisheries	One third fish loss, increased disease infestation
	Forestry	Approx. 30% of nursery seedling damaged
Flood	Agriculture/crop	Two third loss of T. Aman crop and huge damage to preserved food grain storage
	Livestock	Increased disease infestation, large number of deaths of cattle and poultry
	Fisheries	fish washed out from many fish fields
	Forestry	About 20% damage to nursery seedlings and saplings
Drought	Agriculture/crop	Rice crop damage
	Livestock	Scarcity of animal feed, livestock losses
	Fisheries	Approx. 20 % fish loss, increased disease infestation
	Forestry	Lower productivity of coconut and betel nut, increased fruit dropping, plants damaged due to lack of irrigation
Water logging	Agriculture/crop	Increased fallow land area, sweet potato, water melon, potato, gourd, chili, and tomato crop losses,
	Livestock	Increased disease infestation, approx. 20% less production of live stock
	Fisheries	About 15% fish loss
	Forestry	Approx. 15% seedling damaged

The local population in the study area is trying to adapt the changing climatic and other socio-economic conditions. There is, however, scope for identifying new options and for upgrading and/or refining the existing ones to meet the local livelihood needs. many local elite/experienced persons and various government and non-government organizations officials a number of options have been identified as having the potential to improve livelihoods adaptation and the management of climatic risks in the area. They are in below table:

Table 4. Major Problems and proposed Solutions Locally can be adaptation options

Problems	Solution	Recommendations
Fishes are decreasing day by day.	Banning of Trawling in Shallow (less than 40 Bam) water.	- Organize area based organization and aware the local people about the rules and regulation of trawling. - Inform the positive side of banning of trawling in shallow water. - Imposing a system of huge penalty for trawling in shallow water. - Involving Coast Guard and way along with low enforcing agency. - Increase Government Control over trawling. - Involving small fish boat as a source against trawling in shallow sea.
	Banning all types of fishing in the month of June, July & August.	- Form area based organization and make people aware of the positive impact of this solution. - Involving local elites and aged group in awareness campaign. - Measures should be taken so that no trawler & boat can either enter or exit the fishing ghat in these 3 months. - Government should take the responsibilities so that Hatchery cannot collect mother shrimp in these months and trawlers cannot go to sea. - Employ guard along the sea shore to implement the solution.
	Repairing the existing embankments and constructing new one's so that sea water cannot enter in to lakes and canals.	- At first existing embankments, should be repaired and this repair should be made in the month of October - January. - Quality materials must be used in repairing the embankments. For the spots, too close to sea, bolder sand concrete blocks should be used. - After the repair work plantation, should be done with heavily rooted plants species. - Measures should be taken so that lakes and canals cannot be turned to salt field or shrimp farm. - Embankment maintenance and fresh water aquaculture practice should be done by local conservation groups.
Problems in Agriculture.	Dept. of Agriculture should arrange training for the farmers.	- Farmer training program should be arranged through area based group. - Farmers with a small land or no land should be preferred for training. - The topics of the training should include seed bed preparation, using proper fertilizer pesticides etc. - Successful trainees should be facilitated with loan so that they can implement these trainings in field. - Farmers should arrange at least one meeting a month on different Agro- issues and NGO's should represent the meeting.
	Chemical fertilizer should be easily available in the area and should be properly used.	- First of all a list of genuine farmers should be made by the local area organization and should be trained in right use of fertilizer. - The local Agricultural office should test the Farmer's land fertility in free of cost. - The smuggling of fertilizer should be stopped by forming effective committee involving local elites and politicians. - Punishing the involved persons in smuggling. - Farmers should get the fertilizer directly instead via dealers.

Problems	Solution	Recommendations
		- Government should help the local organizations. - Local farmers should be trained and award to use organic fertilizer.
Number of trees has decreased in the area.	Arrange local community through group and involving them in social forestation in the Government Khas land.	- To make aware the local community social forestry. - Through area based group/ organization, Khas land should be leased from UNO/DC/ Forest office. - Beneficiary group selection and completing the agreement draft. - Strong measures should be made so that conflict do not arise in selecting beneficiary group. - Along with plantation the beneficiary group should be well award about the maintenance of the social plantation. If they cannot bear the expense of the maintenance, then Government should help them and it should be noted in the agreement made between Government and beneficiary group. - A committee should be formed from the beneficiary group who will take care of the whole plantation and willmake solution of all conflict. - Upazila level officers should be involved in all steps of agreement. - Along with social plantation, villagers should be trained in alternative income generation activities and they should be trained on "High tech stove.
	Unnecessary deforestation should be protected and trees should be planted in homestead area.	- Villagers should be informed about the financial value of the homestead trees and plants. - Free Government seedlings of timber, medicinal plants should be distributed by locally formed group and this group is also responsible for maintenance of the planted trees. - The seedlings should be distributed without any favoring to anyone. - NGO initiative should be taken to establish nursery in the area. - Government official should visit the plantation to make people interested in plantation.
Embankment related problem.	Embankment should be repaired and highly rooted plantation should be made on the sides of embankment.	- At first locally formed group should take the permission of plantation on embankment from the respective office. - The plants for plantation should be locally available. For this purpose, nursery, should be established. - The villagers should be trained on plantation & plantation management. - The income/ profit from the plantation should be divided 50/50 between the local groups fund and plantation care taker. - The plantation should also be used in other income generative activities. - The repair work of the embankment should be directly supervised by Authority; not the contractors. - Embankment should be repaired through "Atel" types of soil. - Physical survey is must before, construction of new embankment. - Regular visit is essential by Government officials in maintaining

Problems	Solution	Recommendations
		the embankment. - The part of embankment which is too close to sea should turn inside. - Along with Government supervision, a local Committee should be formed to maintain the embankment.
Natural Disaster issues	In case of any devastation like cyclone, Storm etc.	- Build awareness about the natural disaster situation. - Ensure early warning system from all level. - Train people on emergency evacuation, rescue, response & recovery. - Strengthening community by cooperative activity like food bank etc. - Insist community to face any problem collectively. - Increase participation, cooperation and collective effort to the community - NGO's and other Government organization should take initiative to increase capacity and insist community to be united for their better livelihood.

6. Conclusions

In Teknaf Upazila, adverse impacts of climatic risks and its increase in near future the lives of local community are badly affected in many ways. In the four study area unions, increased cyclone is ranked as number one risk, causing huge damage to the crops, water and the entire environment of the area. In addition to Cyclone, the other major risks posing direct threat to livelihoods are salinity and floods, storm surge, droughts, water logging, deforestation and forest degradation, insect and disease infestation and virus infestation in fishes. The vulnerability of different livelihood groups depends on various physical and socio- economic factors, deciding the coping capacity of the individual group to resist the disaster impacts. Among the various livelihood groups, the rural wage laborers are at the highest risks, followed by small / marginal farmers and fishers. Women due to number of socio- cultural and other physical reasons are generally more vulnerable to climatic and non climatic risks. The situation study showed that the local population, vulnerable to various natural and human induced risks is already trying to adapt the changing climatic and other socio- economic conditions. There is however, need for further updating and/ or refining of existing practices. Several adaptation options have been identified which have the potential to improve the livelihoods adaptation capacities and the management of climatic risks in the area. Further research is needed to assess climatic effect on livelihood management of all other disaster prone areas over Bangladesh.

References

[1] C. M. Pickering and W. Hill, "Impacts of recreation and tourism on plant biodiversity and vegetation in protected areas in Australia," *J. Environ. Manage.*, vol. 85, pp. 791–800, 2007.

[1] Asian Disaster Preparedness Centre, 2006: Report on Option Menu for Livelihood Adaptation to Climate Change and Extension Tool Development, prepared by Climate Change Impact assessment and Livelihood Adaptation Options in Drought Prone Areas, DP9/1 BGD/01/004, March, 2006

[2] Ali, A. (1999). Climate Change impact and adaptation assessment in Bangladesh. Space Research and Remote Sensing Organization (SPARSO). Dhaka, Bangladesh.12;109-116. URL:http://v3.weadapt.org/placemarks/files/227/c012p109.pdf

[3] Asian Disaster Preparedness Center (ADPC), (2006). Option Menu for Livelihood Adaptation to Climate Change and Extension Tool Development, Asian Disaster Preparedness Center (ADPC)

Bangkok, Thailand, March, 2006.

[4] BBS.2011. http://www.bbs.gov.bd , accessed December 10, 2014.

[5] BBS.2010. GDP Bangladesh 2004-05 to 2009-10, Accessed on December 10, 2014. http://www.bbs.gov.bd/WebTestApplication/userfiles/Image/SubjectMatterDataIndex/GDP_2008_09.pdf .

[6] Chowdhury, M.S.N., Hossain, M.S. and Barua, P. 2008. Artisanal Fisheries Status and Sustainable Management Options in Teknaf Coast, Bangladesh. Social Change, 1(1): 48-64.

[7] Chowdhury, M.S.N. 2006. Integrated Fisheries Management of the Naaf River Estuary for Livelihood Sustainabilit y of the Coastal Communities. MSc Thesis, Institute of Marine Sciences and Fisheries, University of Chittagong, Bangladesh, 160pp.

[8] Climate Change and Bangladesh, (2007), Climate Change Cell, DOE; "Climate Change and Bangladesh" published by DFID, UNDP and Department of Environment, Bangladesh, September 2007.

[9] Coastal Zone: Integrated Coastal Area Management Planning Project, UARPO, Dhaka, 2005. Daily Samakal 06.07.2008. Weekly, 2000 – 23.05.2008.

[10] Denscombe, M. (2007). The good research guide for small-scale social research projects (3rd ed.). Open University Press.

[11] E, M. K. Siddique; A. Adrika. Jan 2011, Sustainable Coastal Zone Management of Bangladesh, A Scoping Report for Mangroves for the Future, IUCN, Produced by MFF with the financial support of Norad and Sida.

[12] Government of the People's Republic of Bangladesh, 2005: Final Report of National Adaptation Programme of Action (NAPA): Preparing for Future (MOEF: November 2005), Ministry of Environment and Forest, Dhaka.

[13] Government of the people's Republic of Bangladesh, 2007: Improving adaptive capacity to climate variability and change for sustainable food and livelihood security in drought prone areas and in coastal regions of Bangladesh (Subcomponent Plan), Govt of Bangladesh, UNDP and FAO.

[14] Government of the People's Republic of Bangladesh, 2008: National Plan for Disaster Management (2008-2015) (Final Draft), Disaster Management Bureau, Ministry of Food and Disaster Management, Dhaka.

[15] Islam, M.R. and Ahmad, M.2004. Living in the coast: Problems and Opportunities. PDO-ICZMP. WARPO. Dhaka.

[16] M.A. Rouf, M.K. Uddin, S.K. Debsarma and M. Mizanur Rahman, (2011): Climate of Bangladesh: An Analysis of Northwestern and Southwestern Part using AOGCM; The Agriculturist 9(1&2):143-154(2011).

[17] M.A. Rouf, R.A. Feroz, S.K. Debsarma and M.K. Uddin, (2010) : Livilihood Situation, Climate Change Impact and Policy Recommendation of Sapahar&Porsha Upzila of Naogaon District(Drought Prone Area) of Bangladesh; Journal of Environmental Science & Natural Resources, 3(1):131-136, 2010.

[18] M.S.N. Chowdhury, M.S. Hossain, N.G. Das and P. Barua. 2011. Small-scale Fishermen along the Naaf River, Bangladesh in Crisis: A framework for management; Institute of Marine Sciences and Fisheries, University of Chittagong, Chittagong. September2011

[19] PDO-ICZMP, 2001. Resource Use by the Indigenous Communities in the Coastal Zone; a study by Mr. Mesbah Kamal, Dr. Avizit Reaz Quazi & Sanzida Akhter. Program Development Office for Integrated Coastal Zone Management Plan; Water Resources Planning Organization; Ministry of Water Resources, Bangladesh. Dhaka, July 2001.

[20] PDO-ICZMP, 2003c. Coastal livelihoods; situation and context. Program Development Office

for Integrated Coastal Zone Management Plan Project; Working paper WP015. Water Resources Planning Organization; Ministry of Water Resources, Bangladesh. Dhaka, June 2003.

[21] PDO-ICZMP, 2003d. Coastal Livelihoods; an introductory analysis. Program Development Office for Integrated Coastal Zone Management Plan Project; Working paper WP011. Water Resources Planning Organization; Ministry of Water Resources, Bangladesh. Dhaka, January 2003.

[22] PDO-ICZMP, 2004a. Living in the Coast: People and Livelihoods. Program Development Office for Integrated Coastal Zone Management Plan Project; Living in the Coast Series 1. Water Resources Planning Organization; Ministry of Water Resources, Bangladesh. Dhaka, March 2004.

IOP Conf. Series: Earth and Environmental Science **70** (2017) 012045 doi:10.1088/1755-1315/70/1/012045

Rice production model based on the concept of ecological footprint

S A Faiz, A D Wicaksono and D Dinanti

Department of Urban and Regional Planning, Faculty of Engineering, Universitas Brawijaya, Malang, Indonesia

paink.faizasyraf@gmail.com

Abstract. Pursuant to what had been stated in Region Spatial Planning (RTRW) of Malang Regency for period 2010-2030, Malang Regency was considered as the center of agricultural development, including districts bordered with Malang City. To protect the region functioning as the provider of rice production, then the policy of sustainable food farming-land (LP2B) was made which its implementation aims to protect rice-land. In the existing condition, LP2B system was not maximally executed, and it caused a limited extend of rice-land to deliver rice production output. One cause related with the development of settlements and industries due to the effect of Malang City that converted land-function. Location of research focused on 30 villages with direct border with Malang City. Review was conducted to develop a model of relation between farming production output and ecological footprint variables. These variables include rice-land area (X_1), built land percentage (X_2), and number of farmers (X_3). Analysis technique was regression. Result of regression indicated that the model of rice production output $Y=-207{,}983 + 10.246X_1$. Rice-land area (X_1) was the most influential independent variable. It was concluded that of villages directly bordered with Malang City, there were 11 villages with higher production potential because their rice production yield was more than 1,000 tons/year, while 12 villages were threatened with low production output because its rice production yield only attained 500 tons/year. Based on the model and the spatial direction of RTRW, it can be said that the direction for the farming development policy must be redesigned to maintain rice-land area on the regions on which agricultural activity was still dominant. Because rice-land area was the most influential factor to farming production. Therefore, the wider the rice-land is, the higher rice production output is on each village.

1. Introduction

Malang Regency was the center of agricultural development in East Java. Through Region Spatial Planning (RTRW) of Malang Regency for Period 2010-2030, productive districts had been organized to maximize the supply of rice production, including districts bordered with Malang City. But rice-land area usually reduced due to the conversion of agricultural-land to other land-uses and district's development as the consequence of Malang City's growth. The conversion of agricultural-land into non-agricultural-land was evident by the use of farming fertile land for settlement, industry and other usage (Rustiadi, et al, 2011). Rice-land protection policy to reduce land convertion activity, including sustainable food farming-land (LP2B) in Malang Regency, was not maximally implemented. Conversion against farming-land was a factor behind the scarcity of food derived from farming production output (Rahmanto, 2002). As a result, agricultural-land stock for rice production output would be limited.

Ecological footprint concept was about resource supporting capacity. As noted in a theory proposed by Wackernagel and Rees (1996), ecological footprint was a calculation of biological capacity or resource supporting capacity. Resource support in this case, was shown by rice production output because, Ekins in Giljum, et al (2007), ecological footprint did not count all natural assets. Ecological footprint covered only natural environment parts but with irreplaceable important functions. Concerning with their regional characteristic, villages that bordered with Malang City must have huge potentials for supplying rice production.

2. Methods

Type of research was descriptive. Research attempted to investigate and to identify the patterns of development and phenomena that was still undergone or would be happening. This study focused on the villages in the Malang Regency, which bordering the city of Malang. There are 30 villages used as a location for this study and in accordance with the theory of the Roscoe in Sugiyono (2010) decent sample size in the study were between 30-500 samples and Hair et al (2006) which states that the linear regression can be effective with the amount of 20 data or samples. Selection of the 30 villages of the 8 district, also based on criteria such as policy direction, the conditions and characteristics of the village so that the data obtained varies.

Classic assumption test serves to determine and examine the variables that are used in accordance with the criteria variables. Its function is to determine the right variables and variables can generate a good regression model. Classic assumption test to be used are normality test, multicoloniarity test and heteroscedasticity test. Normality Test will be using the Kolmogorov-Smirnov method with significant value that must be larger than 0.05. Multicoloniarity test by looking at the value of tolerance that must be larger than the value of 0.10 and value of VIF (Variance Inflation Factor) should be less than 10. Meanwhile, Park method of heteroscedasticity test that takes into account the value of t-test that must be lower than the value of t-table.

Regression analysis is basically the study of the dependency the dependent variable with one or more of independent variables (Gujarati in Ghozali, 2005). The used variables are variable based on the concept of ecological footprint. This concept is also used in regression analysis with the use of regression model that function to calculate the supply or availability of resources indicated by the dependent variable of rice production. Multiple regression analysis was functioned to see the variables that affect the rice production and regression model can be used in the depiction of the value of agricultural land supply.

Variables for regression analysis were arranged based on ecological footprint. As noted by the concept, resource supply emanated from rice production output of each village. By virtue of ecological footprint concept, then theories were made to support the emergence of variables in rice production model.

Y : rice production (tons)
X1 : rice-land area (hectares)
X2 : built land percentage (percentage)
X3 : number of farmers (people)

The independent variables and the dependent variables will be analyzed on multiple regression using SPSS 16.0 software that simplify the implementation of multiple regression calculation. This calculation also includes the interpretation of model summary and coefficient values. So we get the regression model as follows:

$$Y_1 = a + b_1x_1 + b_2x_2 + b_3x_3 \quad (1)$$

3. Data of The Variables

Datas used in this research was obtained from the secondary survey process. Datas are from government agencies such as BAPPEDA (planning and development agency), PU (infrastructure, settlements and spatial develpment agency), and BPS (statistic agency). Datas obtained in the form of districs in figures contains all data such as rice-land area, rice production, the number of farmers, and settlements area to compare with rice-land area and produce built land percentage.

Table 1. Data of the variables

No	Districts	Villages	Rice production (Tons)	Rice-land area (Hectares)	Bulit land percentage (%)	Number of farmers (people)
1	Dau	Karangwidoro	286,58	46,00	46,97	1.568
2		Kalisongo	118,37	19,00	83,07	1.408
3		Tegalweru	180,67	29,00	9,63	1.073
4		Landungsari	336,42	54,00	63,42	1.032
5		Mulyoagung	249,20	40,00	73,65	170
6	Karangploso	Tegalgondo	1.376,00	179,40	18,53	877
7		Kepuharjo	1.296,23	169,00	21,03	1.014
8		Ngijo	257,71	33,60	53,33	747
9		Ampeldento	880,52	114,80	24,97	454
10	Singosari	Banjararum	2.336,63	114,00	48,55	99
11		Tunjungtirto	3.013,03	147,00	37,95	613
12		Langlang	2.070,17	101,00	26,05	314
13	Pakis	Sekarpuro	556,93	72,00	48,79	94
14		Ampeldento	1.245,36	161,00	22,40	224
15		Sumberkradenan	1.059,72	137,00	10,72	425
16		Kedungrejo	510,52	66,00	20,13	414
17		Mangliawan	1.028,78	133,00	37,81	169
18		Tirtomoyo	154,70	20,00	32,56	202
19	Wagir	Sitirejo	363,09	91,00	22,30	108
20		Sidorehayu	446,88	112,00	14,34	212
21		Jedong	235,41	59,00	16,71	297
22		Pandanlandung	75,81	19,00	18,77	103
23	Tumpang	Ngingit	913,54	90,50	14,18	942
24		Kidal	625,85	62,00	32,40	1.617
25		Kambingan	575,38	57,00	15,02	416
26	Tajinan	Tambaksari	925,05	122,00	37,38	259
27		Tangkilsari	1.139,63	150,30	21,78	203
28		Sumbersuko	947,79	125,00	64,22	919
29	Pakisaji	Kebonagung	1.765,73	184,70	50,36	49
30		Kendalpayak	1.836,47	192,10	48,32	644
Total/Average			26.808,17	2.900,40	34,51	16.666

4. Results

4.1 Result of Classical Asumption Test

Classical assumption test in regression model, including normality test, multicolonearity test, and heterocedasticity test. Normality test on regression model generate significant value of 0,057 and had a value above 0.05. Multicoloniarity test generate all variables in regression model have a value of tolerance and VIF appropriate, so the correlation between variables is not proven. Tolerance and VIF value of rice-land area variable (X1) are 0,938 and 1,066. Built land percentage variabel (X2) had Tolerance and VIF value are 0,939 and 1,065. And the value of Tolerance and VIF of number of farmers variable (X3) are 0,924 and 1,083. All variabels had the value of tolerance are more than 0,10 and VIF value are less than 10.

The value of t-table of heteroscedasticity is 2,776 and all of the variables comply heteroscedastidsitas test because the t-test value is smaller than the value t-table. The t-test of of rice-land area variable (X1) is 2,501, Built land percentage variabel (X2) had the value of t-test is 0,814. And the value of t-test of number of farmers variable (X3) is 0,063. All variabels had the t-test value is less than 2,776. Based on the classic assumption test calculations on the model, all variables have been tested and can be used in the calculation of the regression model.

4.2 Results of Regression Model

The regression model had dependent variable is rice production (Y1) and independent variable are rice area (X1), built land precentage (X2), and the number of farmers (X3). Based on the calculation of regression model, the R-square value by 56,6%, which can be defined independent variables are used in the calculation of regression model can explain the relationship of influence by 56,6% against the dependent variable, and 43,4% are influenced by other variables not included in the independent variables of this study.

Table 2. Result of Model Summary and ANOVA

Model	R	R-square	Adj. R-square	F	Sig.
1	.752	.566	.516	11.302	.000a

The value of F-table on the first regression model was 2,59 and compared with the value F-test, the value of F-table is smaller and can be defined independent variables have a significant effect on the dependent variable.

Table 3. Result of Coefficient value

Model	B	Std. Error	t	Sig.
(Constant)	-207.983	300.865		
Rice-land area (X_1)	10.246	1.810	5.659	.000
Bullt land Perentage (X_2)	4.227	4.942	.855	.400
Number of farmers (X_3)	.063	.213	.295	.770

Based on the calculation of regression model, the variable of rice-land area (X1) is the most influential variable on rice production (Y1) and will form regression model as follows:

$$Y_1 = -207{,}983 + 10{,}246x_1 \quad (2)$$

Rice-land area (X1) had a coefficient of determination of 10,246 with positive effect (+). It means that each hectare of rice-land will have 10,246 tons of rice production. Every increment 1 hectare of rice-land may increase rice production into 10,246 tons. Which means the increasing of rice-land area, will make increasing of rice production.

The constant was -207,983 with negative load (-). The constant itself was a regression value reflecting a determination coefficient rate of independent variable. It means that if rice-land width is 0 or, in the case of a village without rice-land, there will be a deficiancy for rice production output by an average of 207,983 tons. and overall the negaative constant value in this regression model, indicating a shrinking of rice production and rice production is is very dependent on rice-land area.

4.3 Results of Regression Model

The simulation of regression model was aimed to attest the model and also to examine villages with potential for supplying rice production output. Because regression model was created with rice-land area (X1) as the influential variable, then rice production was measured from regression model. Simulation using regression model and existing data of rice-land area in each village. Based on the calculation model, will be obtained results of Rice Production (tons) The area constituted by rice-land in each village. The following findings regression calculation Simulation Model.

IOP Conf. Series: Earth and Environmental Science **70** (2017) 012045 doi:10.1088/1755-1315/70/1/012045

Table 4. Simulation of Regression Model

No	Districts	Villages	Existing data		Result of Regression Model
			Rice-land Area (Ha)	Rice Production (Tons)	Rice Production (Tons)
1	Dau	Karangwidoro	46,00	286,58	263,33
2		Kalisongo	19,00	118,37	-13,31
3		Tegalweru	29,00	180,67	89,15
4		Landungsari	54,00	336,42	345,30
5		Mulyoagung	40,00	249,20	201,86
6	Karangploso	Tegalgondo	179,40	1.376,00	1.630,15
7		Kepuharjo	169,00	1.296,23	1.523,59
8		Ngijo	33,60	257,71	136,28
9		Ampeldento	114,80	880,52	968,26
10	Singosari	Banjararum	114,00	2.336,63	960,06
11		Tunjungtirto	147,00	3.013,03	1.298,18
12		Langlang	101,00	2.070,17	826,86
13	Pakis	Sekarpuro	72,00	556,93	529,73
14		Ampeldento	161,00	1.245,36	1.441,62
15		Sumberkradenan	137,00	1.059,72	1.195,72
16		Kedungrejo	66,00	510,52	468,25
17		Mangliawan	133,00	1.028,78	1.154,74
18		Tirtomoyo	20,00	154,70	-3,06
19	Wagir	Sitirejo	91,00	363,09	724,40
20		Sidorehayu	112,00	446,88	939,57
21		Jedong	59,00	235,41	396,53
22		Pandanlandung	19,00	75,81	-13,31
23	Tumpang	Ngingit	90,50	913,54	719,28
24		Kidal	62,00	625,85	427,27
25		Kambingan	57,00	575,38	376,04
26	Tajinan	Tambaksari	122,00	925,05	1.042,03
27		Tangkilsari	150,30	1.139,63	1.331,99
28		Sumbersuko	125,00	947,79	1.072,77
29	Pakisaji	Kebonagung	184,70	1.765,73	1.684,45
30		Kendalpayak	192,10	1.836,47	1.760,27
Total			2.900,40	26.808,17	23.478,01

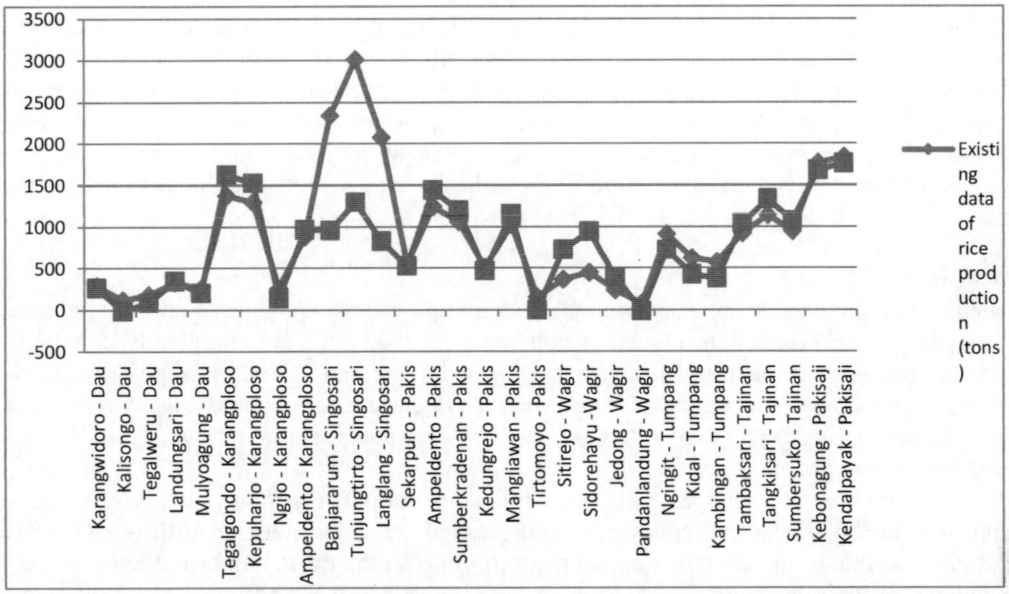

Figure 1. Comparison chart of rice production (tons) based on existing data and result of regression model

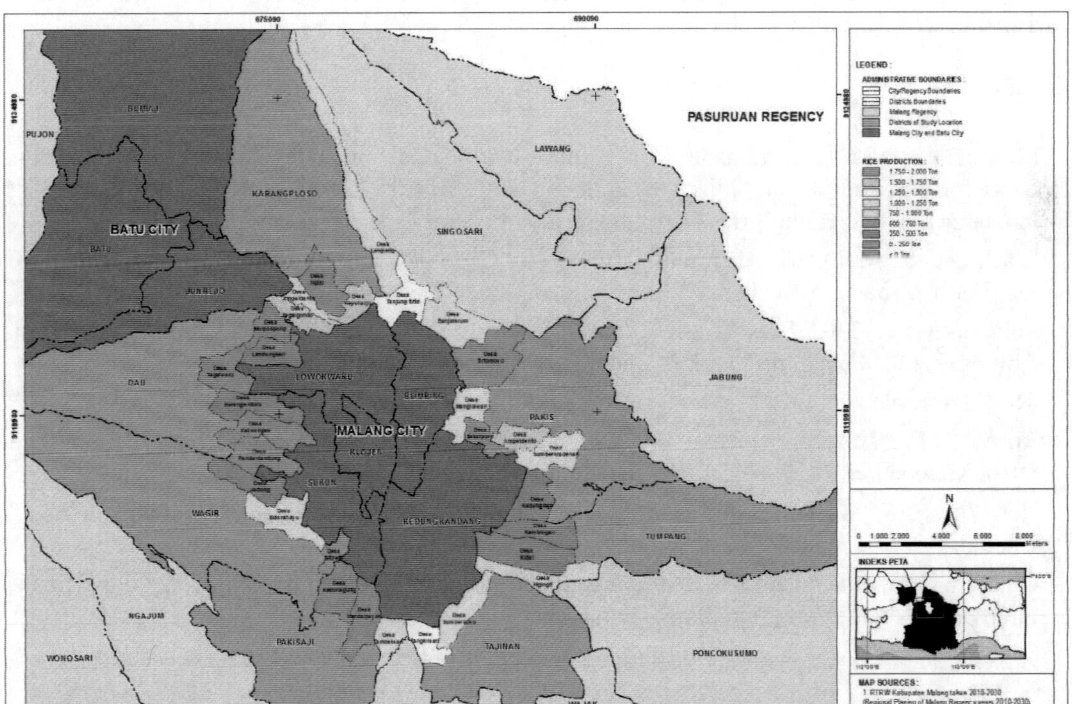

Figure 2. Map of rice production based on result of regression model

Result of regression model's simulation indicted that rice production output for all villages was 23,478.01 tons with rice-land area of 2,900.40 Ha. Based on existing data, rice-land productivity for all villages was 8,09 tons/hectares This productivity was quite high and potential if compared with other regions. Based on the result of regression model's simulation, there were 11 villages with rice production more than 1.000 tons/year. These villages in its existing condition would have quite extensive rice-land. But almost all vilages were influenced by urban effect when land conversion activity must be inescapable. The impact was evident on 12 other villages with rice production output less than 500 tons/year.

5. Conclusions

By taking account the result of regression model calculation, rice-land area was an influential variable for the supply of rice production output. In concern with their regional characteristic, villages that bordered with Malang City have rice-land are that was very potential. Rice-land to produce the rice had a quite higher productivity, precisely 8.09 tons/ha and the rice was a consumable food by few community members. Achieving this productivity, it would need very big and potential rice-land supply to produce the expected rice production output.

However, there were 12 villages with rice production output less than 500 tons/year. It is then recommended that the direction of policy must consider regional characteristic. Villages with extensive rice-land must be able to maintain its rice-land and to improve the quality and productivity of its land. Farming activity should be the dominant work in the village, and the supply of rice production output must be made available for fulfilling the rice demand of the community.

Recomendation on villages that have little rice products due to limited rice-land area, that villages can be used for the development of settlements and economic activity center. So between agriculture and settlements activity, both can able to develop well.

References

[1] Ekins, P et al. 2006. Sustainable Consumption and Production - Development of an Evidence Base: Resource Flows. In Giljum, S, et al. 2007. Scientific Assessment and Evaluation of the Indicator Ecological Footprint. Dessau: German Federal Environment Agency

[2] Wackernagel, Marhis & Rees William E. 1996. Our Logical Footprint: Reducting Impact On The Earth. Canada: New Society.

[3] Rahmanto et al. 2002. Persepsi Mengenai Multifungsi Lahan Sawah dan Implikasinya Terhadap Alih Fungsi Ke Pengguna Non Pertanian. Bogor: Litbang Pertanian.

[4] Rustiadi, Ernan. et.al. 2011. Menuju Desa 2030. Yogyakarta: Pohon Cahaya.

[5] Roscoe, JT. 1975. Fundamental Research Statistic for The Behavior Sciences. In Sugiyono. 2010. Metode Penelitian Kuantitatif, Kualitatif & RND. Bandung. Alfabeta

[6] Hair, Joseph F et al. 2006. Multivariate Data Analysis Fifth Edition. Jakarta. Gramedia Pustaka Utama

[7] Gujarati, Gamodar. 1995. Statistik Ekonometrika. In Ghozali, I. 2005. Aplikasi analisis multivariate dengan program SPSS. Semarang. Dipenogoro University

IOP Conf. Series: Earth and Environmental Science **70** (2017) 012046 doi:10.1088/1755-1315/70/1/012046

Study on integrating disaster risk reduction in Indonesian municipal spatial planning

T A Rachmawati

Department of Urban and Regional Planning, Faculty of Engineering, Universitas Brawijaya, Malang, Indonesia

tyas_49@yahoo.com

Abstract. Indonesia has experienced many major natural disasters. As result, the Spatial Planning Law number 26/2007 and the Disaster Management Law number 24/2007 were enacted. Spatial Planning Law number 26/2007 requires local governments to revise their municipal spatial plans. This paper aims to investigate the obstacles to revising the municipal spatial plans in response to disaster risk reduction and analyze the application of disaster risk reduction by municipalities in relation to the municipal spatial plan. For these purposes, questionnaires were distributed to 106 municipalities (52 responses were obtained) and document analysis was conducted on eight revised municipal spatial plans. The questionnaires, interviews, and document analysis revealed two obstacles to revising the municipal spatial plan (stakeholders with differing interests and a lack of spatial data) and that the eight plans had varying levels of application in disaster risk reduction as a result of the differences in support from government and non-government agencies.

1. Introduction

Indonesia is a nation that has experienced many major disasters, including earthquakes, tsunamis, volcanic eruptions, and mudflows. For this reason, disaster mitigation-based spatial planning is required. The central government amended Spatial Planning Law number 24/1992 (SP 24/2007), to Spatial Planning Law number 26/2007 (SP 26/2007) and enacted Disaster Management Law 24/2007 (DM 24/2007). Both laws require local governments to actively implement disaster risk reduction (DRR) in their spatial planning. After the enactment of those laws, many training sessions and workshops have been conducted by the Public Works Ministry, the National Development Planning Agency, the Indonesian Association of Planners, and other concerned organizations, all to enhance the understanding and knowledge of local government staff and planners regarding spatial planning and the law.

The Indonesian planning system is similar to those of other countries. It has a hierarchical planning structure in which local governments make key decisions within a national policy framework. With the enactment of Laws 32 and 33/2004 on regional administration and regional fiscal balancing (DSZN 32&33/2004), Indonesia is currently experimenting with an ambitious decentralization policy [1]. In line with DSZN 32&33/2004, SP 26/2007 and DM 24/2007 explicitly stipulate the authority and responsibility of local governments in regard to their spatial planning and DRR [2], [3], [4].

This paper aims to investigate the obstacles encountered by the local governments (e.g., regencies and municipalities) in revising the spatial plan, and to analyze the application of DRR with regard to the spatial structure plan, the spatial pattern plan, and the control of spatial utilization. Here, regency and municipal governments have the same level of autonomy, but generally regencies are concerned with rural areas and municipalities with developed areas. Hereafter in this paper, regency and municipality will be termed “municipality” and their spatial plan known as the “municipal spatial plan.”

Questionnaires regarding related obstacles were distributed to the 106 municipality staff who worked on the revision of the municipal spatial plan in Java Island, Kalimantan Island, Sumatera Island, Bali Island, and Lombok Island. Consequently, 52 municipal respondents participated. Additionally, interviews were conducted with nine municipal staff and seven consultants to confirm the questionnaire answers. Furthermore, eight revised municipality spatial plans were selected as samples from the following areas: Sidoarjo Regency, Padang Municipality, and Yogyakarta Municipality (all have experienced a major disaster); and Pasuruan Regency, Malang Regency, Lumajang Regency, Probolinggo Regency, and Sukamara Regency (these areas have not experienced major disaster). The application of DRR in the municipal spatial plans, in terms of its presence in the spatial structure plan, spatial pattern plan, and the control of spatial utilization, was analyzed.

2. Role of DRR in Spatial Planning in Indonesia

2.1. Indonesia Spatial Planning

The Indonesian spatial planning system is complex. It not only contains aspects of spatial development, but also economic, social, political, and environmental development aspects. The Indonesian spatial planning system is similar to land use management, because spatial development there is controlled by rigid zoning regulation (ZR) and codes [5]. Indonesian municipal spatial plan can be formulated either by a self-reliant municipality or by a private consultant under the close supervision of a governmental advisory team (officials from various local agencies, e.g., a development planning agency, public works agency, land administration agency, mining agency, local investment coordinating agency, or forestry agency). Under Ministerial Home Affairs Regulation number 1/2008, municipal spatial planning encompasses a comprehensive plan [6]. A comprehensive plan is advantageous in DRR because it provides powerful tools for a municipality to facilitate and coordinate service delivery (e.g., the central placement of a fire station, hospital units within a residential area, and open space/public space services) and development (spatial development and expanding facilities and infrastructure). **Table 1** shows the general content of Indonesian spatial planning: 1) the goal, policy, and strategy of the spatial plan, 2) the spatial structure plan, 3) the spatial pattern plan, and 4) the control of spatial utilization based on Article 19-32 of the Spatial Planning Law (hereafter SP Article 19-32) [4].

Residential centers and infrastructure networks systems, such as roads, railways, and water supplies, are determined in the spatial structure plan. Conservation and cultivation areas are spatially allocated in the spatial pattern. Conservation areas are classified into land use categories, such as disaster prone areas, natural reserves (wildlife sanctuaries and cultural heritage sites), and locally protected areas (mangroves, rivers, river banks, open spaces, and the seashore). Cultivation areas are classified into the following areas: residential, agriculture, mining, industrial, tourism, and trading and service areas. The determination of the spatial structure and the spatial pattern is a non-structural mitigation countermeasure of DRR, because settlement restrictions, evacuation routes, and evacuation points are outlined in these plans. As previously mentioned, the spatial structure plan and the spatial pattern plan can permit the inclusion of DRR. Building permission has the primary task of structural mitigation within the municipal spatial plan [7]. The control of spatial utilization in the municipal spatial plan can be used as a guideline with regard to building permission regulations. Thus, the building permission itself is to be stipulated in the district spatial plan by the relevant zoning regulation (ZR).

Table 1 shows the hierarchy of the Indonesian spatial plan. As shown in the upper columns of the table, the spatial plan has to follow a nationally set policy framework and that of a higher-level government. SP 26/2007 specifies that the municipal spatial plan must refer to the national plan and the provincial spatial plan with regard to spatial development. In **Table 1**, symbols +, ++ and ○ symbolize the degree of contribution by the various substances to the spatial plans.

DM 24/2007 has resulted in a change in disaster management in Indonesia. It articulates the implementation and enforcement of DRR in the spatial plan and comprises precise enactments for spatial plan regulations, safety standards, and sanctions against violators. Municipalities are required by Article 42.a-b of the Disaster Management Law (hereafter, DM Article 42.a-b) to periodically monitor and evaluate the implementation of the spatial plan [3].

This paper focuses on the municipal spatial plan that is dot-lined in **Table 1**, because the municipal spatial plan has been shown to be the most effective method to achieve the long-term reduction of community vulnerability to multiple hazards [8]. As described above, the municipal spatial plan has a spatial dimension and is the legal basis for the authorization of development location and land administration, and the implementation of physical development. SP 26/2007 stipulates the minimum standards of services for municipal spatial planning (SP Article 58). To preserve the environment, at least 30% of a watershed area must include a designated forest area (SP Article 17), and at least 30% of all urban areas must be open space areas (SP Article 28-30) [4]. SP 26/2007 also provides for new methods of enhancing development control such as zoning, planning permits, the implementation of incentives and disincentives (SP Article 35-40), and imposing sanctions, such as administration and criminal sanctions (SP Article 69-75) [4].

Table 1. Hierarchy and general content of Indonesian spatial law (SP article 19-32) [5]

Focus of research	National spatial plan	Provincial spatial plan	Municipal spatial Plan	District spatial plan	Municipal technical spatial plan
Goal, policy, and strategy	++	++	+	o	–
Spatial structure plan	+	++	++	+	o
Spatial pattren plan	o	+	++	++	+
Control of spatial utilization	–	–	o	+	++

++ = very strong, + = strong, o = moderate, – = low

2.2. Revision of Municipal Spatial Plan under SP 26/2007

This part discusses the obstacles to the revision of the municipal spatial plan in response to DRR under SP 26/2007. The municipality itself begins the revision process of the municipal spatial plan. A draft municipal spatial plan is drawn up and the municipality then has to discuss, consult, and coordinate at a provincial level to ensure accordance with the provincial spatial plan. If the draft is in accord with the provincial plan then the provincial authority gives its approval. After this agreement process, the municipal spatial plan is reviewed by the Kementerian Tata Ruang dan Agraria to ensure integration with the national spatial plan and the various ministerial sectoral plans. This process of discussion, consultation, and coordination requires a long timeframe and sizable budget.

As described above, questionnaires were distributed to 106 municipalities and 52 respondents participated. **Table 2** shows the attributes of the 52 respondents. It can be seen that the respondents can all be considered to have an appropriate level of expertise in spatial planning.

The questionnaire content contained questions regarding: 1) respondent attributes; 2) revision progress of the municipal spatial plan; 3) knowledge regarding SP 26/2007, DM 24/2007, and DSZN 32&33/2004; 4) obstacles to the revision of the municipal spatial plan regarding DRR; 5) the application of DRR in the content of municipal spatial plan; and 6) type of cooperation among municipalities regarding DRR. The questionnaires contained open answered questions except with regard to questions number 3) and 5), where multiple-choice answers were provided. The focus of this paper is on the answers to question 4); the results were classified into seven categories. Figure 1 shows the results for question 4): the obstacles to the revision of the spatial structure plan, the spatial pattern plan, and the control of spatial utilization in the municipal spatial plan with regard to DRR, as answered by the 52 respondents (18 municipalities and 34 regencies). Eleven of the thirty-four regencies answered that the differing interests among the stakeholders were the biggest obstacle. The open answers and the information obtained from the interviews identified the stakeholders as municipal officials, non-government organizations, the private sector, and the community. Within the 18 municipalities, the competing interests among stakeholders were identified as the second obstacle to the revision of municipal spatial plans. With regard to the results in Figure 1, interviews were conducted with the nine municipality staff and seven consultants, as shown in **Table 2**. Interview questions covered: a) the different interests among stakeholders; b) the content of the required spatial data; c) how to obtain spatial data; d) ZR arrangements; e) the existence of hazards maps and their content; and f) evacuation routes and meeting points. The examples given for different interests among stakeholders within the municipalities regarding DRR were: 1) classification of land as either cultivation or conservation areas, and 2) the expansion of residential, industrial, and agriculture areas, and infrastructure in hazard prone areas versus DRR. Respondents also answered that the different interests arose from issues over revenue gathering in the municipality. That is, the taxes paid by companies and individuals (for example, for land acquisition and fix assets) are the main sources of municipal income, and as such competing interests can arise.

Table 2. Attributes of the 52 municipal staff and consultants

	Agency		Work experience			Educational background				Additional knowledge	
			<3	3-6	6<	URP	CE	Arch	Other	Yes	No
Respondent	Local Planning Board Agency	37	16	12	9	22	2	3	10	20	17
	Local Public Works Agency	10	6		2	6	2	1	1	4	6
	Other	5	2	3	-	3	-	1	1	1	4
Inter-view	Local Planning Board Agency	9	3	4	2	7	-	2	-	7	2
	Consultant	7	1	4	2	7	-	-	-	6	1

URP = Urban and Regional Planning, CE = Civil Engineering, Arch = Architecture.

Further obstacles identified in the 34 regencies were: 1) lack of spatial data; 2) ZR arrangements; 3) compliance with regard to 30% of watershed areas to be forested areas; 4) land use planning at border areas between other municipalities; and 5) compliance with regard to 30% of urban areas to be comprised of open spaces. For the 18 municipalities, the main obstacles were identified as being: 1) compliance with regard to 30% of urban areas to be comprised of open spaces; 2) different interests among stakeholders; lack of spatial data; 3) ZR arrangements; compliance with regard to 30% of watershed areas to be forested areas; and land use planning at border areas.

Spatial data are used as a common reference set by various local planning authorities for hazard modification. However, 11 respondents answered that a lack of spatial data was an obstacle to revision. Municipal staff and consultants stated that the absence of up-to-date, accurate, relevant, and reliable

IOP Conf. Series: Earth and Environmental Science **70** (2017) 012046 doi:10.1088/1755-1315/70/1/012046

land-use and other thematic maps hindered the process. These answers suggest that an appropriate governmental authority should be responsible for standardizing the scale of spatial data, classification schemes, and survey methods.

ZR are legal tools for guiding land utilization with consideration to ecological constraints and socioeconomic objectives, and are used to match the operational aspects of spatial pattern plans with municipal spatial plans. As Indonesia is now highly decentralized, the enforcement of zoning, standards, and codes lies in the hands of local governments. The Ministry of Public Works published the latest technical ZR guide in 201o, however the guide did not reflect DRR, and as a result six regencies and two municipalities stated that they have experienced difficulties with regard to ZR arrangements (**Figure 1**).

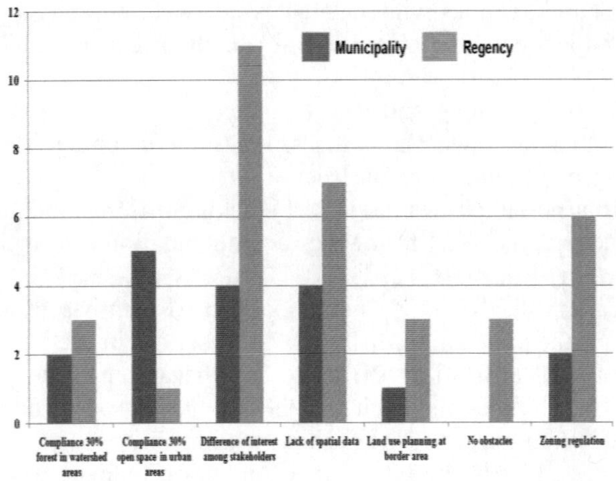

Figure 1. Obstacles revision of municipal spatial plan under SP 26/2007.

As described above, the requirements for watershed areas to be comprised of 30% forested areas and urban areas to include 30% open spaces are new articles in SP 26/2007. The requirement for 30% forested areas will also indirectly prevent flooding and landslides, and the requisite 30% open space area provides an evacuation point and evacuation route plan (SP Article 30) [2]. These requirements are more often seen as obstacles for municipalities than regencies, because a municipality's main activities involve urban residences, government services, social services and economic activities that involve high competition for land utilization. Thus, it is problematic to retain 30% of urban areas as open space. Usually one watershed encompasses a number of municipalities, and to comply with the order that 30% of the watershed be a forested area is a critical issue for three of the respondent regencies and two of the municipalities. A key challenge in implementing forestry decentralization in Indonesia is the increase in requests from municipalities to convert forestland [9]. Negotiations are occurring between the national government and municipalities regarding such conversions. The decision on whether proposals are accepted or rejected is a lengthy process, taking months and often years to complete.

Land use planning at border areas of municipalities is also a crucial issue. Land use planning at administrative boundaries is important because major disasters impact beyond these bounds. Good co-operation among municipalities regarding DRR was demonstrated between Yogyakarta Municipality and Central Java Province at a local DRR forum, which established the Merapi Forum in 2006. The Merapi Forum was a multi-stakeholder forum that addressed a single hazard in a particular area in that instance, volcanic eruption. Merapi Forum members were from the surrounding regencies affected by the Merapi eruption. The forum has fostered mutual understanding among the different stakeholders, and established cross-border and cross-sector collaboration in DRR. This has been maintained through regular communication, information sharing, and knowledge sharing [10]. Government Regulation number 50/2007 regarding co-operation among municipalities articulates that the municipalities in the process of decentralization can still build a suitable governmental institution to optimize the potential of their region, while having regard for DRR.

3. Integrating of DRR in Municipal Spatial Plan

3.1. Application

This chapter analyzes the application of DRR in the eight municipal spatial plans. The eight municipal spatial plans were chosen using the following criteria: 1) the municipal spatial plan was already published as a municipal ordinance, or 2) the revision of the municipal spatial plan was complete and awaiting being made a municipal ordinance. Consequently, three regencies in areas that had experienced a natural disaster and five municipalities without such events were chosen as shown on the left side of **Table 3**. The left of the table shows the characteristics of the regencies and municipalities, and any major disasters.

In the context of DRR, spatial planning and land management must support the following essential functions [10]: 1) early warning systems; 2) risk assessments and mapping; 3) prevention and reduction of disasters; 4) risk management; and 5) reconstruction after disasters. Another expert also indicated that spatial planning has four possible roles in DRR: 1) making differentiated decisions on land use; 2) keeping areas free from development; 3) following recommendations for legally binding land use or zoning plans; and 4) hazard modification [12].

Table 3 shows the application of DRR in the municipal spatial plan based on document analysis. As shown in the upper column, the application of DRR was observed in the following indicators: hazard identification, control of spatial utilization, existence of a hazard map, and existence of evacuation routes. Hazard identification estimates and evaluates the extent, intensity, and magnitude of potentially harmful events caused by certain hazards. The hazard map has to be included in the municipal spatial plan to raise public awareness. It will influence the behaviors of builders, developers, public agencies, and individuals in the location of critical activities and assets. The control of the spatial utilization municipal spatial plan is performed by stipulating the guidelines of ZR, permit incentives, permit disincentives, and sanction imposition. It is a valuable assurance of actual DRR implementation. Evacuation routes and meeting point plans are important issues for the improvement of spatial structures and spatial patterns in the municipal spatial plan, in response to DRR under SP 26/2007.

Table 3 shows the various types of disasters, such as earthquakes, tsunamis, tidal waves, volcanic eruptions, floods, landslides, fires, forest fires, drought, and abrasion. It can be seen from the third column that all the municipalities have already completed their hazard identification of the various disasters, which are considered in response to the various geographical conditions of the municipalities. As shown in the second column, hazard identification was classified into three categories: "AD: Available in detail", "A: Available", and "ND: Not in Detail". For example, there are many "ND" classifications in Sidoarjo Regency, Malang Regency Regency, and Sukamara Regency, representing an insufficient application of DRR in those regencies.

The third column of hazard map in spatial structure plan and spatial pattern plan shows Yogyakarta and Padang Municipalities both having experienced major disasters. All applications of DRR are "AD". In addition, many of the municipalities that have never experienced a major disaster are classified as "ND" with regard to their hazard identifications. This classification indicates a range in levels of DRR among the municipalities. According to the interviewee answers, these situations have arisen because of the different interests among stakeholders, a lack of spatial data, as shown in **Fig. 1**, a lack of understanding regarding the nature of disasters, and little disaster information (as explained below). Tidal waves, droughts, abrasions, and forest fires are disasters for which municipalities cannot provide detailed hazard identification and maps. While Padang Municipality has an "AD" map of an abrasion hazard, Probolinggo Regency's hazard identification of abrasion was "ND", because it is only areas prone to abrasion that are included in a hazard map with other hazards. There are "NDs" with regard to tidal waves in Sidoarjo Regency, drought in Malang Municipality, and forest fire in Sukamara Regency. It is difficult for these municipalities to both identify these types of hazards and to make detailed hazard maps because of a lack of information about these risks.

SP 26/2007 requires an evacuation point and an evacuation route plan. As shown in the third column of **Table 3**, Sidoarjo Regency, Yogyakarta Municipality, and Padang Municipality have hazard maps with an "AD" classification. Yogyakarta Municipality, Padang Municipality, and Pasuruan Municipality have an evacuation point and evacuation route plan, classified as "E: Exist" in the fourth column. **Fig. 2** is an example of the evacuation point and evacuation route plan included in the Yogyakarta Municipal Spatial Plan. The map shows the direction and destination point of any evacuation. Government Regulation number 8/2013 regarding the accuracy of maps for spatial planning has stipulated that the minimum map scale for regency spatial plans is 1:50,000 and the scale for municipality spatial plans is 1:25,000 [13]. A 1:50,000-scale map in a regency spatial plan cannot be said to represent an evacuation point and evacuation route plan for DRR. Therefore, regencies require action plans for evacuation points and evacuation route plans, and this should be stipulated in an article of a municipal ordinance to ensure its implementation.

Table 3 Application of DRR in municipal spatial plans under SP 26/2007 [15], [16], [17], [18], [19], [20], [21], [22].

Municipality		Indicators				
		General description		Spatial structure and pattern		Control of spatial utilization
		Type of disaster	Hazard identification	Hazard map	Evacuation point and evacuation route plan	
Disaster experience	**Sidoarjo Regency** East Java Province (coastal) Area: 634,34 km² Density: 2,441 persons/km²	Flood	ND	ND	NE Scale of map 1:50,000	ND
		Mudflow *	AD	ND		
		Tidal wave	ND	ND		
	Yogyakarta Municipality Yogyakarta Province (inland) Area: 32,5 km² Density: 15,197 persons/km²	Earthquake*	AD	AD	E Scale of map 1:25,000	AD
		Volcanic eruption	AD	AD		
		Landslide	AD	AD		
	Padang Municipality West Sumatera Province (coastal) Area: 69,5 km² Density: 1,101 persons/km²	Tsunami*	AD	AD	E Scale of map 1:25,000	AD
		Earthquake*	AD	AD		
		Landslide	AD	AD		
		Flood	AD	AD		
		Abrasion	AD	AD		
No disaster experience	**Pasuruan Municipality** East Java Province (Coastal) Area: 35,29 km² Density: 4,502 persons/km²	Flood*	AD	A	E Scale of map 1:25,000	ND
		Tidal wave	AD	A		
		Fire	AD	A		
	Malang Regency East Java Province (coastal) Area: 110,06 km²	Tsunami*	ND	ND	NE Scale of map 1:50,000	ND
		Earthquake	ND	ND		

Density:3,459 persons/km^2	Volcanic Eruption	ND	ND		
	Landslide *	AD	AD		
	Flood	AD	AD		
	Drought	ND	ND		
Lumajang Regency East Java Province (coastal) Area: 1,790.90 km^2 Density: 779.82 persons/km^2	Tsunami*	A	A	E Scale of map 1:50,000	A
	Earthquake	A	A		
	Volcanic eruption*	A	A		
	Landslide	A	A		
	Flood	A	A		
Probolinggo Regency East Java Province (coastal) Area: 1.696,17 km^2 Density: 644 persons/km^2	Earthquake	A	A	NE Scale of map 1:50,000	AD
	Volcanic eruption*	A	A		
	Landslide	A	A		
	Flood	A	A		
	Abrasion	ND	A		
Sukamara Regency Central Kalimantan (coastal) Area: 3,827 km2 Density: 11 persons/km^2	Flood	ND	ND	NE Scale of map 1:50.000	ND
	Forest fire*	ND	ND		

*Major disaster

Hazard identification	**Hazard map**	**Evacuation point & evacuation route plan**	**Control of spatial utilization**
AD: Available in detail. Municipal spatial plan states location, estimated impact, and scale of hazards.	**AD**: Available in detail. Hazards map for each hazard are drawn in municipal spatial plan.	**NE:** Non-existent.	**AD**: Available in detail. control of spatial utilization is distinguished by cultivation area and conservation area. Conservation areas are divided into many criteria, such as disaster prone areas, protected areas, etc.
A: Available. Municipal spatial plan states only location and scale of hazard.	**A**: Available. All hazards are drawn in one map.	**E**: Exists.	**A**: Available. Control of spatial utilization is separately applied to conservation area and cultivation area.
ND: Not in detail. Municipal spatial plan states only type of disaster.	**ND**: Not in detail. Only a conservation map is drawn.		**ND**: Not in detail. Control of spatial utilization in a general sense.

In the process of decentralization, the municipality has the authority to develop an appropriate mechanism to act as a spatial utilization permit, taking into account both local conditions and the established national framework [14]. The spatial utilization controls found in a municipal spatial plan consist of four mechanisms: 1) direction of ZR; 2) spatial utilization permit; 3) incentive and disincentive mechanism and; 4) administration and criminal sanctions. The "Controls of spatial utilization" in the fifth column of **Table 3** shows that both the Yogyakarta Municipality Spatial Plan and the Padang Municipality Spatial Plan include detailed information regarding the control of spatial utilization for conservation areas, and are classified as "AD".

In the case of Padang Municipality, the utilization of conservation areas is divided into two categories: protected areas and disaster prone areas. The directions of the ZRs are divided into three categories: "Directed", "Controlled", and "Restricted". Padang Municipality employs a spatial utilization permit to ensure the implementation of the municipal spatial plan. This mechanism is expected to assist in the identification, prediction, and evaluation of the foreseeable environmental consequences of land utilization.

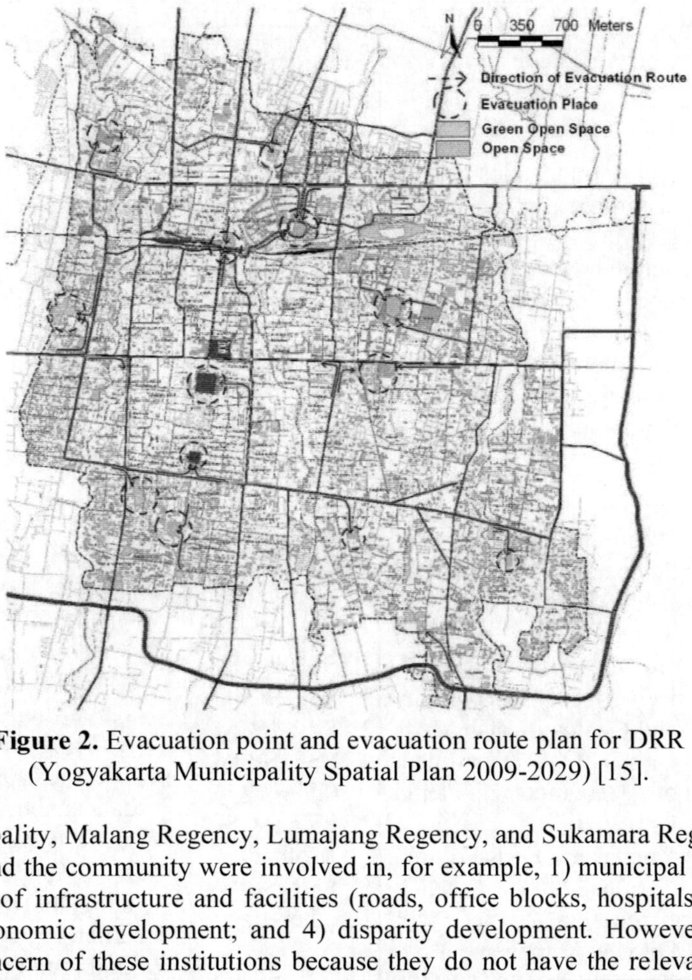

Figure 2. Evacuation point and evacuation route plan for DRR (Yogyakarta Municipality Spatial Plan 2009-2029) [15].

In Pasuruan Municipality, Malang Regency, Lumajang Regency, and Sukamara Regency, local NGOs, the private sector, and the community were involved in, for example, 1) municipal developments such as the development of infrastructure and facilities (roads, office blocks, hospitals, etc.); 2) land use development; 3) economic development; and 4) disparity development. However, DRR is not the responsibility or concern of these institutions because they do not have the relevant experience with major disasters, although SP 26/2007 does state that DRR is an issue to be addressed in municipal spatial plans.

There were a number of institutions in Sidoarjo Regency, Yogyakarta Municipality, and Padang Municipality available to supply spatial data relevant to DRR (and therefore mitigate the lack of data). For example, Deutche Gesselschaft fur Technische (GTZ), a German NGO, developed a guideline on how to integrate DRR criteria into spatial planning. This was then applied to the Yogyakarta Municipality Spatial Plan. Padang Municipality is the area most frequently stricken by earthquakes, and as such it is used as a benchmark for disaster mitigation in Indonesia. Therefore, many DRR plans and

programs have been issued by both national and local governments in this area, such as an evacuation point and evacuation route plan by the Ministry of Marine Affairs and Fisheries.
The Sidoarjo DRR forum, the Yogyakarta DRR forum, and the Padang DRR forum are organizations that exist in three regencies and municipalities. The forums are focused on DRR efforts. The members of the DRR forums include government staff, academics, NGOs, the Indonesian Red Cross, media, and community organizations. The Sidoarjo DDR and Yogyakarta DDR forums have supplied spatial data and risk reduction programs to SPP and CSU to support DRR management. The Padang DDR Forum has supplied data on communities in disaster prone areas and supported evacuation route plans for earthquakes and tsunamis. It is in this manner that the forums play important roles in municipal spatial planning for DRR.

Table 4 Support of government and non-government agencies in the revision of municipal spatial

Municipality	Institution	Support role	Content of supporting and assistance
Sidoarjo Regency	1. BPLS (mudslide disaster task force) 2. Institute Technology Sepuluh November 3. Local NGO; The Indonesian Forum for Environment (WALHI) 4. Sidoarjo DRR Forum 5. Community	1. SSP, SPP & CSU 2. SSP 3. SSP & SPP 4. SPP & CSU 5. SSP & others	1. Data on risk assessment of mudslide 2. a. Data on victims' housing preference b. Suggestions at public hearings 3. a. Data on mudslide victims b. Data on environmental impact of mudslide 4. Supporting spatial data and reducing risks program 5. Input location for resettlements
Yogyakarta Municipality	1. Indonesia Institute of Science (LIPI), BAKOSURTANAL, Meteorological and Geophysical Agency 2. Department of Geography and Geology Gajah Mada University, Department of Geography Veteran University 3. Yogyakarta DRR Forum 4. National NGO: Indonesian Society for Disaster Management 5. International NGO: UNDP, GTZ, CORDAID 6. Community	1. SSP & SPP 2. SSP & SPP 3. SPP & CSU 4. SSP, SPP & CSU 5. SSP, CSU & others 6. SSP, SPP & others	1. Spatial data on disaster prone area 2. Disaster prone areas data (spatial data) 3. Data relating to a community who lives in disaster prone area, supporting data spatial and reducing risks program 4. Technical assistance in spatial data analysis for DRR 5. a. Early warning system design b. Guidelines on how to integrate DRR criteria into spatial planning 6. Community mapping and opinions for rehabilitation process
Padang Municipality	1. Bappenas 2. Ministry of Marine Affairs and Fisheries 3. Reconstruction and Rehabilitation Team Padang Earthquake 2009	1. Others 2. SSP, SPP & CPU 3. Others 4. SSP, SPP & CSU	1. Financial assistance 2. a. Evacuation building shelter plan data b. Strategic disaster mitigation plan data

Municipality	Institution	Support role	Content of supporting and assistance
	4. Padang DRR Forum 5. Community	5. SSP & others	3. a. Manual on building capacity for victims b. Road map of rehabilitation and reconstruction 4. a. Evacuation route plan for tsunamis and earthquakes b. Data on community who lives in a disaster prone area 5. Community mapping for evacuation routes
Pasuruan Municipality	1. Community 2. Local NGO 3. Private sector	1. SSP & SPP 2. SSP & SPP 3. SSP	The stakeholders were concerned with municipality development, such as 1) development of infrastructure and facilities (roads, airports, office blocks, hospitals, etc,; 2) land use development; 3) disparities development, etc.
Malang Regency, Sukamara Regency, Lumajang Regency, and Probolinggo Regency	1. Local NGO 2. Private sector 3. Community		

SSP: Spatial structure plan; **SPP:** Spatial pattern plan; **CSU:** Control of spatial utilization; **Other:** Budget, program, etc

As already shown in **Figure 1**, the differing interests among stakeholders are also obstacles in the revision process of municipal spatial planning. It is desirable for municipalities to obtain external support for the revision of their municipal spatial plans so that the issues relating to DRR and competing interests can be addressed. **Table 4** shows the institutions that supported the revision of the municipal spatial plans and the nature of their assistance. The second column shows the range of institutions and agencies involved, such as governmental institutions, universities, and non-government organizations (NGO), and the third column displays the support and assistance provided with regard to the spatial structure plan, spatial pattern plan (SPP), control of spatial utilization (CSU), and other means such as support for the budget and DRR programs. The numbers in the third and fourth columns correspond to the number of institutions in the second column. Their support and assistance involved the supply of data and assistance with planning, as shown in the fourth column.

Under DM 24/2007, a municipality is required to establish a Municipal Disaster Risk Management Agency (DM Article 180) [3]. This agency must engage in the following key tasks: a) formulate and issue DRR policies; and b) coordinate the implementation of DRR activities. However, none of the eight regencies and municipalities has established such an agency. The municipalities are required to establish the agency in response to DRR in the municipality spatial plan to coordinate the different interests among stakeholders.

4. Conclusion

1. Spatial Planning Law number 26/2007 (SP 26/2007), Disaster Management Law number 24/2007 (DM 24/2007), and Regional Administration and Regional Fiscal Balancing Law

32&33/2004 (DCZN 32&33/2004) explicitly stipulate the authority and responsibility of local governments in regard to their spatial planning and disaster risk reduction (DRR).

2. The questionnaire results from 52 municipality correspondents revealed two main obstacles to the revision of the municipal spatial plans from the viewpoint of DRR. The first was the differing interests among stakeholders, namely municipal officials, NGOs, private sectors, and the community. And second, a lack of spatial data was identified, with regard to 1) the classification of land as either cultivation areas or conservation areas, and 2) the expansion of residential, industrial, and agriculture areas and infrastructure in hazard prone areas versus DRR.
3. A document analysis of the eight municipal spatial plans from Sidoarjo Regency, Yogyakarta Municipality, Padang Municipality, Malang Regency, Lumajang Regency, Probolinggo Regency, and Sukamara Regency indicates that the eight plans were revised in response to DRR with regard to the spatial structure plan, spatial pattern plan, and the control of spatial utilization. Nevertheless, their application to DRR may be limited as a result of a lack of spatial data and a lack of understanding concerning the nature of disasters. Those municipalities that have experienced major disasters, such as Yogyakarta Municipality and Padang Municipality, have designed more detailed hazard maps than those municipalities that have not (such as Malang Regency, Lumajang Regency, Probolinggo Regency, and Sukamara Regency).
4. With regard to the Yogyakarta Municipal Spatial Plan and the Padang Municipal Spatial Plan, the central government, NGOs, and universities were useful in supplying additional data to fill in the gaps concerning spatial data and a lack of understanding of the nature of disasters.
5. DRR forums, which involve government staff, academics, NGOs, the Indonesian Red Cross, media, and community organizations in Padang Municipality, Yogyakarta Municipality and Sidoarjo Regency, play important roles in the revision of the municipal spatial plans and in supporting DRR management by supplying spatial data and DRR programs.
6. Although DM 24/2007 (Article 180) required municipalities to establish a Municipal Disaster Risk Management Agency, none of the eight municipalities has established an agency. The establishment of such an agency is desirable to coordinate the different interests among the stakeholders in DRR.

References

[1] Fahmi F. Z., D. Hudalah, and Firman T. Lessons from Inter-local Government Cooperation Effectiveness in Greater Yogyakarta. Working Paper Regional and Rural Planning Research Group, Institute Technology Bandung-Indonesia 2010; 1–70.

[2] Ministerial of State Secretariat of Republic of Indonesia. Regional Administration and Regional Fiscal Balancing Law 32&33/2004. (In Indonesian)

[3] Ministerial of State Secretariat of Republic of Indonesia. Disaster Management Law 24/2007. (In Indonesian).

[4] Ministerial of State Secretariat of Republic of Indonesia. National Law of Spatial Planning 26/2007. (In English).

[5] Turniningtyas Ayu Rachmawati. Comparison Studies on Integrating of Disaster Risk Reduction (DRR) in Spatial Planning Systems in Indonesia, Japan and European Countries. Jurnal Penanggulangan Bencana. Badan Nasional Penanggulangan Bencana. Jakarta- Indonesia 2013; 13-22.

[6] Ministerial of State Secretariat of Republic of Indonesia. Guidelines for Planning in Urban Areas 1/2008. (in Indonesian)

[7] Fleischhauer Mark and Greiving S. Spatial Planning Response toward Natural and Technological Hazards. Natural and Technological Hazards and Risks Affecting the Spatial Development of European Regions. Geological Survey of Finland 2006; 109-123.

[8] Burdy J. Cooperating With Nature: Confronting Natural Hazards with Land Use Planning for Sustainable Communities. The USA: Joseph Henry Press 1998; 1-29.

[9] Komarudin Heru and Sigiaan L. Yuliana. Power Struggle over Indonesia's Foresland: the Role of Stakeholder Engagement. IASC. 2008.
[10] International Strategy for Disaster Reduction (ISDR) Local Governments and Disaster Risk Reduction. Good Practices and Lessons Learn. United Nations. 2010; 27-29.
[11] Theo Kotter. Prevention of Environmental Disasters by Spatial Planning and Land Management. 2nd FIG Regional Conference, Marrakech, Morocco. 2003.
[12] Fleischhauer Mark. Resilience of Cities to Terrorist and other Threats. The Role of Spatial Planning in Strengthening Urban Resilience. The Netherland: Springer Science. 2008; 273-298.
[13] Ministerial of State Secretariat of Republic of Indonesia. Accuracy of Maps for Spatial Planning 8/2013. (in Indonesian)
[14] Kidokoro Tetsuo, Nguyen Truc Anh, Tran Mai Anh. Improving Spatial Planning System and Development Control Mechanisms toward Sustainable Urban Development in Asia Cities'. The Fourth Urban Research Symposium, Tokyo-Japan. 2010.
[15] Sidoarjo Regency. Municipal Spatial Plan of Sidoarjo Regency 2009-2029. (in Indonesian)
[16] Yogyakarta Municipality. Yogyakarta Municipal Spatial Plan 2009-2029. (In Indonesian).
[17] Padang Municipality. Padang Municipal Spatial Plan 2010-2030. (in Indonesian)
[18] Pasuruan Municipality. Pasuruan Municipal Spatial Plan 2009-2029. (in Indonesian)
[19] Malang Regency. Malang Regency Spatial Plan 2010-2030. (in Indonesia)
[20] Lumajang Regency. Lumajang Regency Spatial Plan 2009-2029. (in Indonesian)
[21] Probolinggo Regency. Probolinggo Regency Spatial Plan 2009-2029. (in Indonesian)
[22] Sukamara Regency. Sukamara Regency Spatial Plan 2009–2029. (in Indonesian)

IOP Conf. Series: Earth and Environmental Science **70** (2017) 012047 doi:10.1088/1755-1315/70/1/012047

The role local initiatives in community based disaster risk management in Kemijen, Semarang City

W Z Fauzie, S Sariffudin

Department of Urban and Regional Planning, Faculty of Engineering, Diponegoro University, Semarang, Indonesia

sariffuddin@live.undip.ac.id

Abstract. Community-based disaster risk reduction is one of the homegrown initiatives efforts and community empowerment oriented in disaster management. This approach is very important because no one can understand the conditions in a region better than the local communities. Therefore, the implementation of CBDRM always emphasize local initiatives in decision making. The existence of local initiative is necessary specially to anticipate the impact of climate change which is increasingly affecting towns in coastal areas, including settlements in Semarang. Kemijen Urban Village is one of the informal settlements in Semarang, which has the highest intensity of flood that is 12 times during 5 years (2011-2015). The research question is how the level of local initiatives in flood disaster management in Kemijen, Semarang? This study aims to assess the level of local initiatives in Kemijen as the community adaptive capacity of flood prevention in pre-disaster, emergency response, and post-disaster. Local initiatives assessed on water supply, sanitation, food, shelter, health, drainage maintenance and waste management. This study shows the level of local initiatives in pre-disaster and post-disaster is almost same and bigger than the response phase. Scoring results showed that pre-disaster is 35.002, 27.9577 for emergency response, and post-disaster is 34.9862 with each category that is independent, empowered, and independent. This study also shows that local initiatives in Kemijen largely formed by individual initiative and only a few were formed by a collective initiative.

1. Introduction

In recent years, the impact of climate change is increasingly affecting the cities in coastal areas. This is a result of unplanned urban growth, especially the increase of informal settlements occurs in large scale in the coastal areas (World Bank, 2012). The growth of informal settlement growth phenomenon occurs in cities in developing countries, including Indonesia (Sariffuddin, Astuti, Farhaeni, & Wahdah, 2017). As unplanned neighborhoods with the heterogeneity of society, the village has problems such as poor quality of the houses, where public space is minimal and the quality of public spaces is low and the legality of land (Kencana & Yuliastuti, 2016). Living in informal settlements makes people more vulnerable to climate change. Public vulnerability is compounded by the very limited resources, inadequate infrastructure, and weak institutional systems weak and ineffective (Ayers, 2011; Danianti & Sariffuddin, 2015). It shows that living in informal settlements is particularly at risk for experiencing a loss in the face of disaster. The ability of communities in disaster risk management is necessary especially in the informal settlements that are vulnerable to disasters. Disaster risk management is a systematic approach to reduce vulnerabilities and risks, and to strengthen disaster resilience (Jörn Birkmann & von Teichman, 2010).

In 2011, extensive tidal inundation in Semarang amounted to 1538.8 ha, it spreads over 8 out of 16 districts. One the district is the Eastern District of Semarang with an area of 44.15 ha reaches tidal inundation, tidal inundation area was just scattered in the Kemijen (Ramadhany, Subardjo, & Suryo, 2012). Kemijen also has the highest intensity of flood in the city of Semarang with 12 times the incidence during the period of 5 years (2011- May 2015) (Nugroho, 2013). However, these conditions do not make public to avoid staying in the Kemijen village and chose to remain in the region for various reasons such as economic constraints. For further information, most people in Kemijen about 45% are as industrial workers whose earn the regional minimum wage (UMR) of Semarang.

Adaptability to reduce disaster risk at the local level is necessary (Sariffuddin & Wijaya, 2014). Since Kemijen community feel and direct observe themselves, so that they can understand the condition of the region especially related to the disaster. It's important to associate the risk of flooding with initiatives related to adaptation to climate change, and the specific issues of urban planning and management, such as the construction of infrastructure and provision of basic services (World Bank, 2012). Local initiative itself is also included in one of the core elements proposed by Fujikake (2008) in an awareness empowerment. All the steps and the activities carried out in PRBBK are based on the process of awareness planned (Nakmofa & Lassa, 2009). Therefore, this local initiative is part of a community-based disaster risk reduction. These local initiatives need to be explored in Kemijen Village community as one of the capital to improve adaptability in disaster management. Taking consideration on those issues, this study was conducted to answer the research question "What level of local initiatives in the prevention of flooding in the Kemijen?" The aim of this study is to assess the level of local initiatives in Kemijen village as the adaptability of the public in flood prevention in three stages of disaster, namely the pre-disaster, emergency responses and post-disaster.

2. Methodology

This study focused on the role of the community in the initiative for disaster management based on aspects of the Sphere Project in three stages of a disaster, according to the issues that have been formulated in this study. The Sphere Project's aspects used as a study variable are namely the water supply, sanitation supply, food supply, provision of shelter/transitional shelters, health care provision and rescue, drainage maintenance and waste management (Sphere, 2011). This research applied quantitative research methods. The sampling technique used was Probability Sampling with random sampling technique. The samples technique was done by assuming one representative family in Kemijen as one respondent. Furthermore, this study used the Slovin formula in determining the number of samples with a tolerance limit of 10% or 90% level of accuracy. So that it can be claimed that this study used 98 families.

Descriptive statistical was used to describe the results of data processing which is then presented in of charts, graphs, and maps. To determine the level of local initiatives, this study used scoring function to assess the criteria listed in each indicator by using Likert scale. The weight of the score for each indicator is based on the number of indicators in each variable, so that the total score is well distributed and balanced.

3. Results and Discussions

3.1 The Analysis of Kemijen Community Characteristics.

This analysis describes the conditions of the community that contains two things: the socio-economic characteristics of society which include level of education, livelihood, income levels and characteristics of local institutions in the form of formal institutions and informal (community) in the local level relating to development and disaster in the Kemijen village. Marfai (2008) and Sariffuddin and Wijaya (2014) described the socio-economic characteristics of the local community as one of the key factors that affect disaster response measures.

3.2 Community Socioeconomic Characteristics.

This aspect is about the level of education and economic conditions as seen from livelihood and income levels in Kemijen. *Income level.* General overview on community's education level in Kemijen is low. It is proved from the population with 36% are primary school graduates, further, there are also the population who do not complete primary school so that the total percentage of both at 44%. The level of education is one of the important indicators which show the quality of the population in Kemijen. The low level of education affects the livelihood and their income, therefore Kemijen village is dominated by the workers. The level of education also influences in disaster relief efforts, because of insight or knowledge affects how people think and take the initiative before they undertake the necessary measures for disaster mitigation in Kemijen.

Economic Conditions. The majority of the population in Kemijen are low-income communities (MBR) or below Rp 2.100.000, - per month with a percentage of 84%. It can be seen that generally people work as laborers in the amount of 51%. It is in line with the level of education of the people. Moreover, based on monographs data, there are 1,340 households included in poor communities. This condition makes people choose to in rural areas. Low income level which indicates poverty in Kemijen makes people vulnerable to floods.

3.3 Analysis of Local Institutions.

The local agency in Kemijen village consists of formal and informal institutions (community). One of the characteristics of the local institution is the existence of collective consciousness, as mentioned by Joern Birkmann, Chang Seng, and Setiadi (2012). The explanation strengthens that community-based disaster risk reduction is related to the concept of community empowerment, in which one of its core elements is awareness.

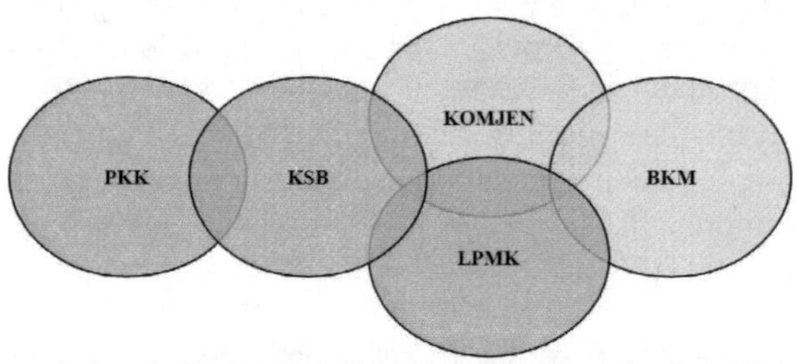

Figure 1. Interrelationship Local Institutions in Sub Kemijen

Figure 1 illustrates the inter-agency coordination mechanism in disaster management. The agency which does not have bound or coordination is directly shown in circles that do not intersect. PKK (Family Welfare Guidance) relates to the KSB Group (Disaster prepared Group) in disaster relief only on emergency response that helped the establishment of a common kitchen. Further, BKM (Community Self-Reliance Agency) is not directly related to KSB, since BKM only concerned on development in Kemijen village, but in fact the construction itself is also part of disaster management. Commissioner General (Kemijen's Community) and LPMK (Urban Village Community Empowerment Organization) is related to both KSB and BKM because they are equally helpful in terms of development as well as contribute to the disaster management program undertaken by KSB. There is a concern issues that the member of KSB, Komjen, LPMK are the same people. They are the ones who are really concerned about the environment in Kemijen.

3.4 Local Initiatives Analysis in Flood Prevention Needs Supply.

This analysis will discuss indirectly how local action and local knowledge contained village Kemijen as part of a local initiative in flood disaster management in Sub Kemijen associated with the provision of clean water, sanitation, food, shelter/ temporary shelter, medical services and rescue, as well as drainage and solid waste management.

3.5 Water Supply.
Clean water sources used by Kemijen's community are same at three stages of disaster. As many as 72% of people in Kemijen are using PDAM (Municipal Waterworks) water and 14% are using wells water. Those who use well water said that they cannot use PDAM water because it is difficult access in the installation, so the installation will be more expensive, especially if the installation is not done at the same time. Seeing these conditions, there is one person in Kemijen initiative in the provision of clean water as well be big opportunity for him becoming a distributor of clean water. The result is amounted 14% of people in Kemijen buy clean water from this distributor. To see the service area, see Figure 2.

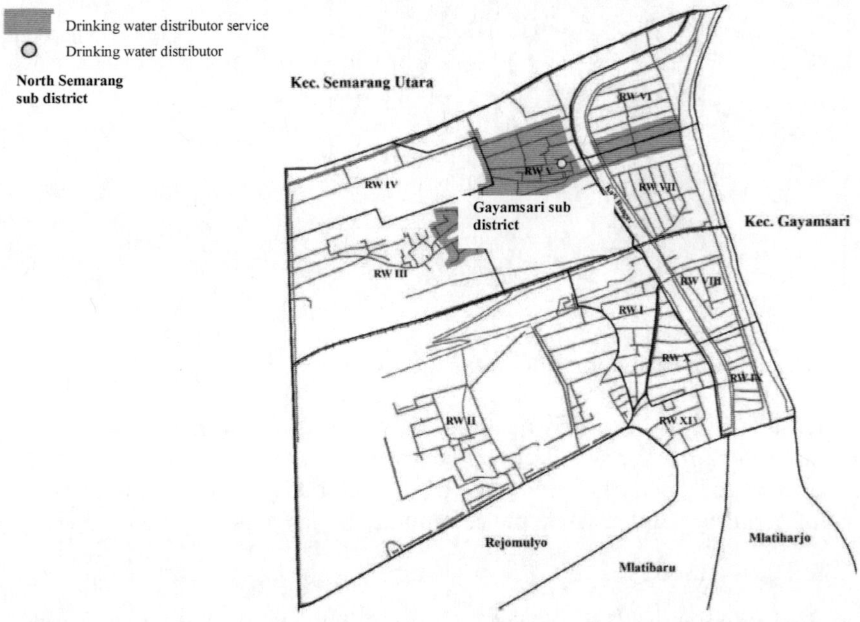

Figure 2. Water supply services from the distributor

Other issues come up related to the clean water supply and the water is turbid. However, people have been thinking over on these issues. They take the water in a container and then wait the dirty go down so that the clean water is in the top, then they consume it. On the other hand, people who are lack of water supply initiate to accumulate the water, so that it can be used whenever they need it. Generally, all the initiatives in this variable are the individual initiative, while the collective initiative has not been seen in the provision of clean water. It is not in line with what Nakmofa and Lassa (2009) said about.

3.6 Provision of Sanitation.
More than 94% of community already have private toilets in their house that, while 6% of the people do not have private toilets. Those who do not have private toilets said that they could not able to build a septic tank and private toilets for sanitation purposes, therefore they use communal toilets. To see the spread of communal toilets can be seen in Figure 3.

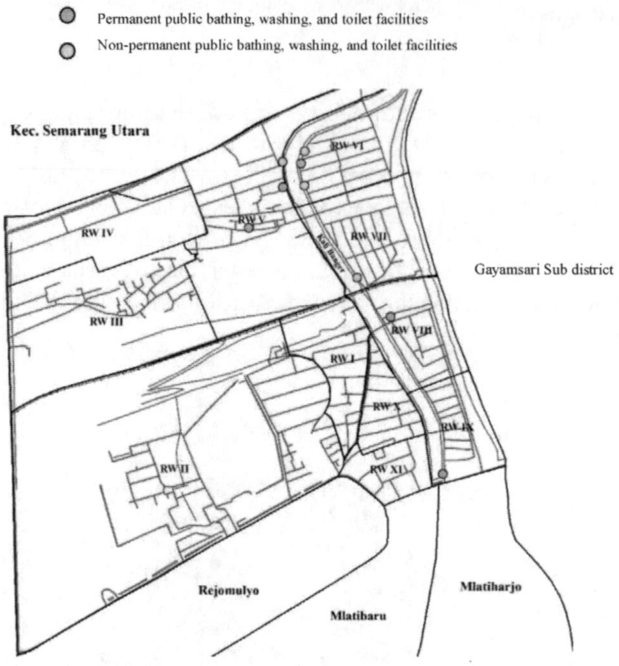

Figure 3. Distribution of communal toilets.

Since there are many people still do not have private toilet, it drives the community to ask the government to build communal toilets for them. All communal toilet construction was done by the people, but the financial resources were from the government, private sector and the community. Usually the government funds a kind of permanent toilet, while the community build semi-permanent toilet.

3.7 Food Supply.

The provision of food in disaster relief is needed, especially in the emergency response phase. Many people died because of the lack of food supply instead of disaster. The availability of food in the Kemijen in pre-disaster stage, disaster response and post-disaster is 100%, ready, so that people feel save.

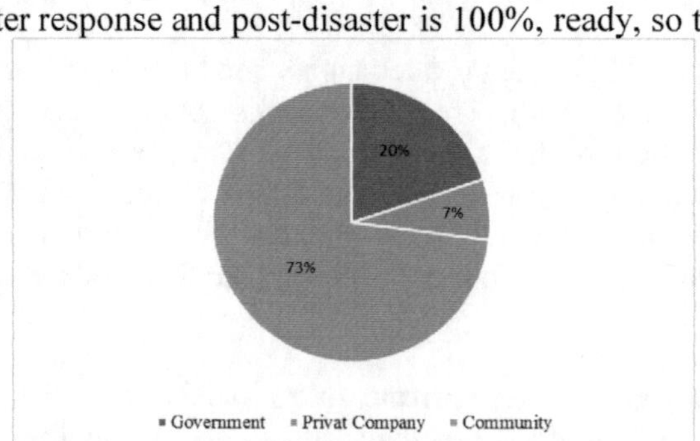

Figure 4. Food Source Provider in Response Stage.

Figure 4 describe some people said when the disaster came, they got help such as rice packs and groceries, some argue the government gave 20% from total assist, and some others argue the private sector (such as the student community) gave 7%, while 73% were from community itself. Furthermore, the community claimed that the food is eatable. In response phase, 38% of the community believes that there is a common kitchen, but most or about 62% of the community did not know these public kitchens.

It is proved that the community has awareness to provide public kitchen in disaster response stage. However, it still needs to tell anyone that public kitchen is available.

3.8 Provision of Housing / Temporary Shelter

Figure 5 show the majority or 53% of the people stated that they did not fine any evacuation place when disaster or emergency response coming, while 47% of people found the evacuation point. However, most people said that they prefer to stay at home when flood coming, because they think that they were save and no need to move to evacuation point. However, at least they have collective initiated to provide evacuation site.

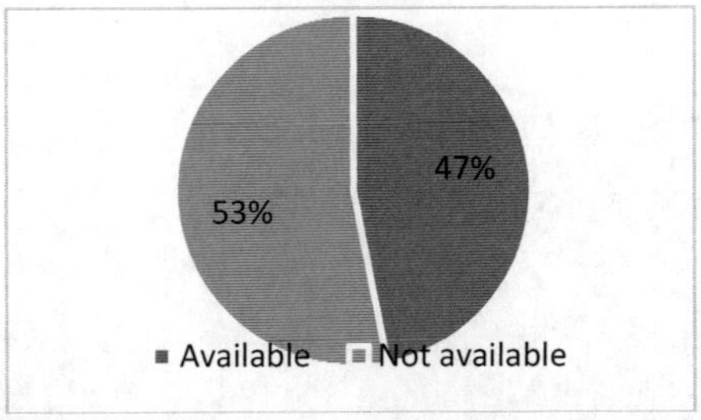

Figure 5. Points Evacuation Response Stage.

However, in 2014 there were houses which affected severely because of the water entered the house and reached 1-1.5 meters. Since this was the only low house that still exist in Kemijen. The evacuation point was selected in other community's houses in Kemijen.

3.9 Provision of Health Care and Rescue

Most people claimed that the disaster volunteers were not available in Kemijen, in fact the KSB also serves as volunteer evacuation, because they were trained directly by BPBD. It indicates that the socialization and information related to KSB is less spread in the community or the possibility that KSB is less active in Kemijen. The community did not know that the emergency kits were available by the BPBD and KSB. Furthermore, the community also did not have any emergency kits for disaster response.

To view the provision of health services can also be viewed on the availability of medical personnel that is provided in Kemijen village. If we see from the third stage of the disaster, the community said that the medical personnels are available most in the response phase, it was about 29%. This is also supported by statements from some respondents who said medical personnel from health centers checked the condition of the people when the floods come. The head of village also still communicate each month to the health center. As for the pre and post disaster medical personnel also played a role in checking larvae associated with jumantik program initiated by the PKK, but jumantik program is less useful related to disaster management.

3.10 Drainage Maintenance.

Most of the people or about 88% of all community still do routine activities for drainage maintenance. 24% of the people do drainage maintenance once a week, while 76% of the people do it every few weeks. In the emergency response stage, community does not do drainage maintenance. Since when flood comes the drainage system can not hold the water flow. However, there is still drainage management during emergency response by using a pump provided by the government. The existence of the government's pump has been very helpful to reduce flood water in Kemijen Village neighborhoods. Thus the public also believes that the floods in Kemijen (although the high level is about 50cm) can recede within 3-4 hours.

In Figure 6, the drainage condition can be quite good if the puddle is not high and clean of trash. The sufficient drainage condition is the inundation moderate to high, and there is garbage, but not too much. As for the poor drainage condition can be described as the inundation is height and a lot of garbage. Drainage condition also describes whether people actually performing maintenance on drainage in their neighborhoods. If you see the previous explanation, the percentage of availability of routines for maintenance of drainage by 88%, but the poor drainage of 16% and a considerable 40%. It indicates that there are some people who do not actually carry out routine maintenance of drainage or they do it unoptimally. It is also likely that there are other factors for their closed drainage makes people difficult to perform drainage maintenance.

Figure 6. Drainage condition.

Figure 6 also shows the image of a closed drainage. Communities whose their settlement environment found a closed drainage said that their closed drainage (the fund is from the government) makes them difficult to clean up the drainage. The purpose of the government or private sector that might be good to use as a drainage well as for the roads and that not a lot of incoming waste. However, it is less good for people, they think that because it is difficult to clean, the deposition of sand or soil in the longer drainage will be higher, it would make drainage more shallow and easily overflow.

3.11 Waste Management.

Garbage disposal system located in the Village Kemijen remained the same at each stage of the disaster. Most people has good enough disposal system with a percentage of 81%. There is even a small portion of the community or 3% who has waste recycled system. It proves that there are some people in Kemijen who have the initiative in waste management. However, there is also a carelessly discarded waste management or about 16%, they do not have a trash can or they have lower awareness to throw garbage properly. Most available trash cans are provided by the community, while 32% of trash cans are given by the government and 9% are assistance from the private sector. The trash cans from the government are quite good, because the trash cans are divided into two, for organic and non-organic garbage.

Figure 7. Waste Systems.

Kemijen society who prefers to transported their garbage must pay a fee of 4000-6000 Rupiah per month. The transportation is done once in three days, three days or twice a week. This system is actually good, but unfortunately the final disposal of the transport of the garbage is dumped on vacant land that distance is still very close to the settlement. This is due to the unavailability of polling stations in Kemijen, so that makes this village became slum garbage as shown in Figure 7. It will not be surprised if Kemijen village established as one of the slums in the city Semarang and frequent flooding. The dustbin aid program is very helpful for Kemijen society, but it would be better if the government provides assistance related to the provision of TPS (polling station), because the dustbin assistance program will be less effective if the polling station in Kemijen is yet available. Waste management by communities, sometime become to a source of social conflict due to unregulated management system.

3.12 Analysis of Local Initiative on Floods.

After doing the calculations scoring for each respondent, then the whole question added up in order to get the total score at each stage of the disaster. The total score will be categorized based on the level of empowerment in the context of disaster management. So it can be formulated whether people in Kemijen seen from initiative at three stages of disaster, namely, pre-disaster, disaster response and post-disaster included in the category of powerless, capable, independent, or civil. Table 1 show the scoring result of the level of local initiatives Kemijen Village community in disaster management.

Table 1 Scoring results of local initiatives Village Kemijen

Variable	Pra Disaster	Emergency Response	Post Disaster
Water Supply	55,1633	54,1224	54,8571
Provision of Sanitation	55,7755	55,3469	55,5306
Food Supply	46,5306	49,8571	46,8367
Provision of Housing / Temporary Shelter	0,0000	13,5612	0,0000
Provision of Health Care and Rescue	4,0408	5,2653	4,6531
Drainage Management	38,5459	0,0000	38,0867
Waste Management	44,9592	17,5510	44,9388

Variable	Pra Disaster	Emergency Response	Post Disaster
Total	35,0022	27,9577	34,9862

The total scores of local initiatives when categorized based on the level of empowerment in the context of disaster management at the three stages of a disaster, the result is Kemijen community is powerless in the stage of the emergency response. Category helpless still considered unfavorable because it signifies that the community as a whole still depend on government and private sector, so that society is less visible initiatives in this stage. As for the pre and post-disaster, community is in independent category. Independent category is assessed quite well in this stage, because it indicates that the ma of people have initiative in making or proposing a program and to apply, besides the public have also been able to access and maximize existing resources.

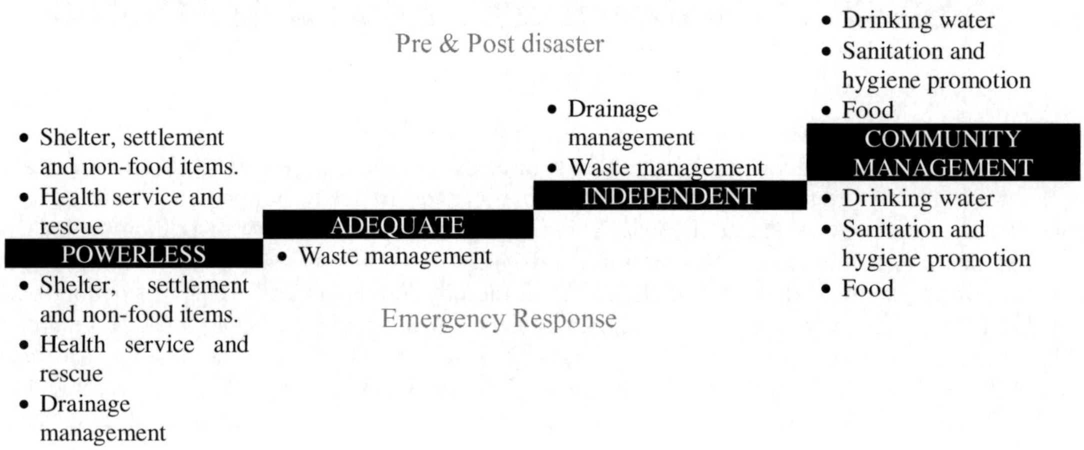

Figure 8. Community Empowerment Study on Pre-disaster, Emergency Response Stage and post-disaster

In Figure 8 tells the categorization of the variables in the response phase which shows that the people in Kemijen with the initiative in the provision of clean water, sanitation, and food in emergency response at the level of civil. The community has been very good at providing these variables. Besides, the local waste management initiatives are still at the level of helplessness. People still need to improve their initiatives in waste management, during this garbage is also a problem that is quite large in Kemijen village. Community initiatives seems weak on providing shelter/temporary shelter, medical services and rescue, and drainage maintenance due to fall into the category of helplessness. When it is linked with the socio-economic characteristics of society, it the level of local community initiatives at the level of civil until helpless is not distinguished by special characteristics. One of the factors affecting the level of local initiatives is society's active engagement in disaster management forums and activities. When they do it collectively, it will indirectly increase the level of initiative, meaning the ability of self-help in disaster response is relatively smaller than the ability of community-help. The knowledge of disaster management is not largely arise from self-help, but community-help, it can be referred to local knowledge. Still in Figure 8 categorization of variable level assessment of local initiatives at pre-disaster and post-disaster stage are put together in one picture because it has almost the same results. It can be seen from the community's mindset about the different stages of a disaster with what is found in the literature, see Figure 9. Some people think that there are only two phase in disaster process, disaster and no disaster. It is supported by the results of the analysis and scoring, it only shows that pre-disaster and post-disaster are merely the same. Therefore, KSB as institutions of local level disaster in Kemijen needs to disseminate to the public related to the disaster, so that people at least know the basics of disaster management.

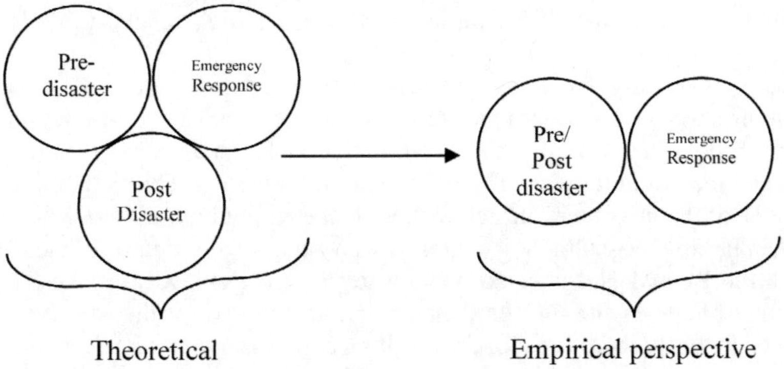

Figure 9. The different Disaster stages between Regulation and Community's opinion

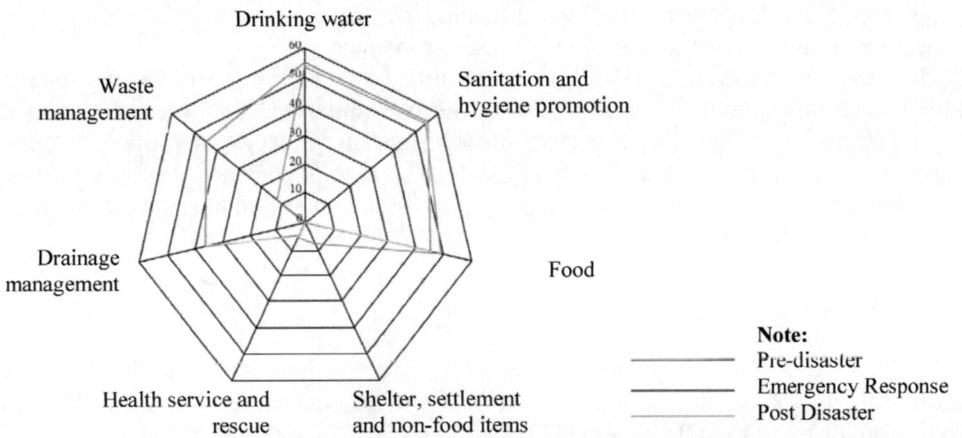

Figure 10. Model Level of Local Initiative in Kemijen

Modeling in Figure 10 makes us easy to view and compare the variables which are being assessed by local-level initiatives, such as pre-disaster and post-disaster stage that has almost the same score. It points out that the majority of local community initiatives more focus on meeting the daily needs, while local knowledge and local action in disaster response is less visible to the public. It can be seen from the variable of disaster response, but it is less related to the fulfillment of their daily needs, the variable provision of shelter/temporary shelters and the provision of health care and rescue. There are indications that the disasters did not significantly affect the people's lives.

4. Conclusions

The overall level of local initiatives in Kemijen included in self-help category at post-and pre-disaster stage, which means generally community can meet their needs in accordance with the initiative by utilizing the existing resources. The community already has initiative in making or proposing a program and applying it, though the government and private sectors still give assistance. On the other hand, response stage of the community included in capable category, which means that the community still needs assistance in responding flood disaster, although the community itself is capable to optimize what they get. Results of assessment on the level of local initiatives at pre-disaster and post-disaster stage have the same results in all the variables. It can be seen from the community's mindset about the different stages of a disaster with what is found in the literature,. Some people think that there are only two phase in disaster process, disaster and no disaster. In addition, a variable which is closely related to the disaster management process has a low yield compared with other variables which has high related to daily living needs.

When referring to the theory used in this study, the study's results supports the statement of Chaskin (2001) on how the interaction aims to promote and maintain the welfare of society emphasizes the action initiative. The results also support Chaskin (2001) argue that the initiative is divided into two types, namely the individual initiative and collective initiatives. The individual initiative is the initiative that the idea comes from individuals and decision-makers in action lies in the individual, although this initiative will have an impact for the public. One example is a person who acts as a distributor of clean water. Costly water supply through the PDAM and unclean wells water is yet provided to make the person take the initiative as a distributor of clean water in Kemijen. Collective initiatives is the collective ideas of several individuals as outlined in the deliberations for the resulting agreement or a joint decision. One example is there are people felt sorrow for access to sanitary, therefore some people apply for aid when (discussion was held) *musrenbang* to the government related to the provision of communal toilet. Collective initiativeint is in line with theory that was purposed by Nakmofa and Lassa (2009), namely the increasing of sense of belonging, sharing the same disaster risks. Geographically, community is in the same location and they are vulnerable to disasters. It means, they have the same catastrophic risks, so that society take the initiative in meeting the needs of disaster management.

However, when we see in the earlier analyzes, local initiatives in Kemijen is largely shaped by individual initiative and only few things is formed by collective initiative. It shows that the flooding problems in Kemijen are not included as collective problem, meaning that the definition of sense of belonging, sharing the same disaster risk has not been established in most societies. Therefore, based on those issues, the accommodation is needed in reducing disaster risk in community-based institutional which is carried out by KSB Kemijen.

References

[1] Ayers, J. (2011). Resolving the adaptation paradox: Exploring the potential for deliberative adaptation policy-making in Bangladesh. Global Environmental Politics, 11(1), 62-88. doi:http://dx.doi.org/10.1162/GLEP_a_00043

[2] Birkmann, J., Chang Seng, D., & Setiadi, N. (2012). Enhancing early warning in the light of migration and environmental shocks. Environmental Science & Policy(0). doi:http://dx.doi.org/10.1016/j.envsci.2012.04.002

[3] Birkmann, J., & von Teichman, K. (2010). Integrating disaster risk reduction and climate change adaptation: key challenges—scales, knowledge, and norms. Sustainability Science, 5(2), 171-184. doi:http://dx.doi.org/10.1007/s11625-010-0108-y

[4] Chaskin, R. J. (2001). Building Community Capacity: A Definitional Framework and Case Studies from a Comprehensive Community Initiative. Urban Affairs Review, 36(3), 291-323. doi:https://doi.org/10.1177/10780870122184876

[5] Danianti, R. P., & Sariffuddin, S. (2015). Tingkat Kerentanan Masyarakat Terhadap Bencana Banjir di Perumnas Tlogosari, Kota Semarang. Jurnal Pengembangan Kota, 3(2), 90-99. doi:http://dx.doi.org/10.14710/jpk.3.2.90-99

[6] Fujikake, Y. (2008). Qualitative Evaluation: Evaluating People's Empowerment. Japanese Journal of Evaluation Studies, 8(2), 25-37. doi:http://doi.org/10.11278/jjoes2001.8.2_25

[7] Kencana, A. D., & Yuliastuti, N. (2016). Penilaian Bentuk Adaptasi Masyarakat Terhadap Hunian di Kawasan Rawan Rob dan Banjir Kelurahan Kemijen. Jurnal Pengembangan Kota, 4(2), 186-196. doi:http://dx.doi.org/10.14710/jpk.4.2.186-196

[8] Marfai, M. A. K., Lorenz; Sartohadi, Junun; Sudrajat, Sudrajat; Budiani, Sri Rahayu; Yulianto, Fajar. (2008). The impact of tidal flooding on a coastal community in Semarang, Indonesia. Environmentalist, 28, 237-248. doi:http://dx.doi.org/10.1007/s10669-007-9134-4

[9] Nakmofa, Y., & Lassa, J. (2009). Community Based Approach to Disaster Risk Reduction and Climate Change Adaptation Towards Sustainable Livelihood: Ten Years Experiences From

PMPB Kupang. Journal of NTT Studies, 1(2), 136-145.

[10] Nugroho, S. H. (2013). Prediksi luas genangan pasang surut (rob) berdasarkan analisis data spasial di Kota Semarang, Indonesia. Jurnal Lingkungan dan Bencana Geologi, 4(1), 71-87.

[11] Ramadhany, A. S., Subardjo, P., & Suryo, A. A. D. (2012). Daerah Rawan Genangan Rob di Wilayah Semarang. Journal of Marine Research, 1(2), 174-180.

[12] Sariffuddin, S., Astuti, K. D., Farhaeni, G., & Wahdah, L. (2017). Vulnerability Assessment: The Role of Coastal Informal Settlement Growth to Social Vulnerability in Genuk Sub-District, Semarang City. IOP Conference Series: Earth and Environmental Science, 55, 012047. doi:https://doi.org/10.1088/1755-1315/55/1/012047

[13] Sariffuddin, S., & Wijaya, A. P. (2014). Pola Adaptasi Masyarakat Pesisir Genuk Kota Semarang. Tataloka, 16(4), 245-253. doi:http://dx.doi.org/10.14710/tataloka.16.4.245-253

[14] Sphere. (2011). Humanitarian Charter and Minimum Standards in Humanitarian Response

[15] World Bank. (2012). Kota dan Banjir: Panduan Pengelolaan Terintegrasi untuk Resiko Banjir Perkotaan di Abad 21. Washington DC: The World Bank.

IOP Conf. Series: Earth and Environmental Science **70** (2017) 012048 doi:10.1088/1755-1315/70/1/012048

Perception Study about Visitors Related to Development of Rowo Bayu Attractions in Kecamatan Songgon Banyuwangi

A G Mahayu

Department of Regional and Urban Planning, Faculty of Engineering, Universitas Brawijaya, Malang 65147, Indonesia

ayundagading11@gmail.com

Abstract. The development of tourism was a process of sustainability, and it was not stand-alone activity because it must involve various sectors. Tourism planning must take into consideration the existing condition and supporting capacity because it should create a long-term mutual interaction in achieving goals such as increasing community welfare and ensuring the sustainability of environmental supporting capacity in the future. Rowo Bayu Tourist Object was greatly potential to be developed into historical and also natural objects of scenery. Some historical heritages of Tawangalun Palace were exposed beautifully by the marsh and this situation could be cultivated into water-based tourism. However, Rowo Bayu Tourist Object still lacked of supporting facilities such as security post, parking lot, cleaning service, prayer house, and others that led only to the inconvenience of the visitors. In this research, the perception of visitors on importance and satisfaction rates of tourist object-related variables was measured. These variables included attraction, accomodation, accessibility, facility, information, and utility, which were then subjected to the analysis technique of IPA. Result of analysis found 14 attributes that were important for tourist object development but in bad condition. These attributes were: the availability of security guard, the availability of food and beverage providers, the availability of transportation modes to the tourist object, the availability of parking lot, the availability of toilet, the availability of garbage can, the availability of information center, the availability of prayer house, the availability of ATM, the availability of fuel-station (SPBU), the availability of tourist object promotion tools, the availability of tour guide, the availability of electricity, and the signal strength of mobile phones. After IPA was finished, it was followed by AHP analysis.

1. Introduction

The development is a capability which is determined by what we can do with what we have in order to to improve our quality of life [1]. The development area is a strategy in utilizing and combining with internal factors (strengths and weaknesses) and external factors (opportunities and challenges) as the potential and opportunities that can be exploited in improving income regions [2]. The development area is also called as an attempt to create an integrated use of resources, improvement of inter-regional harmony and integration between development sectors through a process of spatial planning [3]. The development area is one of the most appropriate way to improve the people welfare in it, one of which is to develop the tourism sector. Pendit [4] defines tourism is a journey that is done temporary from one place to another with the intention to not make a living in the visited places, but simply to enjoy the journey for recreation and leisure or to satisfy the desires from diverse variety. Tourism is process of traveling from one or more people in heading somewhere else outside their homes. The purpose of traveling is due to a variety of interests, either because of economic, social, cultural, political, religious, health interests and other such as simply want to gain knowledge in order to the gain experience or to learn new things. The tourism development is a sustainable process and is an integrated process from many sectors. Tourism planning must be based on the conditions and the carrying capacity in order to create a long-term mutually beneficial between achieving goals, improving the local communities welfare, and the sustainability of environmental carrying capacity in the future [5]. Banyuwangi Regency was famous with an epithet Sunrise of Java. It had many potentials of tourism. One was Rowo Bayu Tourist Object in Bayu Village, Songgon District. This tourist object was located nearby the historical situs of Blambangan Kingdom. However, it was given to improper management, and the number of visitors was quite few. Rowo Bayu Tourist Object had a great potential to be developed into historical and natural objects of tourism. The presence of Tawangalun Palace in the area had been situated by the marsh, and this landscape could be managed into water-based tourism. At the existing condition as recently observed, this tourist object was not equipped with facilities such as security post, parking lot, cleaning service, prayer house, and others that might disturb the convenient visiting. By taking this overview into consideration, the objective of this research was to understand the strategy to develop Rowo Bayu Tourist Object in Songgon District, Banyuwangi Regency, through bottom-up approach, in order to improve community income based on the perception of the visitor and the expert to facilitate the determination of rating of the sub-variables needed to develop.

2. Methods

Method of research was a quantitative description. Focus was given on the perception of visitors at Rowo Bayu Tourist Object. Some variables that were used to assess the perception of importance and satisfaction rates from the visitors were including: attraction, accomodation, accessibility, facility, information, and utility. Sample was taken with purposive sampling by which the subject in the sample was acquired because the subject was representative to the charateristic of population [6]. Number of sample was counted from the number of visitors using Slovin Formula at tolerance rate of 10% and accuracy rate of 90%. Final sample was 98 respondents, but it was then rounded into 100 respondents.

The analysis technique was IPA and it was done to understand visitors' satisfaction rate based on the aspect of importance and of satisfation. Before subjecting the data to the analysis, it derived from the result of questionnaire about community perception on the performance of a certain tourist object based on the predetermined indicators of assessment. Variable "X" was performance rate and variable "Y" was denoted for indicator importance rate. The conformity rate between performance and importance scores would determine the order of priority in how to increase factors influencing the satisfaction of visitors [7]. Result of analysis was set into graphic plot showing that the development of tourist object was set into Quadrant 4 (concentrate here). After IPA was finished, it was followed by AHP analysis. Result of AHP represented expert opinion. The expert in this analysis included community (village chief), academician (lecturer) and government (The Official of Tourism). AHP was an analytical technique used when the decision was made with systemic approach. The decision-making model arranged multi-factor and multi-criteria problems into a hierarchy.

3. Research Result

Tourism development plan in Banyuwangi Regency was focused on three tourism development regions (WPP; wilayah pengembangan wisata), respectively WPP I, WPP II and WPP III. Rowo Bayu Tourist Object belonged to WPP I with Ijen Crate as the leading view. In the last 5 years, Banyuwangi Regency had seen the quite dramatic development of tourism. Tourism potentials had succesfully increased local genuine income and also promoted Banyuwangi Regency as one regency in East Java Province with great diversity of tourist objects. Many tourist objects were situated in Songgon District, including Songgon Pine Forest, Lider Waterfall, Selendang Arum Waterfall, and Badeng River Tourism. Songgon District had been known well by the community of Banyuwangi Regency and also the outsider. All tourist objects mentioned above had produced greater number of visitors if compared to Rowo Bayu

IOP Conf. Series: Earth and Environmental Science **70** (2017) 012048 doi:10.1088/1755-1315/70/1/012048

Tourist Object although this tourist object was the oldest site in Songgon District and containing the historical heritage of Tawangalun Palace. In the other hand, Rowo Bayu Tourist Object could not escape from the War History of Puputan Bayu occuring on 1771. The War of Puputan Bayu was a great clash between the people of Banyuwangi and the Dutch colonial government (VOC). In this war, the Dutch was defeated and suffering great losses. Rowo Bayu Tourist Object in Bayu Village, Songgon District, was quite attractive because besides providing historical tourism object, it was also providing natural scenery, at least through the presence of orchards. It was not suprising because this tourist object was a part of Songgon District Forestry Region under the responsibility of the Forestry Managing Unit of Rogojampi, West Banyuwangi. One potential in Rowo Bayu Tourist Object was a spread of marsh land at 8 meters depth. It resembled to a natural pool with genuine ecosystem. It would be regrettable if this potential was left unmanaged. The fact had showed that the number of visitors at this object was quite few, possibly due to the lacking of visitors' facilities and even the absence of amenities usually existing in the tourist object. The perception of the visitors at Rowo Bayu Tourist Object was used to measure satiscation and importance rates of tourist object-related variables, including attraction, accomodation, accessibility, facility, information, and utility of tourist object. This perception was measured using IPA analysis. Initially, there were 29 attributes that must be assessed to recognize their satisfaction and importance rates based on the perception of visitors. The following was the average rates of satisfaction and importance of these attributes based on the perception of visitors at Rowo Bayu Tourist Object.

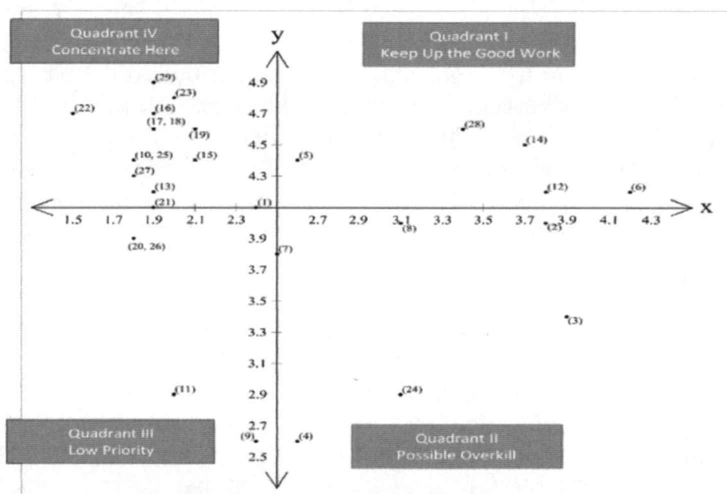

Figure 1. Quadrant IPA Analysis.

Table 1. Attributes Analysis in Quadrant IV.

Variable	Attribute	Analyze
Attraction	The availability of security guard	No security guard was available at Rowo Bayu Tourist Object, and it led visitors to feel incovenient during night tour.
Accommodation	The availability of food and beverage providers	Only one provider of food and beverage existed, and it was rarely standby on site when visitors needed meals and drinks at daylight.
Accessibility	The availability of transportation modes to the tourist object	All visitors must still use personal vehicle due to theabsence of public transport to the tourist object.
Facility	The availability of parking lot	Parking lot for visitors was not available, and visitors found difficulty to park their vehicles.

Variable	Attribute	Analyze
	The availability of toilet	Toilet was not reliable and also in limited numbers. It caused hardiness to the visitors.
	The availability of garbage can	Only few garbage cans were provided and often concentrated. Thus, trash was found everywhere due to the dispersal discard.
	The availability of information center (Tourism Center)	There was no information center at Rowo Bayu Tourist Object. If any, it was only one and situated at the Office of Songgon District that was quite distant from tourist object.
	The availability of prayer house	Prayer house was not available at tourist object, which forced visitors to go to nearby prayer house.
	The availability of ATM and Money Changer	ATM facility was absent at tourist object. This machine was only situated in the capital of Songgon District, and only two machines were available.
	The availability of fuel-station (SPBU)	No fuel-station was found in Songgon District, and thus, visitors must buy fuel from retailer.
Information	The availability of tourist object promotion tools (brochures / pamphlets)	The promotion of Rowo Bayu Tourist Object was not optimum because there was no brochures and pamphlets available informing this tourist object.
	The availability of tour guide (travel agent)	Tour guide was absent. The only reliable informant was ticket seller.
Utility	The availability of electrical network	No electrical network at the tourist object which only induced sense of discomfort among visitors at night.
	The availability of communication network (signal strength)	The signal of mobile phones was weak. Often, it did not exist at all, which caused visitors facing trouble in communication.

These 14 attributes from IPA were subjected to AHP (*Analytic Hierarchy Process*). This follow-up was useful to recognize the most influential sub-variable in the development of Rowo Bayu Tourist Object. AHP was also functional to determine the rating of variables based on the interview with expert to decide which sub-variables on the priority to develop. All sub-variables in AHP were derived from IPA analysis, and these remained at Quadrant IV, which in this case, consisting of important sub-variables but at less satisfying condition. During the development of Rowo Bayu Tourist Object, AHP was a *bottom-up* planning control. In such planning, the government acted as the controller in the determination of priority for development. The conduct of AHP used a Software "Expert Choice" using 3 respondents, respectively: Community (represented by Village Chief), Academician (represented by Lecturer), and the official of tourism.

Table 2. Rating AHP Analysis

Variables	Attributes	Community	Academician	The Official of Tourism	Total	Average	Rating
Attraction	The availability of security guard	0.011	0.057	0.065	0.133	0.0443	7
Accommodation	The availability of food and beverage providers	0.046	0.077	0.010	0.133	0.0443	8
Accessibility	The availability of transportation modes to the tourist object	0.112	0.139	0.115	0.366	0.1220	4
Facility	The availability of parking lot	0.045	0.037	0.094	0.176	0.0587	5
	The availability of toilet	0.051	0.031	0.050	0.132	0.0440	9
	The availability of garbage can	0.026	0.025	0.047	0.098	0.0327	12
	The availability of information center (Tourism Center)	0.122	0.157	0.132	0.411	0.1370	3
	The availability of prayer house	0.042	0.028	0.078	0.148	0.0493	6
	The availability of ATM and Money Changer	0.013	0.017	0.011	0.041	0.0137	14
	The availability of fuel-station (SPBU)	0.015	0.021	0.013	0.049	0.0163	13
Information	The availability of tourist object promotion tools (brochures / pamphlets)	0.232	0.191	0.146	0.569	0.1897	1
	The availability of tour guide (travel agent)	0.172	0.163	0.176	0.511	0.1703	2
Utility	The availability of electrical network	0.045	0.018	0.038	0.101	0.0337	11
	The availability of communication network (signal strength)	0.067	0.039	0.024	0.13	0.0433	10

Based on the average rates given by all respondents, few variables were considered as having the greatest importance rate. One was information variable, represented by the availability of tourist object promotion tools and the availability of tour guide. Other was few sub-variables of facility attribute, involving the availability of information center (tourism center) and the availability of transportation modes to the tourist object.

IOP Conf. Series: Earth and Environmental Science **70** (2017) 012048 doi:10.1088/1755-1315/70/1/012048

4. Conclusions

Pursuant to the result of analysis, conclusion could be inferred as following. Banyuwangi Regency had great potential tourism sector. It was proved by great number of visitors to tourist objects which in turn it helped increasing local income. The development of tourism sector in Banyuwangi Regency might generate the effort to improve the feasibility of facilities at tourist objects, and consequently, would help promoting the existing tourist objects. However, the existing condition in Rowo Bayu Tourist Object was still needing for specific attention because as perceived by visitors, there were 14 important attributes at tourist object but those existed in bad condition (less satisfying state). Specific action must be taken to deal with problematic attributes. These attributes were: the availability of security guard, the availability of food and beverage providers, the availability of transportation modes to the tourist object, the availability of parking lot, the availability of toilet, the availability of garbage can, the availability of information center, the availability of prayer house, the availability of ATM, the availability of fuel station, the availability of tourist object promotion, the availability of tour guide, the availability of electricity, and the signal strength of mobile phones. In regard to result of AHP, the most important attributes for development were found in information variable, especially the availability of tourist object promotion tools and the availability of tour guide, and also in facility attribute, represented by the availability of information center (tourism center) and the availability of transportation modes to the tourist object. It was expected that the result of this research could be used as the base reference for the development of Rowo Bayu Tourist Object.

References

[1] Alkadri, M. (2001). Three Pillars of Development Areas: Natural Resource, Human Resources and Technology. Jakarta: Policy Assessment Center for Technological Development Area.
[2] Friedman, JA (2008). Regional Economic Development and Planning. Massachusetts: MIT Press.
[3] Rustiadi, ES (2011). Planning and Regional Development. Jakarta: Crestpent Pres and Yayasan Pustaka Obor Indonesia.
[4] Pendit, NS (2003). Tourism Studies. Jakarta: Pradya Paramita.
[5] Fandeli. (1995). Management Basics Natural Kapariwisataan. Yogyakarta
[6] Arikunto, S. (2006). Research Procedure A Practical Approach. Jakarta: PT. Rineka Reserved.: Liberty.
[7] Supranto. (1997). Measurement of Customer Satisfaction Levels To Increase Market Share, Jakarta: PT. Renika Notices.

IOP Conf. Series: Earth and Environmental Science **70** (2017) 012049 doi:10.1088/1755-1315/70/1/012049

Changing cultural landscape in post-productivism of rice field in Nyuh Kuning Village Bali

C Maulidi[1, 3], L D Wulandari[2]

[1] Department of Urban and Regional Planning, Faculty of Engineering, Universitas Brawijaya, Malang, Indonesia
[2] Department of Architecture, Faculty of Engineering, Universitas Brawijaya, Malang, Indonesia

c.maulidi@ub.ac.id

Abstract. Natural landscape in developing countries is facing a challenge due to economic growth, a cultural shift, and population dynamics. Farm land where is close to urban areas tending to be converted into more economically valuable spaces. Watershed Pakerisan listed as World Heritage of UNESCO, rich of cultural value on its landscape, especially the Subak, a traditional irrigation system, has a close relationship to the philosophy of Hindu-Bali culture. Nyuh Kuning, a village (local terms is Banjar) located adjacent to the Pakerisan Watershed, and has a spatial pattern in synergic ally connected with tradition, culture, and their religion. Rice field not only for economical but also its place to worship the Goddess (Dewi Sri). Rice Field in Nyuh Kuning declined significantly along past 10 years. The changing landscape of Nyuh Kuning traced through serial of aerial photographs from 2005 until 2015. Along with the broad decline of rice field, villager's attachment on their cultural space is also changing. An economic motive pronounces a winner in the bargaining between the motives of economic value and cultural value in the Nyuh Kuning. Villagers revealed arguments that necessities nowadays prosecute high consumption, both for household and for education. Therefore conversion of rice fields to become more economical is understandable among communities. Villagers rent the rice fields to foreigners (migrants), and then foreigners take rice-fields as personal assets, not for the villagers (ritual activities and the cultural traditions) any longer. In theoritical term, villager's emotional bond to the cultural landscape in post—productivism of rice field, is weakened. Wawedangan Desa and its complex cultural values are not part of their identity anymore. However, place dependence become the reason why the shifting place attachment is happening. Functional economic bond is mentioned as place dependence dominats in villager's attachment. Certainly it's not a sustainable way in conserving cultural landscape. Learning from Nyuh Kuning case, new ideas need to conserve cultural landscape and at the same time increased the economic villagers. Through considering rice fields renters preferences and attachment land in Nyuh Kuning, rice field is an important element for their preferences to stay at Nyuh Kuning. Villas in Nyuh Kuning retaining rice field map as part of the villa's character. Here we can see rice field not only culturally valuable but also as a tourist attraction, which can be sustained if the communities themselves manage it.

1. Introduction

Natural landscapes, particularly in developing countries such as Indonesia, are facing the development challenges due to economic growth, a cultural shift, and of the dynamics demography. Accessible rice field where is close to urban areas tend to be converted into places that more economically valuable, like villa, cafe, and rented to foreigner as a house. Such phenomena in urban studies referred to term 'post-productivity' [1] [2]. Leapfrog migration and urban growth to the suburbs and / or rural farmland; push changes to landscape to be a variety of urban amenities [3]. Moreover, the trend of rural tourist improved greatly. Tourist looking for a destination that is quiet, close to the natural landscape, in order to escape from the bustle of urban life.

Pakerisan Watershed listed as UNESCO World Heritage, rich of cultural value on its landscape. Subak, an irrigation system managed traditionally, based on philosophy of Balinese culture. Nyuh Kuning, a village located nearby to the Pakerisan Watershed, has declined in wide of rice field significantly in the last 10 years. Growing of tourism sector in Ubud Bali has brought an economic sectors shifted, from agriculture to tourism services. These situations drive the villagers to rent their rice filed and convert it into villas, hotels, and house for foreigners. Tha changing landscape in Nyuh Kuning as a pattern of how the development challenges have pressured the landscape character, which reflect phenomenon of post-productivity cultural landscape. Moreover, these phenomenon also faced by other villages, in Pakerisan Watershed. Along with the strengthening of the popularity of Subak System as a world heritage, it invites more and more tourist to Ubud Bali.

Development of urban area and will be succeed when its inline and supported by understanding, acceptance, and participation of the community [4]. In similar way, land use changes in rural areas, occur slowly or rapidly, strongly influenced by the local community perceivement. One of influential factor is the place attachment, which is an emotional bond that grows between an individual or group on the environment [5]. Place attachment as a complex phenomenon that is difficult to model. However the pattern can be observed through how to form people's behavior towards the environment. Attachment at a high level often-tangible form of community commitment is high, positive behavior in protecting and maintaining the environment [6]. If the community is tied to a place, they will tend to participate keep from environmental character changes occur. As an illustration Walker and Ryan in their study of rural communities Norway found that residents of small towns in Maine have a strong attachment to the natural features and rice fields in the middle of their neighborhood [7]. Residents who indicated a level of attachment tends to be high, they are significantly more supportive of preservation strategies in urban development, with a view to protecting the remaining landscape of rice fields.

1.1. The Role of Place Attachment

Attachment to place has a role in how people interpret and react to changes. Place attachment defined as an emotional bond among people and place components [8] [9]. The emotional bond is often referred as place identity, while the functional bond is mentioned as place dependence [10]. Place dependence describes functional bonds that people have to the places. This bond develops when places provide the conditions or features that meet people's needs [11] [6] [12]. Meanwhile, place identity described as symbolic meanings given to a place as, or psychologically people invest value in it, so the place then becomes part of their identity. Place identity develops from memories, ideas and feelings which were associated with a place [13] [14] [12]. Breakwell's Identity Process Theory (IPT) used to explain how places becomes part of people's identities [15]. Refer to IPT, anything that provides people with feelings of distinctiveness, continuity, self-efficacy, and self- esteem can be assimilated into their identities [16]. People can also feel in control of their homes and communities, which is a prerequisite for developing bonds with the place [17]. In order to protect their identities from disruptions, people may engage in efforts to minimise or reduce place change [18].

1.2. Response to Place Change

Relationship between place attachment and place protective actions argued by Devine-Wright about the psychological response to place change [19]. This framework consists of five stages. The first stage people develops awareness to the place change. In stages two and three, people interpret the implications of the change and evaluate the change as positive or negative. Different people within the same place might have differing views on this, depending on their place attachment. Strongly attached people are more likely to keep themselves abreast of place issues in the local area and to take action against changes that they have evaluated negatively. Proposed changes may be evaluated negatively if people believe that the changes threaten their place identity and/or place dependence. People may evaluate change as negative if they feel that it will stop a place from meeting their needs, thus threatening their place dependence. Alternatively, if place change is interpreted as having the potential to enhance a place, strongly attached people may evaluate the change positively. The fourth stage is deciding how to cope with the threat; includes thinking about how to respond to place change. The fifth stage is deciding what can be done and acting. People with stronger place attachment may be more likely to act oppositional behaviour [20].

The theory of planned behaviour (TPB) is used to predict whether people will decide to oppose the change. In this theory, people's behavioural beliefs, normative beliefs and control beliefs affect their behavioural control. This framework is most often applied to predict people's behavioural intentions which are then used to predict both self-reported and observed behaviour [21]. A meta-analysis of the TPB found that attitudes, subjective norms and perceived behavioural control accounted for 39% of the variance in intentions. Behavioural intentions and perceived behavioural control then accounted for 31% of the variance in self-reported behaviour and 30% in observed behaviour [22]. A small number of studies have used the TPB to examine civic engagement. Read et al. (2013) found that, while place attachment was correlated with windfarm opposition, it was the TPB, particularly subjective norms, which predicted intentions to oppose windfarm development [23]. Another study, of an adolescent and youth population, found that attitudes and perceived behavioural control were related to both offline and online civic engagement, but norms were not [23]. The TPB has also been found to predict intentions to engage in women's rights protests, with attitudes appearing to be a much stronger predictor than norms or perceived behavioural control [24]

1.3. Aim and Hypotheses

As initial research, this study was aimed to describe and identify, cultural value on the landscap, its changing, and as well community attachment to the cultural landscape and response to landsape change. Refer to theories above It's hypothesised that people with stronger place attachment would be more likely to interpret place change as negative and threatening.

2. Methods

Place attachment to landscape captured through field survey during the first week of Bali International Field School, 2015. Semi-structured interview techniques conducted along with casual and incidental contacts with villagers, teenagers, and also with foreigners who has live there. Respondents were selected randomly, through snowball sampling technique. Data proceed in content analysis, qualitatively, to interpret variances of landscape changing and construct the pattern of either place attachment and response to the change.

3. Results

3.1. Nyuh Kuning's Cultural Landscape

Nyuh Kuning is a traditional village (Banjar) also tittled as Desa Pakraman, an autonomous village, which has three main major components: a main temple, open public space, and cemetery. Nyuh Kuning is located just south of the most popular tourist destination in Ubud Bali, Monkey Forest. South Gate of Monkey Forest can be accessed by passing Nyuh Kuning village. Whle its location only next to Ubud City Centre. Villagers have string dependence to Ubud Centre for their daily life. Pura Agung (the Great Temple of Ubud) and Ubud Market is located only 2 km from Nyuh Kuning. These strong physical dan funtional connection have made Nyuh Kuning directly affected by the tourism growth in Ubud Center. In the other hand, Nyuh Kuning itself attract tourist through its quiet neighborhood, cheap homestay, traditional Balinese architecture, and ethnical sculpture and traditional ornament along the street village. Nyuh Kuning flanked by two rivers, the Wos River on west, and Tunduh River on east. The village lies from north to south. Formerly Nyuh Kuning identic with extensive landscape of rice field and a row of yellow coconut trees. Irrigation is managed through the Subak system. Subak is a traditional ecologically sustainable irrigation system that has bonded Balinese agrarian society. The water managements are under the authority of a Subak temple priest. Rice is seen as the gift from God. Therefore, Subak System also has significant role in relationship between humans, environment, and the gods. Rituals in Subak temple promote a harmonious relationship between people and their environment through the active engagement of people with ritual concepts that emphasize dependence on the life-sustaining forces of the natural world.

As a Pakraman village, Nyuh Kuning headed by a Bendesa. Nyuh Kuning has *Pura Kahyangan Tiga* (Three Main Temples) and *Setra* (cemetary). This different to common traditional village in Bali, which is consists of hamlets, one temple, and shared cemetary. Therefore Nyuh Kuning also called *Banjar Pakraman* (autonomous village). *Pakraman* derived from two words, *"Pa"* which means that the benchmark and spiritual guide; and *"grama"*, which means living space and its environment as the meaning of words 'village'. So, Pakraman is a living environment, a place for teaching and living, wheres community arranged to follow the spiritual rules, follow the guidance of Hinduism rules.

3.2. Cultural Value on The Landscape

Spatial order of Pakraman village follows Hinduism philosophy. The World is classified into *Bhuwana Agung* (big-world) and *Bhuwana Alit* (small-world). Bhuwana Agung as macrocosm of the village, a man-made environment including buildings, neighborhoods, and so forth. *Bhuwana Alit* is human body who live inside *Bhuwana Agung*. Relationship between the *Bhuwana Agung* and *Bhuwana Alit* illustrate as a fetus inside the womb. Womb as a place for life, provide protection for the growing fetus. Therefore artificial environment *(village)* must be arranged in order to provide protection for people who live inside it.

Tri Hita Karana conception underlying the balance between macro cosmos (universe) and micro cosmos (human being). Tri Hita Karana lead people to keep the environment in harmonize order, among those three elements: *Atma, Prana, and Anga*. *Atma* is Almighty God; *Prana* is the force that drives the universe; *Angga* is the elements of *Panca Maha Bhuta*. *Panca Maha Bhuta* consists of five elements of human body, *Apah* (water or liquid), *Pertiwi* (soil), *Bayu* (wind), *Teja* (heat substance or light), and *Akasa* (space or ether) [25]. Tri Hita Karana philosophy applied on spatial order of the village: *Atma* elements manifestated as temples, *Prana* is the living area (housing), and *Angga* is open area including rice-field. Nyuh Kuning Village's spatial order follows conception of Tri Hita Karana, which is classified into three concentric zones. *Atma* is located in the central part of the village as the most sacred area, characterized by the presence of Temple, sacred banyan tree, and open space as a place to hold traditional and spiritual ceremonies. *Prana* zones intended to shelter villagers termed Karang Ayahan, residential area passed down inheritance according to local tradition. The land in this zone cannot be bought and sold, but only inhibited by family members. *Angga* zone intangibles open land for gardens / forests and fields, where villagers produce goods through gardening and farming. Land in the zone of the latter is commonly referred as Wawedangan, can be owned by individuals and can be traded. Rice field to be a part of this last zone and since the 11th century has become a part of the rural landscape at once part of economic and cultural activity. Rice field irrigation system managed traditionally through the Subak system. There is a Subak Temple, Pura Ulun Carik, at north part of the village, where is farmers used to be pray, put offerings in order to establish harmony communication with God and begging agricultural succeed.

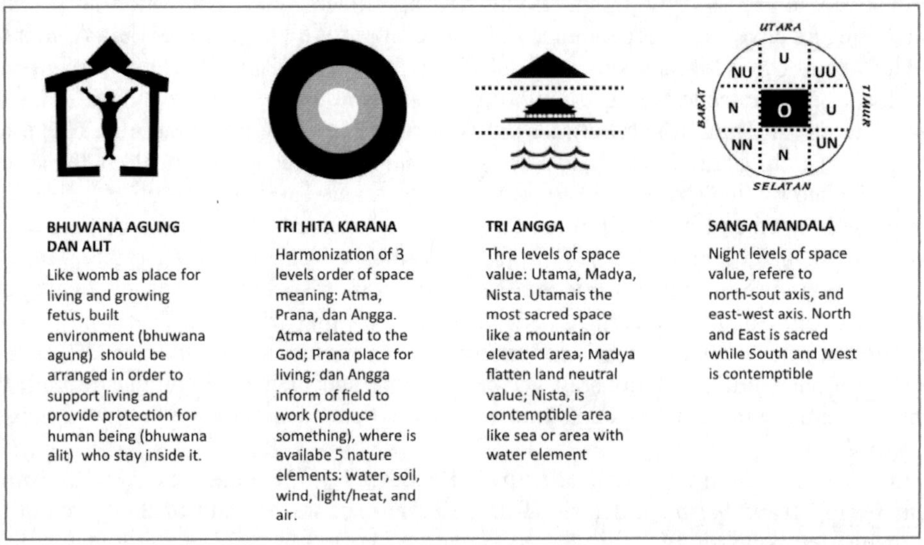

Figure 1. Traditional philosophies which guide spatial order

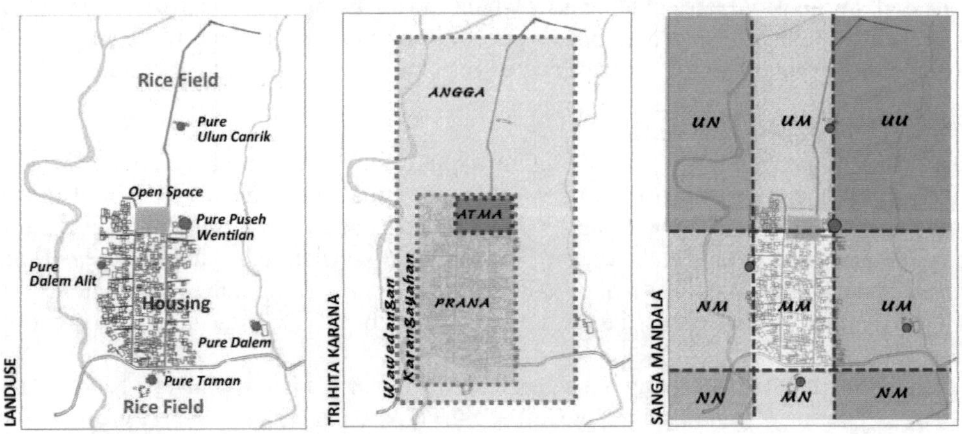

Figure 2. Landuse, spatial zone, and sacred place classification (left to right)

Besides *Tri Hita Karana* conception, also there is *"Tri Anga"* conception than give guidance for village spatial order. *Tri Angga* means three physical values, namely the *Utama Angga* (head); *Madya Angga* (body); and *Nista Angga* (feet) [26]. Main elements of the universe in the form of *Swah Loka Anga*, *Madya or Bhuah Loka Angga*, *Nista Loka or Bhur Loka*. Key elements in a region such as Mountain, associate as *Utama Angga* and Sea as *Nista Angga*. In a village, *Utama Angga* interpreted as place for God, *Madya Angga* for settlements, and the *Nista Angga* for cemetary. Tri Angga in Nyuh Kuning applied as follows: (1) *Utama Anga* elements, there are *Kahyangan Tiga* (three main temples): Pura Desa is located in the northern part of the village; Puseh temple is situated at same location with Pura Desa; Pura Dalem is located at the southeast part of village. (2) *Madya Angga* elements: there are neighborhoods at the center of village. (3) *Nista Angga* elements: a cemetary located adjacent to Pura

IOP Conf. Series: Earth and Environmental Science **70** (2017) 012049 doi:10.1088/1755-1315/70/1/012049

Dalem and Pura Taman, a pond for ceremonial related to purifying soul and heirlooms, whichis located at south part of the village.

There is also Earth's axis conception, *Kaja - Kelod* (North - South) and axis of sunrise - sunset (east - west) which both merged into philosophy *Sanga Mandala* [26]. Implementation on a village, Sanga Mandala concept: put the activities of *Mandala* that is sacred (Pura Desa) in the northern area; *Madya Mandala* form Pura Puseh, neighborhoods to a central area; and *Nistaning Mandala* is a grave and Pura Dalem found in the southern area. According to its function, the concept of the regional division of *Sanga Mandala* of Mandala form of sacred and spiritual activities; Madya Mandala as space worldly activities such as economic, social and housing; *Nistaning Mandala* for activities such dirty-containing waste landfills, disposal channels for rainwater and graves. Sanga Mandala be saw on laying any religious site. Pura Desa and Pura Puseh to worship the god placed in the *Utama Mandala* (north), settlement placed in Madya Mandala (middle of the village), and the graves and a pool to wash the sacred peralatn located in *Nista Mandala* (south). The assertion west as the area marked by the placement *Nista Mandala* Pure Dalem Alit and a small cemetery.

3.3. Changing Landscape

Changing landscape in Nyuh Kuning explored through interpretation of aerial photographs, in 2005, 2009, 2013, and 2015. The interpretation is conducted by classifying visual image (textures and colours) that indicate land cover character. This technique is quite easy to do as a form of land cover can be identified one another based on their color: the residential area is brown and red, garden / forest green with texture grove, and the fields are gray (not embedded) and green textured grain (already planted rice). Land cover image compared between years 2005 - 2015 have showed that landscape changes occurred on lands garden / forest some point turn out to be awake and open field (the garden). Significant changes were observed in rice field were, which is reduced significantly in the last 10 years. Rice field formerly scattered on the north side of the village and longitudinally on the east side of the village. Being on relatively gentle terrain around the Tunduh river flow from the fountain Monkey Forest. West side of the village there is a Wos river, with a larger dimension than another river, but has a steep topography and the river level is below a depth of over 7 meters of elevation settlements Nyuh Kuning. Steep topography may have to be obstacle, so the west side of the village is not used as rice fields but it still left as forest.

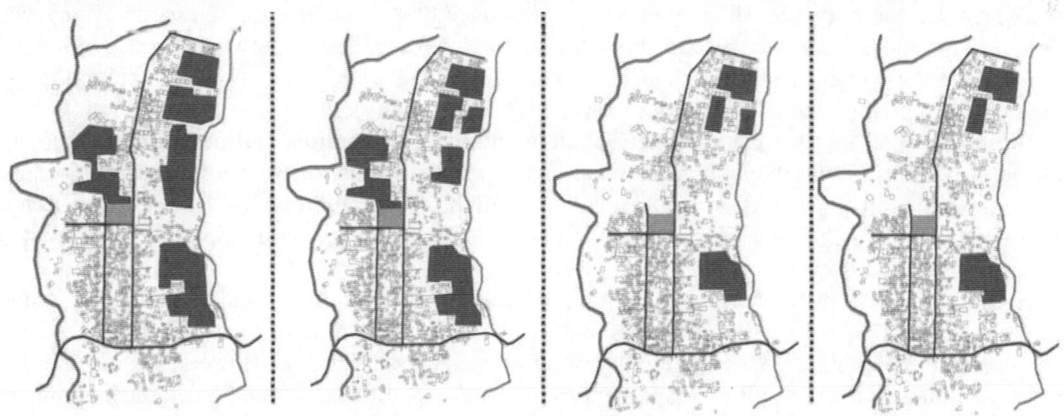

Figure 3. Rice field distribution in 2005, 2009, 2013, and 2015 (left to right)

In 2005 aerial photograph shows there are at least five zones of rice fields in Nyuh Kuning. In 2009 plots of rice fields split by hotel / villa, and finally in 2015 the remaining three zones rice fields two of them no longer as the fields of production, but as an aesthetical garden for villa which offer a quite rural atmosphere for their guests.

3.4. Place Attachment

There are three patterns of place attachment to the landscape which is categorized as follows: (1) Adult Citizens, (2) Youth Citizens (Taruna-Taruni), and (3) Settler who live in Nyuh Kuning. Their place attachment are different each other, have been affected by various factors. Adult citizens have a strong attachment once adhering to the tenets of Tri Hita Karana, living in harmony with God, with nature and with others. Taruni-Taruni have a strong attachment to the landscape of the village was influenced by a childhood memory, about a quiet village and playing around in rice fields. Youth generation affected by contemporary views on environmental sustainability that was gained from formal education. Meanwhile, Settlers or foreign newcomers have a strong attachment to the landscape of rice fields, because of their preference to stay and live in rural scene, close to the green rice fields.

Local residents have grown strong firmness in order philosohpy Pakraman as the principle of Tri Hita Karana. Harmony with God actually pursued through preserving Tria Angga sacred elements in village. Harmony with nature conservation outlined in the order of the elements of the landscape of villages by Sanga Mandala composition. Harmony with human beings embodied in respect personal interest and respects each other's privacy. Deliberation held before agreeing to the collective affairs of the villagers. Their strong attachment looked out of their shared commitment to keep village fabric of space correspond Hindu Bali philosophies. So the composition and layout Pura Karang Ayahan still maintained until now. Meanwhile, the conversion of rice fields, whether, reflects loss of community attachment to the cultural landscape. According to the results of interviews with Bendesa (village head) and some local residents in mind that over the function of rice field is not simple enough to be translated as an attachment disengagement of citizens to the cultural landscape. It is the result of a bargaining process of citizens in applying the concept of Tri Hita Karana in situations of post-productivits of rice field. Villagers strive to maintain harmony and the environment at the same time also seeking harmony with their fellow citizens. As stated by Mesch and Manor (1998) [28] that place attachment is not only a person's attachment to the place, but it could be someone attachment to other people in the place, and the local cultural values. As has been stated previously that Rice field as Wewedangan area is the space for economic activities, in which people utilize this land for production process, and are owned by individuals. Any decision (rent it, sell, and making land villa / hotel) is included in the scope of personal decision. Bendesa indeed involved in any land transactions Wewedangan, however Bendesa has no power to prevent the owner to rent it, because he also must preserve harmony with the neighbors by respecting citizens' private interests and not intervene rice fields transactions.

Taruna-Taruni revealed that their atttachment against childhood memories and the stories about the picturesque of Nyuh Kuning village at past, when many rice fields and quiet still existed. Their place attachment are not much built by economic motives, but still the sophistication ideal Balinese Landscape in many rural landscapes they hear from parents, school subjects, and mass media, also about Subak system on the village landscape.

Foreign resident who live in Nyuh Kuning, to get the experience of living in rural areas in a quiet and close to nature, as well as personal preferences and residential satisfaction as there had been theorized by Kaltenborn and Bjerke [28]. One interesting point of attachment pattern of foreign residents, who live in Nyuh Kuning, is a strong place attachment to the landscape of rice fields. Place attachment they appear on the commitment to protect remaining rice fields. Foreign residents seek to close the access road access to the village into the remaining rice fields, rice fields by hiring down the road access to the fields and then build the houses and private yard. Closed road access to the rice fields would deter other buyers/ tenants to access the remaining rice field. On one side of the action is therefore preserve fields, but on the other hand the access movement and visual villages Nyuh Kuning against paddy is closed, thereby landscape of rice fields can no longer be perceived as part of the Nyuh Kuning landscape, and for cultural rituals related to agriculture.

As Hummon stated that place attachment level will appear on the level of their engagement helped preserve the landscape [27]; Such was evident in the pattern of attachment of foreign nationals in Nyuh Kuning in protecting the remaining rice field. However, they can be regarded as a privatization of rice field landscape in Nyuh Kuning, for the sake of foreigner satisfaction. Certainly these understandable because of their attachment built above personal preferences, without thinking about how the landscape of rice field should be part of thc Nyuh Kuning culture.

Figure 4. Access toward the remaining foce field blocked by private yard (left); Rice field has been made as villa's garden (right).

However in the case of a commitment to the preservation of the rural landscape of rice fields would be more comprehensive if accompanied by large-scale social action and cross-community commitment. Rural communities in general have a tendency to have strong social relationships with one another in the case of Nyuh Kuning more leads to lax protection of rice land conversion. It the conception of the philosophy of Balinese spatial order. Wawedangan village where rice fields are, bearada in the possession of private citizens and other residents with disabilities to intervene affairs of other citizens, especially considered renting rice field is a reasonable for others to boost their economies.. However shifting economic sector to be part of the culture, but in case there Nyuh Kuning cultural values are sacrificed. The tradition of offering to the god of rice, religious rituals related to agriculture, and traditional institutions in the management of Subak be vanished from the village. In addition, Pura Ulun Carik, a Subak Temple in Nyuh Kuning losing its role amongst Nyuh Kuning communities.

4. Conclusion

Nyuh Kuning village cultural landscape facing the challenge of shifting economic sectors of the economic base of agriculture into the tourism sector. These situation makes the value of the agricultural economy lagging behind the tourism sector. As a consequence, citizens releasing rice field switching functions for the more economically valuable, which is sold or leased to be built hotels and villas. Furthermore, this pattern makes the loss of the cultural landscape character Nyuh Kuning which was originally identic as typical rural landscape in Ubud Bali. Nyuh Kuning village developed into a tourist village, but the expense of the natural landscapes of rice fields and traditions which was originally as a part of culture in Nyuh Kuning since the 11th century.

Foreigners have attachments that are specific to the existence of rice fields. However their attachment based on the personal preferences of the resident satisfaction without attachment to make the local cultural landscape views privatize rice field. They preserve rice fields to the personal interests of their villa. Their strong commitment to protect hinders rice paddy fields to become part of the cultural landscape Nyuh Kuning. Local residents Nyuh Kuning village has a strong attachment on the application of the principle of Tri Hita Karana, in order to preserve harmony with the God, Nature and fellow man. Wewedangan, where rice fields located, in cultural settings controlled Bali as personal space and as part of an effort to maintain harmony with each other, the citizens do not participate in providing control over the conversion of rice fields. Bendesa did not take advantage of the authorities to prevent the conversion of rice fields into a hotel and villas. One aspect of the application of Tri Hita Karana in Nyuh Kuning was a gap to let policy for maintaining the rice field as character of Nyuh Kuning Landscape, further compromising the traditional order and Subak system along with the usual tradition in Subak Temple.

Along with the increasingly popular Pakerisan Watershed as UNESCO World Heritage tourists visit more massive, and certainly completeness of tourist facilities will continue to grow. With such a trend, it could be over the function of rice fields, as happened in Nyuh Kuning, will be a challenge for other villages in Pakerisan Watershed. Learning from Nyuh Kuning, can understand the economic motive is to be winning in bargaining between the motives of economic value and cultural value. One of the arguments with residents revealed that the necessaries of life these days demand a high level of consumption, both for household necessities and education. For the conversion of rice fields to become more economical regarded as reasonable fo residents to sell or rent them to foreigners. Foreigners looking at rice field as a personal asset and protects rice field for personal use, not for the village (rituals and cultural traditions of the people). Such a situation results in economic activity and cultural rituals related to agriculture in Nyuh Kuning be disconnected. Foreigners hire and utilize rice field on the basis of place attachment only on their rice field itself, although it is at strong attachment level. But it was not enough to preserve the rice field as a cultural landscape in Nyuh Kuning. Although too debatable to say the attachment of local residents against the cultural landscape is too weak, but it is surely the attachment is defeated by the attachment of citizens on the economic value. In theoritical term, villager's emotional bond to the cultural landscape in post—productivism of rice field, is weakened. *Wawedangan Desa* and its complex cultural values are not part of their identity anymore. However, place dependence become the reason why the shifting place attachment is happening. Functional economic bond is mentioned as place dependence dominats [10] in villager's mind. Learning from this pattern, new ideas need to be built to preserve the culture and at the same time increased the economic citizens. Viewing place dependence of foreigner to the rice-field landscape in Nyuh Kuning. Villas in Nyuh Kuning retaining rice field map as part of the park's villa. Here we can see rice field not only culturally valuable but also valuable to be appointed as a tourist attraction.

If the current state of land simplified rice field successfully utilized by foreigners to a high economic value to the villa, paddy remains as part of the garden villa, and the only thing missing is the rituals and traditions related to farming. The loss of traditions related to agriculture because of land managed by the party that has attachment only on rice field, but not to the cultural landscape. The situation would be different villa and guesthouse managed by the party who has attachment more extensively on cultural values. Villa guest house can be still built as tourist facilities, access to the rice field remains open to residents of the village, rice fields remain as scenery for villa and guest house around him, and rituals and traditions related to farming can continue to be as not only as part of the cultural values, simultaneously also become a tourist attraction. Illustration of this, the cultural landscape will be more sustainable if the fields managed by the Nyuh Kuning, not foreigners. Surely, villager's attachment against the landscape of rice fields need to be reinforced and supported with managerial and business capabilities to manage rice field independently, without rent it to foreigners.

References

[1] G. A. Wilson, "From productivism to post-productivism and back again: Exploring the (un)changed natural and mental landscapes of European agriculture," *Transactions of the Institute of British Geographers,* pp. 77-102, 2001.

[2] J. Holmes, "Impulses towards a multifunctional transition in rural Australia gaps in the research agenda," *Journal of Rural Studies,* vol. 22, no. 2, pp. 142-160, 2006.

[3] K. H. Halfacree, "Talking about rurality: Socia representations of the rural as expressed by residents of six English parishes," *Journal of Rural Studies,* vol. 11, no. 1, pp. 1-20, 1995.

[4] E. Lokocz, "Motivations for land protection and stewardship: Exploring place attachment and rural landscape character in Massachusetts," *Journal of Landscape and Urban Planning,* vol. 99, no. 2, pp. 65-76, 2011.

[5] M. V. Giuliani and R. Feldman, "Place attachment in a developmental and cultural context," *Journal of Environmental Psychology,* vol. 13, pp. 267-274, 1993.

[6] S. A. Shumaker and R. B. Taylor, "Toward a clarification of people-place relationships: A model of attachment to place," in *Environmental psychology: Directions and perspectives*, New York, Praeger Publishers, 1983, pp. 219-251.

[7] A. J. Walker and R. L. Ryan, "Place attachment and landscape preseration in rural New Englan: A Maine case study," *Lanscape and Urban Planning,* vol. 86, no. 2, pp. 141-152, 2008.

[8] G. Brown and C. Raymond, "The relationship between place attacment and landscape values: Toward mapping place attachment," *Applied Geography,* vol. 27, no. 2, pp. 89-111, 2007.

[9] B. S. Jorgensen and R. C. Stedman, "Sense of place as an attitude: Lakeshore owners attitudes toward their properties," *Journal of Environment Psychology,* vol. 21, pp. 233-248, 2001.

[10] D. R. Williams and J. W. Roggenbuck, "Measuring place attachment: Some preliminary results," in *Session on Outdoor lanning and Management NRPA Symposium on Leisure Research*, San Antonio, 1989.

[11] R. L. Moore and A. R. Graefe, "Attachments to recreation settings: The case of rail-trail users," *Leisure Sciences,* vol. 16, no. 1, pp. 17-31, 1994.

[12] D. R. Williams and J. J. Vaske, "The measurement of place attachment: Validity and generalizability of a psychometric approach," *Forest Science,* vol. 49, no. 6, pp. 830-840, 2003.

[13] H. M. Proshansky, A. K. Fabian and R. Kaminoff, "Place-identity: Physical world socialization of the self," *Journal of Environmental Psychology,* vol. 3, pp. 57-83, 1983.

[14] J. J. Vaske and K. C. Kobrin, "Place attachment and environmentally responsible behaviour," *Journal of Environmental Psychology,* vol. 32, pp. 116-121, 2001.

[15] C. Twigger-Ross, M. Bonaiuto and G. M. Breakwell, "Identity theories and environmental psychology," in *Psychological theories for environmental issues*, Aldershot, Ashgate, 2003.

[16] G. Breakwell, "Social representations and social identity," *Papers on Social Representations,* vol. 2, pp. 1-20, 1993.

[17] B. B. Brown, "Territoriality," in *Handbook of Environmental Psychology*, New York, Wiley, 1987.

[18] C. Lin and M. Lockwood, "Forms and sources of place attachments: Evident from two protected areas," *Geoforum,* vol. 53, pp. 74-81, 2014.

[19] P. Devine-Wright, "Rethinking NIMBYism: The role of place attachment and place identity in explaining place-protective action," *Journal of Community & Applied Social Psychology* , vol. 19, no. 6, pp. 426-441, 2009.

[20] P. Devine-Wright and Y. Howes, "Disruption to Place Attachment and the Protection of Restorative Environments: A Wind Energy Case Study," *Journal of Environmental Psychology,* vol. 30, no. 3, pp. 271-280, 2010.

[21] I. Ajzen, "The theory of planned behavior," *Organizational Behavior and Human Decision Processes,* vol. 50, pp. 179-211, 1991.

[22] C. J. Armitage, "Efficacy of the Theory of Planned Behaviour: A meta-analytic review," *British Journal of Social Psychology ,* vol. 40, no. 4, pp. 471-499, 2001.

[23] D. L. Read, R. F. Brown, E. B. Thorsteinsson, M. Morgan and I. Price, "The theory of planned behaviour as a model for predicting public opposition to wind farm developments," *Journal of Environmental Psychology,* vol. 36, pp. 70-76, 2013.

[24] P. Jugert, K. Eckstein, P. Noack, A. Kuhn and A. Benbow, "Offline and Online Civic Engagement Among Adolescents and Young Adults from Three Ethnic Groups," *Journal of Youth and Adolescence,* vol. 42, no. 1, pp. 123-135, 2013.

[25] C. Kelly and S. Breinlinger, "Attitudes, Intentions, and Behavior: A Study of Women's Participation in Collective Action," *Journal of Applied Social Psychology ,* vol. 25, no. 16, pp. 1430-1445, 1995.

[26] Mangku, "Panca Maha Bhuta," 2008. [Online]. Available: http://mang-kuu.blogspot.com. [Accessed 16 June 2011].

[27] N. K. Dwijendra, Arsitetur rumah tradisional Bali, Denpasar: Udayana University Press, 2008.

[28] G. S. Mesch and O. Manor, "Social Ties, Environmental Perception, And Local Attachment," *Environment and Behavior,* vol. 30, no. 4, pp. 504-519, 1998.

[29] B. P. Kaltenborn and T. Bjerke, "Associations between Landscape Preferences and Place Attachment: A study in Røros, Southern Norway," *Landscape Research ,* vol. 27, pp. 381-396, 2002.

[30] D. H. Hummon, "Community attachment: Local sentiment and sense of place," in *Place Attachment*, New York, Plenum Press, 1992, pp. 253-278.

IOP Conf. Series: Earth and Environmental Science **70** (2017) 012050 doi:10.1088/1755-1315/70/1/012050

Reliability analysis for determining performance of barrage based on gates operation

C Adiningrum[1, 3], I K Hadihardaja[2]

[1] Department of Civil Engineering, Universitas Atma Jaya Yogyakarta, Jl. Babarsari 44, Yogyakarta 55281, Indonesia
[2] Department of Civil Engineering, Institut Teknologi Bandung, Jl. Ganesha 10, Bandung 40132, IndonesiaType the author addresses here

cita_adiningrum@staff.uajy.ac.id

Abstract. Some rivers located on a flat slope topography such as Cilemahabang river and Ciherang river in Cilemahabang watershed, Bekasi regency, West Java are susceptible to flooding. The inundation mostly happens near a barrage in the middle and downstream of the Cilemahabang watershed, namely the Cilemahabang and Caringin barrages. Barrages or gated weirs are difficult to exploit since the gate must be kept and operated properly under any circumstances. Therefore, a reliability analysis of the gates operation is necessary to determine the performance of the barrage with respect to the number of gates opened and the gates opening heights. The First Order Second Moment (FOSM) method was used to determine the performance by the reliability index (β) and the probability of failure (risk). It was found that for Cilemahabang Barrage, the number of gates opened with load (L) represents the peak discharge derived from various rainfall (P) respectively one gate with opening height (h=1m) for Preal, two gates (h=1m and h=1,5m) for P50, and three gates (each gate with h=2,5m) for P100. For Caringin Barrage, the results are minimum three gates opened (each gate with h=2,5 m) for Preal, five gates opened (each gate with h=2,5m) for P50, and six gates opened (each gate with h=2,5m) for P100. It can be concluded that a greater load (L) needs greater resistance (R) to counterbalance. Resistance can be added by increasing the number of gates opened and the gate opening height. A higher number of gates opened will lead to the decrease of water level in the upstream of barrage and less risk of overflow.

1. Introduction

Cilemahabang watershed covers two districts in West Java. The headwater is located in Cimahabang (Bogor regency) and flows northward into Ciherang river (Bekasi regency). Cilemahabang barrage is located in the segment of Cilemahabang river and across Cikarang Baru which is an important area in Bekasi. On the other hand, Caringin barrage is located in the downstream in segment of Ciherang river.

A barrage is kind of a weir with a low head consisting of several gates that can be opened or closed to control the amount of water passing through the structure. This is often built in the river on a flat terrain where the flow is tends to be slow and potentially overflow when heavy rainfall occurs. Similiar to a weir, the function of a barrage is to divert the flow for use in irrigation and other systems.

Floods in commercial areas, for example in Cilemahabang watershed, which has inundated roads, housing, and hundreds of factories in Cikarang industrial area, have detrimental effects. The most severe flood, which reached up to 1 meter high, was in Cikarang Baru, upstream of Cilemahabang barrage in the middle of Cilemahabang watershed. Another location was in the downstream of Cilemahabang watershed, near Caringin barrage along the Ciherang river. Because the floodwater inundates several areas near the barrage, the operation of both barrages needs to be evaluated.

This study was undertaken to assess reliability of existing barrages and determine its performance by the reliability index (β) and the probability of failure (risk). Among the existing probabilistic methods, the First Order Second Moment (FOSM) was chosen with corresponding to design flood with various return periods. The reliability analysis will focussed on operational of barrages with respect to the number of gates opened the gates opening heights. At last, the result from this study can become a suggestion for local government (Head of Perum Jasa Tirta II Water Management Division I section Lemahabang) to update the Standard Operating Procedure (SOP) of those two barrages, Cilemahabang and Caringin barrages.

2. Literature Review

Analysis of uncertainties and risks in the hydraulic structures have been developed in recent decades. Reliability of hydraulic design in open channel has also been studied, e.g. the reliability analysis of composite channels was carried out using Advance First Order Second Moment (AFOSM) formulation and Monte Carlo Simulation (MCS) method. Both methods consider the channel design and run off parameters as uncertain variables [1]. The channel design parameters, including friction factors, longitudinal slope, channel width, side slope, and flow depth, while rainfall intensity and drainage as the uncertainty of run off. The study performed a detailed sensitivity study on reliability index for different values of system loading in both static and dynamic environment. It can be concluded from sensitivity analysis that flow depth and bed width were the most influencing parameters that affect the safety factor and reliability.

Another study, an assessment of storm water drain network located in northern part of Bangalore, India has done by considering three failure modes using AFOSM method [2]. There was a need to take into account the uncertainties due to natural and/or inherent randomness for designing a storm water drain network that can effectively convey the discharge. Three parameters were considered to be random variables, namely design flow, conduit dimensions, and roughness of lining material. As a result from reliability analysis, indicate a need to redesign several location of the conduits by the changing of conduit width, slope, or the roughness of conduit lining material.

Failures of hydrosystem infrastructure, as an example in two previous studies, open channel and storm drain sewer, could potentially pose significant threats to public safety and cause enormous damage on properties and the environment. In this study that a barrages or gated weir become an object of study, reliability analysis will be done for determining operation procedure or management strategy. The operation procedure of gates should be checked to minimize the floods and risk of overflow.

2.1. Concepts of Reliability and Risk

With reference to American dictionaries defines risk as the probability of failure to achieve the intended goal. Reliability is defined mathematically as the complement of the risk [3]. Failure of an engineering system can be defined as a situation in which the load L (external forces or demands) on the system exceeds the resistance R (strength, capacity, or supply) of the system. Meanwhile, the reliability is defined as the probability of nonfailure (survive) in which the resistance of the system exceeds the load. In hydrosystem engineering designs, uncertainties associated with the resistance of the hydraulic flow-carrying capacity and the randomness of events, such floods and rainstorms as the load. The mathematical definitions of reliability and failure probability can be seen as,

$$P(survive) = P(L \leq R) \quad and \quad P(failure) = P(L \geq R) \tag{1}$$

In which the resistance R is the designated value of resistance, a deterministic quantity. By considering inherent randomness of annual maximum floods, the annual failure probability for a hydraulic structure designed with a capacity to accommodate a T-year flood, where T-year refers to various return periods.

2.2. First Order Second Moment (FOSM) method
The First Order Second Moment (FOSM) method simplify the functional relationship between independent variables and dependent variable by a truncated Taylor series expansion. The inputs and outputs of these methods are expressed as expected values and standard deviations [1]. This approaches account for uncertainty or randomness in the design parameters by constructing their probability density functions (PDFs). However, such PDFs cannot be derived in most practical applications, owing to complexity of the system and/or data limitations. Hence hydrologists resort to approximate approaches to express the uncertainty based on confidence interval or statistical moments [2].
One of the best-known probability density functions is that forming the familiar bell-shaped curved for the normal distribution, which can be expressed as,

$$f(x) = \frac{1}{\sqrt{2\pi}\sigma}\exp\left[-\frac{(x-\mu)^2}{2\sigma^2}\right] \tag{2}$$

Where x = random variable, μ = mean and σ = standard deviation. If loads and resistances, as well as the parameters determining loads and resistances, are normally distributed, or if their distributions can be approximated by or transformed into normal density functions, the probability of failure of the structure can also be determined from the normal distribution. And if Z (R-L) is a random variable also, the probability of failure can be ilustrated as a shaded area P(Z<0).

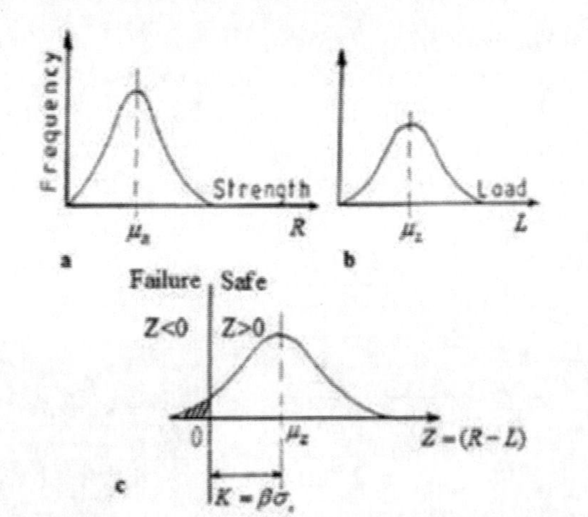

Figure 1. The PDFs normal distribution of resistance/strength (R), load (L), and Z (R-L)

3. Methodology

3.1. Hydrology Analysis using Software HEC-HMS 3.5
This analysis was done to determine the design flood for any return periods (50 years, P_{50} and 100 years, P_{100}) and from extreme rainfall (P_{real}) during the time period 1993-2009. HEC-HMS uses a separate model to represent each component of the runoff process, including models that compute runoff volume/losses, models of direct runoff, models of baseflow, and models of channel flow/routing which method for each model is tabulated below.

Table 1. The method was used in HEC-HMS modelling [4]

No.	Component	Method	Categorization
1.	Precipitation	User hyetograph	The response of a watershed is driven by precipitation. It may be observed rainfall from a historical event, a frequency-based hypothetical rainfall event, or an event that represents the upper limit of precipitation.
2.	Runoff volume	Initial and constant-rate	Address amount about the volume of precipitation that falls on watershed including infiltration on pervious surface and runoff of the impervious surfaces.
3.	Direct runoff	Synthetic Unit Hydrograph: Clark's UH and Snyder's UH	Describe what happens as water that has not infiltrated or been stored on the watershed moves over or just beneath the watershed surface.
4.	Baseflow	Exponential recession	Simulate the slow subsurface drainage of water from the system into the channels.
5.	Routing	Muskingum	Modelling one-dimensional open channel flow.

3.2. Hydraulic Analysis using Software HEC-RAS 4.1.0

The objective of this step is to determine the water surface profiles in the river near the Cilemahabang and Caringin barrages and to calculate the discharge through the gates for various combinations of the gates opened. As the river geometry models, three major tributaries are considered, namely Kali Rasmi, Kali Buntu, and Ci Pegadungan. They will all flow into the main Ciherang river as depicted in Figure 1.

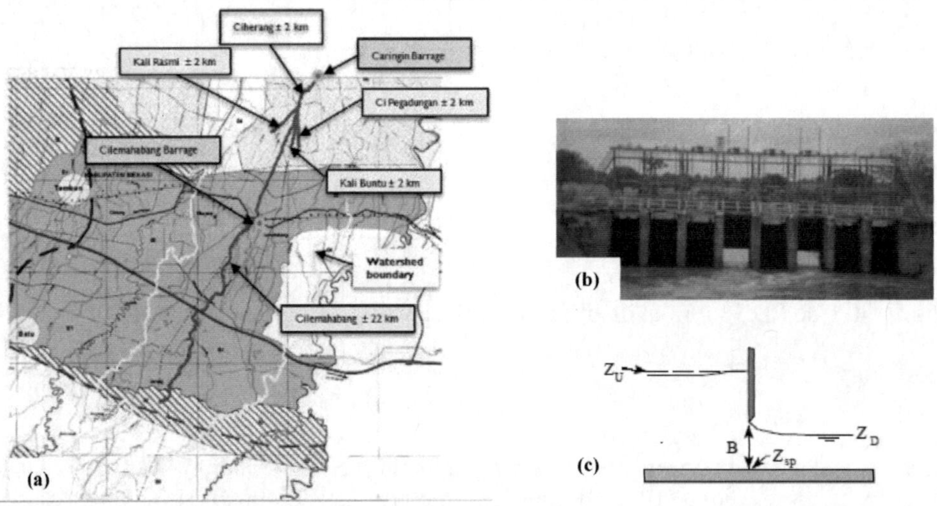

Figure. 2. (a) The layout of Cilemahabang river channel networks (main channel and tributaries); (b) The barrage with low broad crested and several gates which set between flanking piers; (c) example sluice gate with broad crested spillway

Gated weir (barrage) within HEC-RAS was modeled as vertical lift gates (sluices gates). The equations used to model the gates openings that can handle both submerged and unsubmerged conditions at the inlet and outlet of the gates. The spillway crest under the gate was specified as broad crested spillway. The flow through the gate is considered to be "free flow" when the downstream tailwater elevation (Z_D) is not high enough to cause an increase in the upstream headwater elevation for a given flow rate, the equation is as follows [5] :

$$Q = CW B\sqrt{2gH} \tag{3}$$

Where: H = upstream energy head above the spillway crest ($Z_U - Z_{sp}$)
C = coefficient of discharge, typically 0,5 to 0,7
W = weir coefficient

When the downstream tailwater increases to the point at which the gate is no longer flowing freely (downstream submergence is causing a greater upstream headwater for a given flow), the equation becomes:

$$Q = CW B\sqrt{2g3H} \tag{4}$$

Where: H = $Z_U - Z_D$

Submergence begins to occur when the tailwater depth above the spillway divided by the headwater energy above the spillway, is greater than 0,67. Equation (4) is used to transition between free flow and fully submerged flow. This transition is set up so the program will gradually change to the fully submerged orifice equation below when the gates reach a submergence of 0,80.

$$Q = CA\sqrt{2gH} \tag{5}$$

Where: A = area of the gate opening
H = $Z_U - Z_D$
C = discharge coefficient (typically 0,8)

When water is flowing through the top of gate which the gate is closed and at low flow condition (overflow happens through the weir crest under the gate), the flow generally modeled with the standard weir equation:

$$Q = CLH^{3/2} \tag{6}$$

Where: C = Weir flow coefficient, typically 2,6 – 4,0
L = length of the spillway crest
H = upstream energy head above the spillway crest

3.3. Reliability Analysis

This analysis was carried out to determine its performance based on gates operation. Equations below were used to calculate the reliability index (β) and the probability of failure or risk (Tung et.al., 2006).

Reliability index: $$\beta_r = \frac{\mu_z}{\sigma_z} = \frac{\mu_R - \mu_L}{\sqrt{\sigma_R^2 + \sigma_L^2}} \tag{7}$$

Probabilty of failure: $P(failure) = 1 - \Phi(\beta_r)$ (8)

Probability of survive: $P(survive) = \Phi(\beta_r)$ (9)

Where $1-\Phi(\beta_r)$ is risk and $\Phi(\beta_r)$ is reliability with its values are tabulated in PDFs normal distribution tables, and these values may be approximated by the following polynomial [6].

$$B = \frac{1}{2}\left[1 + 0{,}196854|\beta_r| + 0{,}115194|\beta_r|^2 + 0{,}000344|\beta_r|^3 + 0{,}019527|\beta_r|^4\right]^{-4} \tag{10}$$

Where $|\beta_r|$ is the absolute value of reliability index β_r and the standard normal distribution has

$\Phi(\beta_r) = B \quad for\ \beta_r < 0$

$\Phi(\beta_r) = 1 - B \quad for\ \beta_r \geq 0$

The error in $\Phi(\beta_r)$ as evaluated by this formula is less than 0,00025.

4. Results

The design flood was calculated as the Load (L) and the discharge both free flowing and submerged weir flow through the gate openings as the Resistance (R). To avoid a risk of overflow, the water elevation upstream should be kept below the surrounding ground elevation (+11,126 m for Cilemahabang barrage and +5,200 m for Caringin barrage). The result about water elevation upstream of the barrage considering several combinations of the gates operation is summarized in Table 2.

Table 2. The water elevation upstream with several combinations of the gates opened based on various rainfalls Preal, P50, and P100 (CLB = Cilemahabang barrage with 7 gates; CRB = Caringin barrage with 8 gates)

		Water elevation upstream of barrage (m)					
number of gates (pc)	opening height (m)	P_{real}		P_{50}		P_{100}	
		CLB	CRB	CLB	CRB	CLB	CRB
0	-	11,13	5,48	11,21	5,53	11,18	5,5
1	1 m	11,08	5,37	11,2	5,53	11,18	5,5
2	1 m; 2,5 m	11,08	5,24	11,07	5,48	11,17	5,5
3	#gate 2,5 m	10,12	5,16	11,01	5,37	10,96	5,45
4	#gate 2,5 m	9,69	5,16	10,3	5,25	10,96	5,34
5	#gate 2,5 m	9,39	4,11	9,86	5,11	10,04	5,22
6	#gate 2,5 m	9,13	3,84	9,47	5,11	9,59	5,08
7	#gate 2,5 m	8,85	3,63	9,24	4,24	9,48	5,07
8	#gate 2,5 m	-	3,46	-	4,02	-	4,67

Table 2 shows a decrease of water elevation as the effect of increasing number of gates opened. The operating procedure of gates is opened gradually start from no gate opened until all gates opened. Afterward, its respon is monitored due to the change of water elevation upstream of barrage. The opening height is uniform for each opening (2,5 m), which is the maximum height opening.

Furthermore, the performance of barrage can be checked by defining the reliability index and the probability of failure for Cilemahabang and Caringin barrages. Table 3 shows the results of reliability analysis due to gates operation with the design flood was derived from extreme rainfall (P_{real}). The conclusion of these calculation is a higher number of gates opened will lead to minimize risk of overflow.

Two graphics in Figure 3 and Figure 4 give information about relationship between number of gates opened and a risk (%). It compares also about a risk (%) and water elevation upstream on the same graphic.

Table 3. Results of reliability analysis based on gates operation

number of gates (pc)	opening height (m)	Reliability analysis									
		$\mu_Z = \mu_R - \mu_L$		$\sigma_Z = \sqrt{\sigma_R^2 + \sigma_L^2}$		$\beta_r = \mu_Z / \sigma_Z$		$\Phi(\beta_r)$ (%)		$Risk = 1 - \Phi(\beta_r)$ (%)	
		CLB	CRB	CLB	CRB	CLB	CRB	CLB	CRB	CLB	CRB
1	2,5 m	-56,9	-107	17,77	29,40	- 3,20	- 3,63	0,09	0,02	99,91	99,98
2	#gate 2,5 m	-36,9	-70,1	17,83	29,34	- 2,07	- 2,39	1,90	0,85	98,10	99,15
3	#gate 2,5 m	0,003	-35,4	24,46	30,36	0,00	- 1,17	50,01	12,18	49,99	87,82
4	#gate 2,5 m	4,99	-18,9	22,56	29,91	0,22	- 0,63	58,75	26,39	41,25	73,61
5	#gate 2,5 m	10,4	4,16	21,29	30,48	0,49	0,14	68,72	55,41	31,28	44,59
6	#gate 2,5 m	20,55	31,83	19,98	32,54	1,03	0,98	84,79	83,58	15,21	16,42
7	#gate 2,5 m	66,97	50,91	18,22	33,28	3,68	1,53	99,98	93,71	0,022	6,29
8	#gate 2,5 m	-	70,02	-	34,18	-	2,05	-	97,99	-	2,01

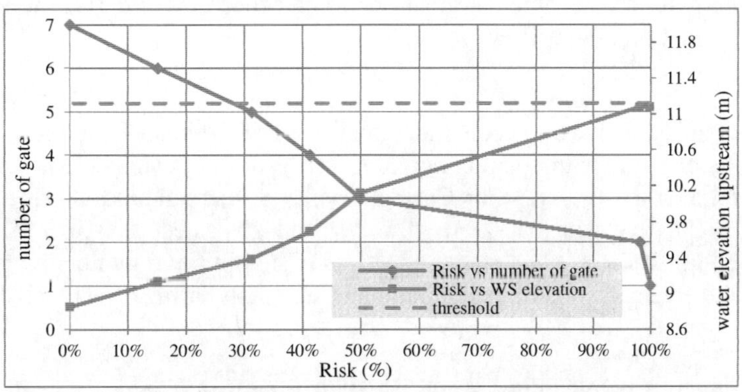

Figure 3. A Graphic relationships between number of gates opened and a risk (%); a risk (%) and water elevation upstream (m) for Cilemahabang barrage

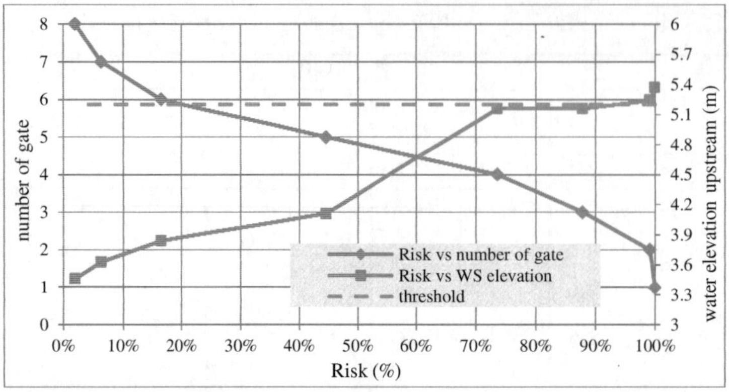

Figure 4. A Graphic relationships between number of gates opened and a risk (%); a risk (%) and water elevation upstream (m) for Caringin barrage

The line graphs in both figures above illustrate changes in a risk of overflow and the water elevation upstream with the combination of different gates operation, which the value of a risk or the probability of failure is showed in Table 3. For both barrages, Cilemahabang and Caringin barrage respectively, a risk declined moderately as the number of gates opened got increased. There was an optimum gate should be opened when the graph of gates opened and a risk overtooked the graph of a risk and water elevation upstream, namely three gates for Cilemahabang barrage and five gates for Caringin barrage. These number of opening gates are become the ideal operating procedure of barrage to maintain a safe condition without any inundation or overflow.

5. Conclusions

From the calculation and analysis, it can be concluded that: (1) For Cilemahabang barrage, the number of gates opened should be one gate with opening height (h=1m) for P_{real}, two gates (h=1m; h=1,5m) for P_{50}, and three gates (h=2,5m) for P_{100}. (2) For Caringin barrage, there should at least be three gates opened (h=2,5 m) for Preal, five gates (h=2,5m) for P_{50}, and six gates (h=2,5m) for P_{100}. (3) The reliability index will increase if more gates were opened. (4) The optimum number of gates should be opened are three gates and five gates for Cilemahabang and Caringin barrages respectively.

References

[1] Reddy M.J., Adarsh S. Reliability Analysis of Composite Channels using First Order Approximation and Monte Carlo Simulations. Stochastic Environmental Research Risk Assessment, 27(2), 477-487; 2013.

[2] Srinivas V.V., Gouri R.L. Reliability Assessment of a Storm Water Drain Network. International Conference on Water Resources, Coastal and Ocean Engineering (ICWRCOE); 2015.

[3] Tung, Y.K., Yen, B.C., dan Melching, C.S. Hydrosystems Engineering Reliability Assessment and Risk Analysis. McGraw-Hill. New York: McGraw-Hill; 2006.

[4] US Army Corps of Engineers, Hydrologic Engineering Center. HEC-HMS Hydrologic Modelling System: Technical Reference Manual. Davis. California; 2000.

[5] US Army Corps of Engineers, Hydrologic Engineering Center. HEC-RAS River Analysis System: Hydraulic Reference Manual. Davis. California; 2010.

[6] Chow, V.T., Maidment, D.R. and Mays, L.W. Applied Hydrology. McGraw-Hill. New York; 1988.

IOP Conf. Series: Earth and Environmental Science **70** (2017) 012051 doi:10.1088/1755-1315/70/1/012051

School services pattern in urban and rural areas: A comparatives study (Case study: Elementary school in Malang City and Malang Regency)

D A Setyono[1], D D Cahyo

Department of Urban and Regional Planning, Faculty of Engineering, Universitas Brawijaya, Malang, Indonesia

deni.setyono08@gmail.com

Abstract. Availability of public facilities are important to support community needs and activities, such as educational facilities (school). Those facilities was needed to endorse the development program implementation which are conducted both of local and national government especially to boost the human resources qualities. This study aims to measures service rates of elementary school in the Malang City and Malang Regency based on supply aspect especially on availability of school unit and also configures the spatial pattern of the school services. Theses study conducted based on the disparity of facility services hypotheses especially on school service provision between urban and rural areas, which are Malang City considered as urban areas and Malang Regency as rural areas. According to the analysis results, rate of elementary school services in the Malang City defined by CGC method about 272% while in Malang the Regency are slightly higher at 319%. The pattern of school services in Malang City relatively similar between its districts, except Klojen District as the growth center of Malang City has the highest rate of services. Meanwhile in the Malang Regency has unique pattern which are high service rates located in the Kepanjen District areas as the growth center of Malang Regency and also several districts that located surrounding the Malang City areas which has impact of city developments. Another district has the lowest service rates due to physical limitations, such as those districts/villages located in the forest areas, coastal areas, or mountainous areas. It is means that students in Malang Regency can access elementary school freely as students in Malang City, they are not only can choose the school in their residential areas but also they can access school everywhere especially from their neighboring areas. It also noticed that there are significant differences of elementary school services between urban center areas and suburban or peripheral areas so that appropriate policy measures are needed to provide equal and balance of educational facilities development throughout each areas. The policy should be arranged appropriately especially in Malang Regency in accordance to the special characteristics of each areas in aims to promote adequate school services and reach all areas equally.

1. Introduction

Access to quality educational facilities considered as a barometer of the socioeconomic development and national development agenda [1]. In these decade, universal primary education development become as part of Millenium Development Goals (MDG's) [2]. However, many developing countries such as Indonesia, providing access to qualified education services are hampered the constraints in educational supply such as education facilities and infrastructures against the quality and efficiency of education systems. In Indonesia, education sector development is the part of indicators national development agenda as it mandated on RPJMN 2015 – 2019. Therefore, in aim to enhance the education sector development there are several programs that can be done and also related to national program "*Wajib Belajar 9 Tahun*" which are qualified and competent teachers and lecturers recruitment, quality of education curriculum improvement and adequate education facilities (school) provision. Those programs are important to improve educational development in each area, and the important program is provision of educational facilities.

Up today, school provision was done based on basic demographical and physical characteristic aspect and it's still need comprehensive and adaptive plan in aim to identify the demand of new facilities based on existed service scales or rates of educational facilities (elementary schools). Research between the effects of competition on achievements and school performance on each area. Those attributes considered as part of facility services and become as important aspect that can increased the quality of services. Moreover, these study is aimed to update the method than can be used to measure the capacities, rates and patterns of education/school facility service in precision. Then those method used to identify areas that has not provided a maximal or optimal facility service and can be used to measure spatial connectivity between regions or areas based on educational facility services, for this research is elementary school.

As mentioned earlier, this research focused to measure educational facilities services especially at primary level that is elementary school which is in Malang City consist of general school and Islamic school. These facilities quite important, especially in supporting the realization of national development priority program in the education field "*Wajib Belajar 9 Tahun*" and currently the management and development of elementary schools are still on the authority of the city and regency government. Until now, there has been no policy on the development and arrangement of educational facility services, ranging from elementary to high school level, so there were competition or accumulation of service occurred related with this educational facilities while on the other hand students have limited choices in accessing the education facility properly especially when student want to choose school near they live. In the end, there is an imbalance or disparity of elementary or primary education services between urban and rural areas. In the Malang City as urban areas and Malang Regency as rural areas, problems related with the elementary school services need to get the best attention especially concentration of school in the certain part of the area. Based on the existing condition, there are areas that have many elementary school that are expected having a high level of services while there are another areas with low availability of elementary school. It is expected that the disparity of elementary school services will be higher in the rural areas than urban areas. Rural communities faced limited services than people in the city. The coverage of elementary school services is as broader as possible so that the capacity of services also expected to be substantial and this certainly affects the level of elementary school services.

2. Configuring the school services pattern

2.1. Literature Review

Several research showed that the school service as part of the school market identification. Garcia-Diaz et.al. (2016) study the relationship between urban school market concentration (lack of competition) and school technical efficiency. Similar with this research, neither efficiency nor competition is not directly observable. In order to measure school efficiency, the stochastic frontier analysis method (SFA) was used to calculate technical efficiency. In other research, Hoxby (2000) measure competition in a school market by using the Herfindahl–Hirschman market concentration index (HHI) as a proxy of the competition level every school market faces. The index employs the application of Geographical Information System (GIS) tools to incorporate the degree of competition each school faces from peer schools spatially located within a radius of 1 km in their local market area. These radius was considered and used in these research to identify the service coverage of elementary school. The consideration also contained on National Standard (SNI) especially to discover public facilities service coverage. Related

IOP Conf. Series: Earth and Environmental Science **70** (2017) 012051 doi:10.1088/1755-1315/70/1/012051

with service coverage, several authors (Barrow and Rouse, 2005; Millimet and Collier, 2008) delimited the school market or school service coverage based on the geographic zones and county sizes, while the others used the school district boundaries in metropolitan areas (Hoxby, 2000; Greene and Kang, 2004). Misra et al. (2012) pointed out that geographic delimitation is subject to an aggregation bias given that sometimes the real competition is not captured. In order to define different school market sizes, the authors drew circles around 1 km radius around each school.

2.2. Methods

These study was intended to compare the calculation method in aim to identify the service rates of facilities between basic/conventional method and the modification ones. Later method is referred as cell/grid based catchment method and it perceived has more accurate/precision result to calculate the service level of facility, especially for the educational facilities (schools) services [Setyono, 2015]. Basically, those method was derived from the Floating Catchment Method (FCM) [Luo, 2004]. The cell/grid based catchment method was applied with the help of GIS software through the buffer and overlay tools. In these research, those method are used to configure the service pattern of educational facility between urban areas (Malang City) and rural areas (Malang Regency). As national standard, the service layer/coverage for elementary school unit takes about 1,200 inhabitants or 1 kilometer radius [6]. The method used to identify the number of school service layers that could be catch in each units (villages and sub-districts) and it will be used as input for the calculation as follows:

$$Rate\ of\ Services = \frac{[\Sigma Layer * 2.400]}{\Sigma\ Population}\ x\ 100\% \qquad (1)$$

Moreover, spatial connectivity analysis was done based on the calculation result in order to find the pattern of elementary school services. Method that can be used to identfy the spatial relationships between are in Malang City and Malang Regency related with school services is Moran Index 'I [5]. Areas with a high index value indicates that the connectivity with the other areas are quite good, especially in terms of accessing elementary school services.

In this research, Malang City and Malang Regency was selected as study area. As it said before, this research focused on elementary school (*sekolah dasar* & *madrasah ibtidaiyah*) as one of main public facilities both for Malang City and Malang Regency in aim to provide qualified, adequate and equitable public education services. By 2015, there are 323 unit of elementary schools that are scattered in various regions in Malang City, in which 271 unit as general school and 52 unit as islamic school [7]. Meanwhile, there are 1.532 unit of elementary schools in Malang Regency consists 1.195 unit of general school and 337 unit of islamic school that are scattered on 33 districts. Development of elementary school will be spread in various part of Malang City and Malang Regency. Development of educational facilities are spread in various districts in Malang City and Malang Regency in order to provide adequate and qualified services to the entire areas and communities, as well in the study areas. Distribution of elementary school in each area/district is not too different in Malang City, contrast with the Malang Regency. In Malang City, each district has more than 60 unit of elementary school except Klojen District. Hence, elementary school number in Malang Regency's district ranging around 30 – 60 unit, except Kasembon and Kromengan District. Purwantoro village has the highest number of elementary school among other villages in Malang City, around 13 unit while the Kepanjen village become area in Malang Regency which highest number of elementary schools. From the distribution of those elementary school, the pattern and service coverages of school will be determined. Detail information of elementary school avalaibility in Malang City and Malang Regency can be seen as follows.

Table 1. Elementary school avalaibility in Malang City and Malang Regency

District	Number of Villages	Population Number	School Number		
			General School	Islamic School	Total
Malang City					
Kedungkandang	12	186.068	55	24	79
Sukun	11	190.053	57	12	69
Klojen	11	104.127	42	6	48
Blimbing	11	177.729	58	5	63
Lowokwaru	12	193.321	59	5	64
Total	**57**	**851.298**	**271**	**52**	**323**
Malang Regency					
Donomulyo	10	62.548	46	2	48
Kalipare	9	60.504	40	11	51
Pagak	8	45.757	30	6	36
Bantur	10	68.816	41	14	55
Gedangan	8	52.938	35	7	42
Sumbermanjing	15	90.350	50	15	65
Dampit	12	119.012	50	14	64
Tirtoyudo	13	60.737	38	4	42
Ampelgading	13	52.622	31	7	38
Poncokusumo	17	92.737	41	23	64
Wajak	13	80.825	40	16	56
Turen	15	114.108	54	18	72
Bululawang	14	71.147	23	19	42
Gondanglegi	14	84.577	32	24	56
Pagelaran	10	67.175	23	16	49
Kepanjen	18	106.668	48	6	54
Sumberpucung	7	53.642	34	2	36
Kromengan	7	38.222	22	2	24
Ngajum	9	49.094	30	7	37
Wonosari	8	41.292	30	3	33
Wagir	12	86.878	36	3	39
Pakisaji	12	88.030	36	6	42
Tajinan	12	53.743	21	15	36
Tumpang	15	75.440	35	14	49
Pakis	15	153.622	33	18	51
Jabung	15	73.850	35	9	44
Lawang	12	109.645	59	5	64
Singosari	17	178.534	60	13	73
Karangploso	9	81.986	25	9	34
Dau	10	74.953	28	2	30
Pujon	13	67.502	32	11	43
Ngantang	13	56.346	38	12	50
Kasembon	6	31.015	19	4	23
Total	**390**	**2.544.315**	**1.195**	**337**	**1.532**

IOP Conf. Series: Earth and Environmental Science **70** (2017) 012051 doi:10.1088/1755-1315/70/1/012051

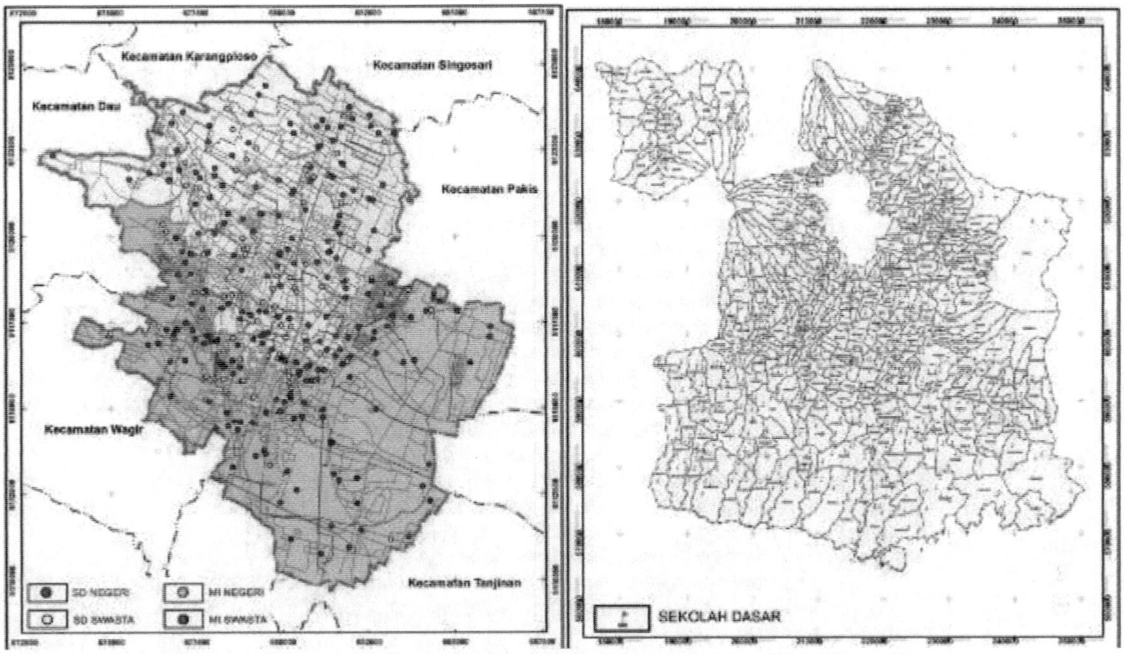

Figure 1. Elementary School Location (a) Malang City; (b) Malang Regency

3. Results

The initial phase in the formulation of the study results is to calculate the level of elementary school services based on the number of existing schools in each village (*desa*) and sub-districts (*kelurahan*) as the smallest administrative units in Malang City and Malang Regency. This calculation is based on the assumption that the communities or students are only able to access elementary schools within the scope of the administrative boundaries of their residential areas. According to the analysis results, it can be seen that the average rates of primary school services in Malang City is lower than the average service rates in Malang Regency which is 98.7% to 140.9% respectively. Those result illustrated that elementary schools in Malang City meet almost all the needs of community as normative standards while elementary school facilities in Malang Regency meet more than the needs of community standards. In the Malang City area, the average of elementary school services in each sub-district is relatively similar which ranged between 90-95%, except Kedungkandang District which has 119.3%. Then in Malang Regency area, almost all villages and sub-districts has a relatively high rate of elementary school services (more than 100%) except Poncokusumo District with only has 56%. Based on this calculation, it can be expected that the peoples and students in Malang City and Malang Regency can only choose elementary school on its local residential boundaries only. In general, the rate of elementary school services based solely on the number of existing schools is quite low compared with the results of other methods. Detailed results of this first/basic method can be seen as in figure 2.

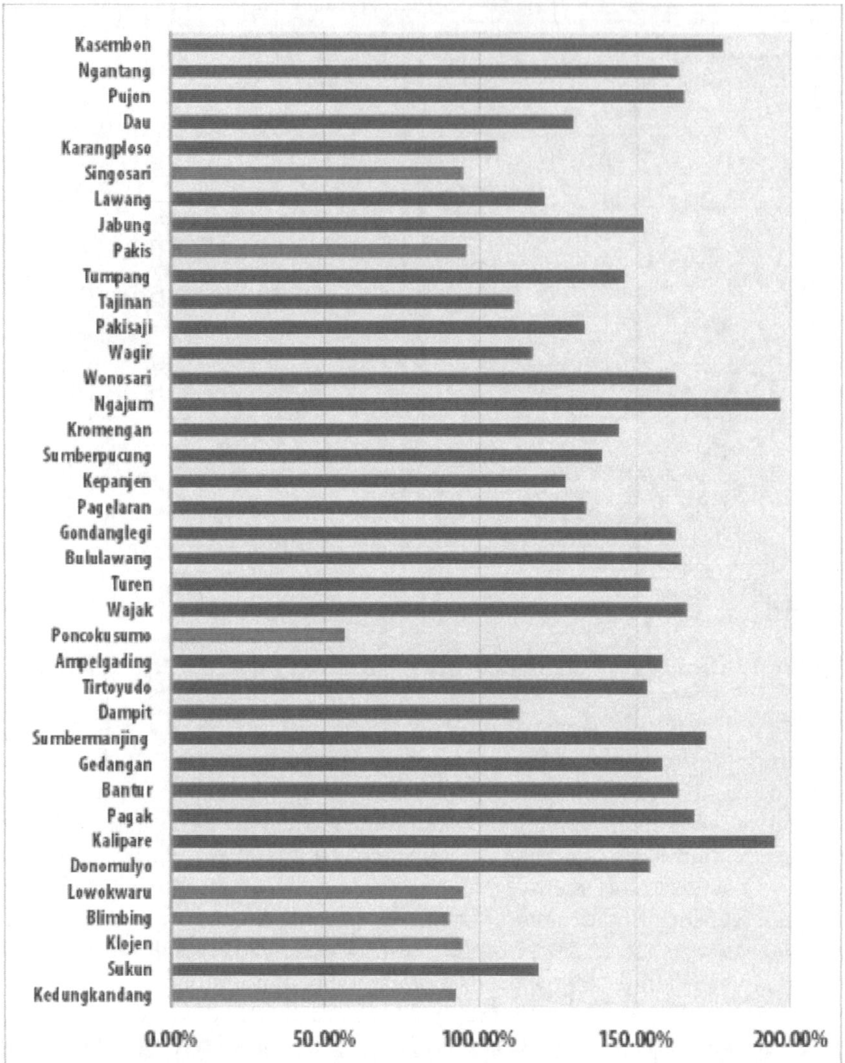

Figure 2. Rate of elementary school services in each district of Malang City & Malang Regency

Another step to determine the rate of elementary school services conducted by using the Administrative Boundary Catchment (ABC) method. As it has been explained, this method was used with the assumption that the communities or students can access not only the elementary school facilities within the scope of their residential areas but also the elementary schools in the neighboring areas while within the radius of elementary school service coverage which is 1 kilometer. These method was used to identify how many or number layers of elementary school services in each villages and sub-districts in Malang City and Malang Regency. Based on the analysis results, it can be seen that the pattern of elementary school (SD) services both in Malang City and Malang Regency showed a tendency rate of services are quite high, especially in urban or growth activity centers. In Malang City area, there are several sub-districts (*kelurahan*) served by more than 50 service layers of elementary school such as Kasin, Sukoharjo, Ciptomulyo, Kesatrian, Polehan and Bumiayu sub-district. Similarly, in the Malang Regency area, villages with more than 50 service layers of elementary school mostly located around the capital area (Kepanjen District), western areas and Malang City border areas such as Singosari, Lawang, Pakis, Dau, Karangploso and Pakisaji District. Those result indicates that the provision of elementary school facilities still concentrated in the several growth center areas. The detailed analysis result of Administrative Boundary Catchment (ABC) method in Malang City and Malang Regency can be seen in the following picture (figure 3).

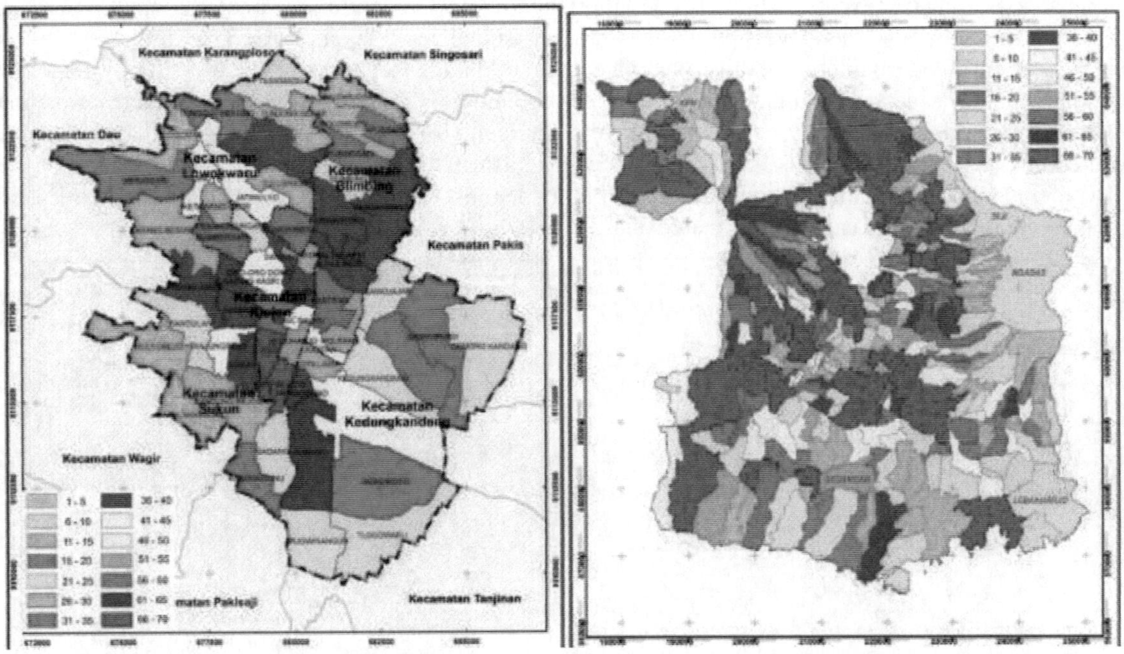

Figure 3. Service layer number based Administrative Boundary Catchment (ABC) method

Those number of school service layers that can be catch in each villages and sub-districts later used as main indicator to calculate the rate of services. Based on the calculation results, it can be obtained that the average rate of elementary school service in Malang City was much lower than the average rate in Malang Regency area which are 560,82% to 3.095,38% respectively. Klojen District became area in Malang City with average rate of elementary school service was more than 930%, while the other districts has average rates range around 400 – 500%. Meanwhile, almost all the districts in the Malang Regency has a high service rates which reaches around 1.000 – 2.500%. Based on those results, it can be estimated that the elementary school service rates in rural areas is much higher than in urban areas. It was also illustrated that elementary school facilities in Malang City were able to serve 4-5 times the population's needs in each district while elementary school facilities in Malang Regency have the higher service condition that reaches 10-25 times the needs of the existed population in these areas. In other words, students in Malang can choose 4-5 times more elementary school facilities surrounding their residential areas. However, the results was considered less good and precise because these method still did not consider the existing land use and conditions in the research areas, especially for non-built land were still included in the calculation of the rate of elementary school services in Malang City and Malang Regency. In general, the results from these method higher than the basic method results.

The last step done in these research was to determine the rate of elementary school services suing the Cell/Grid Based Catchment (CBC) method. Similar to the ABC method, it was used to identify and find the number service layers of elementary school can be catch in each analysis unit (villages & sub-districts). These method also used the assumption that the communities or students can access the primary school which is not only within their residential areas but also the facilities in the surrounding areas as long as it lies within the standard radius of elementary school service coverage of 1 kilometer. However the distinguish variables that made differences from the previous method was this method used the grid system and considered only built-up land use especially settlement areas so that the service rates assessed only in the urban functional areas. The result of Cell/Grid Based Catchment (CBC) analysis

IOP Conf. Series: Earth and Environmental Science **70** (2017) 012051 doi:10.1088/1755-1315/70/1/012051

showed that the average of elementary school service layers that can be catch in each sub district in Malang City was less than 40 layers. Sub-district that high number of elementary school services layers mostly located in the center areas of Malang City, such as Jodipan, Kotalama, Kidul Dalem, and Mergosono Sub-district. While in the sub-district on the Malang City periphery areas has smaller number of elementary school service layers which was ranged around 1 – 5 layers only. The similar condition can be seen from the calculation results of service rates in Malang Regency, whereas most of villages or sub-districts has access elementary school services around 1 – 5 layers. Meanwhile, there are some villages with high elementary school service layers between 16 – 20 layers, especially villages that are located between the main corridors to Malang City. Significant differences related with outcome of the CBC method are due to land use considerations that made most of the layer values in cell of unbuilt land are not considered in the calculation or analysis.

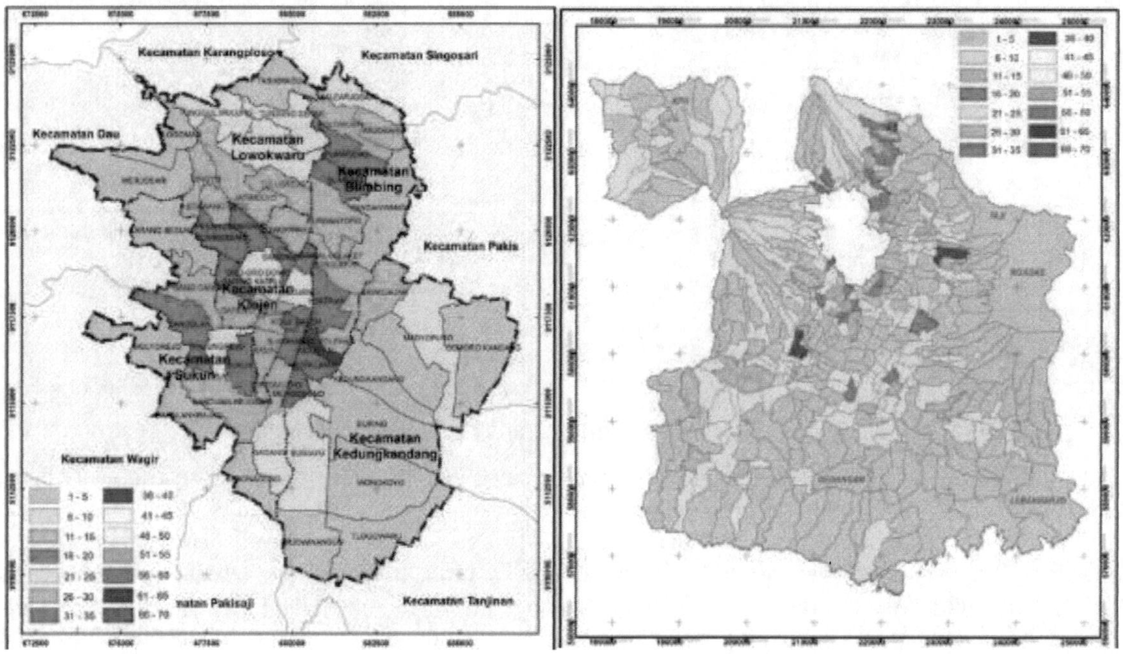

Figure 4. Service layer number based Cell/Grid Based Catchment (CGC) method

Based on the result of using the Cell/Grid Based Catchment (CBC) method, it is known that elementary school service rates has the significant result compared to the result of previous (ABC method). The rate of elementary school services in Malang City about 272% while the rate of elementary school services in Malang Regency are slightly higher at 319%. Related with rate of elementary school services of each district in Malang City, the average rate around 150 – 200%, except Klojen District which reached an average service rates about 569%. Based on the analysis results, it can be seen that there are sub-districts which did not have service rates up to 100% especially those located in suburban or peripheral areas of Malang City such as Pandanwangi, Arjowinangun, Lesanpuro, Merjosari, Tlogomas, Karangbesuki, Mulyorejo and Bakalankrajan sub-district. Then for the rate of elementary school services at the district level in Malang Regency has values with a fairly diverse range, but in general the average service rates around only 200 – 300%. The analysis results also showed that Kepanjen District become the area with the highest service rate reached 723%. Those illustrated that elementary school facilities in Kepanjen District and surrounding areas were able to serve 7 times the population's needs in related areas. All villages and sub-districts in Malang Regency have more than 100% elementary school service rates on averages, but it is noticed that there are villages with rate less than 100%. In general, the average result of the Cell/Grid Based Catchment (CBC) method lower than the Administrative Boundary Catchment (ABC) method but it did not mean poor results and could be considered as a precise result because the consideration in the calculation of the analysis is more specific such as considered only built-up areas.

4. Conclusion

According to the study results, it can be illustrated that calculation using the Cell/Grid Based Catchment (CBC) method can be understood has more effective and accurate results to measure the rate of elementary school services up to now. Then related to the spatial aspect, it can be seen that the average rate of elementary school services in the Malang City are lower than in the Malang Regency. However, there is similar pattern of school services where the central areas or growth center in Malang City and Malang Regency become area with the highest rate of elementary school services compared to other areas which are Klojen District (Malang City) and Kepanjen District (Malang Regency). It can also be figured that there are significant differences of elementary school services between urban center areas and suburban or peripheral areas so that appropriate policy measures are needed to provide equal and balance of educational facilities development throughout each areas.

References

[1] Mizunoya S, Zaw HT. Measuring the holes of the ship: Global cost estimations of internal inefficiency in primary education. Int J of Ed Dev 2017;54(1):8-17.

[2] UNESCO (2015). Education for All 2000–2015: Achievements and Challenges. UNESCO Publishing, Paris, France. Retrieved from: http://unesdoc.unesco.org/images/0023/002322/232205e.pdf.

[3] Garcia-Diaz R, del Castillo E, Cabral R. School competition and efficiency in elementary schools in Mexico. Int J of Ed Dev 2016;46(1):23-34.

[4] Hoxby CM. Does competition among public schools benefit students and taxpayers? Am. Econ. Rev. 2000;90(5):1209–1238.

[5] Millimet DL, Collier T. Efficiency in public schools: does competition matter? J. Econometrics 2008; 145 (1):134–157.

[6] Misra K, Grimes PW, Rogers KE. Does competition improve public school efficiency? A spatial analysis. Econ. Educ. Rev. 2012;31 (6):1177–1190.

[7] Puspitasari R, Susanto I. Spatial Analysis Cases of Dengue Fever in Sukoharjo Central Java Uses Moran Index. Prosiding of Mathematic and Mathematic Education National Conference Yogyakarta 2011.

[8] Minister of National Education RI (2007). Minister of National Education RI Policy No. 24, 2007 about Facilities and Infrastructure of Elementary, Junior High Schools and Senior high School Standards. Republik Indonesia

[9] National Indonesian Standard (SNI) 03-1733-2004: The Procedures for Planning a Residential Neighborhood in Urban Areas.

IOP Conf. Series: Earth and Environmental Science **70** (2017) 012052 doi:10.1088/1755-1315/70/1/012052

Relationship between environmental management with quality of kampong space room (Case study: RW 3 of Sukun Sub District, Malang City)

D K Wardhani[1], D S Azmi, and W D Purnamasari

Department of Regional and Urban Planning, Faculty of Engineering, Universitas Brawijaya, Malang 65147, Indonesia

dkwardhani@yahoo.com

Abstract. RW 3 Sukun Malang was one of kampong that won the competition kampong environment and had managed to maintain the preservation of the kampong. Society of RW 3 Sukun undertake various activities to manage the environment by optimizing the use of kampong space. Despite RW 3 Sukun had conducted environmental management activities, there are several locations in the kampong space that less well maintained. The purpose of this research was to determine the relation of environmental management with the quality of kampong space in RW 3 Sukun. This research used qualitative and quantitative research approaches. Quantitative research conducted by using descriptive statistical analysis in assessing the quality of kampong space with weighting, scoring, and overlay maps. Quantitative research was also conducted on the relation analysis of environmental management with the quality of kampong space by using typology analysis and pearson correlation analysis. Qualitative research conducted on the analysis of environmental management and the relation analysis of environmental management with the quality of kampong space. Result of this research indicates that environmental management in RW 3 Sukun have relation with the quality of kampong space.

1. Introduction

Urban kampong is generally known as settlements with high population density, informally built housing, insufficient facilities and infrastructure, and are very densely inhabited [1]. Inside the urban kampong settlements there are various problems that cause unnoticed environment so that there is an impact increasingly quality of settlements. One of the causes of settlement problems is the result of human behavior for needs of using settlement land [2].

Several local governments have been conducting kampong improvement programs since 1978. The kampong improvement program activities are prioritizing the arrangement of the environment so that people are encouraged to action an active role in improving their environment with the help of physical materials [3]. Kampong improvement program has also been organized by the Government of Malang City trough holding a Kampung Bersinar Competition. Kampung Bersinar Competition has three goals, one of the goal is to increase public awareness of the needs clean and beautiful environment.

RW 3 Sukun is the winner of the environmental competition that has maintained the kampong's sustainability by optimizing the use of kampong space in conducting environmental management activities. However, although RW 3 Sukun has conducted environmental management activities and participated in a competition organized by the Government of Malang City, there are still some locations

IOP Conf. Series: Earth and Environmental Science **70** (2017) 012052 doi:10.1088/1755-1315/70/1/012052

in the kampong space that is less maintained and less clean. So this research is need to know the relation of environmental management to the quality of kampong space in RW 3 Sukun Sub District. The relationship of environmental management with the quality of kampong space can be used as an illustration in managing the kampong environment by utilizing the existing space for another kampong.

2. Contents

2.1. Methods

This research uses qualitative and quantitative research approaches. Quantitative research on this research is done weighting, scoring, map overlay, typology analysis, and correlation analysis by using person correlation test. Qualitative research in this research is conducted on the analysis of environmental management analysis of environmental management relationship with the quality of the kampong space in descriptive qualitative.

This research will divide the kampong space in RW 3 Sukun Sub District based on RT area boundary. The division aims to facilitate the researchers in the analysis process in describing environmental management in each RT and compare the quality of the kampong space between one RT with another RT.

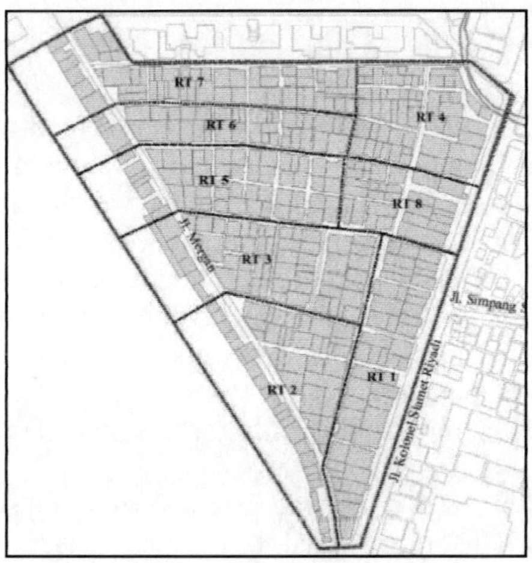

Figure 1 Research Sites.

2.2. Population and Sample

The population in this study is the kampong space in RW 3 Sukun Sub District and the population of RW 3 Sukun Sub District consisting of 1,125 peoples. The sample in this research is divided into two namely the expert sample and the sample of population of RW 3 Sukun Sub District. Sampling of RW 3 Sukun community using probability sampling method with proportional stratified random sampling technique.

$$S = \frac{\lambda^2 \times N \times P \times (1-P)}{(d^2 \times (N-1)) + (\lambda^2 \times P \times (1-P))} = 218 \text{ responden}$$

$$ni = \frac{Ni}{N} x\, n \qquad (1)$$

Table 1. Sample Distribution.

RT	Among Population	Among Sample
RT 1	125	24
RT 2	105	20
RT 3	212	41
RT 4	132	26
RT 5	174	34
RT 6	113	22
RT 7	149	29
RT 8	115	22
	1125	218

Experimental sampling using non-probability sampling method with purposive sampling technique. The number of samples used five people based on the consideration that the respondents are people who understand about the kampong space and environmental management. Expert samples come from BAPPEDA Malang, DKP Malang, academics (Lecture Department, and environmental competition judging).

2.3. Analysis Method

2.3.1. Weighting. Weighing criteria in this study using the pair comparison method to give weight value on the criteria of the quality of the kampong space room. The weight of the criteria for the quality of the kampong space room is obtained from five stakeholders. The weighting steps by the paired comparison method [4] is define criteria, create matrix of pair comparisons, calculate the weight of criteria, make consistency ratio estimation, and create a composite opinion matrix. The criteria that used in this study are taken from the theory of Human Rights [5] which is then divided into two namely physical and nonphysical criteria. Physical criteria consist of environmental criteria, environmental management facilities, greening, settlement layout, and roads. While non-physical criteria consist of container activities criteria that supplement income and social activities related to environmental management.

2.3.2. Scoring. Scoring is a technique in analyzing data by providing scores on existing circumstances based on predetermined criteria. The criteria used in this study based on the criteria for assessment of Kampong Bersinar Competition and literature study on previous research.

1. Physical environmental criteria in assessed from the cleanliness conditions of the kampung environment is road and alley cleanliness, drainage system, and yard of the house.
2. Criteria of environmental management facilities are assessed from facilities used in environmental management consisting of waste collection facilities, drainage system, biopori, and composter.
3. Greening criteria assessed from the presence and maintenance of plants on the road and yard of the house.
4. Criterion of settlement layout assessed from building density and pattern of layout or building regularity.
5. Road criteria are assessed from road construction and road construction conditions.
6. Criteria of activity containers that supplement income are assessed from potential economic activities, side business, and waste utilization in business.
7. Criteria of social activities that support the settlement environment are assessed from social activities related to environmental management that can improve the quality of the village space consisting of socialization related to environmental management, evaluation of environmental management activities, and discussion on environmental management.

2.3.3. Map Overlay. Map overlays are used to get a quality map of the kampong space. In the overlay stage of the map, each criteria score is multiplied by each weight then the whole score is summed to know the value of the quality of the kampong space at each RT in RW 3 Sukun Sub District.

2.3.4. Pearson Correlation Test. Pearson correlation test is used to see the resemblance of the distribution of the value of a variable with the value of other variables so that will know whether there is a relationship between two variables. Pearson correlation is used when two variables X and Y each have a scale of interval and ratio measurement [6]. The steps in the correlation analysis consist of determining the hypothesis, determining the probability value, comparing the significant value, and looking at the value of the correlation coefficient.

3. Results

3.1. Analysis of Environmental Management
The analysis of environmental management consists of the type of activity, actors, location, and time of environmental management activities.

3.1.1. Type of environmental management activities. Types of environmental management activities consist of waste management activities, water treatment management, bio pore management, drainage system management, afforestation, consecration work, and clean and improve the kampong environment.

3.1.2. Person of environmental management activities. Person of environmental management in RW 3 Sukun Sub District can be seen based on the percentage of RT residents involved. Environmental management person in RW 3 Sukun Sub District are dominated by RT 7 peoples as much as 41%.

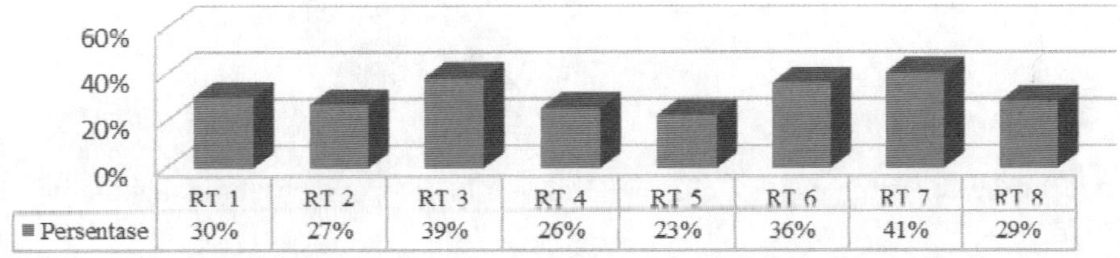

Figure 2. Graph of citizens percentage in environmental management at each RT.

3.1.3. Location of environmental management activities. The location of environmental management in RW 3 Sukun Sub District is viewed based on space that used for environmental management activities. Used for environmental management activities include roads, yard houses, garbage warehouses, BSM unit in RW 3 Sukun Sub District, parks, nurseries, fields, and houses. The most used kampung space is in the kampong road as much as 68%. This is due to the large number of environmental management activities carried out by roads such as waste sorting activities, plant maintenance, management of biopori management, composter management, drainage system improvements, therapeutic stone improvements and village roads, and other activities.

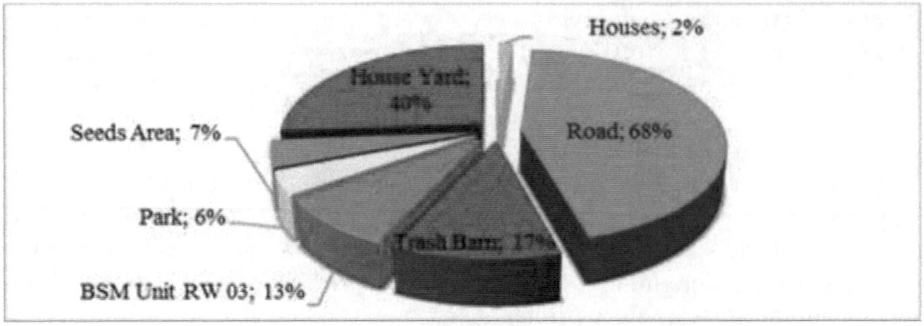

Figure 3. Percentage of space diagram used in environmental management.

3.1.4. Time of environmental management activity. Time of environmental management in RW 3 Sukun Sub District is divided into three in the morning, afternoon, and evening. The environmental management activities of RW 3 Sukun Sub District are mostly done in the morning. Generally, residents RW 3 Sukun Sub District do environmental management activities after doing routines in the morning such as cooking, washing, and others. This condition is caused by the majority of RW 3 Sukun Sub District citizens conducting informal communication or conversation in the afternoon. In addition, many people's social activities conducted in the afternoon such as pray, PKK as organization meeting, informal meeting, istigosah, and other meetings are generally done once to four times a month.

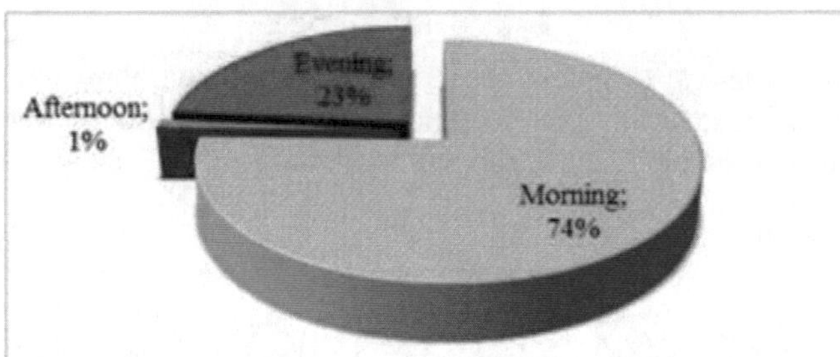

Figure 4. Percentage of citizens diagram in environmental management.

3.2. Quality Analysis of Kampung Space

Quality Analysis of kampong space is an analytical process in knowing the quality of kampong space based on the condition of the kampong space in each RT. The high quality of urban space will be influenced by the overall criteria for the formation of the kampong. The quality of the kampong room is calculated using weighting and scoring.

3.2.1. Weighting. The weighting comparison method is based on the perception of experts / stakeholders from BAPPEDA Malang, Gardening and Hygiene Office of Malang City, academicians, and environmental competition judges. The result of weighting of pair comparison method can be seen in Table 2. The highest quality of space weights are social activities that support the settlement environment with the weight of 0.234. While the value of the lowest weight is a container activity that adds income with a weight of 0.059.

Table 2. Weight Criteria of Kampong Space Quality.

Criteria	Weight	Rating Priority
Physical environment	0,217	3
Environmental management facilities	0,106	4
Container activities to supply income	0,059	7
Greening	0,232	2

Criteria	Weight	Rating Priority
Settlement Layout	0,063	6
Street	0,089	5
Social activities supporting environmental settlements	0,234	1

3.2.2. Scoring. Scoring is based on seven criteria of kampong space quality

1. Physical environment. Scores of physical environmental criteria were obtained from the calculation of average score of road hygiene condition and alley (A), drainage system (B), and home yard (C). RT which has the highest score for environmental physical criteria is RT 3, RT 6, and RT 7 with score 3. Scores of physical environmental criteria can be seen in Table 3.

Table 3. Physical environment Criteria Score.

RT	**Score**			**Criteria Score**
	A	**B**	**C**	
RT 1	3	3	2	2,67
RT 2	2	2	3	2,33
RT 3	2	3	3	3,00
RT 4	2	1	2	1,67
RT 5	2	2	3	2,33
RT 6	3	3	3	3,00
RT 7	3	3	3	3,00
RT 8	2	2	3	2,33

2. Environmental management facilities. Scores of environmental management facility criteria are devided from calculation of average scores of garbage collection (A), drainage system construction (B), drainage system condition (C), number of bio pores (D), and number of composter (E). RT which has the highest score for environmental management facility criteria is RT 3 with score 3. Scores of environmental management facility criteria can be seen in Table 4.

Table 4. Environmental management facilities Criteria Score

RT	**Score**					**Criteria Score**
	A	**B**	**C**	**D**	**E**	
RT 1	3	3	2	3	2	2,60
RT 2	3	3	2	3	2	2,60
RT 3	3	3	3	3	3	3,00
RT 4	3	3	2	2	2	2,40
RT 5	3	3	1	3	3	2,60
RT 6	3	3	2	3	2	2,60
RT 7	3	3	2	3	2	2,60
RT 8	3	3	1	2	2	2,20

3. Reforestation. The score of greening criteria was obtained from the calculation average score of the existence of plants on the road (A), plant maintenance on the road (B), the presence of plants in the yard of the house (C), and the maintenance of the plants in the yard of the house (D). RT with the highest scores for environmental physical criteria were RT 6 and RT 7 with a score of 2.75. The score of greening criteria can be seen in Table 5.

Table 5. Greening Criteria Score.

RT	Score				Criteria Score
	A	B	C	D	
RT 1	1	3	3	3	2,50
RT 2	1	3	3	3	2,50
RT 3	1	3	3	3	2,50
RT 4	1	3	3	3	2,50
RT 5	1	3	3	3	2,50
RT 6	2	3	3	3	2,75
RT 7	2	3	3	3	2,75
RT 8	1	3	2	3	2,25

4. Settlement Layout. Scores of settlement layout criteria are devided from the calculation of the mean score of building density (A) and building layout pattern (B). Scores of settlement layout criteria were dominated by a score of 2-2.5 except on RT 8 which scored 1.5. The road criteria score can be seen in Table 6.

Table 6. Settlement Layout Criteria Score.

RT	Score		Criteria Score
	A	B	
RT 1	1	3	2,00
RT 2	2	3	2,50
RT 3	2	3	2,50
RT 4	1	3	2,00
RT 5	2	3	2,50
RT 6	2	2	2,00
RT 7	2	3	2,50
RT 8	1	2	1,50

5. Street. The street criteria score is divided from the calculation of the average score of road construction (A) and the condition of road construction (B). All RTs get a score of 3 except RT 1 with a score of 2.5. The road criteria score can be seen in Table 7.

Tabel 7. Street Criteria Score.

RT	Score		Criteria Score
	A	B	
RT 1	3	2	2,50
RT 2	3	3	3,00
RT 3	3	3	3,00
RT 4	3	3	3,00
RT 5	3	3	3,00
RT 6	3	3	3,00
RT 7	3	3	3,00
RT 8	3	3	3,00

6. Container activities to supply income. Score criteria container activities that supply the income obtained from the calculation of the average value of potential economic activities, independent side business, and utilization of waste in the business. The RT which has the highest score for the activity container criteria that supplemented the income was RT 1 with a score of 2.33. The score of the income earning activity container criteria can be seen in Table 8.

Tabel 8. Container activities to supply income Criteria Score.

RT	Score			Criteria Score
	A	B	C	
RT 1	3	2	2	2,33
RT 2	1	3	2	2,00
RT 3	1	2	3	2,00
RT 4	1	2	2	1,67
RT 5	1	2	2	1,67
RT 6	1	2	2	1,67
RT 7	1	2	3	2,00
RT 8	1	2	2	1,67

7. Social Activities that Support the Settlement Environment. Criteria score of social activities that support the settlement environment is derived from the calculation of the average community involvement in following social activities. Social activities conducted by RW 3 Sukun Sub District are socialization (A), activity evaluation (B) and discussion (C). The RT with the highest score for the social activity criteria is RT 7 with score 3. The social activity criteria score can be seen in Table 9.

Tabel 9. Social Activities that Support the Settlement Environment.

RT	Social Activity			Score			Criteria Score
	A	B	C	A	B	C	
RT 1	75%	50%	85%	2	1	3	2,00
RT 2	70%	55%	85%	2	1	3	2,00
RT 3	63%	69%	94%	2	2	3	2,33
RT 4	77%	64%	88%	2	2	3	2,33
RT 5	73%	38%	90%	2	1	3	2,00
RT 6	76%	67%	95%	2	2	3	2,33
RT 7	86%	87%	94%	3	3	3	3,00
RT 8	79%	55%	79%	2	1	2	1,67

3.3. The value of the kampong space quality

Determination value of the quality of kampong space by overlaying the map to get the final map result of the quality kampong map. Overlay results indicate that the highest quality of space is in RT 7 with the value of kampung space 2.809 while the lowest space quality is in RT 8 with the value of kampong.

Tabel 10. The value of the kampong space quality.

RT	Value	Space Quality	Large (m^2)	(%)
RT 1	2,389	Good	4629,66	14%
RT 2	2,373	Good	5606,06	17%
RT 3	2,638	Good	4866,70	15%
RT 4	2,233	Medium	3820,83	12%
RT 5	2,353	Good	4308,92	13%
RT 6	2,602	Good	2967,76	9%
RT 7	2,809	Good	4620,89	14%
RT 8	2,111	Medium	2228,33	7%

The criteria for forming quality of each RT space is different. The criteria is seen from the high level of low value between the criteria of the quality of space in each RT. The criteria for the formation of kampung halls in RT 7 as RT with the highest quality of space is the criteria of social activities. The criteria that make the quality of the kampong space in the two RTs with the highest quality of kampong halls after RT 7, RT 3 and RT 6 are the physical environmental criteria. While the criteria quality of kampong space in other RT is the criteria of reforestation. Although the criteria for setting up the quality of kampong space at each RT in RW 3 Sukun Sub District but the criteria of social activities, physical environment, and greening are the three criteria that most influence the quality of kampong. The level of criterion for the quality of space in each RT can be seen in Figure 6.

Figure 5. Level of criteria for the formation of the kampong space quality at each RT.

3.4. Analysis of Environmental Management Relationship with Quality of Kampong Space

3.4.1. Typology Analysis. The analysis of kampong typology was conducted based on the assessment of the percentage of residents conducting environmental management activities and the value of kampong space quality in each RT which was then compared with the respective mean score. A comparison value of spatial quality and environmental management with each mean score will divide the environmental management and the quality of the kampung space into two typologies: low and high. The results of typology analysis show that RW 3 Sukun Sub District consists of two types of typology, RT with low environmental management, low quality of village space and RT with high environmental management, high quality of kampung halls. Both types of typology explain that there is a relationship between environmental management with the kampong space quality.

Table 11. Kampong Typology with Two Interval Class.

RT	Environment Management Clasification	Kampong Space Quality Clasification	Kampong Typology
RT 1	Low	Low	Typology 1
RT 2	Low	Low	Typology 1
RT 3	High	High	Typology 4
RT 4	Low	Low	Typology 1
RT 5	Low	Low	Typology 1
RT 6	High	High	Typology 4
RT 7	High	High	Typology 4
RT 8	Low	Low	Typology 1

Correlation analysis is used to find correlation between environmental management activity conducted by citizen of RW 3 Sukun Sub District with the kampong space quality. Through correlation analysis will be known management of any environment that has a relationship with the kampong space quality. Correlation analysis in this study was conducted using Pearson Correlation test. Based on the results of correlation analysis between environmental management with the kampong space quality, only the persons and spaces of environmental management activities that have a relationship with the kampong space quality while the time of environmental management activities have no relation with the quality of space.
The result of correlation analysis shows that the activity of environmental management has relation with the kampong space quality because the value of Sig. <0.1. Then Ho is rejected, meaning there is a significant relationship between environmental management activities with the kampong space quality. The relationship shows the higher activity of environmental management hence the higher value of kampong space quality. But not all environmental management activities have a significant relationship. Environmental management activities that are related to the kampong space quality include the persons of waste management activities, the persons of bio pore management activities, the person of reforestation activities and the kampong space that used for reforestation activities.

3.4.2. Correlation Analysis

Table 12. Environmental management activity correlation with kampong space quality.

Variable	Sig. (2-tailed)	*Coefficient Correlation*	Relations
waste management activities			
Pearson	.024	.776	Significant, strong, positive
biopori management activities			
Pearson	.035	.743	Significant, strong, positive
reforestation activities			
Pearson	.093	.632	Significant, strong, positive
Space			
• Street	.054	.699	Significant, strong, negative
• House Yard	.089	-,637	Significant, strong, negative

Explanation : Sig. < 0,1 there is a significant relation (Ho rejected)
Sig. > 0,1 there is no a significant relation (Ho accepted)

4. Conclusion

Based on the results of the analysis, the conclusion of the study "Relationship between Environmental Management with Quality of Kampong Space" include:

1. The results of typology analysis indicate that there is a relationship between environmental management and the quality of the kampong space.
2. The result of correlation analysis shows that the activity of environmental management has correlation with the quality of kampong space because the value of Sig <0,1. The value of Sig shows Ho is rejected, the meaning is a significant relationship between environmental management activities with the quality of the kampong space. The relationship shows the higher activity of environmental management hence the higher the value of kampong space quality.
3. Based on the results of correlation analysis between environmental management with the quality of space, only the persons and spaces of environmental management activities that have a relationship with the quality of the kampong space while the time of environmental management activities have no relationship with the quality of space.
4. Environmental management activities that have relationship with the quality of the kampong space are among others the waste management activities, the persons of biopori management activities, the persons of reforestation activities and the space used for reforestation activities.

Overall, environmental management persons have the strongest relationship with the quality of the kampong.

References

[1] Soefaat, Kamus Tata Ruang, Jakarta: Dirjen Cipta Karya Departemen Pekerjaan Umum, Ikatan Ahli Perencanaan Indonesia, 1997.

[2] M. Mayasari and S. Ritohardoyo, "Kualitas permukiman di Kecamatan Pasarkliwon Kota Surakarta," Jurnal Bumi Indonesia, vol. 1, no. 3, pp. 193-201, 2012.

[3] PU-Net, "Kriteria permukiman kumuh harus diperjelas," Kementrian Pekerjaan Umum dan Perumahan Rakyat Republik Indonesia, 2009. [Online]. Available: http://pu.go.id/berita/2190/kriteria-permukiman-kumuh-harus-diperjelas.

[4] G. A. Mendoza, P. Macoun, R. Prabhu, D. Sukadri, H. Purnomo and H. Hartanto, Panduan untuk menerapkan analisis multikriteria dalam menilai kriteria dan indikator, Jakarta: Center for Internationa Forestry Research (CIFOR), 1999.

[5] E. Budihardjo, Arsitektur dan kota di Indonesia, Bandung: PT. Alumni, 1997.

[6] J. Sarwono, Statistik itu mudah: Panduan lengkap untuk belajar komputasi statistik menggunakan SPSS 16, Yogyakarta: Penerbit Universitas Atma Jaya Yogyakarta, 2009.

IOP Conf. Series: Earth and Environmental Science **70** (2017) 012053 doi:10.1088/1755-1315/70/1/012053

The Relationship Between The Availability Of The Supporting Elements Of Pedestrian With Pedestrian Crossing Facility Usage Based On User Preferences (Case Study Corridor of Sumbersari Street, Gajayana Street, MT. Haryono Street, Malang City)

D P Soetrisno

Department of Regional and Urban Planning, Faculty of Engineering, Universitas Brawijaya, Malang 65147, Indonesia

dwimpwkub@gmail.com

Abstract. Pedestrian crossing facilities are effective enough to avoid pedestrians with vehicles, but its utilization is still quite low. It indicated that safety is not the only factor that influences a person to utilize the pedestrian crossing facilities. In addition, the availability of supporting elements of the pedestrian is still not quite attention, which is also became a factor that causes the pedestrians doesn't utilize the pedestrian crossing facilities. Therefore, this research was structured to examine the relationship between the availability of the supporting elements of the pedestrian with pedestrian crossing facility usage based on user preferences. Data collection method used is primary survey consist of observation and the questionnaire. Sampling techniques used is purposive sampling with the number of respondents as many as 211 respondents by using questionnaire with ordinal scales to identify respondents' consideration level of supporting elements pedestrian and crossing facility utilization factors. The survey is done on 15 crossing facilities area in 3 different locations with the same characteristics of land use in the form of higher education area (university area) and trades and services activities area. The analysis technique used is frequency distribution analysis in order to identify preference pedestrian on the availability of supporting elements of pedestrian and pedestrian crossing facility utilization factors, and chi square analysis is used to analyze the relationship between the availability of the supporting elements of the pedestrian with pedestrian crossing facility utilization. Based on the chi square analysis results with significance 5 % obtained the result that there are six supporting elements of pedestrian having correlation to the factors of pedestrian crossing facility utilization consist of the availability of sidewalk, pedestrian lights, Street Lighting Lamps, Pedestrian Crossing Markings Facilities, Sign Crossings Facilities, vegetation, and dustbin. So the result of this research can be considered for the government as main stakehoder especially the local government in preparing policy to provide supporting elements of pedestrian that should be on the area of pedestrian crossing facilities.

1. Introduction

A pedestrian is a person who performs activities on foot and is one of the elements of road users [1]. The term of pedestrian or a pedestrian comes from the Greek pedester/pedestris i.e. people who walk or promenade. Pedestrian also comes from the Greek meaning pedos. Pedestrian are also interpreted as movement or circulation or as the destination on foot (Rubenstein, 1992) [2]. Where pedestrian is part of the transportation system are no less importance compared to other transportation modes. Although the Act of walking will look simple but it has an important role in the transportation system, because if the pedestrian crash then it will affect parts of the transportation system. Therefore, the needs of pedestrians is a part that is integrated with the transportation system. The results of a study from the Institute of Transportation Studies (INSTRAN) get that 65% of traffic accidents caused pedestrians fatalities. [3]. So it can be understood that the pedestrian facilities are often not included in the planning.

If there are any often do not provide comfort for pedestrian who use the facility. With the condition and the facilities inadequate pedestrian often accused one of the causes of traffic congestion. So one of the elements that must be available in the process of engineering traffic is pedestrian facilities (available of pedestrian facility). The importance of the issue of the provision of pedestrian facilities is noteworthy based on the following factors: a). The road plays a role in responding to the problem of urban transportation. b). Pedestrian facilities is an important element in the planning of the city, the pedestrian facilities Set up from a matching vehicle lines would support the potential of the relic in the town center. c). Structuring strategic pedestrian facilities with a high standard of achievement will be able to support the overall circulation in the City Centre (Rhamdani: 1992) [4].

General conditions on the location of research, the existence of pedestrian crossing facilities are adequate is not followed by the availability of supporting elements of the pedestrian, it still is very minimal which forced the most crossing way to not utilize the facilities of the crossing as well as prefer to cross where it's not supposed to. With the high mobility of society requires the availability of pedestrian facilities that are equipped with supporting elements to suit preferred factor crossing the road. So the availability of supporting elements of the pedestrian needs to be examined to see the correlation of each element against the factors that more prioritized by pedestrians.

2. Literature Review

According to (John Fruin 1979), walk is a movement of internal city, the only instrument to meet the needs of interaction face-to-face that is in the commercial activity of and cultural in the city life [5]. The term pedestrian or pedestrians comes from the greek pedester / pedestris who on foot and pedestrians .Pedestrian also derived from a pedos greek which means feet .Pedestrian also interpreted as movement or circulation or as a destination on foot (Rubenstein, 1992) [6].

According to (John Fruin, 1971), planning facility for pedestrians, including crossing facilities must take into the seven main target : safety, security, convenience, comfort, continuity, integration system, attractiveness [7]

According to Anggraini (2009:33), explained that the aspect is included in pedestrian element consist of drainage, sidewalk, pedestrian lights, a seat, guardrail, dustbin, crossing markings facilities, sign crossings facilities, the information board, bus shelter, stalls wait, and the public telephone [8].

3. Methods

Research methods used in this research is quantitative research. Data collection technique used is primary survey consist of observation and the questionnaire. Sampling techniques used is purposive sampling with the number of respondents about 211 respondents using quesioner with ordinal scale (i.e., from 1 = Not Considered to 5 = Very Considered). The survey was carried out on 15 area crossing facilities in three different locations with the same characteristics of land use consist of higher education area (university), trading activities and services. Analysis techniques used frequency distribution analysis in order to identify pedestrian consideration on the availability of supporting elements pedestrian and factors of pedestrian crossing facility utilization and crosstab analysis and chi square analysis to measure association or correlation between availability supporting elements pedestrian with pedestrian crossing facility utilization with 5 percent significance level.

4. Research Results

4.1. The Distribution of Crossing Facility Based On The Kinds Of Pedestrian Crossing Facilities

There are two locations of the point of pedestrian crosing facilities on Sumbersari Street in the form of zebra cross. But seen from the existing condition is not maintained, whereas at that location is one of area with a significant amount of movement. In addition, the lack of adequate supporting elements of pedestrian crossing facilities so that most pedestrians tend to cross in places that are not supposed to.

In Gajayana Street, There are three locations of thc point of pedestrian crosing facilities with the number of crosswalker are quite numerous and intense. The three crossing facilities in the form of two zebra cross and one pelican crossing. However, the condition can be said to be less maintained, it can be observed from marking condition of pedestrian crossing facility that has begun to fade. It is also not supported by supporting elements that can at least make it easier to cross the road in the region.

Pedestrian crossing facilities on Jalan MT. Haryono, there are 10 pedestrian crossing facilities consist of two pelican crossing and 8 zebra cross that existing conditions can be quite good. At 10 points facility, the people movement is very significant especially shopping activities, education, and work. However, the availability of a support element in the area of pedestrian crossing facilities are still very minimal.

Figure 1. Crossing facility in: (a) Sumbersari Street (b) Gajayana Street (c) MT. Haryono Street

Table 1. The Distribution of Crossing Facility Based On The Kinds Of Pedestrian Crossing Facilities

Street Corridor	The Point Of Location	Jenis Fasilitas Penyeberangan
Sumbersari Street	Point I	Zebra Cross
	Point II	Zebra Cross
Gajayana Street	Point III	Zebra Cross
	Point IV	Pelican Crossing
	Point V	Zebra Cross
MT. Haryono Street	Point VI	Zebra Cross
	Point VII	Pelican Crossing
	Point VIII	Pelican Crossing
	Point IX	Zebra Cross
	Point X	Zebra Cross
	Point XI	Zebra Cross
	Point XII	Zebra Cross
	Point XIII	Zebra Cross
	Point XIV	Zebra Cross
	Point XV	Zebra Cross

IOP Conf. Series: Earth and Environmental Science **70** (2017) 012053 doi:10.1088/1755-1315/70/1/012053

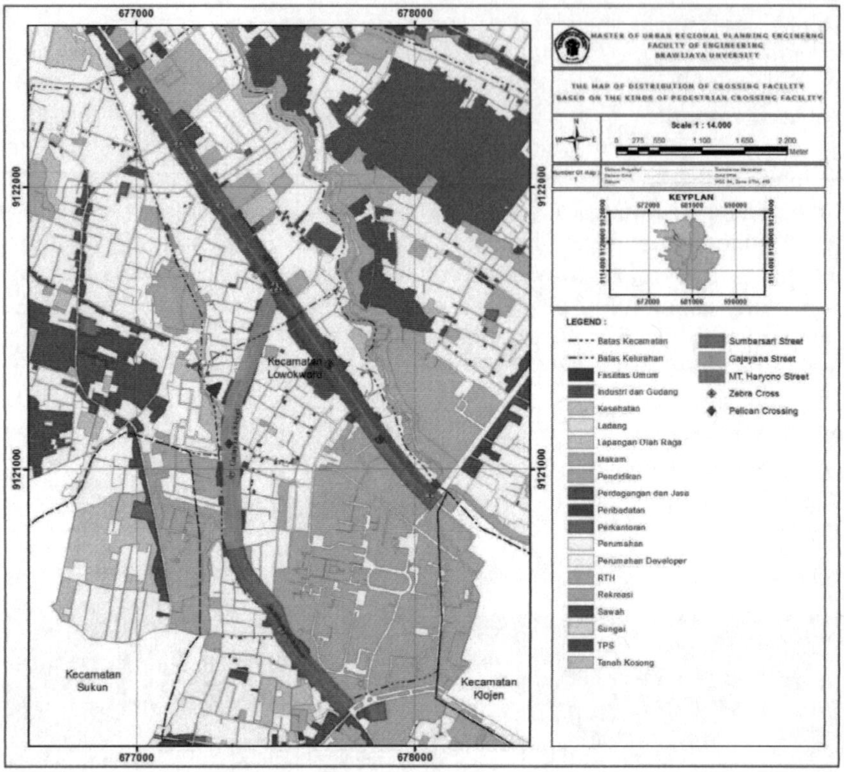

Figure 2. Map of distribution of crossing facilities based on the kinds of pedestrian crossing facilities.

4.2. Analysis of the Pedestrians Preference towards the Availability of Supporting Elements of the Pedestrian

Based on the distribution analysis results, the pedestrian preference describes that the supporting elements of the pedestrian that should be available on the area of pedestrian crossing facilities there are 7 elements, consist of the availability of sidewalks, pedestrian lights, street lighting lamps, pedestrian crossing markings facilities, guardrail, sign crossing facilities, and vegetation that belongs on the category as a whole is very considered and considered.

Table 2. Analysis of the Pedestrians Preference towards the Availability of Supporting Elements of the Pedestrian

No	The Supporting Element Of The Pedestrian	Street Corridor		
		Sumbersari Street	Gajayana Street	MT Haryono Street
1	Availability Of Sidewalks	1	1	1
2	Availability Of Pedestrian Lights	2	2	2
3	Availability Of Street Lighting Lamps	1	2	2
4	Availability Of Stalls Wait	3	3	3
5	Availability Of Pedestrian Crossing Markings Facilities	2	1	1
6	Availability Of Guardrail	2	2	2
7	Availability Of Sign Crossings Facilities	2	2	1
8	Availability Of The Public Telephone	4	5	4

No	The Supporting Element Of The Pedestrian	Street Corridor		
		Sumbersari Street	Gajayana Street	MT Haryono Street
9	Availability Of Vegetation	1	1	1
10	Availability Of Dustbin	4	4	2

Description:
1 = Very considered
2 = Considered
3 = Considered enough
4 = Less considered
5 = Not considere

4.3. Analysis of the Pedestrians Preference towards the Crossing Facilities Utilization Factor

Based on the distribution analysis results, the pedestrian preference describes that overall belongs in two categories consist of very considered and considered. The most dominant factor included by very considered criteria consists of 5 main factors i.e. safety factor, security factor, ease factor, convenience factor, and smooth factor. But, the factor included by considered category consist of 2 factors i.e integration system factors, and atraction factor.

Table 3. Analysis of the Pedestrians Preference towards the Crossing Facilities Utilization Factor

No.	Crossing Facilities Utilization Factor	Street Corridor		
		Sumbersari Street	Gajayana Street	MT Haryono Street
1	Safety Factor	1	1	2
2	Security Factor	1	1	1
3	Ease Factor	1	2	1
4	Convenience Factor	1	1	1
5	Smooth Factor	1	2	1
6	Integration System Factor	2	2	2
7	Attraction Factor	2	2	2

Description:
1 = Very considered
2 = Considered
3 = Considered enough
4 = Less considered
5 = Not considere

4.4. Analysis of the Relationship between the Availability of Supporting Elements of Pedestrians with Pedestrian Crossing Facilities Utilization

Based on the results of the analysis of the relationship between the availability of supporting elements with pedestrian crossing facilities utilization by using the chi square analysis. Chi square analysis mathematically can be calculated with the following formula:

$$x^2 = \frac{(\sum(fo-fe)^2}{fe} \quad (1)$$

Information:
x^2 = The value of the Chi Squared
Fe = Frequency expected
fo = Frequency obtained/observed

The chi square test hypotheses with IE:
H0 = no relationship between availability of the supporting elements of the pedestrian towards the utilization of pedestrian crossing facilities.
H1= There is relationship between availability of the supporting elements of the pedestrian towards the pedestrian crossing facility factors.
Chi square correlation test with significance level of 0.05 as follows:
- If *Asymp. Sig* > 0.05 H_0 is received
- If *Asymp. Sig* < 0.05 H_0 is rejected

Table 4. The Chi Square Test results Correlation between Availability of The Supporting Elements of the Pedestrian Facility Usage Factor towards the Pedestrian Crossing

The Level Of Significance 5% (0.05)	Factor In The Utilization Of Pedestrian Crossing Facilities						
The Supporting Element Of The Pedestrian	Safety Factor	Security Factor	Ease Factor	Convenience Factor	Smooth Factor	Integration System Factor	Attraction Factor
Availability Of Sidewalks	0.373	0.563	0.020	0.095	0.000	0.775	0.062
Availability Of Pedestrian Lights	0.916	0.902	0.020	0.490	0.199	0.010	0.546
Availability Of Street Lighting Lamps	0.712	0037	0.205	0.081	0.673	0.927	0.004
Availability Of Stalls Wait	0.522	0.072	0.467	0.997	0.736	0.185	0.786
Availability Of Pedestrian Crossing Markings Facilities	0.206	0.225	0.017	0.824	0.054	0.989	0.791
Availability Of Guardrail	0.096	0.424	0.675	0.230	0.860	0.247	0.872
Availability Of Sign Crossings Facilities	0.001	0.082	0.567	0.927	0.571	0.840	0.635
Availability Of The Public Telephone	0.195	0.205	0.987	0.607	0.533	0.968	0.835
Availability Of Vegetation	0.537	0.581	0.327	0.033	0.156	0.707	0.379
Availability Of The Trash	0.896	0.892	0.946	0.404	0.633	0.721	0.038

Description:
■ = Have a correlation (significance level <5% /< 0.05)
□ = Do not have a correlation (significance level of >5% /> 0.05)

Based on Chi square analysis the results it can be concluded that there is some element of the pedestrian have a correlation towards utilization of pedestrian crossing facilities.

IOP Conf. Series: Earth and Environmental Science **70** (2017) 012053 doi:10.1088/1755-1315/70/1/012053

- Availability of sign crossings facilities have a correlation with safety factor by significance of 0.001.
- Availability of street lighting lamps have a correlation with security factor by significance of 0.037.
- Availability of sidewalks, pedestrian light availability, and the availability of pedestrian crossing markings facilities have the correlation with ease factor by significance of 0.020 0.020 , and 0.017.
- Availability of vegetation have a correlation with convenience factor by significance of 0.033.
- Availability of sidewalks have correlation with smooth factor by significance of 0.000.
- Availability of pedestrian lights have a correlation with integration system factor by significance of 0.010.
- Availability of street lighting lamps and trash have a correlation with attraction factor by significance levels of 0.004, 0.038.

5. Conclusion

Based on the preferences of pedestrian, there are 5 supporting elements of the pedestrian should be available on the area of pedestrian crossing facilities consist of the availability of sidewalks, pedestrian lights, pedestrian crossing markings facilities, sign crossings facilities, and vegetation.

Based on the pedestrians preference towards the utilization factor of pedestrian crossing facilities belongs in two categories namely very considered and considered consist of. safety factor, security factor, ease factor, convenience factor, and smooth factor.

Based on chi square analysis between the supporting elements of the pedestrian with the pedestrian crossing facility usage. There are 7 supporting elements of pedestrian that has correlation and should be available in the crossing facility area are sign crossings facilities, street lighting lamps, sidewalks, pedestrian light, pedestrian crossing markings facilities, vegetation, and trash.

References

[1] Keputusan Direktorat Jenderal Perhubungan Darat Nomor : SK.43/AJ 007/DRJD/91

[2] Ashadi, 2012,"Analisa Pengaruh Elemen-Elemen Pelengkap Jalur Pedestrian terhadap Kenyamanan Pejalan Kaki", Jurnal Nalars, Volume 11, No.1, hal 79

[3] Mashuri, Muh. Ikbal, 2011,"Studi Karakteristik Pejalan Kaki dan Pemilihan Jenis Fasilitas Penyeberangan Pejalan Kaki di Kota Palu", Jurnal Rekayasa dan Manajemen Transportasi, Volume1, No. 2, hal 70

[4] Tanan, Natalia, 2011, "Fasilitas Pejalan Kaki", Bandung, Kementerian Pekerjaan Umum Badan Penelitian dan Pengembangan Pusat Penelitian dan Pengembangan Jalan dan Jembatan, hal 22

[5] Danoe Iswanto, 2006,"Pengaruh Elemen-Elemen Pelengkap Jalur Pedestrian Terhadap Kenyamanan Pejalan Kaki", Jurnal Ilmiah Perancangan Kota dan Permukiman, Volume 05, No. 1, hal 22

[6] Ashadi, 2012,"Analisa Pengaruh Elemen-Elemen Pelengkap Jalur Pedestrian terhadap Kenyamanan Pejalan Kaki", Jurnal Nalars, Volume 11, No.1, hal 79

[7] Rudi Setiawan, 2006, *Faktor-Faktor Yang Mempengaruhi Pemanfaatan Jembatan Penyeberangan",* Simposium IX FTSPT Universitas Brawijaya, hal. 586

[8] Danoe Iswanto, 2006, "*Pengaruh Elemen-Elemen Pelengkap Jalur Pedestrian Terhadap Kenyamanan Pejalan Kaki*", Ilmiah Perancangan Kota dan Permukiman, Vol. 5, No. 1, Hal. 26 Jurnal

IOP Conf. Series: Earth and Environmental Science **70** (2017) 012054 doi:10.1088/1755-1315/70/1/012054

Building conservation base on assessment of facade quality on Basuki Rachmat Street, Malang

E B Kurniawan, R Y A Putri and D K Wardhani

Department of Urban and Regional Planning, Faculty of Engineering, Universitas Brawijaya, Malang, Indonesia

eddiwawan@gmail.com

Abstract. Visual quality covers aspects of imageability which is associated with visual system and the element of distinction. Within a visual system of specific area, the physical quality may lead to a strong image. Here, the physical quality is one of important that make urban aesthetic. Build a discussion toward visual system of urban area, this paper aim is to identify the influencing factors in defining the façade's visual quality of heritage buildings at Jend. Basuki Rahmat Street, Malang City, East Java-Indonesia. This Street is a main road of Malang city center that was built by Dutch colonial government. It was designed by IR. Thomas Kartsten. It was known as one of Malang area that have good visual quality. In order to idenfity the influencing factors, this paper conducts Multiple linear regression as a tools of analysis. The examined potential factors are resulted from of architecture and urban design expert's assessment to each building's segment at Jend. Basuki Rahmat. Finally, this paper reveals that the influencing factors are color, rhythm, and proportion. This is demonstrated by the results model: Visual quality (Y) = 0.304 + 0.21 Colors(X5) + 0.221 rhythm (X6) + 0.304 proportion (X7). Furthermore, the recommendation of the building facade will be made based on this model and study of historical and typology building in Basuki Rachmat Street.

1. Introduction

Lynch asserts that physical quality provided by a visual system in a region may give rise to a strong image of the region [1]. It further said that the visual quality is a special attribute of a visual system and determined by cultural values and essential physical properties. Vining and Stevens in Smardon, explained that visual quality of aesthetic quality, covers aspects such as proportion, composition, patterns, and structure; imageability is a quality associated with the image of visual system and elements of distinction on a view [2]. While the visual quality is related to how the street can shape positive perception of the pedestrian through visual elements so it will give pedestrian a good psychic impact of leisure and enjoyable visual experience. With the visual quality, the path has an important role in shaping a person's perception of the structure or the image of a city.

The façade is the cultural circumstances it conveying when the building built. The façade is a representation or expression of various aspects that appear and can be observed visually. In the context of the architecture of the city, the façade of the building is not only just but two dimensions are three-dimensional that can represent each of the buildings in the public interest (City) or vice versa [3].
Basuki Rahmat Street is one of the main streets in Malang City which is developed during the Dutch Colonialism era. This street is a part of Kayutangan, the old city centre of Malang. During the colonial

period, the main land use of Kayutangan was of commercial and public service uses. This old city centre is located in strategic area of city centre, especially in the North-South axis.

Basuki Rahmat street is one of the road axis in the Kayutangan city of Malang. The Kayutangan runs from the intersection of the square until the T-junction of Oro-oro Dowo. Before the advent of the colonial city of Malang to the Netherlands, then the Kayutangan and surrounding areas with the ancient village which is in it is part of the Karisidenan town of Pasuruan on the 1100 's which is still led by Hindu and Muslim Kingdoms [4].

Basuki Rahmat Street is the main street in Malang Central which has been used since the existence of the colonial Government of Netherlands.In present, there are several heritage building residing in this location. Basuki Rahmat Street that is part of area named Kayutangan is a special area which was formed at the beginning of the colonial period. During the colonial period, the main function of this area is as a commercial zone and public services. The strategic location of Kayutangan, which is located in the North-South axis and many skipped making Kayutangan as the center of the crowd. At the time of colonial settlement, there are several Office and shopping complex.

Kayutangan developed into a line that is often impassable since the construction of the bridge concrete Tjelaket (now the bridge in front of the Syaiful Anwar Hospital). In addition due to the layout of the Kayutangan which is adjacent to the square with all its activities and the Kayutangan since the beginning of the existence of Malang Regency has become a fast-growing region.

The existence of commercial activity and services advanced enough that also led to the construction of the Kayutangan corridor to get facilities and infrastructure priority compared to other areas such as Chinatown, Kauman, especially when compared with the native settlements tend to be overlooked. The development of the region which resulted in the value of land in the area of Kayutangan and has increased the price when compared to other areas during the time.

The characteristics of land use in Basuki Rahmat street currently dominated by commercial functions in the form of trade. However, some land use also has a signifikat effect on existing activities, namely building which is not used. Architectural characteristic of Jend Basuki Rahmat Street is dominated by image of heritage buildings reside in the corridor. However, due to uncontrolled Indonesian economy development by sufficient urban development system, many historic buildings were changed into the new-style buildings of modern architecture (Johana, 2004).

In 2009, Malang City Spatial Plan (RTRW) defines Kayutangan Area was developed as a tourist and socio-cultural area because of heritage values of the area. Unfortunatelly, this definition gives serious problems for the practical development of the area. By municipal government's policy in spreading the urban economy development, the commercial activities of the city center are declining over the time. In this case, Kayutangan Area are experienced by the increment number of vacant and abandoned buildings, especially the heritage.

Building's physical character of Basuki Rahmat corridor are dominated by the form of a square and rectangle with a flat roof, except for Mega Banks, Telkom Office, herbal medicine shops Kidang. The existing building on the study have a horizontal effect because its existence lining connect into one with a fairly massive nature feels because of the use of the element is not transparent on every building in the form of a wall, aluminum and glass that reflect sunlight with an average altitude of between 6 to 30 metres.

Recently, the building underwent into some changes, such as replacement and addition of some elements. It was done without reducing the colonial's heritage impression. Some of building are experienced with major changes. Therefore defining the actual factors that influence on the building characters changes, particularly the building façade, is important in order to develop theoretical formula in designing the Basuki Rahmat corridor in future as the heritage area of Malang City.

2. Literature Review

Visual quality in difficult enough to be declared objectively but measurable by the human responses. visual quality cannot be mesured with measuring, but it could be a quality result from the human perspective, the measuring should be involed by human being.Visual content refers to the structure and the space of one area giving an infomation to the each individu such a perception in used as purpose of developing the environment [3] Several variables assasment in this research has chosen based on several literature [5] [6] [7] [3] and modified to the needs of this research (table 1).

Table 1. Selected variables.

Variable	Sub variable	Selected Variables	Basic Considerations
Balance	- .	Balance	The balance will show the value of the alignment of the object being studied by involving feelings of each respondent.
Unity	1. Texture 2. Color 3. Tone 4. Direction 5. Proportion 6. Solid and Void 7. Form	Unity	Unity will give a clear image about the visual studies because he explained variable. However, in this study a variable tone sub not used with the consideration that the tone has become part of the assessment component of the sub color variables. In addition to that subsection as well as Solid and Void variables Form also is not used because it is considered by the researcher has been variable and can be explained by other variable sub being used.
Density	-	Density	To see the density of the façade as well as the façade of the components within a segment.
Rhythm	1. Line 2. Form 3. Texture 4. Space 5. Color	Rhythm	Rhythms would indicate a segment. Sub variables used only the lines, shapes and space.

3. Methods

The research method begins with a survey of primary and secondary. where the primary survey covers (1) Observation and documentation; which collects data through field observations to impose limits on each of the segments (1 segment = ± 100 meters) and make a photo montage that will be used as a reference in the assessment of the facade; (2) the interview; interviews were conducted to 5 persons expert in urban design by providing assessment form for each segment based on seven variables assessment, in terms of balance, texture, color, direction, proportion, density and rhythm.

In this research, both sides of the street divided into 16 segments (figure 1). Where the division is based on the division of long segments of approximately 100 meters with the physical limitations of the building. So in the photo montage segments there are no buildings are truncated.

According to Agus [5] to complete a more detailed technical side of design architecture and cities, need input from a range of experts relevant to the design. A range of expert opinion that he should be able to form a single unified draft. This research involve the 5 person of experts as a Experts Group of Urban Heritage (table 2).

Experts were asked to give rate value of each segment using a likert scale. Results of the assessment from experts will be continued using the calculation method of multiple linear regression analysis. It was expected to provide a model equation which will indicate the influential variable. However, prior to the calculation of multiple linear regression, data will be transformed using Sequential Interval Method.

The next process was the analysis of the history of the building to get typology forms of existing buildings along the Basuki Rachmad Street. Based on the model equation and the typology of the building, it will be compiled on facade recommendation for conservation in each segment.

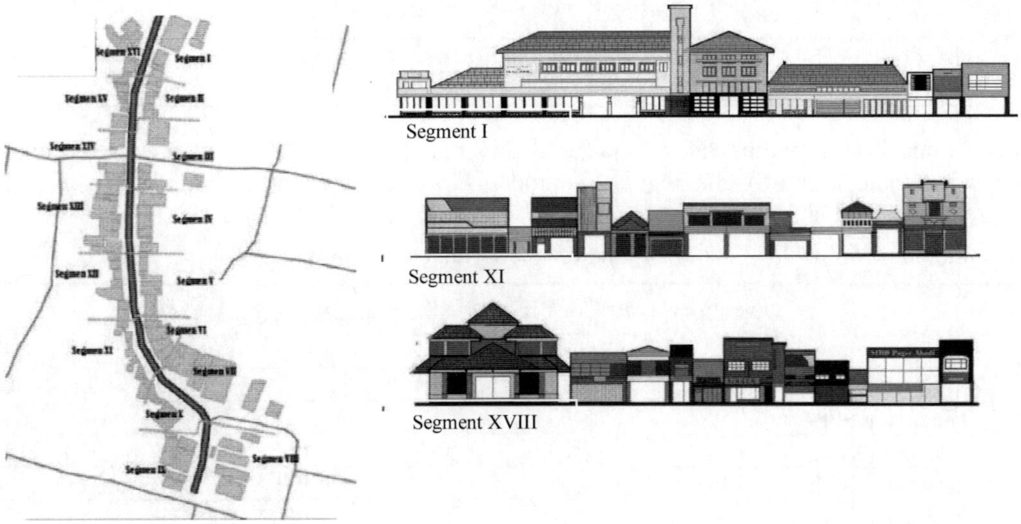

(a) (b)

Figure 1. (a) Divisions of the Segments (b) Building façade at several segment

Table 2. Experts description.

Respondents Name	Description
Prof. Ir. Antariksa Sudikno, M.Eng., Ph.D	Professor Of Department Of Architecture, Faculty Of Engineering University Of Brawijaya
Drs. Ismail Lutfi	Practitioners and observers of cultural heritage Buildings of Malang
Ir. Diah Ayu Kusumadewi MT	Head of Planning Regional Development Of Malang Goverment
Prof. H. Repati Wikantiyoso, ST., MSA., Ph.D	Lecturer Department Of Architecture-Independent University Of Malang – Vice-chancellor Of the Merdeka University Malang
Agung Murti Nugroho ST., MT., Ph.D	Head Of The Department Of Architecture, Faculty Of Engineering University Of Brawijaya

4. Results

The visual assessment done based on segment 6 independent variables (balance, density, direction, texture, color, rhythm, and proportion) and 1 dependent variable (visual quality). From the results of calculation of regression analysis using SPSS 14.0.

Table 3. ANOVA

Model		Sum of Squares	df	Mean Square	F	Sig.
1	Regression	40,708	7	5,815	14,520	,000(a)
	Residual	28,836	72	,401		
	Total	69,544	79			

a Predictors: (Constant), Proportion, Color, Texture, Density, Direction, Balance, Rhythm
b Dependent Variable: Segment's Visual quality

ANOVA results demonstrate that the value of significance 0.05 < which means that the data that will produce a good model. This means Ho accepted (table 3).

IOP Conf. Series: Earth and Environmental Science **70** (2017) 012054 doi:10.1088/1755-1315/70/1/012054

Table 4. Model Summary

Model	R	R Square	Adjusted R Square	Std. Error of the Estimate
1	,765(a)	,585	,545	,63285

Model summary table that the value Adjusted R Square is equal to 54.5% (table 4). Determination of the coefficient indicates how well a regression model is formed by the interaction of the independent variables and dependent variable (table 4).

Table 5. Coefficients

Model		Unstandardized Coefficients		Standardized Coefficients	t	Sig.
		B	Std. Error	Beta		
1	(Constant)	,304	,330		,921	,360
	Balance (X1)	,050	,104	,051	,480	,633
	Density (X2)	,140	,102	,141	1,368	,175
	Direction (X3)	,119	,097	,121	1,226	,224
	Texture (X4)	-,034	,098	-,034	-,344	,732
	Color (X5)	,210	,083	,213	2,522	,014
	Rhythm (X6)	,221	,108	,225	2,053	,044
	Proportion (X7)	,304	,125	,311	2,427	,018

Coefficient tables shows the value of each variable. The resulted value shows that each variable is free of proprietary variable (B), as well as the estimated error rate variable to the model so that the test results be population (Sig). If the value is "Sig" less than 0.05 so it can be assumed that the value of B can describe a model that was later produced with 95% degree of truth (table 5).

From the results above, it can be concluded that the regression equation model is formed are as follows: Visual quality(Y) = 0.304 + 0.21 colors(X5) + 0.221 rhythm(X6) + 0.304 proportion(X7). This Model formed by the variable selection criteria have a calculation result of significance of 0.05, which means < possible errors is 5%. However, in this case the variable Density can be made into consideration in the formulation of directives, because the value of their significance is 17.5% which means it has an 82.5% probability for a valid.

Furthermore, it will perform an analysis of the history and the building typology (Fig. 2). According Hadinoto [4], Basuki Rachmad Street area characterized by colonial building with *neuw bowen* type. The characteristics of the building are:

- Flat roof
- Gevel or long horizontal fasade
- Cubical building mass
- White color
- Tower ornament in the corner of the building
- Used of concrete and steel frame
- Flexible space pattern (wider space, rich of light and open air)
- Have a lot of windows and door
- Color not only for decorative element, but also for element of expresion

After identifying a typology of the façade of the building, then made recommendations for improvement in every segment of the building façade. Based on the model equations that have been generated previously, the repaired item are color, rhythm and proportion (Fig. 3 and Fig. 4) It becomes part of the building conservation directives with the aim of building facade characteristic was maintained.

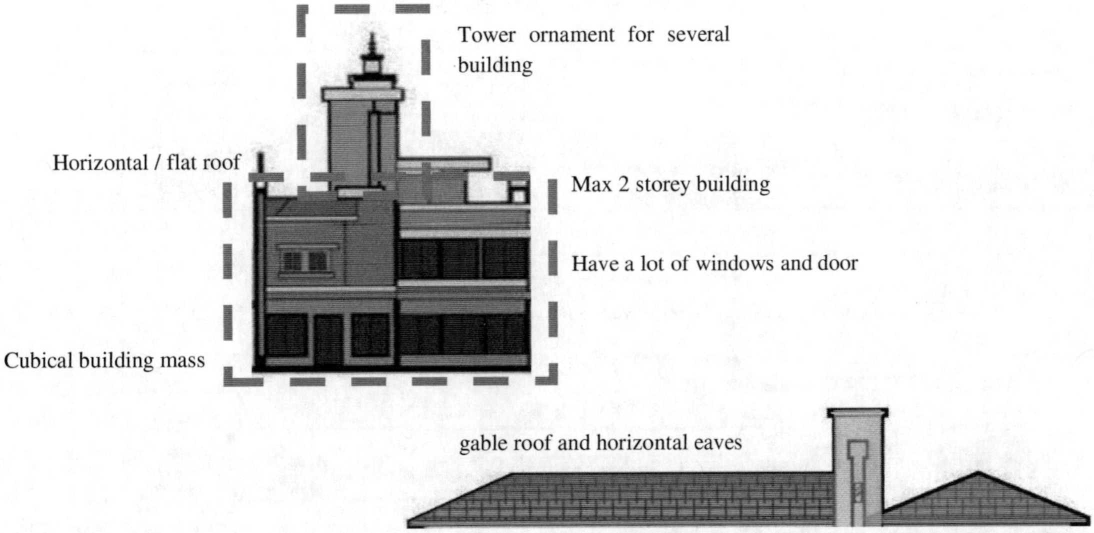

Figure 2. Typology of building façade at Basuki Rachmad Street

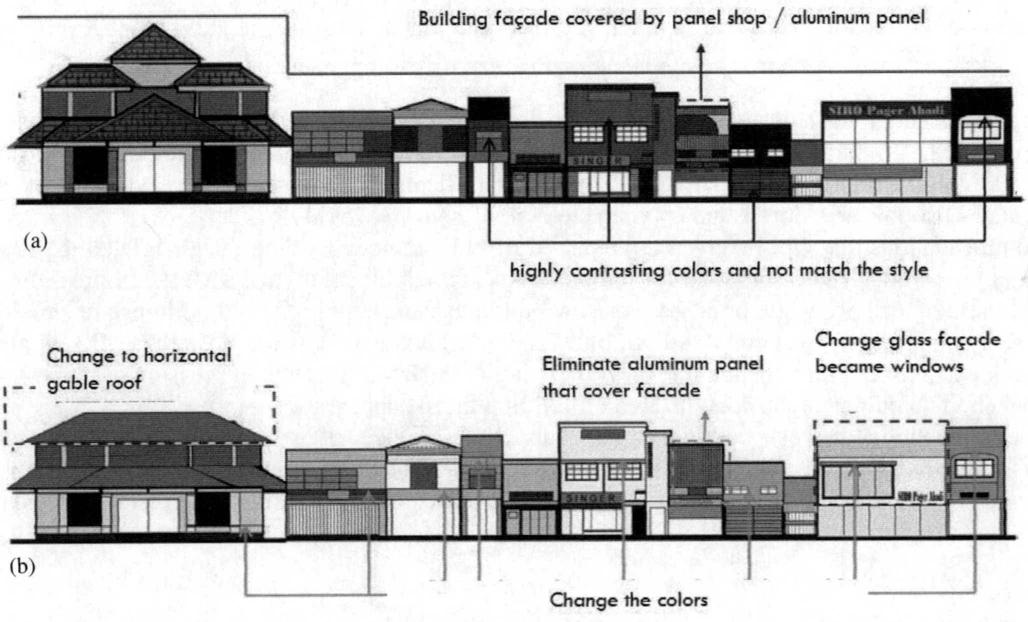

Figure 3. (a) Existing Facade at Segment XI (b) Recommendation Facade at segment XI.

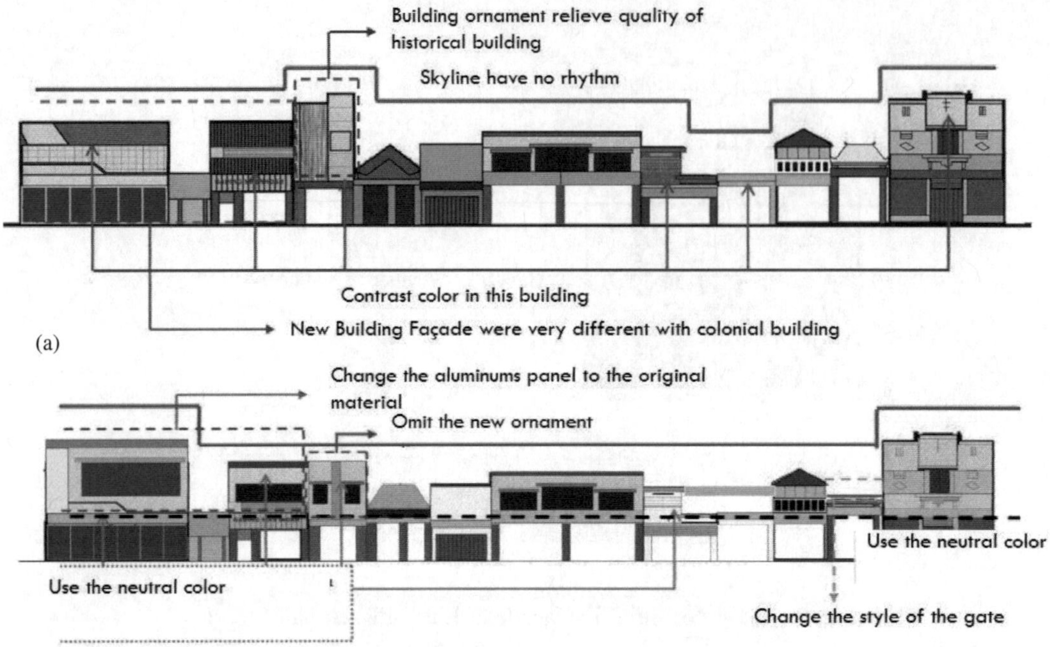

(b) **Figure 4.** (a) Existing facade at segment XV (b) Recommendation facade at segment XV.

5. Conclusion

The results of calculations indicate multiple linear regression models is Visual Quality Color = 0.304 + 0.21Color + 0.221 Rhythm+ 0.304 Proportion. So that the visual factors that affect the building facade is of color, rhythm and proportion. The results of the identification of the history and typology of the building states that this area dominated by characteristic of *neuw bowen* style.

Recommendations that will be given are repaired of old and new building facade related to color, rhythm and proportion. The colors used for old buildings are neutral colors that give the impression of a colonial building that are white or beige. For new buildings can adapt to the old buildings or provide an attractive color but still soft color as light blue, and add glass elements no more than 30%. It also could provide a green color on the top line of the first floor which will give the impression of horizontal line. Panel shop no more than 30% of the area of building facade appearance.

Successful public spaces are characterised by the presence of people, in an often self-reinforcing process. The place would be created by the element of activity, image, and form, Carmona [8]. Research on visual quality directly affect the form elements, and indirectly will affect the image of the area. On perceptual dimension that will create a place.

References

[1] Lynch, Kevin, Image of the City, The Technology Press & Havard University Press, Cambridge 1960

[2] Smardon, R. C., and R. Burton Litton. Development of visual activity classification and advanced testing on visual impact assessment manual procedure. Syracuse, NY. SUNY College of Environmental Science and Forestry, New York 1981

[3] Krier, Rob. Architectural Composition. Erlangga. Jakarta 1996

[4] Hadinoto, Developments of Malang City at Colonial Period 1914-1940. Dimensi. Surabaya. 1996

[5] Purnomo, Agus. Quantitative Techniques for Architects and Urban Design. Rajawali Press. Jakarta. 2009.

[6] Hakim, R. Components of Landscape Architectural Design: Principles, Elements and Application Design. PT. Bumi Aksara. Jakarta. 2012.

[7] Ching, Francis D.K. A visual Dictionary of Architecture. John Wiley & Sons. New York. 2011
[8] Carmona, Matthew, at. al . Public Places-Urban Spaces. Achitectural Press. Oxford. 2003

Malioboro as a Value of Special District of Yogyakarta City

G A Cahya[1], Y K D Mahendra, and I I Damanik

Dept. of Architecture, Faculty of Architecture and Design, DWCU, Yogyakarta, Indonesia

gloria_andida@yahoo.com

Abstract. Yogyakarta is a province that has a status as a Special District. The specialty of Yogyakarta is set based on its history, because it was led by the Noble and has social culture embedded with Javanese tradition. Recently, the development of Yogyakarta has changed, along with the dynamics urban activities and technologies, into a tourist destination. It is a challenge to enhance the status of 'privilege' of Yogyakarta. Malioboro is an important part in the structure of the city of Yogyakarta. Apart from being one of the constituent elements of the north-south axis, Malioboro also serves as a public space that is rich in cultural activities, economic, and social interaction. If it reviewed according to its privileges, Malioboro is one part of the city of Yogyakarta which has an important role in maintaining local values that forms the privilege character of Yogyakarta. As a city that responds to the needs of the society, Yogyakarta provides various supporting infrastructure. Maliboro is located at a strategic area that became the target of investors as site to build public and commercial facilities. Malioboro is an iconic street of Yogyakarta, become one of the tourist destination because of its uniqueness, especially as the affordable shopping area. For that reason, it is necessary to study how Malioboro should maintain its local value to conserve the privilege of Yogyakarta as a special district, but also has ability to adapt nowadays development of Yogyakarta City. This paper will provide the study about the values of Malioboro which conserve the status and the meaning of the privileges for Yogyakarta, when viewed from the historical, social, and cultural. It will continue the discussion about the development of tourism and economic activities that are affected by tourist's demand. The result of the paper will describe (1) the aspects that we have to keep to conserve the privilege of Special District of Yogyakarta and (2) the aspects that we can change or adapt, so that Malioboro continue to develop its role as an iconic street. The study will provide an alternative perspective for Malioboro's development.

1. Introduction

The dynamics of live in the city continue to change with the times that influenced by human needs. In the meantime, the city needs a plan that can be developed with human needs and support human activities. Some experts have given their opinions and the theory about urban design, some of them are Roger Trancik with Finding Lost Space (1986) and Kevin A. Lynch in his book Image of The City (1960).

Yogyakarta City, if it was reviewed through the theory from Finding Lost Space (1986) and Image of The City (1960), will generate a unique urban design and certainly different from the other cities. Administratively, Yogyakarta has its own uniqueness and called as Special Region. The reason is because the governor of the Yogyakarta area is held by hereditary of the Sultanate of Yogyakarta. The

IOP Conf. Series: Earth and Environmental Science **70** (2017) 012055 doi:10.1088/1755-1315/70/1/012055

status shows that Yogyakarta has a role, not only merely administrative privilege but also special character because of the customs, traditions of Javanese culture, and natural beauty.

Malioboro is an important part in the structure of the city of Yogyakarta. Apart from being one of the constituent elements of the north-south axis, Malioboro also serves as a public space that is rich in cultural activities, economic, and social interaction. Many old buildings with specific detail are lined up along the street of Malioboro that is not only as a part of the street, but also as a witness of historical aspect of Yogyakarta urban development [1]. In the terms of the privileges, Malioboro is one part of the city of Yogyakarta which has an important role in maintaining local values and enhance the significant character of the privilege of Yogyakarta.

2. Methodology

The research use primary and secondary data from the site which is collected during survey, observation, and literature. The research use qualitative method and use single case study as research approach [2]. Malioboro is the site, which is investigate as one place that emerge the 'speciality' factor of Special District of Yogyakarta (*Daerah Istimewa Yogyakarta, DIY*), because as a great street of Yogyakarta, it has specific character, not only as a path but also as a linkage and as a place. The paper will focus on the architectural spatial elements of Malioboro that has a dominant impact on the image of Malioboro, so that it could carry on its role as a great street, not only as a street but also as a commercial area, tourism spot, and the most important is as an element that strengthen the character for Special District of Yogyakarta.

3. Analysis

In urban life, the street become one of the important roles. Not only as a physical element, but the street is also a space/place for people to do their activities [3]. In a city, the street can be metaphor as the veins. Because through the street, all the activities of the city flowed. By the street --which is a public space-- there are a lot of social and cultural interaction. Through the street all human activities can be connected and they be able to move the wheels of the economy and social relations in the city.

Streets in urban areas are not only as a current passing for human activities, urban street developed into the economic places, business places, tourist places, commercial places, public places, social place, and even transformed into a symbolic space for a particular area. Streets especially in the Asian region has become part of the culture and tradition of the people; since a long time ago, streets become part of the public space. In urban areas can be seen that along the street, there are a lot of informal sector, such as the emergence of street vendors [3].That's because along the side of the streets always have potential as a gathering place for humans.

In the relation to the urban design, streets can be connected with the theory stated by Roger Trancik (1986) [4] in his book Finding Lost Space, the Linkage theory that explains the dynamics of circulation becoming generator in the city forming. The streets are also have the role as human circulation and helps the movement of people, as well as connecting the dots in a particular city.

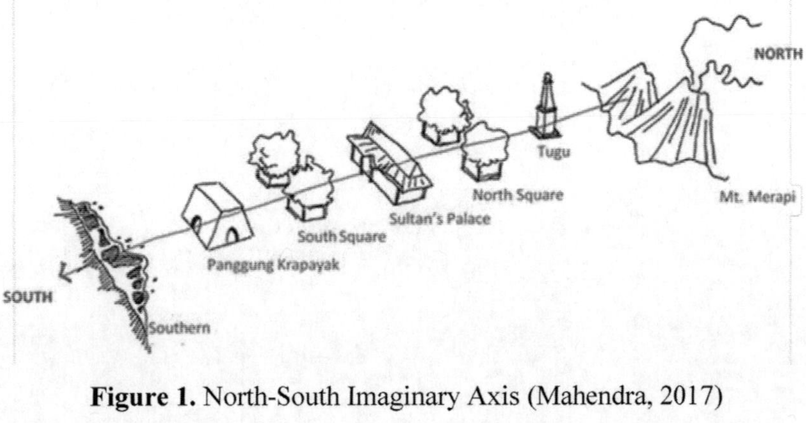

Figure 1. North-South Imaginary Axis (Mahendra, 2017)

IOP Conf. Series: Earth and Environmental Science **70** (2017) 012055 doi:10.1088/1755-1315/70/1/012055

The same thing happened in the Malioboro street, that located in the center of Yogyakarta. Malioboro itself located on the linear configuration between Tugu and Keraton who passed the imaginary axis; a link between the volcano, Keraton and the southern beach. For the people of Yogyakarta, Malioboro street is not just as a connecting street, but also as a place for human activity [3]. Around the Malioboro can be found varieties of economic activity, social, cultural, and tourism.

In 19th century, Malioboro is built as a government and economic center by the Dutch colonial government. Since then, Malioboro street is known as a shopping center and the economy center, which is showed by shophouses that lined up along the street and the traditional market --Pasar Beringharjo-- on the south area of the street. Commodities that are sold on the Malioboro street mostly accessories and souvenir. Nowadays, Malioboro street are also famous for shopping area for tourists. The shopping activities can divided into traditional and modern roles, so the buyers (tourists) have options when they want to enjoy shopping activity at Maliboro. The tourists will know which one is for traditional shopping activity from the way the seller haggling on the stalls of street vendors; and which one is the modern one, that took place in the mall and shop-houses, with fixed price.

In addition to economic activities, on the Malioboro street, there are also take place for social activities, because as a public space, Malioboro is a place of social interaction for people of Yogyakarta with people who come to visit there. With combination of social activities along Malioboro street, people who come to malioboro can recognize the faces of Yogyakarta through the interaction among the people there. Malioboro street is also the place for culture activity, because there exist various cultures. There are Chinatowns, called as Kampung Ketandan on the south-east of Malioboro, there are also buildings with colonial-style architecture called indische buildings, lay along the street from south to north, signifies architectural character of Malioboro. The people of Yogyakarta who uphold the Javanese traditions and also domestic tourists and international tourists who visit there add more various cultures on Malioboro street. Malioboro as a famous street in Yogyakarta, capable of being a symbolic space for the city of Yogyakarta [3], it can be seen from the many cultural events, or other events that concentrate on Malioboro Street and some of the events even become identical with Malioboro street. For the example the event that held in Malioboro street is the Chinese New Year celebration event, carnival, also the other special ceremony.

The number of human activities that occur on Malioboro street with cultural values, social and economic, making the tourism potential of the region Malioboro become stronger and capable to be an attraction for the city of Yogyakarta. Along with the times and technology, especially in the field of transportation and the investors who make their investment in Malioboro area, turned Malioboro to become a commercial and tourist area. High vehicle traffic also makes the area becoming increasingly congested and start to be avoided by the local people, because of the density and traffic jams. Besides that, the construction of facilities for tourists such as hotels around Malioboro area also increased the density in the area. Parking area for motorcycle is took place on the pedestrian area. The density is seen along the Maliboro Street; traffic jam on the street, parking area for motorcycle along pedestrian ways, street vendor along arcade of the building, and people activities around the area.

(a) (b) (c)

Figure 2. (a) Malioboro in the Past (brillio.net, 2017) ; (b) Malioboro 2015 (Mahendra, 2017); (c) Malioboro Nowadays (Theofilia, 2017).

It is changing the habits of interaction on the Malioboro streets, people being oriented into vehicle, no longer enjoying the experience of walking on Malioboro. But now, the area of Malioboro street began to redesign, the most significant change was the side of the road. The arrangement which prioritizes pedestrian, and the vehicle parking area is relocated to the north side of the Malioboro street.

The values and human activities at Malioboro create a special atmosphere of Malioboro Street and the central area of Yogyakarta. The atmosphere cannot be found in other streets or place. The uniqueness of Malioboro Street can be maintained with attention to cultural, social, and economic. One of urban design theory that *place theory* emphasizes the importance of its historical, cultural, and social values on an open area in the city, as well as maintains a sense of urban space [5]. The development on Malioboro needs to pay attention to the cultural aspects as standard rule in the construction, as well as the typologies of the buildings that remain in tune with Malioboro *Indische* Building (Colonial and Chinese style), so that the atmosphere can maintan physically and bring a nostalgic memories from the past. Social aspect give some nodes that Malioboro from time to time is a place for humans, and until noway, the development of Malioboro still able to maintained public spaces for the communities.

The vehicle route on Malioboro street is from north to south (one way). There is no parking area along the street Malioboro anymore, because the motorcycles have a parking building on the northern part of Malioboro. The policy to build parking building on 2016, because the density of Malioboro street is getting high. The activity at Maliobro (table 1):

1. Activites that have same direction with the vehicle (north to south)
   - Jogjakarta Chinese Cultural Week (Culture)
   - Jogjakarta Cultural Festival 2011(Culture)
   - Malioboro Festival (Culture)
   - Jogjakarta Anniversary (Culture)
   - *Gelaran kirab* (politic)
   - Independence day parade (politic)
2. Activites that have different direction with the vehicle (south to north)
   - Royal Wedding of Keraton Jogyakarta (Culture)
   - *Grebeg* (Culture)
3. Activities that take place in 2 direction (south to north and north to south)
   - Economic (tourism)

Table 1. The activity at Maliobro

Activity	Category	Time	Route	Additional Activity
Jogjakarta Chinese Cultural Week	Cultural Festival	1 week after Chinese New Year	Abubakar ali Parking lot– Malioboro – Alun-alun Utara (North Square)	Food festival, Lion dance, several cultural competition at Kampung Ketandan
Grebeg Syawal (Muhammad SAW born celebration) *Grebeg Maulud* (after fasting) *Grebeg Besar* (Eid al-Adha)	Cultural	Once a Year	From Keraton Ngayogyakarta Hadiningrat to 3 different places: Gede Mosque, Kepatihan Office, and Pakualaman Palace	Soldier of Keraton parade complete with special weapon, banner, and traditional music. In the end of the parade, there are several *Gunungan* that prepared for the citizen who watched the ceremony. But the citizen should compete each others to get the food from *Gunungan.*
Jogjakarta Cultural Festival 2011	Cultural Festival	Once a Year (26 June - 5 July 2011)	Tugu – Malioboro, Fort Vredeburg	Performance art and dancing. Local industry and micro economic exhibition at Fort Vredeburg
Malioboro Festival	Cultural Festival	24-26 Juli 2015 09.30-22.00 WIB (tentative)	Malioboro	To strengthen tourism activity
Royal Wedding of Keraton Jogyakarta	Cultural Ceremony	Wednesday, 23 Oktober 2013 09.00 (tentative)	Keraton – Kepatihan (governor office)	Parade of 12 golden chariots, and guarded by 360 Keraton soldiers.
Jogjakarta Anniversary	Culture	Once a year	Abubakar Ali parking lot – zero point (central post office)	Jogjakarta 258 Outdoor Sculpture Exibition 2014
Gelaran kirab (launching the new brand of Jogjakarta City)	Politic	Saturday, 7 March 2016 14.00-16.00 WIB (unconditional)	Abubakar Ali parking lot – zero point (central post office)	Official the new brand of Jogjakarta is reported to governor of special region of Jogjakarta, Sri Sultan HB X and also launched stamp with the new brand image
Independence day parade	Politic	Saturday, 20 August 2016 15.00 WIB (tentative)	Abubakar Ali parking lot – zero point (central post office)	-
Demonstration refused Ahok	Politic	Friday afternoon,	zero point (central post office)	To demand justice about religion defamation

Activity	Category	Time	Route	Additional Activity
		28 October 2016 (unconditional)		
Demonstration (labor)	Politic	Sunday, 1 May 2016 (unconditional)	Malioboro	Commemorate May Day
Demonstration (new airport policy)	Politic	Thursday, 5 November 2015	Malioboro (kepatihan)	They show theatrical as a demonstration
Tourism	Economic	Daily	1. Abubakar Ali parking lot – pedestrian Malioboro – Beringharjo market 2. North Square (Alun-alun utara) –Fort Vredenberg– Beringharjo market – Mirota batik shop, pedestrian – Maliboro	

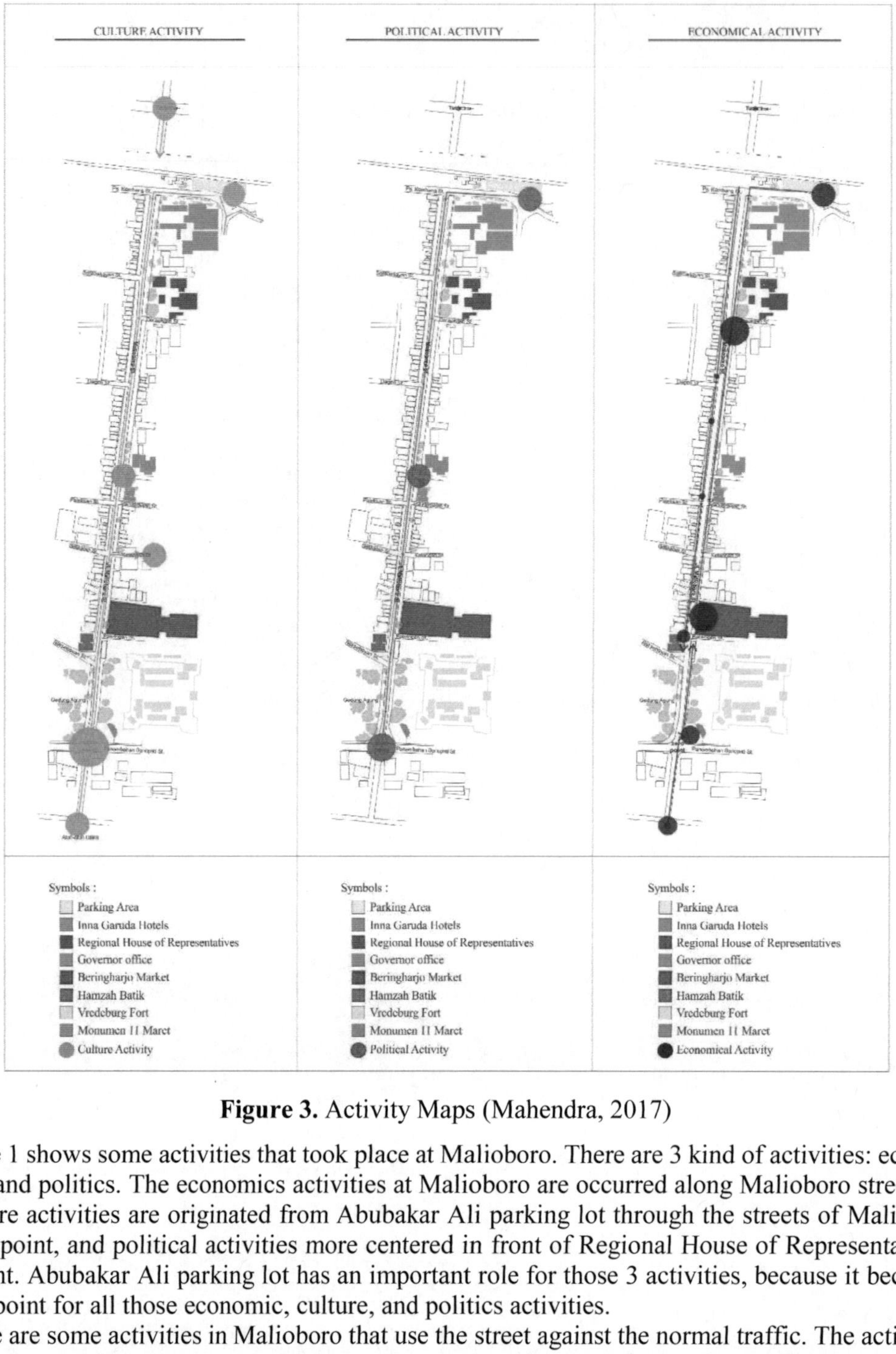

Figure 3. Activity Maps (Mahendra, 2017)

Table 1 shows some activities that took place at Malioboro. There are 3 kind of activities: economic, culture, and politics. The economics activities at Malioboro are occurred along Malioboro street, while for culture activities are originated from Abubakar Ali parking lot through the streets of Malioboro to the zero point, and political activities more centered in front of Regional House of Representative and zero point. Abubakar Ali parking lot has an important role for those 3 activities, because it became the starting point for all those economic, culture, and politics activities.

There are some activities in Malioboro that use the street against the normal traffic. The activity that took the route in opposite direction of the vehicle are activities held by the palace, Keraton Yogyakarta; such as the royal wedding and gerebeg. Because these activities are held by the palace, the sessions start from the palace at Southern area of Malioboro and move to the North, along Malioboro Street and stop at Governor's Office as final destination. For other activities that used Malioboro in opposite direction is economic activity. The majority of economic activity takes place at pedestrian Malioboro, so that economic activity can take place in two directions; move from north to south and otherwise.

The maps (Figure 5) show the order of movement activities at Malioboro Street. The circle on the map shows the central of the activities, and the arrow lines shows the path orientation. Each activity has its own path and activity points scattered along Malioboro street. It shows that the Malioboro street has a lot of value contained therein. Activities that occur also reinforces the values of the privilege of Yogyakarta City. The variety of activities in Malioboro street maintain the manifestation of concept of 'privilege' on Malioboro Street; because it is not just a line of road, but also as a forum for people of Yogyakarta and visitors, as the mixing container, and places for social interaction.

The first map shows the path of cultural activities that take place along Malioboro street. For example is the activity of the Yogyakarta Arts Festival (Festival Kesenian Yogyakarta) in 2011; the activity is an art and culture parade starts from Tugu. The parade moved from north (Tugu) through the Malioboro and ended at south (Fort Vredeburg). Another type of cultural activities is Royal Wedding Procession; the parade also pass through the path of Malioboro street, started from Keraton to Governor Office (from south to the north). There are many other cultural activities took place along Malioboro street which is held to enhance Malioboro as tourism area and the node for Yogyakarta City; such as Gelaran kirab and Yogyakarta Anniversary.

The second map shows the path of political activity on Malioboro street, often used convey aspirations in the form of demonstration. There are several favourite spot for this activity such as the Governor's Office and House of Regional Representatives. The location of political activity often used at in front of the Governor's Office and House of Regional Representatives. The examples of the demonstration are the demonstration in rejecting the construction of the airport in Kulon Progo held in front of the House of Regional Representatives's office. Another political activity is the march of Government Development Work to celebrate the 71st Independence Day; it was started at the parking area at the north and ends at the zero point, at the southern of Malioboro.

The last map shows the path of economic activity that occurred on Malioboro street. Malioboro is well-known as a shopping area for visitors, because Malioboro provides various items of Yogyakarta's products, especially souvenir and art work. The economic activity that occurred on Malioboro street itself scattered at various points on malioboro road, because the road along Malioboro there are many shops, street vendors, shopping malls, hotels, markets. Many other informal sectors also took part in the economic acitivity in Malioboro street. The visitors walk along Malioboro street started from parking lot which consist of (1) buses and motorcycles' parking lot at the North, (2) buses' parking lot at the South, and (3) cars' parking lot on the West and East.

From the description above, there are various activities that takes place in Malioboro, making Malioboro become an important street at Yogyakarta. It is not only important as a street as its original function, but also as a street that has a lot of culture and values which are created by human interaction and the uniqueness of the place. By maintaining the values of local wisdom in Malioboro street, the privilege as Special Province will be sustain within Malioboro street development. Malioboro is an locus to strengthen the image of Yogyakarta, cultural, and tourism city.

4. Conclusion

Street is one elements of the city, where we can find the historical side of the city. Street as a path [4], as a void, as a linkage, and as a place [5]. They give the base point for us to learn how the urban street, like Malioboro, run its role in urban scale.

Malioboro has layers of information, layers of its roles, and layer of its users. So, if there is a new development that applied on Malioboro, it should consider the layers that embedded on it. Learning from the Maliboro that the street is not valued by its dimension and the physical appearance of the old buildings, because people come to Malioboro not because of them. The layer of activities that connecting the street, the building, the pedestrian and the people. We can see the difference of elements of Malioboro from time to time, but we still can see the atmosphere that describe the dynamic of specialty aspects of District of Yogyakarta.

IOP Conf. Series: Earth and Environmental Science **70** (2017) 012055 doi:10.1088/1755-1315/70/1/012055

References

[1] Septirina, Safiera Nur. Takeob, Ozawa. Satoruc, Kaku. (2016) Conservation of Historical Architecture in Malioboro Street, Yogyakarta City, Indonesia, Procedia - Social and Behavioral Sciences, ELSEVIER.

[2] Creswell, John W. (2015), *Penelitian Kualitatif & Desain Riset: Memilih di Antara Lima Pendekatan (edisi ke-3)*, Pustaka Pelajar, Yogyakarta.

[3] Solihah, A. B. (2005), *The Role Of Informal Street Activities In The Context Of Conserving Urban Cultural Entity. Case Study: Malioboro Street, Yogyakarta Indonesia,* Tesis Master of Science FBE Universiti Teknologi Malaysia. http://eprints.utm.my/4210/1/ArifBudiSholihahMBD2005TTT.pdf

[4] Lynch, Kevin. (1960), *The Image of the Cities*, MIT Press.

[5] Tracik, Roger. (1986), *Finding The Lost Space*, John Wiley & Sons. (downloaded: January 2017).

[6] Radar Jogja. (2017, 28 January). *Pekan Budaya Tionghoa Yogyakarta 2017 Digelar Selama 7 Hari*. Downloaded January 2017, from https://www.radarjogja.co.id/pekan-budaya-tionghoa-yogyakarta-2017-digelar-selama-tujuh-hari/

[7] Jogja Tribun News. (2017, 1st Febuary). *Pekan Budaya Tionghoa Yogyakarta 2017 Usung Keragaman Nusantara*. Downloaded Febuary 2017, from http://jogja.tribunnews.com/2017/02/01/pekan-budaya-tionghoa-yogyakarta-2017-usung keragaman-nusantara

[8] Gudeg Net. *Upacara Grebeg Maulud di Yogyakarta*. Downloaded Febuary 2017, from https://gudeg.net/direktori/1882/upacara-grebeg-maulud-di-yogyakarta.html

[9] Regional Kompas. (2016, 12 December). *Liburan Warga Yogyakarta dan Wisatawan Saksikan Grebeg Maulud*. Downloaded Febuari 2017, from http://regional.kompas.com/read/2016/12/12/12485021/liburan.warga.yogyakarta.dan.wisatawan.saksikan.grebeg.maulud

[10] Antara News. (2011, 21st June). *Yogyakarta Malioboro Enviled with Art Festivals*. Downloaded Febuary 2017, from http://www.antaranews.com/en/news/72976/yogyakartas-malioboro-enlivened-with-art-festivals

[11] Bisnis UKM.Com. (2011, 30 June). *Pasar Seni dan UMKM dalam FKY 2011*. Downloaded Febuary 2017, from http://bisnisukm.com/pasar-seni-dan-umkm-dalam-fky-2011.html

[12] Tekno Kompas. (2011, 26 June). *FKY 2011 Resmi Dibuka Sultan HB X*. Downloaded Febuary 2017,from http://tekno.kompas.com/read/2011/06/26/20145973/fky.2011.resmi.dibuka.sultan.hb.x.

[13] Dr Pojokan. (2011, June). *FKY Digelar 26 Juni – 5 Juli 2011*. Downloaded Febuary 2017,from http://www.drpojokan.com/2011/06/fky-di-gelar-26-juni-5-juli-2011-dari.html

[14] Kompasiana. (2015, 7 March). *Hari ini Kawasan Malioboro dan Jalan Menuju Pagelaran Keraton Yogyakarta akan Ditutup*. Downloaded Febuary 2017, from http://www.kompasiana.com/jk.martono/hari-ini-kawasan-malioboro-dan-jalan-menuju-pagelaran-keraton-yogyakarta-akan-ditutup_54fd1d49a333110f1d50f84e
http://jogjapedia.net/acara-budaya/tradisi-grebeg-maulid-keraton-yogyakarta/

[15] Travel Kompas. (2015, 22nd July). *Ingat Festival Malioboro Digelar 24 – 26 Juli*. Downloaded Febuary 2017, from http://travel.kompas.com/read/2015/07/22/134200927/Ingat.Festival.Malioboro.2015.Digelar.24-26.Juli?page=all

[16] Regional Kompas. (2013, 22nd October). Pernikahan Terakhir Keraton Yogyakarta Digelar Lebih Megah. Downloaded Febuary 2017, from http://regional.kompas.com/read/2013/10/22/1249197/Pernikahan.Terakhir.Keraton.Yogyakarta.Digelar.Lebih.Megah

[17] Satu Harapan. (2012, 2nd October). Sambut HUT Kota Seniman pasang Patung Di Sepanjang Malioboro. Downloaded Febuary 2017, from http://www.satuharapan.com/read-detail/read/sambut-hut-kota-seniman-pasang-patung-di-sepanjang-malioboro

[18] Harian Jogja. (2016, 19 August). *Hut Kemerdekaan RI Diikuti 2400 Orang Pawai Pembangunan di Malioboro Meriah*. Downloaded 2017, from

http://www.harianjogja.com/baca/2016/08/19/hut-kemerdekaan-ri-diikuti-2-400-orang-pawai-pembangunan-di-malioboro-meriah-745882

[19] Harian Jogja. (2016, 28 October). *Demo Di Jogja Ribuan Orang Demo Ahok Di Titik Nol Kilometer*. Downloaded Febuary 2017, from http://m.harianjogja.com/baca/2016/10/28/demo-di-jogja-ribuan-orang-demo-ahok-di-titik-nol-kilometer-764460

[20] Detik News. (2016, 1st May). *Buruh Di Yogyakarta Berdemo di Malioboro*. Downloaded Febuary 2017, from https://news.detik.com/berita/3200744/buruh-di-yogyakarta-berdemo-di-malioboro

[21] MerahPutih. (2015, 5 November). *Lagi, Tolak Bandara Kulon ProgoWarga Demo Di Malioboro*. Downloaded febuary 2017, from http://merahputih.com/post/read/lagi-tolak-bandara-kulonprogo-warga-demo-di-malioboro

IOP Conf. Series: Earth and Environmental Science **70** (2017) 012056 doi:10.1088/1755-1315/70/1/012056

Wayfinding Concept in University of Brawijaya

H Firjatullah[1], E B Kurniawan, and W D Purnamasari

Urban and Regional Planning Department, Universitas Brawijaya, Malang 65145, Indonesia

hafizfirjatullah@gmail.com

Abstract. Wayfinding is an activity related to the orientation and motion from first point to point of destination. Benefits of wayfinding in the area of education, namely as a means of helping direct a person to a destination so as to reduce the lostness and assist users in remembering the way through the role of space and objects wayfinding. Around 48% new students of University of Brawijaya (UB) 2015 that was ever lost when entering the campus. This shows the need for wayfinding concept to someone who was unfamiliar with the surrounding area as freshmen. This study uses mental map analysis to find the objects hierarchy wayfinding determination based on the respondents and the space syntax (visual graph analysis) as a hierarchy based on the determination of the configuration space. The overlay result say that hierarchy between both of analysis shows there are several objects which are potentially in wayfinding process on the campus of UB. The concept of wayfinding generate different treatment of the selected object based of wayfinding classification, both in maintaining the function of the object in space and develop the potential of the object wayfinding.

1. Introduction

Wayfinding is an activity related to the orientation and movement from one place to another [1]. In Brawijaya University (UB), the increasing number of new students affect campus' activity movemement such as the addition of infrastructure, widening of road, and new buildings' construction. This affects difficulty in identifying the space to wayfinding in UB, especially new students who are unfamiliar with their surroundings. Based on the survey, 48% of new students UB 2015 ever lost when entering the university. Difficulties and confusion for new students in finding a destination place in the campus became the main idea in this study. Orientation process and wayfinding related to several factors which are capacity of people, processes and cognitive maps awakened in the mind of the individual, and the role of environmental information, includes architectural wayfinding element, signage system, and other sensory information [2]. Architectural wayfinding elements identified by a) path and circulation; b) landmarks or markers; c) nodes; d) edges; and e) districts or zones [1]. In urban design elements, there is one sign or signage elements of architectural design as a form of information and orientation of the city [3]. Then role of wayfinding support in other sensory information, such as the shape of paving sidewalks, street lights, and vegetation can help users in the process of wayfinding [4].

2. Methodology

This study identifying the object condition as the main wayfinding elements in UB and identify wayfinding objects by the experts and respondents whose based on object from the variable architectural wayfinding elements (path, edge, zone, node, and landmark). Wayfinding also needs supportive wayfinding elements, such as signage and other sensory information. Retrieving the data using mental map analysis to determine the hierarchy of objects by the respondents as well as a questionnaire on the supporting wayfinding elements. The hierarchy of mental map is obtained by comparison scores from the result of questionnaire mental map that consist of criteria from each elements. The hierarchy of space syntax is obtained by comparison between averages of imaginary point score from each objects in wayfinding element. After find the hierarchy from both analysis, then do clasification and comparison between the hierarchy of objects by mental map and the spatial configuration hierarchy by space syntax (VGA).

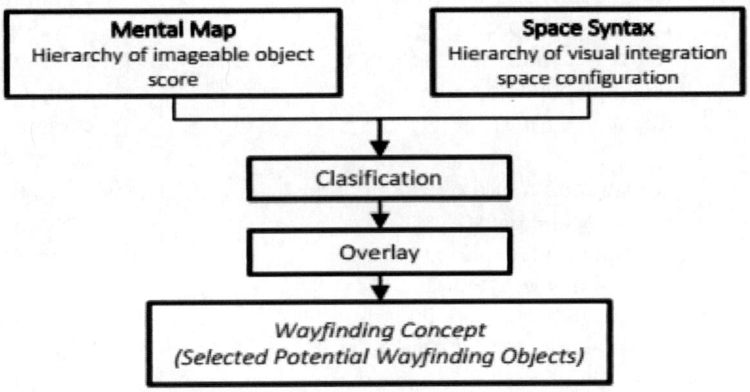

Figure 1. Line of research methodology

The analysis of mental map and space syntax has a difference in the rank or hierarchy on the object wayfinding. Ranking mental map obtained from sequencing scores from high to low, while for space syntax of sorting the average value of an imaginary point of each object architectural wayfinding. The overlay is used on both analysis to determine the potential objects of the overall objects which are selected by the respondents.

In finding the objects that are potential to be the main wayfinding of Universitas Brawijaya, the priority of potential object selection is conducted first. Object selection priorities for space configuration is fixed and is not affected by external influence. Furthermore, the classification is conducted for sorting the eligible object to be included into the draft, thus the object can be classified as low or high. From the selection of the object, the concept of appropriate treatment toward the objects then can be determined, either by merely maintaining the function of the object or developing the potential objects that have the potential to develop yet still have some obstacles in wayfinding process.

Table 1. Priority of Potential Objects for Main Wayfinding

Mental Map	Space Syntax	Information	Priority Objects	Concept
High	High	If the opinions of the respondents as high as spatial configuration, then the object can be a potential cause of the object in terms of the spatial structure, which easily identified and supported by a selection of objects based on the opinions of respondents.	High	• Maintaining the function of the selected object • Doing maintenance on the location of the object
High	Low	If the configuration space is low, it means the space is less integrated, although there are respondents who chose the object	Medium	• Planning the development of space configuration to improve the visual integration with the object of wayfinding.
Low	High	If the opinion of the respondents is low, while the space configuration declare that the object has the potential and is easy to be identified, then there is a problem on that object, both in visual obstacles or caused by other factors.	High	• Strengthen the function of the object as an element designated as the object of wayfinding by arrangement at other sensory information and signage system.
Low	Low	If the objects are equally low, it means that the object is not worthy of being the main wayfinding and only becomes the object complement. As know from the concept of a point of interest, that in a room does not require that all objects have a high value.	Low	• No treatment for low priority objects

The prioritized classification based on criteria are High (mental map) - High (space syntax) and Low (mental map) - High (space syntax) (Tabel 1). The classification object has a different classification in each element based on the average value taken from the analysis (Tabel 2).

Table 2. Clasification Objects from Average of Final Score in Analysis

Objects	Mental Map		Space Syntax	
	Low	High	Low	High
Path	0,18-0,99	0,99-2,43	2,57-4,95	4,95-6,23
Edge	0,09-0,78	0,78-1,45	3,49-3,85	3,85-4,59
Zone	1,14-0,37	0,37-1,14	3,69-4,38	4,38-5,17
Node	1,14-0,27	0,27-1,14	3,69-4,01	5,01-6,57
Landmark	0,03-0,22	0,22-0,98	2,66-4,33	4,33-5,62

The interval of classification consists of two things which are low and high. If the final score result is over the average value, thus the object is classified into high. This is also applied in low classification too.

3. Result

The result of this study is based on the analysis of space syntax, mental map, and other supporting elements, such as signage system and other sensory information. Result forms this study contains of potential wayfinding objects from each elements to support wayfinding activities for new students in Brawijaya University.

3.1. Space Syntax

Researchers use the application Depthmap 0.30, with a base map of UB in dxf format from AutoCAD in the study of wayfinding concept in UB. Results obtained from the analysis is the distribution of color resulting from an imaginary point that describes the visual gradation integration in UB. The red color means that the high value of visual integration, and lower value for blue in gradation.

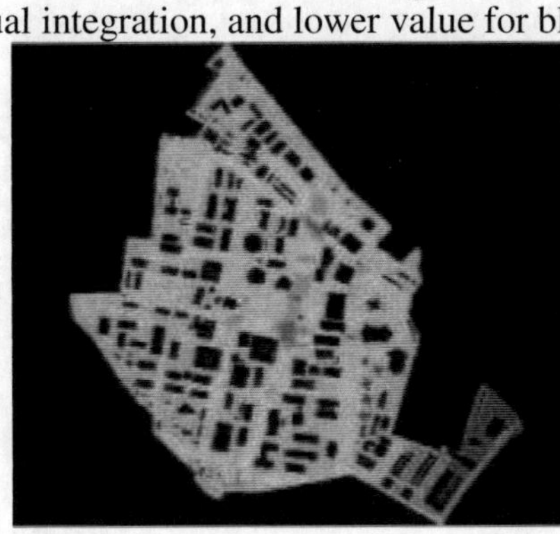

Figure 2. Space Syntax Output (Visual Integration in UB)

3.2. Mental Map

The results of mental map in the form of hierarchy obtained from the questionnaire respondents. The hierarchy result is converted to the map to provide the better visualization based on the users' opinions. The explanation about mental map will be discussed in every elements of wayfinding in UB.

3.2.1. Path. Path is the lane which becomes a potential place to move. It usually takes form as roads, pedestrian pathways, and footpaths [1]. Path is chosen and rated by respondents giving score for each criteria. Overall, respondents give higher scores on each criteria, such as pavement (90%), scale (82%), complementary roads (72%), liaison function with other roads (85%), and historical value (77%).

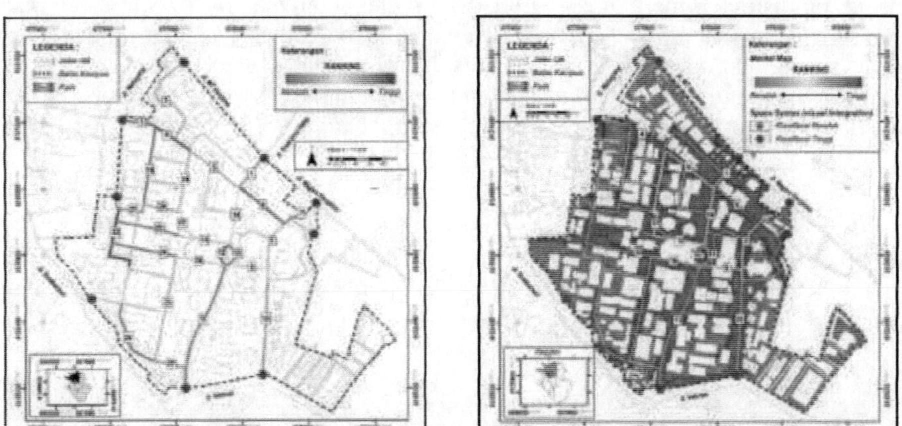

Figure 3. (a) Visualitation Paths of Mental Map ; (b) Overlay Path Mental Map and Space Syntax

Before using the analysis, the element of path has 28 objects obtained from the experts and respondents. After the overlay, path has 15 objects that is potential to become wayfinding objects. The treatment for path elements has 5 objects that only need to be maintained or strengthened, such as Path 4, 10, 11, 12, and 13, and 10 other objects that have the potential to be developed into main wayfinding, such as the Path 1, 5, 6, 8, 9, 14 , 15, 16, 17, and 24.

3.2.2. Edge. Edge is linear element that are not used as a movement path, but it has a function as a barrier between the two adjacent areas [1]. Edge that are selected and rated by the respondents are given levels to determine the ranking of the most chosen and influential objects in wayfinding process. The results shows that respondents give a high score on each criteria such as scale (86%), ease viewed edge (87%), barrier function (85%), the type of construction used (86%), and historical value (73%) of total score per criteria.

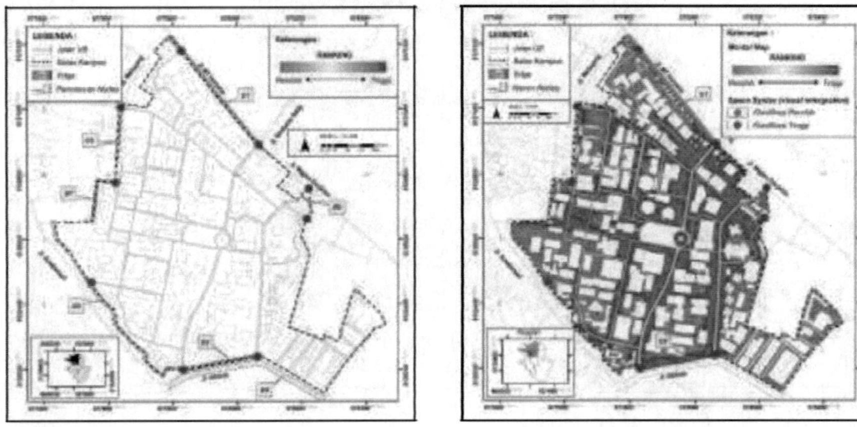

Figure 4. (a) Visualitation Edge of Mental Map; (b) Overlay Edge Mental Map and Space Syntax

The survey shows that there are 7 objects that can be considered as the wayfinding elements. After the overlay process, there are only 2 edges that become the potential object and considered as main wayfinding. The treatment for that edges need to be maintained or strengthened as wayfinding objects in UB space, such as Edge 1 and Edge 2.

3.2.3. Zone. Zone or district is part of a region that is characterized by a specific character and functions so that people can identify and classify different area to another, from the characters and their use, as well as the shapes and materials used [1]. Zone has five criteria used to measure the impact of an object against wayfinding respondents. The percentage of high ratings on those criteria, such as easy to be seen zone (98%), shape (93%), scale (93%), the historical value (84%), and has a special function (93%).

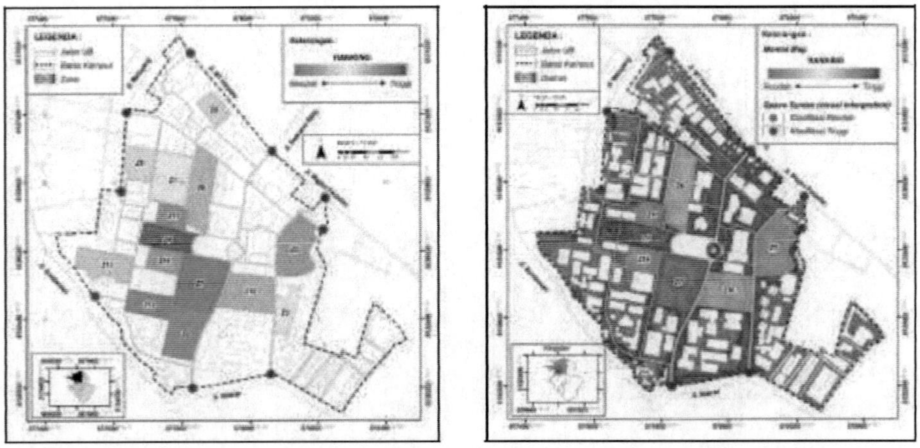

Figure 5. (a) Visualitation Zone of Mental Map; (b) Overlay Zone Mental Map and Space Syntax

Zone element has 14 objects obtained from the experts and respondents. After the overlay, zone has 7 objects that is potential to become wayfinding objects. The treatment for zone elements has 4 objects that need to be maintained or strengthened, such as Zone 4, 5, 6, and 7, and 3 other objects that have the potential to be developed into main wayfinding, the Zone 10, 11, and 14.

3.2.4. Node. Node represent strategic points that can be entered by the observer at the center of activity, such as parks, squares, courts, and others. Node also a meeting point of path, such as intersection or transportation center [1]. The high percentage given by respondents on each criteria are unique shape (93%), worth a look (89%), scale (87%), has a special function (90%), and historical value (83%).

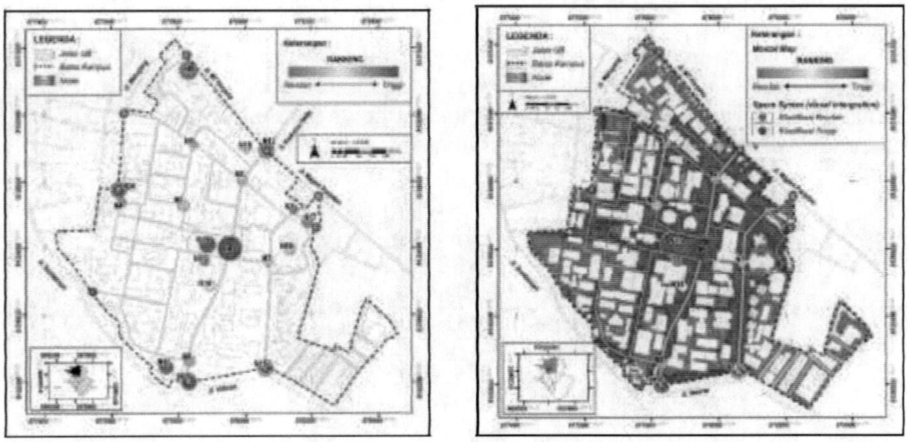

Figure 6. (a) Visualitation Node of Mental Map ; (b) Overlay Node Mental Map and Space Syntax

Node element has 20 objects obtained from the experts and respondents. After the overlay, node has 12 objects that is potential to become wayfinding objects. The treatment for node elements has 3 objects that need to be maintained or strengthened, such as Node 6, 10, and 13, and 9 other objects that have the potential to be developed into main wayfinding, the Node 1, 2, 5, 7, 8, 14, 15, 18 and 19.

3.2.5. Landmark. Landmark can be mentioned as a physical marker that can be easily seen from several angles and distances, such as buildings, direction, gate, field, and others [1]. The high percentage given by respondents on each criteria are unique shape (95%), scale (94%), ease of visits (94%), as a liaison with surrounding objects (92%), and historical value (84%).

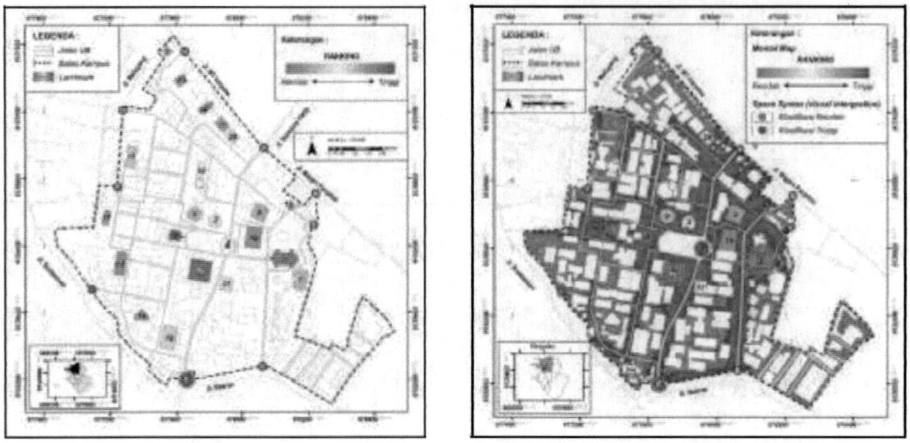

Figure 7. (a) Visualitation Landmark of Mental Map; (b) Overlay Landmark Mental Map and Space Syntax

The survey shows that there are 23 landmark objects that can be considered as the wayfinding objects. After the overlay process, there are 12 objects that become the potential object and considered as the wayfinding. The treatment for landmark elements has 6 objects that need to be maintained or strengthened, such as Landmark 4, 5, 6, 7, 8, and 9, and 6 other objects that have the potential to be developed into main wayfinding, the Landmark 2, 17, 18, 19, 20 and 23.

3.3. Other Sensory Information

Other sensory information is used as a supporting element of wayfinding campus, including paving of sidewalks, street lights, and vegetation. Other sensory objects located near the wayfinding objects and support new student in wayfinding process.

3.3.1. Paving (Sidewalk). Based on a primary survey results, paving on the sidewalk in UB has eight different types of surfaces. It can affect pedestrian's memory about the uniqueness of pavement furniture's visualization. The results of the questionnaire shows that 48 respondents said that the paving is affects someone's wayfinding while 153 respondents said not affected.

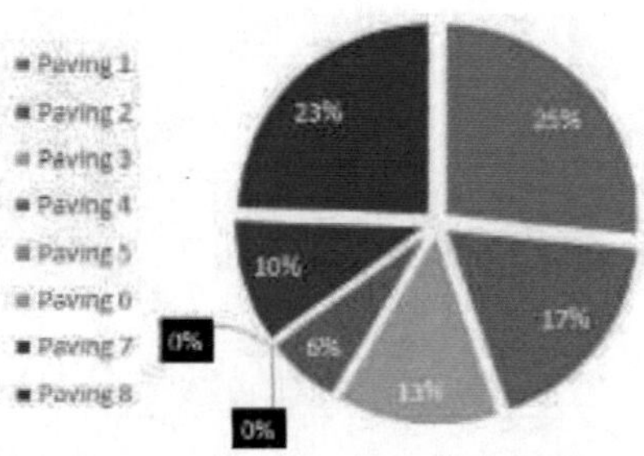

Figure 8. Diagram Percentage Type of Paving Based On Respondents Choices \

3.3.2. Street Lights. The street lights in UB majority are classified into two types which are type A and type B. Street lights is one of street furniture that can be the object that affect the process of wayfinding, especially in the dark situation. In addition to the aesthetic, street lights which follows the contours of the road, can direct people in the night so it can minimize the level of getting lost in campus. Based on the results of the questionnaire, 32 out of 201 respondents said that the street lights affect someone's wayfinding in campus.

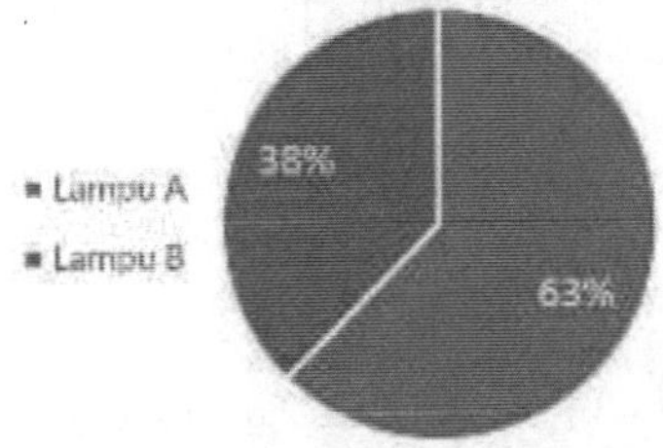

Figure 9. Diagram Percentage Type of Street Lights Based On Respondents Choices

3.3.3. Vegetation. Vegetation has several functions as aesthetics, land cover (ground cover), shade, steering, and the barrier (physical barrier). These function strengthen the objects' mental image into the influential object in wayfinding system. Based on the result of the questionnaire, 60 respondents stated that vegetation influences the wayfinding system while 141 stated it is not influenced.

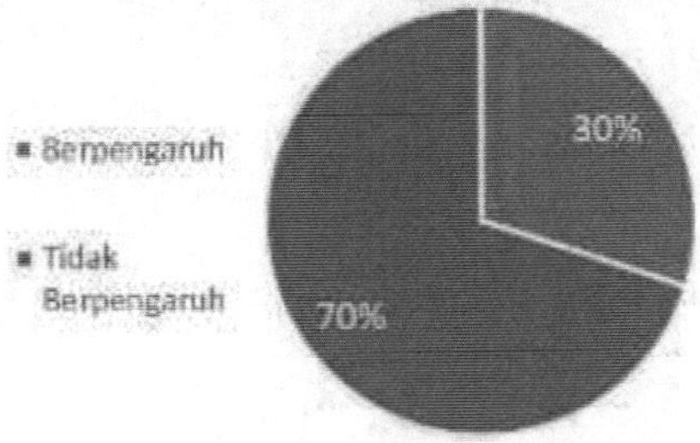

Figure 10. Diagram Percentage Type of Vegetation Based On Respondents Choices

3.4. Signage System

Signage referred in this study is the signage used as a direction and delivers information about the objects in UB. Based on the survey result, there are 10 spots of signage that scattered in the various intersections.

Figure 11. Signage Distribution in University of Brawijaya

References

[1] Lynch, K. 1960. The Image of The City. Cambridge: The MIT Press
[2] Hariyanto, A.D., dkk. 2012. Kualitas Elemen Arsitektur Sebagai Penunjang Kemudahan Wayfinding dan Orientasi. Seminar Nasional Dies Jurusan Arsitektur : Toward Emphatic Architecture University of Kristen Petra.
[3] Shirvani, Hamid. 1985. The Urban Design Process. New York: Van Nostrand Reincold Co.
[4] Apelt, R., dkk. 2007. Wayfinding Design Guidlines. Brisbane: Icon.Net Pty Ltd, Cooperative research Centre (CRC).

IOP Conf. Series: Earth and Environmental Science **70** (2017) 012056 doi:10.1088/1755-1315/70/1/012056

[5] Desyllas, Jake & Elspeth Duxbury. Axial Maps and Visibility Graph Analysis: A Comparison of Their Methodology and Use in Models of Urban Pedestrian Movement. 2001. London: Intelligent Space.
[6] Findlay, C. & Soutwell, K. 2004. 'I just followed my nose' : understanding visitor wayfinding and information needs at forest recreation sites. Edinburge: Routledge (Managing Leisure 9)
[7] Golledge, R.G., (1999). Wayfinding Behaviour : Cognitive Mapping and Other Spatial Processes. London : The Johns Hopkings University Press.
[8] Mohamed, A.A. 2012. Evaluating Way-Finding Ability Within Urban Environtment. Dalam M. Greene, J. Reyes, dan A. Castro (Editor). Proceedings : Eight International Space Syntax Symposium: 1-3moh9. Santiago de Chile: PUC
[9] Porteous, D. 1977. Environtment and Behaviour: Planning and Everyday Urban Life. Massachusetts: Addison-Wesley Publishing
[10] Rubenstein, Havey M. 1992. Pedestrian Malls, Streetscapes, and Urban Space. Canada: John Wiley and Sons Inc.

IOP Conf. Series: Earth and Environmental Science **70** (2017) 012057 doi:10.1088/1755-1315/70/1/012057

Creating kampong as tourist attractions

N Sari[1,3], R Utama[2], A R T Hidayat[1] and A B Zamrony[1]

[1] Department of Urban and Regional Planning, Faculty of Engineering, Universitas Brawijaya, Malang, Indonesia
[2] Department of Architecture, Faculty of Engineering, Universitas Brawijaya, Malang, Indonesia

nindya.s@ub.ac.id

Abstract. Tourism attractions become one of the main components and they drive the tourism activity in a region. The quality of tourism attractions would affect tourists' visits. Tourism power can basically be built on any conditions which can attract people to visit. Towns is full of activities which include their economic, social, cultural and physical features, if they are presented properly, they can be a tourist attraction. Kampung City, as a form of urban settlement, has the potential to be developed as a tourism attraction. Kampung is not only a physical area of housing but it has also productive activities. Even the city's economic activities are also influenced by the productive activities of its Kampung. The shape of Kampung which varies in physical, social, economic and cultural raises special characteristics of each Kampung. When it is linked with the city's tourism activities, these special characteristics of course could be one of the attractions to attract tourists. This paper studies about one of Kampung in the Malang City. Administratively located in the Penanggungan Village Lowokwaru District, but the potential will just be focused on RW 4. Main productive activities of this village are pottery. In contrast to ceramics, pottery is made from clay and its uniqueness in color and shape. Based on the history of pottery in the Malang, it is concentrated in Penanggungan Village. But along with its development, pottery is decreasingly in demand and number of craftsmen is dwindling. Based on these circumstances, a concept is prepared to raise the image of the region as the Kampung of pottery and to repack it as a tourism attraction of the city.

1. Introduction

A city is the concentration of population in the geographical area with a system of human life which characteristics are high and dense population, socio-economic activities as well as heterogeneous cultures, individualistic life, and a land available to be used as public service activities. It is a center of industry, commerce, education, government or in other word covers all of these activities [1]. The city will experience growth and development very rapidly along with activities and human life that keep changing for the sake of improved life quality, as people head into city to look for jobs. A large volume of people in the city will affect development activities in various sectors such as housing, education, commerce, and industry, which result in increases on areas that have urban characteristics.

City as tourism product is formed because of the physical potential, service and experience [2]. It is also the central location of many facilities and attractions which are located within public's reach, in this case is city residents and tourists [3]. In general, tourists are attracted to visit because it offers special function which makes every city is different, unique and peculiar in providing comprehensive facilities [4].

Kampung (village) is one form of urban settlements which are usually built independently by migrants in urban areas [5]. The relationship of a city with its village is very close and it is mutually supportive because the city is not surviving without kampung, while kampong can exist because of the city setting. Many productive activities that support the city's economy are located based in the kampung city. Based on history of the formation, each kampung should have specificity and uniqueness both in terms of social, cultural and economic. Unique and distinctive potentials make kampung's conditions different.

To support tourism activity in a city, the specialty of its kampung can be packed into tourist attraction. Potentials of industrial activities as in Kampung Jayengan which is packed into rural tourism by the support of its potentials in culture, history, facilities, economy and infrastructure [6]. The specificity of a village as historical district also has the potential to be packed into tourist attraction as in Kampung Kauman, Surakarta. Kauman that has oldish-style settlements with a Grand Mosque as community center has been developed into a rural tourism with the packaging of religious tourism and it is integrated with cultural and shopping attractions [7].

Penanggungan Village is one area in Klojen district, Malang City which is known as the hometown of pottery craftsmen. At this moment there are 7 pottery craftsmen who are still actively producing and creating pottery. Forms of produced crafts are changing according to of public needs. In order to pack the Kampung of pottery into a tourist attraction of the city, it is necessary to reform conditions of the village. Kampung's environment must be able to provide comfort and memorable impression for visitors. By structuring the village, it is expected to improve the image of Kampung of pottery thus positive impacts on pottery making and community activities can be achieved, both in economic and social.

2. Methods

The location is a Kampung of pottery is in Penanggungan Village RW 6 Malang City. This area is selected as most craftsmen are concentrated in this region. The process of data collection and drafting are done by the Forum Group Discussion (FGD). People engaged are Chief of Village in Penanggungan and his associated staff, the Head of RW and RT, public figures, pottery craftsmen, youth leaders and representatives of PKK team. The data collection is also done by observation in the field. Based on data collected from public and the results of observation, then they are compiled the potential and problems of the area. The next process concept prepared to accommodate the idea of the forum. We propose the idea of a society that is synergized with the idea of researcher's based on conditions in field into the concept of the development of the kampung as a tourism attraction.

3. Results

3.1. Pottery Village History

Village area of Penanggungan has historical potential as a center for pottery. Pottery is regarded as the forerunner of ceramics which are more developed and more famous nowadays. Pottery activities are started in 1930, with skills inherited from generation to generation. Quality of pottery in the Penanggungan Village is very nice because it is supported quality raw materials obtained from the Village Penanggungan environment. But along with its development, pottery products began to compete with products made from plastic. Number of craftsmen is decreasing since younger generation is not interested in working as a pottery craftsman. Raw materials is also diminishing factor in results of more land is reclassified as residential lots. In the end, needs of raw materials are obtained from the area outside the Penanggungan village, so have to buy them and it increases the cost of transport for the transport of raw materials.

Table 1. History of pottery crafts in the Penanggungan Village.

	1976 In the beginning, a number of craftsmen made and sold *Gentong* in the market
	1987 Craftsmen began to innovate and made mortar for the purposes of the celebration, and it became very popular with the public.
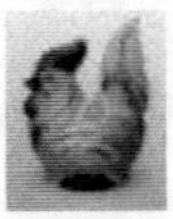	1990 Along the Craftsman movement of money saving, they innovated to create piggy bank in form of animal (rooster)
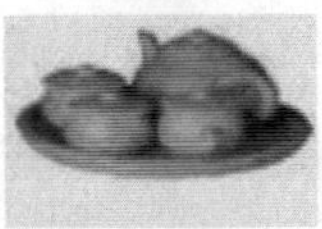	2000 When plastic piggy banks outnumber in the market, craftsmen survived by making pottery household utensils.
	2010 When plastic goods outnumber in the market, craftsmen still survived by making pottery household utensils.
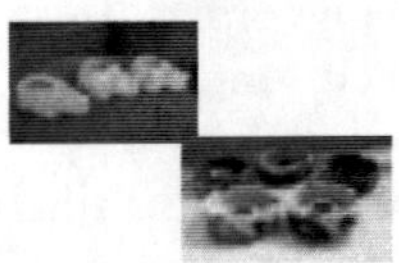	2016 Handicraft products in great demand for wedding gift and souvenirs. So that the shape of the products are varied depending on the order.

3.2. Location

Kampung of pottery is located at strategic locations. Located on near to educational area, it is close to the University of Brawiijaya, State University of Malang, University of Muhammadiyah Malang and ITN, SMAN 8 Malang, SMPN 4 Malang, SMKN 2 Malang. This area is also an area where there are trade and services centers Center (Mall MX and Matos), Hotel Swiss Bell Inn Malang and some other hotels. In certain months, especially during the school holidays, this region becomes one of the attractions and it is tourists' favorite, thus it is expected later to be very easy in guiding tourists to the location of Kampung of pottery.

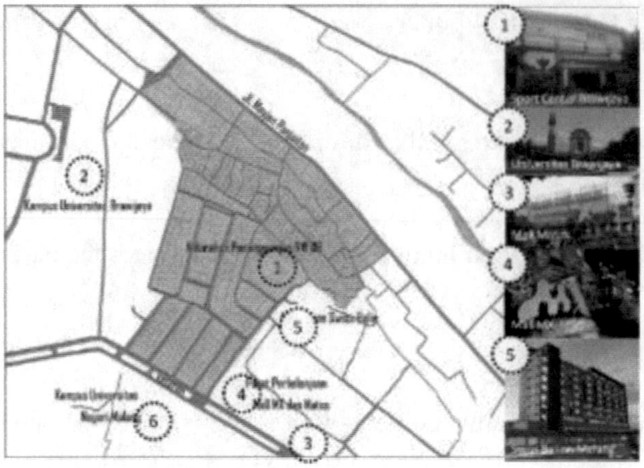

Figure 1. Location of Kampung Pottery.

3.3. Kampung of Pottery Potentials

The main attraction is in the process of pottery making and pottery products. All activities are done manually and traditional. This activity combines art and creativity. There is approximately 15 pottery craftsmen, but there are only 4 craftsmen who processes the manufacture until finishing. Others only do the finishing and sales.

1. Pottery Production Sites *(home industry)*
   There are 4 houses (4 craftsmen) that survive as pottery production house. Activities done there include from pottery making to finishing. The products produced are mostly orders from customers. They produced varies of products depending on the demand of the customer.
2. Sales activities and pottery *finishing*
   There are three houses that work as the pottery finishing and place of sale. Finishing is the final process of pottery making after the firing process. This process is done in by painting the color, defining, carving or sticking to other materials. For coloring, the pottery craftsmen use stencil dyes and oil paint with a capital of ± Rp. 150,000. In the finishing process, the work should not be done directly by pottery craftsmen but it is done by other employees.

Sales activities are done in a *showroom* or a simple *gallery* that is located in front of the house. Most sellers are also a distributor of several craftsmen. There are 5 houses only that work as a *showroom* in form of simple *gallery.* The majority of pottery sales were done in the house that has a strategic location that is located on the main roads.

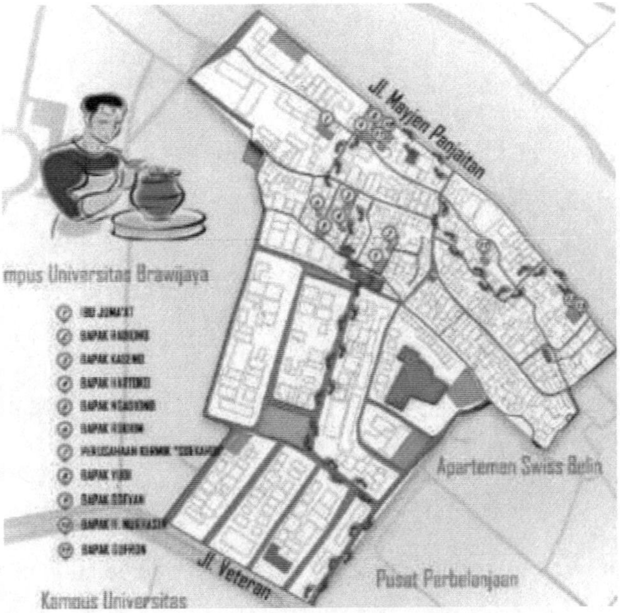

Figure 2. Location of pottery craftsmen.

3.4. Development concept

3.4.1. Zoning concept. To determine the regional zoning is predetermined on point and location has the potential activities to be developed into a tourism attraction. From this zoning will form a travel path from gate to the main attraction, Kampung of pottery. Gate of this region can strengthen the identity of the village and can attract the public attention, especially tourists. Structuring the road corridor that passes through the main gate to the house of pottery as the icon of district and tourists' major destination are also taken into account by providing a consistent concept with the theme of Kampung of pottery. Extra function of building can accommodate artists and craftsmen to show products of their merchandise and arts to attract tourists to visit the Kampung of pottery, Penanggungan. This building at same time will also work to be icons of Kampung of pottery Penanggungan. The park in this should be laid out to support its function as a public space for the society and tourists. The park in this environment is expected to be a place of social interaction of residents of the village, as well as place to display pottery to attract tourists. Arrangement of Village settlement is needed to be cleaner and more attractive at the same time managing the corridors which functioned to guide the visitors to main location of Kampung of pottery. Supporting facilities such as car park is directed by utilizing the MX mall, Matos and Apartment Switzerland Belin parking area, as well as available land in village such as the area near the village official's office and parking area of UB's fitness center. Settings circulation for pedestrians and bicycles can be formed from the main entrance to the home of pottery. Circulation separation is by using different material and specific markers for bicycle and pedestrian paths, thus those additional lanes for pedestrian circulation can indirectly directs visitors towards the main locations of pottery tourism.

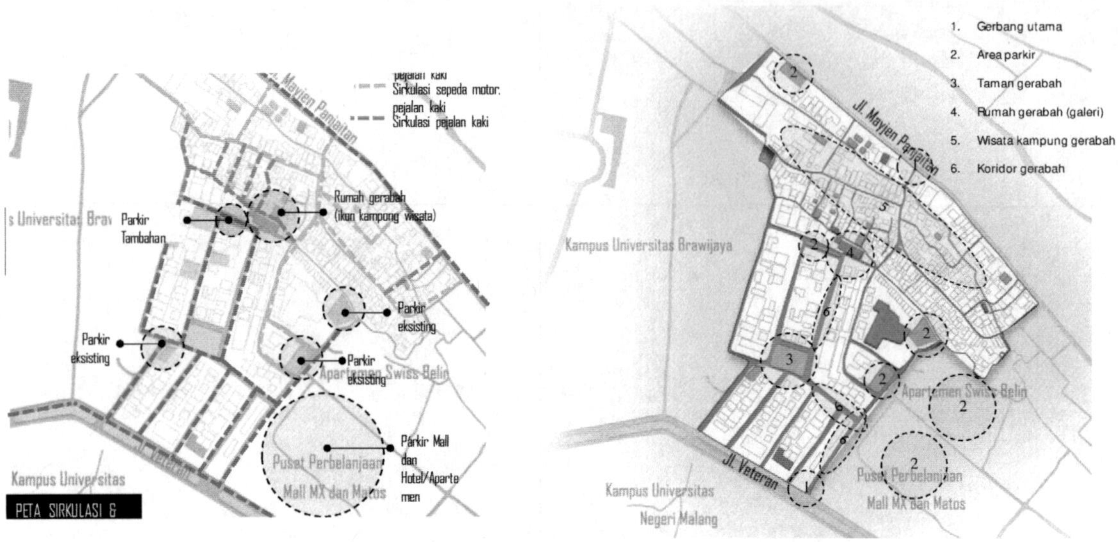

Figure 3. (a) Zoning concept; (b) Activities circulation map.

3.4.2. Tourism Activities Concept. Based on results of zoning, there are some points of potential activities that can be packed as a tourist attraction. These locations include the main gate, pottery garden, house of pottery (gallery), the village environment with craftsmen and people's lifes as well as the corridors in the pottery village. The activities will be offered to tourists is planned in those locations. The main attractions are buildings of house of pottery which setting is created as museums; they are equipped with workshop space and space to present the history of pottery and documentary film pottery making process. Visitors can see the collection of pottery, witness historical film about pottery as well as the manufacturing process and experience training of pottery making. The next fascinated thing is visitors can travel around village area to pottery craftsman's house to see the process directly. Visitors is also offered village atmosphere which is laid out to show comfort and beauty typical pottery village. Purchase of pottery products is centered in gift shop. To attract more visitors, package of activities are packed by adapting the purpose and motivation of visitors. 3 packages are planned with each of the following activities:

Table 2. Tour package.

Tour package 1	• Individual 2-5 people (free / without guide) • Equipped with a road map to location of the pottery craftsmen
Tour Package 2	See collections of pottery that is centered in gallery Purpose : ▪ To give idea to visitor of types of pottery in Penanggungan village ▪ To give visitors the main attraction in kampung of pottery Watching a documentary film about pottery Purpose :

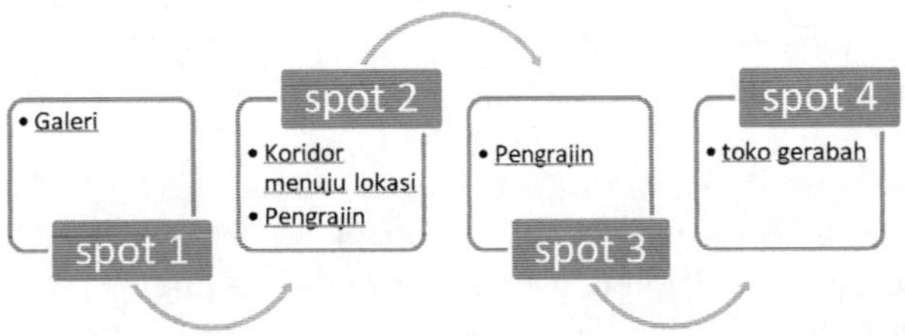	▪ To provide an overview of history of the village and pottery in Penanggungan village to visitors ▪ To provide information on process of pottery making, in order to attract visitors to the craftsmen's location Directed toward pottery craftsmen and passing interesting spots around the environment of kampong of pottery Purpose: ▪ To Introduce kampong of pottery in Penanggungan Village ▪ To accommodate visitors to experience and practice of pottery making Directed toward center of souvenir Purpose : ▪ To buy pottery as souvenir ▪ To provide information to visitors that pottery can be produced into various shapes of souvenirs
Tour Package 3	Special package to learn pottery handicraft

4. Conclusion

Area of RW 6 Penanggungan Village has considerable potential to be packed by the theme kampong of pottery. Different from the ceramic, pottery became one of the main product of Malang. Location of pottery craftsmen basede on t history is concentrated in Penanggungan village. Nowadays, location of pottery-making in Malang can only be found in the Penanggungan Village. This condition is very beneficial for Penanggungan Village, particularly RW 6, to be formed into a unique region with craft and pottery producing activities. Forming the village as planned will certainly need a fairly long process. There are two factors that become the main focus of development which are related to physical and non-physical factors. Physical factor is the implementation of design that has been drafted. This implementation would require considerable expense. Non-physical factors are related to the readiness of local community. Although the drafting process has involved local community, but there should be assistance to provide deeper understanding among local community, particularly people in RW 6 Penanggungan village, so that they could receive and give input, as well as get involved in all the programs that has been planned.

Despite of all the problems in the field conditions which include both physical and non-physical factors, the idea of village establishment with rural-potential based development is very interesting. The goal is to improve local economics as well as changes in village environment to be visually cleaner and more beautiful. The changes would indirectly attract people to come and give comfort to the local community in the village.

IOP Conf. Series: Earth and Environmental Science **70** (2017) 012057 doi:10.1088/1755-1315/70/1/012057

References

[1] A. B. Gallion and S. Eisner, *Pengantar perancangan kota*. Jakarta: Erlangga, 1996.

[2] B. M. Kolb, *Tourism marketing for cities and towns: Using branding and events to attracts tourists*, 10th ed. Burlington: Butterworth-Heinemann, 2006.

[3] C. M. Hall and S. J. Page, *The Geography of Tourism and Recreation: Environment, Place and Space*. New York: Routedge, 1999.

[4] S. Page, *Urban tourism*. New York: Routedge, 1995.

[5] B. Setiawan, "Kampung kota dan kota kampung: Tantangan perencanaan kota di Indonesia." [Online]. Available: http://kampungnesia.org/berita-kampung-kota-dan-kota-kampung--tantangan-perencanaan-kota-di-indonesia--1.html.

[6] W. Astuti, Qomarun, A. Febela, R. A. Putri, and D. W. Astuti, "Identification of Specific Characteristic of Kampung Jayengan as Community-based Industrial Tourism," *Procedia - Soc. Behav. Sci.*, vol. 227, pp. 485–492, Jul. 2016.

[7] W. Setyaningsih, T. Y. Iswati, SriYuliani, W. Nuryanti, B. Prayitno, and A. Sarwadi, "Low-Impact-Development as an Implementation of the Eco-Green-Tourism Concept to Develop Kampung towards Sustainable City," *Procedia - Soc. Behav. Sci.*, vol. 179, pp. 109–117, Apr. 2015.

IOP Conf. Series: Earth and Environmental Science **70** (2017) 012058 doi:10.1088/1755-1315/70/1/012058

Urban greenspace for resilient city in the future: Case study of Yogyakarta City

N M Ni'mah, S Lenonb

Departement of Urban and Regional Planning, Sekolah Tinggi Teknologi Nasional (STTNAS) Yogyakarta, Jl. Babarsari, Catur Tunggal, Depok, Sleman, 55281, Indonesia

novimaulida@sttnas.ac.id

Abstract. The capacity of adaptation is essential elements towards urban resilience. One adaptation that can be done is to consider the provision of open space and public space in the city. Yogyakarta City development which focused on the built area and negates the open space has blurred the characteristics of the city. Efforts in increasing the availability of public space is one of the seven priorities of the programs included in the environmental and the utilization of space in Yogyakarta City. An understanding of the provision of public green open spaces in Yogyakarta is important because the products and processes that take place in a development will determine the successful implementation of the development plan. The objectives of this study arc as follows: (1) to identify the provision green space in Yogyakarta City from the aspects of product and procedure; and (2) to identify the role of green space to build resilient city. This study is used descriptive qualitative approach with in-depth interview, literature review, and triangulation as the method for data collection. Yogyakarta has had instruments for public green open spaces provision called Masterplan Ruang Terbuka Hijau (RTH) Up-Scaling Yogyakarta 2013-2032 which govern the typologies and criteria for green open space development in the city.Public green open spaces development mechanism can be grouped into the planning phase, the utilization phase, and the control phase of each consisting of legal and regulatory aspects, institutional aspects, financial aspects, and technical aspects. The mechanism of green open space provision should regard the need of advocacy for "urban green commons" (UGCs) development as a systematic approach of collective-participatory for urban land management.

1. Introduction

The capacities of adaptation are essential elements towards urban resilience. The provision of open space and public space as a public asset that can minimize negative impacts of climate change in urban areas has to be considered as one of the adaptation [1]. In a resilience framework, one of the criteria for urban resilience is when the city providing green open space as a property of space to maintain the sustainability of socio-cultural, economic, and ecosystem [2]. Investment in urban areas through the provision of diversity biodiversity is believed to be a form of mitigation of natural space to anticipate environmental change phenomenon.

The rapid economic growth in Yogyakarta City, especially in the manufacturing sector, accommodation and restaurants, as well as information and communication led to the city underwent a spatial transformation quite rapidly [3]. The phenomena of the development of hotels, restaurants and cafes, as well as shopping center have ignored the existence of urban public space. The availability of public green open space still very poor with only fulfilled 17% of the 20% minimum requirement of spatial policy [4]. This condition is quite worrying for the fulfillment of public open space, especially public green open space is important to improve the resilience of an urban area through the development of nature-based solutions [5].

Yogyakarta City has made efforts to increase the availability of public space with provision program of green open spaces to support the activities of urban life. The program is one of the seven priorities of the programs included in the environmental and spatial utilization aspect. In 2016, Yogyakarta City should have developed some locations as a green open space [6], [7]. Nonetheless, the challenges for the program implementation is inevitable. The government's action to increase the number of public green open spaces is the right step to improve the quality of life and environment of Yogyakarta. The process of providing public green open spaces are complex activities in which the mechanism is governed by specific policies, involving the various interested parties, and other activities that mutually influence each other. An understanding of the processes in the supply public green open spaces in Yogyakarta City is important to do because the process takes place in the development activity will determine the successful implementation of the development plan. Therefore, efforts to understanding the patterns of integrated and comprehensive in the provision of public green open spaces in Yogyakarta City is performed in this study.

1.1 Objectives

The purpose of this study is to identify the provision of green space in the city of Yogyakarta which plays a role in the realization of the resilience of the city. The objectives are as follows: (1) to identify the provision green space in Yogyakarta City from the aspects of product and procedure; and (2) to identify the role of green space to build resilient city.

1.2 Methods

This study utilized descriptive exploratory qualitative approach to collect the required data and analysis. By using this approach, the researchers had the opportunity to explore and gather information using combined techniques. The methods used to collect data are in-depth interview, literature review, and triangulation. The results will be discussed through description and explanation in accordance with the aims.

2. Result and Discussion

2.1 Plan for City's Green Open Space

Yogyakarta City with an area of 3250.01 Ha has existing green space area of 1028.79 Ha (31.65%) (Fig. 1) with details of public green space covering an area of 399.81 ha (12.30%) and private green space covering an area of 628.98 Ha (19.35%). The overall area of green space and in particular private green space has exceeded the standart that has regulated by the Law No. 26/2007 on Spatial Planning and the Regulation of the Minister of Public Works No. 05/PRT/M/2008 on Guidelines for the Provision and Use of Green Open Space in Urban Area with minimum 30% of provision for green open space with proportion of public green open space is 20% whereas private green open space is 10%. Therefore, in the case study of Yogyakarta City, the provision of public green open space is still lacking to the contribution of green open space availibility in Yogyakarta City. Among 14 districts in the city of Yogyakarta, there are six districts that the existing of green open space has an area of more than 30% of the area of these districts. The sixth districts are District of Gondokusuman, Kotagede, Mantrijeron, Tegalrejo, Umbulharjo and District Wirobrajan. Moreover, District Gondokusuman apart from the compliance with the minimum size of the existing green open space it also has been meeting the public green space area that has more than 20% of the total area. As for the private green open space almost all districts have fulfilled the area of more than 10%, except for three districts that are District Gondokusuman, Kraton and Ngampilan. [8]

Yogyakarta City Government's efforts in improving the quantity and quality of green open space particularly for public green open space has started since 2005 through land acquisition in each district in the city. Until 2012 this program has acquired land 29 units open space with total area of 1,4 Ha which aim for various community activities in the future. To support the procees of urban green open space provision the Masterplan Ruang Terbuka Hijau (RTH) Up-Scaling Yogyakarta 2013-2023 which contains the provision and utilization of green open space plan has been appoved. The plan consists of the stipulation and the plan of green space development in each district. From the document, green open space in Yogyakarta will be developed into some typologies as follows (Fig. 1): (1) Residential Building; (2) Settlement Area; (3) Special Areas; (4) City's Gate Area; and (5) Public Green Open Space in the District. Each typology has criteria of vegetation selection, standard area, and functions for the development of green open space . [8]

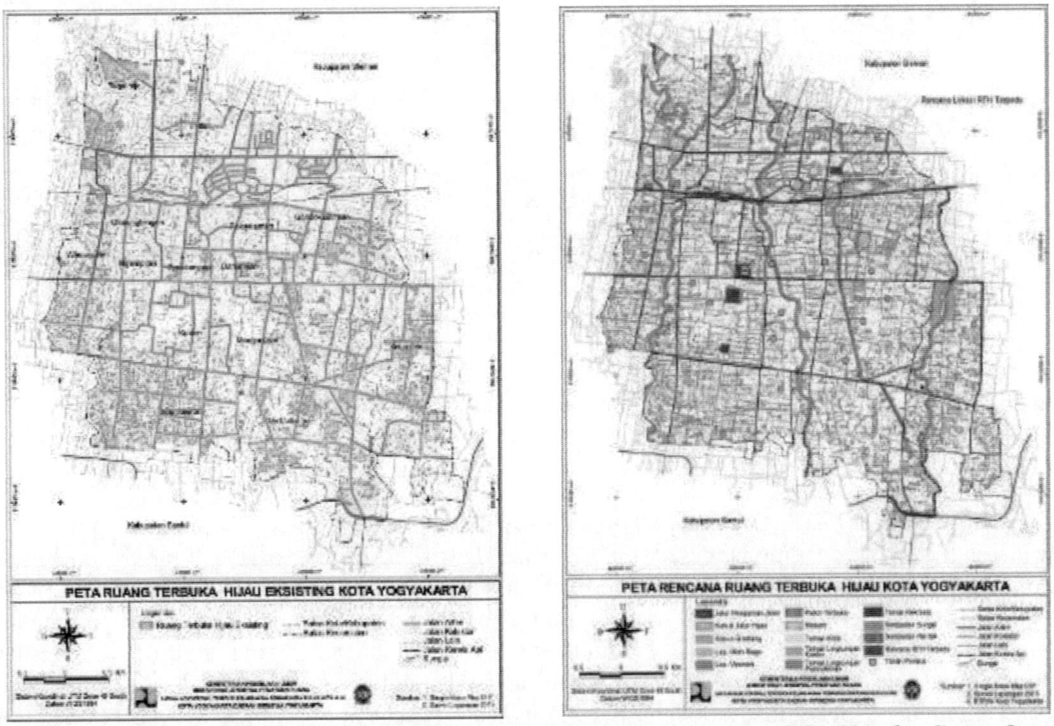

Figure 1. Map of: (a) The Existing Condition of Green Open Space and (b) Plan for Green Open Space in Yogyakarta City [9]

Table 1. Typologies Green Open Space Place in Yogyakarta City

Typology	Details	Vegetation	Area	Functions
Residential Building	Courtyard	Tree with wide canopy Shrub Aestetic vegetation	Depend on Building Coverage Ratio (BCR)	Private
	Offices, shops, and place of business			Public and/or private
	Roof garden	Shrub Ground cover	Depend on the availibility of the roof, terrace, balcony or building canopies	Private

Typology	Details	Vegetation	Area	Functions
Settlement Area	Park/Garden	Tree with wide canopy Fruit Plants Flowering Plants	Depend on the level of neighborhood	Public
	Lane road	Indigenous tree Vines Potted plants	Depend on characteristic of the road and pedestrian The avalibility space under overpass	Public
	Certain Function (riverside, rail line, cemetery)	Tree which suitable for birds ecosystem Productive plants/crops	In accordance with the regulation of the function	Public
Special Areas	Green space for Kraton area	Vegetation that has philosophy and historical feature	Depend on the space that need to be conserve	Public and Private
City's Gate Area	Areas of city's border	Kepel (*Stelechocarpus burahol*) (Decree of the Governor of Yogyakarta Special Region No. 385 / KPTS / 1992 on Establishment of Identity Flora And Fauna Region of Yogyakarta Special Province)	The planned area	Public
Public Green Open Space in the District	Open space in each of district	Depend on the utilization	In accordance with the target of fullfillment of green open space in each district	Public and Private

2.2 The Mechanism of Public Green Open Space Provision

To achieve the ideal conditions based on the spatial aspects of the plan for green open space development, the main mechanism for masterplan implementation, particularly for public green open space provision, tend to emphasize and can be grouped into stages of Planning Phase, Utilization Phase, and Control Phase (Fig. 2). Within the whole process in the mechanisms of green open spaces development, the elements contained in each phase can be distinguished into several aspects include the legal and regulatory aspects, institutional aspects, financial aspects, and technical aspects of operational utilization. The authorized public institutions in the city that run the mechanism namely Regional Planning and Development Agency (Bappeda), Section for Governance and Social Welfare of Regional Secretariat (Sekda), and the Environmental Agency of Yogyakarta (BLH). Bappeda mainly responsible for the planning and control phase whereas Sekda and BLH mainly responsible for the utilization phase, including land acquisition for Sekda and design, development, and maintainance for BLH. [8]

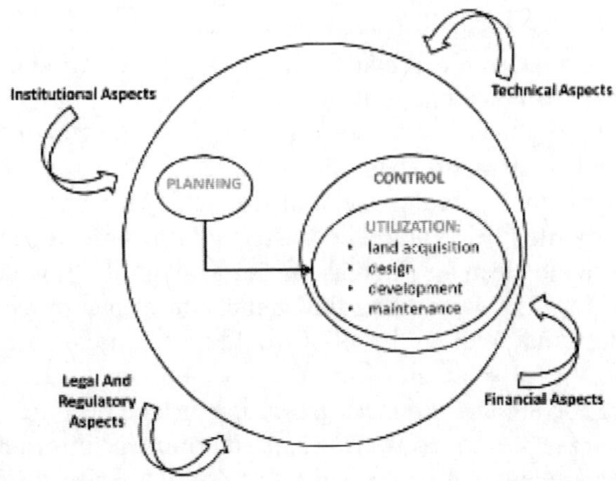

Figure 2. The Mechanism of Green Open Space Development in Yogyakarta City (Adapted) [8]

Table 2. Mechanism Aspect of Green Open Space Provision in Yogyakarta City

Mechanism Aspects	Planning Phase	Utilization Phase	Control Phase
Legal and Regulatory	• Yogyakarta City Regulation No. 2/2010 on Spatial Planning of Yogyakarta • The Minister of Public Works Regulation No. 05 / PRT / M / 2008 on Guidelines for the Provision and Use of Green Open Space In Urban Area	• Yogyakarta Mayor Regulation No. 64/2013 on Request, Procurement and Land Utilization for Public Green Open Space as Supporting Facilities for Community Events (Reference for land acquisition) • The Minister of Public Works Regulation No. 05 / PRT / M / 2008 on Guidelines for the Provision and Use of Green Open Space In Urban Area (Reference for design)	• Yogyakarta Mayor Regulation No. 64/2013 on Request, Procurement and Land Utilization for Public Green Open Space as Supporting Facilities for Community Events • The Minister of Public Works Regulation No. 05 / PRT / M / 2008 on Guidelines for the Provision and Use of Green Open Space In Urban Area • Utilization Permit
Institutional	• Government (Bappeda and other sectoral agency); • Community; and • Private Sector (Planning Consultant)	• Government (Bappeda, BLH, and other sectoral agency); • Community; and • Private Sector (Contractor)	• Government (Bappeda, BLH, and Sekda); • Community; and • Private Sector (Supervisor)
Financial	Indicative budget plan from the regional level (APBD)	Indicative budget plan (APBD) from the regional level for the construction and stimulant fund for the management	Indicative budget plan from the regional level (APBD)
Technical	Masterplan development conducted by planning consultant under contracts with Bappeda.	• The land acquisition process • Technical design process • The development process The utilization and maintenance process	Monitoring and evaluation for the whole process

2.3 Green Open Space for Resilient City in Yogyakarta

Urban green space is part of sustainable urban form typologies which knowns for its aims to bring "nature into the city" [10] and offers a unique landscape that supports biodiversity [5] which will give benefits for urban citizens. From the Green Urbanism perspective, new approaches to urbanism should more incorporate the ecological design to shape the urban form with some of principle among others ecological limits, design for nature, circular metabolism, self-sufficiency, sustainable lifestyles, and neighborhood and community life [11]. Natural ecosystems in the green open space have an important role and benefits such as provide clean air, increase water availability, regulate the urban climate, and create place making in the city [5]. Maintaining the connection of nature with urban environment is a fundamental, therefore understanding of the benefits would be significant to develop sustainable urban development.

As part of major criteria for urban resilience, green open space become one of urban components that have to develop to increase the capacity of urban environment in order to deal with the risk of climate change and disaster. Green open space as the form of ecological diversity in the city is part of strategies of resilience through "nurturing conditions for recovery and renewal after disturbance" [12]. Related to the transition toward resilient city, Yogyakarta City has been in the process of increasing its adapation capacity with the planning and development of green open space. Particularly in the sense of product development, Yogyakarta City with its master plan of green open space has established its foundation for urban green infrastructure development whereas the implementation of policy still has challenges from the mechanism of the provision.

Nonetheless, the provision of green open space still has not developed within the frameworks of urban resilience both for product and the process. The success of green open space provision still only seen by the target of area availaibility with precentage of 30% of total city's area without considering the need to analyze the environment dynamics which could increase the risk of a city to climate change and disaster impacts, to create green open space that can modify for use in a different purposes, and to build management tools of green open space that contribute to the disaster readiness in the future. Therefore, the future green open space development should considering the potential action to build resilience including the planning, the design, and the managament action with principle of flexibility, adaptability, and agility particularly with the global climate change and disaster issues [12].

The insight of common property system within urban management which dealt with what so-called as Urban Green Commons (UGCs) and the potential of the system to build urban resilience is also discussed [13]. The potential of UGCs to manage and to integrate the cultural and biological diversity in the city would be beneficial to develop particularly public green open space. With distinguished institutional characteristic and community participation, UGCs propose a systematic approach of collective-participatory for urban land management that could contribute to the public green open space provision. The UGCs itself rely on the background of how diversity has role to resilience management. Diversity in the urban setting will contribute to the adaptive capacity with its functions to spread the risk, creates buffers, and allow multiple strategies to repurposed and reorganize complex urban system in order to deal with the urban environment changes. Thus, with its fundamental background UGCs could be used as an approach as well as tool to develop product of green open space and to manage the procees of the development with accentuation in the integration of socio-economic function in the context urban resilience. The examples of UGCs, i.e. collectively managed parks, community gardens, and allotment areas, describe how UGCs focus on to build institutional capacity and socio-ecological as well as cultural insight and memory for urban citizens so that there would be massive civic participatory on the public green open space which not only support the ecological system but also socio-cultural and economic system in urban area.

In particular, the finding of this study related to the role of community participation for green open space provision. From the institutional perspective, the Regulation of the Minister of Public Works No. 05/PRT/M/2008 on Guidelines for the Provision and Use of Green Open Space in Urban Area regulate the need of participation of each urban stakeholder. One of which is legal institution as a non-governmental organization that play a major role for connecting between the community, government, and private sector in overcoming the communication gap and information as well as improving people's access to resources [8]. The procedure of green open space development in Yogyakarta has not been involving and optimizing the advocacy aspect that could contribute to the development of green open space. The concept of urban commons property system would be an alternative to bridge the complexity of green open space provision with more collective-participatory of urban land management. With the role of the legal institution which has concerns with the environmental issues there should be more systematic procedure to mediate and to facilitate between relevant stakeholder in terms of communication, conflicts resolution, learning process, space utilization, and law enforcement..

3. Conclusions

Yogyakarta has had instruments for public green open spaces provision called Masterplan Ruang Terbuka Hijau (RTH) Up-Scaling Yogyakarta 2013-2032 which govern the typologies and criteria for green open space development in the city. Public green open spaces development mechanism can be grouped into the planning phase, the utilization phase, and the control phase of each consisting of legal and regulatory aspects, institutional aspects, financial aspects, and technical aspects. Nonetheless, the provision of public green open space still has not appropriately involving the needs of adaptation aspects for the city related to the potential risk of environmental changes in the future. Therefore, urban greenspace development in Yogyakarta City in the future should have more robust framework which include the risk management issues. Moreover, ss some insights have conducted, the mechanism of green open space provision should regard the need of advocacy for "urban green commons" (UGCs) development as a systematic approach of collective-participatory for urban land management.

4. Acknowledgement

Part of this paper is derived from the research of Lenon (2015) while completing bachelor degree in Sekolah Tinggi Teknologi Nasional Yogyakarta. The author express gratitudes to Sekolah Tinggi Teknologi Nasional Yogyakarta for all the support while finishing this paper.

References

[1] C. M. Pickering and W. Hill, "Impacts of recreation and tourism on plant biodiversity and vegetation in protected areas in Australia," *J. Environ. Manage.*, vol. 85, pp. 791–800, 2007.

[1] Vargas-Moreno, J.C.; Meece, B.; Emperador, S. 2014. "A framework for using open green spaces for climate change adaptation and resilience in Barranquilla, Colombia", Proceedings of the Resilient Cities 2014 Congress, 5th Global Forum on Urban Resilience and Adaptation, Bonn, Germany.

[2] Syarifi, Ayyoob and Yamagata, Yoshiki. 2014. "Resilient urban planning: Major principles and criteria", Energy Procedia, 61, 1491 – 1495.

[3] Badan Pusat Statistik (BPS). 2015. Statistik Daerah Kota Yogyakarta 2015, Badan Pusat Statistik Kota Yogyakarta.

[4] http://berita.suaramerdeka.com/ruang-terbuka-hijau-kota-jogja-masih-di-bawah-standar/, acessed 16 Maret 2016.

[5] Van Ham, Chantal. 2014. "Urban green infrastructure: making visible what is valuable", Proceedings of the Resilient Cities 2014 Congress, 5th Global Forum on Urban Resilience and Adaptation, Bonn, Germany.

[6] http://www.harianjogja.com/baca/2015/09/28/penataan-kota-jogja-ruang-terbuka-hijau-ditambah-646738, accessed 16 Maret 2016.

[7] http://www.antaranews.com/berita/528936/yogyakarta-bangun-12-ruang-terbuka-hijau-tahun-depan, acessed 16 Maret 2016.

[8] Lenon, Steven. 2015. "Identifikasi Mekanisme Pengembangan Ruang Terbuka Hijau Publik (RTHP) Studi Kasus Kota Yogyakarta", Skripsi, STNNAS, Yogyakarta.

[9] Anonymous. 2013. Masterplan Ruang Terbuka Hijau Up-Scalling Kota Yogyakarta Tahun 2013-2032, Pemerintah Kota Yogyakarta.

[10] Jabareen, Y. (2012). "Planning the resilient city: concepts and strategies for coping with climate

change and environmental risk", Cities, 31: p.220-229.

[11] Beatley, Timothy. 2000. "Green urbanism: Learning from European cities," in Jabareen, YR. 2006. "Sustainable urban form: Their Typologies, Models, and Concept', Journal of Planning and Education and Research, 26, 38-52.

[12] Desouza, K.C., and Flanery, T.H. (2013). "Designing, planning, and managing resilient cities: A conceptual framework", Cities, 35, 89-99.

[13] Colding, Johan & Barthel, Stephan. 2013. "The potential of 'Urban Green Commons' in the resilience building of cities", Ecological Economics, 86, 156–166.

IOP Conf. Series: Earth and Environmental Science **70** (2017) 012059 doi:10.1088/1755-1315/70/1/012059

Channelling urban modernity to sustainable pro-poor tourism development in Indonesia

R Prasetyanti

Public Administration Department, University of 17 Agustus 1945 Jakarta, Sunter Permai Raya – Jakarta, Indonesia

retnayuprasetyanti@yahoo.com

Abstract. Sustainable urban planning and development requires not only a fast-growing economic growth and modernity, but also social equity and environmental sustainability. Meanwhile, the global goals of sustainable development have fascinatingly set a promising urban development future by enhancing ecology based pro-poor policy program. Apparently, pro-poor development agenda has led to the notion of pro-poor tourism as part of urban development strategies on poverty alleviation. This research presents Jakarta Hidden Tour and Kampung Warna-warni as certain cases of pro-poor tourism in Indonesia. By the emergence of criticism on "pro-growth" paradigm, the critical analysis of this research focuses on the scenario of sustainable pro-poor tourism through eco-cultural based Kampung-Tour development. In accordance, debates and dilemma have been continuously arising as pros and cons regarding the ethical issues of poverty alleviation based Kampung-Tour development. Nevertheless, this paper tries to redefine Slum Kampung as potential; the writer wildly offers a concept of poverty alleviation by reinventing pro-poor tourism strategy; revitalizing slum site to eco-cultural based pro-poor tourism development as an embodiment of a sustainable urban development. By holding system thinking analysis as research method, sustainable pro-poor tourism highlights the urgency community based tourism and eco-tourism so that poverty alleviation based tourism can be tangibly perceived by the poor. In this sense, good local governance and public private partnership must be enhanced, it is due to, like any other development projects; sustainable pro-poor tourism needs a strong political commitment to alleviate urban poverty, as well as to pursue a better future of sustainable nation.

1. Introduction

In the era of uncertainty, Indonesia, as well as the developing nations in Asia is facing two major urban development problems which are poverty and environmental degradation. Indonesia has witnessed a transformational process associated with the drastic changes in economic stability, socio-cultural spheres, administrative and political reform, and environmental quality. In this context, poverty alleviation has become the leading agenda of international development which highly supported by the global deal of Sustainable Development Goals (SDGs) [7].

Tourism is one of the world's largest industries, generating an estimated 11% of global Gross Domestic Product (GDP), employing 200 million people and transporting nearly 700 million international travellers per year – a figure that is expected to double by 2020. International tourism arrivals in developing countries have grown by an average of 9.5% per year since 1990, compared to

4.6% worldwide. The tourism industry makes important contributions to the economies of developing countries, particularly to foreign exchange earnings, employment, and GDP [11].

Tourism sector as one of vital parts of development strategies has increasingly catalysed the growth of national income. The trend of modernization has encouraged the implementation of the principles of Sustainable Development Goals in urban development process. However, the overwhelming growth of economic development has been hectically risking the stability of socio-culture and ecology sectors. Meanwhile, the breakthrough led by eco-tourism and community based tourism has transformed the ideas of urban tourism into a brand new tourism development [3]. These tourism approaches promote eco-friendly tourism activities and community involvement in tourism sector. Yet, eco-tourism, somehow, practically depend on natural beauty/resources as the main tourism attraction, whereas community based tourism development does not always put the ecology aspect as main preference. The major point is whether two types of those tourisms can share benefit to the poor; and how much benefit the poor can tangibly perceive. In addition, due to a globalising and fast-changing world, tourism for the poor meets the need of sustainable urban development in the framework of pro-poor policy [13]. It needs to be noted that, in many cases, urban poverty has gained massive popularity due to the dynamic impacts on the dimensions of economy, socio-culture, politic, and ecology. Moreover, the complexity of urban poverty has turned pro-poor policy into a more creative poverty reduction strategy through pro-poor tourism. Indeed, the connection between pro-poor tourism and Pro-poor policy is close; as a means of poverty alleviation, pro-poor policy has been massively set to overcome the problem of urban poverty – which commonly caused by socio-economy disparity. Nevertheless, debates over pro-poor tourism came up with basic doubt on its effectiveness in reducing poverty and maintaining environmental sustainability.

There has been little consideration of the whole range of impacts on economic sustainability by not incorporating the principle of sustainability into the objectives of poverty elimination. To this extent, the current pro-poor tourism agenda has missed the potential to enhance job opportunities for the poor and sustainability for the environment. As a result, authorities, especially government institutions must promote sustainable pro-poor tourism, defined as tourism that generates net benefits for the poor; encompasses economic, socio-cultural and environmental sustainable benefits, at the same time as aiming to empower the poor rather than merely expanding the profit for the business sector.

Indonesia has experienced pro-poor tourism in some different programs. Jakarta Hidden Tour is one "wild" form of pro-poor tourism developed by private sector. The concept of tourism highlights the idea of social critics on disproportional urban development in the capital city. The attraction ultimately bases on the uniqueness of urban poverty to be exposed as tourism object. Kampung Warna-warni in Malang is another type of pro-poor tourism inspired by Santa Martha and "Favela Painting" in Rio de Janeiro, Brazil. In the perspective of sustainable pro-poor tourism, these types of pro-poor tourism are expected to give economic benefits to the poor and maintain sustainability through community based tourism. Thus, the creation of tourism activities should be sustainably managed by community engagement incorporated with market industry and local authorities. Besides, the uniqueness of each Kampung must highlight indigenous values as local wisdom and heritage. In accordance to the attaining effort of sustainable pro-poor tourism, this research is conducted to display comprehensive analysis on pro-poor tourism in Indonesia by examining practices in Jakarta and Malang.

This paper assumed that pro-poor tourism in Indonesia is growing, and that this trend would show significant progress to achieve sustainable pro-poor tourism. Further, the emphasis is how pro-poor tourism's contribution on sustainable development could be enhanced. The research's problems emphasized on: 1) conceptual and theoretical analysis of pro-poor tourism and pro-poor policy; 2) comparative analysis on pro-poor tourism in Jakarta and Malang (forced field analysis); 3) scenario (system thinking) analysis of the establishment of sustainable pro-poor tourism through Eco-cultural based Thematic Kampung-Tour Development (Causal Loop Diagram, including the input-output diagram).

2. Literature Review

The term of "pro-poor" is embedded into the approach of poverty alleviation. Pro-poor policy is a set of explicit actions which aims to share equity for the poor. It responds to the fact that disproportional benefit of development often perceived by the poor – unemployment, low cost labour; minority group; and women [10], therefore, the unfavourable consequences on economic growth must be resolved by strategic pro-poor policies program including pro-poor tourism. In the late 1999, Department for International Development (DFID) firstly developed research on pro-poor tourism and presented it to the United Nations with ground breaking of "putting poverty at the heart of the tourism agenda". This global controversial approach has led to the shifting paradigm of poverty alleviation as part of tourism activities. Given the complex impacts that pro-poor tourism has on economy; socio-cultural perspectives; and ecology, pro-poor tourism needs to become a central issue of national and local government. Yet, pro-poor tourism is generally viewed as an engine for economic growth rather than as a mechanism for delivering poverty reduction; and that kind of assumption has especially affected the indebted developing nations [9] . The emphasis of "trickle down" benefits is also criticized not to be as good as what it possibly expected. Benefits of the practice of pro-poor tourism come to the rich first, and later, the poor begin to experience the benefits when the rich start spending their gains. Thus, this benefit mechanism implies that the poor benefit from the economic growth only indirectly through vertical flow from the rich [6]. Moreover, the notion of "trickle down" approach in development was aligned with the "pro-growth" paradigm, a mind-set of economic development. Some doubt pro-poor tourism by revealing that in practice, pro-poor tourism is recognised not always to be "community based tourism", the fact is about what some interest parties in tourism sector do for a living by using the brand of pro-poor tourism.

The rising of modernity; the interconnected developed and developing nations; and the linkage of global tourism and international development agendas are the significant points of the theory of Modernization (the liberal perspective) and the theory of dependency (the critical perspective). These platforms have affections at the evaluation of the development perspectives on tourism and poverty. Liberal perspective views modernization paradigm as the best means to achieve the betterment of development in developing countries. Modernization defines development process as an evolutionary of traditional society to a more modern community. The assumption has clearly directed to the massive support of governments in developing nations on pro-poor tourism to solve unemployment, slum settlement, and insufficient infrastructure. With the intention to take part in global governance, a series of pro-poor tourism is set to be well developed by foreign aid. Modernization theory believes that international assistance may contribute to realize the modern way of life for the poor by embedding sharing on pro-poor tourism development. However, theory of dependency sees pro-poor tourism as a product of liberalization which cannot provide instant answer to development. This strategy precisely has impacts on the environmental sustainability because the main goal of pro-poor tourism is "pro-growth" development for the poor.

The newly emerged pro-poor tourism was also influenced by Neo-liberal economy policies, as stated in Structural Adjustment Programs (SAPs). By the emergence of global development reform, gigantic monetary institutions associated with the international development institutions became active in supporting efforts on economic restructuring in developing nations. In this sense, revolutionary ideas of enterprising government and market oriented government have transformed development platform of government's institution. This has linked the network of pro-poor tourism to a greater global scale, an alteration had dynamically occurred at development perspective regarding pro-poor tourism as one of approaches of sustainable development. As a consequence of the era of sustainability, alternative tourism perspective came out to fulfil the empty link to sustainability in tourism. Alternative tourism can be defined as forms of tourism that set out to be consistent with natural, social, and community values and which allow both hosts and guests to enjoy positive and worthwhile interaction and shared experiences [8]. This theoretical reference somehow strengthened the development of the renewed interest of sustainable pro-poor tourism.

3. Methods

A methodological guideline in this research is qualitative approach and specified into the framework of system thinking analysis as the research method. Qualitatively, system thinking can be used to capture the working of a system as a tool of thinking and understanding [12]. System thinking framework is also used as technical guideline to describe and define the complex process and transformation of sustainable pro-poor tourism with various influencing factors (forced field analysis). The data used in this research are primary data through observation and in-depth interview, and also collective secondary data obtained from various sources of local government official websites and the Indonesia Central Bureau of Statistics (relevant literatures from other sources are also included).

STAGE 1	Problem Recognizing (Who cares, and why)
STAGE 2	Problem Understanding and System Description
STAGE 3	Qualitative Analysis (Forced field analysis - Bright ideas and pet theories)
	↓
STAGE 4	Scenario Analysis (Causal Loop Diagram- The input-output diagram)

Figure 1. *System Thinking* Analysis (Qualitative)
Source: Modified by writer (2017), adopted from [2]

4. Results

4.1 Sustainable Pro-poor Tourism in Urban Area

Urban poverty persistently became major concern of urban development due to its consistency of growing (10,34% and 59,35% poor people live in urban area in the late 2016). By the fact of fast-growing urbanization rate, the imbalance of population growth in urban and rural area caused unequal portion of urban spatial development (the unbalanced growth theory) which critically can affect urban sustainability. Urban development is a complex set of technical and political process to manage infrastructure – as part of modernization process; administer equity through public welfare enhancement for the poor; and maintain sustainability of the environment – including water and air management, and land use. Conceptually, urban tourism is core part of urban development, it is simply because tourism sector adds more value to brand image of a city. However, the real challenge of urban tourism development is to improve the sustainability of pro-poor tourism in such a way that pro-poor tourism as part of pro-poor policy fundamentally supports welfare for the poor and protects the environment. Therefore, sustainable pro-poor tourism (particularly in urban area) means not only as tourism for the poor; it is also acknowledged to be the platform of urban development which combines pro-poor tourism and sustainable development – hereinafter known as sustainable pro-poor tourism based urban development.

Like many other industrial developments, sustainable pro-poor tourism based urban development needs strong political attempt supported by green constitution. In Indonesia, urban tourism is latterly growing fast and unique. Whereas pro-poor tourism has to face modernity and culture at the same time, it has (also) to ensure the environmental sustainability and enable ecofeminism. Simply, core value which comes closest to meeting the public welfare in urban pro-poor tourism is set of principles of sustainable development. Principally, tourism relates very closely to sustainability for formerly, tourism always identically recognised as something beautiful, historical, and unique. Somehow, socio-cultural based tourism, like pro-poor tourism, is now globally phenomenal. In 1992, sustainable development became known worldwide by the United Nations Conference on Environment and Development (Rio Earth Summit). This has led to the wider perspective of development – a significant change in the direction of development which facilitates public participation and sustainability.

The development of the pro-poor tourism approach over those ten years is discussed and is differentiated from eco-tourism and community-based tourism [3]. Pro-poor tourism is broadly

IOP Conf. Series: Earth and Environmental Science **70** (2017) 012059 doi:10.1088/1755-1315/70/1/012059

congruent with such ideas; it places greater emphasis on how tourism can benefit the poor. While eco-tourism focuses on issues such how tourism damages nature, culture and the environment, the response of the local community and investment in tourism [1]. However, pro-poor tourism, eco-tourism and community-based tourism rely on community participation.

Despite the difference of pro-poor tourism and eco-tourism, both of those two types of tourism and (also) sustainable development were embedded into "Agenda 21". This prevented negative repercussions in tourism and development which might cause disproportionality by upholding local, national and international partnership. This prevented negative repercussions in tourism and development which might cause disproportionality by upholding local, national and international partnership. Further explanation of pro-poor tourism and eco-tourism is described in the picture below (Figure 2), meanwhile, detailed analysis of sustainable pro-poor tourism explained in Table 1 below:

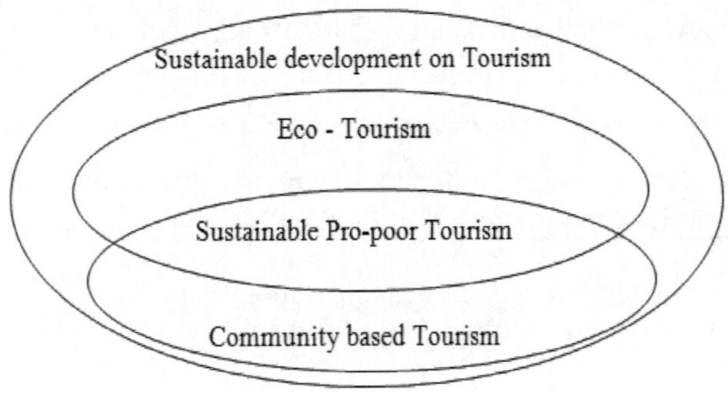

Figure 2. Illustration of Sustainable Pro-poor Tourism Degree
Source: Modified by writer (2017)

Table 1. Analysis of Pro-poor Tourism and Sustainable Pro-poor Tourism

Pro-poor Tourism DOES NOT anti-capitalist; separate from wider tourism systems; a theory or model; a niche type of tourism; specific method; only about the poor; just about hunger and no/ inadequate incomes; only about individual benefits; only for those occupying the "moral high ground".
Pro-poor Tourism DOES focus on incorporating the poor into capitalist markets by increasing job and entrepreneurial opportunities and collective benefits. Like fair trade, it is a form of market intervention, which relies heavily on the private sector; depend on existing tourism structures and markets; orientates research to the net benefits from tourism that can or could accrue to the poor; apply to any kind or type of tourism, including large and small-scale tourism, even if the non- poor also benefit. Can be from regional or national policies or private sector involvement; use numerous methods, none of which are specific to PPT, including value chain analysis, to collect data and show how the poor are and can be further involved in tourism; recognize that the non-poor may also benefit from tourism, even disproportionately. It is less concerned with the relative than the absolute (net) benefits received by the poor; have a broad definition of poverty, including lack of freedom, opportunity, power, skills, and education. It is about development; focus on community benefits; water, sanitation, health, education, infrastructure, (etc.); require wide stakeholder cooperation and commitment, including national and local authorities, planners, the private sectors, (etc.); ideally combining to ensure the poor benefit from tourism
Sustainable Pro-poor Tourism DOES enforces the spirit of Ecocracy (the sovereignty of environment): putting ecology as a core part of governance system; proposes Ecofeminism; promotes local arts and cultures; applies the spirit of Green Economy; balances Economy, Eco-culture, and Empowerment (3E); introduces Poor Public

Partnership and Poor Private Partnership (this indicates that poor people can help providing solution for themselves); redefines the meaning of tourism: a way to share life experience and build empathy; bridging social gaps.

Source: Modified and analyzed by writer (2017); adopted from [4]

4.2 Case Analysis: Pilot Projects in Jakarta (Jakarta Hidden Tour) and Malang, East Java (Kampong Warna-warni)

The model of sustainable pro-poor tourism as poverty reduction program needs to be profound and complex. In accordance with the analysis of Jakarta Hidden Tour and Kampong Warna-warni, there are three models of sustainable pro-poor tourism linked with different impacts for the poor:

Table 2. Analysis of Sustainable Pro-poor Tourism Models on *Jakarta Hidden Tour* and *Kampong Warna-warni*

Direct impacts Model
Jakarta Hidden Tour: Food selling, wages for children who assist and amuse international tourist *Kampong Warna-warni*: service (parking lot); Food selling
Indirect impacts Model
Jakarta Hidden Tour: Tour guides; Translator; Transportation sector *Kampong Warna-warni*: Transportation sector; Local souvenir shop
Dynamic impacts Model
Jakarta Hidden Tour: International tourism development (Trip-Advisor); International aid from group or individual volunteers. Note: *Jakarta Hidden Tour* puts poverty as tourism attraction, therefore, dynamic impact tends to lead to social-economic critics and discourses on development that finally insist local government to reevaluate development and pro-poor tourism strategies. *Kampong Warna-warni*: Environmental preservation (effectively reduced the activity of littering in the river up to 95%, more greenery, clean and healthy environment); International tourism development (Trip-Advisor-one of best tourism destinations promoted by local government); Local economic growth (tourism sector contributes 5,61% to local revenue - it possibly increase to 6-7% per year); Public Private Partnership (local government of Malang and Paint Company *Decofresh*); Community development (community arts group)

The synergy of three models is hardly implemented for most of tourism studies have only emphasized on one model. Yet, the combination of three models might be very effective to mitigate poverty. As a means of sustainable pro-poor tourism, Jakarta Hidden Tour and Kampong Warna-warni need to uphold (firstly) micro strategy in the form of guidance, counseling, stress management, and crisis intervention in order to maximize direct impacts to the poor [5]. Consequently, a future development of Jakarta Hidden Tour and Kampong Warna-warni as pilot project of eco-cultural based Kampong Tour must adopt mezzo strategy with the enhancement of education and training to give more supports on tourism supply – including the advance of community based local economic development. Jakarta Hidden Tour and Kampong Warna-warni also must meet the higher level through macro strategy development which connects the poor to policy-making process and conflict management.

Research result revealed some factual information which is irrelevant with the principles of sustainable pro-poor tourism. Jakarta Hidden Tour might contribute economically, yet, there is no sustainable and tangible impacts perceived by the poor; moreover, governmental support for the program is very poor. Jakarta Hidden Tour meets obstacles to be developed as sustainable Kampong-Tour for there is no active role of community in pro-poor tourism development. However in Malang, Kampong Warna-warni has apparently shared benefits of urban modernity more to the poor. Major principles of sustainable pro-poor tourism are found to be actively implemented in Kampong Warna-warni, thus, a further development of eco-cultural based Kampong Tour can be easily practiced. The achievements of

IOP Conf. Series: Earth and Environmental Science **70** (2017) 012059 doi:10.1088/1755-1315/70/1/012059

sustainable pro-poor tourism elements on Jakarta Hidden Tour and Kampong Warna-warni are briefly explained in Table 3 (below) completed with forced field analysis to describe constraining and supporting factors in Table 4 (below).

Table 3. Applied Principles of Sustainable Pro-poor Tourism on *Jakarta Hidden Tour* and *Kampong Warna-warni*

Sustainable Pro-poor Tourism	***Jakarta Hidden Tour***	***Kampong Warna-warni***
Enforces the spirit of *Ecocracy* (the sovereignty of environment): putting ecology as a core part of governance system	-	✓ (good)
Proposes *Ecofeminism*	-	-
Promotes local arts and cultures	✓ (very poor)	✓ (fair)
Applies the spirit of *Green Economy*	-	✓(poor)
Balances Economy, Eco-culture, and Empowerment (3E)	-	✓(poor)
Introduces *Poor Public Partnership* and *Poor Private Partnership*	✓(poor)	✓ (fair)
Redefines the meaning of tourism: a way to share life experience and build empathy	✓ (good)	✓ (fair)
Bridging social gaps	✓(poor)	✓ (fair)

Table 4. Comparative Forced Field Analysis of Kampong-Tour Development

Jakarta Hidden Tour		*Kampong Warna-warni*	
Driving Forces	Score	Driving Forces	Score
Uniqueness	2 (sufficient)	Political Support	2 (sufficient)
Marketing management	3 (average)	Access	3 (average)
City stories	4 (significant)	Private support	4 (significant)
Private support	5 (most significant)	*Ecocracy* Social capital	5 (most significant)
Restraining Forces	Score	Restraining Forces	Score
Ecocracy	5 (most significant)	Knowledge management	5 (most significant)
Political Support	4 (significant)	Public facility	4 (significant)
Public facility	3 (average)	Marketing management	3 (average)
Social capital	2 (sufficient)	Integrated empowerment system	2 (sufficient)

4.1. *Scenario of Eco-cultural based Kampong Tour Development*

The result of system thinking analysis concluded five subsystems of Kampong-Tour development system, which are (a) Government's role, (b) Community Development, (c) Economic Development, (d) Public Private Partnership, (e) Quality of Environment. All subsystems are displayed in figures below.

IOP Conf. Series: Earth and Environmental Science **70** (2017) 012059 doi:10.1088/1755-1315/70/1/012059

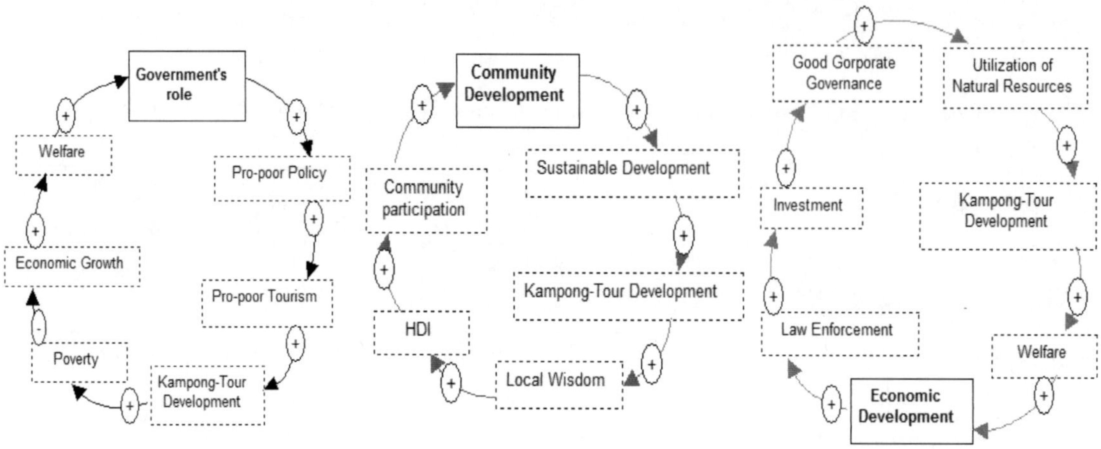

(a) Subsystems of Government's role

(b) Subsystems of Community Development

(c) Subsystems of Economic Development

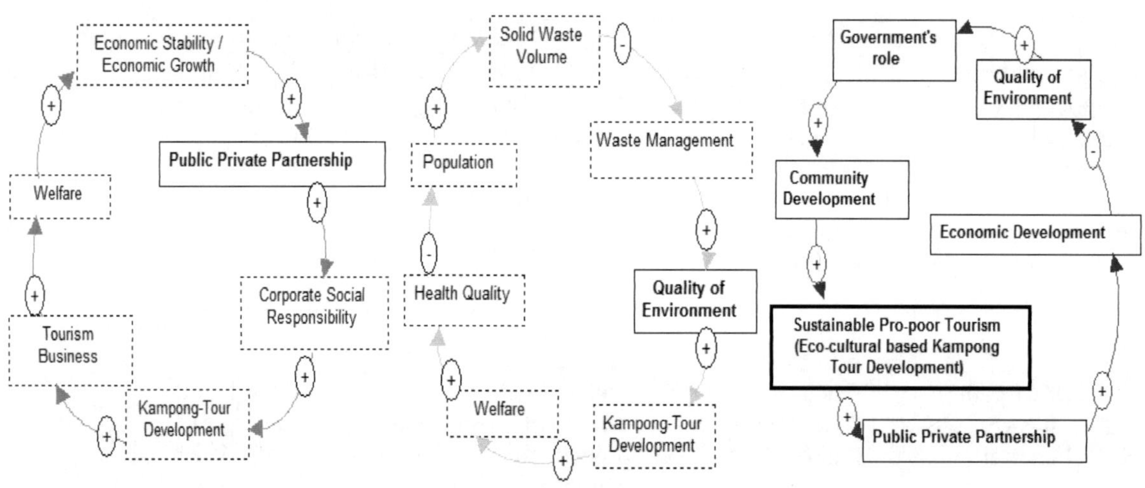

(d) Subsystems of Public Private Partnership

(e) Subsystems of Quality of Environment

(f) System of Kampong-Tour Development System

Figure 3. Subsystems of Kampong-Tour Development System (Causal Loop Diagram)

As a means of sustainable pro-poor tourism strategy, eco-cultural based Kampong Tour development must be commercially realistic, although the private sector cannot be expected to prioritize poverty objective, it must be included in the process of developing pro-poor tourism[1]. Eco-cultural based Kampong Tour development is a concept of urban development in tourism sector which highlights Kampong as tourism site. The idea was inspired from rural development concept "One Village One Product" which thematically promotes local uniqueness. Eco-cultural based Kampong Tour development aims to channel urban modernity to the need of the poor by enhancing pro-poor tourism sector. In addition, by relying on the framework of sustainable development, the principles of ecology and socio-culture are major to be prioritized in Kampong-Tour development. Detailed explanation of inputs and outputs of eco-cultural based Kampong Tour development are explained in Figure 3 (below).

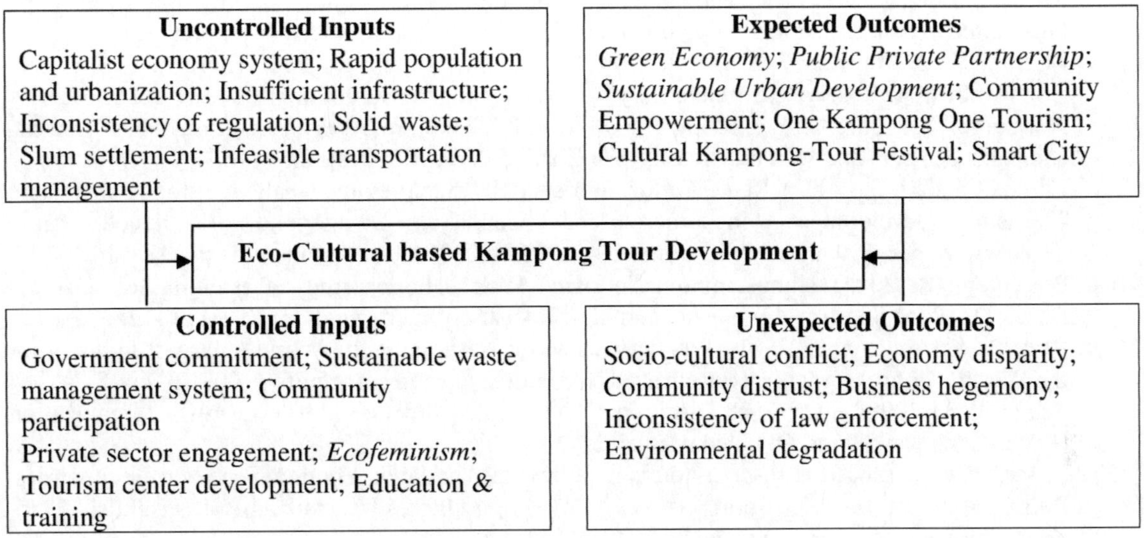

Figure 4. The Input-output Diagram
Source: adapted from [10]

5. Conclusions

Sustainable pro-poor tourism requires the balance of economy, eco-culture, and empowerment (3E) to succeed pro-poor policy program. Sustainable pro-poor tourism considered to be one of effective strategies of pro-poor policy to bridge socio-economy disparity and maintain eco-cultural development in urban area. The analysis of system thinking concludes that sustainable pro-poor tourism may combine community based tourism and eco-tourism. Practically, Kampong Warna-warni is recognized to be more sustainable than Jakarta Hidden Tour. The best supporting factors are social capital and Ecocracy. Direct impacts of Jakarta Hidden Tour are (still) very limited to simple education assistance for children; however Kampong Warna-warni has successfully developed significant benefits on economy and ecology. Research recommendation highlights the enhancement of Good Local Governance and Public Private Partnership in Kampong-Tour development to reinvent sustainable local tourism through community owned business development and Kampong-tour festival. Last, the most important thing in sustainable pro-poor tourism is institutional sustainability which determines the favor of pro-poor policy and urban development future.

References

[1] Ashley, C., Boyd, C., & Goodwin, H. (2000). Pro-Poor Tourism : Putting Poverty At The Heart Of The Tourism Agenda. London, UK.

[2] Coyle, R. G. (1996). System Dynamics Modeling: A Practical Approach. London, UK: Chapman & Hall.

[3] Goodwin, H. (2008). Tourism , Local Economic development , and Poverty Reduction. Applied Research in Economic Development, 5(3), 55–64. Retrieved from http://www.haroldgoodwin.info/uploads/ARED2008goodwin_online_v5n3_propoor_tourism.pdf

[4] Harrison, D. (2008). Pro-poor Tourism: a critique. Third World Quarterly, 29(5 (Tourism and Development in the Global South)), 851–868. https://doi.org/10.1080/01436590802105983

IOP Conf. Series: Earth and Environmental Science **70** (2017) 012059 doi:10.1088/1755-1315/70/1/012059

[5] Ife, J. and Tesoriero, F. (2006). Community Based Alternatives in an Age of Globalization. Australia: Pearson Education.

[6] Kakwani, Nanak, and Ernesto, M. P. (2000). Asian Development Review. Asian Development Review: Studies of Asian and Pacific Economic Issues, 18(1), 1–16. Retrieved from http://citeseerx.ist.psu.edu/viewdoc/download?doi=10.1.1.177.1495&rep=rep1&type=pdf

[7] Manwa, H., & Manwa, F. (2014). Poverty Alleviation through Pro-Poor Tourism: The Role of Botswana Forest Reserves. Sustainability, 6, 5697–5713. https://doi.org/10.3390/su6095697

[8] Newsome, D., Moore, S. A. and Dowling, R., K. (2002). Natural Area Tourism: Ecology, Impacts, and Management (2nd Editio). Bristol, UK: Channel View Publication.

[9] Plüss, C. and M. B. (2002). Red Card for Tourism? 10 Principles and Challenges for a Sustainable Tourism Development in the 21st Century. Switzerland. Retrieved from http://www.unep.fr/shared/publications/cdrom/WEBx0139xPA/statmnts/pdfs/piause.pdf

[10] Prasetyanti, R. (2015). Slum Kampong Tourism " Jakarta Hidden Tour ": Designing Eco -Cultural Based Pro-Poor Tourism. European Journal of Interdisciplinary Studies, 3(3), 111–122.

[11] Roe, D., & Urquhart, P. (2001). Pro-Poor Tourism: Harnessing the World's Largest Industry for the World's Poor. London, UK. Retrieved from http://pubs.iied.org/pdfs/11007IIED.pdf

[12] Senge, P. M. (1994). The Fifth Discipline: The Art and Practice of The Learning Organization (Paperback). New York: Doubleday Business.

[13] Srisantisuk, S. (2015). Pro-poor tourism policy in Thailand. London School of Economics. Retrieved from http://etheses.lse.ac.uk/3146/1/Srisantisuk_Pro-poor_tourism_policy_in_Thailand.pdf

IOP Conf. Series: Earth and Environmental Science **70** (2017) 012060 doi:10.1088/1755-1315/70/1/012060

The Perception of Visitors towards the Level of Satisfaction on Park (Case Study: Singha Merjosari Park Malang)

R Priadaniswari

Department of Urban and Regional Planning, Faculty of Engineering, Universitas Brawijaya, Malang, Indonesia

rpdw.daniswari17@gmail.com

Abstract. Park is one of the public spaces which is used by people to get happiness and comfort. Singha Merjosari Park is one of the parks in Malang that is functioned as a recreational and educational park for the citizen. In weekends and national holidays Singha Merjosari park get visited by so many visitors. But if we see the reality, there are still some problems regarding visitor satisfaction. Also, there are attributes that has performance levels decrease that will become another new problem. The purpose of this study is to analyze the perception of visitors about the level of visitor satisfaction and what attributes that need to be improved and developed by managers in the future. The approach method in this research is descriptive quantitative. Primary data is based on measurement and observation. The number of samples used is 100 respondents referring to the number of samples determination by Slovin formula with the sample selection used is accidental sampling technique. The analysis technique used is Importance Performance Analysis (IPA) and Costumer Satisfaction Index (CSI). Based on the results of IPA analysis, the things that should get important attention and should be improved is the aesthetics of lighting, cleanliness of parking area and toilet, shade in park area, and availability of clean water. While the result of CSI value analysis is 65,30%. This means visitors are satisfied, but visitors are still not satisfied overall. Implications or changes that should be given is the aesthetics of lighting should be more creative and become the identity of the park (for example, lamp lanterns should be suitable with the concept of the park). Also, the change of toilet look so that visitors can enjoy the look and it can be iconic (toilet concept according to local culture of Malang) and the prevalence of lighting in the park area at night.

1. Introduction

The public spaces are important elements in the city that aims to perform several activities with specific interests to interact (Stephen Carr, 1992) [1]. Meanwhile public space according to the Project for Public Spaces in New York in 1984, is a form of space used by human to perform several activities such as roads, pedestrian, parks, plazas and etc.

Park is a fenced plot of land which is used for fun, excitement, and comfort (Laurie, 1986) [2]. Meanwhile, according to (Djamal, 2005) [3], the park is a plot of land that is open for public with a certain area, in which there are trees, shrubs, bushes and grasses that can be combined with the creation of other materials. (Hariyono, 2007) [4] states that the park is part of the public sphere that should be accessible to different kind of citizen. Taman Singha Merjosari is located on Jl. Mertojoyo Selatan, Merjosari Village, Lowokwaru Subdistrict, Malang City.

Malang is the second largest city in the East Java, Indonesia. It is located in the south of Surabaya with 90 km of distance. Malang area is 110.06 km2 and divided into five districts. The population of Malang in late April 2016 reachs 887.443 people (Department of Population and Civil Registration - Malang). Population growth in Malang is about 1.58 percent each year. Malang is one of five cities that have the cleanest air in Asia. This result is released by research institute of non-government organization (NGO) and Cities Clean Air Partnership Program (CCAP). This award is a shared commitment between the government and all citizens of Malang to continue to preserve and beautify the face of Malang with the existence of parks.

Therefore, the assessment of the visitors perceptions is one way to see the level of satisfaction of the park. This assesment is done to improve the quality of the park. This study is conducted to answer the question "What is the level of satisfaction at Singha Merjosari Park?". The purpose of this study is to reveal how the visitor's perception of the satisfaction level at Singha Merjosari Park and also, visitors' desire to Singha Merjosari Park in the future.

2. Methods

The The research method is scientific ways to get the valid data, with the purpose to be found, developed and proven, a certain knowledge that in turn can be used to understand, solve and anticipate problems (Sugiyono, 2008) [5].

In this study, the method used is descriptive quantitative. The method of data collection in this study uses primary data which is a questionnaire to the Singha Merjosari Park Malang visitors.

The use of sampling techniques in this study is accidental sampling. The sampling technique is based on chance, that means anyone who by chance met with researchers can be used as a sample, if the person suits as a source (Sugiyono, 2008) [5]. In this study the number of samples used are 100 respondents. This refers to the determination of the number or size of the sample that uses the Slovin formula: $n = \frac{N}{N(d)^2+1} = \frac{133}{133\,(0.05)^2+1} = 99{,}81$, rounded to 100.

Garden satisfaction parameter reviewed from the factors that affect the level of satisfactory. The factors that affect the satisfactory, which are: aesthetics or beauty, cleanliness, security, circulation, aroma, climate regulation or climatological, and utility.

In analyzing this study, researcher used the Importance Performance Analysis (IPA) and Costomer Satisfaction Index (CSI) analytical techniques. According to (Tjiptono, 2011) [6] IPA technique first introduced by Martilla and James (1977) in their "Importance-Performance Analysis" article. In this technique, the respondents were asked to rate the level of importance and performance of the company, then the value of the average level of interest and performance is analyzed on the Importance-Performance Matrix, which the x-axis represents the perceptions or performance, while the y-axis represents the expectations or interests. There would be results obtained from the value of x, y (of each variable) in four quadrants on a Cartesian diagram. The interpretation of the quadrant is quadrant I (Keep Up the Good Work); Quadrant II (Possible Overkill); quadrant III (Low Priority) and quadrant IV (Concentrate Here).

Customer Satisfaction Index (CSI) is used to determine the level of overall satisfaction with an approach that considers the importance of the attributes of a product or service. According to (Suryawan, 2013) [7], customer satisfaction is determined by customer perception on performance of products or services that meets customer expectations. Customers will be satisfied if they fulfill their expectations thoroughly. According to (Irawan, 2003) [8], a measure of CSI is required because the results of the measurements can be used as a reference to determine the improvement of services goals for respondents overall satisfaction level and it can be seen on the criteria of customer satisfaction levels. CSI obtained from (T divided by 5Y) x 100%. The score 5 (on 5Y) is the maximum value that is used on scale of measurement. CSI formula as follows:

$$CSI = \frac{T}{5Y} x100\%$$

Information:

CSI : Customer Satisfaction Index
T : The total number of variable S
Y : The total number of all the variable P
S : Multiplication of the I and P (each variable)
I : Average Satisfaction or Performance (each variable)
P : Average interest (each variable)

Overall satisfaction level of respondents can be seen in Table 1 below:

Table 1 Criteria of Customer Satisfaction

No.	CSI value	Customer Satisfaction Index
1.	0,81 – 1,00	Very Satisfied
2.	0,66 – 0,80	Satisfied
3.	0,51 – 0,65	Satisfied Enough
4.	0,35 – 0,50	Less Satisfied
5.	0,00 – 0,34	Not Satisfied

3. Results

3.1 Importance Performance Analysis (IPA)

The level of conformity is the result of comparison between the level of visitor satisfaction (x) with the level of interest (y). Here is table of average satisfaction and interest in Singha Merjosari Park Malang as follows:

Table 2. The average of satisfaction and interest in Singha Merjosari Park Malang

No.	Factors	Satisfaction		Interests		Bobot (X)	Bobot (Y)
		Nilai	Rata-rata	Nilai	Rata-rata		
1.	Aesthetics						
	Plants	381	3.8	430	4.3	0.05	0.05
	Supporting facilities – Lights	299	3.0	426	4.3	0.04	0.05
	Supporting facilities – Gazebo	368	3.7	436	4.4	0.05	0.05
2.	Cleanliness						
	Park Area	350	3.5	448	4.5	0.05	0.05
	Parking Area	304	3.0	430	4.3	0.04	0.05
	Trash can	341	3.4	444	4.4	0.05	0.05
	Toilet	274	2.7	423	4.2	0.04	0.05
3.	Security						
	Playground	346	3.5	424	4.2	0.05	0.04
	Activities in park area	367	3.7	420	4.2	0.05	0.04
	Parking area	338	3.4	422	4.2	0.05	0.04
4.	Circulation						
	The ease of access to park area	388	3.9	412	4.1	0.05	0.04
	The ease of surrounding park	390	3.9	410	4.1	0.05	0.04
	The ease of parking vehicles	368	3.7	417	4.2	0.05	0.05
5.	The aroma						
	From waterways (ditch)	336	3.4	410	4.1	0.05	0.04
	Trash can	351	3.5	430	4.3	0.05	0.05
6.	Climate regulator (Climatology)						
	The level of shade in park area	277	2.8	455	4.6	0.04	0.05
	The level of wind circulation in park area	368	3.7	408	4.1	0.05	0.04

No.	Factors	Satisfaction		Interests		Bobot	Bobot
		Nilai	Rata-rata	Nilai	Rata-rata	(X)	(Y)
	The facility to take shelter if it rains in park area	284	2.8	412	4.1	0.04	0.04
7.	Utility						
	Availability of wi-fi	175	1.8	342	3.4	0.03	0.04
	Availibilty of power sources	232	2.3	375	3.8	0.03	0.04
	Availibility of clean water	249	2.5	430	4.3	0.04	0.05
	Availibity of communication network (signal)	350	3.5	430	4.3	0.05	0.05
Total		7136	71.5	9234	92.4	1	1
Average			3.3		4.2		

From Table 2 above shows us the average satisfaction and interest in Singha Merjosari Park in Malang. The calculation result will be displayed in the form of Cartesian diagram, which in the diagram, there are quadrants containing about Keep Up the Good (quadrant 1), redundant (quadrant II), a low priority (quadrant III) and the main priority (quadrant IV). The Cartesian diagram Singha Merjosari Park Malang as follows:

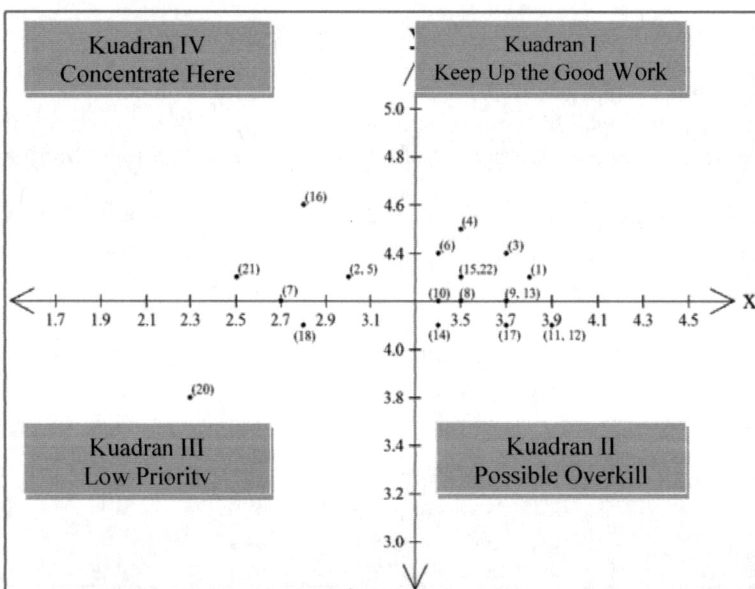

Figure 1. Cartesian Diagram Singha Merjosari Park Malang

In interpreting quadrant on the Cartesian diagram as follows:

1) Keep Up the Good Work

In this quadrant variable is considered as an important factor and should be maintained for consumers to feel satisfied. The variables that must be maintained is:

- Aesthetic factors including: (1) plants or vegetation, and (3) support facility - seat or the gazebo;
- Hygiene factors including (4) area of the park and (6) the trash;
- Safety factors including (8) facility to play, (9) the activities in the park area, and (10) the parking area;
- Circulation factors factors, among other things, (13) the ease to park motor vehicle;
- Aroma or smell factors, among other things, (15) garbage can;

- Utilities factors, among other things, (22) the availability of communications networks - the signal strength.

2) Possible Overkill

In this quadrant variable considered as a factor that is less important, but implementation on the ground is very satisfying. As for the variables are:

- Safety factors including (8) facility to play, (9) the activities in the park area, and (10) the parking area;
- Circulation factors including (11) the ease of access to the location of the park, (12) the ease in the surrounding park, and (13) the ease of parking vehicle;
- Aroma or smell factors, among other things, (14) drains;
- Climate regulation factors including, among other things, (17) the level of air circulation in the garden area.

3) Low Priority

In this quadrant variable considered as a factor that is not too important or not expected by the consumer. As for the variables are:

- Hygiene factors including (7) toilet;
- Climate regulation factors including (18) shelter if it rains in the garden area;
- Utilities factors, among other things, (19) the availability of wi-fi networks, and (20) the availability of electricity network.

4) Concentrate Here

In this quadrant variable is considered as an important factor, but according to the consumer is not satisfying and should be improved. As for the variables are:

- Aesthetic factors including (2) support facility - lighting;
- Hygiene factors including (5) and the parking area (7) toilet;
- Climate regulation factors including (16) the level of shade or shade in the garden area;
- Utilities factors, among other things, (21) the availability of clean water network.

Based on calculations through the use of IPA method, the attributes that need attention are the variables that are in quadrant IV. Based on the results of the analysis, it appears that the variables that need improvement and enhancement to support the people's satisfaction is as follows.

Table 3. Factor analysis of satisfaction and interest of visitor perceptions in Singha Merjosari Park Malang

No.	Variable	Satisfaction Factor	Analysis
1.	Aesthetics	Support facilities – lighting	The design of the lighting in the garden is very monotonous. There needs to be more creative and appropriate lighting design for the park theme.
2.	Cleanliness	Parking area	Cleanliness in the parking area is very dirty because of the large leaves of trees falling. Regular cleaning is required.
		Toilet	Cleanliness in the toilet is very bad because it is less maintanance so that many ceramics are damaged or broken. It is necessary to repair the damaged tile and clean the toilet regularly.
3.	Climate Regulator (Climatology)	The level of shade or shade of the garden	The level of shade or shade of the garden is not enoug sinceh there are many plants. The addition of a shade plant evenly is required
4.	Utility	The availability of clean wate	The availability of clean water in the garden is also needed. The availability of clean water in effective hours is required.
		The lighting	The lighting at night was minimal and less inequality lies the lights throughout the park.

3.2 Customer Satisfaction Index (CSI)

Results of the index calculation of user satisfaction in Singha Merjosari Park Malang can be seen in Table 4.

Table 4. Index of Users Interests and Satisfaction

No.	Factors	Interests (I) Skala 1 - 5	Satisfaction (P) Skala 1 - 5	Score (S) (S) = (I) x (P)
1.	Aesthetics			
	Plants	4.3	3.8	16.34
	Supporting facilities – Lights	4.3	3.0	12.9
	Supporting facilities – Gazebo	4.4	3.7	16.28
2.	Cleanliness			
	Park Area	4.5	3.5	15.75
	Parking Area	4.3	3.0	12.90
	Trash can	4.4	3.4	14.96
	Toilet	4.2	2.7	11.34
3.	Security			
	Playground	4.2	3.5	14.70
	Activities in park area	4.2	3.7	15.54
	Parking area	4.2	3.4	14.28
4.	Circulation			
	The ease of access to park area	4.1	3.9	15.99
	The ease of surrounding park	4.1	3.9	15.99
	The ease of parking vehicles	4.2	3.7	15.54
5.	The aroma			
	From waterways (ditch)	4.1	3.4	13.94
	Trash can	4.3	3.5	15.05
6.	Climate regulator (Climatology)			
	The level of shade in park area	4.6	2.8	12.88
	The level of wind circulation in park area	4.1	3.7	15.17
	The facility to take shelter if it rains in park area	4.1	2.8	11.48
7.	Utility			
	Availability of wi-fi	3.4	1.8	6.12
	Availibilty of power sources	3.8	2.3	8.74
	Availibility of clean water	4.3	2.5	10.75
	Availibity of communication network (signal)	4.3	3.5	15.05
	Total	Y = 92.4		T = 301.69

$$CSI = \frac{301.69}{5 \, x \, 92.4} x100\% = 65{,}30\%$$

Based on the results of data processing and analysis can be concluded that the Singha Merjosari Park visitor satisfaction is achieved, but overall visitors are still "Satisfied Enough". This is proven by the results of the CSI value which is 65.30%. So the improvement is needed in order to increase visitor satisfaction.

4. Conclusions

Based on calculations using the IPA, there are attributes that need to get the attention which are the attributes contained in quadrant IV. Factors contained in quadrant IV are considered important and in poor condition. Based on the results of the analysis, it appears that the attributes that need improvement and people satisfaction enhancement is as follows:

- aesthetic factor which is the variable of support facilities - lighting

IOP Conf. Series: Earth and Environmental Science **70** (2017) 012060 doi:10.1088/1755-1315/70/1/012060

- the cleanliness factor which is the variable of parking area and toilets
- climate regulator factor of which is the variable of level shade or shade in the garden area
- the utility factor is the variable of clean water availability and lighting

Based on the results of data processing and analysis of the use of Customer Satisfaction Index (CSI) that visitor satisfaction Singha Merjosari Park Malang is achieved, but visitors are still not satisfied overall. This is proven by the results of the CSI value which is 65.30%. Improvement in order to increase visitor satisfaction is needed.

Implications or changes that should be given are:

- Lighting:
  The aesthetics or design of lighting should be more creative and become the park iconic identity, for example, the use of lamp lanterns that is suitable with the concept of Singha Merjosari park and the evenness of the park lighting at night.
- Toilet:

It is necessary to fix the floor in the toilet, also the maintenance should be regularly done. It is also important to change or redesign the look of the toilet so that visitors can enjoy the appearance and it can be an icon (the concept of toilet is according to the local culture of Malang), for example the lion statues (male and female) which shows the different type of toilet use (between men and women) so that the appearance can be more interesting.

References

[1] Stephen carr, M. F. (1992). Public Space. Cambridge: Cambridge University Press.

[2] Laurie. (1986). An Introductory Landscape Architecture. Upper Saddle River, United States: Pearson Education (US).

[3] Djamal, Z. (2005). Tantangan Lingkungan Dan Lansekap Hutan Kota. Jakarta: Bumi Aksara.

[4] Hariyono. (2007). Sosiologi Kota Untuk Arsitek. Jakarta: Bumi Aksara.

[5] Sugiyono. (2008). Metode Penelitian Pendidikan: Pendekatan Kuantitatif, Kualitatif dan R&D. Bandung: Alfabeta.

[6] Tjiptono, F. G. (2011). Service, Quality & Satisfaction, Edisi 3. Yogyakarta: Andi.

[7] Suryawan, S. d. (2013). Analisa Hubungan antara Experential Marketing, Customer Satisfaction Dan Customer Loyalty Cafe Nona Manis Grand City Mall Surabaya. Manajemen Pemasaran.

[8] Irawan, H. (2003). Indonesian Customer Satisfaction. Jakarta: PT Gramedia Pustaka Utama.

IOP Conf. Series: Earth and Environmental Science **70** (2017) 012061 doi:10.1088/1755-1315/70/1/012061

Lessons learnt from and sustainability assessment of Indonesian urban kampong

Surjono, Antariksa, and M Ridhoni

Brawijaya University, Jl. M.T. Haryono 167, Malang and 65145, Indonesia

surjono@ub.ac.id

Abstract. The search of good (sustainable) model of development has been evolved from the era of City Beautiful to current era of uncertainty. Many urban planner and designers practice western modern concept and design, which are not entirely suitable for developing countries. It has been identified that local forms and characteristics in the urban context have wisdom and content that sustainable for local condition. This paper supports the argument that local characteristics have the qualities of sustainable development by evaluating sustainable development indicators and promoting method in generating aggregate indicator of sustainable development qualities of urban kampong in Indonesia. Fuzzy logic approach, which is widely used in system control design, is promoted in formulating the aggregate indicator of sustainable kampong in Indonesia. The result show that Indonesian kampongs are good in compactness, density, and access. Other indicators will also show sustainable development quality if the lacks are improved.

1. Introduction

Kampong (Indonesia/Melayu) is often associated with slum within urban areas, similar to favelas in Brazil or ghettos in western countries. A kampong, therefore, is perceived to carry negative aspects and treated as a slum. Kampongs became places for improvement, upgrading, and renewal or even demolition [1][2]. Kampong Improvement Program (KIP) in Indonesia, launched in 1969, was proclaimed by UN-Habitat as the world's first slum upgrading program. Urban designers and planners have not been inspired by kampong and its forms in creating new residential areas, even for low income communities. Imported ideas from western and other 'modern' developments are more popular to them. Positive aspects of kampongs are overridden by the negative perception that kampong is slum.

There are two issues related to urban kampong in Indonesia. Firstly, social and physical characters of urban kampong convey sustainable development principles. Secondly, approaches to measure level of sustainability of urban kampong through an aggregate indicator. Descriptions of Indonesian kampong's characters can counterweigh negative perceptions of the 'Indonesian slums', several authors identify and explore positive aspects of kampong [3][4]. These authors claimed that Indonesian kampongs are successful in managing aspects of sustainable settlement, i.e. vitality and livability. The authors focused the discussion on social characters of kampong, the so called rukun (social harmony) and gotong royong (mutual cooperation) in creating public places. Rahmi et.al., [4] underlined that Indonesian kampongs possess unique characters that cannot be described into other terms such as slum, squatter, marginal or low income settlement. She continuously argues that kampongs best describe Jane

Jacobs' humanistic social environment that she called as 'zoning for diversity', i.e. urban places that characterized by short blocks, mix-uses, many corners and street corridors defining the sidewalks [5].

The physical characters of 'zoning for diversity' are possessed by Indonesian kampongs. Characters that guarantee vibrant and livable environment which are required for urban sustainable development. Recent literatures of sustainable development view that sustainability of urban environment may be achieved through several qualities: (1) urban compactness, including density [6][7][8][9][10][11][12]; (2) diversity, including mixed-use [7][10][11]; (3) accessibility [6][7][9][11]; (4) identity [6]; (5) environment [7][8][9][10][11].

Social and physical character of kampong, in terms of their positive aspects, are actually one of the key poins of Sustainable Development Goals (SDGs), i.e. 'sustainable communities and cities' [13]. Effort to promote sustainable urban dwelling becomes more important since 6 in 10 people will be urban dwellers while today almost one third of urban dwellers in developing countries live in slums [14].

Aggregate indicators to measure social aspects of sustainable development are familiar, such as Human Development Index (HDI) and Multidimensional Poverty Index (MPI) [15]. In terms of community, Putnam's Social Capital Index is perhaps the best example [16]. Sustainable development indicators of physical aspects of design and planning, however, are currently evaluated partially in terms of its diversity, accessibility, and environment. Several methods that combine several indicators, such as spider model, pentagon prism model, Systeme de Evaluation des Logement (SEL), and meta-regression analysis, are not designed to formulate a composite indicator from selected individual indicators.

This research was aimed to respond these two issues, i.e., to identify physical quality of Indonesian Kampong for several sustainable city indicators, and to formulate an aggregate indicator, which is a composite of physical and social indicators, for sustainable settlement criteria. We use Indonesian urban kampong in Malang City as a case.

2. Methods

Major Two steps analysis were utilized in this research. The first step was qualitative descriptive analyses. Qualitative descriptive analysis was used to measure quality of sustainable characters of urban kampong in Klojen District. Six variables were employed in this step: Compactness, accessibility/permeability, diversity, density, poverty, and community participation. The first four represent physical/environmental qualities, the fifth represents economic capabilities and the last represents social capital.

2.1. Environmetal or Physical aspects

Dispersion index or compactness index [17] was used to measure the efficiency in urban utility provision, the scale is ranging from 0 to 1, the closer to 1 the more compact. When the shape is simple (one patch), calculation of compactness index for 2D shapes may use the following formulas:

$$Ci = \frac{Di}{Di'} \text{ and } Di = 2\sqrt[2]{\frac{Ai}{\pi}} \text{ or } CILP = \frac{2\pi\sqrt{S/\pi}}{p} \tag{1}$$

Where

Ci	= compactness index
Di	= diameter of a circle with the same area of area i
Di'	= the longest distance between two points of area i
Ai	= the area of i [18]
Or	
CILP	= compactness index of largest patch
s and p	= area and perimeter of largest patch [19]

To measure connectivity, alpha index is a node-linkage association which measures the ratio of number of circuits relative to maximum number possible in a network [20]. The alpha index (α) is calculated as follows:

$$\alpha = \frac{e - v + 1}{2v - 5}$$

where e = number of edges and v = number of nodes. The α values range from 0 to 1.

Diversity in urban landscape is often measured by the variety of land uses. Recent authors stated that mixed land use, which enabled pedestrians to walk to many destinations of diverse land uses in close proximity, is more desired than segregated land uses which are implicated in health and transportation problems [21]. Diversity of land use is often measured by entropy index (EI).

$$EI = \sum_{i=l}^{N} K_i \, . \log(\frac{1}{Kl})/\log(N)$$

where EI is Entropy Index; K is area of land; N is number of land uses.

Building density, as the fourth indicator of physical settlement quality was simply measured by division of numbers of building by area. The density was clasified into four: slum (more than 100 buildings/Ha; high density (80-100 buildings/Ha); medium density (30-<80 buildings/Ha); low density (less than 30 buildings/Ha). The classification is based on SNI 1733-2004 (Indonesia National Standard) of urban settlement environmental planning mechanism.

2.2. *Economic capabilites*

Poverty and community participation measures were utilized to measure social aspects in the study area. Three basic poverty measures were employed in the research: Poverty head count index (Po), poverty gap index (P1) and poverty severity index (P2). General formula to measure poverty is=

$$P_a = \frac{1}{N} \sum_{i=1}^{q} \left(\frac{z - yi}{z} \right)^a$$

where :

Pa = Poverty head count index (if a=0), or Poverty gap index (if a=1), or Poverty severity index (if a=2); N is total number of population; q is total number of poor population who are living at or below poverty line; z is poverty line; and yi is the income of the poor individual i [22].

2.3. *Social capital*

Social capital comprises of trust, norms, and networks, and in Indonesian case, at the village / kampong level, it is related to gender empowerment [23]. In this research, women were prioritized in the sampling process to evaluate the affiliation network as part of social capital [24]. Social sustainabiliy of urban settlement can be measured from community participation in the community's institution. Rate of participation (RoP) is obtained from sum of diagonal value of adjacency matrix devided by number of institutions.

$$RoP = \frac{sum\ of\ diagonal\ value\ of\ adjacency\ matrix}{number\ of\ institution}$$

where:

Adjacency matrix (A) = I x I' = I(n.m)xI'(m.n). The result of RoP then will be compared to total number of institutions. The closer the value of RoP to total number of institutions the higher community participation to local institutions.

The second step was to formulate a composite index of sustainability of the selected kampong. In this step we inputed discrete indicators obtained from step one. Eventhough the indicators were not so comprehensive but, at least, they represented three legs of sustainable development i.e. environmental, economic, and social aspects. In formulating a composite index from a set of indicators, fuzzy logic approach, using Matlab-Fuzzy Logic Toolbox, was utilized in the research. The framework is described in figure 1.

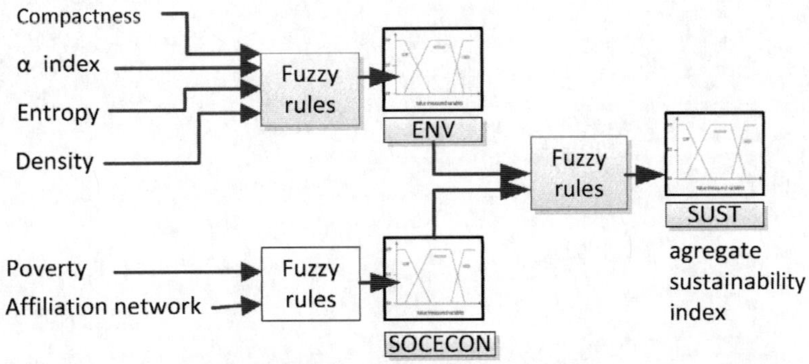

Figure 1. Framework of the model.

Expert judgement was used in the 'if – then' fuzzy rules. Assigned normalization for compactness, α, entropy, poverty, and affiliation network indices were classified into 0 – 0.33, 0.34 – 0.66, 0.67 – 1 as low, medium, and high respectively. Particular for poverty, in contrast to other indicators, high means bad. Density was classified into < 30 houses/Ha, 30 – 80 houses/Ha, and >80 houses/Ha as low, medium, and high respectively. Normalization for final aggregate index classified level of sustainability into five: 0.81–1 (high); 0.61-0.8 (medium high); 0.41-0.6 (medium); 0.21–4 (medium low); and 0-0.2 (low sustainability).

3. Results

The analysis resulted in two type of indicators: discrete indicators of urban kampong which shows each indicator of physical entities found in the three kampong and composite indicators as a result of combination of discrete indicators into single composite indicator.

3.1. Discrete indicators of urban kampong

Three kampongs were evaluated: Arab, Pecinan, and Kebalen Kampongs (figure 1). Two kampongs are ordinary urban kampong: Arab and Pecinan kampongs were named for the majority of population, while the third kampong is Kebalen, whre most of the dwellers are native people. These three Kampongs are located in Klojen District (central district of Malang City).-

Figure 2. The study area: A) Pecinan, B) Arab, C) Kebalen Kampong

The evaluation of discrete indicators of these three *kampongs* are described in table 1.

Table 1: Discrete Indicators for Physical, Economic, and Social Aspects

Kampongs:	compactness	A-index	entropy	density	poverty	affiliation network
Arab	0.73 (good)	0.63 (moderate)	0.76 (good)	0.63 (moderate)	0.05 (good)	0.93 (good)
Pecinan	0.63 (moderate)	0.34 (moderate)	0.17 (bad)	0.58 (moderate)	0.17 (bad)	0.57 (moderate)
Kebalen	0.58 (moderate)	0.36 (moderate)	0.26 (bad)	0.10 (Bad)	0.82 (bad)	0.30 (bad)

In terms of sustainable urban form, table 1 shows that Arab Kampong meet criteria of sustainability. Four indicators of urban form show high indices. Social and economic indicators are also high. Overall evaluation shows that Arab Kampong was the highest, then followed by Pecinan Kampong and Kebalen Kampong. Kebalen kampong was the lowest, majority of indicators were low (bad) and today Kebalen is classified as a slum settlement by the Local Government of Malang. The result shows the case in Klojen District in Malang City. In other places the results may be different. It cannot be inferred that between sustainability indicators and race have relationship. However, overall indices prove that kampongs, to some extent, have met sustainability criteria of urban settlement.

3.2. Composite Indicators

To formulate a composite indicator, inference system based on fuzzy rule was utilized. Fuzzy rules use 'If-then' languages. Based on figure 1, three steps of inference processes were conducted. The first inference system was to formulate a composite indicator of physical aspect. The second one was for a

composite indicator of social-economic aspect, and the last was for an aggregate sustainability indicator of urban kampong.
The inference system for physical aspects was based on 81 permutations of if-then rules (table 2) based on formula nr = 34 where n is the number of objects that can be selected and r is the amount that must be selected.

Table 2: Fuzzy rules for physical aspects (urban form) indicator.

No	IF compactness' is...	AND α-index is...	AND Entropy index is...	AND density index is...	THEN the Physical composite indicator is... (obtained from Delphy analysis)
1	good	good	good	good	good
2	good	good	good	moderate	good
3	good	good	moderate	good	good
etc.	-	-	-	-	-
80	good	bad	bad	bad	bad
81	bad	bad	bad	bad	bad

The inference of 'then' from the result of combination of four indicators (independent variables) was based on Delphi analysis. Defuzzyfying of inputed data resulted in a composite indicator (crisp number) of physical aspects of the form of the three kampongs (table 4).
Composite indicators of social-economic aspects were generated with the same process. Fuzzy rules of social economic aspects consist of 9 possibilities ($n^r=3^2$) (table 3).

Table 3. Fuzzy rules for Social Economic indicator

No	IF poverty is...	AND affiliation network is ...	THEN the social – economy composite indicator is ...(obtained from Delphy analysis)
1	good	good	good
2	good	moderate	good
3	moderate	good	good
4	moderate	moderate	moderate
5	bad	good	moderate
6	good	bad	moderate
7	bad	moderate	bad
8	moderate	bad	bad
9	bad	bad	bad

Defuzzification of the composite indicators of social economic aspects results in crisp number is shown in table 4.

Table 4: Defuzzyfication of Composite indicators for physical aspect (urban form) and social economic aspects)

No	Kampong	Physical aspect (urban form)	Social – economic aspects
1	Arab Kampong	0.520	0.740
2	Pecinan Kampong	0.474	0.612
3	Kebalen Kampong	0.440	0.445

The last process was to formulate an aggregate indicators from the two composite indicators above. The inference system or the rule base to define the level of sustainability was based from Delphi method obtained from key experts in urban development. As in social economic indicators, eight possibilities (table 5).

IOP Conf. Series: Earth and Environmental Science **70** (2017) 012061 doi:10.1088/1755-1315/70/1/012061

Table 5: Rule base of urban kampong sustainability

IF physical sustainability is....	AND social – economic sustainability	THEN the aggregate sustainability indicator is....
good	good	High
good	moderate	Medium – high
moderate	good	Medium – high
moderate	moderate	Medium
good	bad	Medium
bad	good	Medium
bad	moderate	Medium – low
moderate	bad	Medium – low
bad	bad	Low

The determination of high, medium-high, medium, medium –low, and low is based on the set membership. The equation of levels of high sustainability is based on the increasing line of the curve, while for low sustainability is on the opposite way. and the membership of medium- high, medium, and medium-low is based on the equation of triangle area of the curve. The equations of the membership are as follow:

High sustainability = $0.5 \leq X$; $(X – 0.5) / (1\text{-}0.5)$; $x \leq 1$

Medium – high sustainability = $0.5 \leq X$ or $X \leq 1$; $(X – 0.5)/(0.75 – 0)$; $(0.5 – X)/(1\text{-}0.5)$

Medium sustainability = $0.25 \leq X$ or $X \leq 0.75$; $(X – 0.25)/(0.5 – 0)$; $(0.5 – X)/(0.75 – 0.5)$

Medium – low sustainability = $0 – X$ or $X – 0.5$; $(X – 0) / (0.25 – 0)$; $(0.35 – X)/(0.5 – 0.25)$

Low sustainability = $0 \leq X$; $(0.2 – X)/(0.2 – 0)$; $X \leq 0.2$.

Defuzzifying of two composite indicators, physical and social-economic indicators, into an aggregate indicator of sustainable kampong indicates that Arab kampong is more sustainable than the other two kampongs (table 6).

Table 6: Sustainability level of the three kampong in Malang

Kampong	Score	Sustainability level
Arab	0.621	Medium - High
Pecinan	0.548	Medium
Kebalen	0.457	Medium

4. Conclusion

Most of discrete indicators of sustainability in the three kampongs in Malang were good and fair. Bad indicators were found at the entropy of Pecinan and Kebalen Kampongs, while bad indicators of poverty and participation level found in Kebalen Kampong. The evaluation of discrete indicators shows that kampongs are good examples of sustainable settlements in the form of their compactness, density, and access. The composite indicators of their environment were scored in the range of 0.44 to 0.52, while the composite indicators of social-economic aspects were in the range 0.445 to 0.740. Finally, the aggregate of all indicators were in the range from medium to medium – high.

From the results, we promote that this mechanism of assessment can be improved to involve more comprehensive indicators. The good characters of Indonesian urban kampong must be explored further more to provide models for the new local post-modernism planning in the era of new urbanism. These characters of sustainable kampong can be expanded into practices at larger scale in urban context. Of course, more research must be accomplished in determining the 'acceptable zone' of compactness, density, entropy, access and other indicators of sustainable / livable place for all scale of urban development from micro, mezo, to macro settlement.

References

[1] Darrundono, D. A., & Tirtamadja, B. A. (2000). Kampong Improvement Program III. International Workshop on Kampong Improvement Program : Lessons Leading to Strategies for the Future (p. 169 pages). Jakarta: Provincial Government of Jakarta.

[2] Wang, D. D. (2016). Urban Villages in the New China: Case of Shenzhen. NY: Palgrave Macmillan.

[3] Rahmi, D. H., Wibisono, B. H., & Setriawan, B. (2001). Rukun and Gotong Royong: Managing Public Spaces in an Indonesian Kampong. In P. (. Miao, Public Places in Asia Pasific Cities (pp. 119-134). Honolulu: Kluwer Academic Publishers.

[4] Evers, H.-D., & Korff, R. (2003). Southeast Asian Urbanism: The Meaning and Power of Social Space. London: Lit-Verlag.

[5] Jacobs, J. (1961). The Death and Life of Great American Cities. NY: Random House.

[6] Ye, H., He, X. Y., Song, Y., Li, X., Zhang, G., Lin, T., & Xiao, L. (2015). A sustainable urban form: The challenges of compactness from theviewpoint of energy consumption and carbon emission. Energy and Buildings, 90-98.

[7] Wu, J. (2006). An analysis of sustainable urban form in Dalian, China. Thesis. Waterloo, Ontario, Canada: University of Waterloo.

[8] Holden, E. (2004). Ecological footprints and sustainable urban form. Journal of Housing and the Built Environment, 19(1), 91-109. Retrieved 04 12, 2016, from http://www.jstor.org/stable/41107246.

[9] Masoumi, H. E. (2014). A Theoretical Approach to Capabilities of the Traditional Urban Form in Promoting Sustainable Transportation. Theoretical and Empirical Researches in Urban Management, 9(1), 44-60.

[10] Mousavi, S. M., & Nazarian, A. (2013). Urban Form and Sustainable Development: The Case Study of Tehran City. 2nd International Scientific Conference on Economic and Social Development (pp. 592-600). Varazdin: VADEA.

[11] Chen, H., Jia, B., & Lau, S. S. (2008). Sustainable urban form for Chinese compact cities: Challenges of a rapid urbanized economy. Habitat International(32), 28-40.

[12] Jabareen, Y. R. (2006). Sustainable Urban Forms Their Typologies, Models, and Concepts. Journal of Planning Education and Research, 26-38. doi:DOI: 10.1177/0739456X05285119

[13] United Nations. (2015). Transforming Our World : The 2030 Agenda For Sustainable Development. NY: UN. Retrieved from sustainabledevelopment.un.org

[14] United Nations. (2016). The Sustainable Development Goals Report 2016. NY: UN.

[15] De Muro, P., Mazziotta, M., & Pareto , A. (2012). Composite Indices for Multidimensional Development and Poverty: An Application to MDGs Indicators. Social Indicator Research, 1-18. doi:doi:10.1007/s11205-010-9727-z.

[16] Webster, R. (2013). The DImensions of Social Capital. Thesis. Orlando, Florida, USA: University of Central Florida.

[17] Kotharkar , R., Bahadure , P., & Sarda , N. (2014). Measuring Compact Urban Form: A Case of Nagpur City, India. Sustainability, 4246-4272. doi:10.3390/su6074246.

[18] Li, W., Goodchild, M. F., & Church, R. L. (2013). An Efficient Measure of Compactness for 2D Shapes and Its Application in Regionalization Problem. Tempe: Arizona State University.

[19] Gradinaru, S. R., Ioja, C. I., & Patru-Stupariu, I. (2015). Do Post Socialist Urban Areas Maintain Their Sustainable Compact Form? Romanian Urban Areas as A Case Study. Journal of Urban and Regional Analysis, 129-144.

[20] Rodrigue, J.-P. (2013). The Geography of Transport System. New York: Routledge.

[21] Brown, B., Yamada, I., Smith, K., Zick, C., Kowaleski-Jones, L., & Fan, J. (2009). Mixed land use and walkability: Variations in land use measures and relationships with BMI, overweight, and obesity. Health & Place, 1130–1141. doi:10.1016/j.healthplace.2009.06.008.

[22] Haughton, J., & Khandker, S. (2009). Handbook on Equality and Poverty. Washington: The World Bank.

[23] Surjono, Prasisca, Y., & Sutikno, F. (2015). Gender Equality and Social Capital as Rural Development Indicators in Indonesia (Case: Malang Regency, Indonesia). Procedia - Social and Behavioral Sciences, 370-374. doi:doi: 10.1016/j.sbspro.2015.11.048.
[24] Carrington, P., Scott, J., & Wasserman, S. (2004). Models and Methods in Social Network Analysis. New York: Cambridge University Press.

Characteristics utilization of public space in Padang City based on good public space index

T Eriawan

Urban and Regional Planning Department, Faculty of Civil Engineering and Planning, University of Bung Hatta Padang, West Sumatra, Indonesia

tomi.visi@gmail.com

Abstract. Padang is a metropolitan city which has 40 units of open space like plansum park, playground, and sports park with 10,88 hectar of total area. These open spaces are publics', but not all of them are able to use as active public open spaces it caused by some of the parks are dormant parks beacuse they only for planted ground not for public activites.This study conducted to assess the quality of public space that exsist in Padang; Imam Bonjol park (representing of city open space (go-green)), H. Agus Salim sport center (representing of sports park) and Pantai Muaro Padang area (representing of recreation/tourism).The method of this research conducted in several stages, first was the identification toward space's function (Carmona,2008) and assess the space utilization index based approach to public space index (Metha,2007). The spaced quality is measured based on variable in Good Public Space Index which is the intensity of use, social activities, duration of activities, variation and pattern of use. The urgency of this research related to the effort to evaluate the policy of space arrangement and space control through the assessment of the use of existed spaces, until it can be references to Padang goverment in improving and increasing the quality of space thorugh development of spaces that have high quality.The assesment of quality of public open spaces in Padang conducted by assessed five space quality variables, result of analysis, utilization of public space was 0.69. based on the result it can be concluded the quality of public space in some parks in Padang has moderate index quality. This is caused by the time differentiation of the society, there is no interaction among society, and short time duration of the society in visiting the public space.

1. Introduction

Most of public space in padang are dormant, so they become obstacle to the society to do activites in open spaces. Generally, there are 3 public space that used by the society like Iman Bonjol Park, Agus Salim sport center, and Muaro Padang for recreation area which located in downtown include sub-district of west Padang and east Padang.

Open space of Imam Bonjol is in Padang downtown in east Padang sub-district and become an icon of Padang. Based on physycal condition Imam Bonjol space include to soft space which is dominated by plats, grasses then the available facilities such as football field, playground, and parking lot.

Agus Salim sports center located in west Padang sub-district. This is called hard space which one of open space but does not based on go-green areas, the facilities are field, swimming pool,plaza, parking lot, and another sports facilities.

Another area is used by the society as hangout and place to do activites is in Muaro Padang space located in Samudra street (west Padang sub-district). This is recreation/tourism areas in along side of Padang Beach (4 km) then the availabe facilities like parks in the edge of the beach and food streets.

The purpose of this research was to measuring the quality of public space for the people, because many of it's made by government is not used frequently. It's used for passive activity. How to attract the use of public space by increasing the facility diversity so that there is diverse activities in the public space from the different age or different class.

This research used samples of 3 main open spaces in Padang; Imam Bonjol Park (representing open space (go-green) in Padan), Agus Salim sports center (representing sports park) and Pantai Muaro Padang (recreation/tourism).

2. Methodology

Urgency of the related research by evaluating the policy of space arrangement and control thourgh exsisted spaces and as the references for Padang goverment in enhancing the space quality through development of quality spaces.

Based on the function of city space it can be divided into several typologies (Carmona, et al: 2008):

a. Positive space. This is a public space that can be used for positive activites and managed by the goverment. This is natural, semi natural, public space and open public space.
b. Negative space. A public space that cannot be used as a public activities because it has no support for society to do activities comfortly and securely and un managed condition. This is movement space, sevice and abandoned spaces because un managed planned
c. Ambigous space. This is used for transition activities of the society like recreation space, dept stores, cafe, religious site, etc.
d. Private space. This is has a private owner, for instance open private space, backyard, terrace.

The relation with social activities performance in public space, Metha (2007) uses some variables to measure and arrange "Good Public Space Index" as follows:

Good Public Space Index (GPSI) is a method that describes how a public space for the community to use meaningful social activity as well as the characteristics of the outdoor space as an approach (Parlindungan 2013). GPSI variables and measurement method, comprising:

a. *Intensity of use (IU)*

This variable is explained by the number of people engaged in activities in park, by dividing the average number of people with the greatest number of the average person who looks beyond. With formula:

$$IU = \frac{the\ average\ number\ of\ people}{largest\ number\ of\ people} \qquad (1)$$

b. *Intensity of social use (ISU)*

This variable is explained by the existence of the group in park. The group occurs when there are at least two people involved in the same activity. This variable is measured by dividing the number of people involved in the group of the largest number of people involved in park. With formula:

$$ISU = \frac{the\ average\ number\ of\ people\ in\ the\ group}{the\ highest\ number} \qquad (2)$$

c. *People's duration of stay (PDS),*
This variable is explained by the duration of the people to carry out activities in park. This variable is measured by dividing the average time used with the longest time used the space. With formula:

$$\text{PDS} = \frac{\textit{Average time}}{\textit{the longest time}} \quad (3)$$

d. *Temporal diversity of use,*
This variable is measured based on the distribution of activity that occurs in a period of observation. This variable was measured by using the method of Simpson's Diversity Index. Data from this assessment is the number of diversity of activities available at each observation period (morning, noon, evening and night).

e. *Variety of use,*
This variable is measured from the variety of activities. This variable was measured by using the method of Simpson's Diversity Index. Data from this assessment is the number of diversity activities.

f. *Diversity of users,*
This variable is measured from the diversity of the characteristics of the user. This variable was measured by using the method of Simpson's Diversity Index. Data from the assessment is the number of diversity characteristics of the user.

$$\text{Simpson's Diversity Index Formula} = 1 - D$$
$$D = \Sigma n(n-1)/N(N-1) \quad (4)$$

3. Results and Discussion

The public space Imam Bonjol park is a city park with mixed activities such as recreation,sports and ano other social activities. With wide range of public 2.5 Ha and supportive location in Padang downtown make it as the highest visited public space to do activities, like to do sports or only recreation.

Imam Bonjol park is dominated by green field that use for sport activities such as football, ceremonials, for pray (Eid Fitri and Eid Adha). The exsistence of the building inside the park is hall as a tribune for the spectators or dais while ceremonials (60 m2), other supportive facilities are lane for walkers, chairs, playground etc. Some facilities are in less managable for instance broken chairs that made of stone, damange the garden lamp.

The activities around Iman Bonjol park like office, trade and service center. Another activity that can be found is the sellers who sell in the sidewalk, it can be seen on Prof. A Yamin street in front of Policy Deputy Padang.

Pantai Muaro Padang area is in the west of Padang downtown (Pasar Raya), as long as Samudra street in west Padang sub-district. It started from Hayam Wuruk and Muara street (south) until Ir. Juanda street (west and east). It covered beach area along for 4 km.

The activities in Pantai Muaro Padang are dominated by recreation/ tourism activities with some spots of tourism center, service (restaurant and hotel), trading (food stall, store), housing and public facility like education and religious facilities.The area which is dominated by recratetion spots, Pantai Muaro Padang is always become the highest visited by the local or international tourist. The available tourism activites like tour of nature, beautiful scenery, culinary (traditional foods), sports (swimming,surfing) and family recreation. Tourism facilities that available in Pantai Muaro Padang like IORA park,Merpati Perjuangan column, Muaro Lasak park,Lapau Panjang Cimpago, and Rumah Makan Ikan Bakar that availble in east part of Samudra street. The public facilities in Pantai Muaro Padang are parking lot, chairs, sidewalk, sidewalk lamp, trash can, worship place, toilet, food stalls, sports equipment renting (bike).

The street line that accross Pantai Muaro Padang Area is colector prime street, 2 lane and 4 column, the wide rage for each of them is 7 untill 10 m and separated by medians of road with broad is 1.5 untill 3 m. The wide of sidewalk (not all sidewalks are able to passed bt walkers) is 2 untill 3 m.

The existence of west side building are dominated by housing, trading, and general public service. Border of the buildings in this area start from 0 until 12 m. The building which doest has border is in flamboyan Baru, Rimbo Kaluang village, around Pasar Pagi and some in Olo village, meanwhile the buildings with GSB untill 12 m is in Belakang Tangsi village and Purus.

The Agus Salim sport center is in Flamboyan , west Padang sub district 5.5 hectar wide range. The Agus Salim sports center is most visited by the society because there are a lot of sports facilities for instance football stadium, teratai swimming pool, basket ball field (2 units), volley ball field, tenis, parking lot, Pujasera (food stalls park).

Despite of Agus Salim sports center is sport facility, it also become a family recreation, like palyground, music show, circus, exhibition, shopping and many more. The public facilities beside sport facilities are sidewalk, parking lot, playground, sidewalk lamp, and trash can.

Based on identification toward public space in Padang which become the study case in this research, then the typology can be arranged based on varibles; wide, characteristics of land surfaces, activities and etc. Based on comparison analysis between form and characteristic of public space which exist in theoretical or law order, so the public space in Padang can be classified based on some variables, as follows:

Table1. Type of Public Space in Padang based Assessment of Multiple Variables

Variable	Public Space Type		
	Imam Bonjol Park	Padang Beach Park	H. Agus Salim Sports Centre
Area Wide (Typology based on area wide)	2.5 hectares (city park)	10 hectares (region park)	5.5 hectares (region park)
Domination of Land Cover (Typology based on pavement)	Grass and plant (soft space)	Asphalt, paving block, sand (hard space)	Asphalt, paving block, grass and plant (hard space)
Activities	Recreation, sport, celebration	Recreation, sport, celebration	Recreation, sport, celebration
Accessibility	Easy to access	Easy to access	limited to a particular facility
Ownership	Public (local government)	Public (local government)	Public (local government)
Form space	Square	Linier/street	Square

The assessment of public space quality in Padang conducted at 3 public spaces, Imam Bonjol Park, Pantai Muaro Padang, and Agus Salim sport center, they are examined by assessed 5 variables of space quality. Observation and measurement in field were conducted in free day (Sunday), it assumed the intensity of visitors were high. The observation performanced for 4 times for each public space,from morning untill night time. The data was the average of two times observation.

Rate index of the quality of public spaces in the city of Padang conducted by assessing five (5) variable quality of the space, as described in section research methods. Observations and field measurements performed on a weekend (Sunday), where it is assumed on the day of the intensity of use of public space is higher than the other days. Observations were made for 4 (four) times in each public space, with observation time from morning till night. The data used is the average number of such observations. The following are observations and field measurements that have been done.

Table2. Intensity of the use of space

Observation Time	Number of visitors		
	Imam Bonjol Park	Padang Beach Park	H. Agus Salim Sport Centre
Morning (07.00-09.00 am)	122	237	244
Noon (11.00 am-01.00 pm)	365	108	122
Evening (03.00-05.00 pm)	227	285	98
Night (07.00-09.00 pm)	12	41	9
Total number of visitors	726	671	473
Average number of visitors	181.5	167.75	118.25
Highest number of visitors	365	285	244
IU	0.497	0.589	0.485

Table3. Intensity of social use

Observation Time	Number of visitor's group		
	Imam Bonjol Park	Padang Beach Park	H. Agus Salim Sport Centre
Morning (07.00-09.00 am)	98	215	115
Noon (11.00 am-01.00 pm)	201	87	98
Evening (03.00-05.00 pm)	189	241	86
Night (07.00-09.00 pm)	2	22	4
Total number of visitor's group	490	565	303
Average number of visitor's group	122.5	141.25	75.75
Highest number of visitor's group	201	241	115
ISU	0.609	0.586	0.659

Table4. People's duration of stay

Observation Time	Duration of use (minutes)		
	Imam Bonjol Park	Padang Beach Park	H. Agus Salim Sport Centre
Morning (07.00-09.00 am)	120	120	120
Noon (11.00 am-01.00 pm)	45	30	60
Evening (03.00-05.00 pm)	90	60	60
Night (07.00-09.00 pm)	60	90	30
Total Durations (minutes)	315	300	270
Average Duration (minutes)	78.75	75	67.5
Longest Duration (minutes)	120	120	120
PDS	0.656	0.625	0.563

Table5. Temporal diversity of use

Observation Time	Imam Bonjol Park			Padang Beach Park			H. Agus Salim Sport Centre		
	N	n-1	n(n-1)	N	n-1	n(n-1)	N	n-1	n(n-1)
Morning (07.00-09.00 am)	7	6	42	7	6	42	7	6	42
Noon (11.00 am-01.00 pm)	3	2	6	2	1	2	6	5	30
Evening (03.00-05.00 pm)	7	6	42	5	4	20	7	6	42
Night (07.00-09.00 pm)	2	1	2	2	1	2	3	2	6
Total (N)	19	15	92	16	12	66	23	19	120
N-1	18	14	91	15	11	65	22	18	119
N(N-1)	342			240			506		
D		0.266			0.271			0.235	
Simpson Index		0.734			0.729			0.765	

Table 6. Variety of use

Activity	Imam Bonjol Park			Padang Beach Park			H. Agus Salim Sport Centre		
	N	n-1	n(n-1)	N	n-1	n(n-1)	N	n-1	n(n-1)
Jogging	20	19	380	40	39	1560	20	19	380
Walking	20	19	380	20	19	380	20	19	380
Bicycle	5	4	20	10	9	90	5	4	20
Sitting	5	4	20	5	4	20	5	4	20
Eating	5	4	20	5	4	20	5	4	20
Play in groups	20	19	380	10	9	90	15	14	210
Sport in groups	25	24	600	10	9	90	30	29	870
Total (N)	100	93	1800	100	93	2250	100	93	1900
N-1	99	92	1799	99	92	2249	99	92	1899
N(N-1)	9900			9900			9900		
D		0.182			0.227			0.192	
Simpson Index		0.818			0.773			0.808	

Table 7. Diversity of User

Type Of User	Imam Bonjol Park			Padang Beach Park			H. Agus Salim Sport Centre		
	n	n-1	n(n-1)	n	n-1	n(n-1)	n	n-1	n(n-1)
Male									
Toddler (0-5)	0	-1	0	0	-1	0	5	4	20
Children (6-12)	5	4	20	5	4	20	5	4	20
Juvenile (13-18)	15	14	210	10	9	90	5	4	20
Adult (19-45)	20	19	380	10	9	90	20	19	380
Elderly (>45)	10	9	90	20	19	380	10	9	90
Female									
Toddler (0-5)	0	-1	0	0	-1	0	5	4	20
Children (6-12)	5	4	20	5	4	20	5	4	20

Type Of User	Imam Bonjol Park			Padang Beach Park			H. Agus Salim Sport Centre		
	n	n-1	n(n-1)	n	n-1	n(n-1)	n	n-1	n(n-1)
Juvenile (13-18)	20	19	380	15	14	210	10	9	90
Adult (19-45)	15	14	210	10	9	90	25	24	600
Elderly (>45)	10	9	90	25	24	600	10	9	90
Total (N)	100	90	1400	100	90	1500	100	90	1350
N-1	99	89	1399	99	89	1499	99	89	1349
N(N-1)	9900			9900			9900		
D		0.141			0.151			0.136	
Simpson Index		0.859			0.849			0.864	

Table8. Conclusion of Public Space Index Padang City

Variable	Index of Variable		
	Imam Bonjol Park	Muaro Padang Beach	H. Agus Salim Sports Centre
Intensity of Use (IU)	0.497	0.589	0.485
Intensity of Social Use (ISU)	0.609	0.586	0.659
People's Duration of Stay (PDS)	0.656	0.625	0.563
Temporal Diversity of Use (TDU)	0.734	0.729	0.765
Variety of Use (VU)	0.818	0.773	0.808
Diversity of User (DU)	0.859	0.849	0.864
TOTAL GPSI	4.174	4.150	4.143
Average GPSI	0.696	0.692	0.690
Category	**Moderate**	**Moderate**	**Moderate**

4. Conclusions

Based on the research that has been examined, the quality of public spaces in case of Padang have an average index of 0.69 means included in category Moderate quality. The quality of public space that has not reached the Good index, allegedly due to lack of diversity in user activity time, less formation of social interaction between users, and the duration of the average time visitors who are still short in exploiting public space. All public spaces in this case, did not show a significant difference in quality, where the value of the index on each is not far adrift and still be in the same range.

References

[1] C. M. Pickering and W. Hill, "Impacts of recreation and tourism on plant biodiversity and vegetation in protected areas in Australia," *J. Environ. Manage.*, vol. 85, pp. 791–800, 2007.

[1] Carmona M., Magalhaes S, Hammond L. Public Space, the Management Dimension. Routledge, Taylor and francis Group, London and New York. 2008.

[2] Eriawan, Tomi. Prinsip Perancangan Taman Kota dan Taman Bagian Wilayah Kota di Kota Bandung, Thesis, Jurusan Perencanaan Wilayah dan Kota, Institut Teknologi Bandung, Bandung. 2003

[3] Eriawan, Tomi. Effective Land Approach for Determination of Urban Land Use Form, Proceeding International Conference of Construction Industry I. Padang. 2012.

[4] Eriawan, Tomi. Konsep Pengembangan Ruang Terbuka Hijau pada Kawasan yang Memiliki Keterbatasan Lahan Pengembangan, Proceeding Seminar Nasional ASPI. Pekan Baru. 2014

[5] Hariadi dan Setiawan. Arsitektur Lingkungan dan Perilaku. Direktorat Jenderal Pendidikan Tinggi, Departemen Pendidikan dan Kebudayaan Republik Indonesia. 1995.

[6] Mehta V. A Toolkit For Performance Measure of Public Space. 43rd ISOCARP Congress. 2007.

[7] Parlindungan, Johannes. Good Public Space Index, Teori dan Metode. Research Centre of Public Space Laboratory of Urban Design Department of Urban and Regional PlanningUniversity of Brawijaya. 2013.

[8] SNI-1733. Standar Untuk Perencanaan Sarana Lingkungan yang Terdiri Dari Sarana Olahraga dan Daerah Terbuka. 1989.

The development concept of Taman Aloon-Aloon Tulungagung based on visual accessibility, diversity of activities and perception of users

W Sasongko[1], R P Kurniasanti[1] and J P Siregar[1,2]

[1] Department of Urban and Regional Planning, Faculty of Engineering, Universitas Brawijaya, Malang, Indonesia
[2]School of Design, Queensland University of Technology, Brisbane, Australia

sasongkowisnu@yahoo.co.id

Abstract. Tulungagung has a public space that actively used by the community named Taman Aloon-aloon Tulungagung. The importance of research on the Development Concept of Taman Aloon-aloon Tulungagung as open public space is requiring location structuring for appeal aspects of public space in order to spread evenly and can be enjoyed by users of public space. Therefore, this study aimed to assess users' perceptions of the attractiveness of public space by Importance Performance Analysis (IPA) method and assess the level of users' visual accessibility by Visibility Graph Analysis (VGA) method and assess the diversity of users' activity by Behavior Mapping method. As this study shows that the condition of the floor surface, the location of a bird cage, shade conditions of public space and location of lighting lamp become the top priority in the development concept of Taman Aloon-aloon Tulungagung. The location for the prioritized attributes are adjusted to the results of the visual accessibility level evaluation matrix and activities diversity.

1. Introduction

Public open space is an open space owned by all, where people perform functional activities and rituals in a community both everyday life and in the periodic celebration [1]. One of the main functions of public open space is as a place for interactions between communities for a variety of purposes, both individuals and groups. In this case the public open space is part of the public social system whose existence cannot be separated from the social dynamics. Therefore, the existence of public open space should consider a wide range of people's needs class and status that reflects the fulfillment of needs in the community of all levels of society, both top to bottom class, from normal society until people who have different physical abilities with normal people, from children to adults and man or woman.

2. Literature Review

The public spaces are not just limited to green open spaces only, but including the private plots. Distribution and structure of public space is influenced by morphology. Social activities that supported by the existence of a public space, it can be said that the social interaction is also influenced by morphology. Social activity and private need are related to its space and access control. Connectedness each space is determined by the flow of the motion which determined by the structure of land use and street patterns. The high connectivity and flow of motion whether it can create its own public space where humans have the opportunity to meet, engage and interact with each other. Public space is not only seen from the activity or process of communication, but based on access.

The success of creating a public space for people is not only seen in terms of physical public space, but also people's behavior in using it. Therefore, it is necessary to study the utilization pattern of Taman Aloon-Aloon Tulungagung as public open space which is certainly related to user activity and configuration support space. In solving the problems of architectural and town design currently required new approaches that more attention dialectical interaction between humans and the environment, which understands that this interaction process involves decisions of individual human which is not always can be modeled and mathematically structured [2]. Regarding to the principle, the expectation on the relationship between humans and space if all is one, then we will find the level of configuration from space more than activity individual. It is situated on the relationship between configuration from humans and the space configuration [3]. The arguments presented by the Regulation of the Ministry of Domestic Affairs No.2 year 1987 that the need to study all aspects of the problems in town planning including studying the configuration of space for urban design. These considerations are spurring the development of behavioral approaches and develop the concept of space.

Tulungagung has a public space that actively used by community which is Taman Aloon-aloon Tulungagung. Taman Aloon-aloon Tulungagung is a public space with area of 1.3 ha which each day, especially on holidays, used by the community as a sports and leisure facilities. In addition, Taman Aloon-aloon Tulungagung also has historical value as a district center and community center since the beginning of the 20th century. Taman Aloon-aloon Tulungagung located in the center of Tulungagung Town. Taman Aloon-aloon Tulungagung is a public space that has undergone a redesign in 2007. Prior to the redesign, the public space is less well maintained and underused by the public.

At this time, the space layout design of Taman Aloon-aloon Tulungagung forming different activity concentrations in each area so that the impact on the emergence of a crowded area and the area is deserted. The importance of studying about Development Concept of Taman Alun-alun Tulungagung as a public open space is the need of location arrangement to appeal aspects of public space in order to spread evenly and can be enjoyed by users of public space.

3. Methods

Scope of the research areas is Taman Aloon-aloonTulungagung with ± 1.3 Ha wide. Method of data collection uses questionnaires and field observations

3.1. Population & Samples Determination

The population used in this research is all Taman Aloon-aloon Tulungagung users.

3.1.1. Activity mapping sampling. Behavior mapping in this study using the whole user population Taman Aloon-aloonTulungagung that perform activities on public spaces, both static and dynamic activities at 6 a.m. to 7 a.m, 12a.m to 1p.m, and 4 p.m to 5 p.m on holidays (weekend).

3.1.2. Questionnaires sampling. The sampling technique used is Accidental Sampling. Through consideration of the type of research, and the limitations of the study population in terms of time, cost and effort, then used a quota sampling method. Quota sampling is done if the population is not known with certainty either the number or variety of characteristics that make it homogeneous, then it is determined several number of individuals who are considered to represent [4]. Visitor population proportion of Taman Aloon-aloon Tulungagung which not certainty known yet because it is not fixed so that p and q are not known, it can be replaced by 0.25 as the multiplication of 0.5 x 0.5. Total sample calculation using the above equation is:

$$n > 0.5 \times 0.5\ (1{,}96/0{,}1)^2 \quad (1)$$

$$n > 96{,}04 \text{ people} \quad (2)$$

IOP Conf. Series: Earth and Environmental Science **70** (2017) 012063 doi:10.1088/1755-1315/70/1/012063

A sample of 100 subjects classified as essential [5]. Therefore, the number of samples required rounded to 100 people.

Table 1. Sample dividing proportion.

Day	**Users Average**								
	Morning (6 a.m-7 a.m)			**Noon (12 a.m-1 p.m)**			**Afternoon (4 p.m-5 p.m)**		
	R	**D**	**L**	**R**	**D**	**L**	**R**	**D**	**L**
Regular	2	4	1	3	5	2	3	8	3
Holiday	5	13	3	7	10	6	9	13	3
Total	7	17	4	10	15	8	12	21	6

The analytical method used is the analysis of IPA, Behavior Map and Visibility Graph Analysis.

3.2. Analysis of IPA

Evaluative analysis is used to analyze user perception of the public spaces Taman Aloon-aloon Tulungagung attractiveness through analytical methods Importance Performance Analysis (IPA). IPA analysis is an analytical method combination between aspects of the level of interest and perception of the quality or condition of an object into two-dimensional form. Aspects used is space, nature, culture and history, quietness and facilities. There are two parameters in the analysis of IPA, which are represented by the letters x and y, where x is the perception of the quality of public space that can give satisfaction to the user, while y is the level of user interest. The level of interest which meant in this case is interest according to the user toward the public space used.

3.3. Analysis of Behavior Map

Behavior mapping is a systematic observation techniques to document the use of a specific space or a location. Mapping behavior is relatively the most useful way lately in the evaluation process, after the specific location within the local area identified as a place which has a problem, successful or as a place which present opportunities to be redesigned or re-used. This process requires Incentive time on activity observed in a region that has been determined and recorded on maps, with details regarding the activity captured through symbol, notes and photos. Behavior mapping in this study using Place-centered way of mapping.

3.4. Visibility Graph Analysis (VGA)

The purpose of the Visibility Graph Analysis (VGA) is to analyze the extent to which each point in the network spatial visible from any other point. In this research aims to identify the VGA visual access to the garden. VGA can see the value of linkages (connectivity) and integration values (integrity) and intelligibility value that indicates the level of correlation between the measurements of a local scale (connectivity) to the global scale measurements (integrity) [6]. This analysis is a visual approach using open source applications DepthmapX 0.30.

4. Results

4.1. Scope of Research Region

Taman Aloon-aloon Tulungagung placed in Kelurahan Kampungdalem Kecamatan Tulungagung Kabupaten Tulungagung. The limitations of Taman Aloon-aloon Tulungagung are below:
North: Jalan RA. Kartini;
East: Jalan RA. Kartini;
South: Jalan RA. Kartini and Kelurahan Kauman;
West: Jalan RA. Kartini and Kelurahan Kauman

Map of the research region is shown by Figure 1.

IOP Conf. Series: Earth and Environmental Science **70** (2017) 012063 doi:10.1088/1755-1315/70/1/012063

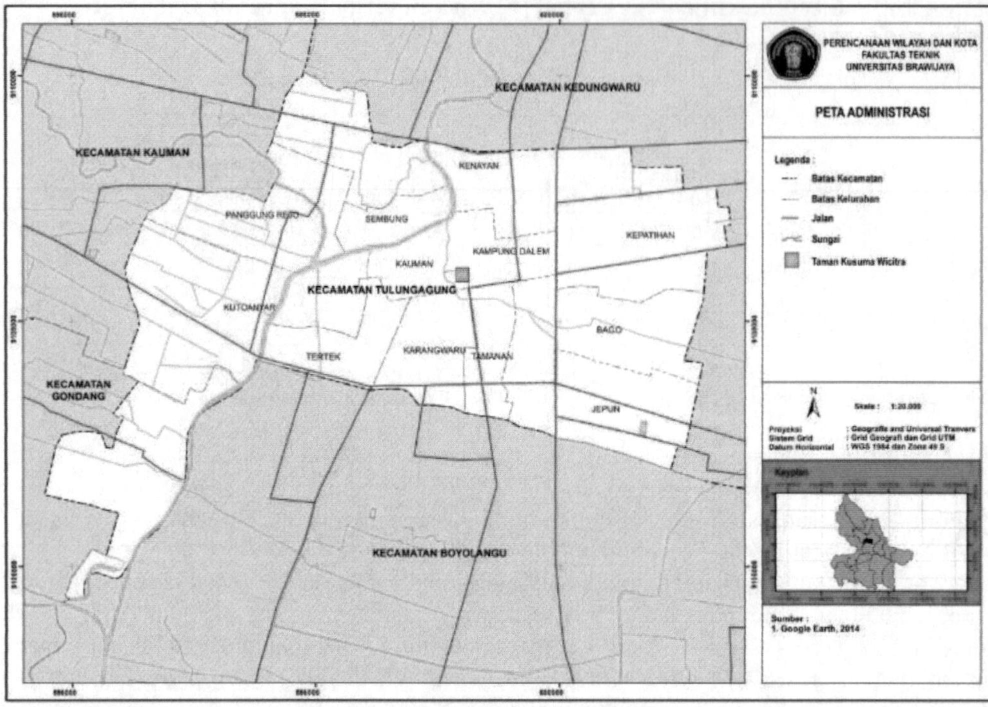

Figure 1. Research region map.

4.2. IPA Analysis

To analyze visitor perceptions about the attractiveness of public spaces Taman Aloon-aloon Tulungagung used IPA analysis. Taman Aloon-aloon Tulungagung condition assessed through 5 variables of the public space attractiveness, namely space, nature, culture and history, quietness, as well as facilities with a total of 22 attributes.

Table 2. Attribute development based on IPA quadrant.

Code	Variable	Level of Compliance (%)	x	y	IPA Quadrant
C4	*Culture and History*	84	364	433	Quadrant I *(Keep Up the Good Work)*
A3	*Space*	92	398	433	
C2	*Culture and History*	94	372	394	
B1	*Nature*	100	399	401	
E8	*Facilities*	100	399	401	
B2	*Nature*	102	381	375	Quadrant II *(Possible Overkill)*
E1	*Facilities*	102	375	366	
E2	*Facilities*	106	376	355	
C1	*Culture and History*	107	393	369	
E3	*Facilities*	108	386	359	
C3	*Culture and History*	109	372	342	
C5	*Culture and History*	109	393	362	
E5	*Facilities*	110	400	363	
E4	*Facilities*	115	408	355	
A2	*Space*	69	263	380	*Quadrant III (Low Priority)*
D1	*Nature*	72	276	384	
A1	*Space*	61	261	426	Quadrant IV

B3	*Nature*	62	266	428	*(Concentrate Here)*
A4	*Space*	65	271	416	
E6	*Facilities*	67	283	420	

4.2.1. Quadrant I. Attributes that go into quadrant keep up the good work is an attribute that is considered as important and was able to meet users' satisfaction. In order of priority, namely: the condition of the monument as a garden center (C4) –location of public spaces shade (A3) – existence of routine activity (C2) – the diversity of vegetation (B1) –existence of PKL(E8).

4.2.2. Quadrant II. Attributes that go into quadrant possible overkill indicating that attributes of the public space are less important to the users but has good quality so that it can be ignored. In order of priority, namely the condition of vegetation shade (B2) – the existence of the entrance / main road (E1) – the location of the entrance / main road (E2) – the existence of diverse visitor activity (C1) - the location of the fountain (C3) - the location of the monument (C5) - the location of the trash can (E5) - the trash can existence (E4).

4.2.3. Quadrant III. Attributes that go into quadrant low priority which indicates that some attributes decreased, due to both the level of interest and the quality is lower than the average value. In order of priority, namely the condition of public spaces guardrail (A2) – condition of trees as a noise reducer (D1).

4.2.4. Quadrant IV. Attributes that go into quadrant concentrate here is an attribute that is considered important and low satisfaction levels. In order of priority, namely: the condition of the floor surface (A1) – the location of bird cage (B1) – conditions of public spaces shade (A4) – location of lighting (E6).

Table 3. Top Priority IPA Result

Quadrant	Priority Order	Level of Compliance	Attribute
IV	1	61%	A1. Concitions of floor surface
	2	62%	B3. Location of bird cage
	3	65%	A4. Conditions of public space shade
	4	67%	E6. Locations of lighting
I	5	84%	C4. Condition of monument as the garden center
	6	92%	A3. Location of public space shade
	7	94%	C2. Existence of routine activity
	8	100%	B1. Diversity of vegetation
	9	100%	E8. Existence of PKL

4.3. Behavior Map

User behavior on holidays in the morning, majority are in the zone 3 while during the noon and afternoon, majority are at zone 5. Here is a map of Behavior Map in the afternoon with the most number of users and their diverse activity value calculation using Simpson's diversity index.

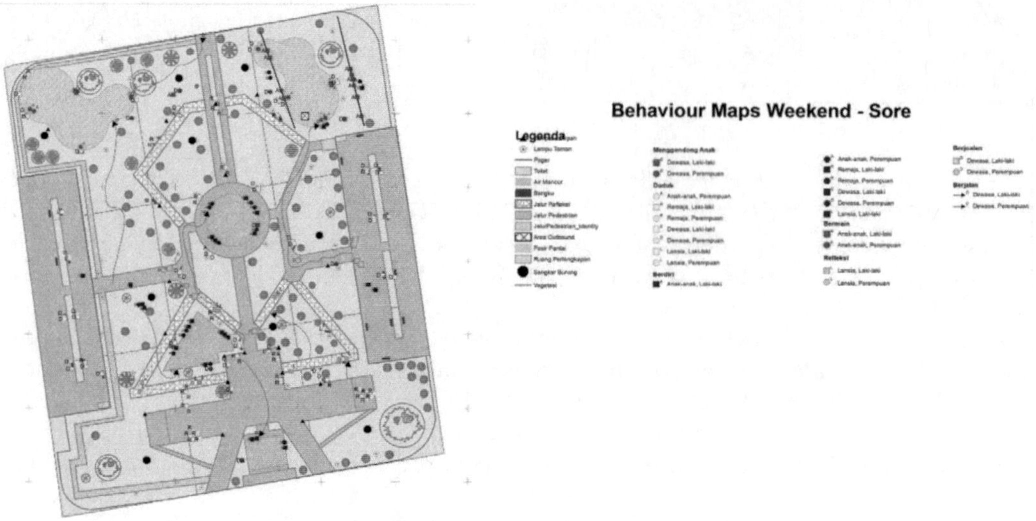

Figure 2. Behavior map weekend – afternoon.

Table 4. Simpson's diversity index value of user activity.

Zone	*Simpson's Diversity Index*	Variety Activity
1.	0.695	Very high
2.	0.426	Very low
3.	0.689	Very high
4.	0.667	Very high
5.	0.701	Very high
6.	0.681	Very high
7.	0.733	Very high
8.	0.609	High
9.	0.457	Very low

Parameter:
Very high : 0.656 – 0.733
High : 0.578 – 0.655
Low : 0.500 – 0.577
Very low : 0.422 – 0.499

4.4. Visibility Graph Analysis (VGA)

By using the Visibility Graph Analysis (VGA), known the extent to which each point in the network spatial visible from any other point so that it can determine the concept location of development direction Taman Aloon-aloon Tulungagung. Connectivity is shown in Figure 3, integrity shown in Figure 4 and intelligibility scatter plot shown in Figure 5.

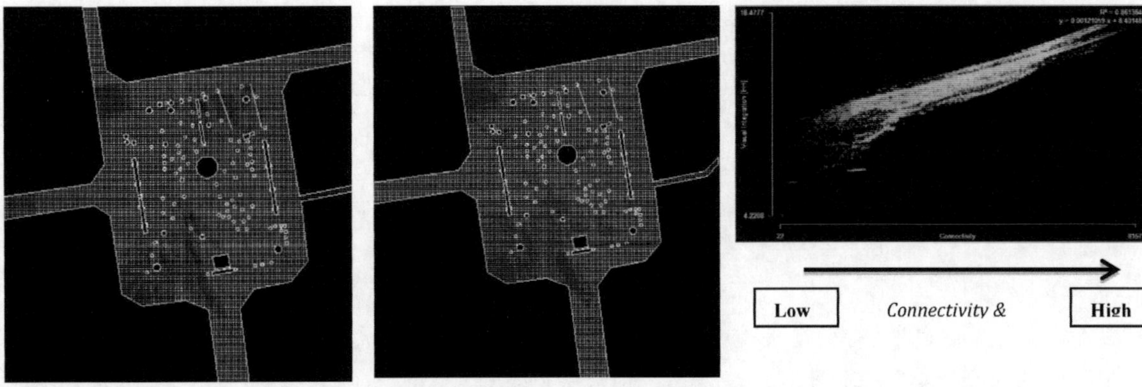

Figure 3. (a) Result of connectivity on VGA (b) Result of integrity on VGA (c) Scatter plot intellibility

4.4.1. Connectivity. Space which has a red color are most common in the zones 7 and 8. This indicates that the space in this zone have visual access to the most excellent in the local scale, meaning that space in this zone has a good visual access to the space around it. Whereas in zone 3, 4 and 6 have a considerable blue. This shows that the space in the zone have a poor visual access within a local scale, meaning that space in this zone has a poor visual access to the space around it.

4.4.2. Integrity. Space which has a red color are most common in the zones 7 and 8. This indicates that the space in this zone has the most excellent visual access on a global scale, meaning that space in this zone has a good visual access to the entire space on the entire zone. Whereas in zone 3 and zone 6 has a blue color which is quite a lot. This shows that the space in the zone have a poor visual access on a global scale, meaning that space in this zone has a poor visual access to the entire space on the entire zone.

Table 5. Integrity value each zone.

Zone	Integrity Average Value	Visual Access
1	13.585	Low
2	12.240	Very low
3	11.440	Very low
4	15.070	High
5	16.491	Very high
6	13.488	Low
7	16.324	Very high
8	17.195	Very high
9	14.982	High

Parameter:
Very High: 15.756 – 17.194
High : 14.316 – 15.755
Low : 12.876 – 14.315
Very Low : 11.436 – 12.875

4.4.3. Intelligibility. The distribution of the connectivity and integrity value produce high intelligibility (R2 = 0.861364) which situate that the correlation is strong. This strong correlation means that connectivity can explain integrity, so that the local position can explain orientation globally.

4.5. Evaluation of Visual Accessibility and Diversity Activities

Once known the value of the diversity of activities from the calculation of Simpson's Diversity Index and the accessibility value of the average room space integrity of each zone, so it can be described by the following diagram.

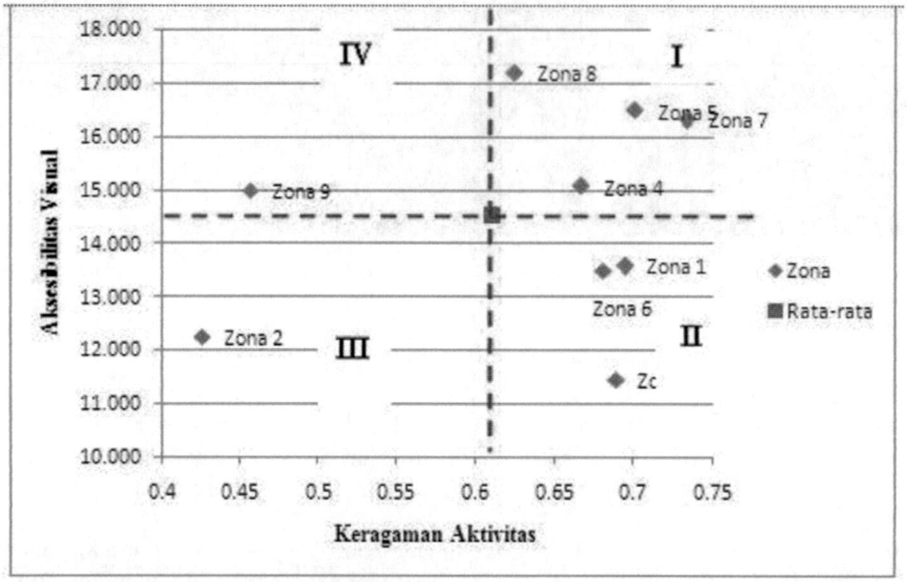

Figure 4. Matrix diagram of evaluation of visual accessibility and diversity activities.

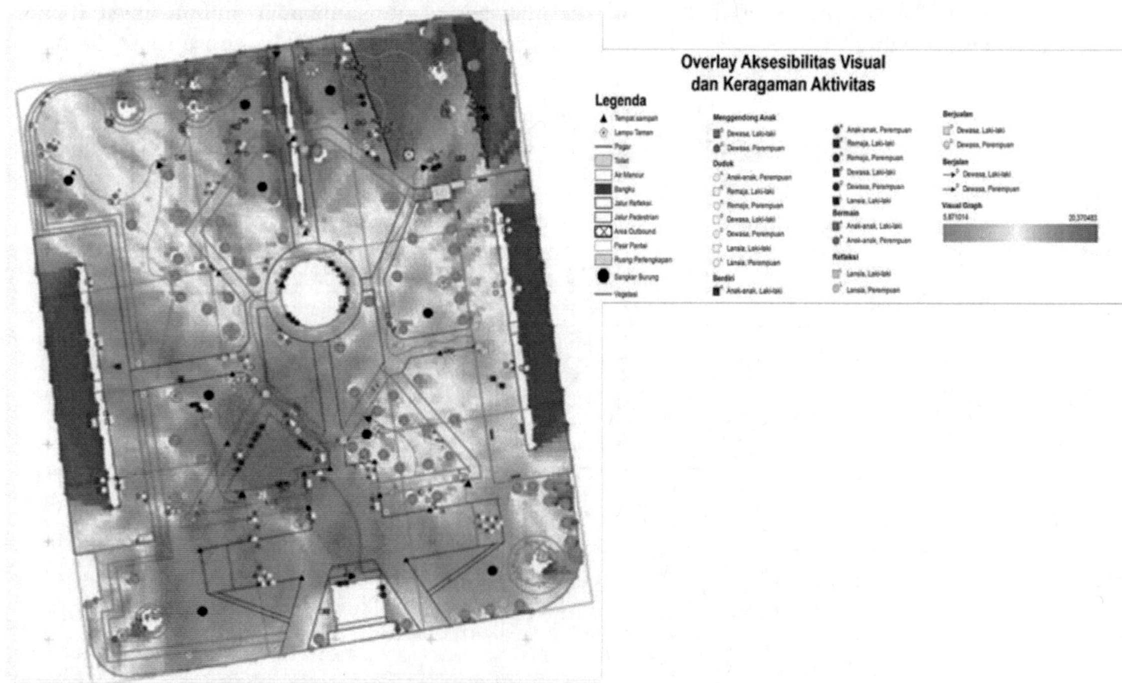

Figure 5. Overlay of visual accessibility – diversity activity.

4.6. Development Concept of Taman Aloon-aloon Tulungagung

1. Garden floor surface pavement Taman Aloon-aloon Tulungagung using grass-block with striking colors, especially in zones that have a high usage patterns of space, especially in zones prone to inundation that zone 7 and 9.
2. Maintenance bird cage conditions so as not to disturb the garden cleanliness and comfort of the user.
3. Location birdhouse transferred to zone 9 which has a low space usage patterns but the visibility value is fair.
4. Improvements of shelter in zone 4 and zone 6 at the entrance. These improvements include the addition of a clump of shade shelter located near a parking lot.
5. Addition massive shade as gazebo directed in zone 7 and zone 9
6. The addition of lighting in zones 1, 3, 7, and 9 who have low visibility level to avoid a negative impression of space. With the direction, the user is expected to be felt safe being in zones located on the corner of Taman Aloon-aloon Tulungagung which has low level of visibility.
7. Maintenance conditions monument located in zone 5, to make it more aesthetically pleasing and able to add value to the attractiveness of the monument as well as the garden center is able to provide value sense of place to the users of Taman Aloon-aloon Tulungagung
8. Location of public spaces shade (trees) focused on space that has a high space usage patterns with low visibility, which is in zone 1 and zone 3.
9. Conducting routine every week to increase the attractiveness of Taman Aloon-aloon Tulungagung directed at the plaza which is located in zone 8.
10. Adding more diversity of vegetation in zone 2 and zone 9. The two zones have low visibility and low usage pattern. With this directive, the expected diversity of vegetation can still be enjoyed by users without disrupting user activity because it is allocated in a zone that has a variety of low activity.
11. PKL arrangement in zones 7 that have high visibility. This will facilitate user access to the location of PKL to trade.

5. Conclusion

Based on the results and discussion, the following conclusions are discussed based on the research objectives:

1. Based on the analysis results obtained through Importance Performance Analysis (IPA) on the user's perception obtained the following results:
   Based on visitor perceptions about the results of the perception of visitors through the analysis of Importance Performance Analysis (IPA) have been the priority attribute of variable attractiveness of public space that will develop as follows:
   a. The condition of the floor surface
   b. Location of birdhouse
   c. The condition of public space shade
   d. Location of lighting
   e. The condition of the monument as a garden center
   f. Location of public space shade
   g. The existence of routine activity
   h. The diversity of vegetation
   i. The existence of PKL
   For attributes that are not mentioned, the development would be done in the last years of planning as the visitors and the community felt that the presence or absence of these attributes is not very significantly effect on Taman Aloon-aloon Tulungagung public space optimization.
2. Based on the analysis results obtained through the space syntax and behavior mappingabout space configuration and user activity patterns obtained the following results:
   a. Results of *space syntax*, regarding the configuration of the room by using the Visibility Graph Analysis (VGA) showed that the zone which has the highest integrity value is zone

5, zone 7, and zone 8. From these results, it can be concluded that zone 5, zone 7, and zone 8 can be more easily accessible from all other zones (globally). It also means that from that space, all kinds of facilities and the attractiveness of the garden can be easily viewed without vertical barriers or it can be said excellent movement and visual access.

b. The results of the behavior mapping, about the user activity patterns of Taman Aloon-aloon Tulungagung can be concluded that the peak user activity most crowded and diverse is in the afternoon at the weekend. From the results of the calculation of user activity value on weekends in the evening using Simpson's Diversity Index is concluded that, a zone that has the highest user activity value is zone 8. This is due to the diversity of facilities that are in the zone 8 so as to support the many activities of the user.

c. The concept of the development direction Taman Aloon-aloon Tulungagung as a public open space is based on result of IPA and combined with Matrix of Evaluation of Visual Accessibility and Diversity Activities. Concept development direction Taman Aloon-aloon Tulungagung as public open space directed by customizing the user's perception of the attractiveness of public spaces and inter-related user activity.

References

[1] S. Carr, M. Francis, L. G. Rivlin, and A. M. Stone, *Public space*. New York: Cambridge University Press, 1992.

[2] Haryadi and B. Setiawan, *Arsitektur, lingkungan dan perilaku: Pengantar ke teori, metodologi dan aplikasi*. Yogyakarta: Gadjah Mada University Press, 2010.

[3] B. Hillier, *Space is the machine: A configurational theory of architecture*. New York: Cambridge University Press, 1996.

[4] A. G. Asapa, “Pemanfaatan ruang publik spontan di perkampungan kota: Studi kasus Kelurahan Samaan, Kecamatan Klojen, Kota Malang,” Universitas Brawijaya, 2006.

[5] J. R. Fraenkel, N. E. Wallen, and H. H. Hyun, *How to design and evaluate research in education*. New York: McGraw-Hill Education, 2015.

[6] M. Batty, J. Desyllas, and E. Duxbury, “The discrete dynamics of small-scale spatial events: agent-based models of mobility in carnivals and street parades,” *Int. J. Geogr. Inf. Sci.*, vol. 17, no. 7, pp. 673–697, Oct. 2003.

IOP Conf. Series: Earth and Environmental Science **70** (2017) 012064 doi:10.1088/1755-1315/70/1/012064

Sustainable and participatory society for the realization of urban settlement (Case study: Green kampong, Malang City)

W D Purnamasari, V Kirana, and D K Wardhani

Department of Regional and Urban Planning, Faculty of Engineering, Universitas Brawijaya, Malang, Indonesia

wulanralando@gmail.com

Abstract. Green Kampong is one of the participatory development to conserve our environment. RW 03 Sukun, Malang City has chosen by researcher become case study because it has be successful to obtain a predicate as "Green Kampong" by following environmental management contests through the participatory society. There was some purpose of this study. The first aim was to find out the level of participation at every kind of the social activities such as waste separation for households, bio-pores project, stone massage therapy on the road, and other greening acts. The second aim was to evaluate the level for sustainable society which consists of: the leadership, the social capital, the internal controls activity, the use of technology, and also the finance of physical development in green kampong. The researcher has used the scoring and schema methods in this study. The result showed that the activity that has reached the maintenance phase are sorting waste and greening, whilst the activity that only reached the stage of control are bio-pores development and stone therapy. The social activities with highest participation was taken by greening act and the lowest was taken by bio-pores project. The sustainability of the community rw 03 Sukun known that of the five components aspects the sustainability of, three of whom were have achieved good progress toward the sustainability of the aspects leadership, social capital, and technology facilities and infrastructure while two aspects of whom were still achieved a good start toward sustainability of the aspects monitoring and evaluation and financing.

1. Introduction

Kampong in the city also has its own characteristics which still appears on the social and cultural system. A form of development that is participatory involving all elements of the community by participating devoting thought and resources owned [1]. The form of participatory development can be exemplified through the middle of a program of green action especially those residing in urban areas. Green programs organized through the green kampong movement aims to improve and preserve the environment of the settlement [2]. While according to Environmental Management Council (BPLH) Jakarta (2013), green kampong movement is a form of environmental management efforts.

RW 03 Sukun is a kampong that got many achievements of race-based environment and preserve sustainability of kampung by applying Green through participation carried out by communities since 2009 to 2015. However, although this RW 03 has been conducting green kampong activities there are still people that have had low participation and will affect the level of sustainability. So that this research conducted to find out how public participation in green activities and discover the level of kampong sustainability community. Participation and sustainability community can be used as an example or picture for other kampongs to actualize the green kampong.

2. Research Methods

This research use the type of qualitative and quantitative research approach. Qualitative research which describe and concern the analysis of the public participation by using scoring. Likewise, to know the level of sustainability of the community is carried out by using the method of scoring.

Figure 1. Research Location RW 03 Sukun, Malang City

2.1. Sampling Methods

The population is the community of RW 03 Sukun with classification 15-65 years old with a population of 750 inhabitants. The sample in this study using a purposive sampling technique. Calculation of the sample using Isaac and Michael through the Krejcie table so that the retrieved sample are 199 respondents.

2.2. Analysis Methods

2.2.1. Community Participation Analysis. The analysis of participation is used to find out the levels of participation by identificate the process of community. The following level is a stage of participation [3], [4].

1. Planning level
   Participation at this stage means a person's involvement in the preparation of plans and budgets on an activity/project. The community participated by giving the proposals, suggestions and criticism through meetings and then do the decision making based on mutual agreement.
2. Doing Level
   Participation in the implementation phase (implementation stage). Participation at this stage means the involvement of the implementation of the work of a project. Community can give you staff or money as one as form of participation.
3. Controlling Level
   Participation consists of monitor and evaluate the development. Monitoring and evaluation of the development was done, not only to achieve a purpose but also to obtain feedback from the problems and constraints that arise in the implementation of development is concerned.
4. Actuating Level
   Participation at this stage means a person's involvement in the maintenance of the project after completing of the project. Public participation at this stage is the form of manpower and money

to operate and maintain the project that has been built from the existence of control by the community in order to be well maintain.

2.2.2. Scoring Analysis. Scoring analysis is a method of analysis by awarding a score in assessing the answers of respondents to obtain a quantitative data needed in the analysis of the perceptions of the respondents based on criteria and indicators [5]. The scoring method is a technique for quantitative analysis of data used to provide value to each characteristic parameters of subcategories a variable so that its value can be calculated and can be rated [6]. The level of public participation is determined based on the total score of the participation of each the respective levels, which are high, medium, and low. So, the minimum score (4 × 1) is 4, maksimum score obtained (4 × 3) is 12. The following is a calculation to determine the class interval:

$$Interval\ Class = \frac{Highest\ Score - Lowest\ Score}{Total\ Class} \quad (1)$$

$$Interval\ Class = \frac{12-4}{3} = \frac{8}{3} = 2{,}6 \quad (2)$$

Thus retrieved the *value* of the classification level of community participation are divided into three classes, namely:
1) Low levels of participation = 4-6.6
2) Medium = 9.2-6.7, and
3) High = 9.3
Valuation and division class to determine the value of the community sustainability aspects referring to the review of Prabasena (2010). The following is a calculation to get the interval class [7]:

$$Interval\ Class = \frac{Highest\ Score - Lowest\ Score}{Total\ Class} \quad (3)$$

$$Interval\ Class = \frac{25-5}{3} = \frac{20}{3} = 6{,}6 \quad (4)$$

The assessment was conducted by five variables, to determine the condition of the sustainability community. Then, five variables summed and averaged to obtain the value of community sustainability in each RT with the following classifications:

Table 1. Classification aspects of community sustainability.

Values	Parameters
18.4 to 25	Show an excellent progress toward sustainability
11.7 to 18.3	Indicate a good start toward sustainability
5 to 11.6	Show a low progress to make a sustainability

2.2.3. Radar Charts. Radar charts are often referred to as diagram of the nest of spiders (Spider Graph). The radar chart is a graph that is described with many axes. Each axis represents a category. This diagram is depicted with existential points placed in the intersection of the lines. These points form a line resembling a Spider Web. Radar charts are used to facilitate the visualization in evaluating the various categories. This chart also makes it easy to observe a logical separation between variables are compared [8]. Research on radar charts, will be used to illustrate the results of the assessment of social aspects of sustainability in RW 03 Sukun at each RT

3. Results and Discussion
Community participation which deliver it to success starting from 2009 to 2015 in which participation is visible from the involved of community in a series of activities of green kampong. In addition to the

carrying out of activities by the community, also coupled with follow the race starting from the city to the national scale.

3.1. Identification of Society Sustainable Aspects

According to Halwatul and Iwan (2012) community sustainability aspects include leadership, social capital, monitoring and evaluation, technology and infrastructure, and financing [9]. The following aspect of community sustainability in RW 03, which are:

1. Leadership
   Leadership in this case defined by the attitude of the leaders in receiving and responding to the aspirations of the community without any limits in delivering it. Leadership by community leaders include Chairman of the RT, RW, and organizations. In addition, based on public perception regarding the support of leaders in innovation as well as the public perception about the influence leader in enhancing community involvement to participate activities.
2. Social Capital
   Social capital can be described of the mutual values, meetings are routinely made through a work program of the Organization, as well as meeting level RT and RW so it can be a very effective communication are created in the execution of a program. On the research of visible from community engagement in the following institutional, community presence in the preparation of the plan of activities and work program, as well as community involvement in activities that overall.
3. Monitoring and Evaluation
   Activities that are performed to see, monitor the operations of the organizations or activities, and assess reached goal. In monitoring can be done with the data collection and analysis results were analyzed, interpreted and explained as input to make improvements. Monitoring and evaluation required in community participation in order to obtain the best results for long term.
4. Technology and infrastructure
   Simple technology that is easily applied by the citizens as well as environment-friendly in order to make the citizens willing to perform and evolve. In this research, technology and infrastructure that's located in the green kampong include availability of technology used in each activity.
5. Finance
   Financing comes from a source of funding comes from dues, cooperation of outside parties, and creativity in the provision of funds. On research, the financing of this relates to the existence of the availability of local banks such as capital and loan-save cooperatives from PKK, as well as the presence of kampong programs related to the development of creativity that can add income as garbage sorting activity results through the development of creativity making crafts.

3.2. Scoring

The value of the classification level of community participation are divided into three classes, namely:
1) Low levels of participation = 4-6.6
2) Being = 9.2-6.7, and
3) High = 9.3

Table 2. Values score of participatory sorting waste

RT	Planning	Implementation	Control	Maintenance	Total
01	1.0	1.2	1.0	1.3	5
02	1.0	1.2	1.1	1.2	5
03	1.4	1.6	1.3	1.6	6
04	1.2	1.4	1.1	1.5	5
05	1.4	1.5	1.3	1.7	6
06	1.0	1.3	1.0	1.4	5
07	1.3	1.7	1.3	1.8	6
08	1.0	1.4	1.1	1.5	5
Total	9.3	11.4	9.1	12.0	42
Classification	High	High	Moderate	High	

Table 3. The calculation value result of participation development activities bio-pores

RT	Planning	Implementation	Control	Maintenance	Total
01	1.2	1.4	1.1	1.0	5
02	1.0	1.5	1.2	1.0	5
03	1.3	1.5	1.4	1.0	5
04	1.2	1.3	1.3	1.0	5
05	1.0	1.5	1.4	1.0	5
06	1.1	1.3	1.2	1.0	5
07	1.3	1.8	1.5	1.0	6
08	1.0	1.3	1.0	1.6	5
Total	9.1	11.6	10.1	8.6	39
Classification	Moderate	High	High	moderate	

Table 4. The calculation value result of participation greening activities

RT	Planning	Implementation	Control	Maintenance	Total
01	1.1	1.2	1.1	1.0	4
02	1.3	1.7	1.4	1.0	6
03	1.4	1.8	1.4	1.0	6
04	1.2	1.5	1.2	1.0	5
05	1.3	1.7	1.5	1.0	6
06	1.1	1.3	1.2	1.0	5
07	1.0	1.5	1.3	1.0	5
08	1.1	1.3	1.1	1.0	5
Total	9.5	12.0	10.4	8.0	40
Classification	moderate	High	High	moderate	

Table 5. The calculation value result of participation greening activities

RT	Planning	Implementation	Control	Maintenance	Total
01	1.2	1.4	1.2	1, 4	5
02	1.2	1.4	1.3	1.5	5
03	1.4	1.7	1.5	1.7	6
04	1.2	1.3	1.2	1.3	5
05	1.4	1.6	1.4	1.7	6
06	1.2	1.4	1.2	1.5	5
07	1.2	1.3	1.3	1.5	5
08	1.2	1.3	1.3	1.4	5
Total	10.1	11.4	10	12 0	44

RT	Planning	Implementation	Control	Maintenance	Total
Classification	High	High	High	High	

Overall of the four activities that have the lowest value is bio-pores activities. This bio-pores development activities gained less attention from the public because these activities do not impact and does not give any sense the community directly so that make the community's lack of enthusiasm to get involved. Furthermore, at the planning stages, the presence and aspirations of society is also very low compared to other activities. In the process of preparing a plan to build this bio-pores done with top-down approach so that the role of the community in the preparation of the plan of this activity is still lacking, especially in the delivery of the aspirations of the community. This activity also doesn't need such a huge space and not much involve the role of the community from various institutional, just the Chairman of RT, RW, as well as a bevy of fathers who was instrumental in this activity so that the community in utilizing the results of development after the project completed is also low. It will certainly have an impact on the treatment and the control carried out by the community against the condition of bio-pores.

3.3. Sustainable Community Scoring Analysis

Based on the results of survey through questionnaire that was given to the community, the results of the recapitulation is then performed using a skoring analysis of the calculation refers to the research of Prabasasena (2010) [7].

The assessment was conducted on five variables, to determine the condition of the sustainability of community, then from five variables result divided by the sample at each of the RT, then summed up the value of all variables and averaged so that the community's sustainability values obtained at each RT.

Table 6. Classification aspects of community sustainability

Values	Parameters
18.4 to 25	Show an excellent progress toward sustainability
11.7 to 18.3	Indicate a good start toward sustainability
5 to 11.6	Show a low progress to make a sustainability

The following below is the result of the calculation of the questionnaire given to the community related to the components of the social aspects of sustainability of community leadership, social capital, monitoring and evaluation, technology and infrastructure, and financing.

Table 7. The calculation value result of participation greening activities

Aspects	RT								Total	Classification
	1	2	3	4	5	6	7	8		
Leadership	2.4	2.5	3.9	2.5	4	2.6	3.6	2.4	23.9	Excellent Progress toward sustainability
social capital	1.9	3.7	3.7	2.4	3.7	2.4	2.8	2.4	23.1	Excellent Progress toward sustainability
Monitoring and evaluation	**2.1**	**2.2**	**2.9**	**2**	**2.9**	**2**	**2.1**	**2.1**	**18.3**	**Good start toward sustainability**
Technology infrastructure	2.4	2.5	3.3	2.6	3.3	2.5	3.3	2.4	22.3	Excellent Progress toward sustainability
Financing	**1.9**	**2.1**	**2.4**	**2.1**	**2.7**	**2.1**	**2.2**	**2**	**17.5**	**Good start toward sustainability**
Total	10,6	13,0	16,2	11,6	16,7	11,6	14,1	11,3		

Based on the analysis of the assessment of the sustainability of the community can be aware that all the components of community sustainability has been toward sustainability. There are three of these two components that shown good progress towards the sustainability of the community, which are the components of leadership, social capital, as well as the technology infrastructure while the two of them achieved a good start toward sustainability, namely monitoring and evaluation as well as financing. Besides, it can also be noted that the highest rated RT is RT 05 with value 16.7 and RT who get the lowest value is RT 01 with 10.6 value. The results of the analysis of the sustainability of the community is also depicted with a radar graph that is as follows below;

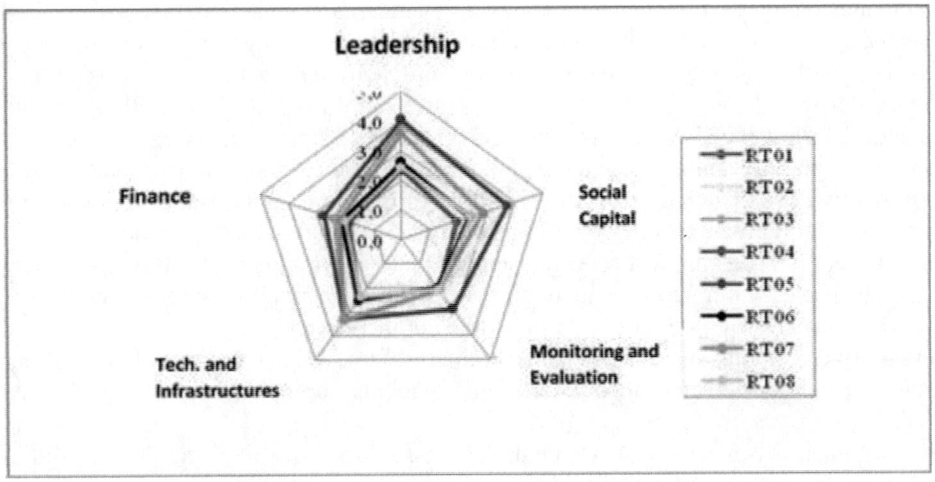

Figure 2. Graphic Radar of Results Analysis sustainable society aspect

The highest value on the five components of community sustainability aspect is leadership. Leadership styles reflect the effectivity of leader that is able to create a sense of trust from the community, persistence (unanimous determination) to achieve the goal, the ability to communicate, willingness to hear, and objective in receiving and running the aspirations of the community. In addition, with the implementation of innovation as a method or manner that can enhance the progress of repair green kampong. In this case, the value of which RT has most highv value are RT 03, 05, and 07 while the RT that had a lower value when compared to the other RT.

Social capital also described mutual value as well as a large number of joint activities on a regular basis, such as the community service, RT and RW gathering of thus nurtures communication. Communications society looks upon society as well as the large number of Community institutional following is involved in an activity that's located in the green kampong. In this case RT that has the highest value are RT 02, 03, and 05. While the RT that have the lowest value is RT 01.

In addition, due to the lack of a society incorporated in the institutional community, the percentage on the RT in attending gatherings as well as meetings and community service still less on any activities of the green kampong.

Public participation also relates to how the control and evaluation of community in follow up the activities that have been held up. Monitoring and evaluation in RW 03 was performed by the entire community itself without any oversight of government agencies. Communities are more likely to be involved in the activities of the evaluation compared the monitoring activities. Although in terms of knowledge and community cooperation is already good, but in terms of the monitoring and evaluation of community involvement in average on the whole less except on RT RT 03 and 05. That is because most of the people depend on the apparatus RW mainly on the president of organization.

The technology and infrastructure that can make the public wants to make green kampong support. Look on the main activities of green kampong organic waste segregation, i.e. construction of bio-pores, stone therapy development and greening.

Technology and infrastructure other than intended so that the community wants to do activities, also so that the community can inprove their knowledge in managing and preserving the activities that have been implemented as there are mowers organic waste, the presence of bio-pores as catchment area so that it can reduce surface run off water on rainy days because previously there was the problem of flooding in the RW 03 so as solution the community build the bio-pores, stone therapy development, utilizing part of the middle of the road as one of the efforts of maintaining health with reflection stone therapy, as well as the provision of water treatment by utilizing river water for watering the plants. In this component, RT who has highest value is RT 03, 05, and 07. The third society at RT this often make use of technology and infrastructure mowers organic waste to produce fertilizer as the fulfillment of a need for the plants in our daily lives.

The Sustainable community in carrying out activities of green kampong is not only seen from how big the community is willing to get involved from all existing activities, but also in terms of financing. As a whole the existence of local banks (cooperative), help from third parties, as well as dues community is already adequate. However, help from third parties also had a limit time, so as community to be able to keep running the activity and is not dependent in terms of financing to third parties, society also needs to be independently develop activities of the creativity of the community associated with the economy so that it can be a source of financing for the next activity. Creativity development activities already found in RW 03 but people's interest to follow institutional and matters related to the development of the independent business community is also still low. Moreover, interests of the society which is still low, it is also due to a lack of motivation, training, and yet there is a clear brand product to be developed by the community. So, the results of sales from products that have been created by the public is limited and the results of their sales were not so much. This financing component that has the highest value is RT 05.

Community sustainability aspects which include leadership, social capital, monitoring and evaluation, technology and infrastructure, and financing for all components already shows that there is a good progress toward sustainability except on aspects of monitoring and evaluation as well as financing. If community sustainability percentage showed RW 03 by 60%. So, it can be known that for 40% of the components are still not achieving good progress towards sustainability but has achieved a good start toward sustainability. Thus, to achieve sustainability of the community, the need for community efforts to improve the community's component is still lacking.

4. Recommendation

Recommendations for communities based on the results of the analysis that has been done to increase participation and community sustainability through activities that have been implemented. These recommendations are (a) Increasing the awareness of people from greening activities through empowering the youth organization, (b) Socialization related to the manufacture handicraft products processed from trash to woman organization in green kampong, (c) Procurement partnerships with others as one as source of financial aspect in order to make all programs can be realized.

References

[1] D. Hernawati, C. Saleh, and Suwondo, "Partisipasi Masyarakat Dalam Pengelolaan Sampah Berbasis 3R (Reduce, Reuse Dan Recycle) (Studi Pada Tempat Pengelolaan Sampah Terpadu Di Desa Mulyoagung Kecamatan Dau Kabupaten Malang)," *J. Adm. Publik*, vol. 1, no. 2, pp. 181–187, 2005.

[2] K. Mega and S. Paulla, "Analisis peran gerakan kampung hijau sebagai upaya pengelolaan di Kelurahan Rajawati Kecamatan Pancoran Jakarta Selatan," *J. Ruang*, vol. 1, no. 1, pp. 71–80, 2013.

[3] B. Repi, B. O. L. S, O. Porajow, and J. F. J. Timban, "Partisipasi masyarakat terhadap pembangunan infrastruktur jalan melalui PNPM-PPIP di Desa Munte Kecamatan Tumpaan Kabupaten Minahasa Selatan," *ASE*, vol. 11, no. 1, pp. 38–50, 2015.

[4] S. Y. Wardhani, "Partisipasi kelompok tani dalam pengelolaan hutan mangrove di Kelurahan Wonorejo Kecamatan Rungkut Kota Surabaya," *J. Ilmu Sos.*, vol. 1, no. 1, pp. 1–10, 2012.

[5] A. Sumaryadi, "Kesesuaian rencana pengembangan pariwisata kawasan kaki Jembatan Suramadu sisi Madura dengan persepsi wisatawan terhadap kebutuhan pariwisata," *J. Perenc.*

Wil. dan Kota, vol. II, no. 1, pp. 34–43, 2013.

[6] D. W. Gunawan, S. J. Hartati, and Y. M. Maulana, "Jurnal Sistem Informasi Jurnal Sistem Informasi," *J. Sist. Inf.*, vol. 3, no. 2, pp. 97–103, 2014.

[7] D. Prabasasena, "Studi persepsi masyarakat terhadap tingkat keberlanjutan lanskap Kampung Sindang Barang, Desa Pasireurih, Kecamatan Tamansari, Kabupaten Bogor," Institut Pertanian Bogor, 2010.

[8] A. Nurdiana, "Aplikasi manajemen resiko dari persepsi para stakeholders: Studi kasus proyek pembangunan jalan tol Semarang-Solo seksi I ruas Tembalang-Gedawang," Universitas Diponegoro Semarang, 2011.

[9] Halwatul and Iwan, "Keberlanjutan pengelolaan sampah berbasis masyarakat, Kelurahan Taman Sari," *J. Perenc. Wil. dan Kota*, vol. 1, pp. 283–290, 2012.

IOP Conf. Series: Earth and Environmental Science **70** (2017) 012065 doi:10.1088/1755-1315/70/1/012065

Redefining smart city concept with resilience approach

Y Arafah[1 2 3], **H Winarso**[1]

[1] Urban and Regional Planning, School of Architecture, Planning, and Policy Development, Institut Teknologi Bandung, Jl. Ganesha 10, Bandung 40132, Indonesia.
[2]Architecture Department, Faculty of Engineering, Syiah Kuala University, Jl. T. Abdurrauf, No.7 Darussalam, Banda Aceh 23111, Indonesia.

yunitaarafah@gmail.com

Abstract. The smart city concept originally aimed at dealing with various urban problems, in particular, those related to the urban environment and infrastructure, such as modeling transport flow in a city. As it developed, the concept is now widely used to accelerate the process of urban management by using IT technology and by the availability of big data. However, the smart city discourses are still debated. There is a number of critical literature on the discourses; some are more concerned with the use and development of information communication technology (ICT). ICT and modern technology are considered the key aspect of the smart city concept. Meanwhile, others emphasize the importance of the people who operate the technology. Very few, if any, literature emphasizes the importance of resilience in the smart city discourse. The city as a complex system should have the ability to be resilient, especially when technology fails either due to technical/man-made or natural disasters. This paper aims to redefine the smart city concept in urban planning through a literature study in the context of planning using a resilience approach. This paper describes and defines what the smart city concept is, what it means, as well as explains the relation and linkage of the importance of using resilience approach in defining the smart city. Factors of resilience will lead to a soft infrastructure approach, such as enhancement in many aspects, e.g. community capacity, social and human capital, knowledge inclusion, participation, social innovation, and social equity. Discussion and analysis are conducted through a deep literature study using systematic literature review methodology.

1. Introduction

Cities continue to develop with various types of innovations and progress. According to the United Nation (UN), cities generate 75% of the global carbon emissions and consume two third of world energy. Moreover, due to urbanization and economic growth, energy consumption is expected to rise to three-fourth of the world energy consumption by 2030. Moreover, the UN estimates that 66% of the world population will reside in urban areas by 2050 [1]. This will certainly increase the amount of carbon dioxide, greenhouse emissions, congestion, and waste disposal that will affect public health. This will also create greater challenges in dealing with air pollution, population density, waste management and human health [2].

During the last years of the 20th century, two major phenomena emerged, namely urbanization and information technology [3]. In the 1990s the "smart growth" concept began to spread in an effort to seek solutions to problems such as population density, congestion, air pollution, open space running out, and the soaring cost of public facilities [4]. The idea of the smart city evolved in an attempt to save the earth and human health. It is an idea or concept people had hoped would be able to solve urban problems while considering the environment [3]. Internationally, the smart city discourse developed as a way to meet the Kyoto Protocol's goals that call for reducing emission levels that had increased on an international scale. From there, the smart city concept is now widely used, supported and adopted by many institutions such as the European Commission Setis-EU, OECD, etc., along with a rapid increase in the number of research and publications that are conducted on the topic ([5] & [6]).

As the concept is more widely used, there is still unclarity in understanding its definition ([7]; [8]; [9]; [10]). Based on mutual understanding by the European Commission, the smart city is a concept that could achieve sustainability through using a wide range technology [11]. Therefore, the main objective of the smart city is to increase sustainability with the use of technology, hence, "smart sustainable cities" is seen as more suitable and accurate compared to the term smart cities [12]. In addition, terminology related to the smart city has developed in accordance with different definitions as well, such as the virtual city [13]; digital city [14]; knowledge city [15]; wired city [16]; ubiquitous city [17], and many more.

Currently, literature explaining smart city is plentiful and continues to expand; some of the most prominent focus is on the development of modern technology and information communication technology. Meanwhile definitions are not limited to only those, but in fact extend to other areas. This theory is supported by Caragliu et al., 2011 [18] who stated that a city is categorized as smart when it has invested in human capital, social capital, transportation, and modern ICT. It should also be capable of creating economically sustainable growth and a high quality of life with good management of its natural resources through participatory governance. Adding to that, Goldsmith and Crawford, 2014 [19], also discussed the importance of community involvement and good leadership to ensure proper implementation of the smart city concept. In addition, there is also need for community empowerment and related factors [20].

Various attempts are made at conceptualizing the smart city concept in planning, one of which is by developing several methods of benchmarking and indicators to evaluate and measure impacts. Benchmarking has the purpose of comparing smart cities based on various factors and ideas or concepts [21]. The benchmarking method is used to evaluate smart cities from different perspectives, e.g. its sustainability, global city performance, resilience, local government effectiveness, urban competitiveness, and good urban governance [21].

Benchmarking smart city planning now considers the perspective of resilience. Desouza and Flanery, 2013 [22], have even created an instrument or guideline in benchmarking of resilience based on an evaluation of the city's resilience, as well as a framework for its implementation. Their implementation framework is divided into two urban components, i.e. resources and process (physically) and communities, institutions, and activities (socially). Therefore, the factor of a resilience approach is needed in the development of smart city planning.

In the context of Indonesia, many cities, in particular, big cities such as Jakarta, Bogor, Bandung, and Surabaya, claim that their cities have implemented the smart city concept [23]. However, it is unconfirmed whether urban residents are 'smart' or intelligent enough or whether there is equality and fairness among all levels of society in both accessing and using ICT as well as in the modern technology provided? These sorts of factors have been a particular challenge in development and in implementing the smart city concept. The concept of resilience, which planners attempt to include as one of the factors or indicators in smart city planning, is a concept that is hoped to increase the resistance and resilience of smart cities in dealing with unexpected predicaments, such as natural disasters, terrorism, energy crises, and climate change. When a city planned to be smart, it must also be prepared to be resilient at all times.

This paper will focus on the smart city concept with a resilience approach. The analysis is conducted through literature study and aims to redefine the smart city concept in the context of planning using a resilience approach. This paper will also describe and define the smart city concept and the meaning of recently used concepts, as well as describe the relevance and linkage of the importance of a resilience approach in defining the smart city concept. This paper consists of four major parts:

IOP Conf. Series: Earth and Environmental Science **70** (2017) 012065 doi:10.1088/1755-1315/70/1/012065

1) Introduction, which discusses the factors and background that underlie this research, a brief introduction to the definition and history of smart cities, and its linkage with resilience.
2) Methodology, which will define the research methodology used to identify and organize relevant literature into themes in order to establish the conceptual framework of the analysis.
3) Discussion, which will further explain the concept. This chapter is divided into several sub-topics, i.e. the current smart city definitions; classifying themes from the definitions found; discussing smart city plan positioning in the context of planning in the era of uncertainty; explaining the smart city concept and resilience in terms of their definition, relation, and its relevance in planning a city; and lastly, synergizing the smart city concept with a resilience approach into a new concept based on the discussion and literature.
4) Result and conclusion, this part presents a conclusion of the analysis aimed at finding the definition of a smart city concept based on resilience.

2. Research Methodology

The literature review on smart cities and resilience uses a systematic literature review methodology. This methodology is carried out by compiling previous studies that have presented facts comprehensively and balanced. The results are synthesized using narrative or qualitative techniques. This qualitative approach is used to synthesize the earlier descriptive qualitative analysis. This method is also called meta-synthesis technique, which is a technique that integrates data for a deeper and more thorough level of understanding of new concepts [24]. Data collection in systematic literature review methodology uses previous journals, internet searches, and recently published studies. Qualitative systematic literature review methodology [25] comprises the following steps:

1. Formulating the review question.
2. Conducting a systematic literature search.
3. Screening and selecting appropriate research articles.
4. Analyzing and synthesizing qualitative findings.
5. Maintaining quality control.
6. Presenting findings.

Research using a meta-synthesis method or also called qualitative data synthesis can use two approaches, i.e. meta-aggregation and meta-ethnography [26]. In the meta-aggregation approach, synthesis aims to compile previous studies in order to answer the research question. Meanwhile, in meta-ethnography, the synthesis has the goal of developing a new theory to complete the existing ones. This paper will use the first approach of meta-aggregation, to find solutions to answer the research questions about the role of the factor of resilience in the smart city concept.

In the meta-aggregation approach, the topic of analysis is elaborated into certain themes to build a conceptual framework. In these themes, articles will be searched, compared with one another and summarized. In meta-aggregation, the synthesis is an "aggregate" of various studies with the same theme. Synthesis is carried out using the following five steps: (1) extracting relevant themes and concepts from previous material; (2) results of extraction will then be organized into primary/main findings; [3] the findings will then be classified into categories; [4] each category will then be synthesized into one particular theme (adjusted with the conceptual framework compiled).

The explanation above shows that this analysis uses systematic literature review with a meta-aggregation methodology. Meta-aggregation is carried out to aggregate relevant previous studies. Therefore, in the meta-aggregation synthesis, the conceptual framework that shows the relevancy must be made in advance. Then primary study results are plotted into previously identified themes. In other words, presented in the aggregate (descriptive) model.

3. Analysis and Discussion

3.1 Defining Smart City Concept

Intrinsically, smart city is actually not a new concept, nevertheless the discussion is still warmly debate. Its definition has also spread to other areas, thus, a mutual understanding and agreement of its exact definition has not been found yet. Based on current studies, it is talked that a cities whom used smart city concept are valued to be able to increase the use of city services, hence will nevertheless enhance the level of urban management eficiency and support social, environmental, and economic sustainability. Innovative approaches are also necessary to advance the development of smart city as well as its citizen oriented urban service management for the city in order to achieve higher living standard [27].

Paralleling with the widespread of its definition, the usage and its implementation experience the same phase. However, researchers view the definition has not found a firm consistency ([16]; [9]; [10]). Analysis conducted as yet are stil debated and vastly developed, however, the fundamental problem lies in defining the real definition of smart city. There is still no mutual understanding on the word "smart" or "intelligent" of a city. The existing definition is still viewed multi interpretated, not universal, and solutions taken in problem solving are not those related to previous case problem solving ([8]; [16]; [28]; [4]).

In the history of its development, the roots of smart city idea has long begun, it was when Singapore develop long-term planning agenda by implementing the concept The Intelligent City" (Toh, M. H., & Low, L.,1993, in [29], followed by Adelaide City, whom in 1994 applied the Smart City Adelaide (SCA), the concept which uses technology and communication infrastructures to provide electronic service in the field of commercial in its utilization by the local residents. Infrastructures developed by using software applications through telecommunication providers. Others alike projects that has also showed up during those times are, Singapore's IT 2000 " Intelligent Island" I implemented by Singapore's National Computing Board (NCB), and Silicon Valley (JVSV BLUW 93), the "Joint Venture : Silicon Valley " (JVSV) project [23].

Based from the above various project examples, identified that the idea of smart city originally aim to advanced city quality to be "smart" through the improvements on technology infrastructure, ICT in particular, by developing software applications while seeking assistance from private sector vendor for ideas development and implementation, but, over time, the development of the concept had increased quite rapidly, both in terms of publications and application, especially since the EU in 2010, introduced the smart city project, a project that aims to develop intelligent solutions for future challenges, like urbanization, globalisation, dan weather changes ([6] & [5]).

Some other reasons supporting the development of smart city are, the extensive use of technology devices, such as: mobile devices and the internet in all level of community, the circumstances in which urban dimension is constantly getting higher and the need to protect the environment from pollution and high energy consumption [30]. Whereas, in fact, understanding this smart city concept, is not limited to the use of ICT and modern technology only, roles of other aspects also plays an important part. The encroachment of this concept to different areas are still warmly debated. This theory supported by Caragliu et al., 2011 [18], of whom said that a city is modified into smart city when it has invested on human capital, social capital, transportations, and modern ICT, along with creating a sustainable economic and high quality of life, with good natural resources management through participatory governance. Goldsmith and Crawford, 2014 [19], had also mentioned about the importance of community participation and leadership factor in an effort to implement smart city concept. Other than that, community empowerment and linkages between the actors are seen needed [20].

The development of smart city concept, the implementation can be phased into three main group, starting from focusing only to hardware or so called smart city as digital city, then shifted to software only, known as social inclusive city, and lasted with, focusing on both, which formed smart city as City with high quality of life [31].

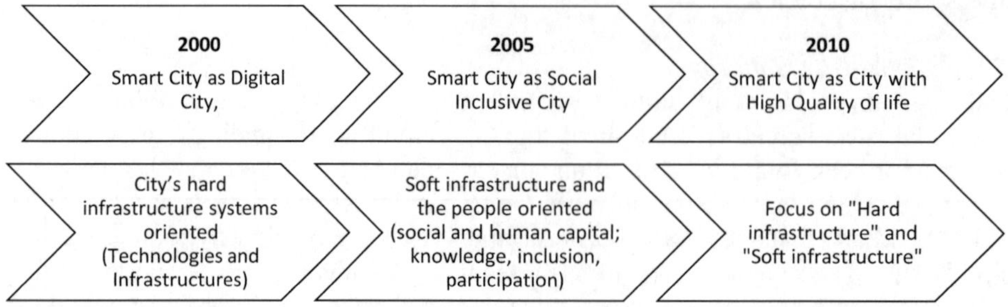

Figure. 1 Evolution of the smart city concept.

There are four preferred approaches in the strategic planning of smart cities. One approach comprises planning which focuses on hard infrastructure, e.g. the management of transportation, water, sewerage, waste, and energy. Another planning approach focuses on soft infrastructure, which comprises a focus on communities, community capacity building, human resources, knowledge improvement, community participation, social innovation, social justice, etc. Every approach in strategic planning has its own advantages and disadvantages. Therefore, if a city plans to use the smart concept, it must first identify its local characteristics, potential, priorities, and its needs in terms of creating a global market and availability of technology [32].

Some definitions of the smart city emphasize the use of modern technology in daily urban life, which enables the city to produce an innovative transportation and infrastructure system and an efficient energy saving system [33]. However, quite a few definitions refer to factors that improve the city's quality of life. Figures 1 and 2 give an overview of smart city definitions based on their orientation of the concept.

Table 1. Smart city definition oriented and focusing on 'hardware'.

No	Definition	Reference
1.	A smart city is a city that monitors and integrates the condition of every critical infrastructure, such as roads, bridges, tunnels, railway, subway lines, airports, ports, communications, water, electricity, and major buildings, A smart city must conduct preventive maintenance, surveillance on security aspects, and maximize service for the community [34].	Hall, 2000.
2.	IBM defines the smart city as the skill of a city to use information and communication technology to sense, analyze, and integrate key information of the core systems in running a city [35].	IBM, 2015.
3.	The smart city is a concept that aims for the residents, object, utilities, and other city elements to be well connected by using new and advanced technologies provided anywhere, in order to significantly increased the living experience in the 21st-century urban environment [6].	Northstream, 2010, in Cocchia, 2014.
4.	The smart city is a digitalized city product combined with the 'internet of things' [36].	Sue et al., 2011.
5.	A smart city is a city that is able to harmonize many types of technologies as diverse as water recycling and advanced energy grids and also uses mobile communication, aiming to reduce environmental impacts to offer residents a better quality of life [11].	Setis-EU, 2012.
6.	A smart city is where the advantages of technology products are available within reach to make living in the big city much more convenient [37].	Townsend, 2013.

No	Definition	Reference
7.	ICT and technology utilization as an effort to restore and improve the quality of life and degradation of environmental qualities [38].	IEEE, 2014.

Apart from an orientation on hardware, the smart city concept also focuses on software aspects, or better yet, on both.

Table 2. Smart city definitions focusing on people and 'software'.

No	Definition	Reference
1.	A smart society is a community that consciously makes effort to use information technology in improving their quality of life by cooperating with the government and other factors linked as a whole in significant and fundamental ways, not separately [6].	California Institute, 2001, in Cocchia, 2014.
2.	A smart city is a city with a good appearance and is built with a smart combination of its qualities, independent and with awareness of its residents [39].	Giffinger, 2007.
3.	A city transforms into a smart city when it has invested in human and social capital, as well as in traditional and modern ICT, creating sustainable economic growth and a better quality of life, with good management of human resources through participatory governance [18].	Caragliu et al., 2009.
4.	Smart cities are defined as geographic regions that possess advanced technologies such as ICT, logistic, energy production, which can benefit its residents in terms of their wellbeing, inclusion, and participation, improving their quality of life and the development of intelligence [29].	Dameri, 2013.
5.	A smart city utilizes ICT and technology to achieve welfare, effectiveness, and competitiveness of the city [32].	Angelidou, 2014.

Based on the internet searches, sourced from various smart city articles, the definition of smart city that focuses on hardware are relatively more that those found on software focuses, in fact, some of those seems vague and bias. The harmonization of both focus, are viewed as best step to establish a new smart city definition.

3.2 Smart City Definition

Definitions of the smart city are grouped based on various perspectives in several research publications. One focus is on 'hardware' and ICT. The discussion can be classified into four parts: (1) ICT (Information and Communications Technology) and modern technology is seen as the key driver in the smart city concept ([40]; [41]; [42]; [5]; [43]; [44]); (2) the role of ICT and technology in actualizing the welfare, effectiveness, and competitiveness of its residents [32]; (3) the utilization of advanced technology products to make living in a big city much more convenient [37]; (4) a broader understanding; the utilization of ICT and technology also constitutes an effort to improve the quality of life and prevent the degradation of environmental quality [38].

From the grouping of definitions, it is evident that the focus on 'hardware' is more dominant than the 'software' sides. Group 1, 3, and 4 emphasize the 'hardware' sides, and group 2 focuses more on the software side. In practice, cities mostly apply the concept proposed by EU, in which implementation is aimed at improving the environment. This is based on the desire to obtain public funding for projects undertaken to create a better city form. This is different to researchers in an academic environment, who do not only focus on the environment but also on other issues such as improving the quality of life, which relies on the level of culture, data information, sharing of knowledge and its relation to some aspects of community life, such as human capital, communication between citizens, etc. [6].

The grouping of smart city concepts is also carried out by other experts, one of which groups smart city concepts into four themes [45],: (1) innovation, economy, competitiveness, and added value on the global level, with a focus on added value locally in a global context, as well as a focus on innovation so people will have global talent and a global network to increase competitiveness; (2) a focus on developing technology and ICT; (3) attention to stakeholders of development, which emphasizes the importance of the private sector and smart social community capital, the importance of community involvement and the leadership factor in implementing smart cities, along with community empowerment; (4) the process towards a smart city, both in the process of analysis of the smart city as a model or as a development process, top down as well as bottom up. Cocchia, 2014 [6] also grouped some definitions based on their field and focus, taken from frequently cited literature: (1) the role of public services and the private sector in improving the quality of life; (2) the importance of attention to the damage caused by urban activities; (3) the importance of the concept of good natural resource management; and (4) sustainable economic growth.

These two groupings show that concepts with a focus on hardware are not too dominant; instead, there is more focus on aspects of environmental improvement, innovation, process, and the improvement of community's quality of life through smart city concepts.

3.3 Resilience Approach in Smart City Concept

The increasing urban population creates significant challenges for the city, such as urbanization, climate change, terrorism, and the increased risk of natural disasters. To face these circumstances, the city must learn to adapt in dealing with these challenges; cities must learn to develop its resilience in facing the problems of this era of unpredictability and uncertainty. The concept of resilience is a means to deal with these challenges. In the context of urban planning, the strategies that have been implemented mirror the philosophy of sustainable development, which focuses on managing resources to ensure the welfare of future generations. The study of resilience in sustainable development initially focuses on ecological safety. However, as the concept developed it was viewed as an important step in building sustainability [46].

The concept of resilience was initially introduced by a well-known ecologist, CS. Holling, who in 1973 suggested two general approaches, i.e. first, man and nature are closely linked and evolve together and must thus be conceived as one social ecology system; second, the responses of this system towards changes are unpredictable, but not proven [47]. After that, the focus resilience changed to disasters ([48] & [49] and then to social study ([50]; [51]; [52]), social-ecology studies [53] [54], and entered the field of planning ([55]; [56]; [57]; [58]).

In planning, the concept of resilience is developed by Wildazsky in 1988. It has evolved from the context of extremely powerful disasters, terrorism, energy crises, and climate change ([56];[59]); social-ecological resilience in a context of communicative planning [57]; and collaborative planning for governance in addressing resilience [58]. Following is a scheme of the three main stages of evolution of the concept of resilience.

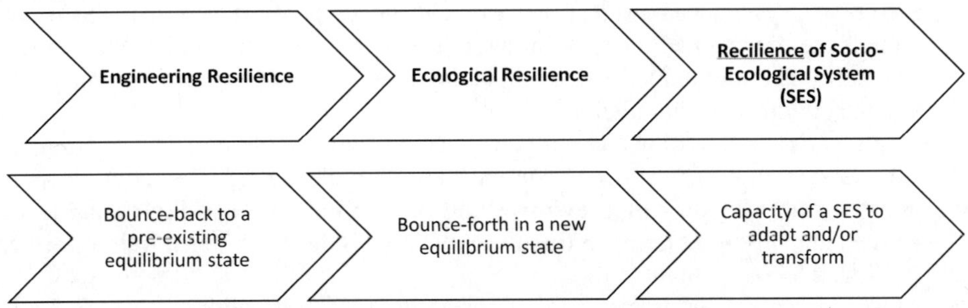

Figure. 2 Evolution of the concept of resilience.

Table 3. Definitions and understanding of the concept of resilience, particularly in the science of urban planning.

No	Definition	Reference
1.	Planning urban resilience is conducted to analyze the impact of pressures that a city faces. Therefore, its planning requires an evaluation of the city's vulnerabilities, understanding of the process, procedures, interaction, and capacity building to develop several components structures and their interaction with the main purpose achieving resilience [19].	Desouza and Flanery, 2013.
2.	The concept of resilience is used as a critique towards sustainability, which gained popularity in a different field of policy [60].	Davoudi, 2012.
3.	Resilience has 3 features, i.e. persistence, adaptability, and transformability which each interact from the local to the global scale [61].	Folke, 2006.
4.	Resilience refers to the capacity of individuals, households, groups in society, or systems to absorb and recover from the impact and danger of climate change and other shocks and stresses in the long term [62].	Turnbull et al., 2013
5.	Factors that affect resilience comprise institutional, political, cultural, social, environmental, human, economic, and physical factors. All factors are expected to become a unity and fair [62].	Turnbull et al., 2013
6.	A resilient city mirrors urban capacity through its individuals, communities, institutions, companies, and system to survive, adapt and develop, no matter how hard or great the shock faced [63].	Arup, 2013.
7.	A resilient city is a city that is capable of enduring and absorbing the impact of hazards, shocks, and pressures through adaptation or transformation in order to ensure long-term sustainability as well as basic functions, characteristics and structure [64].	UNISDR, 2012.
8.	Resilience is the capacity of a social-ecological system and its components to overcome hazardous shocks and pressures at a precise and efficient time to respond, adapt, and change ways to recover, maintain, and improve the main functions, structures, and identities in preserving the capacity to growth and changes [65].	IPCC, Arup, in ICLEI, 2016.
9.	A vision of a resilient city reflects the efforts to enhance the city's ability to respond to heterogenic pressure factors (climate, environment, energy and economy) with targets to ensure a higher quality of life and sustainable urban development [31].	Papa et al., 2015.

Understanding the smart city planning concept by incorporating elements of resilience can be re-interpreted to find a new definition. Based on the previous discussion, it can be concluded that the existing definitions of the smart city are insufficient and an additional definition is required. In the context of urban planning, particularly for smart city concepts, the resilient city plays an important role.

A city as a complex system and its resilient capacity needs to be developed; the multi-component capacity, which is the process and interaction that raises it above the city's physical limits. When a city is considered smart it is important to be resilient at any time [22]. Therefore, the concept of resilience is one of the key factors in smart city planning.

Based on various definitions about smart cities and resilience, the existing concepts and definitions of the smart city need to be redefined and complemented. To redefine the concept, the following steps conducted: gathering previous smart city definitions focusing on hardware aspects, to deduct the most frequently used terms in literature, which will be a reference in the re-interpretation of the smart city with a focus on 'hardware'.

The most frequently used terms in literature on the smart city with a focus on hardware consist of: (1) first, the utilizations of modern or advanced technology in information and communication; (2) second, the importance of integrated infrastructure, utilities, and urban systems; and (3) finally, the goal of the smart city concept, i.e. to improve the quality of life and to restore degraded environmental quality. From the analysis, it is can be concluded that the definition of the smart city with a hardware orientation is a concept of a city that utilizes modern and advanced information and communication technology in order to realize city planning that offers a better quality of life and better environment quality in the city. The re-definition process of software aspects also uses the same steps, the software-focused smart city concept also refers to the most frequently mentioned terms in current definitions. From the selected literature, the most frequently discussed material consists of (1) the role of using information and communication technology; (2) the awareness and intelligence, well-being, participation, effectiveness, and competitiveness of the city's community; (3) the objective of the smart city to create higher quality of life and to improve environment quality. From the analysis, it can be concluded, that the software-focused smart city concept relates to the utilization of information and communication technology as a means to obtain high quality of life and improve environmental quality.

3.4 Analysis

Analysis of the above definitions shows that the hardware-oriented and software-oriented definitions commonly mention the utilization of information and communication as well as the objectives, which are improved quality of life and of the environment, while the difference is the object of the smart city concept. The hardware-oriented concept focuses on the city's infrastructure, utilities, and urban system, whilst the software-oriented concept emphasizes the community, e.g. intelligence, wellbeing, participation, effectiveness and competitiveness.

In this paper, the factor of resilience will only be included in the definition of the smart city concept with a software-oriented approach. The combination of the definitions of the smart city concept and resilience is expected to establish a new definition that will represent both concepts.

The definition of resilience is based on the most commonly discusses aspects in literature. Based on the analysis, the most commonly appearing and discussed aspects are: (1) the city's capacity in dealing with pressure, shocks, and hazards; (2) the ability to survive and adapt, being resilient and able to transform; (3) the process of dealing with changes in the era of uncertainty; (4) toward sustainability. Based on the literature, it can be analyzed that resilience is the capacity of a city to deal with pressures, shocks, and hazards, in an effort to survive, adapt, be resilient, and able to transform through a process in facing changes in the uncertainty era towards city's sustainability.

Through re-definition of the concept mentioned above, the re-definition of the smart city concept based on a resilience approach should, thus, at least comprise some of the following factors: (1) the role of information communication technology utilization; (2) the awareness and intelligence, well-being, participation, effectiveness and competitiveness of the community; (3) the ability to deal with pressure, shocks and hazards; (4) to survive, adapt, have determination, and able to transform; (5) the process of facing changes in the era of uncertainty; (6) to create a high quality of life as well as improve environmental quality; (7) towards sustainability. Overall, it can be concluded that a smart city with a resilience approach is a concept that utilizes ICT in enhancing awareness, intelligence, well-being, and community participation in dealing with pressures, shocks, and hazards in order to survive, adapt, be tough, and be able to transform to have a future with a better environment and higher quality of life, which is sustainable in facing the era of uncertainty.

4. Conclusions

The phenomena of urbanization and the development of information technology lead to the emergence of the smart city idea. Initially, this concept was expected to overcome city issues while considering the environment. As the concept developed, it is widely used in urban planning and is supported and adopted by many institutions, such as the European Commission, Setis-EU, OECD, etc. Since then, publications and analysis related to smart cities experienced a fast development. However, since the concept has become widespread, its original meaning and definition are still vaguely defined and there is still inconsistency in defining the smart city concept. Moreover, several new terminologies are used in the smart city concept, e.g. ubiquitous city, wired city, knowledge city, digital city, etc.

Currently, the most prominent literature on the smart city concept focuses on the development of modern technology and information communication technology. ICT and modern technology are considered as key to the smart city concept and all planning processes are conducted based on ICT for efficiency, whereas in fact, this is not the main factor. A city as a complex system needs to build its resilience capacity. Unpredictable events and natural disasters might happen anywhere and anytime. Such issues should be considered in a smart city as a system. In urban planning, when a city is planned to be smart, it is important to be resilient at any time.

This paper aims to redefine the smart city concept through literature study, in the context of the current uncertainty era, with a resilience approach. Literature review on the smart city and resilience is conducted using a systematic literature review methodology. The methodology consists of summarizing previous research and carry out with a more comprehensive and balanced analysis. Synthesis uses a narrative technique with a meta-aggregation approach. This aim to aggregate relevant analysis and research, which is then re-interpreted to gain a new understanding concerning the definition and concept of the smart city.

From the above discussion, it can be concluded that the smart city concept through a resilience approach can be redefined as a concept of a city that utilizes ICT to increase citizen's awareness, intelligence, wellbeing as well as community participation in facing pressures, shocks, and hazard in order to be able to survive, adapt, be tough, and able to transform, in order to that the community achieve a higher quality of life and environment, which is sustainable in facing the future era of uncertainties.

References

[1] United Nations, Department of Economic a, nd Social Affairs, Population Division, and United Nations, Department of Economic a, World Urbanization Prospects The 2014 Revision. New York, 2015.

[2] OECD, "OECD Environmental Outlook to 2050. The Consequences of inaction," OCD Publ., 2012.

[3] A. Cocchia, "Smart and Digital City: A Systematic Literature Review," in Smart City, R. P. Dameri and C. Rosenthal-Sabroux, Eds. Cham: Springer International Publishing, 2014, pp. 13–43.

[4] T. Nam and T. A. Pardo, "Conceptualizing smart city with dimensions of technology, people, and institutions," 2011, p. 282.

[5] R. Jucevičius, I. Patašienė, and M. Patašius, "Digital Dimension of Smart City: Critical Analysis," Procedia - Soc. Behav. Sci., vol. 156, pp. 146–150, Nov. 2014.

[6] A. Cocchia, "Smart and Digital City: A Systematic Literature Review," in Smart City, R. P. Dameri and C. Rosenthal-Sabroux, Eds. Cham: Springer International Publishing, 2014, pp. 13–43.

[7] M. Angelidou, "Smart cities: A conjuncture of four forces," Cities, vol. 47, pp. 95–106, Sep.

2015.

[8] H. Chourabi et al., "Understanding Smart Cities: An Integrative Framework," 2012, pp. 2289–2297.

[9] M.-L. Marsal-Llacuna, J. Colomer-Llinàs, and J. Meléndez-Frigola, "Lessons in urban monitoring taken from sustainable and livable cities to better address the Smart Cities initiative," Technol. Forecast. Soc. Change, vol. 90, pp. 611–622, Jan. 2015.

[10] R. S. Wall and S. Stavropoulos, "Smart cities within world city networks," Appl. Econ. Lett., vol. 23, no. 12, pp. 875–879, Aug. 2016.

[11] European Commission, "Communication from the commission. Smart cities and communities – European innovation partnership," Brussels, 2012.

[12] H. Ahvenniemi, A. Huovila, I. Pinto-Seppä, and M. Airaksinen, "What are the differences between sustainable and smart cities?," Cities, vol. 60, pp. 234–245, Sep. 2016.

[13] D. Schuler, "Digital Cities and Digital Citizens," presented at the Revised Papers from the Second Kyoto Workshop on Digital Cities II, Computational and Sociological Approaches, 2001, pp. 71–85.

[14] H. Couclelis, "The Construction of the Digital City," Environ. Plan. B Plan. Des., vol. 31, no. 1, pp. 5–19, Feb. 2004.

[15] K. Ergazakis, K. Metaxiotis, and J. Psarras, "Towards knowledge cities: conceptual analysis and success stories," J. Knowl. Manag., vol. 8, no. 5, pp. 5–15, Oct. 2004.

[16] R. G. Hollands, "Will the real smart city please stand up?: Intelligent, progressive or entrepreneurial?," City, vol. 12, no. 3, pp. 303–320, Dec. 2008.

[17] L. Anthopoulos and P. Fitsilis, "From Digital to Ubiquitous Cities: Defining a Common Architecture for Urban Development," 2010, pp. 301–306.

[18] A. Caragliu, C. Del Bo, and P. Nijkamp, "Smart Cities in Europe," J. Urban Technol., vol. 18, no. 2, pp. 65–82, Apr. 2011.

[19] S. Goldsmith and S. Crawford, The Responsive City: Engaging Communities Through Data-Smart Governance, 1 edition. San Francisco, CA: Jossey-Bass, 2014.

[20] S. Kumar, "Institutional Arrangements for Managing Large-Scale Recovery: Key Lessons from 2004 Tsunami," in Recovery from the Indian Ocean Tsunami, R. Shaw, Ed. Springer Japan, 2015, pp. 103–119.

[21] L. Anthopoulos, M. Janssen, and V. Weerakkody, "Comparing Smart Cities with Different Modeling Approaches," presented at the International World Wide Web Conference Committee (IW3C2), Florence, Italy., 2015.

[22] K. C. Desouza and T. H. Flanery, "Designing, planning, and managing resilient cities: A conceptual framework," Cities, vol. 35, pp. 89–99, Dec. 2013.

[23] Direktur Perkotaan dan Perdesaan,, Kementerian PPN/ Bappenas, "Pengembangan Kota Cerdas di Indonesia," presented at the Konferensi e-Indonesia Initiative (eII) dan Smart Indonesia Initiatives (SII) Forum ke-1, Bandung, Indonesia., Oktober-2015.

[24] A. Perry and N. Hammond, "Systematic Review: The Experience of a PhD Student.," Psychology Learning and Teaching, vol. 2 (1), 2002.

[25] C. Francis and Baldesari, "Systematic Reviews of Qualitative Literature," Oxford: UK Cochrane Centre, 2006.

[26] S. Lewin, Methods to Synthesise Qualitative Evidence Alongside a Cochrane Intervention Review. London: London School of Hygiene and Tropical Medicine, 2008.

[27] D. Belanche, L. V. Casaló, and C. Orús, "City attachment and use of urban services: Benefits for smart cities," Cities, vol. 50, pp. 75–81, Feb. 2016.

[28] N. Komninos, P. Tsarchopoulos, and C. Kakderi, "New services design for smart cities: a planning roadmap for user-driven innovation," 2014, pp. 29–38.

[29] R. P. Dameri and C. Rosenthal-Sabroux, Eds., "Smart City and Value Creation," in Smart City, Cham: Springer International Publishing, 2014, pp. 1–12.

[30] I. Oberti and A. S. Pavesi, "The triumph of the smart city," TECHNE - J. Technol. Archit. Environ., vol. 0, no. 5, pp. 117–122, 2013.

[31] R. Papa, A. Galderisi, M. C. Vigo Majello, and E. Saretta, "Smart and Resilient Cities. A Systemic Approach for Developing Cross-sectoral Strategies in the Face of Climate Change," TeMA - J. Land Use Mobil. Environ., vol. 8(1):19-49, 2015.

[32] M. Angelidou, "Smart city policies: A spatial approach," Cities, vol. 41, pp. S3–S11, Jul. 2014.

[33] P. Lombardi, "New challenges in the evaluation of Smart Cities," Netw. Ind. Q., vol. 13, no. 3, pp. 8–10, 2011.

[34] P. Hall, "Creative cities and economic development," Urban Stud., vol. 37, no. 4, pp. 639–649, 2000.

[35] "IBM - United States," 01-Oct-2015. [Online]. Available: http://www.ibm.com/smarterplanet/us/en/.

[36] K. Su, J. Li, and H. Fu, "Smart city and the applications," 2011, pp. 1028–1031.

[37] A. M. Townsend, Smart Cities: Big Data, Civic Hackers, and the Quest for a New Utopia, 1 edition. New York: W. W. Norton & Company, 2013.

[38] IEEE (the Institute of Electrical and Electronics Engineers), "IEEE Smart Cities," 2014. .

[39] G. Rudolf, C. Fertner, H. Kramar, R. Kalasek, N. Pichler-Milanovic, and E. Meijers, "Smart cities-ranking of european medium-sized cities," Rapp. Tech. Vienna Cent. Reg. Sci., 2007.

[40] J. A. A. González and A. Rossi, "New trends for smart cities, open innovation mechanisms in smart cities. European commission with the ICT policy support programme.," vol. 270896, 2011.

[41] C. Harrison and I. A. Donnelly, "A Theory of Smart Cities," Proc. 55th Annu. Meet. ISSS - 2011 Hull UK, vol. 55, no. 1, Sep. 2011.

[42] H.-N. Hsieh, C.-Y. Chou, C.-C. Chen, and Y.-Y. Chen, "The evaluating indices and promoting strategies of intelligent city in Taiwan," 2011, pp. 6704–6709.

[43] S. Paroutis, M. Bennett, and L. Heracleous, "A strategic view on smart city technology: The case of IBM Smarter Cities during a recession," Technol. Forecast. Soc. Change, vol. 89, pp. 262–272, Nov. 2014.

[44] D. Washburn, U. Sindhu, S. Balaouras, R. A. Dines, N. M. Hayes, and L. E. Nelson, "Helping CIOs Understand 'Smart City' Initiatives De ning the Smart City, Its Drivers, and the role Of the CIO," MA: Forrester Research, Inc., vol. Cambridge, 2010.

[45] R. Sutriadi, Perspektif Perencana: Smart City, Inovasi, Kota Komunikatif, dan Kota Berkeadilan. Bandung, Indonesia.: Inside Publisher, 2015.

[46] Levin, "Science and Sustainability," Ecol. Appl., vol. 3, no. 4, pp. 759–762, 1993.

[47] C. S. Holling, "Resilience and Stability of Ecological Systems," Annu. Rev. Ecol. Syst., vol. 4, pp. 1–23, 1973.

[48] P. Timmerman, Vulnerability, resilience and the collapse of society: a review of models and possible climatic applications. Toronto: institute for Environmental Studies, University of Toronto, 1981.

[49] G. A. Tobin and B. E. Montz, Natural Hazards: Explanation and Integration, 1 edition. New York: The Guilford Press, 1997.

[50] W. N. Adger, "Social-Ecological Resilience to Coastal Disasters," Science, vol. 309, no. 5737, pp. 1036–1039, Aug. 2005.

[51] B. Obrist, C. Pfeiffer, and R. Henley, "Multi-layered social resilience: a new approach in mitigation research," Prog. Dev. Stud., vol. 10, no. 4, pp. 283–293, Oct. 2010.

[52] R. K. Larsen, E. Calgaro, and F. Thomalla, "Governing resilience building in Thailand's tourism-dependent coastal communities: Conceptualising stakeholder agency in social–ecological systems," Glob. Environ. Change, vol. 21, no. 2, pp. 481–491, May 2011.

[53] C. S. Holling, "Understanding the Complexity of Economic, Ecological, and Social Systems," Ecosystems, vol. 4, no. 5, pp. 390–405, Aug. 2001.

[54] B. Starzomski, "Berkes, F., J. Colding, and C. Folke. 2003. Navigating social-ecological systems: Building resilience for complexity and change. Cambridge University Press, Cambridge, UK.," Ecol. Soc., vol. 9, no. 1, Jan. 2004.

[55] A. Wildavsky, Searching for Safety. Transaction Publ., 1988.

[56] D. R. Godschalk, "Urban Hazard Mitigation: Creating Resilient Cities," Nat. Hazards Rev., vol. 4, no. 3, pp. 136–143, Aug. 2003.

[57] B. Goldstein, "Resilience to surprises through communicative planning," Ecol. Soc., vol. 14, no. 2, p. 33, 2009.

[58] J. E. Innes and D. E. Booher, Planning with Complexity: An Introduction to Collaborative Rationality for Public Policy, 1 edition. Milton Park, Abingdon, Oxon ; New York, NY: Routledge, 2010.

[59] L. J. Vale and T. J. Campanella, Eds., The Resilient City: How Modern Cities Recover from Disaster, 1 edition. New York: Oxford University Press, 2005.

[60] S. Davoudi, "Resilience: A Bridging Concept or a Dead End? "," Plan. Theory Pract., vol. 13, no. 2, pp. 299–333, Jun. 2012.

[61] C. Folke, "Resilience: The emergence of a perspective for social–ecological systems analyses," Glob. Environ. Change, vol. 16, no. 3, pp. 253–267, Aug. 2006.

[62] M. Turnbull, C. L. Sterrett, and A. Hilleboe, Toward Resilience: A Guide to Disaster Risk Reduction and Climate Change Adaptation. Practical Action Pub, 2013.

[63] ARUP, "Arup | Publications | City Resilience Index," 2013. [Online]. Available: http://publications.arup.com/publications/c/city_resilience_index. [Accessed: 19-Feb-2017].

[64] UNISDR, How To Make Cities More Resilient A Handbook For Local Government Leaders. Geneva: United Nations, 2012.

[65] ICLEI, "Resilient Cities - ICLEI: Glossary of key terms," 2016. [Online]. Available: http://resilient-cities.iclei.org/resilient-cities-hub-site/resilience-resource-point/glossary-of-key-terms/. [Accessed: 18-Feb-2017].

Built urban heritage conservation in Islamic societies: Study case in Banda Aceh, Indonesia

Z D Meutia

Department of Urban and Regional Planning School of Architecture, Planning and Policy Development (SAPPD), Intitute of Technology Bandung, Indonesia-Department of Architecture, University of Muhamaddiyah Aceh

zyadyena@s.itb.ac.id

Abstract. This paper aims to find the concept of the built heritage related to Islamic societies with a case study in the city of Banda Aceh through study literature review, with the context of the planning in the era of uncertainty. In this paper will be elaborated and described what it was about heritage and urban heritage and conservation Islamic values in heritage, as well as explain the concept of the built heritage conservation in Islamic societies. Discussion and analysis will be done through its study literature. Literature reviews about built urban heritage conservation and perspective of the Islamic societies in Banda aceh was done using systematic methodology literature review. This methodology summarizes research results earlier that presents the fact that a more comprehensive and balanced. The synthesis of the results conducted using narrative techniques or technique of qualitative. The discovery paper in this paper is to understand the relationship the built heritage conservation of Islamic societies perspective that consider shari'a aspect and local tradition in built urban heritage that can affect to heritage planning.

1. Introduction

The concept of conserved and preserved historic cities in the Islamicate world is a recent development [1]. Heritage planning in historic cities, like the ones in Islamic societies, have to be considered carefully. They often do not represent the conditions that gave rise to the traditional urban conservation and preservation in planning approaches. This emphasis on a discontinuity between the present and the past still is a part of modern practical problem as those included in the Athens and Venice Charters [[2]-[3]] and are often transferred to, or even asserted on, many other parts of the world, where in most cases the past is still very much alive in the present [4]. How to deal with these changes of the historic cities in Islamic societies, therefore, is concern for all those interested in conserving urban heritage and identity.

The differential to what many people in the West might believe, Islamic societies is not monolithic [5]. Many Islamic societies in the world today while some agreement may lot across cultures, classes, religious and continents of the Muslim world, the diversity is at least as glaring [6]. This is because of a diversity of interpretations and practices of Islam in different parts of the world. The term 'Islamic societies' has been used in this paper to indicate Muslim majority countries or regions where religion-based principles and local traditions have a central role in everyday life. The plural thing has been used to identify the diversity of Islamic practices and interpretations across the world.

Now days, the heritage planning of historic site and cities has become significance as amount of designated urban World Heritage Sites (WHSs) has developed across the world over the last two decades

[7]. Banda Aceh has developed with historical city. That historic city has through in the kingdom, colonial, and independence period. Banda Aceh City as the capital of Aceh Province is one of the priorities in the program heritage city of Indonesia. The reason for establish Banda Aceh as one of the ten cities in the priorities of the program because the city has a rich culture, local wisdom and history both heritage tangible and intangible. They also decide zone of the priority area, but it is not comprehensive espescially from perspective Islamic societies for built heritage conservation in Banda Aceh. Aceh which was once the largest Islamic center in Southeast Asia is currently one of the provinces in Indonesia, which regulate Islamic law. Realization of this spesific law into his own challenge for the capital of Aceh in general and the city of Banda Aceh specifically scope to conserved heritage in Islamic societies. In addition to the application of sharia law in Aceh is also known as the starting point of the history of the spread of Islam in Southeast Asia, the beautiful landscape of nature, the wealth of Islamic history and culture, and heritage sites are also tragedi tsunami 2004.

Geerts argument that Islam may driven as a layer over existing indigenous local and belief systems also must be considered here [8]. In Indonesia, he argues, inspiration is specific to a societies, whereas Islam provided a constant belief system for move societies coming from Arabia. Yet there is no doubt, better both the culture and sense of the Verandah of Mecca is construed as being meaningful in an Islamic values of heritage. A view of Aceh in the 1930s offers a context in which heritage can be problematized, Siegel has argued that "the Acehnese see their past as periods of either rationality or irrationality. They think of history as cyclical, not however as the repetition of certain events but as the emergence and retreat of "akal" [9]. In the reconstruction of society, it is only need to bring about the drive control of inner life, and a good society will automatically develop. This calls for in depth study of the role of Islam in creating a special conservation philosophy.

The concept of built heritage conservation is very complicated in terms of planning approaches, actors involved, technical methods and implementation processes. The literature review of discussion presented in this paper is divided into several sections: firstly, introducing the terminology of heritage and urban heritage and secondly finding relation of built heritage conservation and Islamic societies in Banda Aceh. The paper used the literature review linked the evolution of the built heritage conservation concept both intellectually and practically with case study.

2. Literature Review

This paper also described about heritage, urban heritage and built heritage in Indonesia from literature review that related with perspective Islamic societies.

2.1 Terminology of Heritage and Urban Heritage

Heritage has many meanings and connotations, which vary with the perspective from which we look at it. [10] explained that heritage is a symbolic resource, strongly related between of collective memory and identity [10]. Furthermore, [11] claims that: "The content of heritage is commonly seen as embracing both the tangible (monuments, natural landscapes, buildings and the like) and the intangible (traditions, faiths, myths, folklore, rituals, and others)"[12]. Whether the heritage is tangible or intangible, it acquires a symbolic significance from its architectural elements, its historial value, or its contemporary importance [13]. The Open University [14] described defining heritage as: "Heritage have to be seen as separate part from the search of historic knowledge, as it is concerned with the re-packaging of the past for some purpose in the present and future." [15] said that most urban planners and architect disposed to define urban heritage as "monuments". He also states [13] that "this comprehension expel historic residential areas and historic city centres, which equally describe the urban heritage." Whether it buildings authenthicity had culture, colonial, social, military, religious or economic value, they are all part of the national heritage in this perspective, and the assets of urban heritage must be preserved so that people in the present and the future can get benefit from them.

Built heritage conservation is a "positive", "dynamic", and "wide-ranging", concept exceeding the earlier, narrow sense of "preservation"; the latter means preventing further decrease, whereas conservation actually requires of change. [16] expresses this with firmly: conservation is "part of the process of change" and it requires the past to be upgraded, modernised, reshaped, and even fabricated to meet contemporary expectations [17]. However, [18] said that built heritage conservation has two opposing side, declare the past in a positive ray as inherently valuable and worth cultivating, but also managing and detaining the future by congealing the heritage site in its new form [19]. Thus, the present can changeable between a tendency to preserve the past as it was and the forward momentum of a projected future [20]. From another perspective, heritage has a value in educating present and future generations as to the historical significance of their built heritage and the necessary of its conservation [21]. Increased concern over the value and management of heritage is reflected in the literature on dissonance heritage [22], integrated heritage management [[19]-[20]], heritage visitor attraction [20], clarifying the core heritage [23], heritage and postmodern heritage [24] and consuming heritage [25].

The materialization and acknowledment of heritage is because of the continuing maintenance and use of them by the societies that inhabit and utilize them [26]. [24] refers to link between material culture and living culture, where the local societies who use their built heritage are agree more valuable as a cultural asset than the buildings themselves. In this way, the principles set forth by the Venice Charter are unique inadequate. While the Charter values built heritage as the most effective server of meaning from the past, living cultures also have the responsibility and capability to effectively illness values from the past, values that may not be able to be informed uniquely through the built heritage. Cowherd describes that "to the expand it the Venice Charter concern on the material objects of conservation to the disimissal of their material local customs and tradition, it needs to be considered and modified to the discrimination of a valuable physical culture from its living culture" [27]. Indonesia, like other Asian nations, has rich history and culture of challenging the nature of the global heritage discourse. This endurance must be seen as an authenticity way to prevent the destruction of heritage values that would come at the hands not enough global discourse. Such opposition emerges therefore as reality non-Western understanding of cultural heritage.

2.2 Conservation and Islamic values

It is not clear why there is a void in debates about the religious context of conservation principles, particularly when it comes to Islamic Heritage and values. Still to be over taked is an interrogation of how the dominant heritage discourse and world heritage model understanding Islamic Values and whether there is an emerging resistance to suggest alternative ways of preserving the value of materility in this context. In addition, the phsycal and rhetorical construction and destruction of cultural heritage in the context of religion has not been discussed. Do research how heritage is constructed within the spiritual values of Islam demands that we depart from an assumption that the attendance of islam is identify in the records of material culture. Islamic heritage acknowledges that the completeness and variety of islamic material culture may be recognizabled and used to construe Muslim societies in the past [28]. In Indonesia, it has been argued that Islam did not construct a civilization, it appropriated one, resulting in a lack of uniformity in the forms that Islam took [8]. In Geertz's view, the "struggle for the real" [30], between secular and religious interpretations of the world in this region is chronic and intense rendering the history of islam unintelligible. It could be argued, however that moving away from the idea of Islam as a visibly clearly different cultural practice and further seeking to construct it as an essentially unique Islamic conservation ethos, is simply another way for the West to produce the oriental in the cultural world [29]. In this sense, the use of an "Islamic" label may be seen as arbitatry [5], it should be noted that an Oriental dichotomy is already in the making with an earlier proposal to define an "East and west" approach to conservation [31] a proposal that assumes more or less clear boundaries in an Asian approach to materiality.

What would a conservation philosophy that was intrinsically Islamic entail? An identification of islamic heritage needs to depart from a consideration of the "social imaginary" of Muslims [32], the way that Islam or other cultures visualize themselves, which has been heavily influenced by romantic and a historical visions of the past. To comprehend the built of heritage, one must also differentiate between the world of the religious, the secular and the sacred within islamic cultures and nature. In

IOP Conf. Series: Earth and Environmental Science **70** (2017) 012066 doi:10.1088/1755-1315/70/1/012066

Islamic law (shari'a), does not offer any admonition on whether preserving Islamic heritage is required, recommended, discouraged, prohibited, or permitted but morally indifferent. Facing a shortage of explicit discourse of materiality and authenticity in the Quran or the hadith, we are left, with allotment commentaries on how to approach material culture from the most significance of Islamic symbols. From this history, it is possible to infer an Islamic philosophy of conservation and ideas of originality and "best practices", but only at a high cost of speculation. There is no doubt, however a suitable way to care for the monuments of Islam. Notions of pollution and determined purifications associated with its materiality and spirituality is one central concern of conservation practices. That is, a religious authority establish the conditions where these objects may be handled for conservation. There are rules in trace, for the handling of manuscripts of the Quran such as the curatorial regulation manual in place in the Islamic Arts Museum in Malaysia, that decide how the collection will be curated [32]. However, there is no research to help construct a positive or negative critique of rules for handling Islamic Heritage, and how these rules are embedded in larger global world of conservation and management.

A Result but significance considerations in the meaningful of islamic heritage is the way that the significance of its power is conserved and transmited through the legitimate text of the Quran. The transmission of the text was and is strengthened by a strong tradition of recitation that consolidate its written transmission and its precision. Possible it can stand as an example of the relative competence that material culture has particularly as the vehicle for the perpetuation os Islamic traditions and beliefs. This opinion would turn any essentially Islamic conservation philosophy into an artificial creation and traffict my opinions back to the question of legitimizing communication of heritage expertise.

3. Methods

The methodological framework developed for this paper is summarizes on a review of the international academic literature, collected through a scoping review of reviews to get new information related topic in this paper. Study literature review about built urban heritage in an era of uncertainty especially in Indonesia was conducted using methodology systematic literature review. This methodology was done in summarizes results research that previous presenting the fact a more comprehensive result and balanced. Synthesis result is effected using a technique narrative or technique qualitative. A qualitative approach in methodology systematic literature review used to synthesize research results previous is descriptive qualitative. A method of summarizes the results of the study this qualitative called by meta-sintesis, that is technique in doing integration data for have the theory and a new concept in tiers a profound understanding and through [33]. The data in literature methodology used systematic review done through searching journals via internet and results of the recent research who has posted. Steps to be performed in to do this research, through methodology systematic literature review qualitatively [34]. Research through a method of meta-sintesis or also called with the methods synthesis qualitative data can be done with two approach. The first approach that is use meta-aggregation and meta-ethnography [35]. On the approach of meta aggregation, this synthesis aims to answer questions research by means of summarizes of various research results. While approach meta-etnografy the synthesis aimed at to develop the new theory in order to furnish the theory existing.

4. Results

Relation of Built Urban Heritage Conservation and Islamic Societies in Banda Aceh, Indonesia

Islam in Indonesia has some unique characteristics. Viewed within this context, the key question that often appears to us is the way the authenticity of Islam should be interpreted, especially when the daily practices of the religion encounter local tradition. Historically speaking, a dispute over Islamic authenticity has existed for ages among Muslims in Indonesia. Nowadays, the direction of the development of urban development in Banda Aceh, especially in the old area, is weak controlled so that are likely to die, stopped and concealed, less productive and disorderly. Restoration Ulee lheue Mosque, for example, have deprive of the character primary of dome originally the graceful. Another example, the old area around Mosque and Keraton who conserved covering broad 100 ha [36]. Besides Mosque and Keraton, old town is also has some of objects and a historic landmark, as Taman Sari, Gunongan, Pinto Khop, Pendopo, Grave Sultan Iskandar Muda, Kandang XII, Kerkhof, Bells cakra donya, and the Museum Aceh. Various efforts the local government in determine policy conserved heritage in Islam Community and consider also the concept of a line direction (Kiblat) Mosque Baiturrahman need to be considered also in planning conservation and revitalisation of historical city in Banda Aceh. This conservation must be considered about local identity and shari'a as aspect preservation heritage tangible and intangible in Banda Aceh also how to built urban heritage in Islam communities.

5. Conclusions

The paper discuss about terminology heritage, urban heritage and built urban heritage conservation in the era of uncertainty with consider Islamic communities to get new perspective in heritage planning. Banda Aceh have strong religious character. The kingdom period is the Islamic civilization in Banda Aceh. Up to the present time, historical and cultural against the influential Islamic of Banda Aceh. Historical of Banda Aceh is not just in a grand Islam course, but also influence to the heritage value from perspective social and spiritual Islamic communities that establish local policy based on shari'a in heritage planning. The discovery paper in this is to understand the relationship the built heritage conservation of Islamic communities perspective that consider shari'a aspect and local tradition in built urban heritage that can affect to heritage planning in the era of uncertainty.

References

[1] Nasser,N.(2002). Islamicate urbanism: The state of heart, Built Environment, 28, pp.173.

[2] League of Nations. (1931). The Piagam of Venice Charter and Athens Charter.

[3] ICOMOS (1994), The Venice Charter, ICOMOS,Venice.

[4] Byrne,D.(1991).Western Hegemony in Archaeological Heritage Management. History and Anthropology 5:269-276.

[5] Said, E. (1978). Orientalism. New York. Pantheon.

[6] Shaheed, F. (1995), "Networking for Change : the role's of women's groups in iniating dialogue on woman's issues, in M. Afkhami (ed). Faith and Freedom ; Women's Human Rights in the Muslim World, London : I.B.Tauris, pp. 78-103.

[7] Pendlebury, J., et al. (2009) Urban World Heritage Sites and the problem of authenticity, Cities Journal, issue No.26, pp.349–358.

[8] Geertz, C. (1968). Islam Observed: Religious Development in Morocco and Indonesia.

[9] Siegel, J. T. (1969) The Rope of God. University of California Press, Berkeley and Los Angeles.

[10] Behiri,A. (2011), Heritage rehabilitation in sustainable development policy for a better environment quality in small historical coastal cities: The case of Cherchell in Algeria, International Conference on Green Buildings and Sustainable Cities, pp.753-759.

[11] Graham, B. and P. Howard (editors). (2008) The Ashgate Research Companion to Heritage and Identity. Ashgate Publishing Limited, Aldershot (UK) and Burlington (USA).

[12] Boukas et al. (2013). Olympic legacy and cultural tourism: exploring the facets of Athens' Olympic heritage, International Journal of Heritage Studies,Volume19, Issue 2, pp. 203-228

[13] Open University. (2016). http://www.open.edu/openlearn/history-the-arts/history.

[14] Steinberg, F. (1994). Conservation and rehabilitation of urban heritage in developing countries, Habitat International Volume 20, Issue3, September 1996, Pages 463–475.

[15] Lowenthal, D. (1985), The Past is a Foreign Country, Cambridge: Cambridge University Press,University Press, New Haven ,CT, 180-188.

[16] Carter, G. (1994), Heritage interpretation and environmental education, in Harrison R. (ed),

Manual of Heritage Management, Oxford : Butterworth Heinemann, pp 359.

[17] Radoine, H. (2003), Conservation-Based Cultural, Environmental, and Economic Development : The Case of the Walled City of Fez, In The Human Sustainable City: Challenges and Perspectives from the Habitat Agenda, ed. Luigi Fusco Girard. Burlington, Ashgate.

[18] Garrod,B., and A.Fyall. (2000). Managing Heritage Tourism, Annals of Tourism Research 21 :682–708/.

[19] Hall, C. M. and McArthur, S. (1998), Integrated heritage management: Principles and practice, London : Stationery Office.

[20] Leask, A. and Yeoman, I. (1999), Heritage visitor attractions: An operation management Perspective (eds.), London: Cassell.

[21] Tunbridge, E. and Ashworth, J. (1996), Dissonant heritage: The management of the past as are source in conflict,West Sussex:John Wiley and Sons Limited.

[22] Nuryanti, W. (1996), Heritage and post modern tourism, Annals of Tourism Research, 23(2): 249-260.

[23] Waitt, G. (2000), Consuming heritage perceived historical authenticity, Annals of Tourism Research,27:835-862.

[24] Cowherd, R. (1999) Indonesia and the new conservation. In Monuments and Sites Indonesia, edited by ICOMOS Indonesia, pp.1-8.PF Book, Bandung.

[25] Rossi, Aldo. (1982). The Architecture of The City. Cambridge : MITPress.

[26] Arif, K. (2009). Ragam Citra Kota: Banda Aceh: Interpretasi Sejarah, Memori Kolektif dan Arketipe Arsitekturnya.Pustaka Bustanussalatin.

[27] Waitt, G. (2000), Consuming heritage perceived historical authenticity, Annals of Tourism Research,27 : 835-862.

[28] Insoll, T. (1999). The Archaelogy of Islam. Blackwell Publishers, Oxford, Malden Mass.

[29] Cooperson, M. (2010). Culture. In Key Themes for the Study of Islam, edited by J. J. Elias, pp. 104-122. Oneworld Book, Oxford.

[30] Wei, C. and A. Aass. (1989). Heritage Conservation: East and West. ICOMOS Information 3:3-8.

[31] Arkoun, M. (1973). Essais sur la pensee Islamique. Maisonneuve et Larose, Paris.

[32] Zekrgoo, A. H. and M. Barkeshli. (2005). Collection management of Islamic heritage in accordance with the worldview and Shari´a of Islam. In Conservation of Living Religious Heritage : papers from the ICCROM 2003 Forum on living religious heritage : conserving the sacred edited by H. Stovel, N. Stanley-Price and R. Killick, pp. 94-101. ICCROM Conservation Studies vol. 3. ICCROM, Rome.

[33] Perry, A. & Hammond, N. (2002). Systematic Review: The Experience of a PhD Student. Psychology Learning and Teaching, 2(1), 32–35.

[34] Francis C. & Baldesari. (2006). Systematic Reviews of Qualitative Literature. Oxford: UK Cochrane Centre.

[35] Lewin, S. (2008). Methods to Synthesise Qualitative Evidence Along side a Cochrane Intervention Review. London : London School of Hygiene and Tropical Medicine.

[36] Ditjen Cipta Karya, Departemen Pekerjaan Umum. (2002). Laporan Akhir, Rencana Tata Ruang Kawasan Banda Aceh. Jakarta : Proyek Penataan Ruang Wilayah Nasional

Institute of Physics
Dirac House, Temple Back
Bristol BS1 6BE UK

ISSN: 1755-1307
ISBN 978-1-5108-4293-9

CURRAN ASSOCIATES INC.
proceedings.com